W. Greiner · S. Schramm · E. Stein

QUANTUM CHROMODYNAMICS

Springer
Berlin
Heidelberg
New York
Barcelona
Hong Kong
London
Milan
Paris
Tokyo

Greiner
Quantum Mechanics
An Introduction 4th Edition

Greiner
Quantum Mechanics
Special Chapters

Greiner · Müller
Quantum Mechanics
Symmetries 2nd Edition

Greiner
Relativistic Quantum Mechanics
Wave Equations 3rd Edition

Greiner · Reinhardt
Field Quantization

Greiner · Reinhardt
Quantum Electrodynamics
3rd Edition

Greiner · Schramm · Stein
Quantum Chromodynamics
2nd Edition

Greiner · Maruhn
Nuclear Models

Greiner · Müller
Gauge Theory of Weak Interactions
3rd Edition

Greiner
Classical Mechanics I
(in preparation)

Greiner
Classical Mechanics II

Greiner
Classical Electrodynamics

Greiner · Neise · Stöcker
**Thermodynamics
and Statistical Mechanics**

Walter Greiner · Stefan Schramm
Eckart Stein

QUANTUM CHROMODYNAMICS

With a Foreword by
D. A. Bromley

Second Revised and Enlarged Edition
With 161 Figures,
and 66 Worked Examples and Exercises

Springer

Professor Dr. Walter Greiner
Institut für Theoretische Physik der
Johann Wolfgang Goethe-Universität Frankfurt
Postfach 111932
60054 Frankfurt am Main
Germany

Street address:

Robert-Mayer-Strasse 8–10
60325 Frankfurt am Main
Germany

email: greiner@th.physik.uni-frankfurt.de

Dr. Eckart Stein
Physics Department
Maharishi University of Management
6063 NP Vlodrop
The Netherlands

email: EStein@maharishi.net

Dr. Stefan Schramm
Argonne National Laboratory
9700 S. Cass Avenue
Argonne, IL 60439
USA

email: schramm@theory.phy.anl.gov

Title of the original German edition: *Theoretische Physik,* Ein Lehr- und Übungsbuch, Band 10: *Quantenchromodynamik* © Verlag Harri Deutsch, Thun, 1984, 1989

Library of Congress Cataloging-in-Publication Data applied for.

Die Deutsche Bibliothek – CIP-Einheitsaufnahme

Greiner, Walter:
Quantum chromodynamics/Walter Greiner; Stefan Schramm; Eckart Stein. –
Berlin; Heidelberg; New York; Barcelona; Hongkong; London; Mailand;
Paris; Singapur; Tokio: Springer, 2002
ISBN 3-540-66610-9

ISBN 3-540-66610-9 2. Auflage Springer-Verlag Berlin Heidelberg New York

ISBN 3-540-57103-5 1. Auflage Springer-Verlag Berlin Heidelberg New York

Springer-Verlag Berlin Heidelberg New York
a member of BertelsmannSpringer Science+Business Media GmbH

http://www.springer.de

Printed in Germany

Typesetting: Data conversion by LE-TEX, Leipzig
Cover design: Design Concept, Emil Smejkal, Heidelberg
Copy Editor: V. Wicks, Heidelberg
Production Editor: P. Treiber, Heidelberg
Printed on acid-free paper SPIN 10646206 56/3141/tr - 5 4 3 2 1 0

Foreword to Earlier Series Editions

More than a generation of German-speaking students around the world have worked their way to an understanding and appreciation of the power and beauty of modern theoretical physics – with mathematics, the most fundamental of sciences – using Walter Greiner's textbooks as their guide.

The idea of developing a coherent, complete presentation of an entire field of science in a series of closely related textbooks is not a new one. Many older physicists remember with real pleasure their sense of adventure and discovery as they worked their ways through the classic series by Sommerfeld, by Planck and by Landau and Lifshitz. From the students' viewpoint, there are a great many obvious advantages to be gained through use of consistent notation, logical ordering of topics and coherence of presentation; beyond this, the complete coverage of the science provides a unique opportunity for the author to convey his personal enthusiasm and love for his subject.

The present five volume set, *Theoretical Physics*, is in fact only that part of the complete set of textbooks developed by Greiner and his students that presents the quantum theory. I have long urged him to make the remaining volumes on classical mechanics and dynamics, on electromagnetism, on nuclear and particle physics, and on special topics available to an English-speaking audience as well, and we can hope for these companion volumes covering all of theoretical physics some time in the future.

What makes Greiner's volumes of particular value to the student and professor alike is their completeness. Greiner avoids the all too common "it follows that ... " which conceals several pages of mathematical manipulation and confounds the student. He does not hesitate to include experimental data to illuminate or illustrate a theoretical point and these data, like the theoretical content, have been kept up to date and topical through frequent revision and expansion of the lecture notes upon which these volumes are based.

Moreover, Greiner greatly increases the value of his presentation by including something like one hundred completely worked examples in each volume. Nothing is of greater importance to the student than seeing, in detail, how the theoretical concepts and tools under study are applied to actual problems of interest to a working physicist. And, finally, Greiner adds brief biographical sketches to each chapter covering the people responsible for the development of the theoretical ideas and/or the experimental data presented. It was Auguste Comte (1798–1857) in his *Positive Philosophy* who noted, "To understand a science it is necessary to know its history". This is all too often forgotten in modern physics teaching and the bridges that Greiner builds to the pioneering figures of our science upon whose work we build are welcome ones.

Greiner's lectures, which underlie these volumes, are internationally noted for their clarity, their completeness and for the effort that he has devoted to making physics an integral whole; his enthusiasm for his science is contagious and shines through almost every page.

These volumes represent only a part of a unique and Herculean effort to make all of theoretical physics accessible to the interested student. Beyond that, they are of enormous value to the professional physicist and to all others working with quantum phenomena. Again and again the reader will find that, after dipping into a particular volume to review a specific topic, he will end up browsing, caught up by often fascinating new insights and developments with which he had not previously been familiar.

Having used a number of Greiner's volumes in their original German in my teaching and research at Yale, I welcome these new and revised English translations and would recommend them enthusiastically to anyone searching for a coherent overview of physics.

Yale University
New Haven, CT, USA
1989

D. Allan Bromley
Henry Ford II Professor of Physics

Preface to the Second Edition

The theory of strong interactions, quantum chromodynamics (QCD), was formulated 30 years ago and has since been a very active field of research. The underlying equations of motion for the gauge degrees of freedom are nonlinear and minimally coupled to fermions with global and local SU(3) charges. This leads to spectacular problems compared with those of QED since the gauge bosons themselves interact with each other. On the other hand, it is exactly the self-interaction of the gluons which leads to asymptotic freedom and the possibility to calcuate quark–gluon interaction at small distances in the framework of perturbation theory. We discover one of the most complicated but most beautiful gauge theories which poses extremely challenging problems on modern theoretical and experimental physics today.

Quantum chromodynamics is the quantum field theory that allows us to calculate the propagation and interaction of colored quarks and gluons at small distances. Today's experiments do not allow these colored objects to be detected directly; instead one deals with colorless hadrons: mesons and baryons seen far away from the actual interaction point. The hadronization itself is a complicated process and not yet understood from first principles. Therefore one may wonder how the signature of quark and gluon interactions can be traced through the process of hadronization.

Very advanced analytical and numerical techniques have been developed in order to analyze the world of hadrons on the grounds of fundamental QCD. Together with a much improved experimental situation we seem to be ready to answer the question wether QCD is the correct theory of strong interactions at all scales or just an effective high-energy line of a yet undiscovered theory.

In this book, we try to give a self-consistent treatment of QCD, stressing the practitioners point of view. For pedagogical reasons we review quantum electrodynamics (QED) (Chap. 2) after an elementary introduction. In Chap. 3 we study scattering reactions with emphasis on lepton–nucleon scattering and introduce the language for describing the internal structure of hadrons. Also the MIT bag model is introduced, which serves as an illustrative and successful example for QCD-inspired models.

In Chap. 4 the general framework of gauge theories is described on the basis of the famous Standard Model of particle physics. We then concentrate on the gauge theory of quark–gluon interaction and derive the Feynman rules of QCD, which are very useful for pertubative calculations. In particular, we show explicitily how QCD is renormalized and how the often-quoted running coupling is obtained.

Chapter 5 is devoted to the application of QCD to lepton–hadron scattering and therefore to the state-of-the-art description of the internal structure of hadrons. We start by presenting two derivations of the Dokshitzer–Gribov–Lipatov–Altarelli–Parisi equations. The main focus of this chapter is on the indispensable operator product expansion and its application to deep inelastic lepton–hadron scattering. We show in great detail how to perform this expansion and calculate the Wilson coefficients. Furthermore, we discuss perturbative corrections to structure functions and perturbation theory at large orders, i.e. renormalons.

After analyzing lepton–hadron scattering we switch in Chap. 6 to the case of hadron–hadron scattering as described by the Drell–Yan processes. We then turn to the kinematical sector where the so-called leading-log approximation is no longer sufficient. The physics on these scales is called small-x physics.

Chapter 7 is devoted to two promising nonperturbative approaches, namely QCD on the lattice and the very powerful analytical tool called the QCD sum rule technique. We show explicitely how to formulate QCD on a lattice and discuss the relevant algorithms needed for practical numerical calculations, including the lattice at finite temperature. This is very important for the physics of hot and dense elementary matter as it appears, for example, in high-energy heavy ion physics. The QCD sum rule technique is explained and applied to the calculation of hadron masses.

Our presentation ends with some remarks on the nontrivial QCD vacuum and its modification at high temperature and/or baryon density, including a sketch of current developments concerning the so-called quark–gluon plasma in Chap. 8. Modern high-energy heavy ion physics is concerned with these issues.

We have tried to give a pedagogical introduction to the concepts and techniques of QCD. In particular, we have supplied over 70 examples and exercises worked out in great detail. Working through these may help the practitioner in perfoming complicated calculations in this challenging field of theoretical physics.

In writing this book we profited substantially from a number of existing textbooks, most notably J.J.R. Aitchison and A.J.G. Hey: 'Gauge Theories in Particle Physics', O. Nachtmann: 'Elementarteilchenphysik', B. Müller: 'The Physics of the Quark–Gluon Plasma', P. Becher, M. Boehm and H. Joos: 'Eichtheorien', J. Collins: 'Renormalization', R.D. Field: 'Application of Perturbative QCD', and M. Creutz: 'Quarks, Gluons and Lattices', and several review articles, especially: L.V. Gribov, E.M. Levin and M.G. Ryskin: 'Semihard processes in QCD', Phys. Rep. 100 (1983) 1, Badelek, Charchula, Krawczyk, Kwiecinski, ' Small x physics in deep inelastic lepton hadron scattering', Rev. Mod. Phys. 927 (1992), L.S. Reinders, H. Rubinstein, S. Yazaki: 'Hadron properties from QCD sumrules', Phys. Rep. 127 (1985) 1.

We thank many members of the Institute of Theoretical Physics in Frankfurt who added in one way or another to this book, namely Dr. J. Augustin, M. Bender, Ch. Best, S. Bernard, A. Bischoff, M. Bleicher, A. Diener, A. Dumitru, B. Ehrnsperger, U. Eichmann S. Graf, N. Hammon C. Hofmann, A. Jahns, J. Klenner, O. Martin, M. Massoth, G. Plunien, D. Rischke, and A. Scherdin.

Special thanks go to Mrs. A. Steidel who drew the figures, Dr. Henning Weber for supervising and solving numerous technical problems and, in particularly to Dr. Stefan Hofmann who supervised the editing process and gave hints for many improvements.

We hope that our readers will enjoy learning about quantum chromodynamics.

Frankfurt am Main,
May 2002

Walter Greiner
Stefan Schramm
Eckart Stein

Contents

Contents of Examples and Exercises

1. The Introduction of Quarks

About 70 years ago, only a small number of "elementary particles",[1] thought to be the basic building blocks of matter, were known: the proton, the electron, and the photon as the quantum of radiation. All these particles are stable (the neutron is stable only in nuclear matter, the free neutron decays by beta decay: $n \rightarrow p + e^- + \bar{\nu}$). Owing to the availability of large accelerators, this picture of a few elementary particles has profoundly changed: today, the standard reference *Review of Particle Properties*[2] lists more than 100 particles. The number is still growing as the energies and luminosities of accelerators are increased.

1.1 The Hadron Spectrum

The symmetries known from classical and quantum mechanics can be utilized to classify the "elementary-particle zoo". The simplest baryons are p and n; the simplest leptons e^- and μ^-. Obviously there are many other particles that must be classified as baryons or leptons.

The symmetries are linked to conserved quantum numbers such as the baryon number B, isospin T with z component T_3, strangeness S, hypercharge $Y = B + S$, charge $Q = T_3 + Y/2$, spin I with z component I_z, parity π, and charge conjugation parity π_c. Conservation laws for such quantum numbers manifest themselves by the absence of certain processes. For example, the hydrogen atom does not decay into two photons: $e^- + p \rightarrow \gamma + \gamma$, although this process is not forbidden either by energy–momentum conservation or by charge conservation. Since our world is built mainly out of hydrogen, we know from our existence that there must be at least one other conservation law that is as fundamental as charge conservation. The nonexistence of the decays $n \rightarrow p + e^-$ and $n \rightarrow \gamma + \gamma$ also indicates the presence of a new quantum number. The proton and neutron are given a baryonic charge $B = 1$, the electron $B = 0$. Similarly the electron is assigned leptonic charge $L = 1$, the nucleons $L = 0$. From the principle of simplicity it appears very unsatisfactory to regard all observed particles as elementary. To give an impression of the huge number of hadrons known today, we have collected together the baryon resonances in Fig. 1.1. The data are

[1] For a detailed discussion of the content of this chapter see W. Greiner and B. Müller: *Symmetries* (Springer, Berlin, Heidelberg 1994).

[2] See *Review of Particle Properties*: Phys. Rev. D **45** (1992).

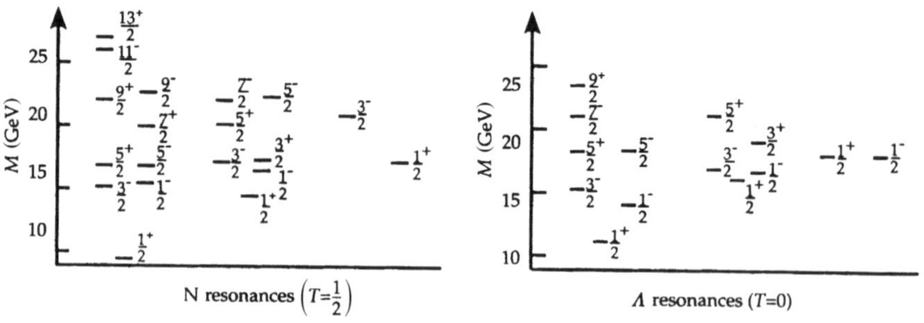

Fig. 1.1. The mass spectra of baryons. Plotted are the average masses of the multiplets. For example, the state $N_{5/2^+}$ at 1.68 MeV stands for two particles, one protonlike and one neutronlike, both with spin 5/2 and positive internal parity. The figure contains 140 particle states in total

taken from the "Review of Particle Properties". Particles for which there is only weak evidence or for which the spin I and internal parity P have not been determined have been left out. Note that each state represents a full multiplet. The number of members in a multiplet is $N = 2T + 1$ with isospin T. Thus the 13 Δ resonances shown correspond to a total of 52 different baryons.

When looking at these particle spectra, one immediately recognizes the similarity to atomic or nuclear spectra. One would like, for example, to classify the nucleon resonances (N resonances) in analogy to the levels of a hydrogen atom. The $\frac{1}{2}^+$ ground state (i.e., the ordinary proton and neutron) would then correspond to the $1s_{\frac{1}{2}}$ state, the states $\frac{3}{2}^-$, $\frac{1}{2}^-$, and $\frac{1}{2}^+$ at approximately 1.5 GeV to the hydrogen levels $2p_{\frac{3}{2}}$, $2p_{\frac{1}{2}}$, and $2s_{\frac{1}{2}}$, the states $\frac{5}{2}^+$, $\frac{3}{2}^+$, $\frac{3}{2}^-$, $\frac{1}{2}^-$, $\frac{1}{2}^+$ to the sublevels of the third main shell $3d_{\frac{5}{2}}$, $3d_{\frac{3}{2}}$, $3p_{\frac{3}{2}}$, $3p_{\frac{1}{2}}$, $3s_{\frac{1}{2}}$, and so on.

Although one should not take this analogy too seriously, it clearly shows that a model in which the baryons are built from spin-$\frac{1}{2}$ particles almost automatically leads to the states depicted in Fig. 1.1. The quality of any such model is measured by its ability to predict the correct energies. We shall discuss specific models in Sect. 3.1.

We therefore interpret the particle spectra in Fig. 1.1 as strong evidence that the baryons are composed of several more fundamental particles and that most of the observable baryon resonances are excitations of a few ground states. In this way the excited states $\frac{3}{2}^-$ and $\frac{1}{2}^-$ are reached from the nucleon ground state N(938 MeV) $\frac{1}{2}^+$ by increasing the angular momentum of one postulated component particle by one. ($\frac{1}{2}^+$ can be coupled with 1^- to give $\frac{1}{2}^-$ or $\frac{3}{2}^-$.) As the energy of the baryon resonances increases with higher spin (i.e., total angular momentum of all component particles), one can deduce that all relative orbital angular momenta vanish in the ground states.

To investigate this idea further, one must solve a purely combinatorial problem: How many component particles (called *quarks* in the following) are needed, and what properties are required for them to correctly describe the ground states of the hadron spectrum? It turns out that the existence of several quarks must be postulated. The quantum numbers given in Table 1.1 must be given to them.

The three light quarks u, d, s can be identified with the three states in the fundamental representation of SU(3). This is initially a purely formal act. It gains importance only as one shows that the *branching ratios* of particle reactions and

Table 1.1. Quark charge (Q), isospin (T, T_3), and strangeness (S)

	Q	T	T_3	S
u	$\frac{2}{3}$	$\frac{1}{2}$	$\frac{1}{2}$	0
d	$-\frac{1}{3}$	$\frac{1}{2}$	$-\frac{1}{2}$	0
s	$-\frac{1}{3}$	0	0	-1
c	$\frac{2}{3}$	0	0	0
t	$\frac{2}{3}$	0	0	0
b	$-\frac{1}{3}$	0	0	0

the *mass differences* between stable baryons show – at least approximately – the same symmetries. This means that the so-called *flavor* SU(3) can be interpreted as the symmetry group of a more fundamental interaction.

Hadrons are therefore constructed as flavor SU(3) states. As the spin of the quarks must also be taken into account, the total symmetry group becomes SU(3) $\times$ SU(2). As an example we give the decomposition of the neutron into quark states:

$$\begin{aligned}|\mathrm{n}\!\uparrow\rangle = \frac{1}{\sqrt{18}}\Big(&2\,|\mathrm{d}\!\uparrow\rangle\,|\mathrm{d}\!\uparrow\rangle\,|\mathrm{u}\!\downarrow\rangle - |\mathrm{d}\!\uparrow\rangle\,|\mathrm{d}\!\downarrow\rangle\,|\mathrm{u}\!\uparrow\rangle - |\mathrm{d}\!\downarrow\rangle\,|\mathrm{d}\!\uparrow\rangle\,|\mathrm{u}\!\uparrow\rangle \\ &- |\mathrm{d}\!\uparrow\rangle\,|\mathrm{u}\!\uparrow\rangle\,|\mathrm{d}\!\downarrow\rangle + 2\,|\mathrm{d}\!\uparrow\rangle\,|\mathrm{u}\!\downarrow\rangle\,|\mathrm{d}\!\uparrow\rangle - |\mathrm{d}\!\downarrow\rangle\,|\mathrm{u}\!\uparrow\rangle\,|\mathrm{d}\!\uparrow\rangle \\ &- |\mathrm{u}\!\uparrow\rangle\,|\mathrm{d}\!\uparrow\rangle\,|\mathrm{d}\!\downarrow\rangle - |\mathrm{u}\!\uparrow\rangle\,|\mathrm{d}\!\downarrow\rangle\,|\mathrm{d}\!\uparrow\rangle + 2\,|\mathrm{u}\!\downarrow\rangle\,|\mathrm{d}\!\uparrow\rangle\,|\mathrm{d}\!\uparrow\rangle\Big) \ . \end{aligned} \quad (1.1)$$

Particularly interesting for the topic of this volume are the corresponding decompositions of the states Ω^-, Δ^{++}, and Δ^-:

$$\begin{aligned}\left|\Omega^-\right\rangle &= |\mathrm{s}\!\uparrow\rangle\,|\mathrm{s}\!\uparrow\rangle\,|\mathrm{s}\!\uparrow\rangle \ , \\ \left|\Delta^{++}\right\rangle &= |\mathrm{u}\!\uparrow\rangle\,|\mathrm{u}\!\uparrow\rangle\,|\mathrm{u}\!\uparrow\rangle \ , \\ \left|\Delta^-\right\rangle &= |\mathrm{d}\!\uparrow\rangle\,|\mathrm{d}\!\uparrow\rangle\,|\mathrm{d}\!\uparrow\rangle \ . \end{aligned} \quad (1.2)$$

To obtain the spin quantum numbers of hadrons, one must assume that the quarks have spin $\frac{1}{2}$. This poses a problem: spin-$\frac{1}{2}$ particles should obey Fermi statistics, i.e., no two quarks can occupy the same state. So the three quarks in Ω^-, Δ^{++}, and Δ^- must differ in at least one quantum number, as we shall discuss in Chapt. 4. Before proceeding to the composition of baryons from quarks, we shall first repeat the most important properties of the symmetry groups SU(2) and SU(3).

SU(2) and SU(3) are special cases of the group SU(N) the special unitary group in N dimensions. Any unitary square matrix $\hat{U}$ with N rows and N columns can be written as

$$\hat{U} = \mathrm{e}^{\mathrm{i}\hat{H}} \ , \quad (1.3)$$

where $\hat{H}$ is a Hermitian matrix. The matrices $\hat{U}$ form the group SU(N) of unitary matrices in N dimensions. $\hat{H}$ is Hermitian, i.e.,

$$\hat{H}^*_{ij} = \hat{H}_{ji} \ . \quad (1.4)$$

Of the N^2 complex parameters (elements of the matrices), N^2 real parameters for $\hat{H}$ and hence for $\hat{U}$ remain, owing to the auxiliary conditions (1.4). Since $\hat{U}$ is unitary, i.e. $\hat{U}^\dagger\hat{U} = 1$, $\det\hat{U}^\dagger \det\hat{U} = (\det\hat{U})^* \det\hat{U} = 1$ and thus

$$\left|\det\hat{U}\right| = 1 \ . \quad (1.5)$$

Owing to (1.4), $\mathrm{tr}\left\{\hat{H}\right\} = \alpha$ (α real) and

$$\det\hat{U} = \det\left(\mathrm{e}^{\mathrm{i}\hat{H}}\right) = \mathrm{e}^{\mathrm{i\,tr}\hat{H}} = \mathrm{e}^{\mathrm{i}\alpha} \ . \quad (1.6)$$

If we additionally demand the condition

$$\det \hat{U} = +1 \ , \tag{1.7}$$

i.e., $\alpha = 0 \mod 2\pi$, only $N^2 - 1$ parameters remain. This group is called the special unitary group in N dimensions (SU(N)).

Let us now consider a group element $\hat{U}$ of U(N) as a function of N^2 parameters ϕ_μ ($\mu = 1, \dots, n$). To this end, we write (1.3) as

$$\hat{U}(\phi_1, \dots, \phi_n) = \exp\left(-\mathrm{i}\sum_\mu \phi_\mu \hat{L}_\mu\right) \ , \tag{1.8}$$

where $\hat{L}_\mu$ are for the time being unknown operators:

$$-\mathrm{i}\hat{L}_\mu = \left.\frac{\partial \hat{U}(\boldsymbol{\phi})}{\partial \phi_\mu}\right|_{\boldsymbol{\phi}=0} \tag{1.9}$$

($\boldsymbol{\phi} = (\phi_1, \dots, \phi_n)$). For small $\phi_\mu (\delta\phi_\mu)$ we can expand $\hat{U}$ in a series ($\mathbb{1}$ is the unit matrix):

$$\hat{U}(\boldsymbol{\phi}) \approx \mathbb{1} - \mathrm{i}\sum_{\mu=1}^{n} \delta\phi_\mu \hat{L}_\mu - \frac{1}{2}\sum_{\mu,\nu} \delta\phi_\mu \delta\phi_\nu \hat{L}_\mu \hat{L}_\mu + \dots \ . \tag{1.10}$$

Boundary conditions (1.4) and (1.5) imply after some calculation that the operators $\hat{L}_i$ must satisfy the commutation relations

$$\left[\hat{L}_i, \hat{L}_j\right] = c_{ijk}\hat{L}_k \ . \tag{1.11}$$

Equation (1.11) defines an algebra, the Lie algebra of the group U(N).

The operators $\hat{L}_i$ generate the group by means of (1.10) and are thus called *generators*. Obviously there are as many generators as the group has parameters, i.e., the group U(N) has N^2 generators and the group SU(N) has $N^2 - 1$. The quantities c_{ijk} are called *structure constants* of the group. They contain all the information about the group. In the Lie algebra of the group (i.e., the $\hat{L}_k$), there is a maximal number R of commutating elements $\hat{L}_i$ ($i = 1, \dots, R$)

$$\left[\hat{L}_i, \hat{L}_j\right] = 0 \quad (i = 1, \dots, R) \ . \tag{1.12}$$

R is called the *rank* of the group. The eigenvalues of the $\hat{L}_i$ are, as we shall see, used to classify elementary-particle spectra. We shall now discuss the concepts introduced here using the actual examples of the spin and isospin group SU(2) and the group SU(3).

SU(2). U(2) is the group of lineary independent Hermitian 2×2 matrices. A well-known representation of it is given by the Pauli matrices and the unit matrix

$$\hat{\sigma}_1 = \begin{pmatrix} 0 & 1 \\ 1 & 0 \end{pmatrix}, \ \hat{\sigma}_2 = \begin{pmatrix} 0 & -\mathrm{i} \\ \mathrm{i} & 0 \end{pmatrix}, \ \hat{\sigma}_3 = \begin{pmatrix} 1 & 0 \\ 0 & -1 \end{pmatrix}, \ \mathbb{1} = \begin{pmatrix} 1 & 0 \\ 0 & 1 \end{pmatrix}. \tag{1.13}$$

These span the space of Hermitian 2×2 matrices, i.e., they are linearly independent. SU(2) has only three generators; the unit matrix is not used. From (1.3) we can write a general group element of the group SU(2) as

$$\hat{U}(\boldsymbol{\phi}) = \exp\left(-\mathrm{i}\sum_{i=1}^{3} \phi_i \hat{\sigma}_i\right) \qquad (1.14)$$

(or, using the summation convention, $\exp(-\mathrm{i}\phi_i\hat{\sigma}_i)$). Here $\boldsymbol{\phi} = (\phi_1, \phi_2, \phi_3)$ is a shorthand for the parameter of the transformation. The Pauli matrices satisfy the commutation relations

$$\left[\hat{\sigma}_i, \hat{\sigma}_j\right] = 2\mathrm{i}\varepsilon_{ijk}\hat{\sigma}_k \ , \qquad (1.15)$$

with

$$\varepsilon_{ijk} = \begin{cases} 0 & \text{for two equal indices,} \\ 1 & \text{for even permutations of the indices,} \\ -1 & \text{for odd permutations of the indices.} \end{cases}$$

Usually, instead of $\hat{\sigma}_i$, the $\hat{S}_i = \frac{1}{2}\,\hat{\sigma}_i$ are used as generators, i.e.

$$\left[\hat{S}_i, \hat{S}_j\right] = \mathrm{i}\varepsilon_{ijk}\hat{S}_k \ .$$

According to (1.11), $\mathrm{i}\varepsilon_{ijk}$ are the structure constants of SU(2). Equation (1.15) shows that no generator commutes with any other, i.e., the rank of SU(2) is 1. According to the *Racah theorem*, the rank of a group is equal to the number of Casimir operators (i.e., those operators are polynomials in the generators and commute with all generators). Thus there is one Casimir operator for SU(2), namely the square of the well-known angular momentum (spin) operator:

$$\hat{C}_{\mathrm{SU(2)}} = \sum_{i=1}^{3} \hat{S}_i^2 \ . \qquad (1.16)$$

The representation of SU(2) given in (1.13) (and generally of SU(N)) by 2×2 matrices (generally $N \times N$ matrices) is called the fundamental representation of SU(2) (SU(N)). It is the smallest nontrivial representation of SU(2) (SU(N)). It is a 2×2 representation for SU(2), a 3×3 representation for SU(3), and so on. From Schur's first lemma the Casimir operators in the fundamental representation are multiples of the unit matrix (see Exercise 1.1):

$$\hat{C}_{\mathrm{SU(2)}} = \sum_{i=1}^{3}\left(\frac{\hat{\sigma}_i}{2}\right)^2 = \frac{3}{4}\mathbb{1} \ . \qquad (1.17)$$

SU(3). The special unitary group in three dimensions has $3^2 - 1 = 8$ generators. In the fundamental representation they can be expressed by the

Gell-Mann matrices $\hat{\lambda}_1, \dots, \hat{\lambda}_8$:

$$\hat{\lambda}_1 = \begin{pmatrix} 0 & 1 & 0 \\ 1 & 0 & 0 \\ 0 & 0 & 0 \end{pmatrix}, \quad \hat{\lambda}_2 = \begin{pmatrix} 0 & -\mathrm{i} & 0 \\ \mathrm{i} & 0 & 0 \\ 0 & 0 & 0 \end{pmatrix}, \quad \hat{\lambda}_3 = \begin{pmatrix} 1 & 0 & 0 \\ 0 & -1 & 0 \\ 0 & 0 & 0 \end{pmatrix},$$

$$\hat{\lambda}_4 = \begin{pmatrix} 0 & 0 & 1 \\ 0 & 0 & 0 \\ 1 & 0 & 0 \end{pmatrix}, \quad \hat{\lambda}_5 = \begin{pmatrix} 0 & 0 & -\mathrm{i} \\ 0 & 0 & 0 \\ \mathrm{i} & 0 & 0 \end{pmatrix}, \quad \hat{\lambda}_6 = \begin{pmatrix} 0 & 0 & 0 \\ 0 & 0 & 1 \\ 0 & 1 & 0 \end{pmatrix},$$

$$\hat{\lambda}_7 = \begin{pmatrix} 0 & 0 & 0 \\ 0 & 0 & -\mathrm{i} \\ 0 & \mathrm{i} & 0 \end{pmatrix}, \quad \hat{\lambda}_8 = \frac{1}{\sqrt{3}} \begin{pmatrix} 1 & 0 & 0 \\ 0 & 1 & 0 \\ 0 & 0 & -2 \end{pmatrix} . \tag{1.18}$$

The Gell-Mann matrices are Hermitian,

$$\hat{\lambda}_i^\dagger = \hat{\lambda}_i \ , \tag{1.19}$$

and their trace vanishes,

$$\mathrm{tr}\left\{\hat{\lambda}_i\right\} = 0 \ . \tag{1.20}$$

They define the Lie algebra of SU(3) by the commutation relations

$$\left[\hat{\lambda}_i, \hat{\lambda}_j\right] = 2\mathrm{i}\, f_{ijk}\hat{\lambda}_k \ , \tag{1.21}$$

where the structure constants f_{ijk} are, like the ε_{ijk} in SU(2), completely antisymmetric, i.e.,

$$f_{ijk} = -f_{jik} = -f_{ikj} \ . \tag{1.22}$$

The anticommutation relations of the $\hat{\lambda}_i$ are written as

$$\left\{\hat{\lambda}_i, \hat{\lambda}_j\right\} = \frac{4}{3}\delta_{ij}\mathbb{1} + 2d_{ijk}\hat{\lambda}_k \ . \tag{1.23}$$

The constants d_{ijk} are completely symmetric:

$$d_{ijk} = d_{jik} = d_{ikj} \ . \tag{1.24}$$

The nonvanishing structure constants are given in Table 1.2.

As in SU(2), generators $\hat{F}_i = \frac{1}{2}\hat{\lambda}_i$ ("hyperspin") are used instead of the $\hat{\lambda}_i$ with the commutation relations

$$\left[\hat{F}_i, \hat{F}_j\right] = \mathrm{i}\, f_{ijk}\hat{F}_k \ . \tag{1.25}$$

One can easily check that among the $\hat{F}_i$ only the commutators $\left[\hat{F}_1, \hat{F}_8\right] = \left[\hat{F}_2, \hat{F}_8\right] = \left[\hat{F}_3, \hat{F}_8\right] = 0$ vanish. As the $\hat{F}_i$, $i = 1, 2, 3$, do not commute with each other, there are at most two commuting generators, i.e., SU(3) has rank

Table 1.2. The nonvanishing, completely antisymmetric structure constants f_{ijk} and the symmetric constants d_{ijk}

ijk	f_{ijk}	ijk	d_{ijk}
123	1	118	$\frac{1}{\sqrt{3}}$
147	1/2	146	1/2
156	−1/2	157	1/2
246	1/2	228	$\frac{1}{\sqrt{3}}$
257	1/2	247	−1/2
345	1/2	256	1/2
367	−1/2	338	$\frac{1}{\sqrt{3}}$
458	$\frac{\sqrt{3}}{2}$	344	1/2
678	$\frac{\sqrt{3}}{2}$	355	1/2
		366	−1/2
		377	−1/2
		448	$-\frac{1}{2\sqrt{3}}$
		558	$-\frac{1}{2\sqrt{3}}$
		668	$-\frac{1}{2\sqrt{3}}$
		778	$-\frac{1}{2\sqrt{3}}$
		888	$-\frac{1}{\sqrt{3}}$

two (in general SU(N) has rank $N-1$), and hence two Casimir operators, one of which is simply

$$\hat{C}_1 = \sum_{i=1}^{8} \hat{F}_i^2 = -\frac{2\mathrm{i}}{3} \sum_{i,j,k} f_{ijk}\, \hat{F}_i\, \hat{F}_j\, \hat{F}_k \ . \tag{1.26}$$

In the fundamental representation

$$\left(\hat{C}_1\right)_{j\ell} = \frac{1}{4} \sum_{i=1}^{8} \sum_{k=1}^{3} \left(\hat{\lambda}_i\right)_{jk} \left(\hat{\lambda}_i\right)_{k\ell} = \frac{4}{3}\delta_{j\ell} \ . \tag{1.27}$$

From the structure constants f_{ijk}, new matrices $\hat{U}_i$ can be constructed according to

$$\left(\hat{U}_i\right)_{jk} = -\mathrm{i}\, f_{ijk} \ , \tag{1.28}$$

which also satisfy the commutation relations

$$\left[\hat{U}_i, \hat{U}_j\right] = \mathrm{i}\, f_{ijk} \hat{U}_k \ . \tag{1.29}$$

This representation of the Lie algebra of SU(3) is called *adjoint* (or *regular*). In it (see Exercise 1.2)

$$\begin{aligned} (\hat{C}_l)_{kl} &= \sum_{i=1}^{8} (\hat{U}_i^2)_{kl} = \sum_{i,j} (\hat{U}_i)_{kj} (\hat{U}_i)_{jl} \\ &= -\sum_i \sum_j f_{ikj} f_{ijl} = \sum_{i,j} f_{ijk} f_{ijl} \\ &= 3\delta_{kl} \ . \end{aligned} \tag{1.30}$$

A form of the complete SU(3) group element according to (1.3) is ($\hat{U}(0)$ designates in contrast to $\hat{U}_i$ the transformation matrix from (1.3))

$$\hat{U}(\theta) = \mathrm{e}^{-\mathrm{i}\boldsymbol{\theta}\cdot\hat{\boldsymbol{F}}} \ , \tag{1.31}$$

where $\hat{\boldsymbol{F}}$ is the vector of eight generators and $\boldsymbol{\theta}$ the vector of eight parameters.

After this short digression into the group structure of SU(2) and SU(3), we return to the classification of elementary particles. As indicated above, the eigenvalues of commuting generators of the group serve to classify the hadrons. For SU(3) there is only one such operator among the $\hat{T}_i$ $(i = 1, 2, 3)$, usually chosen to be $\hat{T}_3$ (the z component). The structure of SU(2) multiplets is thus one-dimensional and characterized by a number T_3. In the framework of QCD the most important application of SU(2) is as the isospin group (with generators $\hat{T}_i$) and as the angular momentum group with, for examples, the spin operator $\hat{S}_i$. The small mass difference between neutron and proton (0.14% of the total mass) leads to the thought that both can be treated as states of a single particle, the nucleon. According to the matrix representation

$$\hat{T}_3 = \frac{1}{2}\begin{pmatrix} 1 & 0 \\ 0 & -1 \end{pmatrix} = \frac{1}{2}\hat{\tau}_3 \ , \tag{1.32}$$

one assigns the isospin vector $\Psi_\mathrm{p} = \begin{pmatrix} 1 \\ 0 \end{pmatrix}$ to the proton and $\Psi_\mathrm{n} = \begin{pmatrix} 0 \\ 1 \end{pmatrix}$ to the neutron, so that the isospin eigenvalues $T_3 = \pm\frac{1}{2}$ are assigned to the nucleons:

$$\hat{T}_3 \begin{pmatrix} 1 \\ 0 \end{pmatrix} = +\frac{1}{2}\begin{pmatrix} 1 \\ 0 \end{pmatrix} \ , \tag{1.33}$$

$$\hat{T}_3 \begin{pmatrix} 0 \\ 1 \end{pmatrix} = -\frac{1}{2}\begin{pmatrix} 0 \\ 1 \end{pmatrix} \ . \tag{1.34}$$

Analogously one introduces

$$\hat{\tau}_1 = \begin{pmatrix} 0 & 1 \\ 1 & 0 \end{pmatrix} \qquad \text{and} \qquad \hat{\tau}_2 = \begin{pmatrix} 0 & -\mathrm{i} \\ \mathrm{i} & 0 \end{pmatrix} \tag{1.35}$$

such that the

$$\hat{T}_k = \frac{1}{2}\,\hat{\tau}_k \qquad (k = 1, 2, 3) \tag{1.36}$$

satisfy the same commutation relations as the spin operators. One can check by direct calculation that raising and lowering operators can be constructed from the $\hat{\tau}_i$:

$$\hat{\tau}_+ = \frac{1}{2}(\hat{\tau}_1 + \mathrm{i}\hat{\tau}_2) = \begin{pmatrix} 0 & 1 \\ 0 & 0 \end{pmatrix} \ ,$$

$$\hat{\tau}_- = \frac{1}{2}(\hat{\tau}_1 - \mathrm{i}\hat{\tau}_2) = \begin{pmatrix} 0 & 0 \\ 1 & 0 \end{pmatrix} \ . \tag{1.37}$$

They have the following well-known action on nucleon states:

$$\begin{aligned} &\hat{\tau}_+\Psi_\mathrm{p} = 0 \ , \qquad &&\hat{\tau}_+\Psi_\mathrm{n} = \Psi_\mathrm{p} \ , \\ &\hat{\tau}_-\Psi_\mathrm{p} = \Psi_\mathrm{n} \ , \qquad &&\hat{\tau}_-\Psi_\mathrm{n} = 0 \ , \end{aligned} \tag{1.38}$$

i.e., the operators change nucleon states into each other (they are also called *ladder operators*). From (1.3), we can give the general transformation in the abstract three-dimensional isospin space

$$\hat{U}(\boldsymbol{\phi}) = \hat{U}(\phi_1, \phi_2, \phi_3) = \mathrm{e}^{-\mathrm{i}\phi_\mu \hat{T}_\mu} \ , \tag{1.39}$$

where the ϕ_μ represent the rotation angles in isospin space. The Casimir operator of isospin SU(2) is

$$\hat{T}^2 = \hat{T}_1^2 + \hat{T}_2^2 + \hat{T}_3^2 \qquad . \tag{1.40}$$

We can now describe each particle state by an abstract vector $|TT_3\rangle$ (analogously to the spin, as the isospin SU(2) is isomorphic to the spin SU(2)), where the following relations hold:

$$\hat{T}^2 \, |TT_3\rangle = T(T+1) \, |TT_3\rangle \quad , \tag{1.41}$$

$$\hat{T}_3 \, |TT_3\rangle = T_3 \, |TT_3\rangle \quad . \tag{1.42}$$

Thus the nucleons represent an isodoublet with $T = \frac{1}{2}$ and $T_3 = \pm\frac{1}{2}$. The pions $(\pi^\pm, \pi^0)$ (masses $m(\pi^0) = 135\,\mathrm{MeV}/c^2$ and $m(\pi^\pm) = 139.6\,\mathrm{MeV}/c^2$, i.e. a mass difference of $4.6\,\mathrm{MeV}/c^2$) constitute an isotriplet with $T = 1$ and $T_3 = -1, 0, 1$. Obviously there is a relation between isospin and the electric charge of a particle. For the nucleons the charge operator is immediately obvious:

$$\hat{Q} = \hat{T}_3 + \frac{1}{2}\mathbb{1} \tag{1.43}$$

in units of the elementary charge e, while one finds in a similarly simple way for the pions

$$\hat{Q} = \hat{T}_3 \quad . \tag{1.44}$$

To unify both relations, one can introduce an additional quantum number Y (the so-called hypercharge) and describe any state by T_3 and Y:

$$\hat{Y} \, |YT_3\rangle = Y \, |YT_3\rangle \quad , \tag{1.45}$$

$$\hat{T}_3 \, |YT_3\rangle = T_3 \, |YT_3\rangle \quad . \tag{1.46}$$

In this way the nucleon is assigned $Y = 1$ and the pion $Y = 0$, so that (1.42) and (1.43) can be written as

$$\hat{Q} = \frac{1}{2}\hat{Y} + \hat{T}_3 \quad . \tag{1.47}$$

Relation (1.45) is the *Gell-Mann–Nishijima relation*. The hypercharge characterizes the center of charge. It is often customary to express Y by the strangeness S and the baryon number B using $Y = B + S$. Here $B = +1$ for all baryons, $B = -1$ for antibaryons, and $B = 0$ otherwise (in particular for mesons). Thus $Y = S$ for mesons. To classify elementary particles in the framework of SU(3), it is customary to display them in a T_3–Y diagram. The baryons with spin $\frac{1}{2}$ constitute an octet in this diagram (see Fig. 1.2).

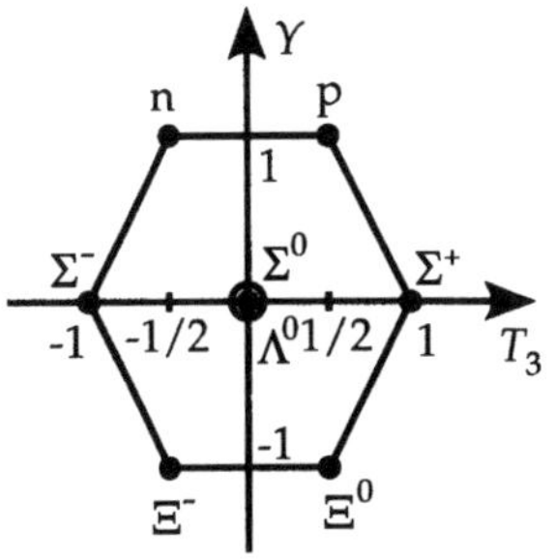

Fig. 1.2. An octet of spin-$\frac{1}{2}$ baryons

The spectrum of antiparticles is obtained from this by reflecting the expression with respect to the Y and T_3 axes. The heavier baryons and the mesons can be classified analogously. We introduced the hypercharge by means of the charge and have thus added another quantum number. SU(2) has rank 1, i.e., it provides only one such quantum number. SU(3), however, has rank 2 and thus two commuting generators, $\hat{F}_3$ and $\hat{F}_8$. We can therefore make the identification $\hat{T}_3 = \hat{F}_3$ and $\hat{Y} = 2/\sqrt{3}\hat{F}_8$ and interpret the multiplets as SU(3) multiplets. The SU(3)-multiplet classification was introduced by M. Gell-Mann and is initially purely schematic. There are no small nontrivial representations among these multiplets (with the exception of the singlet, interpreted as the Λ^* hyperon with mass 1405 MeV$/c^2$ and spin $\frac{1}{2}$). The smallest nontrivial representation of SU(3) is the triplet. This reasoning led Gell-Mann and others to the assumption that physical particles are connected to this triplet, the quarks (from James Joyce's Finnegan's Wake: "Three quarks for Muster Mark"). Today we know that there are five quarks. They are called up, down, strange, charm, and bottom quarks. A sixth quark, the top, most probably exists but it has not yet been possible to detect it directly owing to its large mass. The different kinds of quarks are called "flavors". The original SU(3) flavor symmetry is therefore only important for low energies, where c, b, and t quarks do not play a role owing to their large mass. It is, also still relevant for hadronic ground-state properties.

All particles physically observed at this time are combinations of three quarks (baryons) or a quark and an antiquark (mesons) plus, in each case, an arbitrary number of quark–antiquark pairs and gluons. This requires that quarks have

(1) baryon number $\frac{1}{3}$

(2) electric charges in multiples of $\pm\frac{1}{3}$.

Uneven multiples of charge $\frac{1}{3}$ have never been conclusively observed in nature, and there therefore seems to exist some principle assuring that quarks can exist in bound states in elementary particles but never free. This is the problem of quark confinement, which we shall discuss later. Up to now, we have considered the SU(3) symmetry connected with the flavor of elementary particles. Until the early 1970s it was commonly believed that this symmetry was the basis of the strong interaction. Today the true strong interaction is widely acknowledged to be connected with another quark quantum number, the *color*. The dynamics of color (chromodynamics) determines the interaction of the quarks (which is, as we shall see, flavor-blind).

Quantum electrodynamics is reviewed in the following chapter. Readers familiar with it are advised to continue on page 75 with Chap. 3.

EXERCISE

1.1 The Fundamental Representation of a Lie Algebra

Problem. (a) What are the fundamental representations of the group SU(N)? (b) Show that according to Schur's lemma the Casimir operators in these fundamental representations are multiples of the unit matrix.

Solution. (a) The fundamental representations are those nontrivial representations of a group that have the lowest dimension. All higher-dimensional representations can be constructed from them. We shall demonstrate this using the special unitary groups SU(N).

SU(2). As we have learned, its representation is characterized by the angular-momentum quantum number $j = 0, \frac{1}{2}, 1, \frac{3}{2}, \ldots$, and states are classified by

Exercise 1.1

$(j) \equiv |jm\rangle$, $m = -j, \ldots, -j$. The scalar representation is $j = 0$. The lowest-dimensional representation with $j \neq 0$ would then be $j = \frac{1}{2}$. From it we can construct all others by simply coupling one to another:

$$\left[\frac{1}{2}\right] \times \left[\frac{1}{2}\right] = \left[\frac{1}{2}\right]^2 = [1] + [0] \ , \tag{1a}$$

$$\left[\frac{1}{2}\right] \times \left[\frac{1}{2}\right] \times \left[\frac{1}{2}\right] = \left[\frac{1}{2}\right]^3 = \left[\frac{3}{2}\right] + \left[\frac{1}{2}\right] + \left[\frac{1}{2}\right] \ . \tag{1b}$$

"$\times$" indicates the direct product, "+" the direct sum. The first two $j = \frac{1}{2}$ representations can be coupled to $j = 0, 1$. Adding another $j = \frac{1}{2}$, it couples with $j = 1$ to give $j = \frac{3}{2}, \frac{1}{2}$ and with $j = 0$ to give only $j = \frac{1}{2}$. In total, $\left[\frac{1}{2}\right]^3$ contains the representations $\left[\frac{3}{2}\right]$, $\left[\frac{1}{2}\right]$, $\left[\frac{1}{2}\right]$. Figure 1.3 depicts this angular momentum coupling graphically. It must be noted that a representation can appear more than once, e.g., $\left[\frac{1}{2}\right]$ appears twice in $\left[\frac{1}{2}\right]^3$ and [1] thrice in $\left[\frac{1}{2}\right]^4$.

In the next example, an alternative representation according to "maximal weight" is of interest. For this, all operators in the algebra that commute with each other are considered (*Cartan subalgebra*). Their eigenvalues classify states in a representation. In the case of SU(2) there is only one operator commuting with itself. This can be chosen to be any of the j_i, usualy one takes j_3, the third component of the angular momentum vector. Its eigenvalues are $m = -j, \ldots, +j$. The "maximal weight" is $m_{\text{max}} = j$. In direct products $\left[\frac{1}{2}\right]^n$ the maximal weight is $m_{\text{max}} = \frac{n}{2}$, which is the "maximal weight" of the "straight coupling" (see Fig. 1.3).

SU(3). Its representations (multiplets) are classified by the eigenvalues of the Casimir operators. These give us, in the case of SU(3), two numbers $[p, q]$. These are in turn connected to the rank of the algebra, i.e., the number of commuting generators in the algebra. In general, the representations of SU(N) are characterized by $N - 1$ numbers. Another possibility would be to classify representations by their "maximal weight". As is known, each state in a representation

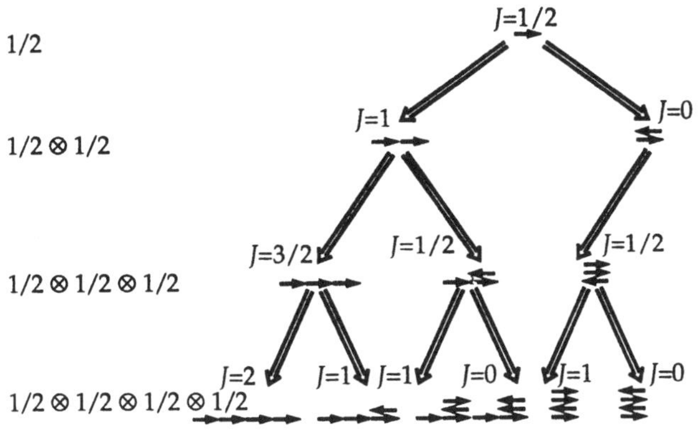

Fig. 1.3. Multiple coupling of spins $\frac{1}{2}$ to various total spins J

Exercise 1.1

of SU(3) (a multiplet) is labeled by the eigenvalues of the third component of isospin $\hat{T}_3$ and hypercharge $\hat{Y}$. The weight is given by the tuple (T_3, Y). A weight (T_3, Y) is higher than (T_3', Y') if

$$T_3 > T_3' \quad \text{or} \quad T_3 = T_3' \quad \text{and} \quad Y > Y' \ . \tag{2}$$

The highest weight in a representation is given by the maximal value of T_3, and, if there is more than one, by the maximal value of Y. This is demonstrated by the following examples:

(1) $[p, q] = [1, 0]$.

This is the representation whose "weight diagram" is depicted in Fig. 1.4. The states carry the weights

$$(T_3, Y) = \left(\frac{1}{2}, \frac{1}{3}\right), \left(-\frac{1}{2}, \frac{1}{3}\right), \left(0, -\frac{2}{3}\right) \ .$$

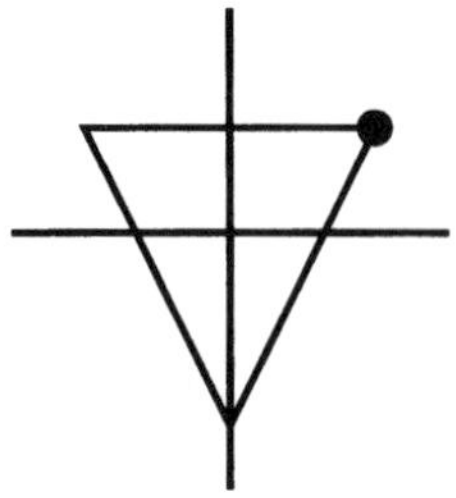

Fig. 1.4. The quark weight diagram

The tuple $\left(\frac{1}{2}, \frac{1}{3}\right)$ is the maximal weight.

(2) $[p, q] = [0, 1]$.

This is the representation of antiquarks with the "weight diagram" in Fig. 1.5. The states carry the weights

$$(T_3, Y) = \left(0, \frac{2}{3}\right), \left(\frac{1}{2}, -\frac{1}{3}\right), \left(-\frac{1}{2}, -\frac{1}{3}\right) \ .$$

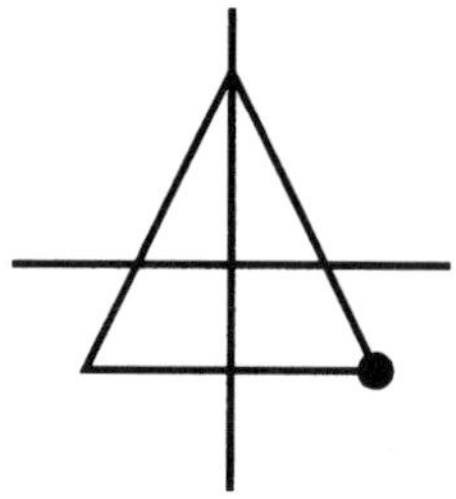

Fig. 1.5. The antiquark weight diagram

The state of maximal weight is $\left(\frac{1}{2}, -\frac{1}{3}\right)$.

In the case of SU(3), the trivial (scalar) representation is $[p, q] = [0, 0]$. The first nontrivial representations would be (1,0) and (0,1) of the same lowest dimension. By definition we thus have two fundamental representations. All others can be constructed from these two representations! To do so, we must construct the direct product of states

$$\begin{aligned} &(1, 0)^p (0, 1)^q \to \\ &|T_3(1)Y(1)\rangle\, |T_3(2)Y(2)\rangle \cdots \\ &|T_3(p)Y(p)\rangle\, \left|\overline{T}_3(1)\overline{Y}(1)\right\rangle \left|\overline{T}_3(2)\overline{Y}(2)\right\rangle \cdots \left|\overline{T}_3(q)\overline{Y}(q)\right\rangle \ . \end{aligned} \tag{3}$$

Owing to the additivity of the isospin component $\hat{T}_3$ and the hypercharge $\hat{Y}$, it holds that

$$\hat{T}_3 = \sum_i \hat{T}_3(i) \ , \tag{4a}$$

$$\hat{Y} = \sum_i \hat{Y}(i) \ . \tag{4b}$$

Thus many-quark states have T_3 and Y eigenvalues

$$(T_3, Y) = \left(\sum_{i=1}^{p} T_3(i) + \sum_{i=1}^{q} \overline{T}_3(i), \sum_{i=1}^{p} Y(i) + \sum_{i=1}^{q} \overline{Y}(i)\right) \ . \tag{5}$$

Exercise 1.1

In these, there is one state of maximal weight, namely the one that is composed of p quarks of maximal weight $\left(\frac{1}{2}, \frac{1}{3}\right)$ and q antiquarks of maximal weight $\left(\frac{1}{2}, -\frac{1}{3}\right)$, i.e.,

$$(T_3)_{\max} = \frac{p+q}{2} \ , \quad (Y)_{\max} = \frac{p-q}{3} \ . \tag{6}$$

It characterizes a representation contained in (5). If we subtract it, there is a remainder. Within this there is another state (or several states) of maximal weight. They are analogously given tuples $[p, q]$, i.e., a multiplet. We repeat the above steps until nothing is left, i.e., the direct product is completely reduced. In this way we can construct all SU(3) decompositions.

We consider $[p_1, q_1] \times [p_2, q_2] = [1, 0] \times [0, 1]$ and first add the two weight diagrams, i.e., at each point of the one diagram, we add the other diagram (see Fig. 1.6).

We are thus led to a weight diagram whose center is occupied three times! The maximal weight appearing there is

$$(T_3, Y)_{\max} = (1, 0) \ . \tag{7}$$

For $[p, q]$, it follows from (6) that

$$[p, q] = [1, 1] \ , \tag{8}$$

corresponding to an octet with dimension 8. On subtracting the octet which is twice degenerate at the center, only the singlet remains

$$(T_3, Y)_{\max} = (0, 0) \ , \tag{9a}$$

that is,

$$[p, q] = [0, 0] \ . \tag{9b}$$

We thus obtain the following result:

$$[1, 0] \times [0, 1] = [1, 1] + [0, 0] \ . \tag{10}$$

Note: Constructing $[1, 0] \times [0, 1]$ with this method, we obtain

$$[1, 0] \times [1, 0] = [2, 0] + [0, 1] \ . \tag{11}$$

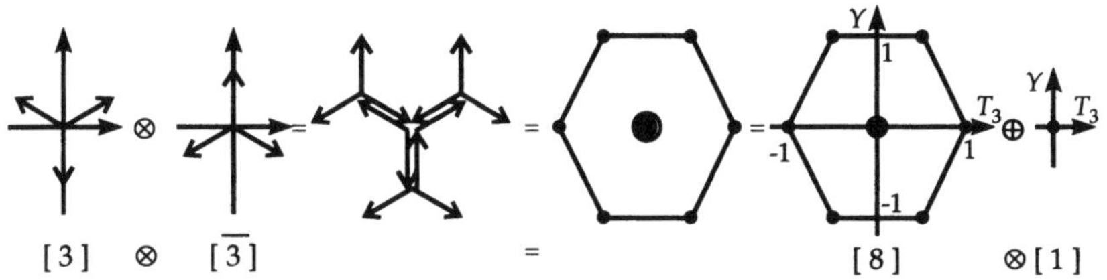

Fig. 1.6. Adding $[1, 0]$ and $[0, 1]$ weight diagrams

Exercise 1.1

On the right-hand side, [0, 1] appears. Mathematically, we can construct [0, 1] from [1, 0]. Thus one is inclined to call only [1, 0] the fundamental representation. Physically, however, the right-hand side describes two-quark states and not, as [0, 1] does, antiquark states.

SU(N). Its solutions are classified by $N-1$ numbers:

$$[h_1, \cdots, h_{N-1}] \ . \tag{12}$$

Analogously to SU(3), there is the scalar (trivial) representation

$$[0, \cdots, 0] \tag{13}$$

and $N-1$ fundamental representations

$$\begin{aligned} &[1, 0, \cdots, 0] \ , \\ &[0, 1, \cdots, 0] \ , \\ &\quad\vdots \\ &[0, \cdots, 0, 1] \ . \end{aligned} \tag{14}$$

From these, all other multiplets in (12) can be constructed by direct products.

Solution. (b) Schur's lemma indicates that any operator $\hat{H}$ commuting with all operators $\hat{U}(\boldsymbol{\alpha})$ ($\boldsymbol{\alpha} =$ group parameter), in particular with the generators $\hat{L}_i$,

$$\left[\hat{H}, \hat{U}(\boldsymbol{\alpha})\right] = 0 \Leftrightarrow \quad \left[\hat{H}, \hat{L}_i\right] = 0 \quad \Rightarrow \quad \left[\hat{H}, \hat{C}(\lambda)\right] = 0 \ ,$$

has the property that every state in a multiplet of the group is an eigenvector and that all states in a multiplet are degenerate. $\hat{C}(\lambda)$ is a Casimir operator of the group in the irreducible representation λ.

Since $\hat{C}(\lambda)$ commutes with $\hat{H}$, $\hat{C}(\lambda)$ and $\hat{H}$ can be simultaneously diagonalized, i.e., $\hat{C}(\lambda)$, too, is diagonal with respect to any state of the irreducible representation (multiplet) of the group. Calling $C(\lambda)$ the eigenvalues of $\hat{C}(\lambda)$, $\hat{C}(\lambda)$ has the following form with respect to the irreducible representation of the group:

$$\hat{C}(\lambda) = C(\lambda)\mathbb{1}(\lambda) \ , \tag{15}$$

where $\mathbb{1}_\lambda$ is the unit matrix with the multiplet's dimension. As the fundamental representation is by construction irreducible, (15) holds. In matrix representation, the Casimir operator has the following form:

$$\begin{pmatrix} C(\lambda_1)\mathbb{1}(\lambda_1) & 0 & 0 & \cdots \\ 0 & C(\lambda_2)\mathbb{1}(\lambda_2) & 0 & \cdots \\ 0 & 0 & C(\lambda_3)\mathbb{1}(\lambda_3) & \cdots \\ \vdots & \vdots & \vdots & \ddots \end{pmatrix} .$$

Each diagonal submatrix appearing in it is of the form $C(\lambda)\mathbb{1}(\lambda)$ and characterizes a representation (multiplet) of the same dimension as this multiplet.

EXERCISE

1.2 Casimir Operators of SU(3)

Problem. The regular (adjoint) representation of SU(3) is given by the eight generators $\hat{U}_i$, $i = 1, \ldots, 8$ with

$$(\hat{U}_i)_{jk} = -\mathrm{i} f_{ijk} \qquad (1)$$

($\hat{U}_i$ are 8×8 matrices). Show that for $\hat{C}_1$, one of the two Casimir operators of SU(3) in the regular representation, it holds that

$$\hat{C}_1 = \sum_{i=1}^{8} \hat{U}_i^2 = 3\mathbb{1}_{8\times 8} \ . \qquad (2)$$

Table 1.3. The eigenvalues of the Casimir operator $\hat{C}_1$ for the regular representation

m	$\sum_{ij} f_{ijm}^2$
1	$2f_{123}^2 + 2f_{147}^2 + 2f_{156}^2 = 3$
2	$2f_{123}^2 + 2f_{246}^2 + 2f_{257}^2 = 3$
3	$2f_{123}^2 + 2f_{345}^2 + 2f_{367}^2 = 3$
4	$2f_{246}^2 + 2f_{345}^2 + 2f_{147}^2 + 2f_{458} = 3$
5	$2f_{156}^2 + 2f_{257}^2 + 2f_{345}^2 + 2f_{458} = 3$
6	$2f_{156}^2 + 2f_{246}^2 + 2f_{367}^2 + 2f_{678} = 3$
7	$2f_{147}^2 + 2f_{257}^2 + 2f_{367}^2 + 2f_{678} = 3$
8	$2f_{458}^2 + 2f_{678}^2 = 3$

Solution. Each irreducible representation of SU(3) is uniquely determined by the eigenvalues of its Casimir operators. Each state in a multiplet has the same eignvalues with respect to $\hat{C}_1$. Thus this operator must be proportional to the unit matrix. This is checked here using an example. Using (1) it follows for $\hat{C}_1$ that

$$(\hat{C}_1)_{lm} = -\sum_{i,j} f_{ilj} f_{ijm} \ . \qquad (3)$$

From the table of f_{ijk}, one recognizes that $f_{ilj} \neq 0$ and $f_{ijm} \neq 0$, which implies that $l = m$:

$$(\hat{C}_1)_{lm} = +\sum_{i,j} f_{ijm}^2 \delta_{lm} = 3\delta_{lm} \ . \qquad (4)$$

This proves (2).

2. Review of Relativistic Field Theory

2.1 Spinor Quantum Electrodynamics

As a general introduction, this section reviews the basics of spinor quantum electrodynamics that are referred to in the following text.[1] Section 2.2 will give a similar review of scalar quantum electrodynamics. Readers who are familiar with this material should continue with Chap. 3.

2.1.1 The Free Dirac Equation and Its Solution

The equation of motion for the free spinor field Ψ is the free Dirac equation (we use natural units, $\hbar = c = 1$):

$$\mathrm{i}\frac{\partial}{\partial t}\Psi = (-\mathrm{i}\hat{\boldsymbol{\alpha}} \cdot \nabla - \beta m_0)\Psi \ . \tag{2.1}$$

Its solution is of the form

$$\Psi = w\mathrm{e}^{-\mathrm{i}p \cdot x} \ , \tag{2.2}$$

where

$$p \equiv p^\mu = (p^0, \boldsymbol{p}) \tag{2.3}$$

is the momentum four-vector and w a four-component Dirac spinor. The spinor w is usually decomposed into the two two-component spinors φ and χ:

$$w = \begin{pmatrix} \varphi \\ \chi \end{pmatrix} \ . \tag{2.4}$$

With this, the Dirac equation becomes a coupled system of equations for φ and χ:

$$p^0 \begin{pmatrix} \varphi \\ \chi \end{pmatrix} = \begin{pmatrix} m_0 \mathbb{1} & \hat{\boldsymbol{\sigma}} \cdot \boldsymbol{p} \\ \hat{\boldsymbol{\sigma}} \cdot \boldsymbol{p} & -m_0 \mathbb{1} \end{pmatrix} \begin{pmatrix} \varphi \\ \chi \end{pmatrix} \ , \tag{2.5}$$

[1] For a detailed discussion see W. Greiner: *Relativistic Quantum Mechanics – Wave Equations*, 3rd ed. (Springer, Berlin, Heidelberg 2000)

where $\mathbb{1} = \begin{pmatrix} 1 & 0 \\ 0 & 1 \end{pmatrix}$ is the 2×2 unit matrix and $\hat{\boldsymbol{\sigma}}$ the vector of the 2×2 Pauli matrices. Equation (2.5) possesses solutions of positive and negative energy.

Plane Waves of Positive Energy. In this case

$$p^0 \equiv E = +\sqrt{\boldsymbol{p}^2 + m_0} > 0 \ . \tag{2.6}$$

Exploiting the covariance of the Dirac equation, we first give the solutions for a particle at rest for which

$$p^0 = m_0 \ , \quad \boldsymbol{p} = \boldsymbol{0} \tag{2.7}$$

holds. The system of equations (2.5) then has the form

$$m_0 \begin{pmatrix} \varphi \\ \chi \end{pmatrix} = \begin{pmatrix} m_0 \mathbb{1} & 0 \\ 0 & -m_0 \mathbb{1} \end{pmatrix} \begin{pmatrix} \varphi \\ \chi \end{pmatrix} \tag{2.8}$$

and leads to

$$\chi = 0 \tag{2.9}$$

and

$$w(p^0 = m_0) = \begin{pmatrix} \varphi \\ 0 \end{pmatrix} \ . \tag{2.10}$$

The two linearly independent solutions for the two-spinor φ are

$$\begin{aligned} \varphi^1 &= \begin{pmatrix} 1 \\ 0 \end{pmatrix} \quad (\text{spin } \uparrow) \ , \\ \varphi^2 &= \begin{pmatrix} 0 \\ 1 \end{pmatrix} \quad (\text{spin } \downarrow) \ , \end{aligned} \tag{2.11}$$

which clearly shows that the Dirac equation describes particles of spin $\frac{1}{2}$. For nonvanishing spatial momentum

$$\boldsymbol{p} \neq \boldsymbol{0} \ , \quad p^0 = E = \sqrt{\boldsymbol{p}^2 + m_0^2} \ , \tag{2.12}$$

χ can be expressed by φ using (2.5), and one obtains the spinors of positive energy in the form

$$w^s = \begin{pmatrix} \varphi^s \\ \frac{\hat{\boldsymbol{\sigma}} \cdot \boldsymbol{p}}{E + m_0} \varphi^s \end{pmatrix} \ , \quad s = 1, 2 \ . \tag{2.13}$$

Plane Waves of Negative Energy. These are characterized by

$$p^0 \equiv -E = -\sqrt{\boldsymbol{p}^2 + m_0} \ , \tag{2.14}$$

where E always indicates the positive square root $\sqrt{\boldsymbol{p}^2 + m_0^2}$, i.e., in this notation $E > 0$. To construct the solutions we proceed as above. For a particle at rest with $p^\mu = (p^0 = -m_0, \boldsymbol{p} = \boldsymbol{0})$ a system analogous to (2.8) leads to

$$\varphi = 0 \tag{2.15}$$

and to the four-spinor

$$w(p^0 = -m_0) = \begin{pmatrix} 0 \\ \chi \end{pmatrix} \ , \tag{2.16}$$

respectively. For nonvanishing spatial momentum, i.e., for the four-momentum $p^\mu = (-E, \boldsymbol{p})$, φ can now be eliminated and one obtains

$$w = \begin{pmatrix} -\frac{\hat{\boldsymbol{\sigma}}\cdot\boldsymbol{p}}{E+m_0}\chi \\ \chi \end{pmatrix} \ . \tag{2.17}$$

We give the following important definition, which can be understood from hole theory. A particle (electron) is identified with a solution of positive energy and positive momentum $\boldsymbol{p}$, i.e.,

$$\Psi \sim \mathrm{e}^{-\mathrm{i}p\cdot x} \ , \quad p = (E, \boldsymbol{p}) \ , \tag{2.18}$$

and an antiparticle (positron) with the solution of negative energy and negative momentum, i.e.,

$$\Psi \sim \mathrm{e}^{\mathrm{i}p\cdot x} = \mathrm{e}^{-\mathrm{i}(-p\cdot x)} \equiv \mathrm{e}^{-\mathrm{i}p'\cdot x} \ , \quad p' = (-E, -\boldsymbol{p}) \ . \tag{2.19}$$

The particle and antiparticle solutions are therefore connected by the transformation

$$p^\mu \to -p^\mu \ . \tag{2.20}$$

Similarly, one expects for the spin that a missing particle of spin $\uparrow$ corresponds to an antiparticle of spin $\downarrow$. In other words, electron solutions of negative energy, negative momentum, and spin $\downarrow$ correspond to positron solutions of positive energy, positive momentum, and spin $\uparrow$. For this reason one puts

$$\chi^1 = \begin{pmatrix} 0 \\ 1 \end{pmatrix} \quad \text{and} \quad \chi^2 = \begin{pmatrix} 1 \\ 0 \end{pmatrix} \ . \tag{2.21}$$

The four-spinors w representing particles and antiparticles are now

$$w^s = \begin{pmatrix} \frac{\hat{\boldsymbol{\sigma}}\cdot\boldsymbol{p}}{E+m_0}\chi^s \\ \chi^s \end{pmatrix} \ , \quad s = 1, 2 \ . \tag{2.22}$$

With definitions (2.18)–(2.21) it is guaranteed that the quantities E and $\boldsymbol{p}$, as well as the basis spinors χ^1 and χ^2 that appear in the solutions (2.22), always denote energy, momentum, and spin $\uparrow$ or spin $\downarrow$ of the (physically observed) antiparticle.

2.1.2 Density and Current Density

The density ϱ and current density $\boldsymbol{j}$ of the Dirac field are, independent of the sign of the energy, given by

$$\varrho = \Psi^\dagger \Psi \ , \tag{2.23a}$$

$$\boldsymbol{j} = \Psi^\dagger \hat{\boldsymbol{\alpha}} \Psi \ , \tag{2.23b}$$

and satisfy the continuity equation

$$\frac{\partial}{\partial t}\varrho + \nabla \cdot \boldsymbol{j} = 0 \ . \tag{2.24}$$

For any spinor of the form

$$\Psi = w\, u(\boldsymbol{x}, t) \tag{2.25}$$

it follows that

$$\varrho = w^\dagger w |u(\boldsymbol{x}, t)|^2 \ , \tag{2.26a}$$

$$\boldsymbol{j} = w^\dagger \boldsymbol{\alpha} w |u(\boldsymbol{x}, t)|^2 \ . \tag{2.26b}$$

Obviously, $\varrho \geq 0$ always holds, and this is independent of (2.25) designating a particle or an antiparticle solution. Similarly, $\boldsymbol{j}$ does not change its sign when moving from particle to antiparticle solutions. This is most quickly verified for the z component:

$$(j_z)_{\mathrm{e}^-\uparrow} = \begin{pmatrix} \begin{pmatrix} 1 \\ 0 \end{pmatrix} \\ \frac{p_z}{E+m_0}\begin{pmatrix} 1 \\ 0 \end{pmatrix} \end{pmatrix}^\dagger \begin{pmatrix} 0 & \hat{\sigma}_z \\ \hat{\sigma}_z & 0 \end{pmatrix} \begin{pmatrix} \begin{pmatrix} 1 \\ 0 \end{pmatrix} \\ \frac{p_z}{E+m_0}\begin{pmatrix} 1 \\ 0 \end{pmatrix} \end{pmatrix} = \frac{2p_z}{E+m_0} \ ,$$

$$(j_z)_{\mathrm{e}^+\uparrow} = \begin{pmatrix} -\frac{p_z}{E+m_0}\begin{pmatrix} 0 \\ 1 \end{pmatrix} \\ \begin{pmatrix} 0 \\ 1 \end{pmatrix} \end{pmatrix}^\dagger \begin{pmatrix} 0 & \hat{\sigma}_z \\ \hat{\sigma}_z & 0 \end{pmatrix} \begin{pmatrix} -\frac{p_z}{E+m_0}\begin{pmatrix} 0 \\ 1 \end{pmatrix} \\ \begin{pmatrix} 0 \\ 1 \end{pmatrix} \end{pmatrix} = \frac{2p_z}{E+m_0} \ . \tag{2.27}$$

A sign change of charge and current density is, however, desired. To put it in by hand one inserts an extra minus sign whenever an electron (fermion) of negative energy appears in a final state. This rule is very naturally included in the definition of the Feynman propagator.

2.1.3 Covariant Notation

It is customary to introduce γ matrices, which replace $\hat{\boldsymbol{\alpha}}$ and $\hat{\beta}$ (we leave out the operator hats for γ matrices in the following):

$$\begin{aligned} \gamma^0 &= \hat{\beta}\ , \quad (\gamma^0)^2 = \mathbb{1}\ , \\ \gamma^i &= \hat{\beta}\hat{\alpha}_i\ , \quad (\gamma^i)^2 = -\mathbb{1}\ , \quad i = 1,2,3\ , \\ \gamma^\mu\gamma^\nu + \gamma^\nu\gamma^\mu &= g^{\mu\nu}\mathbb{1}\ . \end{aligned} \tag{2.28}$$

Here $\mathbb{1}$ designates the unit matrix. The free Dirac equation (2.1) takes the form

$$\left(\mathrm{i}\gamma^\mu \frac{\partial}{\partial x^\mu} - m_0\right)\Psi = 0\ ,$$

or

$$(\mathrm{i}\not{\partial} - m_0)\Psi = 0\ , \tag{2.29}$$

using the (Feynman) dagger notation ($\not{a} \equiv \gamma^\mu a_\mu$). Density ϱ and current density $\boldsymbol{j}$ can be combined to form a four-current density:

$$j^\mu = \bar{\Psi}\gamma^\mu\Psi\ , \tag{2.30}$$

and the continuity equation (2.24) can be written as a four-divergence:

$$\frac{\partial}{\partial x^\mu} j^\mu = 0\ . \tag{2.31}$$

Here

$$\bar{\Psi} = \Psi^\dagger\gamma^0 \tag{2.32}$$

designates the adjoint spinor.

2.1.4 Normalization of Dirac Spinors

It is useful to consider again the normalization of spinor wave functions. Let us first consider plane waves of positive energy,

$$\Psi^{1,2} = N' w^{1,2}\,\mathrm{e}^{-\mathrm{i}p\cdot x}\ . \tag{2.33}$$

We normalize such a wave in a box of volume V in such a way that

$$\int \mathrm{d}^3x\,\varrho = 2E \tag{2.34}$$

holds. This normalization differs from the usual normalization to unity of quantum mechanics but is often used in field theory. Using the explicit form of the spinors $\omega^{1,2}$ of positive energy yields the normalization factor

$$N' = \sqrt{\frac{E + m_0}{V}}\ . \tag{2.35}$$

Usually one absorbs the factor $\sqrt{E-m_0}$ in the definition of the spinor w^s and designates the spinors for positive energy by $u(p,s)$:

$$u(p,s) = \sqrt{E+m_0}\begin{pmatrix} \varphi^1 \\ \frac{\hat{\boldsymbol{\sigma}}\cdot\boldsymbol{p}}{E+m_0}\varphi^2 \end{pmatrix} , \qquad s = 1,2 ,$$
$$\varphi^1 = \begin{pmatrix} 1 \\ 0 \end{pmatrix} , \quad \varphi^2 = \begin{pmatrix} 0 \\ 1 \end{pmatrix} . \tag{2.36}$$

One similarly introduces spinors $v(p,s)$ for negative energy:

$$v(p,s) = \sqrt{E+m_0}\begin{pmatrix} \frac{\hat{\boldsymbol{\sigma}}\cdot\boldsymbol{p}}{E+m_0}\chi^1 \\ \chi^2 \end{pmatrix} , \qquad s = 1,2 ,$$
$$\chi^1 = \begin{pmatrix} 0 \\ 1 \end{pmatrix} , \quad \chi^2 = \begin{pmatrix} 1 \\ 0 \end{pmatrix} . \tag{2.37}$$

We again emphasize that the spinors $v(p,s)$ are constructed such that E, $\boldsymbol{p}$, and $s = 1,2$ in (2.37) correspond to energy, momentum, and spin projection $\uparrow$ or $\downarrow$ of the positron. It is easy to check that

$$u^+u = v^+v = 2E \tag{2.38}$$

and

$$(\not{p} - m_0)u = 0 , \tag{2.39a}$$
$$(\not{p} + m_0)v = 0 \tag{2.39b}$$

hold. Equations (2.39) are the momentum-space Dirac equation for the (free) solutions of positive and negative energy, respectively. Correspondingly one finds the Dirac equations for the adjoint spinors $\bar{u}$ and $\bar{v}$:

$$\bar{u}(\not{p} - m_0) = 0 , \tag{2.40a}$$
$$\bar{v}(\not{p} + m_0) = 0 . \tag{2.40b}$$

The normalization conditions for the spinors can be summarized as

$$\bar{u}(p,s)u(p',s') = 2m_0\delta_{ss'} , \tag{2.41a}$$
$$\bar{v}(p,s)v(p',s') = -2m_0\delta_{ss'} \tag{2.41b}$$

or

$$\overline{w}^r(p)w^r(p') = 2m_0\varepsilon_r\delta_{rr'} , \qquad \varepsilon_r = \begin{cases} 1 & \text{for} \quad r = 1,2 \\ -1 & \text{for} \quad r = 3,4 \end{cases} . \tag{2.42}$$

We want to emphasize that another normalization of $u(p,s)$ and $v(p,s)$ can also quite often be found:

$$u' = \frac{1}{\sqrt{2m_0}}u , \quad v' = \frac{1}{\sqrt{2m_0}}v . \tag{2.43}$$

The advantage of the normalization used here is its covariance. This is understandable from (2.38): the densities u^+u and v^+v are proportional to the energy E and transform as 0 components of a four-vector. When we compute cross sections, this and the corresponding transformation properties of phase space and flow factors make Lorentz invariance obvious.

With these conventions we can write electron and positron wave functions as

$$\Psi(\mathrm{e}^-) = Nu(p,s)\mathrm{e}^{-\mathrm{i}p\cdot x} \ , \tag{2.44a}$$

$$\Psi(\mathrm{e}^+) = Nv(p,s)\mathrm{e}^{\mathrm{i}p\cdot x} \ , \qquad N = \frac{1}{\sqrt{V}} \ . \tag{2.44b}$$

These explicit expressions enable us to write down directly the transition currents, as we shall see below.

2.1.5 Interaction with a Four-Potential A^μ

The interaction of the field with an electromagnetic potential A^μ is introduced by the so-called "minimal" coupling to preserve gauge invariance. For an electron (of charge $-e$), the minimal substitution is

$$\frac{\partial}{\partial x_\mu} \equiv \partial_\mu \to \partial_\mu - \mathrm{i}eA_\mu \ . \tag{2.45}$$

Thus the free Dirac equation (2.1) is changed into (written in noncovariant form)

$$\mathrm{i}\frac{\partial}{\partial t}\psi = (-\mathrm{i}\hat{\boldsymbol{\alpha}}\cdot\boldsymbol{\nabla} + \hat{\beta}m_0 + \hat{V})\psi \ , \tag{2.46}$$

where the interaction $\hat{V}$ is given by

$$\hat{V} = -eA^0\mathbb{1} + e\,\hat{\boldsymbol{\alpha}}\cdot\boldsymbol{A} \ . \tag{2.47}$$

In covariant form, the Dirac equation with interaction (substitution of (2.49) into (2.29)) reads

$$\left[\mathrm{i}\gamma^\mu(\partial_\mu - \mathrm{i}eA_\mu) - m_0\right]\Psi = 0 \tag{2.48}$$

and the interaction is written as

$$\gamma^0\hat{V} = -e\gamma^\mu A_\mu \ . \tag{2.49}$$

2.1.6 Transition Amplitudes

The transition amplitude (S-matrix element) of an initial electron state $\Psi_\mathrm{i}(\mathrm{e}^-; p, s)$ with four-momentum p and spin projection s into a final electron state $\Psi_\mathrm{f}(\mathrm{e}^-; p', s')$ characterized by momentum p' and spin s' is, in first-order perturbation theory,

$$\begin{aligned} S^{(1)}_{\mathrm{fi}} &= -\mathrm{i} \int \mathrm{d}^4x\, \Psi^\dagger_\mathrm{f}(\mathrm{e}^-; p', s') \hat{V}\, \Psi_\mathrm{i}(\mathrm{e}^-; p, s) \\ &= \mathrm{i} \int \mathrm{d}^4x\, \Psi^\dagger_\mathrm{f}(-e\gamma^\mu A_\mu)\Psi_\mathrm{i} \\ &= \mathrm{i} \int \mathrm{d}^4x\, J^\mu(\mathrm{e}^-) A_\mu \,, \end{aligned} \tag{2.50}$$

where

$$J^\mu(\mathrm{e}^-) = (-e)\bar{\Psi}_\mathrm{f}\, \gamma^\mu\, \Psi_\mathrm{i} \tag{2.51}$$

are the electron (fermion) transition current densities. Using the plane waves (2.44a) this becomes explicitly

$$J^\mu(\mathrm{e}^-) = \frac{(-e)}{V} \bar{u}_\mathrm{f}(p', s')\, \gamma^\mu\, u_\mathrm{i}(p, s)\, \mathrm{e}^{\mathrm{i}(p'-p)\cdot x} \,, \quad N_\mathrm{i} = N_\mathrm{f} = \frac{1}{\sqrt{V}} \tag{2.52}$$

and now allows the calculation of scattering process in lowest order according to (2.50).

2.1.7 Discrete Symmetries

We restrict ourselves here to a summarizing and tabulating the properties of the discrete symmetry transformations parity $\hat{P}$, charge conjugation $\hat{C}$, and time reversal $\hat{T}$.

Dirac fields can be combined into the following bilinear forms (currents), distinguished by their tensor character:

$$S(x) = \bar{\Psi}(x)\Psi(x) \quad \text{scalar} \,, \tag{2.53a}$$

$$V^\mu(x) = \bar{\Psi}(x)\gamma^\mu\Psi(x) \quad \text{vector} \,, \tag{2.53b}$$

$$T^{\mu\nu}(x) = \bar{\Psi}(x)\sigma^{\mu\nu}\Psi(x) \quad \text{tensor} \,, \tag{2.53c}$$

$$P(x) = \mathrm{i}\bar{\Psi}(x)\gamma_5\Psi(x) \quad \text{pseudoscalar} \,, \tag{2.53d}$$

$$A^\mu(x) = \bar{\Psi}(x)\gamma_5\gamma^\mu\Psi(x) \quad \text{pseudovector} \,. \tag{2.53e}$$

The behavior of these currents under the transformations $\hat{P}$, $\hat{C}$, $\hat{T}$, as well as under the mixed symmetry $\hat{O} = \hat{P}\hat{C}\hat{T}$, is given in Table 2.1 where $\tilde{x} = (t, -\boldsymbol{x})$.

We also give the corresponding transformations for the electromagnetic four-potential A_μ:

$$\begin{aligned} \hat{P} A^\mu(x) \hat{P}^+ &= A^\mu(x) \,, \quad \hat{C} A^\mu(x) \hat{C}^+ = -A^\mu(x) \,, \\ \hat{T} A^\mu(x) \hat{T}^+ &= A^\mu(-x) \,, \quad \hat{O} A^\mu(x) \hat{O}^+ = -A^\mu(-x) \,. \end{aligned} \tag{2.54}$$

Table 2.1. The behavior of the currents (2.53) under the transformations $\hat{P}$, $\hat{C}$, $\hat{T}$, and $\hat{O} = \hat{P}\hat{C}\hat{T}$

	$S(x)$	$V^\mu(x)$	$T^{\mu\nu}(x)$	$P(x)$	$A^\mu(x)$
$\hat{P}$	$S(\tilde{x})$	$V^\mu(\tilde{x})$	$T^{\mu\nu}(\tilde{x})$	$-P(\tilde{x})$	$-A^\mu(\tilde{x})$
$\hat{C}$	$S(x)$	$-V^\mu(x)$	$-T^{\mu\nu}(x)$	$P(x)$	$A^\mu(x)$
$\hat{T}$	$S(-\tilde{x})$	$V^\mu(-\tilde{x})$	$-T^{\mu\nu}(-\tilde{x})$	$-P(-\tilde{x})$	$A^\mu(-\tilde{x})$
$\hat{O}$	$S(-x)$	$-V^\mu(-x)$	$T^{\mu\nu}(-x)$	$P(-x)$	$-A^\mu(-x)$

2.2 Scalar Quantum Electrodynamics

2.2.1 The Free Klein–Gordon Equation and its Solutions

It is known that pions as spin-0 particles satisfy the Klein–Gordon equation. Here we compile the main results of pion quantum electrodynamics. Starting from the four-momentum vector and relativistic energy conservation

$$p^\mu = (E, \boldsymbol{p}) \ , \tag{2.55}$$

$$p_\mu p^\mu = E^2 - \boldsymbol{p}^2 = m_0^2 \tag{2.56}$$

and the correspondence between momentum and momentum operator

$$p^\mu \to \hat{p}^\mu = \mathrm{i}\partial^\mu \ , \tag{2.57}$$

the free Klein–Gordon equation follows:

$$\left(\hat{p}_\mu \hat{p}^\mu - m_0^2\right) \phi(\boldsymbol{x}, t) = 0 \ , \tag{2.58a}$$

which can be written as

$$\left(\Box + m_0^2\right) \phi(\boldsymbol{x}, t) = 0 \tag{2.58b}$$

using the *d'Alembertian operator (quabla operator)*, which is defined by

$$\Box \equiv \partial_\mu \partial^\mu = \frac{\partial^2}{\partial t^2} - \boldsymbol{\nabla}^2 \ . \tag{2.59}$$

Plane waves of the form

$$\phi(\boldsymbol{x}, t) = N\mathrm{e}^{-\mathrm{i}p\cdot x} = N\mathrm{e}^{-\mathrm{i}(Et - \boldsymbol{p}\cdot\boldsymbol{x})} \tag{2.60}$$

are solutions of (2.58) if condition (2.56) is satisfied. Therefore we also have solutions of positive and negative energy:

$$E = \pm\sqrt{p^2 + m_0^2} \ . \tag{2.61}$$

The question of their interpretation is thus raised. To answer it, we shall derive expressions for the probability density ϱ and the probability current density $\boldsymbol{j}$ by multiplying (2.58) by ϕ^* and its complex conjugate-equation by ϕ and subtracting each from the other. This leads to the continuity equation ($\partial_t \equiv \frac{\partial}{\partial t}$)

$$\partial_t \varrho + \nabla \cdot \boldsymbol{j} = 0 \ , \tag{2.62}$$

where

$$\varrho = \mathrm{i}\left[\phi^*(\partial_t \phi) - (\partial_t \phi^*)\phi\right] \tag{2.63a}$$

and

$$\boldsymbol{j} = -\mathrm{i}\left[\phi^*(\nabla\phi) - (\nabla\phi^*)\phi\right] \ . \tag{2.63b}$$

In four-dimensional notation, this is concisely written as

$$\partial_\mu j^\mu = 0 \ , \tag{2.64}$$

with the four-current density

$$j^\mu = (\varrho, \boldsymbol{j}) = \mathrm{i}\left[\phi^* \left(\partial^\mu \phi\right) - \left(\partial^\mu \phi^*\right) \phi\right] \ . \tag{2.65}$$

The three-current density $\boldsymbol{j}$ in (2.63b) is formally identical with that known from the Schrödinger equation. However, the probability density ϱ contains, in contrast to the Schrödinger density, additional time derivatives. This has the consequence that ϱ is not positive definite, which can be immediately checked using plane waves (2.60), taking into account (2.61). In this way it follows for (2.63a) that

$$\varrho = 2\,|N|^2\, E \ . \tag{2.66}$$

The probability current density (2.63b) is obtained as

$$\boldsymbol{j} = 2\,|N|^2\, \boldsymbol{p} \ . \tag{2.67}$$

Since E can be positive or negative owing to (2.61), the above statement about ϱ is obvious. To interpret ϱ nonetheless as a probability, one must make use of the particle–antiparticle interpretation. By the Feynman–Stückelberg prescription, it holds that:

A solution of negative energy for a particle propagating backward in time corresponds to a solution of positive energy for an antiparticle propagating forward in time.

The Klein–Gordon equation describes both neutral and charged mesons. In the case of charged scalar particles we not only have to analyze their spatial propagation but also the assignment of charges as discussed next for the charged pions π^+ and π^-. To this end we consider the scattering of a particle (e.g., a π^+) off a potential in second-order perturbation theory. The space–time diagram of such a process is shown in Fig. 2.1. An incoming π^+ scatters off the potential at position x_1 and time t_1 and propagates to position x_2, where it scatters again at a later time $t_2 \gg t_1$ and then moves on freely. According to Stückelberg and Feynman there must be the possibility that particles are scattered backward in time (Fig. 2.2). Thus one must allow in relativistic quantum field theory for the processes shown in the these figures.

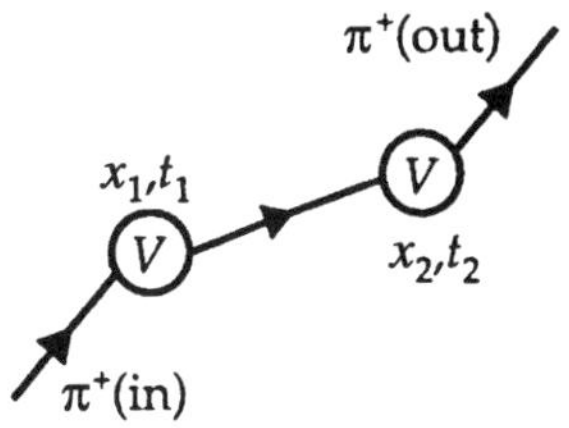

Fig. 2.1. Double scattering of a particle (π^+) off a potential

We interpret this second process according to Feynman in such a way that particle solutions of positive energy propagate exclusively forward and particle solutions of negative energy exclusively backward in time. The π^+ moving backward in time between t_2 and t_1 must have negative energy. It is equivalent to a π^- (antiparticle) moving forward in time. This is obviously implied by charge conservation: only particle–antiparticle pairs can be created or annihilated. Figure 2.3 illustrates this reinterpretation of Fig. 2.2. At t_2, a $\pi^+\pi^-$ pair is created whose π^- – which is identical to the originally incoming π^+ – is annihilated at t_1 and whose π^+ propagates on.

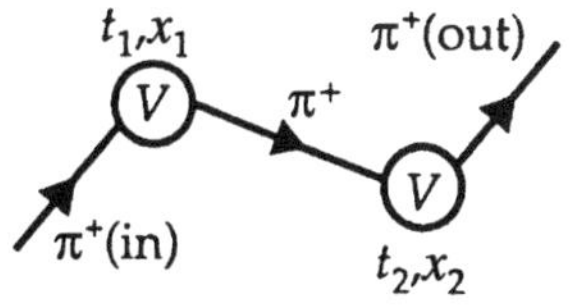

Fig. 2.2. Scattering backward in time: The particle here has, according to Stückelberg and Feynman, negative energy

There is also another way to demonstrate the concept of a charged Klein–Gordon field. The charged currents for π^+ and π^- at positive energy are obtained by multiplying the charge density (2.11), calculated for waves by positive energy, by the positive and negative unit charge ($e > 0$), respectively, that is

$$j^\mu(\pi^\pm) = (\pm e) \times \text{probability current density for a } \pi^\pm \text{ at positive energy.} \tag{2.68}$$

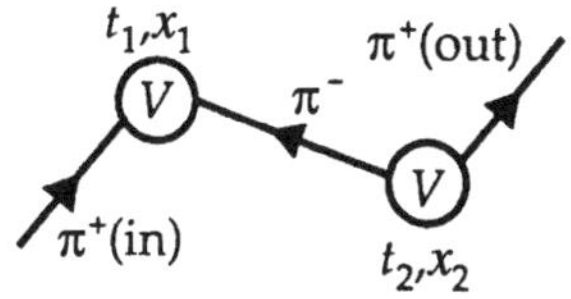

Fig. 2.3. Feynman's reinterpretation of the scattering process shown in Fig. 2.2

Inserting the plane wave (2.60) into (2.63), we have

$$j^\mu(\pi^+) = (+e)2\,|N|^2\left(\sqrt{p^2+m_0^2},\, \boldsymbol{p}\right) \tag{2.69}$$

and

$$j^\mu(\pi^-) = (-e)2\,|N|^2\left(-\sqrt{p^2+m_0^2},\, \boldsymbol{p}\right) . \tag{2.70}$$

Comparing (2.70) with (2.69), we see that it is obvious that (2.70) can also be written as

$$j^\mu(\pi^-) = (+e)2\,|N|^2\left(-\sqrt{p^2+m^2},\, -\boldsymbol{p}\right) , \tag{2.71}$$

which equals the current density of a π^+ with negative energy and negative momentum. In other words, a π^- thus corresponds to a π^+ with inverse four-momentum.

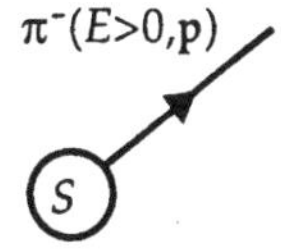

Fig. 2.4. Emission of a π^- with $(E > 0, \boldsymbol{p})$ by the system S

This correspondence can be expressed more precisely: if a system S emits a π^- of positive energy $E > 0$ and momentum $\boldsymbol{p}$ (see Fig. 2.4), the energy of S

is reduced by E, its momentum by $\boldsymbol{p}$, and its charge by $(-e)$. But all this is equivalent to the absorption of a π^+ with negative four-momentum $(-E, -\boldsymbol{p})$, as demonstrated in Fig. 2.5. We summarize this with the following statement:

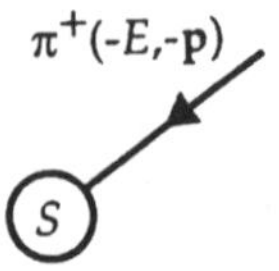

Fig. 2.5. Reinterpretation of the emission process in Fig. 2.4 as an absorption process

The emission (absorption) of an antiparticle with four-momentum p^μ is physically equivalent to the absorption (emission) of a particle with four-momentum $-p^\mu$.

2.2.2 Interaction of a π^+ with a Potential A^μ

Just as in the case of the Dirac equation, the electromagnetic potential A^μ is coupled in by the minimal-coupling prescription (π^+ has the charge $+e$)

$$\partial^\mu \to \partial^\mu + \mathrm{i}eA^\mu \tag{2.72}$$

to preserve gauge invariance. If this is inserted into (2.58b), one obtains the Klein–Gordon equation with electromagnetic interaction

$$(\Box + m_0^2)\phi = -\mathrm{i}e(\partial_\mu A^\mu + A^\mu \partial_\mu)\phi + e^2 A^2 \phi \equiv -\hat{V}\phi \ . \tag{2.73}$$

In contrast to the Dirac theory, a coupling term quadratic in A^μ appears. However, we shall neglect it whenever scattering processes are considered in lowest order. In this approximation the coupling potential reduces to

$$\hat{V}(x) = \mathrm{i}e(\partial_\mu A^\mu(x) + A^\mu(x)\partial_\mu) \ . \tag{2.74}$$

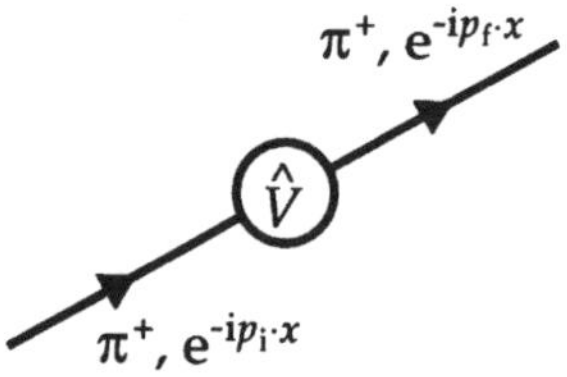

Fig. 2.6. π^+ scattering off a potential V to lowest order. The potential is denoted by the vertex x

To calculate scattering processes, we also need the scattering amplitude. This is for scattering in first order of a potential $\hat{V}$, as before, given by

$$S^{(1)}_{\mathrm{fi}} = -\mathrm{i}\int \mathrm{d}^4x\, \phi^*_{\mathrm{f}} \hat{V} \phi_{\mathrm{i}} \ . \tag{2.75}$$

It is displayed by the graph in Fig. 2.6.

We shall now calculate the transition amplitude (2.75). Incoming and outgoing π^+ states are described by plane waves

$$\phi_{\mathrm{i}} = N_{\mathrm{i}}\, \mathrm{e}^{-\mathrm{i}p_{\mathrm{i}}\cdot x} \ , \tag{2.76a}$$

$$\phi_{\mathrm{f}} = N_{\mathrm{f}}\, \mathrm{e}^{-\mathrm{i}p_{\mathrm{f}}\cdot x} \ . \tag{2.76b}$$

Together with (2.74), (2.75) becomes

$$\begin{aligned} S^{(1)}_{\mathrm{fi}} &= -\mathrm{i}N_{\mathrm{i}}N_{\mathrm{f}} \int \mathrm{d}^4x\, \mathrm{e}^{\mathrm{i}p_{\mathrm{f}}\cdot x} (\mathrm{i}e)(\partial_\mu A^\mu + A^\mu \partial_\mu) \mathrm{e}^{-\mathrm{i}p_{\mathrm{i}}\cdot x} \\ &= -\mathrm{i}N_{\mathrm{i}}N_{\mathrm{f}}(p_{\mathrm{i}} + p_{\mathrm{f}})_\mu \int \mathrm{d}^4x\, \mathrm{e}^{-\mathrm{i}q\cdot x} A^\mu(x) \\ &= -\mathrm{i}eN_{\mathrm{i}}N_{\mathrm{f}}(p_{\mathrm{i}} + p_{\mathrm{f}})_\mu A^\mu(q) \ , \end{aligned} \tag{2.77}$$

where the four-momentum transfer

$$q = p_\mathrm{i} - p_\mathrm{f} \tag{2.78}$$

has been introduced. Also, in calculating (2.77), a partial integration of the type

$$\int_{-\infty}^{+\infty} \mathrm{d}x^\mu \, f \frac{\mathrm{d}g}{\mathrm{d}x^\mu} = fg\Big|_{-\infty}^{\infty} - \int_{-\infty}^{\infty} \mathrm{d}x^\mu \, \frac{\mathrm{d}f}{\mathrm{d}x^\mu} g \tag{2.79}$$

has been performed twice. Here the assumption has been made that contributions of the form gf vanish at infinity, i.e.,

$$fg\Big|_{-\infty}^{\infty} = 0 \; . \tag{2.80}$$

This requires that either the potentials A_μ or the wave amplitudes decay fast enough at infinity. Exact plane waves do not have this property. But, strictly speaking, any particle is always represented by a wave packet. Even if it can be, for large distances from the scattering center and for large times before or after the scattering, arbitrarily delocalized; it will, however, decay asymptotically. Taking into account that we use plane waves (2.76) just to simplify calculations, the surface contributions can be neglected and the S-matrix element (2.75) can be written in a more convenient form:

$$\begin{aligned} S_{\mathrm{fi}}^{(1)} &= e \int \mathrm{d}^4x \, \phi_\mathrm{f}^*(\partial_\mu A^\mu + A^\mu \partial_\mu)\phi_\mathrm{i} \\ &= e \int \mathrm{d}^4x \left[-(\partial_\mu \phi_\mathrm{f}^*)\phi_\mathrm{i} + \phi_\mathrm{f}^*(\partial_\mu \phi_\mathrm{i}) \right] A^\mu \\ &= -\mathrm{i} \int \mathrm{d}^4x \, j_\mu(\pi^+) A^\mu \; . \end{aligned} \tag{2.81}$$

Here

$$j_\mu(\pi^+) = \mathrm{i}e \left[\phi_\mathrm{f}^* \left(\partial_\mu \phi_\mathrm{i}\right) - \left(\partial_\mu \phi_\mathrm{f}^*\right) \phi_\mathrm{i} \right] \tag{2.82}$$

is the transition current density for the π^+ meson. For plane waves (2.76) this is particularly simple:

$$j_\mu(\pi^+) = e N_\mathrm{i} N_\mathrm{f} (p_\mathrm{i} + p_\mathrm{f})_\mu \, \mathrm{e}^{\mathrm{i}(p_\mathrm{f} - p_\mathrm{i}) \cdot x} \; , \tag{2.83}$$

which also appears in (2.77). In the following problem (2.1), the steps discussed here are illustrated once more.

EXERCISE

2.1 The Matrix Element for a Pion Scattered by a Potential

Problem. Consider the matrix element

$$M_{\mathrm{fi}} = \int \mathrm{d}^3x \int \mathrm{d}t\, \mathrm{e}^{\mathrm{i}p_\mathrm{f}\cdot x}\left(\partial_\mu A^\mu(x) + A^\mu(x)\partial_\mu\right)\mathrm{e}^{-\mathrm{i}p_\mathrm{i}\cdot x} \; . \tag{1}$$

Assume that the four-potential fulfills the conditions

$$A^0(\boldsymbol{x}, t) \to 0 \;\text{ for }\; t \to \pm\infty \; , \tag{2a}$$

$$|\boldsymbol{A}(\boldsymbol{x}, t)| \to 0 \;\text{ for }\; |\boldsymbol{x}| \to \infty \; , \tag{2b}$$

and show that

$$\text{(a)} \int \mathrm{d}t\, \mathrm{e}^{\mathrm{i}p_\mathrm{f}\cdot x}\partial_t\left(A^0\mathrm{e}^{-\mathrm{i}p_\mathrm{i}\cdot x}\right) = \left(-\mathrm{i}(p_\mathrm{f})_0\right)\int \mathrm{d}t\, \mathrm{e}^{\mathrm{i}p_\mathrm{f}\cdot x}A^0\mathrm{e}^{-\mathrm{i}p_\mathrm{i}\cdot x} \; , \tag{3}$$

$$\text{(b)} \int \mathrm{d}^3x\, \mathrm{e}^{\mathrm{i}p_\mathrm{f}\cdot x}\boldsymbol{\nabla}\cdot\left(\boldsymbol{A}\mathrm{e}^{-\mathrm{i}p_\mathrm{i}\cdot x}\right) = \mathrm{i}\boldsymbol{p}_\mathrm{f}\cdot\int \mathrm{d}^3x\, \mathrm{e}^{\mathrm{i}p_\mathrm{f}\cdot x}\boldsymbol{A}\mathrm{e}^{-\mathrm{i}p_\mathrm{i}\cdot x} \; , \tag{4}$$

and therefore also

$$\text{(c)} \int \mathrm{d}^4x\, \mathrm{e}^{\mathrm{i}p_\mathrm{f}\cdot x}\left(\partial_\mu A^\mu + A^\mu\partial_\mu\right)\mathrm{e}^{-\mathrm{i}p_\mathrm{i}\cdot x} = -\mathrm{i}(p_\mathrm{f}+p_\mathrm{i})_\mu\cdot\int \mathrm{d}^4x\, \mathrm{e}^{\mathrm{i}p_\mathrm{f}\cdot x}A^\mu\mathrm{e}^{-\mathrm{i}p_\mathrm{i}\cdot x} \tag{5}$$

hold.

Solution. (a) A partial integration of the time integral yields

$$\begin{aligned}\int_{-\infty}^{\infty} \mathrm{d}t\, \mathrm{e}^{\mathrm{i}p_\mathrm{f}\cdot x}\partial_t A^0\mathrm{e}^{-\mathrm{i}p_\mathrm{i}\cdot x} &= \left[A^0\mathrm{e}^{\mathrm{i}(p_\mathrm{f}-p_\mathrm{i})\cdot x}\right]_{t=-\infty}^{t=+\infty} - \int_{-\infty}^{+\infty}\mathrm{d}t\, A^0\mathrm{e}^{-\mathrm{i}p_\mathrm{i}\cdot x}\partial_t\,\mathrm{e}^{\mathrm{i}p_\mathrm{f}\cdot x}\\ &= -\mathrm{i}(p_\mathrm{f})_0\int_{-\infty}^{+\infty}\mathrm{d}t\, \mathrm{e}^{\mathrm{i}p_\mathrm{f}\cdot x}A^0\mathrm{e}^{-\mathrm{i}p_\mathrm{i}\cdot x} \; .\end{aligned} \tag{6}$$

The surface term vanishes because of the boundary condition (2a).

(b) Analogously, a partial integration over the spacial coordinates leads to (Gauss's theorem)

$$\begin{aligned}&\int \mathrm{d}^3x\, \mathrm{e}^{\mathrm{i}p_\mathrm{f}\cdot x}\boldsymbol{\nabla}\cdot\left(\boldsymbol{A}\mathrm{e}^{-\mathrm{i}p_\mathrm{i}\cdot x}\right)\\ &\quad= \int_{\text{surface, } |\boldsymbol{x}|\to\infty} \mathrm{d}\boldsymbol{F}\cdot\boldsymbol{A}\,\mathrm{e}^{\mathrm{i}(p_\mathrm{f}-p_\mathrm{i})\cdot x} - \int \mathrm{d}^3x\, \mathrm{e}^{-\mathrm{i}p_\mathrm{i}\cdot x}\boldsymbol{A}\cdot\boldsymbol{\nabla}\,\mathrm{e}^{\mathrm{i}p_\mathrm{f}\cdot x}\\ &\quad= \mathrm{i}\boldsymbol{p}_\mathrm{f}\cdot\int \mathrm{d}^3x\, \mathrm{e}^{\mathrm{i}p_\mathrm{f}\cdot x}\boldsymbol{A}\,\mathrm{e}^{-\mathrm{i}p_\mathrm{i}\cdot x} \; .\end{aligned} \tag{7}$$

Exercise 2.1

Owing to boundary condition (2b), the surface integral again vanishes.

(c) Summarizing (a) and (b) we obtain

$$\int \mathrm{d}^4x\, \mathrm{e}^{\mathrm{i}p_\mathrm{f}\cdot x}\partial_\mu \left(A^\mu \mathrm{e}^{-\mathrm{i}p_\mathrm{i}\cdot x}\right) = -\mathrm{i}(p_\mathrm{f})_\mu \int \mathrm{d}^4x\, \mathrm{e}^{\mathrm{i}p_\mathrm{f}\cdot x} A^\mu \mathrm{e}^{-\mathrm{i}p_\mathrm{i}\cdot x} \ . \tag{8}$$

On the other hand we have

$$\partial_\mu \mathrm{e}^{-\mathrm{i}p_\mathrm{i}\cdot x} = -\mathrm{i}(p_\mathrm{i})_\mu \mathrm{e}^{-\mathrm{i}p_\mathrm{i}\cdot x} \ , \tag{9}$$

i.e.,

$$\begin{aligned} M_{\mathrm{fi}} &= \int \mathrm{d}^4x\, \mathrm{e}^{\mathrm{i}p_\mathrm{f}\cdot x} \left(\partial_\mu A^\mu + A^\mu \partial_\mu\right) \mathrm{e}^{-\mathrm{i}p_\mathrm{i}\cdot x} \\ &= -\mathrm{i}(p_\mathrm{f}+p_\mathrm{i})_\mu \int \mathrm{d}^4x\, \mathrm{e}^{\mathrm{i}p_\mathrm{f}\cdot x} A^\mu\, \mathrm{e}^{-\mathrm{i}p_\mathrm{i}\cdot x} \ . \end{aligned} \tag{10}$$

2.2.3 π^+K^+ Scattering

As a further example we now consider π^+K^+scattering and again evaluate the transition matrix element $S^{(1)}_{\pi^+K^+}$. Being a spin-0 particle, the K^+ meson obeys the same wave equation as the pion. Since π^+ and K^+ are distinguishable particles, they need not be symmetrized and exchange amplitudes do not have to be taken into accout. The scattering reaction can be described in the following manner. The electric charge of the K^+ creates a vector potential by which the π^+ is scattered. First we have to determine this vector potential, because it enters the scattering amplitude (2.75).

A^μ obeys Maxwell's equations

$$\Box A^\mu - \partial^\mu(\partial_\nu A^\nu) = j^\mu_{\mathrm{em}} \ . \tag{2.84}$$

Here j^μ_{em} denotes an electromagnetic current density, which will be further specified later. It is well known that (2.84) can be simplified by choosing a specific gauge. One should remember that (2.84) remains invariant under gauge transformations of the form

$$A^\mu = A'^\mu - \partial^\mu \Lambda \ , \tag{2.85}$$

with an arbitrary scalar function $\Lambda(x)$, i.e., A'^μ obeys the same equations (2.84) as A^μ. One can therefore always choose the gauge $\Lambda(x)$ in such a way that

$$\partial_\mu A^\mu = 0 \tag{2.86}$$

holds. Equation (2.86) is referred to as the *Lorentz condition.* By requiring condition (2.86) we have fixed a certain gauge and are now able to determine A^μ. In this Lorentz gauge the Maxwell equations reduce to

$$\Box A^\mu = j^\mu_{\text{em}} \ . \tag{2.87}$$

In order to derive the vector potential A^μ of K^+ mesons, the K^+ transition current has to be specified and inserted into the right-hand side of (2.87). As already mentioned the K^+ is a Klein–Gordon particle just like the π^+ and we can therefore construct $j^\mu(\mathrm{K}^+)$ in complete analogy to the pion current (2.82) or (2.83):

$$\begin{aligned} j^\mu(\mathrm{K}^+) &= \mathrm{i}e\left[\varphi_4^*(\partial^\mu\varphi_2) - (\partial^\mu\varphi_4^*)\varphi_2\right] \\ &= eN_2N_4(p_2+p_4)^\mu \mathrm{e}^{\mathrm{i}(p_4-p_2)\cdot x} \ . \end{aligned} \tag{2.88}$$

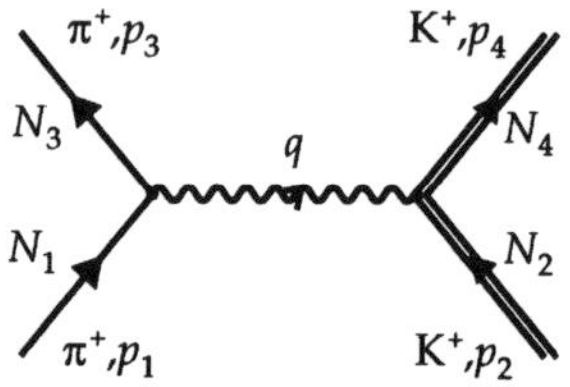

Fig. 2.7. A Feynman diagram for $\pi^+\mathrm{K}^+$ scattering to lowest order (one-photon exchange). The normalization factors are also shown

The notation is explained in Fig. 2.7, which represents $\pi^+\mathrm{K}^+$ scattering to lowest order. The formal solutions of (2.87) corresponding to the K^+ transition current (2.88) are

$$A^\mu = \Box^{-1} j^\mu(\mathrm{K}^+) \ , \tag{2.89}$$

where the inverse quabla operator is defined by

$$\Box^{-1}\Box = \mathbb{1} \ . \tag{2.90a}$$

$\Box^{-1}$ can be identified by its action on a plane wave:

$$\Box^{-1}\left(\Box \mathrm{e}^{-\mathrm{i}q\cdot x}\right) = \Box^{-1}\left(-q^2\mathrm{e}^{-\mathrm{i}q\cdot x}\right) = \mathrm{e}^{-\mathrm{i}q\cdot x} \ , \tag{2.90b}$$

$$\Box^{-1}\mathrm{e}^{-\mathrm{i}q\cdot x} = -\frac{1}{q^2}\mathrm{e}^{-\mathrm{i}q\cdot x} \ . \tag{2.90c}$$

Now the four-potential (2.89), which is created by the transition current (2.88), is readily obtained:

$$A^\mu(\mathrm{K}^+) = -\frac{1}{q^2} j^\mu(\mathrm{K}^+) = -\frac{1}{q^2} eN_2N_4(p_2+p_4)^\mu \mathrm{e}^{\mathrm{i}q\cdot x} \ . \tag{2.91}$$

Here the transferred four-momentum is

$$q^\mu = (p_4-p_2)^\mu = (p_1-p_3)^\mu \ . \tag{2.92}$$

Inserting this result into the scattering amplitude of (2.81) leads to

$$
\begin{aligned}
S^{(1)}_{\mathrm{fi}}(\pi^+\mathrm{K}^+) &= -\mathrm{i}\int \mathrm{d}^4x\, j_\mu(\pi^+)A^\mu(\mathrm{K}^+) \\
&= \mathrm{i}\int \mathrm{d}^4x\, j_\mu(\pi^+)\frac{1}{q^2}\, j^\mu(\mathrm{K}^+) \\
&= \mathrm{i}e^2 N_1N_2N_3N_4(p_1+p_3)_\mu \frac{1}{q^2}(p_2+p_4)^\mu \\
&\quad\times \int \mathrm{d}^4x\, \mathrm{e}^{\mathrm{i}(p_3-p_1)\cdot x}\mathrm{e}^{\mathrm{i}(p_4-p_2)\cdot x} \\
&= -N_1N_2N_3N_4\,(2\pi)^4\,\delta^4(p_3+p_4-p_1-p_2) \\
&\quad\times e(p_1+p_3)_\mu \left(-\frac{\mathrm{i}g^{\mu\nu}}{q^2}\right) e(p_2+p_4)_\nu \\
&= -\mathrm{i}N_1N_2N_3N_4\,(2\pi)^4\,\delta^4(p_3+p_4-p_1-p_2)F_{\mathrm{fi}}\ .
\end{aligned}
\tag{2.93}
$$

The last step includes the definition of the *reduced scattering amplitude* F_{fi}, which is mainly given by the current–current coupling connected with the photon propagator. Now we can interpret the factors occurring in this result (2.93):

1. Every external line in a Feynman graph yields a normalization factor N_i.
2. The mesons interact via exchange of a virtual photon, which is represented in the graph by a wavy line. This line corresponds to the photon propagator

$$
D_{\mu\nu}(q) = g_{\mu\nu}D(q^2) = -\mathrm{i}g_{\mu\nu}\frac{1}{q^2} \tag{2.94}
$$

in (2.93). The last step of (2.94) shows how the photon propagator is represented in graphs (diagrams). The square of the momentum transfer q^2 is often referred to as the squared mass of the virtual photon. This is completely analogous to $q^2 = m_0^2$, which holds for every four-momentum of a particle with rest mass m_0. A free photon obeys the homogeneous Maxwell equations

$$
\Box A^\mu = 0\ . \tag{2.95}
$$

This equation is only solved by a plane wave of the form $\exp(-\mathrm{i}qx)$ if

$$
q^2 = 0 \tag{2.96}
$$

holds. But this condition shows that real photons are massless ($m_0 = 0$). Virtual photons, however, which are exchanged by electromagnetically interacting particles, are characterized by $q^2 \neq 0$ and referred to as *off mass shell*. We should emphasize that the form (2.94) of the photon propagator is only valid within the Lorentz gauge. It is defined by

$$
\Box D_{\mu\nu}(x-x') = -\mathrm{i}g_{\mu\nu}\delta^4(x-x') \tag{2.97}
$$

according to the Maxwell equations in Lorentz gauge. We shall later discuss the form of the photon propagator in different gauges.

3. There are two vertices in the graph for π^+K^+ scattering. Since the virtual photon propagator $D_{\mu\nu}$ is a tensor with respect to Lorentz indices, there must be four-vectors on the left- and the right-hand sides, in order to produce a scalar. In the case of spinless mesons, however, there is only one characterizing four-vector, which is the four-momentum. This fact and the symmetry of the initial and final lines at the vertex explain the factors $e(p_1+p_3)$ and $e(p_2+p_4)$ in (2.93). The tensor character of the photon propagator is due to the photon being a spin-1 particle.

4. The transition currents in momentum space (vertices) as well as the photon propagator have been defined with factors $\pm\mathrm{i}$ in a way that yields the correct sign also at higher orders. The main advantage of this convention is that scattering amplitudes for arbitrary graphs can immediately be constructed.

5. The four-momenta of the incoming and outgoing particles (the external lines in the graph) are subject to four-momentum conservation, which is taken into account by the factor $(2\pi)^4\,\delta^4(p_1+p_2-p_3-p_4)$.

2.2.4 The Cross Section

We have already mentioned that plane waves (2.60) with probability density (2.63a) are not normalized as usual to one particle per volume V but rather to $2E_i$ particles per volume V, i.e.,

$$\varrho_i = |N_i|^2\, 2E_i \ , \tag{2.98}$$

where we have made use of (2.66) once again. This makes sense because both ϱ and E are the zero components of four-vectors. We have already become acquainted with the covariant normalization of Dirac spinors and its advantages in describing meson–meson scattering processes. Of course, one has to choose flux and phase-space factors correspondingly, since the cross section must not depend on the normalization scheme. Here we employ the normalization

$$\int_V \mathrm{d}^3x\, \varrho_i = 2E_i \ , \tag{2.99}$$

which leads to $N_i = 1/\sqrt{V}$. The transition probability per unit volume and unit time is given by

$$P_{\mathrm{fi}} = |S_{\mathrm{fi}}|^2 / VT \ . \tag{2.100}$$

Making use of the final result (2.93) and taking into account that the square of momentum functions, δ^4, can be expressed as

$$\begin{aligned} \left[(2\pi)^4\delta^4(p_3+p_4-p_1-p_2)\right]^2 &= (2\pi)^4\delta^4(p_3+p_4-p_1-p_2)(2\pi)^4\delta^4(0) \\ &= (2\pi)^4\delta^4(p_3+p_4-p_1-p_2)VT \ , \end{aligned} \tag{2.101}$$

we obtain

$$P_{\mathrm{fi}}^{(1)} = (2\pi)^4\,\delta^4(p_3+p_4-p_1-p_2)(N_1N_2N_3N_4)^2\,|F_{\mathrm{fi}}|^2 \ . \tag{2.102}$$

This transition rate is still proportional to the flux of incoming particles and the density of target particles. In order to eliminate this dependence we have to divide by these two quantities. The beam particle flux (projectile quantities carry index 1, here, for example, π^+ particles) is defined as the number of incoming particles per unit area that can reach the target (K^+ particles, index 2) per unit time. The velocity of such beam particles is $\boldsymbol{v} = \boldsymbol{v}_1 - \boldsymbol{v}_2$. The densities are normalized to $2E/V$ particles per unit volume (see (2.99)) and therefore the flux factor is

$$|\boldsymbol{j}_1| = |\boldsymbol{v}|\,\frac{2E_1}{V} \ . \tag{2.103}$$

For the density of target particles we obtain

$$\varrho_2 = \frac{2E_2}{V} \ . \tag{2.104}$$

If the target particles are at rest, (2.104) assumes the value $2m_2/V$. All together (2.102) has to be multiplied by the factor

$$\frac{1}{|\boldsymbol{v}|}\,\frac{V}{2E_1}\,\frac{V}{2E_2} \ . \tag{2.105}$$

In order to derive the cross section, one has to sum over all two-particle final states. If the volume V contains a particle, this yields an integration over the two-particle phase space with the voulume element

$$\frac{V}{(2\pi)^3}\mathrm{d}^3p_3\frac{V}{(2\pi)^3}\mathrm{d}^3p_4 \ . \tag{2.106}$$

According to the normalization employed above, the phase space factor is

$$\frac{V}{(2\pi)^3}\,\frac{\mathrm{d}^3p_3}{2E_3}\,\frac{V}{(2\pi)^3}\,\frac{\mathrm{d}^3p_4}{2E_4} \ , \tag{2.107}$$

which then yields the following cross section:

$$\begin{aligned} \mathrm{d}\sigma &= P_{\mathrm{fi}}\frac{V^2}{2E_12E_2\,|\boldsymbol{v}|}\,\frac{V}{(2\pi)^3}\,\frac{\mathrm{d}^3p_3}{2E_3}\,\frac{V}{(2\pi)^3}\,\frac{\mathrm{d}^3p_4}{2E_4} \\ &= \frac{|F|^2}{2E_12E_2\,|\boldsymbol{v}|}(2\pi)^4\,\delta^4(p_3+p_4-p_1-p_2)\frac{\mathrm{d}^3p_3}{(2\pi)^32E_3}\,\frac{\mathrm{d}^3p_4}{(2\pi)^32E_4} \ . \end{aligned} \tag{2.108}$$

Since $N_i = 1/\sqrt{V}$, all factors V cancel and the cross section does not depend on the normalization volume. The Lorentz invariant form of the flux factor (see Exercise 2.2) is

$$E_1E_2\,|\boldsymbol{v}| = \sqrt{(p_1\cdot p_2)^2 - m_1^2m_2^2} \ . \tag{2.109}$$

This identity holds only for collinear collisions. For noncollinear collisions only the relativistically invariant expression on the right-hand side of (2.109) remains valid. We therefore introduce the Lorentz-invariant phase-space factor

$$\begin{aligned} &\mathrm{d}\,\mathrm{Lips}(s;\, p_3, p_4) \\ &\quad = (2\pi)^4\, \delta^4(p_3 + p_4 - p_1 - p_2) \frac{1}{(2\pi)^3} \frac{\mathrm{d}^3 p_3}{2E_3} \frac{1}{(2\pi)^3} \frac{\mathrm{d}^3 p_4}{2E_4}\ , \end{aligned} \tag{2.110}$$

where

$$s = (p_1 + p_2)^2 \tag{2.111}$$

denotes one of the so-called Mandelstam variables. Equation (2.108) then assumes the form

$$\mathrm{d}\sigma = \frac{|\, F\,|^2}{4\sqrt{(p_1 \cdot p_2)^2 - m_1^2 m_2^2}}\, \mathrm{d}\,\mathrm{Lips}(s;\, p_3, p_4)\ . \tag{2.112}$$

EXERCISE

2.2 The Flux Factor

Problem. Verify that in the center-of-mass system as well as in the laboratory system the flux factor $4E_1E_2|\boldsymbol{v}|$ is given by the invariant expression

$$4E_1E_2|\boldsymbol{v}| = 4\sqrt{(p_1\, p_2)^2 - m_1^2 m_2^2}\ .$$

Solution. In the center-of-mass system the momenta of the projectile and target point in opposite directions, i.e.,

$$\begin{aligned} E_1 &= \sqrt{m_1^2 + \boldsymbol{p}^2}\ , \qquad \boldsymbol{p}_1 = +\boldsymbol{p}\ , \\ E_2 &= \sqrt{m_2^2 + \boldsymbol{p}^2}\ , \qquad \boldsymbol{p}_2 = -\boldsymbol{p}\ . \end{aligned} \tag{1}$$

Therefore $p_1\, p_2 = E_1E_2 + \boldsymbol{p}^2$, and we obtain

$$\begin{aligned} (p_1\, p_2)^2 - m_1^2 m_2^2 &= (E_1E_2 + \boldsymbol{p}^2)^2 - m_1^2 m_2^2 \\ &= E_1^2(E_2^2 - m_2^2) + 2E_1E_2\boldsymbol{p}^2 + \boldsymbol{p}^4 + (E_1^2 - m_1^2)m_2^2 \\ &= \boldsymbol{p}^2(E_1^2 + 2E_1E_2 + E_2^2) \\ &= \boldsymbol{p}^2(E_1 + E_2)^2 \\ &= (E_1E_2)^2 \left| \frac{\boldsymbol{p}_1}{E_1} + \frac{\boldsymbol{p}_2}{E_2} \right|^2\ , \end{aligned} \tag{2}$$

which immediately leads to

Exercise 2.2

$$
\begin{aligned}
4\sqrt{(p_1\,p_2)^2 - m_1^2 m_2^2} &= 4E_1E_2 \left| \frac{\boldsymbol{p}_1}{E_1} - \frac{\boldsymbol{p}_2}{E_2} \right| \\
&= 4E_1E_2\,|\boldsymbol{v}_1 - \boldsymbol{v}_2| \ . \qquad (3)
\end{aligned}
$$

In the laboratory system the target particle is at rest and we have

$$
\begin{aligned}
E_1 &= \frac{m_1}{\sqrt{1-\boldsymbol{v}^2}} \quad , \quad \boldsymbol{p}_1 = \frac{m_1\boldsymbol{v}}{\sqrt{1-\boldsymbol{v}^2}} \\
E_2 &= m_2 \quad , \quad \boldsymbol{p}_2 = \boldsymbol{0} \ . \qquad (4)
\end{aligned}
$$

Then the scalar product $p_1 \cdot p_2$ is simply equal to $m_1m_2/\sqrt{1-\boldsymbol{v}^2}$ and we obtain

$$
\begin{aligned}
4\sqrt{(p_1 \cdot p_2)^2 - m_1^2 m_2^2} &= 4m_1m_2\sqrt{\frac{1}{1-\boldsymbol{v}^2} - 1} \\
&= 4\frac{m_1m_2}{\sqrt{1-\boldsymbol{v}^2}}|\boldsymbol{v}| \\
&= 4E_1E_2|\boldsymbol{v}| \ . \qquad (5)
\end{aligned}
$$

EXERCISE

2.3 The Mandelstam Variable *s*

Problem. Introduce the Mandelstam variable $s = (p_1 + p_2)^2$ and show that

$$
4\left[(p_1 \cdot p_2)^2 - m_1^2 m_2^2\right] = \left[s - (m_1+m_2)^2\right]\left[s - (m_1-m_2)^2\right]
$$

holds.

Solution. The Mandelstam variable s provides an invariant measure for the energy of the particles participating in the reaction. In the center-of-mass system, where one has $\boldsymbol{p}_1 + \boldsymbol{p}_2 = \boldsymbol{0}$, $\sqrt{s}$ is equal to the sum of all particle energies:

$$
\begin{aligned}
s = (p_1+p_2)^2 &= p_1^2 + 2\,p_1 \cdot p_2 + p_2^2 \\
&= m_1^2 + 2\,p_1 \cdot p_2 + m_2^2 \ ,
\end{aligned}
$$

i.e.,

$$
2\,p_1 \cdot p_2 = s - m_1^2 - m_2^2 \ . \qquad (1)
$$

The square of the flux factor then assumes the form

$$
\begin{aligned}
&(2\,p_1 \cdot p_2 - 2m_1m_2)(2\,p_1 \cdot p_2 + 2m_1m_2) \\
&\quad = \left[s - (m_1+m_2)^2\right]\left[s - (m_1-m_2)^2\right] \ .
\end{aligned}
$$

EXERCISE

2.4 The Lorentz-Invariant Phase-Space Factor

Problem. Show that the phase-space factor

$$\begin{aligned} &\mathrm{d}\,\mathrm{Lips}(s;\, p_3,\, p_4) \\ &\quad = (2\pi)^4\, \delta^4(p_3 + p_4 - p_1 - p_2) \frac{1}{(2\pi)^3} \frac{\mathrm{d}^3 p_3}{2E_3} \frac{1}{(2\pi)^3} \frac{\mathrm{d}^3 p_4}{2E_4} \end{aligned} \tag{1}$$

is Lorentz invariant.

Solution. The phase-space factor is apparently invariant under spacial rotations. We must therefore investigate its behavior under proper Lorentz transformations, which are induced by the matrix $\Lambda(\boldsymbol{v}_\mathrm{b})$ with a *boost velocity* $\boldsymbol{v}_\mathrm{b}$. Owing to rotational invariance one can, without loss of generality, put $\boldsymbol{v}_\mathrm{b}$ parallel to the z axis. This considerably simplifies the dependence of the particle coordinates in the system at rest p_n $(n = 1, 2, 3, 4)$ on the new coordinates p'_n in the moving reference system. In general this dependence is

$$p'_n = \Lambda(\boldsymbol{v}_\mathrm{b})\, p_n \ , \tag{2}$$

with the inversion

$$p_n = \Lambda^{-1}(\boldsymbol{v}_\mathrm{b})\, p'_n = \Lambda(-\boldsymbol{v}_\mathrm{b})\, p'_n \ . \tag{3}$$

For the differentials $\mathrm{d}p^x_{n'}$, $\mathrm{d}p^y_{n'}$, $\mathrm{d}p^z_{n'}$ we obtain $\left(\gamma = 1/\sqrt{1 - \boldsymbol{v}_\mathrm{b}^2}\right)$:

$$\begin{aligned} \mathrm{d}p^x_n &= \mathrm{d}p^{x\prime}_n \ , \\ \mathrm{d}p^y_n &= \mathrm{d}p^{y\prime}_n \ , \\ \mathrm{d}p^z_n &= \gamma \left(\mathrm{d}p^{z\prime}_n + |\boldsymbol{v}_\mathrm{b}\mathrm{d}E'_n\right) \\ &= \gamma \mathrm{d}p^{z\prime}_n \left(1 + |\boldsymbol{v}_\mathrm{b}| \frac{p^{z\prime}_n}{E'_n}\right) \\ &= \mathrm{d}p^{z\prime}_n \frac{E_n}{E'_n} \ . \end{aligned} \tag{4}$$

Here we have employed the relations

$$E_n \equiv p^0_n = \left(\boldsymbol{p}^2_n + m^2_n\right)^{\frac{1}{2}} = \gamma(E'_n + |\boldsymbol{v}_\mathrm{b}|\, p^{z\prime}_n) \tag{5}$$

and

$$\frac{\mathrm{d}E'_n}{\mathrm{d}p^{z\prime}_n} = \frac{\mathrm{d}}{\mathrm{d}p^{z\prime}_n} \left[(p^{z\prime}_n)^2 + (p^{y\prime}_n)^2 + (p^{x\prime}_n)^2\right]^{\frac{1}{2}} = \frac{p^{z\prime}_n}{E'_n} \ . \tag{6}$$

Exercise 2.4

For the volume element $\mathrm{d}^3 p_n = \mathrm{d}p_n^x \mathrm{d}p_n^y \mathrm{d}p_n^z$ we therefore have

$$\frac{\mathrm{d}^3 p_n}{E_n} = \frac{\mathrm{d}^3 p'_n}{E'_n} \ . \tag{7}$$

It remains to prove that the four-delta function is a Lorentz scalar. By definition we have

$$\int \mathrm{d}^4 p \delta^4(p) = 1 \tag{8}$$

in any reference frame. Owing to the properties of the matrix $\Lambda(\boldsymbol{v})$ the volume element is a Lorentz scalar too:

$$\mathrm{d}^4 p = \left| \frac{\partial p^\mu}{\partial p'^\nu} \right| \mathrm{d}^4 p' = |\det(\lambda(-\boldsymbol{v}_\mathrm{b}))| \mathrm{d}^4 p' = \mathrm{d}^4 p' \ . \tag{9}$$

EXAMPLE

2.5 $\pi^+\pi^+$ and $\pi^+\pi^-$ Scattering

As an example of the scattering of identical particles we study $\pi^+\pi^+$ scattering. The main modification compared with the π^+K^+ scattering discussed above is the symmetrization of initial and final states. The total scattering amplitude has to be symmetric under exchange of the incoming or outgoing identical bosons. The direct graph of $\pi^+\pi^+$ scattering is depicted in Fig. 2.8.

Substituting the final state p_3 for p_4 and vice versa leads to the corresponding exchange graph (Fig. 2.9). The complete scattering amplitude, is then

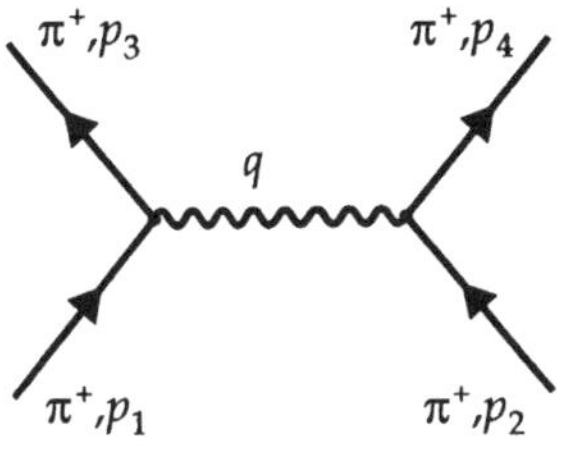

Fig. 2.8. A Feynman diagram for the direct amplitude of $\pi^+\pi^+$ scattering in the one-photon exchange approximation

$$S_{\mathrm{fi}}^{(1)}(\pi^+\pi^+) = S_{\mathrm{fi}}^{(1)}(\text{direct}) + S_{\mathrm{fi}}^{(1)}(\text{exchange}) \ .$$

Employing (2.93), $S_{\mathrm{fi}}^{(1)}$ assumes the form

$$\begin{aligned} S_{\mathrm{fi}}^{(1)}(\pi^+\pi^+) &= -\mathrm{i}(2\pi)^4 \delta^4(p_3 + p_4 - p_1 - p_2) N_1 N_2 N_3 N_4 \\ &\quad \times \left[\frac{-e^2(p_1+p_3)_\mu (p_2+p_4)^\mu}{(p_2-p_4)^2} + \frac{-e^2(p_1+p_4)_\mu (p_2+p_3)^\mu}{(p_2-p_3)^2} \right] \\ &\equiv -\mathrm{i}(2\pi)^4 \delta^4(p_3+p_4-p_1-p_2) \\ &\quad \times N_1 N_2 N_3 N_4 F_{\pi^+\pi^+}(p_1 p_2; p_3 p_4) \ . \end{aligned} \tag{1}$$

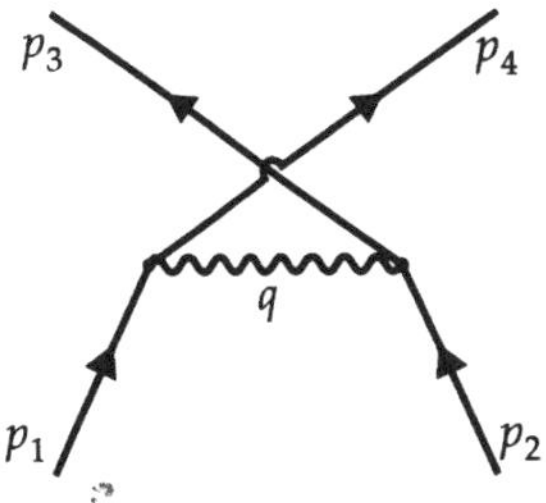

Fig. 2.9. The exchange amplitude of $\pi^+\pi^+$ scattering

Again $F_{\pi^+\pi^+}$ denotes the invariant scattering amplitude. All further steps can be performed in complete analogy to π^+K^+ scattering.

A similar argument holds for $\pi^+\pi^-$ scattering, if we take the antiparticle interpretation into account (cf. the discussion connected with (2.14)–(2.17)). We can therefore make the identification shown in Fig. 2.10. Owing to the

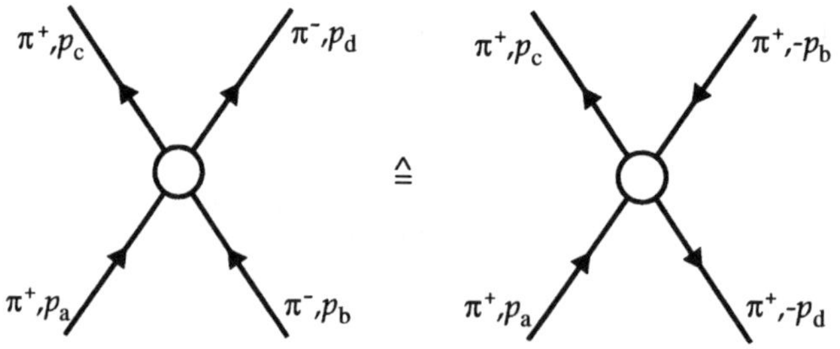

Fig. 2.10. The interrelation between $\pi^+\pi^+$ and $\pi^+\pi^-$ scattering

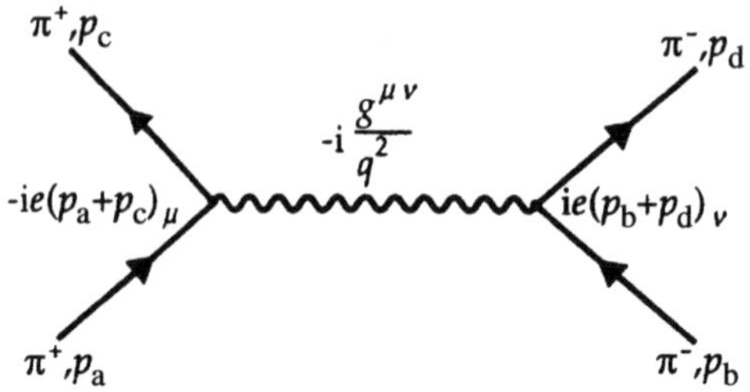

Fig. 2.11. The one-photon exchange amplitude for $\pi^+\pi^-$ scattering in detail

antiparticle concept disussed earlier we can interpret an incoming π^- with four-momentum p_b as an outgoing π^+ with four-momentum $-p_\mathrm{b}$ and an outgoing π^- with four-momentum p_d as an incoming π^+ with four-momentum $-p_\mathrm{d}$. Hence we can immediately write down the invariant scattering amplitude for $\pi^+\pi^-$ scattering:

$$F_{\pi^+\pi^-}(p_\mathrm{a}p_\mathrm{b};\, p_\mathrm{c}p_\mathrm{d}) = F_{\pi^+\pi^+}(p_\mathrm{a}-p_\mathrm{d};\, p_\mathrm{c}-p_\mathrm{b}) \ , \tag{2}$$

which explicitly is

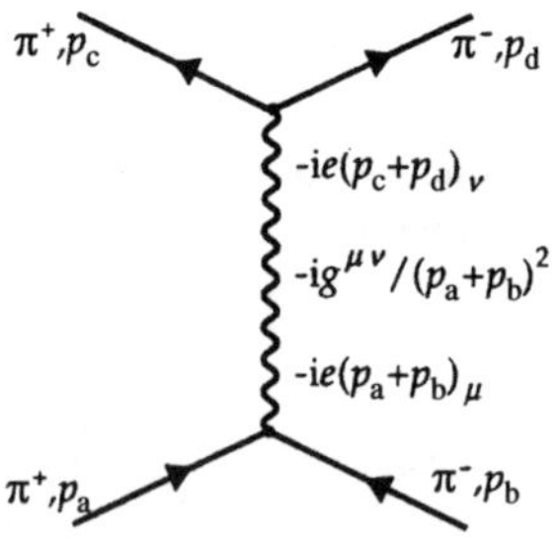

Fig. 2.12. The detailed exchange graph for $\pi^+\pi^-$ scattering. Obviously this amplitude can also be understood as one-photon $\pi^+\pi^-$ annihilation connected with $\pi^+\pi^-$ pair creation

$$F_{\pi^+\pi^-}(p_\mathrm{a}p_\mathrm{b};\, p_\mathrm{c}p_\mathrm{d}) = \left[\frac{-e^2(p_\mathrm{a}+p_\mathrm{c})_\mu(-p_\mathrm{d}-p_\mathrm{b})^\mu}{(-p_\mathrm{d}+p_\mathrm{b})^2}\right.$$
$$\left. + \frac{-e^2(p_\mathrm{a}-p_\mathrm{b})_\mu(+p_\mathrm{c}-p_\mathrm{d})^\mu}{(p_\mathrm{a}+p_\mathrm{b})^2}\right] \ . \tag{3}$$

A graphical representation of the two contributions more detailed than that given above is given in Fig. 2.11. The difference between the direct parts of the amplitudes for $\pi^+\pi^-$ scattering (or π^+K^+ scattering) and $\pi^+\pi^+$ scattering is only that of sign if the corresponding four-momenta are considered equal (because of the different masses of K^+ and π^- they are in general not equal). The different signs of these two processes correspond to attractive and repulsive interaction, respectively.

EXERCISE

2.6 The Cross Section for Pion–Kaon Scattering

Problem. Derive the explicit form of the differential cross section for electromagnetic π^+K^+ scattering in the center-of-momentum system (cm system).

Solution. According to (2.112) the cross section for π^+K^+ scattering is

$$\mathrm{d}\sigma = \frac{|F|^2}{4\sqrt{(p_1 p_2)^2 - m_1^2 m_2^2}}\mathrm{d\,Lips}(s;\, p_3 p_4)\ , \tag{1}$$

where

$$\begin{aligned} F &= -\frac{e^2}{q^2}(p_1+p_3)\cdot(p_2+p_4)\ , \\ q^\mu &= (p_3-p_1)^\mu = -(p_4-p_2)^\mu\ , \end{aligned} \tag{2}$$

and

$$\begin{aligned} &\mathrm{d\,Lips}(s;\, p_3 p_4) \\ &\quad = (2\pi)^4\,\delta^4(p_3+p_4-p_1-p_2)\frac{1}{(2\pi)^3}\frac{\mathrm{d}^3\mathrm{p}_3}{2\mathrm{E}_3}\frac{1}{(2\pi)^3}\frac{\mathrm{d}^3\mathrm{p}_4}{2\mathrm{E}_4} \end{aligned} \tag{3}$$

denote the invariant scattering amplitude and the Lorentz-invariant phase-space factor, respectively. The cm system is defined by

$$\boldsymbol{p}_1+\boldsymbol{p}_2=\boldsymbol{p}_3+\boldsymbol{p}_4=\boldsymbol{0}\ . \tag{4}$$

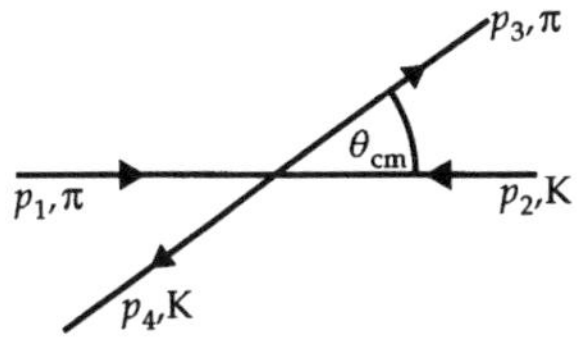

Fig. 2.13. π^+K^+ scattering in the cm system

In this system the scattering process is described by the scattering angle θ_{cms} (see Fig. 2.13). Now we transform the four-momenta to the cm system,

$$\begin{aligned} p_1^\mu &= (E_1, \boldsymbol{p})\ , \quad p_2^\mu = (E_2, -\boldsymbol{p})\ , \\ p_3^\mu &= (E_3, \boldsymbol{p}')\ , \quad p_4^\mu = (E_4, -\boldsymbol{p}')\ , \end{aligned} \tag{5}$$

which leads to the total energy

$$\begin{aligned} E_{\mathrm{cms}} &= E_1+E_2 = E_3+E_4 \\ &= \sqrt{\boldsymbol{p}^2+m_1^2}+\sqrt{\boldsymbol{p}^2+m_2^2} \\ &= \sqrt{\boldsymbol{p}'^2+m_1^2}+\sqrt{\boldsymbol{p}'^2+m_2^2} \\ &= \sqrt{p^2+m_1^2}+\sqrt{p^2+m_2^2}\ . \end{aligned} \tag{6}$$

Here and in the following we denote the absolute value of the spatial momentum by

$$|\boldsymbol{p}| = |\boldsymbol{p}'| \equiv p\ . \tag{7}$$

Exercise 2.6

By integrating the invariant phase-space factor $\mathrm{d\,Lips}(s;\, p_3 p_4)$ over the spatial momenta $\mathrm{d}^3 p_4$ we obtain

$$\int \frac{\mathrm{d}^3 p_4}{E_4}\, \delta^4(p_3 + p_4 - p_1 - p_2) = \frac{1}{E_4}\delta(E_3 + E_4 - E_1 - E_2) \ . \tag{8}$$

The right-hand side of (8) contains E_4 as well as $|\boldsymbol{p}_4|$. These two variables, however, are not independent of each other: they are connected by

$$\boldsymbol{p}_4 = \boldsymbol{p}_1 + \boldsymbol{p}_2 - \boldsymbol{p}_3 \ , \quad E_4 = \sqrt{\boldsymbol{p}_4 + m_2^2} \ . \tag{9}$$

Next we transform $\mathrm{d}^3 p_3$ into spherical coordinates

$$\mathrm{d}^3 p_3 = p_3^2 \mathrm{d}p_3 \mathrm{d}\Omega \ , \tag{10}$$

where $p_3 = |\boldsymbol{p}_3|$ and $\mathrm{d}\Omega$ denotes the spherical angle into which the π^+ is scattered. Because of

$$E_3^2 = p_3^2 + m_1^2 \tag{11}$$

we have

$$E_3 \mathrm{d}E_3 = p_3 \mathrm{d}p_3 \tag{12}$$

and

$$\mathrm{d\,Lips}(s;\, p_3 p_4) = \frac{1}{(4\pi)^2}\delta(E_3 + E_4 - E_1 - E_2)\frac{p_3 \mathrm{d}E_3}{E_4}\mathrm{d}\Omega \ . \tag{13}$$

This formula is valid in any Lorentz system. Now we go into the cm system by making use of the relations

$$E_3^2 = p_2 + m_1^2 \ , \quad E_4^2 = p^2 + m_2^2 \ , \tag{14}$$

and

$$E_3 \mathrm{d}E_3 = E_4 \mathrm{d}E_4 = p \mathrm{d}p \ . \tag{15}$$

Introducing the free variable

$$E' = E_3 + E_4 \ , \tag{16}$$

we also have, according to (15),

$$\mathrm{d}E' = \frac{p}{E_3}\mathrm{d}p + \frac{p}{E_4}\mathrm{d}p = \frac{E'}{E_3 E_4} p \mathrm{d}p = \frac{E'}{E_4}\mathrm{d}E_3 \ . \tag{17}$$

Now (13) assumes the form

$$\mathrm{d\,Lips}(s;\, p_3 p_4)\bigg|_{\mathrm{cms}} = \frac{1}{(4\pi)^2}\delta(E_{\mathrm{cms}} - E')\frac{p}{E'}\mathrm{d}E'\mathrm{d}\Omega \ . \tag{18}$$

Exercise 2.6

An integration over E' then yields

$$\mathrm{d}\,\mathrm{Lips}(s;\, p_3 p_4)\bigg|_{\mathrm{cms}} = \frac{1}{(4\pi)^2}\frac{p}{E_{\mathrm{cms}}}\mathrm{d}\Omega \tag{19}$$

for the two-particle phase-space factor in the cm system. Finally the flux factor has to be rewritten in terms of cm variables. With the help of (5) and (6) we immediately find that

$$\begin{aligned}\sqrt{(\boldsymbol{p}_1\cdot \boldsymbol{p}_2)^2 - m_1^2 m_2^2} &= \sqrt{(E_1E_2+p^2)^2-(E_1^2-p^2)(E_2^2-p^2)}\\ &= p\,E_{\mathrm{cms}}\ .\end{aligned} \tag{20}$$

The result for the differential cross section is then

$$\frac{\mathrm{d}\sigma}{\mathrm{d}\Omega}\bigg|_{\mathrm{cms}} = \frac{1}{(8\pi E_{\mathrm{cms}})^2}|F|^2\ . \tag{21}$$

By introducing the second Mandelstam variable t the result (21) can be brought into an invariant form. t is simply defined as the square of the four-momentum transfer:

$$t = q^2 = (p_1 - p_3)^2 = (p_2 - p_4)^2\ . \tag{22}$$

In the cm system we consequently have

$$\begin{aligned} t &= (0, \boldsymbol{p}-\boldsymbol{p}')^2 = -(\boldsymbol{p}-\boldsymbol{p}')^2\\ &= -\left(\boldsymbol{p}^2 - 2\,\boldsymbol{p}\cdot\boldsymbol{p}' + \boldsymbol{p}'^2\right)\\ &= -2p^2\,(1-\cos\theta_{\mathrm{cms}})\ ,\end{aligned} \tag{23}$$

which leads to

$$\mathrm{d}t = 2p^2\mathrm{d}\cos\left(\theta_{\mathrm{cms}}\right)\ . \tag{24}$$

Since the cross section for spinless particles is cylindrically symmetric around the beam axis, i.e.

$$\mathrm{d}\Omega_{\mathrm{cms}} = 2\pi\mathrm{d}\cos\left(\theta_{\mathrm{cms}}\right)\ , \tag{25}$$

we obtain the relation

$$\frac{\mathrm{d}}{\mathrm{d}t} = \frac{\pi}{p^2}\frac{\mathrm{d}}{\mathrm{d}\Omega_{\mathrm{cms}}}\ . \tag{26}$$

Therefore the two-particle cross section is, in invariant form

$$\begin{aligned}\frac{\mathrm{d}\sigma}{\mathrm{d}t} &= \frac{1}{64\pi}\frac{1}{(pE_{\mathrm{cms}})^2}|F|^2\\ &= \frac{1}{64\pi}\frac{|F|^2}{(p_1\cdot p_2)^2 - m_1^2m_2^2}\ ,\end{aligned} \tag{27}$$

Exercise 2.6

where we have inserted (20). Finally we introduce the Mandelstam variable s (see Exercise 2.3) and take $p_1^2 = m_1^2$ and $p_2^2 = m_2^2$ into account. Equation (27) then assumes the form

$$\frac{\mathrm{d}\sigma}{\mathrm{d}t} = \frac{1}{16\pi} \frac{|F|^2}{\left[s-(m_1+m_2)^2\right]\left[s-(m_1-m_2)^2\right]} \ . \tag{28}$$

As already mentioned these expressions are valid for any scattering reaction with two unpolarized particles in both initial and final states. Now we want to express $|F|^2$ in the case of π^+K^+ scattering by invariant Mandelstam variables. From the definitions

$$\begin{aligned} s &= (p_1+p_2)^2 = (p_3+p_4)^2 \ , \\ u &= (p_1-p_4)^2 = (p_2-p_3)^2 \end{aligned} \tag{29}$$

we derive the relations

$$\begin{aligned} 2\, p_1 \cdot p_2 &= 2\, p_3 \cdot p_4 = s - m_1^2 - m_2^2 \ , \\ 2\, p_1 \cdot p_4 &= m_1^2 + m_2^2 - u \ . \end{aligned} \tag{30}$$

With the help of these relations the invariant scattering amplitude (2) assumes the form

$$|F|^2 = \left(\frac{4\pi\alpha}{t}\right)(s-u)^2 \ , \tag{31}$$

where we have transformed the fine-structure constant

$$\alpha = \frac{e^2}{4\pi} \simeq \frac{1}{137} \tag{32}$$

to so-called Heaviside–Lorentz units. In these units Gauss's law reads $\nabla \cdot \boldsymbol{E} = p$. It is particularly simple to transform (31) into the cm system, since we have, according to (29) and (5),

$$\begin{aligned} s &= (E_1+E_2, \mathbf{0})^2 = E_{\text{cms}}^2 \\ u &= (E_1-E_4, \boldsymbol{p}+\boldsymbol{p}')^2 = (E_1-E_4)^2 - (\boldsymbol{p}+\boldsymbol{p}')^2 \\ &= E_1^2 + E_4^2 - 2E_1E_4 - (E_1^2 - m_1^2) - (E_4^2 - m_2^2) - 2\,\boldsymbol{p}\cdot\boldsymbol{p}' \\ &= m_1^2 + m_2^2 - 2\,p^2\cos(\theta_{\text{cms}}) - 2E_1E_4 \ . \end{aligned} \tag{33}$$

The variable t has already been given in (23).

2.2.5 Spin-1 Particles and Their Polarization

In the preceding sections on the basic elements of scalar QED we discussed the scattering of charged spin-0 mesons. Their mutual interaction is mediated by the exchange of virtual photons (massless spin-1 vector bosons). The kind of virtual quanta exchanged is of course specific for each interaction. For example in pion Compton scattering, virtual pions are exchanged (see Example 2.8), and the weak interaction in lepton scattering is mediated by vector bosons (Z^0, $W^{\pm}$). One difference between such particles is the number of internal degrees of freedom, which depends on their spin (or polarization).

Massive Spin-1 Particles. Massive spin-1 bosons are described in the framework of the Proca theory. From the Lagrangian density of the classical four-vector field $\phi^\mu(x)$

$$L = -\frac{1}{4}F_{\mu\nu}F^{\mu\nu} + \frac{1}{2}M^2\phi_\mu\phi^\mu \ , \tag{2.113a}$$

$$F^{\mu\nu} = \partial^\mu\phi^\nu - \partial^\nu\phi^\mu \ , \tag{2.113b}$$

the wave equation follows:

$$\partial_\alpha F^{\alpha\mu} + M^2\phi^\mu = 0 \ . \tag{2.114}$$

Taking the four-divergence of this equation, we find that

$$M^2\partial_\mu\phi^\mu = 0 \ . \tag{2.115}$$

As it is assumed that $M^2 \neq 0$ here, the divergence of ϕ^μ vanishes, and (2.114) is reduced to the Proca equation

$$\left(\Box + M^2\right)\phi^\mu = 0 \ , \quad \partial_\mu\phi^\mu = 0 \ . \tag{2.116}$$

We first consider the polarization vectors of these massive vector bosons. In the rest system of such particles there are three possible positions for spin 1, i.e., three spin vectors, which can generally be chosen as

$$\begin{aligned} \boldsymbol{\varepsilon}^{(1)} &= (1,0,0) \ , \\ \boldsymbol{\varepsilon}^{(2)} &= (0,1,0) \ , \\ \boldsymbol{\varepsilon}^{(3)} &= (0,0,1) \ . \end{aligned} \tag{2.117a}$$

These obviously satisfy the orthogonality relations

$$\boldsymbol{\varepsilon}^{(i)} \cdot \boldsymbol{\varepsilon}^{(j)} = \delta^{ij} \ . \tag{2.117b}$$

It is more useful to use the spherical representation $\boldsymbol{\varepsilon}(\lambda)$ with

$$\begin{aligned} \boldsymbol{\varepsilon}(\lambda=1) &= -\frac{1}{\sqrt{2}}(1,\mathrm{i},0) \ , \\ \boldsymbol{\varepsilon}(\lambda=0) &= (0,0,1) \ , \\ \boldsymbol{\varepsilon}(\lambda=-1) &= \frac{1}{\sqrt{2}}(1,-\mathrm{i},0) \ , \end{aligned} \tag{2.118a}$$

and

$$\boldsymbol{\varepsilon}^*(\lambda) \cdot \boldsymbol{\varepsilon}(\lambda') = \delta_{\lambda\lambda'} \tag{2.118b}$$

instead of the Cartesian representation (2.117a).

In the massless case, i.e., for photons, this change of basis vectors corresponds to the transition from linearly to circularly polarized light. As is known from nonrelativistic quantum mechanics, $\lambda = \pm 1, 0$ is the projection of the particle's spin on, for example, the z axis.

We shall now formulate (2.118) in a manifestly covariant way. For a spin-1 particle in motion, four-vectors $\varepsilon^\mu(\lambda)$ must be found that transform into (2.118) in the rest frame. Setting in the rest frame

$$\varepsilon^0(\lambda) = 0 \ , \qquad \text{for} \quad \lambda = 0, \pm 1 \tag{2.119}$$

and thus defining (in the rest frame)

$$\varepsilon^\mu(\lambda) = \left(\varepsilon^0(\lambda), \boldsymbol{\varepsilon}(\lambda)\right) \ , \tag{2.120}$$

the polarization vector in any other inertial system can be found by the Lorentz transformation. Since the four-momentum in the rest frame is given by

$$p^\mu = (M, \mathbf{0}) \ , \tag{2.121}$$

it follows that

$$p \cdot \varepsilon(\lambda) = 0 \ . \tag{2.122}$$

Equation (2.122) is basically a direct consequence of the condition $\partial_\mu \phi^\mu = 0$, since it reduces the number of relevant degrees of freedom to 3. Since a general free solution of the wave equation (2.116) can always be written as a superposition of linearly independent solutions in the form

$$\phi^\mu(x) = \sum_{\lambda=0,\pm 1} \varepsilon^\mu(p;\lambda)\,\mathrm{e}^{-\mathrm{i}p\cdot x} \ , \qquad p^2 = M^2 \ ,$$

(2.122) becomes evident. The normalization of polarization vectors is given by

$$\varepsilon^{*\mu}(p;\lambda)\varepsilon_\mu(p;\lambda') = -\delta_{\lambda\lambda'} \ . \tag{2.123}$$

Let us consider, for example, a system in which the particle is moving along the z axis with momentum $\boldsymbol{p}$. Hence its four-momentum is

$$p^\mu = (E, 0, 0, p) \ , \quad |\boldsymbol{p}| \equiv p \ , \tag{2.124}$$

and we recognize immediately from (2.122) that the tranverse polarization vectors in this system are the same as those in the rest system,

$$\varepsilon^\mu(p; \lambda = \pm 1) \ , \tag{2.125}$$

but that the longitudinal polarization vector must be given by

$$\varepsilon^{\mu}(p;\lambda=0)=\frac{1}{M}(p,0,0,E)\ , \tag{2.126}$$

in order to satisfy (2.122) and the normalization (2.123). It turns out that the p and E dependence of the longitudinal polarization vector have very interesting consequences for massive spin-1 particles such as the Z and $W^{\pm}$ bosons mediating the weak interaction. The polarization vectors $\varepsilon^{\mu}(p;\lambda)$ satisfy the completeness relation

$$\sum_{\lambda}\varepsilon^{\mu^*}(p;\lambda)\varepsilon^{\nu}(p;\lambda)=-g^{\mu\nu}+\frac{p^{\mu}p^{\nu}}{M^2}\ , \tag{2.127}$$

which we shall prove in Exercise 2.7. The factor on the right-hand side projects out the physical states and appears, as we shall see, in the propagator of virtual spin-1 particles.

EXERCISE

2.7 Polarization States of a Massive Spin-1 Particle

Problem. The polarization vectors of a massive spin-1 particle with four-momentum p^{μ} and helicity λ are denoted by $\varepsilon^{\mu}(p;\lambda)$. It holds that

$$\varepsilon^{\mu^*}(p;\lambda)\varepsilon_{\mu}(p;\lambda')=-\delta_{\lambda\lambda'}\ . \tag{1}$$

The minus sign occurs because these vectors are spacelike. Owing to Lorentz covariance, the sum over the polarization states

$$\sum_{\lambda}\varepsilon^*_{\mu}(p;\lambda)\varepsilon_{\nu}(p;\lambda)\equiv\eta_{\mu\nu}(p) \tag{2}$$

has to be of the form

$$\eta_{\mu\nu}(p)=A\,g_{\mu\nu}+Bp_{\mu}p_{\nu}\ . \tag{3}$$

Find arguments for this fact and determine the constants A and B. Make use of scalar multiplications by p^{μ}, p^{ν}, and $g^{\mu\nu}$. Note that $g^{\mu\nu}g_{\mu\nu}=4$.

Solution. The condition $\partial_{\mu}\phi^{\mu}=0$ for the wave function of a spin-1 particle leads to $p^{\mu}\varepsilon_{\mu}(p;\lambda)=0$. Here p^{μ} and $\varepsilon^{\mu}(p;\lambda)$ are a system of four linearly independent, orthogonal vectors; i.e., any four-vector a^{μ} can be represented as a linear combination

$$a^{\mu}=a_p p^{\mu}+\sum_{\lambda}a_{\lambda}\varepsilon^{\mu}(p;\lambda)\ . \tag{4}$$

Exercise 2.7

Now we evaluate

$$
\begin{aligned}
a^{\mu}\eta_{\mu\nu}(p) &= \sum_{\lambda}\left[a_p p^{\mu}\varepsilon^*_{\mu}(p;\lambda) + \sum_{\lambda'} a_{\lambda'}\varepsilon^{*\mu}(p;\lambda)\varepsilon_{\mu}(p;\lambda')\right]\varepsilon_{\nu}(p;\lambda)\\
&= \sum_{\lambda}\left[a_p\cdot 0 + \sum_{\lambda'}(-\delta_{\lambda\lambda'})a_{\lambda'}\right]\varepsilon_{\nu}(p;\lambda)\\
&= -\sum_{\lambda} a_{\lambda}\varepsilon_{\nu}(p;\lambda)\ .
\end{aligned} \tag{5}
$$

This result does not depend on the specific choice of the $\varepsilon_{\mu}(p;\lambda)$ and shows that $-\eta_{\mu\nu}(p)$ eliminates the part of a given four-vector that is proportional to p^{μ}. Therefore $\eta_{\mu\nu}(p)$ can only depend on p^{μ}, i.e., $a^{\perp}_{\nu}(p) = -a^{\mu}\eta_{\mu\nu}(p)$. Since the $\varepsilon_{\mu}(p;\lambda)$ are four-vectors, the polarization sum $\eta_{\mu\nu}(p)$ is transformed like a second-rank tensor. Any symmetric tensor of second rank that is built by a four-vector p^{μ} is of the general form

$$
\eta_{\mu\nu}(p) = A(p^2)g_{\mu\nu} + B(p^2)p_{\mu}p_{\nu}\ . \tag{6}
$$

There are no other possibilities, because the only Lorentz-covariant quantities available are $g_{\mu\nu}$, p_{μ}, and p^2. Here we have $p^2 = M^2 = \text{const.}$, and therefore A and B must be constants. Multiplying the polarization sum by p^{μ} yields

$$
\begin{aligned}
& p^{\mu}\eta_{\mu\nu} = \sum_{\lambda} p\cdot\varepsilon^*(p;\lambda)\varepsilon_{\nu}(p;\lambda) = 0\\
& \to p^{\mu}(A\, g_{\mu\nu} + Bp_{\mu}p_{\nu}) = (A + Bp^2)p_{\nu} = 0\\
& \to B = -\frac{A}{M^2}\ .
\end{aligned} \tag{7}
$$

Hence the polarization sum assumes the form $A(g_{\mu\nu} - p_{\mu}p_{\nu}/M^2)$. A contraction with $g^{\mu\nu}$ leads to

$$
\begin{aligned}
g^{\mu\nu}\eta_{\mu\nu}(p) &= \eta^{\mu}{}_{\mu}(p) = \sum_{\lambda}\varepsilon^{\mu^*}(p;\lambda)\varepsilon_{\mu}(p;\lambda)\\
&= \sum_{\lambda}(-\delta_{\lambda\lambda}) = -3
\end{aligned} \tag{8}
$$

$$
\begin{aligned}
\to \eta^{\mu}{}_{\mu}(p) &= A\, g^{\mu}{}_{\mu} + Bp^2 = A\left(g^{\mu}{}_{\mu} - \frac{p^2}{M^2}\right)\\
&= A(4-1) = 3A\\
\to A &= -1\ .
\end{aligned} \tag{9}
$$

The final result is then

$$
\sum_{\lambda}\varepsilon^*_{\mu}(p;\lambda)\varepsilon_{\nu}(p;\lambda) = -\left(g_{\mu\nu} + \frac{p_{\mu}p_{\nu}}{M^2}\right)\ . \tag{10}
$$

Massless Spin-1 Particles: Photons. Photons do not possess a rest system as massive vector bosons do, since from $p^2 = 0 \leftrightarrow |E| > 0$ we immediately get $\boldsymbol{p} \neq 0$ in any inertial frame. It is therefore impossible to proceed as for massive spin-1 particles, formulating the polarization vectors in the rest frame and then obtaining them in any frame by a Lorentz transformation, which was the method just discussed. To find the number of relevant internal degrees of freedom of the photon field we will use gauge invariance, which holds for massless, but not for massive, vector bosons. It will turn out that real photons have only two transverse polarization degrees of freedom.

We first summarize. For a free photon field, the wave equation and the Lorentz condition are

$$\Box A^{\mu} = 0 \tag{2.128a}$$
$$\partial_{\mu} A^{\mu} = 0 \tag{2.128b}$$

This auxiliary condition can be always satisfied in a special gauge – the Lorentz gauge – and only in this gauge does the wave equation (2.128) have this simple form. For $p^2 = 0$ (real photons) its solutions are plane waves

$$A^{\mu} = N\varepsilon^{\mu}\, \mathrm{e}^{-\mathrm{i}p\cdot x} \;, \tag{2.129}$$

where N is a normalization factor and ε^{μ} the polarization vector of the photon. The Lorentz condition (2.128) leads immediately to the condition

$$p \cdot \varepsilon = 0 \tag{2.130}$$

for the polarization vector. Equations (2.128) and (2.130) are analogous to (2.116) and (2.122), derived for massive vector bosons. There $\partial_{\mu}\phi^{\mu}$ followed directly from the field equations (2.114) and $p \cdot \varepsilon = 0$ was first derived in the rest frame and then recognized to be covariant in general. Now (2.128) and (2.130) follow from an arbitrary choice of gauge for the photon field. The Lorentz condition (2.128) reduces the number of internal degrees of freedom to three. But as we shall see in the following, the gauge condition reduces this number to two.

In the Lorentz gauge, one can still perform the symmetry transformation

$$A^{\mu} \to A^{\mu} - \partial^{\mu} \Lambda = A'^{\mu} \;, \tag{2.131}$$

provided Λ satisfies the Klein–Gordon equation for the massless scalar field,

$$\Box \Lambda = 0 \;. \tag{2.132}$$

Such regauging obviously does not change the Lorentz condition (2.128). For the plane waves (2.129), the regauging amounts to changing the polarization vector ε^{μ} by a multiple of p^{μ}:

$$\begin{aligned} A^{\mu} - \partial^{\mu} \Lambda &= N\varepsilon^{\mu}\, \mathrm{e}^{-\mathrm{i}p\cdot x} - \alpha \partial^{\mu}\, \mathrm{e}^{-\mathrm{i}p\cdot x} \\ &= N\left(\varepsilon^{\mu} + \beta p^{\mu}\right) \mathrm{e}^{-\mathrm{i}p\cdot x} \;, \end{aligned} \tag{2.133}$$

that is,

$$\varepsilon^\mu \to \varepsilon'^\mu = \varepsilon^\mu + \beta p^\mu \ . \tag{2.134}$$

Moreover, the condition

$$\varepsilon' \cdot p = 0 \tag{2.135}$$

still holds since $\partial_\mu A'^\mu = 0$ holds as well. The freedom expressed by (2.134) has profound consequences. To illustrate this, we consider a photon with the four-momentum

$$p^\mu = \left(p^0, \boldsymbol{p}\right) \tag{2.136}$$

and the polarization vector

$$\varepsilon^\mu = \left(\varepsilon^0, \boldsymbol{\varepsilon}\right) \ , \tag{2.137}$$

which satisfy the Lorentz condition $\varepsilon \cdot p = 0$. Gauge invariance now allows us to add, according to (2.134), any multiple of p^μ to ε^μ, still obtaining admittable polarization vectors. We can therefore always choose the gauge such that the time component of ε^μ vanishes in (2.137) and the four-dimensional equation (2.135) is reduced to the three-dimensional equation

$$\boldsymbol{\varepsilon} \cdot \boldsymbol{p} = 0 \ . \tag{2.138}$$

This means that there are only two linearly independent polarization vectors for photons (massless bosons). For a plane photon wave propagating in the z direction, we can choose $\boldsymbol{\varepsilon}^{(1)}$ and $\boldsymbol{\varepsilon}^{(2)}$ from (2.117a) for linearly polarized photons and $\boldsymbol{\varepsilon}(\lambda = +1)$ and $\boldsymbol{\varepsilon}(\lambda = -1)$ for circularly polarized photons.

This is the result – known from classical electrodynamics – that the electric and magnetic field strengths

$$F_{\mu\nu} = \partial_\mu A_\nu - \partial_\nu A_\mu$$

are purely transverse. Although photons are described by a vector field A^μ and thus must have spin 1, there are only two independent spin projections and not three, as one might naively expect. As we have seen, this result is rooted in the photons having no mass. If the wave equation (2.128) had a mass term, as in the Proca equation (2.116), the theory would no longer be gauge invariant and gauge freedom would be lost.

We thus describe incoming photons with four-momentum p and polarization state λ by the wave function

$$A^\mu = N\varepsilon^\mu(\lambda)\,\mathrm{e}^{-\mathrm{i}p\cdot x} \ , \quad \lambda = \pm 1 \ , \tag{2.139}$$

and outgoing photons by

$$A^\mu = N\varepsilon^{*\mu}(\lambda)\,\mathrm{e}^{\mathrm{i}p\cdot x} \ , \quad \lambda = \pm 1 \ . \tag{2.140}$$

As in (2.123) for massive spin-1 particles, the orthogonality

$$\varepsilon^*(\lambda)\cdot\varepsilon(\lambda') = -\delta_{\lambda\lambda'} \tag{2.141}$$

holds. Normalizing in the same way as with the pion wave function, we obtain

$$N = \frac{1}{\sqrt{V}} \tag{2.142}$$

as the normalization factor.

In a certain sense we can associate the gauge system where (2.138) holds, i.e., where photons are tranverse, to the rest system (2.118), (2.119) for particles with a mass. In this rest system $\boldsymbol{\varepsilon}(\lambda)\cdot\boldsymbol{p} = 0$ also holds, since the momentum $\boldsymbol{p}$ does vanish there.

2.2.6 The Propagator for Virtual Photons

In view of the different fermion–boson scattering processes that will be derived in the next section, we now proceed to derive the pion propagator. This derivation is completely analogous to that for the photon propagator. The equation corresponding to (2.87) is

$$\left(\Box + m_0^2\right)\phi(x) \equiv -J(x)\ . \tag{2.143}$$

The Green function for this inhomogeneous Klein–Gordon equation is

$$\left(\Box + m_0^2\right)G(x-x') = -\mathrm{i}\delta^4(x-x')\ . \tag{2.144}$$

By Fourier transformation, we obtain

$$G(p^2) = \frac{\mathrm{i}}{p^2 - m_0^2 + \mathrm{i}\varepsilon} \tag{2.145}$$

where $p^2 \neq m_0^2$ has been assumed for virtual pions. We refer to standard QED textbooks for the detailed proof that the propagator is given by (2.145) and that the Feynman interpretation of waves with positive and negative energy corresponds to the $+\mathrm{i}\varepsilon$ prescription for treating the poles in (2.145).

EXAMPLE

2.8 Compton Scattering by Pions

Elastic photon scattering by a charged particle is called Compton scattering. In the case of a pion this process is written as

$$\gamma + \pi^+ \to \gamma + \pi^+ \ .$$

First we consider the direct graph (symmetrization is necessary!), which is depicted in Fig. 2.14a. The lower vertex, which is separately shown in Fig. 2.14b, corresponds to the absorption of a photon with four-momentum k_1 and polarization ε_1 and to the transition of the pion state from p_1 to q. The corresponding scattering amplitude is proportional to (cf. (2.81))

$$\sim \mathrm{i}e \int \mathrm{d}^4x\, \phi_\mathrm{f}^*(\pi^+, q)(\partial_\mu A^\mu + A^\mu \partial_\mu)\phi_\mathrm{i}(\pi^+, p_1) \ . \tag{1}$$

Inserting the plane waves

$$\begin{aligned} \phi_\mathrm{i} &\approx \mathrm{e}^{-\mathrm{i}p_1\cdot x} \ , \\ \phi_\mathrm{f}^* &\approx \mathrm{e}^{\mathrm{i}q\cdot x} \ , \\ A^\mu &\approx \varepsilon_1^\mu\, \mathrm{e}^{-\mathrm{i}k_1\cdot x} \end{aligned} \tag{2}$$

into (1) leads to a vertex amplitude proportional to

$$F_1 \approx e(p_1+q)_\mu \varepsilon_1^\mu \ , \tag{3}$$

where

$$q = p_1 + k_1 \tag{4}$$

denotes the four-momentum of the virtual photon. Energy conservation is ensured by the delta function

$$\delta^4(q - p_1 - k_1) \ . \tag{5}$$

An analogous procedure for the upper vertex in Fig. 2.14a (cf. Fig. 2.14c) leads to the vertex amplitude

$$F_2 \approx e(q+p_2)_\nu \varepsilon_2^{\nu *} \tag{6}$$

with four-momentum conservation,

$$q = p_2 + k_2 \ , \quad \delta^4(p_2 + k_2 - q) \ . \tag{7}$$

Putting graphs 2.14b and 2.14c together we obtain the complete direct graph 2.14a. Formally this connection is achieved by inserting the virtual photon propagator

$$G(q^2) = \frac{\mathrm{i}}{q^2 - m_0^2} \tag{8}$$

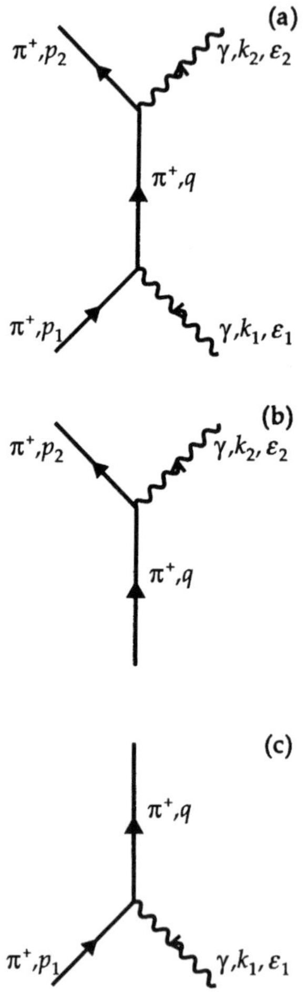

Fig. 2.14a–c. The direct graph of $\pi^+\gamma$ Compton scattering. The vertices of $\pi^+\gamma$ Compton scattering

Example 2.8

between the scattering amplitudes (3) and (6). The invariant scattering amplitude for the direct graph of Compton scattering by a pion is then

$$F^{(d)}_{\gamma\pi} = e^2 \varepsilon_1 \cdot (p_1 + q) \frac{1}{q^2 - m_0^2} \varepsilon_2^* \cdot (q + p_2) \ , \tag{9}$$

where energy conservation is ensured by $\delta^4(p_2 + k_2 - p_1 - k_1)$. Taking into account the Lorentz-gauge condition

$$\varepsilon_1 \cdot k_1 = 0 \ , \qquad \varepsilon_2 \cdot k_2 = 0 \tag{10}$$

and introducing the Mandelstam variable

$$s = q^2 = (p_1 + k_1)^2 = (p_2 + k_2)^2 \ , \tag{11}$$

we can easily transform (9) into

$$F^{(d)}_{\pi\gamma} = \frac{4e^2(\varepsilon_1 \cdot p_1)(\varepsilon_2^* \cdot p_2)}{s - m_0^2} \ . \tag{12}$$

Now we have to symmetrize with respect to the two photons (or the two pions). The resulting exchange graph is shown in Fig. 2.15. First the outgoing photon k_2, ε_2 is emitted and later the incoming photon k_1, ε_1 is absorbed. In analogy to (9) the scattering amplitude for this process is

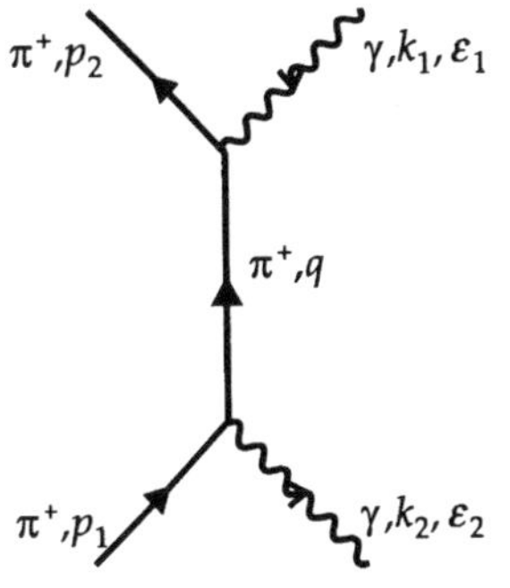

Fig. 2.15. The exchange graph for $\pi^+\gamma$ Compton scattering

$$F^{(e)}_{\gamma\pi} = e^2 \varepsilon_1 \cdot (q' + p_2) \frac{1}{q'^2 - m_0^2} \varepsilon_2^* \cdot (q' + p_1) \ , \tag{13}$$

with

$$q' = p_1 - k_2 = p_2 - k_1 \ . \tag{14}$$

If we employ the Mandelstam variable

$$u = (p_1 - k_2)^2 = (p_2 - k_1)^2 = q'^2 \ , \tag{15}$$

which represents the squared mass of the virtual pion in the exchange graph, and take (10) into account, the amplitude (13) assumes the form

$$F^{(e)}_{\gamma\pi} = \frac{4e^2(\varepsilon_1 \cdot p_2)(\varepsilon_2^* \cdot p_1)}{u - m_0^2} \ . \tag{16}$$

This exchange amplitude is sometimes called the u-channel contribution (amplitude).

Now the question arises whether there are contributions to Compton scattering that are of order e^2 caused by the squared interaction term

$$-e^2 A_\mu A^\mu \ . \tag{17}$$

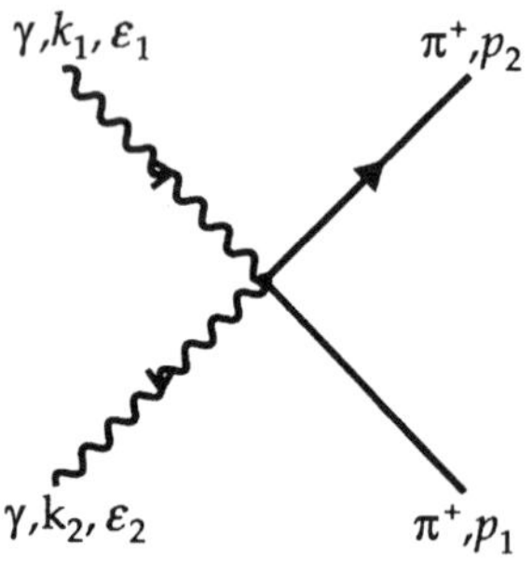

Fig. 2.16. The four-point contact graph for $\gamma\pi^+ \to \gamma\pi^+$

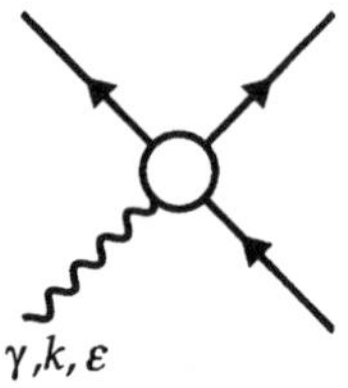

Fig. 2.17. The graph of a general one-photon process

This interaction will directly lead to graphs of the form shown in Fig. 2.16, i.e., to vertices with two photon lines. Such contributions are called contact terms. They can be interpreted using gauge invariance arguments. To this end we first consider the process shown in Fig. 2.17, where the initial state consists of one photon and one particle and the final state of two particles. One photon is absorbed and two particles are emitted in the final channel. Of course, one of the outgoing particles can again be a photon, i.e., the Compton scattering treated above is included in such a process. The corresponding amplitude (see Fig. 2.17) must be linear in the polarization ε^μ and can therefore be factorized into

$$A = \varepsilon^\mu T_\mu \ , \tag{18}$$

where T_μ contains all the details of the process. In the Lorentz gauge, the condition

$$\varepsilon \cdot k = 0$$

must hold. As we have already discussed in Sect. 2.2.5 an additional gauge transformation

$$\varepsilon'^\mu = \varepsilon^\mu + \beta k^\mu$$

may be performed without changing any physical results. This transformation corresponds to an additional gauge transformation within the Lorentz gauge. Since the amplitude (18) must be gauge invariant, we are lead to the condition

$$k_\mu T^\mu = 0 \ . \tag{19}$$

The total scattering amplitude for $\gamma\pi^+$ Compton scattering derived above can be written as

$$\begin{aligned} F_{\gamma\pi} &= F^{(\mathrm{d})}_{\gamma\pi} + F^{(\mathrm{e})}_{\gamma\pi} \\ &= 4e^2 \varepsilon_1^\mu \varepsilon_2^{\nu *} \left(\frac{p_{1\mu} p_{2\nu}}{s - m_0^2} + \frac{p_{2\mu} p_{1\nu}}{u - m_0^2} \right) \\ &\equiv \varepsilon_1^\mu \varepsilon_2^{\nu *} T_{\mu\nu} \ . \end{aligned} \tag{20}$$

According to the transition from (18) to (19) we have to replace ε_1 by k_1 and evaluate

$$k_1^\mu \varepsilon_2^{\nu *} T_{\mu\nu} \ .$$

The scalar products that occur,

$$2p_1 \cdot k_1 = s - m_0^2 \ , \quad 2p_2 \cdot k_1 = -(u - m_0^2) \ , \tag{21}$$

then give

$$\begin{aligned} k_1^\mu \varepsilon_2^{\nu *} T_{\mu\nu} &= 2e^2 \varepsilon_2^{\nu *} (p_2 - p_1)_\nu \\ &= 2e^2 \varepsilon_2^* \cdot k_1 \neq 0 \ . \end{aligned} \tag{22}$$

Example 2.8

This result is surprising, because the total scattering amplitude (20) derived so far explicitly violates condition (19). If we also replace ε_2 by k_2 in (22), we obtain

$$k_1^\mu k_2^\nu T_{\mu\nu} = 2e^2 k_1 \cdot k_2 \neq 0 \ . \tag{23}$$

Again this result is not equal to zero. The only possible explanation for this observation is that the amplitude (20) is not gauge invariant. But where did we make a mistake? We wanted to evaluate a process of second order in e, but so far we have not taken into account the contact graphs according to the interaction (17), which are of the same order. In order to restore gauge invariance and fulfill conditions (19), T_μ (or $T_{\mu\nu}$ in (20)) must contain all the interactions of a given order. If the coupling constant e is interpreted as a variable quantity (which then assumes some fixed value), gauge invariance must separately be fulfilled in every order in e. We therefore expect an additional term $F_{\gamma\pi}^{(c)}$ for the complete Compton scattering amplitude, i.e.,

$$F_{\gamma\pi} = F_{\gamma\pi}^{(e)} + F_{\gamma\pi}^{(c)} \ , \tag{24}$$

where the superscript (c) denotes the contact term. Also $F_{\gamma\pi}^{(c)}$ must be linear in ε_1 and ε_2^* and by means of the replacements

$$\varepsilon_1 \to k_1 \ , \quad \varepsilon_2^* \to k_2 \tag{25}$$

it must yield (22) or (23) with the opposite sign. Only in this way can the gauge invariance of the scattering amplitude (24) be ensured. Apparently,

$$F_{\gamma\pi}^{(c)} = -2e^2 \varepsilon_1 \cdot \varepsilon_2^* \tag{26}$$

must hold. This scattering amplitude is caused by the quadratic interaction. One must understand that if we had in general ignored the interaction $-e^2 A^2$, gauge invariance in second order would have demanded its existence. This is a first example of the power of gauge symmetry. A further comment on (26): this term is the only one that is linear in ε_1 and ε_2^* and up to a sign equal to (22) and (23). There are no other terms fulfilling these conditions!

Taking into account (20), (24), and (26), we get for the total invariant scattering amplitude for Compton scattering by a pion

$$F_{\gamma\pi} = e^2 \varepsilon_1^\mu \varepsilon_2^{\nu *} \left(\frac{4p_{1\mu} p_{2\nu}}{s - m_0^2} + \frac{4p_{2\mu} p_{1\nu}}{u - m_0^2} - 2g_{\mu\nu} \right) \ . \tag{27}$$

The factor 2 in front of the $g_{\mu\nu}$ term is plausible, because each factor A_μ in A^2 can represent one absorption and one emission process. Multiplying (27) by the four-momentum conservation $(2\pi)^4 \delta^4(p_2 + k_2 - p_1 - k_1)$ and normalization factors then yields the complete Compton S-matrix element.

2.3 Fermion–Boson and Fermion–Fermion Scattering

In this section we discuss a number of problems and examples to review and deepen our knowledge of QED. We shall encounter well-known subjects in new forms and also gain new insights. Our notation will get closer to that employed in high-energy physics.

EXERCISE

2.9 Elastic $e^-\pi^+$ Scattering (I)

Problem. Determine the scattering amplitude and explain the formal steps necessary to evaluate the cross section.

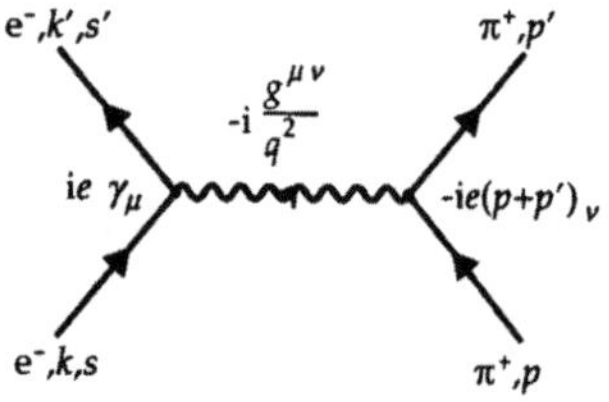

Fig. 2.18. The Feynman diagram for $e^-\pi^+$ scattering in the one-photon exchange approximation

Solution. The graph for $e^-\pi^+$ scattering is of the following form (see Fig. 2.18). Most of the above notation is readily understood. Only the factor $+\mathrm{i}$ in the transition current at the electron vertex, which has been denoted by $(\mathrm{i}e\gamma_\mu)$, needs additional explanation. Reviewing our knowledge of QED, we start with the scattering amplitude (2.50), i.e., with

$$S^{(1)}_{\mathrm{fi}} = -\mathrm{i}\int \mathrm{d}^4x\, j_\mu(\mathrm{e}^-)A^\mu \ , \tag{1}$$

where according to (2.52)

$$j_\mu(\mathrm{e}^-) = (-e)NN'\,\bar{u}(k',s')\,\gamma_\mu\,u(k,s)\,\mathrm{e}^{\mathrm{i}(k'-k)\cdot x} \tag{2}$$

denotes the electron transition current density. The electromagnetic four-potential A^μ in (1) is created by the pion transition current density

$$j^\mu(\pi^+) = (+e)\bar{N}\bar{N}'(p+p')^\mu\,\mathrm{e}^{\mathrm{i}(p'-p)\cdot x} \ . \tag{3}$$

According to (2.91) we have

$$A^\mu = -\frac{1}{q^2}\,j^\mu(\pi^+) \ , \tag{4}$$

with the four-momentum transfer

$$q = p' - p = k - k' \ . \tag{5}$$

Exercise 2.9

The scattering amplitude (1) in detail is then

$$
\begin{aligned}
S_{\mathrm{fi}}^{(1)} &= -\mathrm{i}\int \mathrm{d}^4x\, j_\mu(e^-)\left(-\frac{1}{q^2}\right) j^\mu(\pi^+) \\
&= \mathrm{i}NN'\bar{N}\bar{N}'\int \mathrm{d}^4x\left[\bar{u}(k',s')\gamma_\mu u(k,s)\,\mathrm{e}^{\mathrm{i}(k'-k)\cdot x}\right] \\
&\quad\times\left(-\frac{e^2}{q^2}\right)\left[(p+p')^\mu\,\mathrm{e}^{\mathrm{i}(p-p')\cdot x}\right] \\
&= -\mathrm{i}NN'\bar{N}\bar{N}'(2\pi)^4\,\delta^4(k'+p'-k-p)(-e)\bar{u}(k',s')\gamma_\mu u(k,s) \\
&\quad\times\left(-\frac{g^{\mu\nu}}{q^2}\right)(+e)(p+p')_\nu \\
&= -\mathrm{i}NN'\bar{N}\bar{N}'(2\pi)^4\,\delta^4(p'+k'-p-k)F_{ss'}(kp;k'p')\ .
\end{aligned}
\tag{6}
$$

In the last step we have introduced the invariant scattering amplitude

$$
F_{ss'}(kp;k'p') = (-e)\left[\bar{u}(k',s')\gamma_\mu u(k,s)\right]\left(-\frac{g^{\mu\nu}}{q^2}\right)(+e)\left[(p+p')_\nu\right]\ . \tag{7}
$$

Only the spin variables occur in addition. Now we see from (6) that the above Feynman rules yield the correct total sign for the scattering amplitude if a factor $+\mathrm{i}$ is assigned to the vertex of the leptonic transition current. Figure 2.18 already contains this factor. Note that the spinor combinations $(\bar{u}\gamma_\mu u)$ are the components of a four-vector. Contracting this vector with $(p+p')_\nu$ yields a Lorentz scalar and therefore a Lorentz-invariant scattering amplitude. The derivation of the cross section consists in the same steps, which have been discussed in detail for π^+K^+ scattering. Employing the four-vectors

$$
\begin{aligned}
k^\mu &= (\omega,\boldsymbol{k})\ ,\quad k'^\mu = (\omega',\boldsymbol{k}')\ ,\\
p^\mu &= (E,\boldsymbol{p})\ ,\quad p'^\mu = (E',\boldsymbol{p}')\ ,
\end{aligned}
\tag{8}
$$

we find the differential cross section to be

$$
\mathrm{d}\sigma_{ss'} = (2\pi)^4\delta(k'+p'-k-p)\frac{|F_{ss'}|^2}{2E2\omega|\boldsymbol{v}|}\frac{1}{(2\pi)^3}\frac{\mathrm{d}^3k'}{2\omega'}\frac{1}{(2\pi)^3}\frac{\mathrm{d}^3p'}{2E'}\ . \tag{9}
$$

The scattering amplitude $F_{ss'}$ can be easily evaluated if we insert the spinors (see (2.36))

$$
u(k,s) = \sqrt{w+m_0}\begin{pmatrix}\phi^s\\ \frac{\hat{\boldsymbol{\sigma}}\cdot\boldsymbol{k}}{\omega+m_0}\phi^s\end{pmatrix} \tag{10}
$$

into (7). This procedure is quite tedious and, more importantly, does not yield the quantity observed in most of the actual experiments. The interesting quantity is the so-called nonpolarized cross section, which is obtained from (9) by averaging over the initial spins and summing over the final spins, i.e.,

$$
\begin{aligned}
\mathrm{d}\bar{\sigma} &= \frac{1}{2}(d\sigma_{\uparrow\uparrow}+\mathrm{d}\sigma_{\uparrow\downarrow}+\mathrm{d}\sigma_{\uparrow\downarrow}+\mathrm{d}\sigma_{\downarrow\downarrow})\\
&= \frac{1}{2}\sum_{ss'}\mathrm{d}\sigma_{ss}\ .
\end{aligned}
\tag{11}
$$

Exercise 2.9

These lengthy summations need not be performed explicitly. Instead, employing Feynman's trace techniques enables us to drastically simplify the spin summations (11).

2.3.1 Traces and Spin Summations

Let us briefly review the basic facts about trace techniques. The trace of a matrix is the sum of its diagonal elements, i.e.,

$$\mathrm{tr}\{A\} = \sum_i A_{ii} \ . \tag{2.146}$$

Cyclic permutability holds under the trace

$$\begin{aligned} \mathrm{tr}\{AB\} &= \mathrm{tr}\{BA\} \ , \\ \mathrm{tr}\{ABC\} &= \mathrm{tr}\{CAB\} = \mathrm{tr}\{BCA\} \ . \end{aligned} \tag{2.147}$$

The most important relations for traces over products of γ matrices and Feynman "daggers" needed in this context are as follows. By using the anticommutation relation

$$\{\gamma^\mu, \gamma^\nu\} = 2g^{\mu\nu}\mathbb{1}$$

the following traces can be easily evaluated:

$$\begin{aligned} \mathrm{tr}\{\mathbb{1}\} &= 4 \ , \\ \mathrm{tr}\{\gamma_\mu\gamma_\nu\} &= 4\, g_{\mu\nu} \ , \\ \mathrm{tr}\{\not{a}\not{b}\} &= 4\, a\cdot b \ , \\ \mathrm{tr}\{\not{a}\not{b}\not{c}\not{d}\} &= 4\big[(a\cdot b)(c\cdot d) + (a\cdot d)(b\cdot c) - (a\cdot c)(b\cdot d)\big] \ , \end{aligned} \tag{2.148}$$

where use has been made of the fact that

$$\not{a}\not{b} = -\not{b}\not{a} + 2\, a\cdot b \tag{2.149}$$

holds (see Exercise 2.10). An expression for the scattering cross section in lepton–pion scattering was derived in the Exercise 2.9. There the square of the scattering cross section appeared:

$$\begin{aligned} |F_{ss'}|^2 = \left(\frac{e^2}{q^2}\right)^2 & \Big\{\big[\bar{u}(k',s')\gamma_\mu u(k,s)\big][p+p']^\mu\Big\} \\ & \times \Big\{\big[\bar{u}(k',s')\gamma_\nu u(k,s)\big][p+p']^\nu\Big\}^* \ . \end{aligned} \tag{2.150}$$

Since $F_{ss'}$ is a complex number (all matrix indices are summed over), we can replace the conjugate term $\{\dots\}^*$ by the Hermitian conjugate; we find that

$$\begin{aligned}\left\{\left[\bar{u}(k',s')\gamma_\nu u(k,s)\right][p+p']^\nu\right\}^\dagger &= \left\{\left[u^\dagger(k,s)\gamma_\nu^\dagger\gamma_0 u(k',s')\right][p+p']^\nu\right\}\\ &= \left\{\left[\bar{u}(k,s)\gamma_\nu u(k',s')\right][p+p']^\nu\right\}, \end{aligned} \tag{2.151}$$

since

$$\gamma_0\gamma_\nu^\dagger\gamma_0 = \gamma_\nu , \quad \gamma_0^\dagger = \gamma_0 \tag{2.152}$$

holds. The four-vector $(p+p')^\nu$ has real components and is therefore not changed by the operations $(\dots)^*$ and $(\dots)^\dagger$. In performing the spin summation according to (11) in Exercise 2.9 we must calculate the following expression:

$$\begin{aligned}\frac{1}{2}\sum_{s,s'}|F_{ss'}|^2 &= \frac{1}{2}\left(\frac{e^2}{q^2}\right)^2\sum_{s,s'}\left[\bar{u}(k',s')\gamma_\mu u(k,s)\right]\\ &\quad\times\left[\bar{u}(k,s)\gamma_\nu u(k',s')\right](p+p')^\mu(p+p')^\nu\\ &\equiv \left(\frac{e^2}{q^2}\right)^2 L_{\mu\nu}T^{\mu\nu} , \end{aligned} \tag{2.153}$$

where we have introduced the so-called lepton tensor

$$L_{\mu\nu} = \frac{1}{2}\sum_{s,s'}\bar{u}(k',s')\gamma_\mu u(k,s)\bar{u}(k,s)\gamma_\nu u(k',s') \tag{2.154}$$

and the so-called hadron tensor

$$T^{\mu\nu} = (p+p')^\mu(p+p')^\nu . \tag{2.155}$$

The factorization of the scattering cross section in a leptonic and a hadronic part stems from the one-photon approximation. To higher order, the situation is more complicated. The middle term (the direct product of two spinors)

$$\sum_s u(k,s)\otimes\bar{u}(k,s) = \sum_s u_\beta(k,s)\bar{u}_\gamma(k,s) , \tag{2.156}$$

which no longer depends on s (s is summed over), is a 4×4 matrix. Making use of the explicit form $u(k,s)$ of the spinor and the fact that for two-component spinors ϕ^s

$$\begin{aligned}\sum_s \phi^s\phi^{s\dagger} &= \phi^1\phi^{1\dagger}+\phi^2\phi^{2\dagger}\\ &= \begin{pmatrix}1\\0\end{pmatrix}(1,0)+\begin{pmatrix}0\\1\end{pmatrix}(0,1)\\ &= \begin{pmatrix}1&0\\0&0\end{pmatrix}+\begin{pmatrix}0&0\\0&1\end{pmatrix} = \mathbb{1}\end{aligned} \tag{2.157}$$

holds, we can deduce that

$$\sum_s u(k,s)\bar{u}(k,s) = u(k,\uparrow)\bar{u}(k,\downarrow) + u(k,\downarrow)\bar{u}(k,\downarrow)$$
$$= (\omega + m_0)\begin{pmatrix} 1\!\!1 & -\frac{\hat{\sigma}\cdot\boldsymbol{k}}{\omega+m_0} \\ \frac{\hat{\sigma}\cdot\boldsymbol{k}}{\omega+m_0} & -\frac{\omega-m_0}{\omega+m_0}1\!\!1 \end{pmatrix}$$
$$= (\not{k} + m_0) \ . \tag{2.158}$$

In Exercise 2.10, we shall perform this calculation in detail. Using this result, (2.156) can be expressed by (2.158) and inserted into (2.154), which leads to

$$L_{\mu\nu} = \frac{1}{2}\sum_{s'}\sum_{\alpha,\beta,\gamma,\delta} \bar{u}_\alpha(k',s')(\gamma_\mu)_{\alpha\beta}(\not{k}+m_0)_{\beta\gamma}(\gamma_\nu)_{\gamma\delta}u_\delta(k',s') \tag{2.159}$$

written in expanded matrix notation. Each single factor in this term represents a c number, and the factors can therefore be reordered arbitrarily. In particular, the sum

$$\sum_{s'} u_\delta(k',s')\,\bar{u}_\alpha(k',s') = (\not{k}' + m_0)_{\delta\alpha} \tag{2.160}$$

can be performed using (2.158), and the lepton tensor is thus reduced to the form

$$L_{\mu\nu} = \frac{1}{2}\sum_{\alpha,\beta,\gamma,\delta} (\not{k}' + m_0)_{\delta\alpha}(\gamma_\mu)_{\alpha\beta}(\not{k}+m_0)_{\beta\gamma}(\gamma_\nu)_{\gamma\delta}$$
$$= \frac{1}{2}\mathrm{tr}\{(\not{k}' + m_0)\gamma_\mu(\not{k}+m_0)\gamma_\nu\} \ . \tag{2.161}$$

Using the trace formulas from (2.148) we obtain

$$\mathrm{tr}\{(\not{k}' + m_0)\gamma_\mu(\not{k}+m_0)\gamma_\nu\} = \mathrm{tr}\{\not{k}'\gamma_\mu\not{k}\gamma_\nu\} + m_0^2\mathrm{tr}\{\gamma_\mu\gamma_\nu\}$$
$$= 4(k'_\mu k_\nu + k'_\nu k_\mu - k\cdot k' g_{\mu\nu}) + 4m_0^2 g_{\mu\nu} \ . \tag{2.162}$$

If we take into account that the mass squared is

$$q^2 = (k-k')^2 = k^2 + k'^2 - 2k\cdot k' = 2(m_0^2 - k\cdot k') \ , \tag{2.163}$$

this can be inserted into (2.152), and the lepton tensor can finally be written as

$$L_{\mu\nu} = 2\left(k'_\mu k_\nu + k'_\nu k_\mu + \frac{1}{2}q^2 g_{\mu\nu}\right) \ . \tag{2.164}$$

This tensor not only is important for elastic scattering, but also plays a major role in most quark–parton calculations. We shall perform the contraction of $L_{\mu\nu}$ with the hadron tensor $T^{\mu\nu}$ in the pion rest system ($p^\mu = (M,\mathbf{0})$) according to (2.153) in Exercise 2.11 to calculate the cross section. Neglecting the electron mass (the ultrarelativistic limit), the result is

$$\frac{d\bar{\sigma}}{d\Omega} = \frac{\alpha^2}{4k^2 \sin^4(\theta/2)} \frac{k'}{k} \cos^2\left(\frac{\theta}{2}\right)$$
$$= \frac{\alpha^2}{4\omega^2 \sin^4(\theta/2)} \frac{\omega'}{\omega} \cos^2\left(\frac{\theta}{2}\right) , \quad (2.165)$$

where

$$k = |\boldsymbol{k}| = \omega ,$$
$$k' = |\boldsymbol{k}'| = \omega' , \quad (2.166)$$

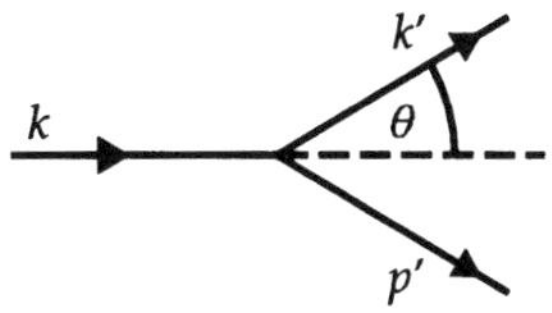

Fig. 2.19. The definition of the e^- scattering angle in $e^-\pi^+$ scattering

and

$$q^2 = -4\,kk' \sin^2\left(\frac{\theta}{2}\right) . \quad (2.167)$$

θ is the scattering angle of the electron (see Fig. 2.19).

The preceding considerations that culminate in the scattering cross section (2.165) are based on treating pions and electrons as pointlike. We thus label the cross section "n.s." (meaning "no structure"):

$$\left(\frac{d\bar{\sigma}}{d\Omega}\right)_{\text{n.s.}} \quad (2.168)$$

Today it is accepted that pions have an internal structure, being composed of quarks, antiquarks, and gluons. Leptons, on the other hand, are still considered to be elementary, i.e., without internal structure. This conclusion is drawn mainly from lepton–lepton scattering.

EXERCISE

2.10 Features of Dirac Matrices

Problem. Start with the anticommutator

$$\{\gamma^\mu, \gamma^\nu\} = 2g^{\mu\nu}\mathbb{1} \quad (1)$$

and show that the following relations hold:

(a) $\not{a}\not{b} = -\not{b}\not{a} + 2a \cdot b$. (2)

(b) $(\not{k} - m_0)(\not{k} + m_0) = (\not{k} + m_0)(\not{k} - m_0)$
$= 0 , \quad \text{for} \quad k^2 = m_0^2$. (3)

Exercise 2.10

(c) $\hat{\Lambda}_+(k) = (\not{k} + m_0)$, (4)

which eliminates the negative energy parts of an arbitrary spinor, and

$$\hat{\Lambda}_-(k) = (\not{k} - m_0) \ , \tag{5}$$

which eliminates the positive energy parts of an arbitrary spinor.
(d) Employ the explicit forms

$$u(k,s) = \sqrt{\omega + m_0}\begin{pmatrix} \phi^s \\ \frac{\hat{\boldsymbol{\sigma}}\cdot\boldsymbol{k}}{\omega+m_0}\phi^s \end{pmatrix} \ ,$$
$$\phi^1 = \begin{pmatrix} 1 \\ 0 \end{pmatrix} \ , \quad \phi^2 = \begin{pmatrix} 0 \\ 1 \end{pmatrix} \ , \tag{6}$$

$$v(k,s) = \sqrt{\omega + m_0}\begin{pmatrix} \frac{\hat{\boldsymbol{\sigma}}\cdot\boldsymbol{k}}{\omega+m_0}\chi^s \\ \chi^s \end{pmatrix} \ ,$$
$$\chi^1 = \begin{pmatrix} 0 \\ 1 \end{pmatrix} \ , \quad \chi^2 = \begin{pmatrix} 1 \\ 0 \end{pmatrix} \tag{7}$$

and show how the 4×4 matrices $\sum_s u(k,s)\bar{u}(k,s)$ and $\sum_s v(k,s)\bar{v}(k,s)$ depend on $\hat{\Lambda}_+(k)$ and $\hat{\Lambda}_-(k)$, respectively.

Solution.

(a)
$$\begin{aligned}\not{a}\not{b} + \not{b}\not{a} &= a_\mu\gamma^\mu b_\nu\gamma^\nu + b_\nu\gamma^\nu a_\mu\gamma^\mu = a_\mu b_\nu\left\{\gamma^\mu, \gamma^\nu\right\} \\ &= 2a_\mu b_\nu g^{\mu\nu}\mathbb{1} = 2a\cdot b\mathbb{1} \ ;\end{aligned} \tag{8}$$

from this equation follows in particular that $\not{a}^2 = a^2\mathbb{1}$.

(b)
$$\begin{aligned}(\not{k} - m_0\mathbb{1})(\not{k} + m_0\mathbb{1}) &= \left(\not{k}^2 - m_0^2\right)\mathbb{1} \\ &= \left(k^2 - m_0^2\right)\mathbb{1} = 0 \ .\end{aligned} \tag{9}$$

(c) The wave function of a particle with positive energy and four-momentum k_μ can be written as

$$\Psi^{(+)}(x) = \sum_s b_s u(k,s)\,\mathrm{e}^{-\mathrm{i}k\cdot x} \ . \tag{10}$$

Correspondingly a solution with negative energy and four-momentum $k_{-\mu}$ is

$$\Psi^{(-)}(x) = \sum_s d_s v(k,s)\,\mathrm{e}^{\mathrm{i}k\cdot x} \ . \tag{11}$$

Both wave functions must obey the Dirac equation, i.e.,

$$\begin{aligned}(\mathrm{i}\not{\nabla} - m_0)\,\Psi^{(+)}(x) &= (\not{k} - m_0)\,\Psi^{(+)}(x) \\ &= \hat{\Lambda}_{-(k)}\Psi^{(+)}(x) = 0\end{aligned} \tag{12}$$

Exercise 2.10

and

$$(\mathrm{i}\nabla\!\!\!/ - m_0)\,\Psi^{(-)}(x) = -\,(k\!\!\!/ + m_0)\,\Psi^{(-)}(x)$$
$$= -\hat{\Lambda}_{+(k)}\Psi^{(-)}(x) = 0\ . \qquad (13)$$

Equations (12) and (13) can also be put into the form

$$k\!\!\!/\Psi^{(+)}(x) = m_0\Psi^{(+)}(x)\ ,$$
$$k\!\!\!/\Psi^{(-)}(x) = -m_0\Psi^{(-)}(x)\ . \qquad (14)$$

From (14) we immediately obtain

$$\hat{\Lambda}_+\Psi = 2m_0\Psi^{(+)}\ ,$$
$$\hat{\Lambda}_-\Psi = -2m_0\Psi^{(-)}\ . \qquad (15)$$

(d) $$\sum_s u(k,s)\bar{u}(k,s) = \sum_s u(k,s)u^\dagger(k,s)\gamma^0$$
$$= (\omega+m_0)\sum_s \begin{pmatrix} \phi^s \\ \frac{\hat{\boldsymbol{\sigma}}\cdot\boldsymbol{k}}{\omega+m_0}\phi^s \end{pmatrix} \left(\phi^{s\dagger}, \phi^{s\dagger}\frac{\hat{\boldsymbol{\sigma}}^\dagger\cdot\boldsymbol{k}}{\omega+m_0}\right) \begin{pmatrix} 1\!\!1 & 0 \\ 0 & -1\!\!1 \end{pmatrix}$$
$$= (\omega+m_0)\sum_s \begin{pmatrix} \phi^s\phi^{s\dagger} & -\phi^s\phi^{s\dagger}\frac{\hat{\boldsymbol{\sigma}}^\dagger\cdot\boldsymbol{k}}{\omega+m_0} \\ \frac{\hat{\boldsymbol{\sigma}}\cdot\boldsymbol{k}}{\omega+m_0}\phi^s\phi^{s\dagger} & -\frac{\hat{\boldsymbol{\sigma}}\cdot\boldsymbol{k}}{\omega+m_0}\phi^s\phi^{s\dagger}\frac{\hat{\boldsymbol{\sigma}}^\dagger\cdot\boldsymbol{k}}{\omega+m_0} \end{pmatrix}\ . \qquad (16)$$

Now we employ the Hermiticity of the Pauli matrices, $\hat{\boldsymbol{\sigma}}^\dagger = \hat{\boldsymbol{\sigma}}$, and make use of the identity $\sum_s \phi^s\phi^{s\dagger} = 1\!\!1$ (cf. (2.157)). Equation (16) then assumes the form

$$(\omega+m_0)\begin{pmatrix} 1\!\!1 & -\frac{\hat{\boldsymbol{\sigma}}\cdot\boldsymbol{k}}{\omega+m_0} \\ \frac{\hat{\boldsymbol{\sigma}}\cdot\boldsymbol{k}}{\omega+m_0} & -\frac{(\hat{\boldsymbol{\sigma}}\cdot\boldsymbol{k})^2}{(\omega+m_0)^2} \end{pmatrix}\ . \qquad (17)$$

But since we have

$$(\hat{\boldsymbol{\sigma}}\cdot\boldsymbol{k})^2 = \frac{1}{2}k_ik_j\left(\hat{\sigma}_i\hat{\sigma}_j + \hat{\sigma}_j\hat{\sigma}_i\right) = k_ik_j\sigma_{ij}$$
$$= \boldsymbol{k}^2 = \omega^2 - m_0^2\ ,$$

the result becomes

$$\begin{pmatrix} (\omega+m_0)1\!\!1 & -\hat{\boldsymbol{\sigma}}\cdot\boldsymbol{k} \\ \hat{\boldsymbol{\sigma}}\cdot\boldsymbol{k} & (-\omega+m_0)1\!\!1 \end{pmatrix}\ . \qquad (18)$$

Employing the γ-matrix representation

$$\gamma^0 = \begin{pmatrix} 1\!\!1 & 0 \\ 0 & -1\!\!1 \end{pmatrix}\ , \quad \boldsymbol{\gamma} = \begin{pmatrix} 0 & \boldsymbol{\sigma} \\ -\boldsymbol{\sigma} & 0 \end{pmatrix}\ , \qquad (19)$$

Exercise 2.10

we obtain the result

$$\sum_s u(k,s)\bar{u}(k,s) = \omega\gamma^0 - \boldsymbol{k}\cdot\boldsymbol{\gamma} + m_0$$
$$= (\not{k} + m_0) = \hat{\Lambda}_+(k) \ .$$

An analogous calculation for negative solutions yields

$$\begin{aligned}
&\sum_s v(k,s)\bar{v}(k,s) \\
&= (\omega+m_0)\sum_s \begin{pmatrix} \frac{\hat{\boldsymbol{\sigma}}\cdot\boldsymbol{k}}{\omega+m_0}\chi^s \\ \chi^s \end{pmatrix} \left(\chi^{s\dagger}\frac{\hat{\boldsymbol{\sigma}}\cdot\boldsymbol{k}}{\omega+m_0}, -\chi^s\right) \\
&= (\omega+m_0)\begin{pmatrix} \frac{\omega^2-m_0^2}{(\omega+m_0)^2} & -\frac{\hat{\boldsymbol{\sigma}}\cdot\boldsymbol{k}}{\omega+m_0} \\ \frac{\hat{\boldsymbol{\sigma}}\cdot\boldsymbol{k}}{\omega+m_0} & -\mathbb{1} \end{pmatrix} \\
&= \begin{pmatrix} (\omega-m_0)\mathbb{1} & -\hat{\boldsymbol{\sigma}}\cdot\boldsymbol{k} \\ \hat{\boldsymbol{\sigma}}\cdot\boldsymbol{k} & (-\omega-m_0)\mathbb{1} \end{pmatrix} \\
&= \omega\gamma^0 - \boldsymbol{k}\cdot\boldsymbol{\gamma} - m_0 \\
&= (\not{k} - m_0) = -\hat{\Lambda}_-(k) \ .
\end{aligned} \tag{20}$$

EXERCISE

2.11 Electron–Pion Scattering (II)

Problem. Evaluate in detail the nonpolarized π^+e^- cross section. Start with the previously discussed formula

$$\mathrm{d}\bar{\sigma} = \frac{1}{4E\omega|\boldsymbol{v}|}\frac{1}{2}\sum_{s,s'}|F_{ss'}|^2\mathrm{d}\,\mathrm{Lips}(s;k'p') \ .$$

Determine $\mathrm{d}\bar{\sigma}/\mathrm{d}\Omega$ in the rest system of the pion ($p^\mu = (M,\boldsymbol{0})$).

Solution. We denote the four-momenta of the pion before and after the collision by $p^\mu = (M,\boldsymbol{0})$ and $p'^\mu = (E',\boldsymbol{p}')$, respectively. $k^\mu = (\omega,\boldsymbol{k})$ and $k'^\mu = (\omega',\boldsymbol{k}')$ are the corresponding four-momenta of the incoming and outgoing electrons. The scattering angle θ is the angle between the directions $\boldsymbol{k}$ and $\boldsymbol{k}'$ and $q = k' - k$ denotes the momentum transfer. In the following we only consider high electron energies, i.e., since $\omega, \omega' \gg m$, the electron rest mass can be neglected. Therefore we have $\omega = |\boldsymbol{k}|$ and $\omega' = |\boldsymbol{k}'|$ and the invariant flux factor is simply

$$4\sqrt{(k\cdot p)^2 - m^2M^2} \simeq 4M\omega \ . \tag{1}$$

Exercise 2.11

According to the Feynman rules the spin average of the squared scattering amplitude describing the exchange of one photon is given by the contraction of lepton and hadron tensor:

$$\frac{1}{2}\sum_{s,s'}|F_{ss'}|^2 = \left(\frac{4\pi\alpha}{q^2}\right)^2 L_{\mu\nu}T^{\mu\nu} \ . \tag{2}$$

Employing (2.155) and (2.164) we obtain

$$L_{\mu\nu}T^{\mu\nu} = 8\left(2(k\cdot p)(k'\cdot p) + \frac{q^2}{2M^2}\right) \ . \tag{3}$$

In the ultrarelativistic limit, q^2 becomes

$$\begin{aligned} q^2 &= (k-k')^2 = (\omega-\omega')^2 - (\boldsymbol{k}-\boldsymbol{k}')^2 \\ &\approx -2\omega\omega'(1-\cos\theta) = -4\omega\omega'\sin^2\left(\frac{\theta}{2}\right) \ . \end{aligned}$$

Hence

$$\left(\frac{4\pi\alpha}{q^2}\right)^2 = \left(\frac{\pi\alpha}{\omega\omega'}\right)^2 \frac{1}{\sin^4\left(\frac{\theta}{2}\right)} \tag{4}$$

and

$$L_{\mu\nu}T^{\mu\nu} = 16M^2\omega\omega'\cos^2\left(\frac{\theta}{2}\right) \ . \tag{5}$$

Now we have to evaluate the invariant phase-space factor:

$$\begin{aligned} \mathrm{d}\,\mathrm{Lips}(s;k'p') &= (2\pi)^4\delta^4(k'+p'-k-p)\frac{1}{(2\pi)^3}\frac{\mathrm{d}^3p'}{2E'}\frac{1}{(2\pi)^3}\frac{\mathrm{d}^3k'}{2\omega'} \\ &= \frac{1}{(4\pi)^2}\delta^3(\boldsymbol{k}'+\boldsymbol{p}'-\boldsymbol{k})\delta(E'-M+\omega'-\omega)\frac{\mathrm{d}^3p'}{E'}\frac{\mathrm{d}^3k'}{\omega'} \ . \end{aligned} \tag{6}$$

In order to evaluate the cross section for the electron to be scattered into a given final state, we have to integrate over all final states of the pion. This is readily achieved by means of the δ^3 function in (6) (owing to momentum conservation only one final pion state is possible for a given final state of the electron). In the remaining integrand $\boldsymbol{p}'$ must then be replaced by $\boldsymbol{k}-\boldsymbol{k}' = -\boldsymbol{q}$. Because of the identity

$$\begin{aligned} E' &= \sqrt{M^2+\boldsymbol{p}'^2} = \sqrt{M^2+\boldsymbol{q}^2} \\ &\approx \sqrt{M^2+\omega^2+\omega'^2-2\omega\omega'\cos\theta} \end{aligned} \tag{7}$$

E' is not an independent quantity either. Except for this factor there are no further dependences of the integrand on $\boldsymbol{p}'$, including the scattering amplitude and flux factor. Finally $\omega' = |\boldsymbol{k}'|$ leads to

$$\frac{\mathrm{d}^3k'}{\omega'} = \omega'\,\mathrm{d}\omega'\,\mathrm{d}\Omega \ , \tag{8}$$

Exercise 2.11

where $\mathrm{d}\Omega$ denotes the spherical angle into which the electron is scattered. The ω' integration is performed by using the remaining energy δ function, i.e., for a given scattering angle the absolute value of the electron momentum is fixed by kinematic arguments. Note that in the argument of the δ function E' depends on ω' as well (cf. (7)). Therefore we have to employ the relation

$$\delta\left(f(x)\right) = \sum_i \frac{\delta(x-x_i)}{f'(x)|_{x=x_i}} \,, \tag{9}$$

where the x_i denote the zeros of the function $f(x)$. According to (7) we have

$$\frac{\mathrm{d}(E'-M+\omega'-\omega)}{\mathrm{d}\omega'} = \frac{\omega'-\omega\cos\theta+E'}{E'} \,. \tag{10}$$

Inserting now the identity $E' = M+\omega-\omega'$, which has been derived by integrating over the δ function, into (7) and (10), we obtain

$$M(\omega-\omega') = \omega\omega'(1-\cos\theta) \,, \tag{11}$$

$$\frac{\mathrm{d}}{\mathrm{d}\omega'}(E'-M+\omega-\omega') = \frac{\omega(1-\cos\theta)+M}{E'} \,, \tag{12}$$

or

$$\frac{\mathrm{d}}{\mathrm{d}\omega'}(E'+\omega') = \frac{M\omega}{E'\omega'} \,. \tag{13}$$

The (partially integrated) Lorentz-invariant phase-space factor now assumes the form

$$\mathrm{dLips} = \frac{1}{(4\pi)^2}\frac{\omega'^2}{M\omega}\mathrm{d}\Omega \,, \tag{14}$$

where ω' is fixed by (11). Summarizing equations (1), (4), (5), and (14) finally yields the nonpolarized cross section

$$\left(\frac{\mathrm{d}\bar{\sigma}}{\mathrm{d}\Omega}\right)_{\mathrm{n.s.}} = \frac{\alpha^2}{4\omega^2\sin^4\left(\frac{\theta}{2}\right)}\frac{\omega'}{\omega}\cos^2\left(\frac{\theta}{2}\right) \,. \tag{15}$$

EXERCISE

2.12 Positron–Pion Scattering

Problem. Show that the cross section for $e^+\pi^+$ scattering is, in the one-photon-exchange approximation, equal to that for $e^-\pi^+$ scattering.

Solution. The graph for $e^+\pi^+$ scattering is of the following form. The incoming positron with four-momentum k and spin s is described as an outgoing electron with four-momentum $-k$ and spin $-s$. Correspondingly the outgoing positron with k', s' can be interpreted as an incoming electron with $-k'$, $-s'$. Therefore we have for the positron transition current

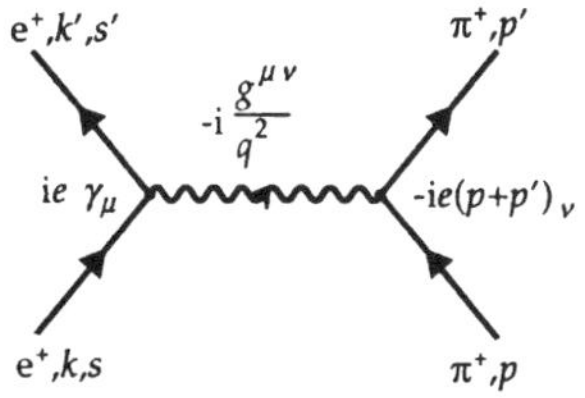

Fig. 2.20. $e^+\pi^+$ scattering in the one-photon- exchange approximation

$$j_\mu(e^+) = (-e)NN'\bar{v}(k,s)\gamma_\mu v(k',s')\,e^{-i(-k'+k)\cdot x} \;, \tag{1}$$

where $v(k',s')$ and $\bar{v}(k',s')$ represent an incoming electron wave $-k'$, $-s'$ and an outgoing electron wave $-k$, $-s$, respectively. If we construct the scattering matrix element with $j_\mu(e^+)$, an additional minus sign is inserted according to the previously introduced rules. Everything else remains the same as in the $e^-\pi^+$ scattering discussed in Exercises 2.9 and 2.11. The evaluation of the cross section, too, is analogous to that of $e^-\pi^+$ scattering except for the spin average, where the sum

$$\sum_s u(k,s)\bar{u}(k,s) = (\not{k}+m_0) \tag{2}$$

in the lepton tensor must be replaced by (cf. (2.154))

$$\sum_s v(k,s)\bar{v}(k,s) = (\not{k}-m_0) \;. \tag{3}$$

This calculation was made in Exercise 2.10 and leads to the trace

$$\mathrm{tr}\left\{(\not{k}'-m_0)\gamma_\mu(\not{k}-m_0)\gamma_\nu\right\} \tag{4}$$

instead of

$$\mathrm{tr}\left\{(\not{k}'+m_0)\gamma_\mu(\not{k}+m_0)\gamma_\nu\right\} \;. \tag{5}$$

Because of rules (2.148) we immediately recognize that these two traces are equal. Therefore the $e^+\pi^+$ cross section is to lowest order exactly equal to the $e^-\pi^+$ cross section. This result does not surprise us at all, since only the lepton charge changed its sign and the lowest-order cross section contains only the square of this charge.

2.3.2 The Structure of the Form Factors from Invariance Considerations

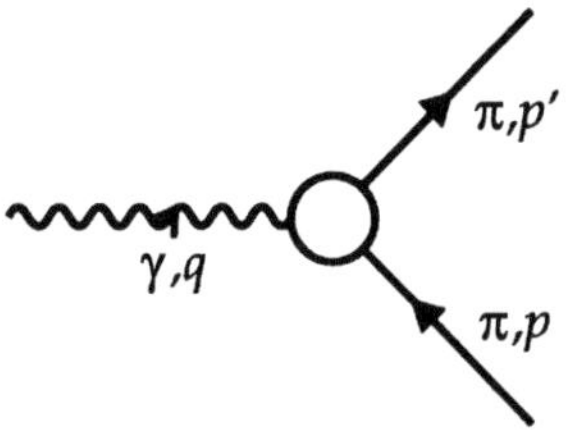

Fig. 2.21. The general photon–pion vertex. The circle indicates the internal structure of the pion

We now assume that the pion has an internal structure that we do not know exactly but that can be parametrized in some rather general way, as we shall demonstrate shortly. The photon–pion vertex with an internal pion structure is drawn in Fig. 2.21. In place of some simple vertex, a circle is drawn, symbolizing the internal strcuture of the pion, while a simple junction indicates the vertex of a pointlike pion (without internal structure).

To exemplify which kind of processes can be contained in the circle, we mention the two following graphs, drawn in the conventional framework without imagining a quark structure of the pion.

Owing to the strong interaction (large coupling constant) such virtual processes can contribute considerably to the total scattering amplitude. As we shall see, even without detailed knowledge of the internal structure of the pion, general statements about the form of the required modifications of the vertex can be obtained from the Lorentz and gauge invariance of the theory. These two general requirements fix the form of the scattering amplitude for any internal structure imaginable.

For a pointlike pion, the transition amplitude is

$$j^\mu(\pi^+) = eNN'(p+p')^\mu \, \mathrm{e}^{\mathrm{i}(p'-p)\cdot x} \ . \tag{2.169}$$

We now consider the current (2.169) as a matrix element of an electromagnetic current operator (Heisenberg operator)

$$\hat{j}^\mu_{\mathrm{em}}(x) \tag{2.170}$$

and write

$$j^\mu(\pi^+) = \langle \pi^+ p' | \hat{j}^\mu_{\mathrm{em}}(x) | \pi^+ p \rangle \ . \tag{2.171}$$

If (2.171) describes the special transition current of pointlike pions, we identify

$$\langle \pi^+ p' | \hat{j}^\mu_{\mathrm{em}}(0) | \pi^+ p \rangle = eNN'(p+p')^\mu \ . \tag{2.172}$$

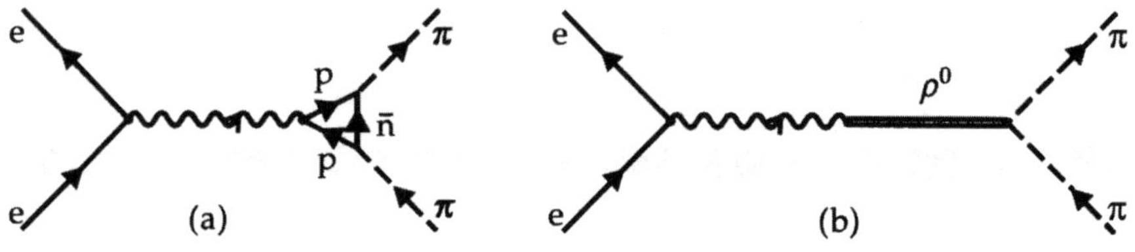

Fig. 2.22a,b. Complex virtual processes founded on strong interactions and contributing to the structure of the $\gamma\pi^+$ vertex: (**a**) a virtual nucleon loop; (**b**) a virtual ϱ^0 meson

Since the matrix element (2.171) is taken with plane pion waves, its x dependence is given by the exponential factor in (2.169). This will also hold unmodified for pions with internal structure, but we expect that the strong interaction modifies the right-hand side of (2.172), i.e., the four-momentum dependence:

$$\langle \pi^+ p' | \hat{j}^\mu_{\text{em}}(0) | \pi^+ p \rangle \equiv N N' \Gamma^\mu(p, p', q) \ , \tag{2.173}$$

where the momentum transfer or four-momentum of the virtual photon q can also enter.

Taking a rather pragmatic point of view, we simply try to parametrize the four-current or Γ^μ. As Γ^μ must remain a four-vector, we shall first discuss which general four-vector Γ^μ can be constructed from the available four-vectors p, p', and q. Owing to four-momentum conservation at the vertex

$$p' = p + q \ , \tag{2.174}$$

only two independent four-vectors are available, which can be chosen to be

$$(p' + p)^\mu \quad \text{and} \quad (p' - p)^\mu = q^\mu \ . \tag{2.175}$$

Both can be utilized in the construction of Γ^μ. Additionally, both four-vectors (2.175) can be multiplied with an unknown scalar function. From the relations

$$\begin{aligned} &p^2 = p'^2 = M^2 \ , \\ &p'^2 = p^2 + 2\,p\cdot q + q^2 \ \Leftrightarrow\ 2\,p\cdot q = -q^2 \ , \end{aligned} \tag{2.176}$$

and

$$q^2 = 2M^2 - 2\,p\cdot p' \tag{2.177}$$

it can be seen that only one nontrivial scalar can be composed of the four-vectors (2.175), namely $p\cdot p'$ or, equivalently, q^2, the square of the momentum transfer at the vertex. Using this, we can write the vertex function most generally using Lorentz invariance as

$$\Gamma^\mu(p, p', q) = e\left[F(q^2)(p+p')^\mu + G(q^2)q^\mu\right] \ . \tag{2.178}$$

The scalar functions $F(q^2)$ and $G(q^2)$ are called form factors.

To discover what statements can be made about the form factors from gauge invariance, we remember that the Maxwell equations are left invariant by the gauge transformation $A^\mu \to A'^\mu = A^\mu - \partial^\mu \Lambda$. The Maxwell equations in the Lorentz gauge are

$$\Box A^\mu = \Box A'^\mu = j^\mu_{\text{em}} \ . \tag{2.179}$$

Since, owing to current conservation, only those Λ satisfying $\Box \Lambda = 0$ can be considered, this gauge condition implies current conservation:

$$\partial_\mu j^\mu_{\text{em}} = 0 \ . \tag{2.180}$$

Charge conservation must obviously hold both for the transition current (2.169) of a pointlike pion,

$$-\mathrm{i}\partial_\mu j^\mu(\pi^+) = q_\mu j^\mu(\pi^+) = 0 \ , \quad q\cdot(p+p') = 0 \ , \tag{2.181}$$

and for the current

$$q_\mu \langle \pi^+ p' | \hat{j}^\mu(0) | \pi^+ p \rangle = 0 \ . \tag{2.182}$$

Both conditions thus demand in general that

$$q_\mu \Gamma^\mu = q_\mu e \left[F(q^2)(p+p')^\mu + G(q^2) q^\mu \right] = 0 \ . \tag{2.183}$$

The first term always vanishes since $q\cdot(p+p') = 0$ owing to (2.175). Only the second term will not always vanish, if $q^2 \neq 0$, so that (2.183) can only be satisfied if

$$G(q^2) = 0 \ . \tag{2.184}$$

In other words, gauge invariance of the theory leads automatically to the statement that all structural effects of the pion (mainly caused by the strong interaction) can be described by a form factor as a function of the photon mass q^2. We thus find that

$$(\Gamma^\mu)_{\text{n.s.}} = e(p'+p)^\mu \ , \quad (\Gamma^\mu)_{\text{w.s.}} = eF(q^2)(p'+p)^\mu \ , \tag{2.185}$$

where the subscript "w.s." indicates "with structure". Since the charge e appears as a prefactor, (2.185) also contains the definition of charge in the sense that the form factor at vanishing four-momentum transfer ($q^2 = 0$) must be unity:

$$F(0) = 1 \ . \tag{2.186}$$

The form factor is measured in scattering processes for all values of q^2, both for $q^2 \leq 0$ (spacelike four-momenta measured in $\mathrm{e}^-\phi^+$ scattering) and $q^2 \geq 4M^2$ (timelike four-momenta in the "crossing reaction" $\mathrm{e}^+\mathrm{e}^- \to \pi^+\pi^-$). Naturally these measurements ask new questions of the theory in order to explain the form factor $F(q^2)$. For physical reasons, we are led to expect that $F(q^2)$ diminishes when $|q^2|$ is increased, since it becomes increasingly difficult to transfer momentum to the various constituents of the pion in order that it stays intact (elastic scattering as opposed to inelastic scattering).

EXAMPLE

2.13 Electron–Muon Scattering

As an example of lepton–lepton scattering we briefly discuss $e^-\mu^-$ scattering, which is represented in the one-photon-exchange approximation by the following graph. Using the experiences of previous problems we can immediately write down the invariant scattering amplitude, which now depends on four spin indices:

$$F_{sr;s'r'} = (-e)^2 \overline{u}(k',s')\gamma_\mu u(k,s)\left(-\frac{g^{\mu\nu}}{q^2}\right)\overline{u}(p',r')\gamma_\mu u(p,r) \ . \tag{1}$$

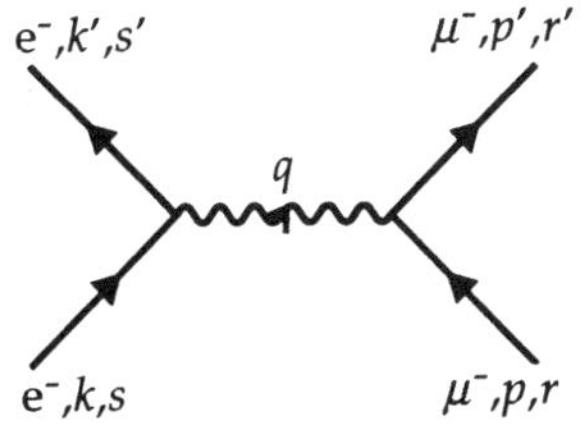

Fig. 2.23. The one-photon-exchange amplitude for $e^-\mu^-$ scattering

Again the nonpolarized cross section is proportional to the square of this amplitude averaged over initial spins and summed over final spins, i.e.,

$$d\overline{\sigma} = \frac{1}{4}\sum_{r,s,r',s'} \left|F_{sr;s'r'}\right|^2 \ . \tag{2}$$

We can perform the same steps as for $e^-\pi^+$ scattering (cf. (2.153)–(2.164)) for each transition current separately. This simplification is caused by the factorization of the currents in the one-photon-exchange approximation. We obtain

$$\begin{aligned}\frac{1}{4}\sum_{r,s,r',s'} &= \left(\frac{e^2}{q^2}\right)^2 \left[\frac{1}{2}\mathrm{tr}\left\{(\not{k}'+m_0)\gamma_\mu(\not{k}+m_0)\gamma_\nu\right\}\right] \\ &\quad\times\left[\frac{1}{2}\mathrm{tr}\left\{(\not{p}'+M_0)\gamma^\mu(\not{p}+M_0)\gamma^\nu\right\}\right] \\ &\equiv \left(\frac{e^2}{q^2}\right)^2 L_{\mu\nu}M^{\mu\nu} \ .\end{aligned} \tag{3}$$

Employing our previous results (2.164) we can immediately write down the electron tensor

$$L_{\mu\nu} = 2\left[k'_\mu k_\nu + k'_\nu k_\mu + \frac{q^2}{2}g_{\mu\nu}\right] \tag{4}$$

and the muon tensor

$$M^{\mu\nu} = 2\left[p'^\mu p^\nu + p'^\nu p^\mu + \frac{q^2}{2}g^{\mu\nu}\right] \ . \tag{5}$$

In order to obtain the cross section we have to evaluate the contraction of these two tensors $L_{\mu\nu}M^{\mu\nu}$. The direct evaluation is straightforward but quite lengthy. We prefer to employ the following trick, which follows from the current conservation. Because $q^\mu = k'^\mu - k^\mu$, the electron current conservation

$$\begin{aligned}\partial^\mu j_\mu(e^-) &= 0 \ , \\ q^\mu\left[\overline{u}(k',s')\gamma_\mu u(k,s)\right] &= 0\end{aligned} \tag{6}$$

Example 2.13

can be written as

$$\overline{u}(k', s') \left(\not{k}' - \not{k}\right) u(k, s) = 0 \; . \tag{7}$$

Equation (7) can be explicitly obtained from the corresponding Dirac equations for $\overline{u}\left(k', s'\right)$ and $u\,(k, s)$, respectively, and is valid for all possible spin projections. But since $L_{\mu\nu}$ is a product of two transition currents, we immediately get

$$q^{\mu} L_{\mu\nu} = q^{\nu} L_{\mu\nu} = 0 \; . \tag{8}$$

This result is very useful because in evaluating the contraction $L_{\mu\nu} M^{\mu\nu}$ we can omit all terms proportional to q. Therefore we are able to simplify the quantities $p' = p + q$ and to consider the so-called *effective muon tensor*

$$M_{\text{eff}}^{\mu\nu} = 2\left[2p^{\mu} p^{\nu} + \frac{q^2}{2} g^{\mu\nu}\right] , \tag{9}$$

which yields the same result for the contraction to be calculated, i.e.,

$$L_{\mu\nu} M^{\mu\nu} = L_{\mu\nu} M_{\text{eff}}^{\mu\nu} \; . \tag{10}$$

A straightforward but cumbersome calculation yields the following result for the nonpolarized cross section in the rest frame of the muon ($p^{\mu} = (M, 0, 0, 0)$):

$$\frac{\mathrm{d}\overline{\sigma}}{\mathrm{d}\Omega} = \left(\frac{\mathrm{d}\overline{\sigma}}{\mathrm{d}\Omega}\right)_{\text{n.s.}} \left(1 - \frac{q^2 \tan^2\left(\frac{\theta}{2}\right)}{2M^2}\right) , \tag{11}$$

where θ denotes the angle between $\boldsymbol{k}$ and $\boldsymbol{k}'$. The following remarks should be noted:

1. $(\mathrm{d}\overline{\sigma}/\mathrm{d}\Omega)_{\text{n.s.}}$ is the *no-structure* cross section known from $\mathrm{e}^-\pi^+$ scattering (cf. (2.165) and (2.168)). It is modified by an additional term proportional to $\tan^2(\theta/2)$. This effect is caused by the spin-$\frac{1}{2}$ nature of the muon. The muon has not only a charge but also a magnetic moment. The latter is automatically taken into account by the Dirac equation. In other words, compared with $\mathrm{e}^-\pi^+$ scattering, we observe an additional scattering by the normal magnetic moment in the case of $\mathrm{e}^-\mu^-$ scattering.

2. The electron rest mass was neglected in the kinematics of (11), i.e., we considered only the ultrarelativistic limit.

3. We wrote down the $\mathrm{e}^-\pi^+$ as well as the $\mathrm{e}^-\mu^-$ cross sections in the rest system of the π^+ and μ^-, respectively, which can hardly be realized in experimental setups. Later this kind of cross section for structureless particles will be useful in the discussion of the quark–parton model. One has to understand these cross sections in order to acknowledge the physical content of parton dynamics.

4. The *crossed reaction* $\mathrm{e}^+\mathrm{e}^- \to \mu^+\mu^-$ is frequently investigated in electron–positron collisions in the context of so-called *colliding-beam experiments*. It is also important for testing the quark–parton model if compared with the reaction

$$\mathrm{e}^+\mathrm{e}^- \to \text{hadrons} \; .$$

Example 2.13

An analogous calculation leads to the cross section

$$\frac{\mathrm{d}\overline{\sigma}}{\mathrm{d}\Omega} = \frac{\alpha^2}{4q^2}\left(1+\cos^2\theta\right) , \tag{12}$$

where all variables are defined in the center-of-mass system of the e^+e^- pair and all masses are neglected (ultrarelativistic limit). θ denotes the angle between the axis of the incoming and the axis of the outgoing particles.

3. Scattering Reactions and the Internal Structure of Baryons

3.1 Simple Quark Models Compared

In the first chapter we showed that the baryon spectrum by itself already suggests that baryons are composed of quarks. However, the interaction between quarks cannot be easily deduced from the energies of the states, since different models yield nearly identical and well-fitting descriptions of the mass spectrum. Such models include the flavor SU(6) model, the MIT bag model, the Skyrmion bag model, and potential models with nonrelativistic quarks. Luckily lattice calculations are now good enough to demonstrate that the *correct* model, namely QCD, gives equally good results.

The flavor SU(6) model postulates that the up, down, and strange quark species are eigenvalues of an internal symmetry group, namely SU(3). To include spin and thus the splitting between the spin-$\frac{1}{2}$ and spin-$\frac{3}{2}$ multiplets, this group is extended to SU(6) $\supset$ SU(2) $\otimes$ SU(3). This symmetry is then broken in such a manner that the mass terms that appear depend only on operators that can be diagonalized simultaneously. In this way we obtain mass formulas that describe mass differences in a multiplet. The simplest expression is the Gursey–Radicati mass formula:

$$M = a + bY + c\left[T(T+1) - \frac{1}{4}Y^2\right] + dS(S+1) \; ; \tag{3.1}$$

Y, T, and S are the hypercharge, isospin, and spin of the baryon. Using the four parameters a to d we can fit the lowest baryon resonances very well. With

$$a = 1065.5\,\text{MeV}, \; b = -193\,\text{MeV}, \; c = 32.5\,\text{MeV}, \; \text{and } d = 67.5\,\text{MeV} \; , \tag{3.2}$$

for example, the numbers given in Table 3.1 are obtained. To describe the other baryons as well an internal angular momentum must be introduced, i.e., we suppose that quarks inside baryons can fill states with any angular momentum. As the rotation group is O(3), we are thus led to the SU(6) $\otimes$ O(3) symmetry group.

Indeed we can describe the full baryon spectrum starting from the SU(6) $\otimes$ O(3) mass formulas. Owing to the size of the symmetry group, many different contributions appear, making the procedure rather tedious.[1] Also the discovery of every new quark, such as the charm and bottom quarks, demands an

[1] For a discussion, see M. Jones, R.H. Dalitz, and R.R. Hougan: Nucl. Phys. B **129**, 45 (1977).

Table 3.1. The Gursey–Radicati mass formula

Particles	Mass from (3.1)	Experimental data
$N1/2^+$	939 MeV	939 MeV
$\Lambda 1/2^+$	1116 MeV	1116 MeV
$\Sigma 1/2^+$	1181 MeV	1189 MeV
$\Xi 1/2^+$	1325 MeV	1318 MeV
$\Delta 3/2^+$	1239 MeV	1230–1234 MeV
$\Sigma 3/2^+$	1384 MeV	1385 MeV
$\Xi 3/2^+$	1528 MeV	1533 MeV
$\Omega 3/2^+$	1672 MeV	1672 MeV

extension of the flavor SU(6) model, leading to yet more complicated and unsatisfactory models. The extremely large masses of the heavy quarks signal that flavor symmetry is heavily broken, rendering such models much less attractive.

In the MIT bag model, presented in Sect. 3.3, quarks can occupy all states satisfying the specific boundary conditions. Such states exist for any angular momentum, i.e., the single-particle spectrum of the quarks contains all states known from atomic physics: $s_{1/2}, p_{1/2}, p_{3/2}, d_{3/2}, d_{5/2}, \ldots$. In principle, many-particle states with definite spin and parity could be constructed from this, thus deriving the corresponding masses from the bag boundary conditions. However, this procedure gives disastrously bad results. To improve these, additional residual interactions (like the one-gluon exchange) and other corrections can be taken into account, but satisfactory baryon spectra are obtained only after introducing a sufficiently large number of parameters.[2]

Also the Skyrmion bag model, which is based on totally different assumptions, yields similar results. There the baryon number is regarded as a topological quantum number. We shall not investigate this model further but shall illustrate the basic idea with a simple example. As we remarked in Sect. 1.1, spin and isospin are isomorphic. In particular, the regular representation of isospin, e.g., pions, with its three isospin unit vectors $|\pi^0\rangle$, $\frac{1}{\sqrt{2}}(|\pi^+\rangle + |\pi^-\rangle)$ and $\frac{1}{\sqrt{2}\mathrm{i}}(|\pi^+\rangle - |\pi^-\rangle)$ is isomorphic to angular momentum. As angular momenta can be represented by vectors in three-dimensional space, pions also can be interpreted as vectors in three-dimensional isospin space. A very interesting construction is now obtained by coupling the direction of this isospin vector to the position vector, for example by demanding that the isospin vector $\boldsymbol{t}$ at position $\boldsymbol{x}$ points in the direction $\boldsymbol{t} = \boldsymbol{x}/|\boldsymbol{x}|$ in isospin space. In this way we obtain a pion field consisting purely of π^0s along the z axis and of a mixture of π^0, π^+, and π^- at other positions (see Fig. 3.1). This construction is termed the *hedgehog* solution (because the isospin vectors point outwards like the spikes of a hedgehog). To reverse this orientation of the pion field, one would have to change $\boldsymbol{\pi}(\boldsymbol{x})$ in

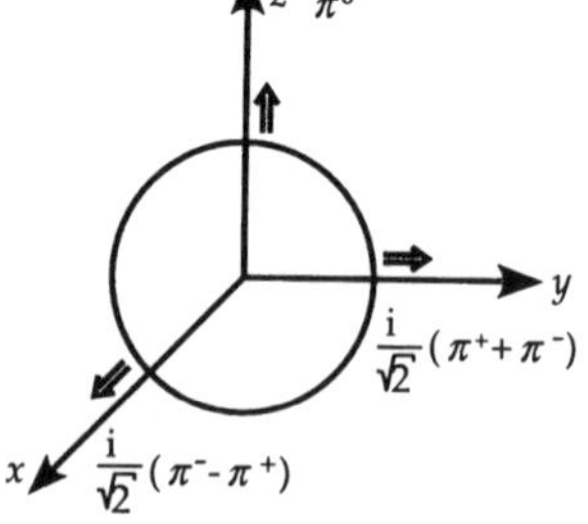

Fig. 3.1. The schematic form of the *hedgehog* solution

[2] See, for example, T.A. De Grand and R.L. Jaffe: Ann. Phys. **100**, 425 (1976) and T.A. De Grand: Ann. Phys. **101**, 496 (1976).

Table 3.2. A test of the potential model

Particles	Potential model	Experiment
$N1/2^-$	1490 MeV	1520–1555 MeV
$N1/2^-$	1655 MeV	1640–1680 MeV
$N3/2^-$	1535 MeV	1515–1530 MeV
$N3/2^-$	1745 MeV	1650–1750 MeV
$\Lambda1/2^-$	1490 MeV	1407 ± 4 MeV
$\Lambda1/2^-$	1650 MeV	1660–1680 MeV
$\Lambda1/2^-$	1800 MeV	1720–1850 MeV
$\Lambda3/2^-$	1490 MeV	1519.5 ± 1 MeV
$\Lambda3/2^-$	1690 MeV	1685–1695 MeV
$\Lambda3/2^-$	1880 MeV	?

an infinite spatial domain (at $|\boldsymbol{x}| \to \infty$), which would require infinite energy. Thus a single hedgehog is stable, and the number of hedgehogs can be identified with the baryon number. More precisely, a *topological quantum number* is defined that specifies how often $\boldsymbol{\pi}(|\boldsymbol{x}|)$ covers all isospin values for $|\boldsymbol{x}| \to \infty$. The different states with topological quantum number 1 are then identified with the different baryons.

Potential models simply solve the Schrödinger equation for nonrelativistic quarks including a spin–spin and spin–tensor interaction. The basic Hamiltonian is

$$\begin{aligned} H = & \sum_i \frac{\boldsymbol{p}_i^2}{2m_i} + \frac{1}{2}K \sum_{i>j} V_{ij}^2 \\ & + \sum_{i>j} \frac{2\alpha_s}{3m_i m_j} \left\{ \frac{8\pi}{3} \boldsymbol{s}_i \cdot \boldsymbol{s}_j \, \delta^3(\boldsymbol{r}) + \frac{1}{r^3}\left[3(\boldsymbol{s}_i \cdot \boldsymbol{r})(\boldsymbol{s}_j \cdot \boldsymbol{r}) - \boldsymbol{s}_i \cdot \boldsymbol{s}_j\right] \right\} \\ & + U(r_{ij}) \ . \end{aligned} \tag{3.3}$$

The coupling constants K, α_s, the masses m_i, and the (weak) residual interaction are fitted to the baryon ground states,[3] and excited states are then predicted. As Table 3.2 shows, the predictions obtained in this way coincide rather well with experimental values.[4]

In conclusion, completely different models describe the mass spectrum equally well, which implies that nothing can be learned about the underlying interaction from baryon masses alone. Also other parameters, such as magnetic moments, do not give more information. However, there are experimental results which are really sensitive. These are the so-called *structure functions* deduced

[3] See, for example, N. Isgur and G. Karl: Phys. Rev. D **19**, 2653 (1979a).

[4] See A.J.G. Hey and R.L. Kelly: Phys. Rep. **96**, 72 (1983).

from scattering reactions. Their definition, measurement, and meaning will be discussed in detail in this chapter. Structure functions are sensitive to the details of the interaction to such an extent that, contrary to the situation with the mass formulas, no current model yields a really satisfactory description. Only a complete solution of quantum chromodynamics could achieve this.

3.2 The Description of Scattering Reactions

To learn about the internal structure of nucleons, we must consider the scattering of particles as pointlike as possible, such as the scattering of high-energy electrons, muons, or neutrinos off nucleons:

$$\mathrm{e}^-(E \gg 1\,\mathrm{GeV}) + \mathrm{N} \to \mathrm{e}^- \ldots \;, \tag{3.4}$$

$$\nu_\mathrm{e}(E \gg 1\,\mathrm{GeV}) + \mathrm{N} \to \mathrm{e}^- \ldots \;. \tag{3.5}$$

Since highly energetic leptons have a very small wavelength, namely $\lambda \approx 1/E < 0.2\,\mathrm{fm}$, and do not possess a resolvable internal structure, the cross sections of these reactions depend solely on the internal structure of the nucleon. As electron scattering takes place mainly by photon exchange, it senses the electromagnetic charge distribution, whereas reaction (3.5) occurs through the weak interaction and gives information about the corresponding distribution of "weak charge". By comparing the results of different scattering reactions, we thus obtain a nearly complete description of the internal structure of the nucleon. The internal structure of baryon resonances and heavy mesons cannot, of course, be determined in this way because of the small lifetime of these particles. Although some information can be obtained from their decay properties, only the structure functions of the proton, neutron, and pion are known.

We shall now discuss the scattering of an electron off a nucleon. This is often discussed in textbooks on *quantum electrodynamics*, leading to the Rosenbluth formula. We shall shortly repeat this discussion and introduce a new, more practical notation for the process of Fig. 3.2. Since QED is Lorentz-covariant, the vertex function Γ_μ (or, more precisely, the matrix element $\bar{u}(P', S')\, \Gamma_\mu\, u(P, S)$) must be a Lorentz vector. The most general structure of Γ_μ is thus

$$\Gamma_\mu = A\gamma_\mu + BP'_\mu + CP_\mu + \mathrm{i}DP'^\nu\sigma_{\mu\nu} + \mathrm{i}EP^\nu\sigma_{\mu\nu} \;, \tag{3.6}$$

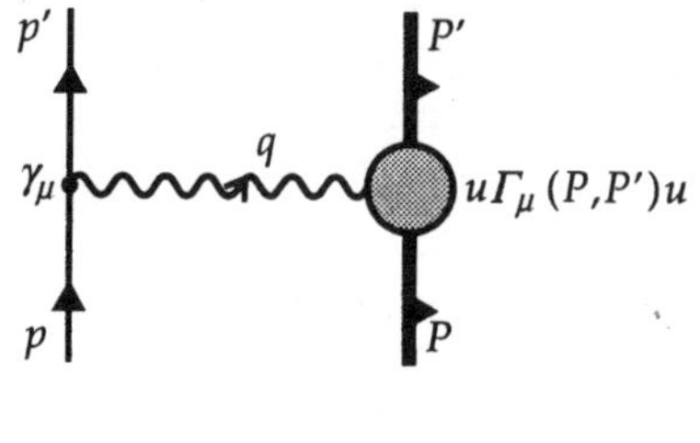

Fig. 3.2. Elastic electron-nucleon scattering

where the quantities A, B, ..., E depend only on Lorentz-invariant quantities. Since all these invariants can be expressed in terms of M_N^2 and q^2,

$$\begin{aligned}
P\cdot P &= P'\cdot P' = M_\mathrm{N}^2 \;, \\
P\cdot P' &= -\frac{1}{2}(P-P')^2 + M_\mathrm{N}^2 = -\frac{1}{2}q^2 + M_\mathrm{N}^2 \;, \\
P\cdot q &= P\cdot P' - P^2 = -\frac{1}{2}q^2 \;, \\
P'\cdot q &= P'^2 - P'\cdot P = \frac{1}{2}q^2 \;,
\end{aligned} \tag{3.7}$$

$A = A(q^2)$, $B = B(q^2)$, etc. hold. From the demand for gauge invariance, it follows, on the other hand, that

$$q^\mu\,\bar{u}(P')\,\Gamma_\mu\,u(P) = 0 \;. \tag{3.8}$$

Substituting from (3.6) yields $D = -E$ and $C = B$ and thus

$$\begin{aligned}
\overline{u}(P')\,\Gamma_\mu(P',P)\,u(P) = \overline{u}(P')\Big[&A(q^2)\gamma_\mu + B(q^2)(P'+P)_\mu \\
&+\mathrm{i}D(q^2)(P'-P)^\nu\sigma_{\mu\nu}\Big]u(P) \;.
\end{aligned} \tag{3.9}$$

On physical grounds we demand that the transition current must be Hermitian. For (3.9) to be invariant under the transformation $(\dots)^+|P_\mu \to P'_\mu$, A, B, and D must be real (see Exercise 3.3). Using the Gordon decomposition, [5]

$$\overline{u}(P')\,\gamma_\mu\,u(P) = \left(P+P'\right)_\mu\overline{u}(P')\,u(P) + i\left(P'-P\right)^\nu\overline{u}(P')\,\sigma_{\mu\nu}\,u(P)$$

the second term on the right-hand side can be expressed by the first and the third terms, and we get

$$\overline{u}(P')\,\Gamma_\mu(P',P)\,u(P) = \overline{u}(P')\Big[A(q^2)\gamma_\mu + \mathrm{i}B(q^2)q^\nu\sigma_{\mu\nu}\Big]u(P) \;. \tag{3.10}$$

The absolute square of this expression enters in the cross section. Hence one is led to the expression

$$\begin{aligned}
W_{\mu\nu} &= \frac{1}{2}\sum_{\text{spin}}\left[\overline{u}(P')\,\Gamma_\mu\,u(P)\right]^*\left[\overline{u}(P')\,\Gamma_\nu\,u(P)\right] \\
&= \frac{1}{2}\mathrm{tr}\Big\{(A\gamma_\mu - \mathrm{i}Bq^\lambda\sigma_{\mu\lambda})(\not{P}' + M_\mathrm{N})(A\gamma_\nu + \mathrm{i}Bq^\varrho\sigma_{\nu\varrho})(\not{P} + M_\mathrm{N})\Big\} \;.
\end{aligned} \tag{3.11}$$

where we have utilized the relation $\gamma^0\hat{\Gamma}_\mu^\dagger\gamma^0 = \hat{\Gamma}_\mu$, which can be directly verified in the standard representation of the γ_μ. Note that we have have used here the normalization convention for the spinors u and v expressed in (2.43). Therefore the projection operator is $\sum_s u(PS)\overline{u}(PS) = \not{P} + M_\mathrm{N}$. Remember that in

[5] See W. Greiner and J. Reinhardt: *Quantum Electrodynamics*, 2nd ed. (Springer, Berlin, Heidelberg 1996)

the spinor normalization used standardly in QED, the right-hand side would be $(\not{P}+M_\mathrm{N})/2M_\mathrm{N}$.

After some lengthy calculation we find that

$$\begin{aligned} W_{\mu\nu} &= 2(A+2M_\mathrm{N}B)^2\Big[P_\mu P'_\nu + P'_\mu P_\nu - (P\cdot P' - M_\mathrm{N}^2)g_{\mu\nu}\Big] \\ &\quad + \left[-4(A+2M_\mathrm{N}B)M_\mathrm{N}B + 2M_\mathrm{N}^2B^2\left(\frac{P\cdot P'}{M_\mathrm{N}^2}+1\right)\right] \\ &\quad \times (P_\mu + P'_\mu)(P_\nu + P'_\nu)\ . \end{aligned} \tag{3.12}$$

Replacing P'_μ by $q_\mu + P_\mu$ yields

$$\begin{aligned} W_{\mu\nu} &= 2(A+2M_\mathrm{N}B)^2\left[P_\mu q_\nu + P_\nu q_\mu + 2P_\mu P_\nu + \frac{q^2}{2}g_{\mu\nu}\right] \\ &\quad + \left[-4(A+2M_\mathrm{N}B)M_\mathrm{N}B + 2M_\mathrm{N}^2B^2\left(-\frac{q^2}{2M_N^2}+2\right)\right] \\ &\quad \times (q_\mu + 2P_\mu)(q_\nu + 2P_\nu)\ , \end{aligned} \tag{3.13}$$

and using

$$P_\mu q_\nu + P_\nu q_\mu + 2P_\mu P_\nu = \frac{1}{2}(q_\mu + 2P_\mu)(q_\nu + 2P_\nu) - \frac{1}{2}q_\mu q_\nu \tag{3.14}$$

we get for (3.13)

$$\begin{aligned} W_{\mu\nu} &= -(A+2M_\mathrm{N}B)^2q^2\left(\frac{q_\mu q_\nu}{q^2} - g_{\mu\nu}\right) \\ &\quad + (A^2 - B^2q^2)4\left(P_\mu - q_\mu\frac{P\cdot q}{q^2}\right)\left(P_\nu - q_\nu\frac{P\cdot q}{q^2}\right)\ . \end{aligned} \tag{3.15}$$

In the last step, we have used the fact that, owing to (3.7), $P\cdot q/q^2 = \frac{1}{2}$. We now introduce two new functions W_1, W_2 and the variable $Q^2 = -q^2$ and write the elastic scattering tensor as

$$\begin{aligned} W_{\mu\nu}^{\mathrm{elastic}} &= \left(-g_{\mu\nu} + \frac{q_\mu q_\nu}{q^2}\right)W_1(Q^2) \\ &\quad + \left(P_\mu - q_\mu\frac{P\cdot q}{q^2}\right)\left(P_\nu - q_\nu\frac{P\cdot q}{q^2}\right)\frac{W_2(Q^2)}{M_\mathrm{N}^2}\ . \end{aligned} \tag{3.16}$$

This structure is immediately evident if one considers that, owing to gauge invariance,

$$q^\mu W_{\mu\nu}^{\mathrm{elastic}} = q^\nu W_{\mu\nu}^{\mathrm{elastic}} = 0\ . \tag{3.17}$$

If we consider instead of elastic scattering special *inelastic processes* like $\mathrm{e}+\mathrm{N}\to\mathrm{e}+\mathrm{N}+\pi$, more momentum vectors can be combined and the general structure of $W_{\mu\nu}$ becomes more complicated. However, a simple expression can again be obtained if one sums over all possible processes or, more precisely, over all possible final hadron states, since this sum can again only depend on P and q. The only change is that the *Lorentz invariants q^2 and $q \cdot P$ are now independent*. One thus obtains the following general form for the inclusive inelastic scattering tensor:

$$W^{\text{incl.}}_{\mu\nu} = \left(-g_{\mu\nu} + \frac{q_\mu q_\nu}{q^2}\right) W_1(Q^2, \nu) + \left(P_\mu - q_\mu \frac{P \cdot q}{q^2}\right)\left(P_\nu - q_\nu \frac{P \cdot q}{q^2}\right) \frac{W_2(Q^2, \nu)}{M_{\mathrm{N}}^2} \tag{3.18}$$

with the *inelasticity variable* [6]

$$\nu = P \cdot q \ . \tag{3.19}$$

In the rest frame of the proton $\nu = M_{\mathrm{N}}(E' - E)$, i.e. it equals the energy loss of the electron (see (3.28) below). At low energies, the energy loss of the electron will end up in a recoil energy of the nucleon. At high energies, however, most of the energy will go into production of other particles, mostly pions. It is therefore justified to call ν inelasticity variable.

The two functions $W_1(Q^2, \nu)$ and $W_2(Q^2, \nu)$ are called *structure functions* for inclusive electron–nucleon scattering. They are most important because they precisely exhibit the impact of the structure of the nucleon on the inclusive cross section. All the rest in (3.18) is relativistic kinematics! To obtain the differential cross section, we must multiply $W_{\mu\nu}$ with the corresponding tensor for the electrons

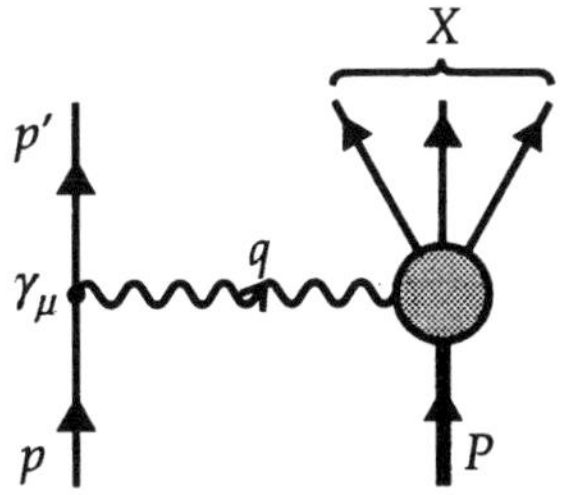

Fig. 3.3. Inclusive inelastic electron–nucleon scattering

$$\begin{aligned} L_{\mu\nu} &= \frac{1}{2}\mathrm{tr}\left\{(\not p + m)\gamma_\mu(\not p' + m)\gamma_\nu\right\} \\ &= 2\left[p_\mu p'_\nu + p_\nu p'_\mu - g_{\mu\nu}\, p \cdot p' + g_{\mu\nu} m^2\right] \ . \end{aligned} \tag{3.20}$$

Equation (3.20) can be obtained by setting $A = 1$, $B = 0$ in (3.12). (The factor $\frac{1}{2}$ comes from averaging over the spin directions of the incoming electrons.) The other factors appearing are the coupling constant and the photon propagator

$$\left(\frac{e^2}{q^2}\right)^2 = \frac{e^4}{Q^4} \ . \tag{3.21}$$

Finally, this must be multiplied with normalization and phase-space factors (cf. Example 3.1). The final expression for unpolarized electron–nucleon scattering in the laboratory system is

$$\frac{\mathrm{d}^2\sigma}{\mathrm{d}E'\mathrm{d}\Omega} = \frac{E'\alpha^2}{EQ^4} L^{\mu\nu} W_{\mu\nu} \ . \tag{3.22}$$

[6] In the literature one may also find the following definition for ν: $\nu = P \cdot q / M_{\mathrm{N}}$. It differs from our definition by the factor $1/M_{\mathrm{N}}$. We shall use the definition (3.19) throughout this book. It is the one standardly used in connection with the operator product expansion and the DGLAP equations (see Chap. 5).

Here E and E' are defined by

$$p_\mu = (E, \boldsymbol{p}) \ ,$$
$$p'_\mu = (E', \boldsymbol{p}') \ , \tag{3.23}$$

and the electron mass has been neglected (the extremely relativistic approximation).

Substituting (3.18) and (3.20) into (3.22), this yields

$$\frac{d^2\sigma}{dE'd\Omega} = \frac{E'\alpha^2}{EQ^4} 2(2p^\mu p^\nu - g^{\mu\nu} p\cdot p') \tag{3.24}$$
$$\times \left[\left(-g_{\mu\nu} + \frac{q_\mu q_\nu}{q^2} \right) W_1 + \left(P_\mu - q_\mu \frac{P\cdot q}{q^2} \right) \left(P_\nu - q_\nu \frac{P\cdot q}{q^2} \right) \frac{W_2}{M_N^2} \right] .$$

Because $q^\mu W_{\mu\nu} = 0$, we were able to replace $p'_\mu = p_\mu - q_\mu$ by p_μ, and we have again neglected the electron mass in comparison with p_μ and p'_μ. We shall set it to zero in the following. Taking into account that

$$-Q^2 = q^2 \approx -2\, p\cdot p' \tag{3.25}$$

and

$$2\,p\cdot q = 2\,p\cdot(p-p') = 2m^2 - 2\,p\cdot p' \approx -Q^2 \ , \tag{3.26}$$

we obtain

$$\begin{aligned} \frac{d^2\sigma}{dE'd\Omega} &= \frac{E'}{E}\frac{\alpha^2}{Q^4} 2 \left[\left(2\frac{(p\cdot q)^2}{q^2} + 3\,p\cdot p' \right) W_1 + 2\left(P\cdot p - \frac{q\cdot p}{q^2} q\cdot P \right)^2 \frac{W_2}{M_N^2} \right. \\ &\quad \left. - \left(P_\mu - q_\mu \frac{q\cdot P}{q^2} \right)^2 p\cdot p' \frac{W_2}{M_N^2} \right] \\ &= \frac{E'}{E}\frac{2\alpha^2}{Q^4} \left[Q^2 W_1 + 2\left(P\cdot p - \frac{1}{2} P\cdot q \right)^2 \frac{W_2}{M_N^2} \right. \\ &\quad \left. - \frac{Q^2}{2}\left(M_N^2 - \frac{(P\cdot q)^2}{q^2} \right) \frac{W_2}{M_N^2} \right] . \end{aligned} \tag{3.27}$$

Let us look at the scattering in the laboratory system, in which the nucleon is at rest before the collision, i.e.,

$$P_\mu = (M_N, 0), \quad \nu = P\cdot q = P\cdot(p-p') = M_N(E-E') \ . \tag{3.28}$$

In addition we introduce the scattering angle θ of the electron in the laboratory system; thus

$$\boldsymbol{p}\cdot\boldsymbol{p}' = |\boldsymbol{p}|\,|\boldsymbol{p}'|\cos(\theta) = EE'\cos(\theta) \tag{3.29}$$

and

$$Q^2 = 2\,p \cdot p' = 2(EE' - \boldsymbol{p} \cdot \boldsymbol{p}') = 4EE' \sin^2\left(\frac{\theta}{2}\right) \ . \tag{3.30}$$

Finally we use definition (3.19):

$$\begin{aligned}
\frac{\mathrm{d}^2\sigma}{\mathrm{d}E'\mathrm{d}\Omega} =& \frac{E'}{E}\frac{\alpha^2}{Q^4}2\left\{4EE'\sin^2\left(\frac{\theta}{2}\right)W_1 + 2M_\mathrm{N}^2\left(E - \frac{\nu}{2M_\mathrm{N}}\right)^2\frac{W_2}{M_\mathrm{N}^2}\right.\\
&\left. - \left[2EE'\sin^2\left(\frac{\theta}{2}\right) + \frac{\nu^2}{2M_\mathrm{N}^2}\right]W_2\right\}\\
=& \frac{E'}{E}\frac{\alpha^2}{Q^4}2\left\{4EE'\sin^2\left(\frac{\theta}{2}\right)W_1\right.\\
&\left. + \left[2E^2 - 2E(E-E') - 2EE'\sin^2\left(\frac{\theta}{2}\right)\right]W_2\right\}.
\end{aligned} \tag{3.31}$$

Because $\nu = (P \cdot q) = (P \cdot p - P \cdot p') = M_\mathrm{N}(E - E')$, the last term simplifies to give

$$2M_\mathrm{N}^2 EE'\left[1 - \sin^2\left(\frac{\theta}{2}\right)\right]\frac{W_2}{M_\mathrm{N}^2} = 2EE'\cos^2\left(\frac{\theta}{2}\right)W_2 \ , \tag{3.31a}$$

and we obtain

$$\left.\frac{\mathrm{d}^2\sigma}{\mathrm{d}E'\mathrm{d}\Omega}\right|_{\mathrm{eN}} = 4E'^2\frac{\alpha^2}{Q}\left[2\sin^2\left(\frac{\theta}{2}\right)W_1\left(Q^2,\nu\right) + \cos^2\left(\frac{\theta}{2}\right)W_2\left(Q^2,\nu\right)\right]. \tag{3.32}$$

This is the final expression for the inclusive unpolarized electron–nucleon scattering cross section. As mentioned before, the functions $W_1\left(Q^2,\nu\right)$ and $W_2\left(Q^2,\nu\right)$ are called the *structure functions* of inclusive electron–nucleon scattering. To avoid confusion, we write them in what follows as $W_1^{\mathrm{eN}}\left(Q^2,\nu\right)$ and $W_2^{\mathrm{eN}}\left(Q^2,\nu\right)$. Experimentally one measures for a definite electron beam energy the direction and energy of the scattered electrons and sums the total reaction cross section for each (θ, E') bin. From this one obtains $W_1^{\mathrm{eN}}\left(Q^2,\nu\right)$ and $W_2^{\mathrm{eN}}\left(Q^2,\nu\right)$.

What now is the advantage of (3.32)? So far we have only managed to eliminate one of the three parameters E, E', and θ. While $\mathrm{d}^2\sigma/\mathrm{d}E'\mathrm{d}\Omega$ can in general be any function of E, E', and θ, we have to express it by arbitrary functions of Q^2 and ν. The importance in (3.32) lies mainly in the fact that the *structure functions* $W_1\left(Q^2,\nu\right)$ and $W_2\left(Q^2,\nu\right)$ can be calculated from the microscopic properties of the quark model and that then the variables $x = Q^2/2\nu$ (see (3.39) below) and Q^2 are the relevant ones. In the leading order of α_s the structure functions turn out to be Q^2 independent. Only a very tiny Q^2 dependence is observed. This residual Q^2 dependence can be used, however, as a most sensitive test for the quark interaction, i.e., for QCD.

$W_{\mu\nu}^{\mathrm{eN}}$ can be expressed by

$$W_{\mu\nu}^{\mathrm{eN}}(P,q) = \frac{1}{2\pi}\int \mathrm{d}^4x\; \mathrm{e}^{\mathrm{i}q\cdot x}\frac{1}{2}\sum_{\mathrm{pol.}}\langle \mathrm{N}(P)|\hat{J}_\mu(x)\hat{J}_\nu(0)|\mathrm{N}(P)\rangle \ , \tag{3.33}$$

where $|\mathrm{N}(P)\rangle$ designates the *state vector of a nucleon* with momentum P and $\hat{J}$ the *electromagnetic current operator*. The derivation of this relation will be given in Example 3.2.

In the laboratory frame $P_\mu = (M, 0)$. Also, we can orient the coordinate system such that $\boldsymbol{q}$ points in the z direction. Then from (3.18) it holds that

$$W_{00}^{\mathrm{eN}} = \left(\frac{q_0 q_0}{q^2} - 1\right) W_1^{\mathrm{eN}}\left(Q^2,\nu\right) + \left(1 + \frac{\nu}{Q^2}q_0\right)^2 W_2^{\mathrm{eN}}\left(Q^2,\nu\right) \tag{3.34}$$

and

$$W_{11}^{\mathrm{eN}} = W_1^{\mathrm{eN}}\left(Q^2,\nu\right) \tag{3.34a}$$

with

$$q_0 = E - E' = \nu/M_\mathrm{N} \ . \tag{3.35}$$

For $W_1^{\mathrm{eN}}\left(Q^2,\nu\right)$ one has according to (3.33):

$$W_1^{\mathrm{eN}}\left(Q^2,\nu\right) = \frac{1}{2\pi}\int \mathrm{d}^4x\; \mathrm{e}^{\mathrm{i}q\cdot x}\frac{1}{2}\sum_{\mathrm{pol.}}\langle \mathrm{N}(P)|\hat{J}_1(x)\hat{J}_1(0)|\mathrm{N}(P)\rangle \ . \tag{3.36}$$

The right-hand side can (at least in principle) be computed for any quark model, and its correctness can thus be tested by comparing the result with the experimental values for W_1^{eN} and W_2^{eN}.

We shall illustrate the calculation and the meaning of the nucleon structure functions for a very simple, but inadequate, model in Exercise 3.9. This model is indeed so simple that we do not need to use (3.36) but can choose a simpler way to calculate $W_1^{\mathrm{eN}}\left(Q^2,\nu\right)$.

Before we discuss the experimentally determined properties of structure functions and their meaning, we give the analogous result for *neutrino–nucleon* and *antineutrino–nucleon scattering*. These formulas again hold in the rest system of the nucleon, i.e., in the laboratory system:

$$\begin{aligned}\left.\frac{\mathrm{d}^2\sigma}{\mathrm{d}E'\mathrm{d}\Omega}\right|_{\nu\mathrm{N}} = \frac{G_\mathrm{F}^2}{2\pi^2}E'^2\Big[&2\sin^2\left(\frac{\theta}{2}\right)W_1^{\nu\mathrm{N}}\left(Q^2,\nu\right) + \cos^2\left(\frac{\theta}{2}\right)W_2^{\nu\mathrm{N}}\left(Q^2,\nu\right)\\ &- \frac{E+E'}{M_\mathrm{N}}\sin^2\left(\frac{\theta}{2}\right)W_3^{\nu\mathrm{N}}\left(Q^2,\nu\right)\Big] \ ,\end{aligned} \tag{3.37}$$

$$\begin{aligned}\left.\frac{\mathrm{d}^2\sigma}{\mathrm{d}E'\mathrm{d}\Omega}\right|_{\bar{\nu}\mathrm{N}} = \frac{G_\mathrm{F}^2}{2\pi^2}E'^2\Big[&2\sin^2\left(\frac{\theta}{2}\right)W_1^{\bar{\nu}\mathrm{N}}\left(Q^2,\nu\right) + \cos^2\left(\frac{\theta}{2}\right)W_2^{\bar{\nu}\mathrm{N}}\left(Q^2,\nu\right)\\ &+ \frac{E+E'}{M_\mathrm{N}}\sin^2\left(\frac{\theta}{2}\right)W_3^{\bar{\nu}\mathrm{N}}\left(Q^2,\nu\right)\Big] \ .\end{aligned} \tag{3.38}$$

Here G_F is the Fermi coupling constant of weak interactions. The derivation of these relations is to be found in Exercises 3.3 and 3.4.

An important assumption we have made tacitly up to now is that single-photon and single-W exchange dominate the cross section. This assumption is safe because of the smallness of the coupling constants. For strongly interacting particles the situation becomes more complicated. Such reactions will be analyzed in detail in Chap 6.

EXAMPLE

3.1 Normalization and Phase Space Factors

We start with the relation (see (2.108))

$$\mathrm{d}\sigma = \frac{1}{4E\omega|\boldsymbol{v}|}\frac{1}{2}\sum_{s,s'}|F_{ss'}|^2\,(2\pi)^4\,\delta^4(k'+p'-k-p)\frac{\mathrm{d}^3\boldsymbol{k}'}{(2\pi)^3 2\omega'}\frac{\mathrm{d}^3\boldsymbol{p}'}{(2\pi)^3 2E'}\,, \tag{1}$$

where $(E,\boldsymbol{p})$, $(E',\boldsymbol{p}')$, $(\omega,\boldsymbol{k})$ and $(\omega',\boldsymbol{k}')$ denote the four-momenta of the incoming electron, the outgoing electron, the incoming nucleon, and the outgoing multihadron state, respectively. Now we have to integrate this expression over $\boldsymbol{k}'$ and, in contrast to elastic electron–pion scattering (see Exercise 2.9), also over ω', since ω' is no longer fixed by $\boldsymbol{k}'$. In elastic e–N scattering $\boldsymbol{k}'^2 = \mathrm{m}_N^2$. In inelastic e–N scattering, however,

$$\boldsymbol{k}'^2 = (\boldsymbol{k}+(\boldsymbol{p}-\boldsymbol{p}'))^2 = (\boldsymbol{k}+\boldsymbol{q})^2 \geq \mathrm{m}_N^2\,,$$

i.e., the outgoing nucleon is no longer on the mass shell. Therefore, the energy ω' of the outgoing nucleon becomes an independent variable. Inelasticity opens, so to speak, a new degree of freedom (see Figure 3.3a). Hence (1) becomes

$$\mathrm{d}\sigma = \frac{\mathrm{d}^3p'}{4EE'\,|\boldsymbol{v}|}\int \mathrm{d}^4k'\left(\frac{1}{2}\frac{1}{4\omega\omega'}\sum_{s,s'}|F_{ss'}|^2\right)(2\pi)^{-2}\,\delta^4(k'+p'-k-p)\,. \tag{2}$$

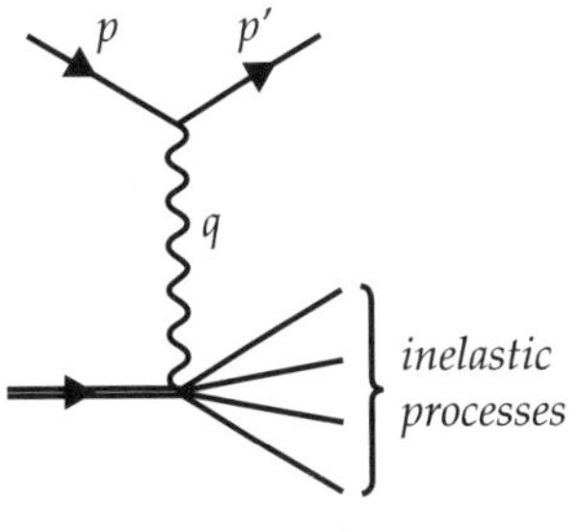

Fig. 3.3a. Inelastic scattering process. The nucleon fragments into a baryon and a number of pions.

Obviously, the dimension of the integrals has changed by one, implying that the structure function has got an additional dimension of 1/energy. This is why we shall later on define the dimensionless structure function F_1 as $F_1 = W_1 M_N$. The integral on the right-hand side is equal to $e^4 L^{\mu\nu}W_{\mu\nu}/Q^4$. Note that the nucleon normalization factors 2ω and $2\omega'$ are included in $W_{\mu\nu}$ (see (3.11)) and that $L_{\mu\nu}$ contains the factor 1/2, which is due to spin averaging (see (3.20)). Thus (2) can be written as

$$\mathrm{d}\sigma = \frac{\mathrm{d}^3p'}{4EE'\,|\boldsymbol{v}|}\frac{e^4}{(2\pi)^2 Q^4}L^{\mu\nu}W_{\mu\nu}\,. \tag{3}$$

Example 3.1

For our case (a massless electron incident on a resting nucleon)

$$|\boldsymbol{v}| \approx c = 1 \ . \tag{4}$$

Furthermore we employ

$$\mathrm{d}^3 p' = (\boldsymbol{p}')^2 \,\mathrm{d}\left|\boldsymbol{p}'\right| \mathrm{d}\Omega' \approx E'^2 \mathrm{d}E' \,\mathrm{d}\Omega \tag{5}$$

and in this way obtain (3.22):

$$\frac{\mathrm{d}^2\sigma}{\mathrm{d}E'\mathrm{d}\Omega} = \frac{E'}{E}\frac{e^4}{(4\pi)^2 Q^4} L^{\mu\nu} W_{\mu\nu} = \frac{E'}{E}\frac{\alpha^2}{Q^4} L^{\mu\nu} W_{\mu\nu} \ . \tag{6}$$

Comparing with (2) we find

$$L_{\mu\nu} W^{\mu\nu} = \int \frac{\mathrm{d}^4 k}{(2\omega')(2\pi)^3} \left(\frac{1}{4\omega}\right) \sum_{s,s'} \left|\tilde{F}_{s,s'}\right|^2 (2\pi)^3 \delta^4(k' + p' - k - p) \ , \tag{7}$$

where we factored out the photon propagator and the elementary charges $\left|F_{s,s'}\right|^2 = \frac{e^4}{Q^4}\left|\tilde{F}_{s,s'}\right|^2$. The integral $\int \mathrm{d}^4 k/(2\omega')(2\pi)^3$ corresponds to a complete sum $\sum_{X,P_X}$ over the outgoing multihadron state, and the delta function will be absorbed in the definition of the hadronic tensor (see (4) of Example 3.2).

EXAMPLE

3.2 Representation of $W_{\mu\nu}$ by Electromagnetic Current Operators

The coupling of the exchanged photon to the nucleon can be written in a general way as

$$\hat{J}_\mu(x) A^\mu(x) \ , \tag{1}$$

where for the moment no assumption about the structure of $\hat{J}_\mu(x)$ has been made. Now we consider the corresponding current matrix element between the initial nucleon state with momentum P_μ and an arbitrary hadronic final state X with four-momentum $P_{\mathrm{X},\mu}$:

$$\left\langle \mathrm{X}(P_\mathrm{X}) \left| \hat{J}_\mu(x) \right| \mathrm{N}(P) \right\rangle = \left\langle \mathrm{X}(P_\mathrm{X}) \left| \hat{J}_\mu(0) \right| \mathrm{N}(P) \right\rangle \mathrm{e}^{-\mathrm{i}(P-P_\mathrm{X})\cdot x} \ . \tag{2}$$

In order to evaluate the inclusive cross section, (2) is squared and summed over all X states:

$$\sum_{\mathrm{X},P_\mathrm{X}} \left\langle \mathrm{N}(P) \left| \hat{J}_\mu(0) \right| \mathrm{X}(P_\mathrm{X}) \right\rangle \left\langle \mathrm{X}(P_\mathrm{X}) \left| \hat{J}_\nu(0) \right| \mathrm{N}(P) \right\rangle \ . \tag{3}$$

Example 3.2

In addition we have to average the spin of the incoming nucleon, since this quantity is not observed. Finally energy-momentum conservation is ensured by a δ function, which is due to the integration over x:

$$W_{\mu\nu} = \frac{1}{2}\sum_{\text{pol.}}\sum_{X,P_X}\left\langle N(P)\left|\hat{J}_\mu(0)\right|X(P_X)\right\rangle\left\langle X(P_X)\left|\hat{J}_\nu(0)\right|N(P)\right\rangle \times (2\pi)^3\,\delta^4(P_X - P - q)\ . \tag{4}$$

q denotes the momentum of the virtual photon (see Fig. 3.3a). The δ function arises from the transition amplitude for the scattering process, which is proportional to

$$\int d^4y d^4x \hat{J}_e^\mu(y)\frac{e^{-iq\cdot(x-y)}}{q^2}\hat{J}_N^\mu(x)\ ,$$

where $\hat{J}_e^\mu(y)$ and $\hat{J}_N^\mu(x)$ represent the electron and nucleon transition currents, respectively. The latter one, which is in fact explicitly given by (2), also reveals its x dependence. The x integration then yields

$$\int d^4x e^{-iq\cdot x}e^{-i(P-P_X)\cdot x} = (2\pi)^4\delta^4(P_X - P - q)\ .$$

Squaring the amplitude implies also a squaring of the δ function, which yields in the well-known way $\left[(2\pi)^4\delta^4(P_X - P - q)\right]^2 = (2\pi)^4\delta^4(P_X - P - q)VT$, where VT is the space–time normalization volume which drops out when the transition rate and cross section are finally calculated. At the end of Example 3.1 above we indicated that the factor $(2\pi)^3$ occurs instead of the usual factor $(2\pi)^4$ in the definition of the structure function. The summation of the projection operator $|X(P_X)\rangle\langle X(P_X)|$ in (4) has the form of a complete projection operator. But the completeness relation cannot be employed yet, since the argument of the δ function depends on P_X .

Using the integral representation

$$\delta^4(P_X - P - q) = \int \frac{d^4y}{(2\pi)^4}\, e^{-i(P_X-P-q)\cdot y}\ , \tag{5}$$

we can write (4) in the form

$$\begin{aligned} W_{\mu\nu} &= \frac{1}{2}\sum_{\text{pol.}}\sum_{X,P_X}\frac{1}{2\pi}\int d^4y\,\left\langle N(P)\left|\hat{J}_\mu(0)\right|X(P_X)\right\rangle \\ &\quad\times\left\langle X(P_X)\left|\hat{J}_\nu(0)\right|N(P)\right\rangle e^{-i(P_X-P)\cdot y}\,e^{iq\cdot y} \\ &= \frac{1}{4\pi}\sum_{\text{pol.}}\sum_{X,P_X}\int d^4y\left\langle N(P)\left|\hat{J}_\mu(y)\right|X(P_X)\right\rangle\left\langle X(P_X)\left|\hat{J}_\nu(0)\right|N(P)\right\rangle e^{iq\cdot y}, \end{aligned} \tag{6}$$

Example 3.2

where $\hat{J}_\mu(y)$ has been obtained by means of (2). Now we are able to employ the completeness relation

$$\sum_{\mathrm{X}, P_\mathrm{X}} |\mathrm{X}(P_\mathrm{X})\rangle\langle \mathrm{X}(P_\mathrm{X})| = 1 \ , \tag{7}$$

which yields the expression for $W_{\mu\nu}(P, q)$ given in (3.33):

$$W_{\mu\nu}(P,q) = \frac{1}{4\pi} \sum_{\mathrm{pol.}} \int \mathrm{d}^4y \ \mathrm{e}^{\mathrm{i}q\cdot y} \left\langle \mathrm{N}(P) \left| \hat{J}_\mu(y)\hat{J}_\nu(0) \right| \mathrm{N}(P) \right\rangle \ . \tag{8}$$

Now it is interesting to note that instead of the operator product $\hat{J}_\mu(y)\hat{J}_\nu(0)$ also the expectation value of the commutator of the current operators could be used in (8), i.e.,

$$W_{\mu\nu} = \frac{1}{4\pi} \sum_{\mathrm{pol.}} \int \mathrm{d}^4y \ \mathrm{e}^{\mathrm{i}q\cdot y} \left\langle \mathrm{N}(P) \left| \left[\hat{J}_\mu(y)\hat{J}_\nu(0) \right] \right| \mathrm{N}(P) \right\rangle \ . \tag{9}$$

In order to show the equivalence of (8) and (9) one has to prove that the following expression vanishes:

$$\begin{aligned}
&\int \mathrm{d}^4y \ \mathrm{e}^{\mathrm{i}q\cdot y} \left\langle \mathrm{N}(P) \left| \hat{J}_\nu(0)\hat{J}_\mu(y) \right| \mathrm{N}(P) \right\rangle \\
&= \sum_{\mathrm{X}, P_\mathrm{X}} \int \mathrm{d}^4y \ \mathrm{e}^{\mathrm{i}q\cdot y} \left\langle \mathrm{N}(P) \left| \hat{J}_\nu(0) \right| \mathrm{X}(P_\mathrm{X}) \right\rangle \left\langle \mathrm{X}(P_\mathrm{X}) \left| \hat{J}_\mu(y) \right| \mathrm{N}(P) \right\rangle \\
&= \sum_{\mathrm{X}, P_\mathrm{X}} \int \mathrm{d}^4y \ \mathrm{e}^{\mathrm{i}(q+P_\mathrm{X}-P)\cdot y} \left\langle \mathrm{N}(P) \left| \hat{J}_\nu(0) \right| \mathrm{X}(P_\mathrm{X}) \right\rangle \left\langle \mathrm{X}(P_\mathrm{X}) \left| \hat{J}_\mu(0) \right| \mathrm{N}(P) \right\rangle \\
&= (2\pi)^4 \sum_{\mathrm{X}, P_\mathrm{X}} \delta^4(P_\mathrm{X} - P + q) \left\langle \mathrm{N}(P) \left| \hat{J}_\nu(0) \right| \mathrm{X}(P_\mathrm{X}) \right\rangle \left\langle \mathrm{X}(P_\mathrm{X}) \left| \hat{J}_\mu(0) \right| \mathrm{N}(P) \right\rangle .
\end{aligned} \tag{10}$$

The energy δ function requires that $E_\mathrm{X} = P_0 - q_0 = M_\mathrm{N} - (E - E') < M_\mathrm{N}$ holds in the lab system. However, such a state X does not exist, since the nucleon is the state of lowest energy with baryon number 1. Therefore (10) vanishes and (9) is proved.

EXERCISE

3.3 The Nucleonic Scattering Tensor with Weak Interaction

Problem. Repeat the discussion leading from (3.6) to (3.18) assuming that parity is not conserved, i.e., that Γ_μ consists of Lorentz vectors and Lorentz axial vectors. Take into account that time-reversal invariance still holds.

Exercise 3.3

Solution. Have a look at Fig. 3.2. We are again discussing the elastic process described by that figure, but allow for non-parity-conserving currents. In this case the transition current is of the general form

$$\begin{aligned}\bar{u}(P')\Gamma_\mu u(P) = {} & B(P'_\mu + P_\mu)S + \mathrm{i}C(P'_\mu - P_\mu)S + A\, V_\mu + D(P'^\nu + P^\nu)T_{\mu\nu} \\ & + \mathrm{i}E(P'^\nu - P^\nu)T_{\mu\nu} + B'(P'_\mu + P_\mu)P + \mathrm{i}C'(P'_\mu - P_\mu)P + A'\, A_\mu \\ & + D'(P'^\nu + P^\nu)\,\varepsilon_{\mu\nu\alpha\beta}\, T^{\alpha\beta} + \mathrm{i}E'(P'^\nu - P^\nu)\,\varepsilon_{\mu\nu\alpha\beta}\, T^{\alpha\beta} \qquad (1)\end{aligned}$$

with real functions A, B, C, ... and with

$$\begin{aligned} S &= \bar{u}(P')\,u(P)\ , \\ P &= \bar{u}(P')\,\mathrm{i}\,\gamma_5\,u(P)\ , \\ V_\mu &= \bar{u}(P')\,\gamma_\mu\,u(P)\ , \\ A_\mu &= \bar{u}(P')\,\gamma_\mu\,\gamma_5\,u(P)\ , \\ T_{\mu\nu} &= \bar{u}(P')\,\sigma_{\mu\nu}\,u(P)\ . \qquad (2)\end{aligned}$$

In (1) we have assumed that the transition current is real, i.e.

$$\left(\bar{u}(P')\,\Gamma_\mu\,u(P)\right)^\dagger = \bar{u}(P')\,\Gamma_\mu\,u(P)$$

and therefore

$$u^\dagger(P)\gamma_0\,\gamma_0\Gamma^\dagger_\mu\gamma_0 u(P') = \bar{u}(P)\Gamma_\mu u(P')\ .$$

Here the relation $\gamma^0\Gamma^\dagger_\mu\gamma^0 = \Gamma_\mu$ was used, which can easily be verified in the standard representation of the γ_μ and $\sigma_{\mu\nu} = \frac{\mathrm{i}}{2}\left[\gamma_\mu, \gamma_\nu\right]$. Obviously, the exchange $P \leftrightarrow P'$ should have the same effect as complex conjugation. In other words, the right-hand side must therefore be invariant under the transformation

$$(\cdots)^*\big|_{P'_\mu \leftrightarrow P_\mu}\ . \qquad (3)$$

Time inversion yields (see Table 2.1)

$$S \to S\ , \quad P \to -P\ , \quad V_\mu \to V^\mu\ , \quad A_\mu \to A^\mu\ , \quad T_{\mu\nu} \to -T^{\mu\nu}\ , \qquad (4)$$

and

$$\boldsymbol{P} \leftrightarrow -\boldsymbol{P}'\ , \quad P^0 \leftrightarrow P'^0\ , \quad \text{i.e.} \quad P_\mu \leftrightarrow P'^\mu\ . \qquad (5)$$

The transition $\boldsymbol{P} \leftrightarrow -\boldsymbol{P}'$, $P^0 \leftrightarrow P'^0$ is due to the complex conjugation of $\bar{u}(P')\,\Gamma_\mu\,u(P)$, which replaces the momentum P'_μ by the negative value of P_μ and vice versa, i.e., $P'_\mu \leftrightarrow -P_\mu$. Because $t \to -t$ there is an additional change in the sign of the zero component. This is easily understood, because under time reversal the direction of motion changes and initial and final states are exchanged. The energies, however, remain positive.

Exercise 3.3

Under combined transformations (4) and (5), (1) assumes the form

$$\begin{aligned}\bar{u}(P')\Gamma_\mu u(P) \rightarrow\ & B(P'^\mu + P^\mu)S - \mathrm{i}C(P'^\mu - P^\mu)S + A\,V^\mu - D(P'_\nu + P_\nu)T^{\mu\nu} \\ & + \mathrm{i}E(P'_\nu - P_\nu)T^{\mu\nu} - B'(P'^\mu + P^\mu)P + \mathrm{i}C'(P'^\mu - P^\mu)P + A'\,A^\mu \\ & - D'(P'_\nu + P_\nu)\,\varepsilon_{\mu\nu\alpha\beta}\,T_{\alpha\beta} + \mathrm{i}E'(P'_\nu - P_\nu)\,\varepsilon_{\mu\nu\alpha\beta}\,T_{\alpha\beta}\ .\end{aligned} \tag{6}$$

Only the spatial components of the current vector should change sign under time reversal. In order to conserve T invariance (6) must therefore be equal to $\bar{u}(P')\,\Gamma^\mu\,u(P)$. For the ε tensor the relation $\varepsilon_{\mu\nu\alpha\beta} = -\varepsilon^{\mu\nu\alpha\beta}$ holds. This can be verified from the definition $-4\mathrm{i}\varepsilon^{\mu\nu\alpha\beta} = \mathrm{tr}\left[\gamma_5\gamma_\mu\gamma_\nu\gamma_\alpha\gamma_\beta\right]$. The indices μ, ν, α, β have to be 0, 1, 2, 3, and different from each other. Now $-4\mathrm{i}\varepsilon^{0,1,2,3} = \mathrm{tr}\left[\gamma_5\gamma^0\gamma^1\gamma^2\gamma^3\right] = -\mathrm{tr}\left[\gamma_5\gamma_0\gamma_1\gamma_2\gamma_3\right] = +4\mathrm{i}\varepsilon_{0,1,2,3}$. For different permutations of 0, 1, 2, 3 the analogous relation holds. Employing now $\varepsilon_{\mu\nu\alpha\beta} = -\varepsilon^{\mu\nu\alpha\beta}$ we have

$$C = D = B' = E' = 0\ . \tag{7}$$

Because of the Gordon decomposition (see earlier text and (3.10)), E can again be replaced by A and B. In an analogous way D' can also be eliminated using A' and C'. In order to derive this identity we consider the expression

$$\begin{aligned}&\bar{u}(P')\left(-2M_\mathrm{N}\not{a}\gamma_5 + \left\{\frac{\not{P}' - \not{P}}{2}, \not{a}\right\}_+ \gamma_5 + \left[\frac{\not{P}' + \not{P}}{2}, \not{a}\right]_- \gamma_5\right)u(P) \\ &\quad = \bar{u}(P')\left(-2M_\mathrm{N}\not{a}\gamma_5 + \not{P}'\not{a}\gamma_5 - \not{a}\not{P}\gamma_5\right)u(P) \\ &\quad = \bar{u}(P')\left(-2M_\mathrm{N}\not{a}\gamma_5 + M_\mathrm{N}\not{a}\gamma_5 + \not{a}\gamma_5 M_\mathrm{N}\right)u(P) \\ &\quad = 0\ .\end{aligned} \tag{8}$$

Differentiating this relation with respect to a^μ yields

$$\frac{\partial}{\partial a^\mu}a^\alpha\gamma_\alpha = \delta^\alpha_\mu\gamma_\alpha = \gamma_\mu$$

and we obtain

$$\begin{aligned}&\bar{u}(P')\left(-2M_\mathrm{N}\gamma_\mu\gamma_5 + (P' - P)^\lambda\frac{1}{2}\left\{\gamma_\lambda, \gamma_\mu\right\}\gamma_5\right. \\ &\qquad \left. + (P' + P)^\lambda\frac{1}{2}\left[\gamma_\lambda, \gamma_\mu\right]\gamma_5\right)u(P)\ .\end{aligned} \tag{9}$$

With $(1/2)\left\{\gamma_\lambda, \gamma_\mu\right\} = g_{\lambda,\mu}$ and $(1/2)\left[\gamma_\lambda, \gamma_\mu\right] = -\mathrm{i}\sigma_{\lambda,\mu}$ and with (8) we therefore conclude that

$$\begin{aligned}0 &= -2M_\mathrm{N}A_\mu + \bar{u}(P')\Big[(P'_\mu - P_\mu)\gamma_5 - \mathrm{i}(P'^\nu + P^\nu)\sigma_{\nu\mu}\,\gamma_5\Big]u(P) \\ &= -2M_\mathrm{N}A_\mu - \mathrm{i}(P'_\mu - P_\mu)P - \frac{1}{2}\,\varepsilon_{\mu\nu\alpha\beta}\,T^{\alpha\beta}(P'^\nu + P^\nu)\ ,\end{aligned} \tag{9b}$$

which represents the desired result. Consequently we have reduced (1) to

$$\bar{u}(P')\Gamma_\mu u(P) = B(P'_\mu + P_\mu)S + A\,V_\mu + \mathrm{i}C'(P'_\mu - P_\mu)P + A'A_\mu\ . \tag{10}$$

Exercise 3.3

Here the letters B, A, C', and A' denote constants while S, V_μ, P, and A_μ stand for the various currents denoted in (2). Similarly, as in (3.11), $W_{\mu\nu}$ then becomes

$$
\begin{aligned}
W_{\mu\nu} = \mathrm{tr}\Big\{\Big[& A\gamma_\mu + B(P'_\mu + P_\mu) \\
&+ A'\gamma_\mu\gamma_5 + C'(P'_\mu - P_\mu)\gamma_5)\Big](\not{P}' + M_\mathrm{N})\Big[A\gamma_\nu + B(P'_\nu + P_\nu) \\
&+ A'\gamma_\nu\gamma_5 - C'(P'_\nu - P_\nu)\gamma_5\Big](\not{P} + M_\mathrm{N})\Big\} \\
= \mathrm{tr}&\Big\{\Big[A\gamma_\mu + B(P'_\mu + P_\mu)\Big](\not{P}' + M_\mathrm{N})\Big[A\gamma_\nu + B(P'_\nu + P_\nu)\Big](\not{P} + M_\mathrm{N})\Big\} \\
&+ \mathrm{tr}\Big\{\left[A'\gamma_\mu + C'(P'_\mu + P_\mu)\right](-\not{P}' + M_\mathrm{N})\left[-A'\gamma_\nu - C'(P'_\nu - P_\nu)\right](\not{P} + M_\mathrm{N})\Big\} \\
&+ \mathrm{tr}\left\{A'\gamma_\mu\gamma_5\not{P}'A\gamma_\nu\not{P} + A\gamma_\mu\not{P}'A'\gamma_\nu\gamma_5\not{P}\right\} . \qquad (11)
\end{aligned}
$$

Note again that we have used the normalization for the u and v spinors expressed in (2.43). The first two traces do not have to be evaluated explicitly. It is sufficient to know that because $P'_\mu = q_\mu + P_\mu$ their contribution can only be of the form

$$
g_{\mu\nu}V_1 + P_\mu P_\nu V_2 + (P_\mu q_\nu + P_\nu q_\mu)V_3 + q_\mu q_\nu V_4 \ . \qquad (12)
$$

In the case of the last trace the identity

$$
\mathrm{tr}\left\{\gamma_5 \not{a}\not{b}\not{c}\not{d}\right\} = 4\mathrm{i}\varepsilon_{\mu\nu\alpha\beta}\, a^\mu\, b^\nu\, c^\alpha\, d^\beta
$$

holds. Therefore the hadronic tensor is of the form

$$
W_{\mu\nu} = g_{\mu\nu}V_1 + P_\mu P_\nu V_2 + (P_\mu q_\nu + P_\nu q_\mu)V_3 + q_\mu q_\nu V_4 + \mathrm{i}\varepsilon_{\mu\nu\alpha\beta}P^\alpha q^\beta V_5 \ . \qquad (13)
$$

We have derived (13) for elastic lepton–nucleon scattering. In the case of inelastic processes, including particle creation, $q_\mu = P'_\mu - P_\mu$ is no longer valid. $q \cdot P'$, q^2 and $q \cdot P$ are then independent quantities. However, the results again are quite simple if one sums over all possible final states, i.e., over the final states of the reactions

$$
\begin{aligned}
&\nu_\mathrm{e}(p) + \mathrm{N}(P) \to \mathrm{e}^-(p-q) + \mathrm{X} \ , \\
&\bar{\nu}_\mathrm{e}(p) + \mathrm{N}(P) \to \mathrm{e}^+(p-q) + \mathrm{X} \ . \qquad (14)
\end{aligned}
$$

Therefore the inclusive inelastic scattering tensor for parity-violating lepton–nucleon scattering is

$$
\begin{aligned}
W_{\mu\nu} = g_{\mu\nu}&V_1(Q^2,\nu) + P_\mu P_\nu V_2(Q^2,\nu) + (P_\mu q_\nu + P_\nu q_\mu)V_3(Q^2,\nu) \\
&+ q_\mu q_\nu V_4(Q^2,\nu) + \mathrm{i}\varepsilon_{\mu\nu\alpha\beta}P^\alpha q^\beta V_5(Q^2,\nu) \ . \qquad (15)
\end{aligned}
$$

EXERCISE

3.4 The Inclusive Weak Lepton–Nucleon Scattering

Problem. The coupling of the charged vector bosons $W_\mu^\pm$ of the weak interaction to leptons is described by the interaction Lagrangean (4.23),

$$L_{\text{int}} = \frac{+g}{\sqrt{2}}\bar{\Psi}_{\text{f}}(\hat{T}_- W_\mu^+ + \hat{T}_+ W_\mu^-)\gamma^\mu \Psi_{\text{f}} \ , \tag{1}$$

where $\Psi_{\text{f}} = \binom{\nu_{\text{e}}}{\text{e}}_{\text{L}}$ denotes the doublet of the left-handed electron–neutrino field: $e_{\text{L}} = \frac{1-\gamma_5}{2}e_{\text{L}}$. Writing (1) we have used $\hat{T}_\pm = \frac{1}{\sqrt{2}}\left(\frac{\lambda_1}{2} \pm \text{i}\frac{\lambda_2}{2}\right)$. With $\hat{T}_- = \begin{pmatrix} 0 & 0 \\ 1 & 0 \end{pmatrix}$ and $\hat{T}_+ = \begin{pmatrix} 0 & 1 \\ 0 & 0 \end{pmatrix}$ we get for the fundamental $(e, \bar{\nu})$ interaction

$$\frac{+g}{\sqrt{2}}\bar{\nu}(x)\gamma_\mu \frac{1-\gamma_5}{2}\text{e}(x)W^\mu(x) + \text{h.c.} \ . \tag{1a}$$

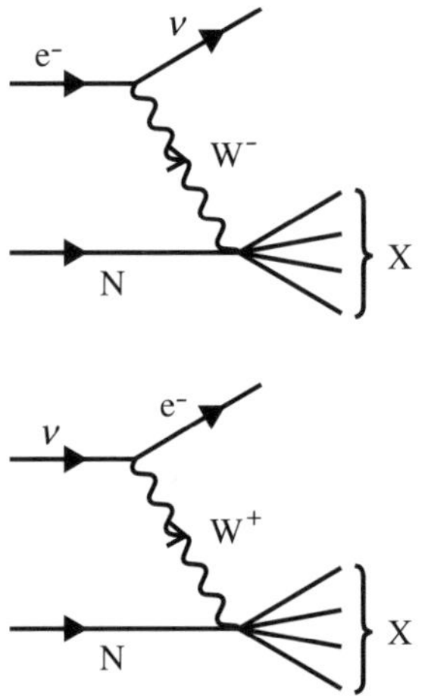

Fig. 3.3b. Deep inelastic scattering with charged vector bosons.

The corresponding processes are depicted in the figure. Derive from this equation the leptonic scattering tensor $L_{\mu\nu}$. Employ the result of Exercise 3.3 to evaluate the differential cross section for the inclusive weak lepton–nucleon scattering.

Solution. Since the neutrino is massless and the electron mass can be neglected, it is helpful to choose the following spinor normalization.

$$\begin{aligned} &u_\alpha^\dagger(p)u_\beta(p) = 2E\delta_{\alpha\beta} \ , \\ &\sum_{\text{spin}} u(p)\bar{u}(p) = \not{p} + m = \not{p} \ . \end{aligned} \tag{2}$$

The leptonic scattering tensor is then

$$\begin{aligned} L_{\mu\nu} &= \frac{g^2}{8}\sum_s \text{tr}\left\{\bar{\nu}(p)\gamma_\mu(1-\gamma_5)\text{e}(p', s)\bar{\text{e}}(p', s)\gamma_\nu(1-\gamma_5)\nu(p)\right\} \\ &= \frac{g^2}{8}\text{tr}\left\{\not{p}\gamma_\mu(1-\gamma_5)\not{p}'\gamma_\nu(1-\gamma_5)\right\} \quad \text{(utilizing (2))} \\ &= \frac{g^2}{4}\left(\text{tr}\left\{\not{p}\gamma_\mu\not{p}'\gamma_\nu\right\} - \text{tr}\left\{\gamma_5\not{p}'\gamma_\nu\not{p}\gamma_\mu\right\}\right) \\ &= g^2\left(p'_\mu p_\nu + p_\mu p'_\nu - p' \cdot p\, g_{\mu\nu} - \text{i}\varepsilon_{\mu\nu\alpha\beta}p^\alpha p'^\beta\right) \ . \end{aligned} \tag{3}$$

Since the neutrino is always left-handed, there is no averaging of the initial spin. Nevertheless, in order to obtain the trace in the second line of (3) we have summed over all neutrino spins and made use of (2). This procedure is correct, because the operator $(1-\gamma_5)$ cancels the contribution of the nonphysical right-handed neutrino state. In analogy to (3.22) the differential cross section then becomes *Exercise 3.4*

$$\frac{d^2\sigma}{dE'\,d\Omega} = \frac{E'}{(4\pi)^2 E}\left(\frac{1}{q^2 - M_W^2}\right)^2 L^{\mu\nu}\left(\frac{g}{\sqrt{2}}\right)^2 W_{\mu\nu} \ . \tag{4}$$

The differences can be explained as follows.

(1) The propagator for massive W bosons differs from that for massless photons. For $|q^2| \ll M_W^2$ we can use

$$\left(\frac{1}{q^2 - M_W^2}\right)^2 \approx \frac{1}{M_W^4} \ . \tag{5}$$

This yields Fermi's theory of weak interaction.
(2) The coupling constant of the W bosons to the nucleon is effectively chosen to be $G_F/\sqrt{2}$ where $G_F = g^2/(4\sqrt{2}M_W^2)$ is the Fermi constant of weak interaction. Since we have already absorbed a factor of $g^2/8$ (see (3)) in the definition of the leptonic tensor we are left with a factor of $g^2/(4\cdot 2)$ in front of the hadronic tensor $W_{\mu\nu}$, which according to (3.11) does not contain any coupling constant.
(3) The factor $1/(4\pi)^2$ arises by replacing $\alpha^2 = e^4/(4\pi)^2 \to g^4/(4\pi)^2$. One factor g^2 is contained in $L_{\mu\nu}$, the other one appears explicitly in (4).

In order to evaluate (4) we first show that $q^\mu L_{\mu\nu}$ and $q^\nu L_{\mu\nu}$ vanish:

$$\begin{aligned} q^\mu L_{\mu\nu} &= g^2\big(q\cdot p'\, p_\nu + q\cdot p\, p'_\nu - p'\cdot p\, q_\nu - \mathrm{i}\varepsilon_{\mu\nu\alpha\beta}\, q^\mu p^\alpha p'^\beta\big) \\ &= g^2\left[-\frac{q^2}{2}p_\nu + \frac{q^2}{2}p'_\nu + \frac{q^2}{2}(p_\nu - p'_\nu)\right] = 0 \ . \end{aligned} \tag{6}$$

Here we have employed

$$\begin{aligned} &q_\mu = p_\mu - p'_\mu \quad \Rightarrow \quad p\cdot p' = -\frac{q^2}{2} \ , \\ &p\cdot q = \frac{q^2}{2} \quad \text{and} \quad p'\cdot q = -\frac{q^2}{2} \ , \end{aligned} \tag{7}$$

which all hold in the high-energy limit $m^2 \ll q^2$. With help of 15 from Exercise 3.3 it therefore follows that

$$\begin{aligned} L^{\mu\nu}W_{\mu\nu} &= L^{\mu\nu}\big(g_{\mu\nu}V_1 + P_\mu P_\nu V_2 + \mathrm{i}\varepsilon_{\mu\nu\alpha\beta}P^\alpha q^\beta V_5\big) \\ &= g^2\Big[(2\,p'\cdot p - 4\,p'\cdot p)V_1 + \big(2\,p'\cdot P\, p\cdot P - M_N^2\, p\cdot p'\big)V_2 \\ &\quad + \varepsilon^{\mu\nu}_{\alpha\beta}\,\varepsilon_{\mu\nu\delta\gamma}\, p^\delta q^\gamma P^\alpha q^\beta V_5\Big] \ . \end{aligned} \tag{8}$$

With

$$p = (E, \boldsymbol{p}) \ , \quad p' = (E', \boldsymbol{p}') \ , \quad P = (M, \boldsymbol{0}) \ , \tag{9}$$

Exercise 3.4

and[7]

$$\varepsilon^{\mu\nu}_{\alpha\beta}\,\varepsilon_{\mu\nu\delta\gamma} = -2\left(g_{\alpha\delta}\,g_{\beta\gamma} - g_{\alpha\gamma}\,g_{\beta\delta}\right) , \tag{10}$$

and, also, utilizing relations (7) this assumes the form

$$\begin{aligned} L^{\mu\nu}W_{\mu\nu} &= g^2\left[q^2V_1 + M_\mathrm{N}^2\left(2EE' + \frac{1}{2}q^2\right)V_2 - 2\left(p\cdot P\,q^2 - P\cdot q\;p\cdot q\right)V_5\right] \\ &= g^2\left\{q^2V_1 + 2M_\mathrm{N}^2\left(EE' + \frac{1}{4}q^2\right)V_2 - 2M_\mathrm{N}\left[Eq^2 - (E-E')\frac{q^2}{2}\right]V_5\right\} . \end{aligned} \tag{11}$$

Again, using (3.30), we find that

$$q^2 = -Q^2 = -4EE'\sin^2\left(\frac{\vartheta}{2}\right) \tag{12}$$

leads to

$$\begin{aligned} L^{\mu\nu}W_{\mu\nu} = g^2 2EE'\bigg[&-2\sin^2\left(\frac{\vartheta}{2}\right)V_1 + M_\mathrm{N}^2\cos^2\left(\frac{\vartheta}{2}\right)V_2 \\ &+2M_\mathrm{N}\sin^2\left(\frac{\vartheta}{2}\right)(E+E')V_5\bigg] . \end{aligned} \tag{13}$$

In order to provide all structure functions with the same dimension we define

$$W_1 = -V_1 , \quad W_2 = V_2M_\mathrm{N}^2 , \quad W_3 = -2V_5M_\mathrm{N}^2 \tag{14}$$

and obtain

$$\begin{aligned} \frac{\mathrm{d}^2\sigma}{\mathrm{d}E'\mathrm{d}\Omega} = \frac{E'^2g^4}{4M_\mathrm{W}^4(4\pi)^2}\bigg[&2\sin^2\left(\frac{\vartheta}{2}\right)W_1(Q^2,\nu) + \cos^2\left(\frac{\vartheta}{2}\right)W_2(Q^2,\nu) \\ &-\frac{E+E'}{M_\mathrm{N}}\sin^2\left(\frac{\vartheta}{2}\right)W_3(Q^2,\nu)\bigg] . \end{aligned} \tag{15}$$

Finally we substitute the so-called Fermi constant of weak interaction:

$$G_\mathrm{F} = \frac{g^2}{4\sqrt{2}M_\mathrm{W}^2} ; \tag{16}$$

$$\begin{aligned} \frac{\mathrm{d}^2\sigma}{\mathrm{d}E'\mathrm{d}\Omega} = \frac{G_\mathrm{F}^2E'^2}{2\pi^2}\bigg[&2\sin^2\left(\frac{\vartheta}{2}\right)W_1(Q^2,\nu) + \cos^2\left(\frac{\vartheta}{2}\right)W_2(Q^2,\nu) \\ &-\frac{E+E'}{M_\mathrm{N}}\sin^2\left(\frac{\vartheta}{2}\right)W_3(Q^2,\nu)\bigg] . \end{aligned} \tag{17}$$

[7] The proof of this relation can be found in Exercise 2.4 of W. Greiner and B. Müller: *Gauge Theory of Weak Interactions*, 3rd ed. (Springer, Berlin, Heidelberg 2000).

This is identical to (3.37). In the reaction $\bar{\nu} + \mathrm{N} \to \mathrm{e}^+ + \mathrm{X}$ the incoming antineutrino corresponds to the outgoing electron in the reaction $\nu + \mathrm{N} \to \mathrm{e}^- + \mathrm{X}$, i.e., p and p' are exchanged. Since the third term in $L_{\mu\nu}$ is antisymmetric under this exchange, the sign of W_3 changes.

We shall now discuss the experimental structure functions of electron–nucleon scattering. These functions will have a particularly simple form when ν is replaced by $x = Q^2/2\nu$. This quantity is called the *Bjorken* or *scaling variable*. We shall first consider some of its properties. For elastic scattering, (3.19) and (3.7) give

$$x = \frac{Q^2}{2\nu} = \frac{-q^2}{2P\cdot q} = \frac{-q^2}{-q^2} = 1 \ . \tag{3.39}$$

Since the nucleon is the lightest state with baryon number 1, we know in addition that the invariant mass of the final state

$$M_\mathrm{X} = \sqrt{(P+q)^2} \tag{3.40}$$

must be larger than M_N. If this were not the case, there would be a state X (of arbitrary structure) with baryon number 1 into which the proton could decay, by emitting γ quanta, for example.

From (3.40) it follows that x is always positive and less than 1:

$$\begin{aligned} x &= \frac{Q^2}{(P+q)^2 - P^2 + Q^2} = \frac{Q^2}{M_\mathrm{X}^2 - M_\mathrm{N}^2 + Q^2} \\ &= \frac{1}{1 + (M_\mathrm{X}^2 - M_\mathrm{N}^2)/Q^2} \ ; \end{aligned} \tag{3.41}$$

Q^2 is always positive, as can be seen from (3.30), and

$$0 < x \leq 1 \ . \tag{3.42}$$

Additionally, *it is customary to include a factor of* ν *in the definition of* W_2 *and* W_3 *and a factor of* M_N *in the definition of* W_1. Thus we are lead to "new" form factors F_1, F_2, F_3 instead of W_1, W_2, and W_3, respectively:

$$\begin{aligned} M_\mathrm{N} W_1^{\mathrm{eN}}(Q^2,\nu) &= F_1^{\mathrm{eN}}(Q^2,x) \ , \\ \frac{\nu}{M_\mathrm{N}} W_2^{\mathrm{eN}}(Q^2,\nu) &= F_2^{\mathrm{eN}}(Q^2,x) \ , \\ \frac{\nu}{M_\mathrm{N}} W_3^{\nu\mathrm{N}}(Q^2,\nu) &= F_3^{\mathrm{eN}}(Q^2,x) \ . \end{aligned} \tag{3.43}$$

Frequently, also the parameter Q^2 is replaced by a new quantity, the inelasticity parameter y:

$$y = \frac{\nu}{M_\mathrm{N} E} = \frac{E - E'}{E} \ . \tag{3.44}$$

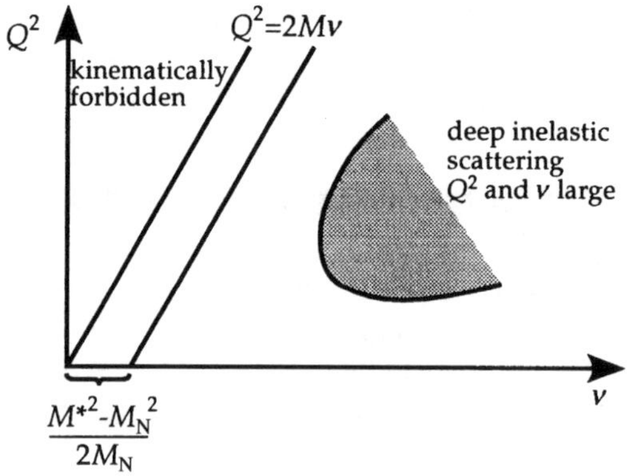

Fig. 3.4. The physical meaning of the different domains of x, Q^2, and ν. $Q^2 = 2M_\nu$ holds for elastic scattering only, otherwise Q^2 and ν are independent

y is dimensionless and specifies which fraction of the lepton energy in the laboratory system is transferred to the nucleon. The value of y thus lies in the range $0 \leq y \leq 1$. The meaning of Q^2, ν, and x is also illustrated in Fig. 3.4. It can easily be checked (see Exercise 3.5) that

$$\frac{\mathrm{d}^2\sigma}{\mathrm{d}E'\mathrm{d}\Omega} = \frac{E'}{2\pi M_\mathrm{N} E y} \frac{\mathrm{d}^2\sigma}{\mathrm{d}x\mathrm{d}y} . \tag{3.45}$$

In this notation, (3.32) becomes (see Exercise 3.5):

$$\left.\frac{\mathrm{d}^2\sigma}{\mathrm{d}x\,\mathrm{d}y}\right|_{\mathrm{eN}} = \frac{8\pi M_\mathrm{N} E\alpha^2}{Q^4} \times \left[xy^2 F_1^{\mathrm{eN}}(Q^2, x) + \left(1 - y - \frac{M_\mathrm{N} xy}{2E}\right) F_2^{\mathrm{eN}}(Q^2, x) \right] . \tag{3.46}$$

The first measurements of inelastic electron–proton scattering at high energies were performed in 1968 at the Stanford Linear Accelerator (SLAC). These and subsequent experiments have yielded two important results:

(1) For sufficiently large values of Q^2, i.e., for $Q^2 \geq 1\mathrm{GeV}^2$, F_2^{eN} and F_1^{eN} do not depend significantly on Q^2, but only on x. This behavior is termed *Bjorken scaling*.

(2) In this domain (i.e., for $Q^2 \geq 1\mathrm{GeV}^2$) it holds to good approximation that

$$F_2^{\mathrm{eN}} \approx 2x F_1^{\mathrm{eN}} .$$

EXERCISE

3.5 The Cross Section as a Function of x and y

Problem. Derive (3.45) and (3.46).

Solution. Equations (3.44), (3.39), and (3.30) yield

$$y = \frac{E - E'}{E} = 1 - \frac{E'}{E} \,,$$
$$x = \frac{Q^2}{2(E - E')M_N} = \frac{2EE'}{(E - E')M_N} \sin^2\left(\frac{\theta}{2}\right) \,. \tag{1}$$

Owing to the cylindrical symmetry of the problem we have in addition

$$\frac{d^2\sigma}{dE'd\Omega} = \frac{d^2\sigma}{dE' 2\pi \sin(\theta) d\theta} \,. \tag{2}$$

Therefore we only have to evaluate the Jacobi determinant

$$\begin{aligned}
\frac{d^2\sigma}{dE'd\Omega} &= \frac{1}{2\pi \sin(\theta)} \left| \frac{\partial(x, y)}{\partial(E', \theta)} \right| \frac{d^2\sigma}{dx dy} \\
&= \begin{vmatrix} \frac{2E \sin^2\left(\frac{\theta}{2}\right)}{M_N} \left(\frac{E' + (E - E')}{(E - E')^2}\right) & \frac{2EE'}{(E - E')M_N} \sin\left(\frac{\theta}{2}\right) \cos\left(\frac{\theta}{2}\right) \\ -\frac{1}{E} & 0 \end{vmatrix} \frac{1}{2\pi \sin(\theta)} \frac{d^2\sigma}{dx dy} \\
&= \frac{1}{2\pi \sin(\theta)} \frac{2E'}{(E - E')M_N} \frac{\sin(\theta)}{2} \frac{d^2\sigma}{dx\, dy} \\
&= \frac{E'}{2\pi M_N E y} \frac{d^2\sigma}{dx\, dy} \,.
\end{aligned} \tag{3}$$

This is identical to (3.45). With the help of the definitions introduced in (3.39)–(3.44) the double-differential cross section for electron–nucleon scattering consequently becomes

$$\frac{d^2\sigma}{dx dy} = \frac{8\pi M_N y E E' \alpha^2}{Q^4} \left[\frac{2}{M_N} \sin^2\left(\frac{\theta}{2}\right) F_1^{eN}(Q^2, x) + \frac{\cos^2\left(\frac{\theta}{2}\right)}{yE} F_2^{eN}(Q^2, x) \right] . \tag{4}$$

Here we have employed (3.32) and (3.43). Now we replace θ by x and y, using (1):

$$\sin^2\left(\frac{\theta}{2}\right) = \frac{(E - E')M_N}{2EE'} x = \frac{M_N}{2E'} xy \,, \tag{5}$$

Exercise 3.5

$$\frac{d^2\sigma}{dxdy} = \frac{8\pi M_N yEE'\alpha^2}{Q^4}\left[\frac{xy}{E'}F_1^{eN}(Q^2,x) + \frac{1-\frac{M_N}{2E'}xy}{yE}F_2^{eN}(Q^2,x)\right]$$
$$= \frac{8\pi M_N E\alpha^2}{Q^4}\left[xy^2F_1^{eN}(Q^2,x) + \left(1-y-\frac{M_N}{2E}xy\right)F_2^{eN}(Q^2,x)\right], \quad (6)$$

which yields (3.46).

These results are illustrated in Figs. 3.5 and 3.6. In Fig. 3.5 it can be seen that, at least for medium values of x, the measured points are practically horizontal. In the following we shall discuss the physical meaning of this. In Chap. 5 we shall discuss how the Q^2 dependence of the structure functions for large and small x can be understood in the framework of quantum chromodynamics.

These two unusual properties can be easily reduced to the common physical cause that the scattering is off pointlike spin-$\frac{1}{2}$ constituents of the nucleon. These constituents are termed, following R.P. Feynman, *partons*. What do we know about this kind of scattering?

The scattering cross section $d\sigma_i/d\Omega$ for the scattering of electrons off a pointlike parton of mass M_i and charge $q_i e$ can be constructed from the re-

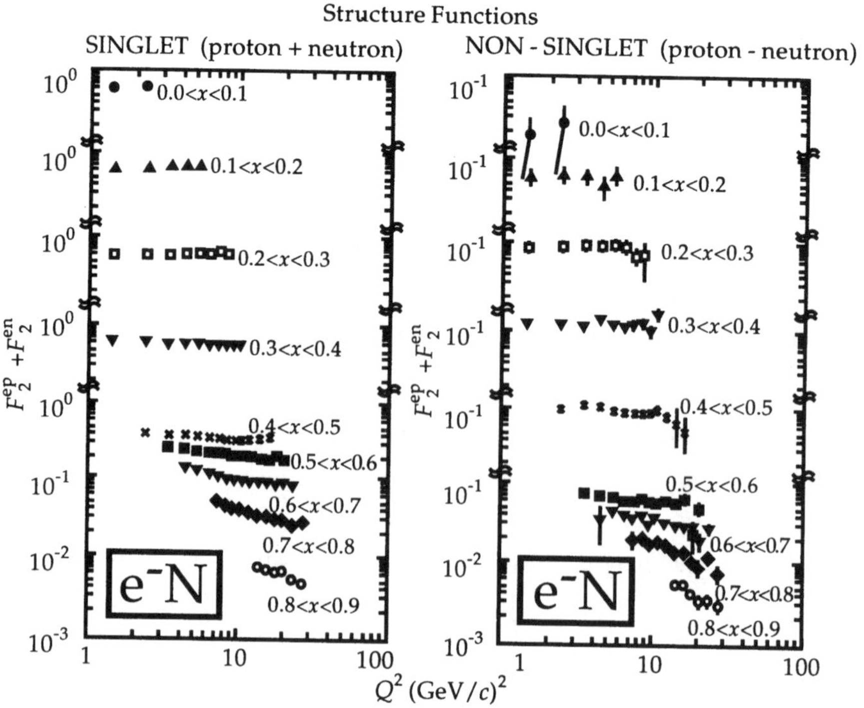

Fig. 3.5. The F_2 structure functions for electron–proton and electron–neutron scattering have been combined into $F_2^{ep} \pm F_2^{en}$ and plotted against Q^2. The Bjorken variable x is defined as $x = Q^2/2\nu M_N$ (see Review of Particle Properties: Phys. Lett. **170** (1986))

sults for electron–muon scattering, namely (2.165), (see in particular 11 in Exercise 2.13), which now reads

$$\begin{aligned}\frac{\mathrm{d}\sigma_i}{\mathrm{d}\Omega} &= \frac{\alpha^2 q_i^2}{4E^2\sin^4\theta/2}\frac{E'}{E}\cos^2\theta/2\left(1-\frac{q^2\tan^2\theta/2}{2M_i^2}\right)\\ &= \frac{\alpha^2 q_i^2}{4E^2}\frac{E'}{E}\frac{1}{\sin^4\theta/2}\left[\cos^2\theta/2+\frac{Q^2}{4M_i^2}\sin^2\theta/2\right] .\end{aligned}$$

From

$$2\left(E-E'\right)M = 2\nu = Q^2 = +4EE'\sin^2\left(\frac{\theta}{2}\right) ,$$

one finds

$$\frac{E}{E'} = 1+\frac{2E}{M}\sin^2\left(\frac{\theta}{2}\right) ,$$

and therefore

$$\begin{aligned}\frac{\mathrm{d}\sigma_i}{\mathrm{d}\Omega} &= \frac{\alpha^2 q_i^2}{4E^2\sin^4\left(\frac{\theta}{2}\right)}\frac{1}{1+\frac{2E}{M_i}\sin^2\left(\frac{\theta}{2}\right)}\cos^2\left(\frac{\theta}{2}\right)\left[1-\frac{q^2\tan^2\left(\frac{\theta}{2}\right)}{2M_i^2}\right]\\ &= \frac{\alpha^2 q_i^2}{4E^2}\frac{1}{\sin^4\left(\frac{\theta}{2}\right)}\frac{\left[\cos^2\left(\frac{\theta}{2}\right)-\frac{q^2}{2M_i^2}\sin^2\left(\frac{\theta}{2}\right)\right]}{1+\frac{2E}{M_i}\sin^2\left(\frac{\theta}{2}\right)} .\end{aligned} \tag{3.47}$$

This cross section is computed in the laboratory system as the rest system of the parton. The laboratory system, however, is quite unsuitable for our needs since here partons cannot be treated as free particles because they are held together by the interaction to which this volume is dedicated. Thus we do not know at all the initial momenta of the partons; they are definitely not at rest. There are, however, frames of reference where the initial momentum is known. These are all frames of reference in which the nucleon is moving very fast. The momentum of the ith parton is then

$$P_{i,\mu} = \xi_i\cdot P_\mu + \Delta P_{i\mu} , \quad \max_\mu(\Delta P_\mu) \ll \max_\mu(P_\mu) . \tag{3.48}$$

Here $\max_\mu(\Delta P_\mu)$ stands for the maximum of ΔP_μ for the various space–time components μ. The unknown momentum fraction $\Delta P_{i\mu'}$, which comes from the interactions of the partons with each other, is much smaller than the momentum from the collective motion. To describe electron–parton scattering sensibly, (3.47) must be transformed into a different Lorentz system.[8] This is done by rewriting it using only the invariant variables

$$\begin{aligned}&(p-p')^2 =: t = -Q^2 = (P-P')^2 ,\\ &(p+P)^2 =: s = (p'+P')^2 ,\\ &(p'-P)^2 =: u = (p-P')^2 .\end{aligned} \tag{3.49}$$

[8] Here and in the following we omit the parton index i on the mass and momentum variables.

These quantities are called *Mandelstam variables*. In the laboratory system

$$Q^2 = 4EE' \sin^2\left(\frac{\theta}{2}\right) \ , \tag{3.50a}$$

$$s = 2\,p\cdot P + M^2 = M(2E+M) \ , \tag{3.50b}$$

$$u = -2\,p'\cdot P + M^2 = -M(2E'-M) \ . \tag{3.50c}$$

The small mass of the electron ($m^2 \ll M^2,\ p\cdot P$) has been neglected here. Q^2 is solely a function of θ, because the E' appearing in (3.50a) can be expressed in terms of E and θ (see (3.53) below). Therefore we may use (note the cylindrical symmetry)

$$\frac{\mathrm{d}\sigma}{\mathrm{d}\Omega} = \frac{1}{2\pi}\frac{\mathrm{d}\sigma}{\sin(\theta)\mathrm{d}\theta} = \frac{1}{2\pi\sin(\theta)}\frac{\mathrm{d}Q^2}{\mathrm{d}\theta}\frac{\mathrm{d}\sigma}{\mathrm{d}Q^2} \ . \tag{3.51}$$

From $P'_\mu = P_\mu + p_\mu - p'_\mu$, it follows that

$$M^2 = P'^2_\mu = (P_\mu + p_\mu - p'_\mu)^2 = M^2 + (2E-E')M - Q^2 \ , \tag{3.52}$$

or, after solving for E',

$$E' = E\left[1 + 2\frac{E}{M}\sin^2\left(\frac{\theta}{2}\right)\right]^{-1} \ . \tag{3.53}$$

For the derivative $\mathrm{d}Q^2/\mathrm{d}\theta$ we obtain, with (3.50a) and (3.53),

$$\begin{aligned}\frac{\mathrm{d}Q^2}{\mathrm{d}\theta} &= 4EE'\sin\left(\frac{\theta}{2}\right)\cos\left(\frac{\theta}{2}\right) - \frac{E\,2\frac{E}{M}\sin\left(\frac{\theta}{2}\right)\cos\left(\frac{\theta}{2}\right)}{\left[1+2\frac{E}{M}\sin^2\left(\frac{\theta}{2}\right)\right]^2}4E\sin^2\left(\frac{\theta}{2}\right)\\ &= 2EE'\sin\theta\left\{1 - \frac{2E\sin^2\left(\frac{\theta}{2}\right)}{M\left[1+2\frac{E}{M}\sin^2\left(\frac{\theta}{2}\right)\right]}\right\}\\ &= \frac{2EE'\sin\theta}{\left[1+2\frac{E}{M}\sin^2\left(\frac{\theta}{2}\right)\right]} \ .\end{aligned} \tag{3.54}$$

For $\mathrm{d}\sigma/\mathrm{d}Q^2$ we therefore find according to (3.51) and (3.47) that

$$\begin{aligned}\frac{\mathrm{d}\sigma}{\mathrm{d}Q^2} &= \frac{2\pi\sin^2}{\frac{\mathrm{d}Q^2}{\mathrm{d}\theta}}\frac{\mathrm{d}\sigma}{\mathrm{d}\omega}\\ &= \frac{\pi}{EE'}\frac{\alpha^2 q_i^2}{4E^2}\frac{1}{\sin^4\left(\frac{\theta}{2}\right)}\left[\cos^2\left(\frac{\theta}{2}\right) + \frac{Q^2}{2M_i^2}\sin^2\left(\frac{\theta}{2}\right)\right] \ .\end{aligned} \tag{3.55}$$

We further insert, according to (3.50),

$$
\begin{aligned}
E' &= \frac{u-M_i^2}{-2M_i} \ , \quad E = \frac{s-M_i^2}{2M_i} \ , \\
\sin^2\left(\frac{\theta}{2}\right) &= \frac{Q^2}{4EE'} = -\frac{Q^2 M_i^2}{\left(u-M^2\right)\left(s-M_i^2\right)} \ , \\
\cos^2\left(\frac{\theta}{2}\right) &= 1+\frac{Q^2M_i^2}{\left(u-M_i^2\right)\left(s-M_i^2\right)} \ ,
\end{aligned}
\tag{3.56}
$$

so that after appropriate substitutions, (3.55) becomes

$$
\begin{aligned}
\frac{\mathrm{d}\sigma_i}{\mathrm{d}Q^2} &= \frac{4\pi\alpha^2 q_i^2}{Q^4} \\
&\quad\times\left(-\frac{u_i-M_i^2}{s_i-M_i^2}\right)\left[1+\left(1-\frac{Q^2}{2M_i^2}\right)\frac{Q^2M_i^2}{\left(u_i-M_i^2\right)\left(s_i-M_i^2\right)}\right] .
\end{aligned}
\tag{3.57}
$$

The quantities u_i and s_i refer to the ith parton. With this, we have succeeded in the first step. We must now choose an appropriate laboratory system and substitute the parton masses. It is now a major statement of the parton model that *one obtains a good description of experimental results when one assumes that partons are massless*. For $M_i \to 0$, (3.57) simplifies to

$$
\frac{\mathrm{d}\sigma_i}{\mathrm{d}Q^2} = \frac{4\pi\alpha^2 q_i^2}{Q^4}\left(\frac{Q^4}{2u_is_i}-1\right)\frac{u_i}{s_i} \ . \tag{3.58a}
$$

In addition, from (3.49) one deduces

$$
\begin{aligned}
Q^2 - s_i - u_i &= 2p\cdot p' - 2p\cdot P_i + 2p\cdot P_i{}' \\
&= 2p\cdot\left(p'-P_i+P_i{}'\right) = 2p^2 = 0
\end{aligned}
\tag{3.59}
$$

and therefore $Q^4 = s_i^2 + u_i^2 + 2u_is_i$, so that

$$
\left(\frac{Q^4}{2u_is_i}-1\right)\frac{u_i}{s_i} = \frac{s_i^2+u_i^2}{2s_i^2} \ . \tag{3.60}
$$

Thus (3.58a) becomes

$$
\frac{\mathrm{d}\sigma_i}{\mathrm{d}Q^2} = \frac{4\pi\alpha^2 q_i^2}{Q^4}\left(\frac{s_i^2+u_i^2}{2s_i^2}\right) . \tag{3.58b}
$$

For our frame of reference, we now choose the so-called *Breit system*. This is characterized by the property that q_μ is purely spacelike and points in the z direction (see Exercise 3.6):

$$
\begin{aligned}
q_\mu &= \left(0,0,0,\sqrt{-q^2}\right) = \left(0,0,0,\sqrt{Q^2}\right) \ , \\
P_\mu &= \left(\tilde{P},0,0,-\tilde{P}\right) \ , \quad \tilde{P} \gg M_{\mathrm{N}} \ , \quad P^2 = \tilde{P}^2 - \tilde{P}^2 = 0 \approx M_{\mathrm{N}}^2 \ .
\end{aligned}
\tag{3.61}
$$

Since we neglect the momentum fraction of the partons stemming from the parton–parton interaction, for consistency we must also neglect the nucleon masses:

$$P_{i,\mu} = \xi_i \left(\tilde{P}, 0, 0, -\tilde{P}\right) \ . \tag{3.62}$$

The cross section for scattering off a parton in the Breit system thus becomes

$$\begin{aligned}
\frac{\mathrm{d}\sigma_i}{\mathrm{d}Q^2} &= \frac{4\pi\alpha^2 q_i^2}{Q^4}\left(\frac{(p+P_i)^4+(p'-P_i)^4}{2(p+P_i)^4}\right) \\
&= \frac{4\pi\alpha^2 q_i^2}{Q^4}\left(\frac{(\xi_i 2p\cdot P)^2+(\xi_i 2p'\cdot P)^2}{2(2\xi_i p\cdot P)^2}\right) \\
&= \frac{4\pi\alpha^2 q_i^2}{Q^4}\left(\frac{(p+P)^4+(p'-P)^4}{2(p+P)^4}\right) \\
&= \frac{4\pi\alpha^2 q_i^2}{Q^4}\left(\frac{s^2+u^2}{2s^2}\right) \ .
\end{aligned} \tag{3.63}$$

s and u are now Mandelstam variables for the whole nucleon. We recognize that $\mathrm{d}\sigma/\mathrm{d}Q^2$ is completely independent of ξ_i. To compare this expression with (3.46), we must write in the Breit system not only the elastic scattering cross section for pointlike particles, i.e.(3.63) above, but also the inelastic scattering cross section for particles with an internal structure (see (3.46)). To this end, (3.46) is rewritten such that only Lorentz-invariant quantities appear. For $M_\mathrm{N}^2 \ll u, s, Q^2$ (see (3.56))

$$\begin{aligned}
E' &= -\frac{u}{2M_\mathrm{N}} \ , \qquad E = \frac{s}{2M_\mathrm{N}} \ , \\
y &= \frac{E-E'}{E} \equiv \frac{s+u}{s}
\end{aligned} \tag{3.64}$$

and

$$x = \frac{Q^2}{2\nu} = \frac{Q^2}{2P\cdot q} = \frac{Q^2}{s+u} \ . \tag{3.65}$$

Equation (3.46) is thus nearly in invariant form. One only has to substitute s for E:

$$\left.\frac{\mathrm{d}^2\sigma}{\mathrm{d}x\,\mathrm{d}y}\right|_\mathrm{eN} = \frac{4\pi s\alpha^2}{Q^4}\left[xy^2 F_1^\mathrm{eN}(Q^2,x) + \left(1-y-\frac{xyM_\mathrm{N}^2}{s}\right) F_2^\mathrm{eN}(Q^2,x)\right] . \tag{3.66}$$

Because $M_\mathrm{N} \ll s$ the second term can be further simplified. This expression must now be compared to the sum over partons i of (3.63). To this end, we must first rewrite (3.63) as a double-differential cross section.

To do this, we start from the momentum conservation $P_i' = P_i + q$ for the scattering parton, i.e.,

$$0 \approx P_i'^2 = (P_i+q)^2 = (\xi_i P+q)^2 = \xi_i^2 M_i^2 + 2\xi_i P\cdot q + q^2 \ . \tag{3.67}$$

When the parton mass vanishes, this implies that

$$\xi_i = \frac{Q^2}{2P\cdot q} = x \ . \tag{3.68}$$

This is an interesting relation, which allows us to interprete the variable x in a new way: If we observe scattering characterized by a Bjorken variable x, this means in the parton model that the interacting parton carried a fraction $\xi_i = x$ of the nucleon's total momentum! Hence the variable x has a simple meaning (but only in reference frames with very large nucleon momentum, e.g. in the Breit frame).

Owing to (3.68), (3.63) can immediately be written as a double differential:

$$\frac{\mathrm{d}^2\sigma}{\mathrm{d}x\,\mathrm{d}Q^2} = \frac{4\pi\alpha^2 q_i^2}{Q^4}\frac{s^2+u^2}{2s^2}\delta(\xi_i - x) \ . \tag{3.69}$$

The transition from Q^2 to the desired variable y is achieved with the aid of (3.39), (3.44), and (3.56), according to which

$$Q^2 = 2\,P\cdot q\,x = 2M_\mathrm{N}(E-E')x = 2M_\mathrm{N}Eyx = syx \ , \tag{3.70}$$

and we thus obtain

$$\frac{\mathrm{d}^2\sigma}{\mathrm{d}x\,\mathrm{d}y} = \left.\frac{\mathrm{d}Q^2}{\mathrm{d}y}\right|_x \frac{\mathrm{d}^2\sigma}{\mathrm{d}x\,\mathrm{d}Q^2} = \frac{2\pi\alpha^2 q_i^2}{Q^4}\frac{s^2+u^2}{s^2}\delta(\xi_i - x)\frac{Q^2 s}{s+u} \ . \tag{3.71}$$

Here, in the last step, we have put $x = Q^2/(s+u)$, according to (3.65). To obtain the total cross section in the parton model, we must sum (3.71) over all partons i and all possible momentum fractions ξ_i. The single terms of the sum must be weighted with their proper probabilities $f_i(\xi_i)$:

$$\begin{aligned}\left.\frac{\mathrm{d}^2\sigma}{\mathrm{d}x\,\mathrm{d}y}\right|_{\mathrm{eN}} &= \sum_i \int_0^1 \mathrm{d}\xi_i\, f_i(\xi_i)\frac{2\pi\alpha^2 q_i^2}{Q^4}\frac{s^2+u^2}{s^2}\delta(\xi_i - x)\frac{Q^2 s}{s+u} \\ &= \frac{2\pi\alpha^2 s}{Q^4}\frac{s^2+u^2}{s^2}\sum_i f_i(x)q_i^2 x \qquad \left(\text{remember } x = \frac{Q^2}{s+u}\right) \\ &= \frac{2\pi\alpha^2 s}{Q^4}\left[(y-1)^2+1\right]\sum_i f_i(x)q_i^2 x \end{aligned} \tag{3.72}$$

with

$$\begin{aligned}(y-1)^2+1 = y^2-2y+2 &= \left(\frac{s+u}{s}\right)^2 - 2\frac{s+u}{s} + 2 \\ &= \frac{s^2+u^2}{s^2} + \frac{2su}{s^2} - 2\frac{s+u}{s} + 2 \\ &= \frac{s^2+u^2}{s^2}\end{aligned}$$

according to (3.64). For the momentum distribution functions $f_i(\xi_i)$ of the partons the normalization

$$\int_0^1 \mathrm{d}\xi_i\, f_i(\xi_i) = 1 \tag{3.73}$$

holds. Comparing (3.72) with (3.66) shows that in the parton model the following identification must hold:

$$2\left[xy^2 F_1^{\mathrm{eN}}(Q^2,x) + (1-y)F_2^{\mathrm{eN}}(Q^2,x)\right] = \left[(y-1)^2+1\right]\sum_i f_i(x) q_i^2 x \ , \qquad s = \mathrm{const} \ . \tag{3.74}$$

This equation (3.74) holds for all scattering processes, i.e., for any value of E, E', and θ. We use this fact by letting the parameter $s = 2M_{\mathrm{N}}E$, which was fixed up to now, vary:

$$s \to s' = \frac{s'}{s} s \ . \tag{3.75}$$

Simultaneously, we substitute

$$u \to \frac{s'}{s} u = u' \ , \quad Q^2 \to \frac{s'}{s} Q^2 = Q'^2 \ . \tag{3.76}$$

x and y are not changed under the simultaneous transformations

$$y \to \frac{s'+u'}{s'} = \frac{s+u}{s} \ , \quad x \to \frac{Q'^2}{s'+u'} = \frac{Q^2}{s+u} \ , \tag{3.77}$$

or

$$y' = y \ , \quad x' = x \ . \tag{3.78}$$

Since the right-hand side of (3.74), computed in the parton model, depends only on the variables x and y, and is therefore invariant under the scaling transformations (3.75)–(3.77), it follows for the left-hand side that

$$xy^2 F_1^{\mathrm{eN}}(Q^2,x) + (1-y)F_2^{\mathrm{eN}}(Q^2,x) = \mathrm{const} \quad \text{for all } Q^2 = Q'^2 \ , \tag{3.79}$$

which implies that the functions $F_1^{\mathrm{eN}}(Q^2,x)$ and $F_2^{\mathrm{eN}}(Q^2,x)$ do not depend on Q^2 at all, i.e.

$$F_1^{\mathrm{eN}}(Q^2,x) = F_1^{\mathrm{eN}}(x) \ , \quad F_2^{\mathrm{eN}}(Q^2,x) = F_2^{\mathrm{eN}}(x) \ . \tag{3.80}$$

The structure functions in the parton picture thus no longer depend on Q^2. Since we have neglected here both the momenta coming from the binding of the partons and the nucleon mass, (3.79) should hold for sufficiently large values for Q^2. This has indeed been experimentally observed as discussed above. It is the so-called *Bjorken scaling*, which we now do understand within the parton model. Equation (3.74) then becomes

$$2\left[xy^2F_1^{\mathrm{eN}}(x)+(1-y)F_2^{\mathrm{eN}}(x)\right]=\left[y^2+2(1-y)\right]\sum_i f_i(x)q_i^2x\ . \qquad (3.81)$$

By comparing powers of y, we find that

$$F_1^{\mathrm{eN}}(x)=\frac{1}{2}\sum_i f_i(x)q_i^2\ , \quad F_2^{\mathrm{eN}}(x)=\sum_i f_i(x)q_i^2x\ , \qquad (3.82)$$

and

$$F_2^{\mathrm{eN}}(x)=2xF_1^{\mathrm{eN}}(x)\ . \qquad (3.83)$$

This is the *Callan–Gross relation*. Moreover, the nucleon form factors are now – within the parton model – simply related to the parton distribution functions $f_i(x)$, describing the distribution of the fraction of the total nucleon momentum shared by the ith parton.

We have thus seen that deep inelastic scattering processes can be explained by the basic assumptions of the parton model: the hadron is built of pointlike constituents whose energy of interaction with each other and whose mass is small compared to $\sqrt{Q^2}>1\mathrm{GeV}$.

The physical foundations of (3.81) and (3.82) can also be illustrated from a different point of view: we can assume that partons are massless. There are no dimensional quantities in electron–parton scattering except for the momenta. Hence, the only Lorentz-invariant quantities are q^2, P^2, and $P\cdot q$. Now, $P^2=M^2=0$, because of the vanishing parton masses. From $Q^2=-q^2$ and $P\cdot q$ one can only have Q^2 and the dimensionless $x=Q^2/2P\cdot q$. Since $F_1^{\mathrm{eN}}(Q^2,x)$ is dimensionless, it can only depend on x alone, i.e. $F_1^{\mathrm{eN}}(Q^2,x)=F_1^{\mathrm{eN}}(x)$.

We have already seen that three structure functions are present in a parity-violating interaction (see (3.37) and (3.38)). On the other hand, we obtain only one structure function for scalar particles, i.e., $F_1=0$ (see Exercise 3.7). In addition to deep inelastic electron–nucleon scattering, photon–nucleon scattering also yields information about structure functions. In fact, in Exercise 3.8, we shall show that the relationship between the cross sections for transverse and for scalar photons, in particular, is of interest, since it tests the Callan–Gross relation (3.82). As can be seen from Fig. 3.6, it holds very well experimentally. Therefore it suffices to investigate scattering reactions as a function of $F_2(x)$. For this reason, Fig. 3.5 shows only $F_2(x)=\frac{1}{2}(F_2^{\mathrm{ep}}+F_2^{\mathrm{en}})$. In Exercise 3.9 we shall finally compute the structure functions following from a simple constituent-quark model. Independent of the special assumption of each model, all such computations yield the following results.

(1) If one assumes that the nucleon is composed of three quarks, the one-particle distributions $f(x)$ are typically maximal at $x=1/3$ and, accordingly, the structure function $F_2(x)$ at somewhat higher values of x.

Fig. 3.6. The ratio of the structure functions F_1 and F_2 within the scaling region provides a test of the so-called Callan–Gross relations (see (3.82))

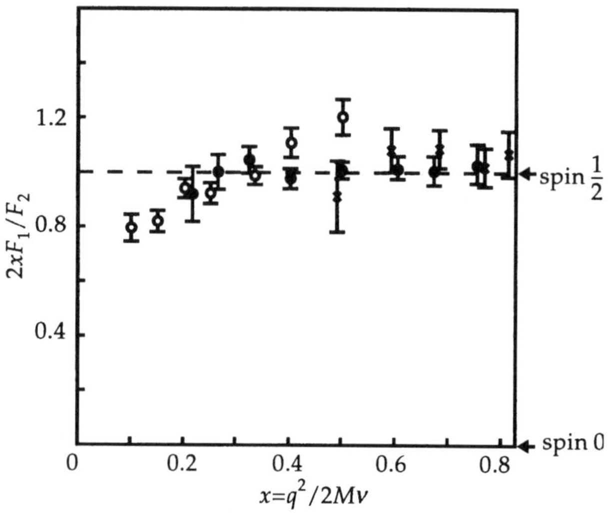

(2) The structure functions averaged over protons and neutrons are designated by $F_2(x) = \frac{1}{2}[F_2^{\mathrm{ep}}(x) + F_2^{\mathrm{en}}(x)]$, where

$$F_2^{\mathrm{eN}}(x) = \sum_i f_i(x) q_i^2 x = f(x) \cdot x \sum_i q_i^2 \tag{3.84}$$

is the structure function for the nucleon (N = p, n). Here it is assumed that all three quarks in the nucleon have, owing to their negligible masses, the same wave function.

We proceed now somewhat differently then before and introduce momentum distribution functions for quark flavors. Let us call the momentum distribution function for an u quark $u(x)$, the one for a d quark $d(x)$, and those for the corresponding antiquarks $\bar{u}(x)$, $\bar{d}(x)$, etc. Then

$$\begin{aligned} F_2^{\mathrm{ep}}(x) &= \frac{4}{9}\left(u^p(x) + \bar{u}^p(x)\right) x + \frac{1}{9}\left(d^p(x) + \bar{d}^p(x)\right) x \ , \\ F_2^{\mathrm{en}}(x) &= \frac{4}{9}\left(u^n(x) + \bar{u}^n(x)\right) x + \frac{1}{9}\left(d^n(x) + \bar{d}^n(x)\right) x \ . \end{aligned} \tag{3.85}$$

Note that $u(x)$ contains now all u quarks in the nucleon. The same is true for $\bar{u}(x)$, $d(x)$, $\bar{d}(x)$, etc. This is different from the $f_i(x)$ introduced in e.g. (3.81). There each individual quark is counted separately with its own distribution function $f_i(x)$. Here, only the distributions of the flavors are considered.

Assuming isospin symmetry, one concludes that the $u^n(x)$ distribution is the same as the $d^p(x)$ distribution, and similarly $u^p(x) = d^n(x)$. Therefore we set

$$\begin{aligned} u^p(x) &= d^n(x) \equiv u(x) \ , \\ d^p(x) &= u^n(x) \equiv d(x) \ . \end{aligned}$$

and similarly for the antiquarks. Hence (3.85) becomes

$$F_2^{\mathrm{ep}}(x) = \frac{4}{9}\left(u(x)+\bar{u}(x)\right)x + \frac{1}{9}\left(d(x)+\bar{d}(x)\right)x \ ,$$
$$F_2^{\mathrm{en}}(x) = \frac{1}{9}\left(u(x)+\bar{u}(x)\right)x + \frac{4}{9}\left(d(x)+\bar{d}(x)\right)x \ , \tag{3.86}$$

and it follows that

$$\begin{aligned} F_2(x) &= \frac{1}{2}\left(F_2^{\mathrm{ep}}(x)+F_2^{\mathrm{en}}(x)\right) \\ &= \frac{1}{2}\left[\frac{5}{9}\left(u(x)+\bar{u}(x)\right)+\frac{5}{9}\left(d(x)+\bar{d}(x)\right)\right]x \\ &= \frac{5}{18}\left[\left(u(x)+\bar{u}(x)\right)+\left(d(x)+\bar{d}(x)\right)\right]x \ . \end{aligned} \tag{3.87}$$

If we sum over the momenta of all possible partons in a nucleus, the total momentum P of the nucleon should be obtained, i.e.

$$\begin{aligned} &\int_0^1 \mathrm{d}x\,\left(xP_\mu\right)\left[u(x)+\bar{u}(x)+d(x)+\bar{d}(x)+s(x)+\bar{s}(x)+\dots\right] \\ &\quad = P_\mu - \left(P_\mu\right)_g \ . \end{aligned} \tag{3.88}$$

Here $\left(P_\mu\right)_g$ stands for the momentum carried by partons not taken into account in the sum of the left-hand side. The index g in $\left(P_\mu\right)_g$ shall indicate that this missing momentum is possibly due to gluons. Neglecting gluons, one deduces from (3.88) that

$$\int_0^1 \mathrm{d}x\; x\left[(u(x)+\bar{u}(x)+d(x)+\bar{d}(x)+s(x)+\bar{s}(x)+\dots\right] = 1 \ , \tag{3.89}$$

and it follows from (3.87) that

$$\frac{18}{5}\int_0^1 \mathrm{d}x F_2(x) = 1 \ . \tag{3.90}$$

This relation can be tested with experiments (see Fig. 3.12 below). One finds:

(1) $F_2(x)$ is maximal for small x, i.e., there must be charged partons carrying only a small momentum fraction. It is natural to identify them with quark–antiquark pairs, which are – due to the interaction between the quarks (gluon exchange) – also created out of the vacuum (ground state). One might better call them vacuum ground-state correlation. A careful analysis of electromagnetic and electroweak structure functions indeed justifies this assumption, as we shall see below.

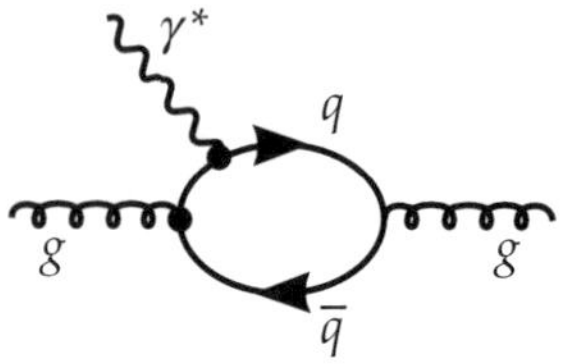

Fig. 3.7. The quark–antiquark pairs act effectively as electrically neutral objects, as do the gluons, of course. However, there is an important difference between quark–antiquark pairs and gluons, namely that electrons with sufficiently high momentum transfer can always scatter from the former, but never from gluons

(2) The integral $\int \mathrm{d}x F_2(x)$ is experimentally much smaller than unity, namely approximately 0.45. Half of the nucleon momentum is thus carried by electrically neutral particles. This is an important point since, according to our consideration above, it is the most direct evidence for the existence of gluons. The valence quarks interact with other quark–antiquark pairs via gluons, which leads to vacuum ground-state correlations. They can be graphically depicted as in Fig. 3.7. We shall return to this point in Sect. 4.2.

It may be added that more-refined models are also not able to give a satisfying description of the experimental data in Fig. 3.12 starting from the quark–quark interaction without ad hoc assumptions. It is possible, however, to describe the relationship between two structure functions. One can, for example, calculate quite well the relative change in F_2 for different values of Q^2. We shall discuss this in Chap. 5. The reason for this only partial success is that the calculation of structure functions itself is a completely nonperturbative problem and thus very difficult, while the Q^2 dependence is calculable by summing a few classes of graphs, i.e., in perturbation theory. As mentioned above, one introduces distribution functions for the various quarks and antiquarks:

$$u(x)\,, \quad \bar{u}(x)\,, \quad \text{etc.}$$

Then (3.86) can be generalized as

$$\begin{aligned} F_2^{\mathrm{ep}}(x) = x \bigg\{ &\frac{4}{9}\left[u(x)+\bar{u}(x)\right] + \frac{1}{9}\left[d(x)+\bar{d}(x)\right] \\ &+ \frac{4}{9}\left[c(x)+\bar{c}(x)\right] + \frac{1}{9}\left[s(x)+\bar{s}(x)\right] \bigg\} \,. \end{aligned} \tag{3.86a}$$

The crucial point is now that the structure functions for the reactions $\nu_\mathrm{e} + \mathrm{n} \rightarrow \mathrm{e}^- + \mathrm{p}$, $\nu_\mathrm{e} + \mathrm{n} \rightarrow \nu_\mathrm{e} + \mathrm{n}$, etc. involve different combinations of $\mathrm{u}(x)$, $\bar{u}(x)$, $\mathrm{d}(x)$, and $\bar{d}(x)$. (The weak charges of the quarks are not proportional to their electric charges.) Therefore quark distributions can be deduced from the different structure functions.

We will examplify this idea by considering the neutrino–nucleon structure functions that are measured in the reactions

$$\begin{aligned} &\nu\mathrm{p} \rightarrow \mathrm{e}^-\mathrm{X}, \quad \bar{\nu}\mathrm{p} \rightarrow \mathrm{e}^+\mathrm{X}\,, \\ &\nu\mathrm{n} \rightarrow \mathrm{e}^-\mathrm{X}, \quad \bar{\nu}\mathrm{n} \rightarrow \mathrm{e}^+\mathrm{X}. \end{aligned} \tag{3.91}$$

Due to charge conservation in these reactions the following parton distribution functions are measured:

$$\begin{aligned} F_2^{\nu\mathrm{p}}(x) &= 2x\left[d(x)+\bar{u}(x)\right]\,, \\ F_2^{\bar{\nu}\mathrm{p}}(x) &= 2x\left[u(x)+\bar{d}(x)\right]\,, \\ F_2^{\nu\mathrm{n}}(x) &= 2x\left[u(x)+\bar{d}u(x)\right]\,, \\ F_2^{\bar{\nu}\mathrm{n}}(x) &= 2x\left[d(x)+\bar{u}(x)\right]\,. \end{aligned} \tag{3.92}$$

The factor of 2 reflects the presence of both vector and axial vector parts in the weak currents. From that we get for the combination $\nu p \to e^- X$ and $\nu n \to e^- X$

$$F_2^{\nu p}(x) + F_2^{\bar{\nu} p}(x) = 2x(u + \bar{u} + d + \bar{d}) \ .$$

Combining this with (3.87) gives

$$F_2^{\mathrm{e\,p}}(x) + F_2^{\mathrm{e\,n}} = x\frac{5}{9}(u + \bar{u} + d + \bar{d}) + x\frac{2}{9}(s + \bar{s})$$

we can extract the strange quarks content of the nucleon

$$F_2^{\mathrm{e\,p}}(x) + F_2^{\mathrm{e\,n}} - \frac{5}{18}\left[F_2^{\nu \mathrm{p}}(x) + F_2^{\bar{\nu} \mathrm{p}}(x)\right] = \frac{2}{9}x(s + \bar{s}) \ . \qquad (3.93)$$

The experimental data indicate a nonvanishing right-hand side only for the small $x \leq 0.2$ region.

In Fig. 3.8 we have sketched approximately what the parton distributions look like in a proton. Most interesting is the fact that quark and antiquark distributions coincide for $x \to 0$. This can be understood if both are vacuum excitations (so-called sea quarks but one should better call them vacuum correlations (ground-state correlations) of quark–antiquark pairs), generated by the quark–quark interaction. The obvious divergence of the functions $u(x)$, $d(x)$, and $\bar{q}(x)$ for $x \to 0$ thus indicates that the interaction is large for small momentum transfers.

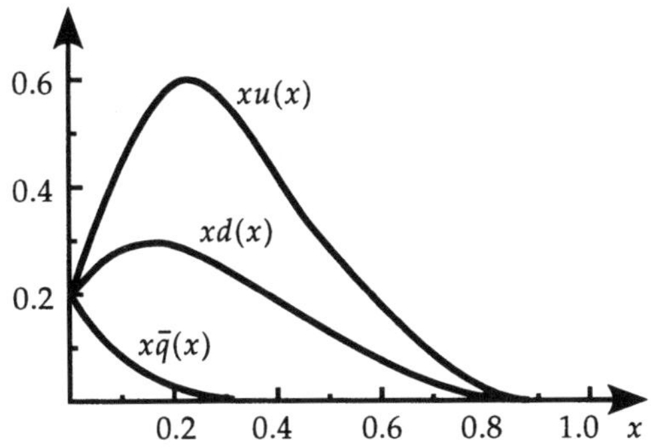

Fig. 3.8. A sketch of the quark distribution functions for protons with $\overline{q}(x) = \frac{1}{2}(\overline{u}(x) + \overline{d}(x))$. It can be recognized that the contributions of the valence quarks $x(u(x) - \overline{q}(x))$ and $x(d(x) - \overline{q}(x))$ vanish for $x \to 0$. Hence vacuum excitations in form of quark–antiquark pairs dominate for small values of x

EXERCISE

3.6 The Breit System

Problem. Prove the existence of a Lorentz reference frame where (3.61) holds.

Solution. We start with the laboratory system. Here

$$\begin{aligned} q_\mu &= (q_0, \boldsymbol{q}) \ , \quad q_0 > 0 \ , \\ P_\mu &= (M, \boldsymbol{0}) \ . \end{aligned} \tag{1}$$

First we perform a rotation in such a way that $\boldsymbol{q}$ is parallel to the z direction.

$$\begin{aligned} q'_\mu &= (q'_0, 0, 0, q'_3) \ , \quad q'_0, q'_3 > 0 \\ P'_\mu &= (M_\mathrm{N}, 0, 0, 0) \ . \end{aligned} \tag{2}$$

Now we boost the system in the z direction, i.e., we transform to a reference frame moving with the velocity

$$v_z = \frac{q'_0}{q'_3} c = \beta c \ . \tag{3}$$

This is possible, because the momentum transfer q_μ is spacelike, i.e.,

$$Q^2 = q_3'^2 - q_0'^2 > 0 \to q'_3 > q'_0 \ . \tag{4}$$

This follows easily from (3.50a), according to which, $Q^2 = 4EE' \sin^2\theta/2 > 0$. Then in the new reference frame

$$\begin{aligned} q''_\mu &= \gamma(q'_0 - \beta q'_3, 0, 0, q'_3 - \beta q'_0) \\ &= \gamma\left(0, 0, 0, q'_3 - \beta^2 q'_3\right) \ , \beta = \frac{q'_0}{q'_3} \\ &= \left(0, 0, 0, q'_3 \sqrt{1 - \frac{q_0'^2}{q_3'^2}}\right) = \left(0, 0, 0, \sqrt{Q^2}\right) \ , \end{aligned} \tag{5}$$

with

$$\beta = \frac{v_z}{c} = \frac{q'_0}{q'_3} \quad \text{and} \quad \gamma = \left(1 - \beta^2\right)^{-1/2} \ . \tag{6}$$

Under this transformation the nucleon momentum becomes

$$P''_\mu = \gamma(M_\mathrm{N}, 0, 0, -\beta M_\mathrm{N}) \ . \tag{7}$$

Since we consider reactions where Q^2 and

$$\nu = q' \cdot P' = q'_0 = q'' \cdot P'' \tag{8}$$

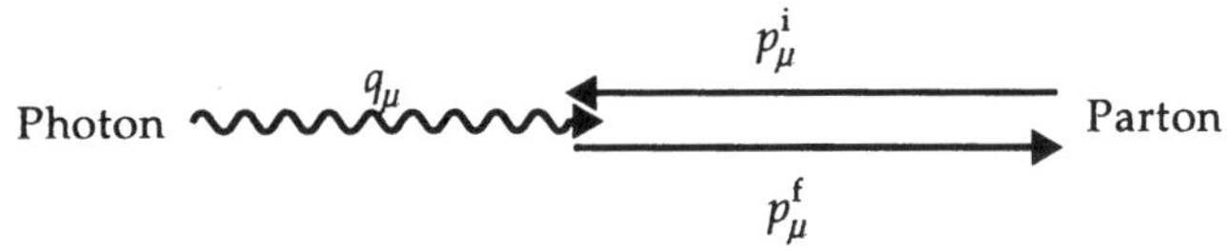

Fig. 3.9. Photon–parton interaction in the Breit system. The initial photon and parton carry the four-momenta $q_\mu = (0,0,0,\sqrt{Q^2})$ and $P^{\rm f}_\mu = \frac{1}{2}(\sqrt{Q^2},0,0,-\sqrt{Q^2})$, respectively. The final parton then has the momentum $P^{\rm f}_\mu = \frac{1}{2}(\sqrt{Q^2},0,0,\sqrt{Q^2})$

get very large, but their ratio $x = Q^2/2\nu$ remains constant,

$$\frac{\nu}{M_{\rm N}} = \frac{q''\cdot P''}{M_{\rm N}} = \beta\gamma\sqrt{Q^2} = \frac{Q^2}{2M_{\rm N}x} \Rightarrow \beta\gamma = \frac{\nu/M_{\rm N}}{\sqrt{Q^2}} = \frac{\sqrt{Q^2}}{2M_{\rm N}x} \tag{9}$$

must hold. Since $\sqrt{Q^2} \gg M_{\rm N}$ and $x \le 1$, equation (9) can be fulfilled only for $\beta \approx 1$, $\gamma \gg 1$. Hence (7) becomes

$$P''_\mu \approx (\tilde{P},0,0,-\tilde{P}), \quad \text{where} \quad \tilde{P} = \beta\gamma M_{\rm N} = \frac{\sqrt{Q^2}}{2x}. \tag{10}$$

This is (3.62). In the Breit system the nucleon (and with it all partons) and the photon move in the z direction towards each other. In the final state the scattered parton carries the momentum (x denotes the initial momentum fraction of the parton, i.e., $P^{\rm i}_\mu = xP''_\mu$)

$$\begin{aligned} P^{\rm f}_\mu &= xP''_\mu + q''_\mu = x(\tilde{P},0,0,-\tilde{P}) + q''_\mu = \left(\frac{\sqrt{Q^2}}{2},0,0,-\frac{\sqrt{Q^2}}{2}\right) + q''_\mu \\ &= \left(\frac{\sqrt{Q^2}}{2},0,0,\frac{\sqrt{Q^2}}{2}\right) . \end{aligned} \tag{11}$$

Obviously the spatial momentum of the parton is flipped to its opposite direction by the reaction (see Fig. 3.9). In the Breit system the parton is simply reflected. We shall come back to this in Example 3.8.

EXERCISE

3.7 The Scattering Tensor for Scalar Particles

Problem. Repeat the steps leading to (3.18) for scalar particles.

Solution. We start with (3.6). In the case of scalar particles all terms containing γ matrices vanish:

$$\Gamma_\mu = BP'_\mu + CP_\mu \ . \tag{1}$$

The requirement of gauge invariance, (3.8), then yields

$$\begin{aligned} q^\mu\Gamma_\mu &= (P'^\mu - P^\mu)\Gamma_\mu = B(P'^\mu - P^\mu)P'_\mu + C(P'^\mu - P^\mu)P_\mu \\ &= BP'^\mu P'_\mu - CP^\mu P_\mu = (B-C)M_{\rm N}^2 = 0 \quad \Rightarrow C = B \ , \end{aligned} \tag{2}$$

Exercise 3.7

$$\Gamma_\mu = B(P'_\mu + P_\mu) \ . \tag{3}$$

This is rewritten by replacing P'_μ by P_μ and q_μ:

$$\begin{aligned}\Gamma_\mu &= B(P_\mu + q_\mu + P_\mu) \\ &= 2B\left(P_\mu + \frac{1}{2}q_\mu\right) \\ &= 2B\left(P_\mu - q_\mu\frac{-q^2/2}{q^2}\right) \\ &= 2\left(P_\mu - q_\mu\frac{P\cdot q}{q^2}\right)B(Q^2) \ ,\end{aligned} \tag{4}$$

where relation (3.7) has been employed in the last step. For a free scalar field one simply has

$$u^*(P)u(P) = \text{const} \ . \tag{5}$$

Therefore $W_{\mu\nu}$ follows directly from Γ_μ as

$$\begin{aligned}W_{\mu\nu} &= \text{const} \times \Gamma_\mu\Gamma_\nu \\ &= \text{const} \times \left(P_\mu - q_\mu\frac{P\cdot q}{q^2}\right)\left(P_\nu - q_\nu\frac{P\cdot q}{q^2}\right)B^2(Q^2) \ .\end{aligned} \tag{6}$$

Up to now we have considered the elastic process. To obtain the equation equivalent to (6) for the inclusive inelastic case, we again have to replace $B^2(Q^2)$ by $B^2(Q^2,\nu) =: W_2(Q^2,\nu)$.

$$W_{\mu\nu} = \text{const} \times \left(P_\mu - q_\mu\frac{P\cdot q}{q^2}\right)\left(P_\nu - q_\nu\frac{P\cdot q}{q^2}\right)W_2(Q^2,\nu) \ . \tag{7}$$

Comparing this result with (3.18) shows that there is only a W_2 function and no W_1 function for scalar particles. Consequently (3.66) then becomes

$$\frac{\mathrm{d}^2\sigma}{\mathrm{d}x\mathrm{d}y} = \frac{4\pi s\alpha^2}{Q^4}(1-y)F_2(Q^2,x) \ , \tag{8}$$

with a corresponding structure function $F_2(Q^2,x)$. Here, in the last step, the term xyM_N^2/s has been neglected for large s.

EXAMPLE

3.8 Photon–Nucleon Scattering Cross Sections for Scalar and Transverse Photon Polarization

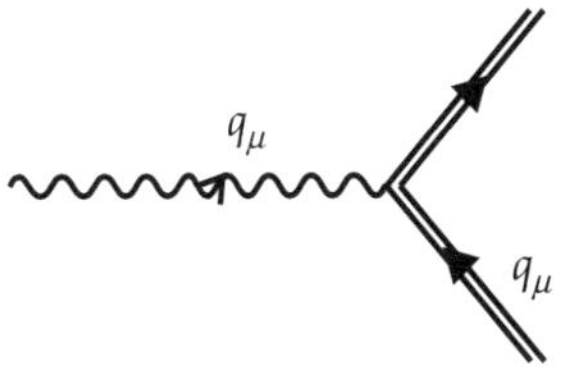

Fig. 3.10. Photon-nucleon vertex

Electron–nucleon scattering can be viewed as the scattering of high-energy transverse and longitudinal photons off the nucleon. These photons are created in the scattering process of the electron. Note that these are virtual photons, i.e. they are not on the mass shell. These photons have the effective mass $q^2 = q_\mu q^\mu = -Q^2$ and carry the energy $\nu = M_\mathrm{N}(E' - E)$. Thus their energy is exactly given by the inelasticity ν known from (3.19) and (3.28). For scalar as well as for transverse polarized photons the spin 4-vector is perpendicular to the photon momentum. With

$$q_\mu = (\nu/M_\mathrm{N}, 0, 0, q_3), \quad Q^2 = -q_\mu q^\mu = q_3 - \frac{\nu^2}{M_\mathrm{N}^2} \tag{1}$$

the spin unit vectors can, in the laboratory system, be chosen as follows. For the transverse case

$$\varepsilon_{1,\mu} = (0, 1, 0, 0) \ , \quad \varepsilon_{2,\mu} = (0, 0, 1, 0) \ . \tag{2}$$

Scalar polarization is defined by

$$\varepsilon_{0,\mu} = (q_3, 0, 0, \nu/M_\mathrm{N}) \frac{1}{\sqrt{Q^2}} \ . \tag{3}$$

Clearly

$$\varepsilon_{\lambda,\mu}\, q^\mu = 0 \tag{4}$$

and

$$\begin{aligned} \varepsilon_{0,\mu}\varepsilon_0^\mu &= 1 \ , \\ \varepsilon_{1,\mu}\varepsilon_1^\mu &= -1 \ , \\ \varepsilon_{2,\mu}\varepsilon_2^\mu &= -1 \ , \end{aligned}$$

or simply

$$\left|\varepsilon_{\lambda,\mu}\varepsilon^{\lambda,\mu}\right| = 1 \quad \text{for} \quad \lambda = 0, 1, 2 \ . \tag{5}$$

In the case of photon–nucleon scattering we get

$$\sigma(\gamma\mathrm{N} \to \mathrm{X}) = \mathrm{const} \times \varepsilon_{\lambda,\mu}\, \varepsilon_{\lambda,\nu}\, W^{\mu\nu} \tag{6}$$

for the inclusive cross section, where $\varepsilon_{\lambda,\mu}$ denotes the polarization vector of the incoming photon. By means of (3.18) and

$$P_\mu = (M_\mathrm{N}, 0, 0, 0) \ , \tag{7}$$

Example 3.8

it follows from the general structure (6) and from (2) that the cross section for transverse photons is proportional to W^{11}, W^{22}, W^{12}, and W^{21}. By inspection of (3.18) it is clear that only W^{11} and W^{22} contribute and both are proportional to $W_1\left(Q^2, \nu\right)$. Therefore

$$\sigma_{\mathrm{T}}(\gamma\mathrm{N} \to \mathrm{X}) = \mathrm{const} \times W_1\left(Q^2, \nu\right) \tag{8}$$

can be derived in the laboratory system for transversely polarized photons. For scalar photons one has

$$\begin{aligned}
\varepsilon_0 \cdot P &= q_3 M_{\mathrm{N}} \frac{1}{\sqrt{Q^2}} \ , \\
\varepsilon_0 \cdot q &= \left(\frac{\nu}{M_{\mathrm{N}}} q_3 - q_3 \frac{\nu}{M_{\mathrm{N}}}\right) = 0 \ , \\
\varepsilon_0 \cdot \varepsilon_0 &= \left(q_3^2 - \frac{\nu^2}{M_{\mathrm{N}}^2}\right) \frac{1}{Q^2} = \frac{-Q^2}{Q^2} = -1 \ .
\end{aligned}$$

Therefore

$$\begin{aligned}
&\sigma_{\mathrm{S}}(\gamma\mathrm{N} \to \mathrm{X}) \\
&= \mathrm{const} \times \left[\left(\frac{-q_3^2 + \nu^2/M_{\mathrm{N}}^2}{Q^2} + 0\right) W_1\left(Q^2, \nu\right) + \left(\frac{M_{\mathrm{N}} q_3}{\sqrt{Q^2}}\right)^2 \frac{W_2\left(Q^2, \nu\right)}{M_{\mathrm{N}}^2}\right] \\
&= \mathrm{const} \times \left[-W_1\left(Q^2, \nu\right) + \frac{q_3^2}{Q^2} W_2\left(Q^2, \nu\right)\right] \ .
\end{aligned} \tag{9}$$

Inserting (3.43) now yields

$$\begin{aligned}
\sigma_{\mathrm{T}}(\gamma\mathrm{N} \to \mathrm{X}) &= \mathrm{const} \times \frac{1}{M_{\mathrm{N}}} F_1^{\mathrm{eN}}\left(Q^2, x\right) \ , \\
\sigma_{\mathrm{S}}(\gamma\mathrm{N} \to \mathrm{X}) &= \mathrm{const} \times \frac{1}{M_{\mathrm{N}}} \left[-F_1^{\mathrm{eN}}\left(Q^2, \nu\right) + \frac{q_3^2 M_{\mathrm{N}}^2}{Q^2 \nu} F_2^{\mathrm{eN}}\left(Q^2, \nu\right)\right] \ .
\end{aligned} \tag{10}$$

Because

$$Q^2 = q_3^2 - q_0^2 = q_3^2 - \nu^2/M_{\mathrm{N}}^2 \ , \tag{11}$$

the factor in front of $F_2^{\mathrm{eN}}\left(Q^2, \nu\right)$ can be simplified to

$$\frac{q_3^2 M_{\mathrm{N}}^2}{Q^2 \nu} = \frac{\left(Q^2 + \nu^2\right) M_{\mathrm{N}}^2}{Q^2 \nu} = \frac{M_{\mathrm{N}}^2}{\nu} + \frac{1}{2x} \ , \tag{12}$$

with $x = Q^2/2\nu$, as usual. In the scaling region $\nu \gg M_{\mathrm{N}}^2$ we therefore obtain

$$\sigma_{\mathrm{T}}(\gamma\mathrm{N} \to \mathrm{X})|_{\nu^2, Q^2 \gg (1\mathrm{GeV})^2} = \mathrm{const} \times \frac{1}{M_{\mathrm{N}}} F_1^{\mathrm{eN}}(x) \ , \tag{13}$$

$$\sigma_{\mathrm{S}}(\gamma\mathrm{N} \to \mathrm{X})|_{\nu^2, Q^2 \gg (1\mathrm{GeV})^2} = \mathrm{const} \times \frac{1}{M_{\mathrm{N}}} \left[\frac{1}{2x} F_2^{\mathrm{eN}}(x) - F_1^{\mathrm{eN}}(x)\right] \ . \tag{14}$$

Example 3.8

Because of the Callan–Gross relation (3.83), σ_S should therefore vanish. See Fig. 3.11.

In fact, Fig. 3.6 was obtained by analyzing photon–nucleon scattering. It should be noted that on the other hand $F_1(x) = 0$ holds for scalar particles (see Exercise 3.7). We would therefore expect $\sigma_T/\sigma_S \to 0$ for scalar partons. But this contradicts experimental observations, i.e., all models that represent the charged constituents of the nucleon by scalar particles are nonphysical. These results can easily be understood within the Breit system. To that end one only has to re-

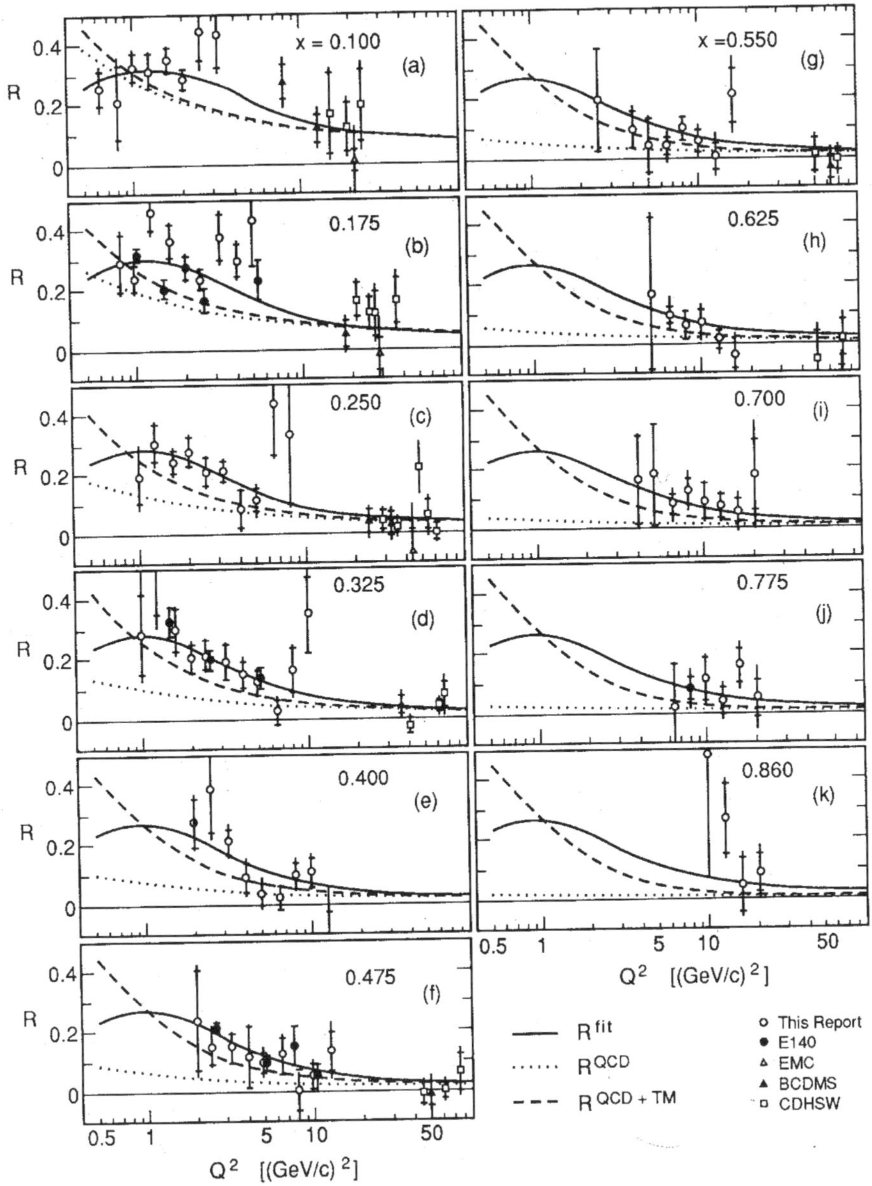

Fig. 3.11a–k. Measured values of $R(x, Q^2) = \sigma_s/\sigma_T$ as analyzed by Whitlow et al. (Phys. Lett. B **250**, 193 (1990)). The dotted and dashed curves represent perturbative QCD predictions with and without target mass corrections. The solid line represents a best-fit model to all available data

Example 3.8

member that for transverse photon polarization the spin is either parallel to the direction of motion or opposite to it (right or left circular polarization). More precisely, this means that

$$\varepsilon_\mu^+ = -(\varepsilon_{1,\mu} + \mathrm{i}\varepsilon_{2,\mu})\frac{1}{\sqrt{2}} , \quad \boldsymbol{S}\cdot\boldsymbol{q} = +|\boldsymbol{q}| , \tag{15}$$

and

$$\varepsilon_\mu^- = (\varepsilon_{1,\mu} - \mathrm{i}\varepsilon_{2,\mu})\frac{1}{\sqrt{2}} , \quad \boldsymbol{S}\cdot\boldsymbol{q} = -|\boldsymbol{q}| . \tag{16}$$

Here $\boldsymbol{S}$ denotes the photon spin. According to Fig. 3.11 the partons do not carry angular momentum and spin and consequently the photon can only be absorbed by the parton if the spin component in the z direction of the latter particle is changed by 1. But this is impossible for scalar particles and leads to $\sigma_\mathrm{T} = 0$. Massless spin-$\frac{1}{2}$ particles, however, encounter a completely different situation. For these particles the spin component parallel to the direction of motion can only assume the values $+1/2$ and $-1/2$. The corresponding spin states are known as positive and negative helicity or as right-handed and left-handed particles. Helicity states are defined by the projection operator $(1 \pm \gamma_5)/2$ applied to the wave function (spinor), left-handed and right-handed states by the projection of the spin onto the momentum axis. For ultrarelativistic particles, positive helicity corresponds to right-handed particles, that is the spin points in the direction of motion, and negative helicity corresponds to the spin pointing in the opposite direction. The vector γ_μ conserves the helicity, i.e., left-handed particles, for example, couple only to other left-handed particles (see Sect. 4.1). Since the direction of motion of a parton is changed into its opposite in the Breit system, the spin must consequently be flipped at the same time (for sufficiently fast partons, i.e., for sufficiently large Q^2). Therefore spin-$\frac{1}{2}$ partons are only able to absorb a photon if $\boldsymbol{S}_z$ is equal to ± 1, i.e., if the photon is transverse. In this case σ_T is nonzero (see Fig. 3.11).

Scalar photons (the Coulomb-field, for example, consists of such photons) have zero spin projection, i.e., here $\sigma_\mathrm{S} = 0$ for spin-$\frac{1}{2}$ partons and $\sigma_\mathrm{S} \neq 0$ for spin-0 partons, which is the opposite of the situation encountered by their transverse counterparts.

EXAMPLE

3.9 A Simple Model Calculation for the Structure Functions of Electron–Nucleon Scattering

As already stated, the structure functions $W_1(Q^2, \nu)$ and $W_2(Q^2, \nu)$ of (3.36) can be derived from any microscopic model. This procedure, however, is quite cumbersome, but it allows predictions for values of Q^2 and ν^2 smaller than $(1\ \mathrm{GeV})^2$, i.e., for values beyond the scaling region. If we are interested only in

Example 3.9

the restricted information provided by $F_2(x)$, then there is an easier way. In this case we can start directly from (3.81):

$$F_2^{\mathrm{eN}}(x) = \sum_i f_i(x) q_i^2 x \ . \tag{1}$$

The only input needed is the functions $f_i(x)$, i.e., the probabilities that the parton with index i carries a fraction x of the total momentum. We are going to evelu-ate these probabilities in an extremely simple model by making the following assumptions:

(1) The nucleon consists of quarks.

(2) The wave functions of up and down quarks are identical, which yields

$$F_2^{\mathrm{eN}}(x) = \left(\sum_i q_i^2\right) f(x) x \ . \tag{2}$$

Averaging the structure function over the proton and neutron gives

$$\begin{aligned} F_2^{\mathrm{eN}}(x) &= \frac{1}{2}\left[F_2^{\mathrm{ep}}(x) + F_2^{\mathrm{en}}(x)\right] \\ &= \frac{3}{2}\left[\left(\frac{2}{3}\right)^2 + \left(\frac{1}{3}\right)^2\right] f(x) x \\ &= \frac{5}{6} f(x) x \ . \end{aligned} \tag{3}$$

(3) We employ Gaussian distributions for the internal wave functions in position space, i.e., we set

$$\begin{aligned} \Psi_P(r,s) = \frac{\alpha^2}{3\pi^2} \exp\left\{\frac{\alpha}{2}\left[\frac{1}{2}\left(r^2 - 2\left(\frac{P\cdot r}{M}\right)^2\right)\right.\right. \\ \left.\left. + \frac{2}{3}\left(s^2 - 2\left(\frac{P\cdot s}{M}\right)^2\right)\right]\right\} \end{aligned} \tag{4}$$

with four-dimensional (i.e. space–time) Jacobi coordinates

$$\begin{aligned} r_\mu &= (z_2 - z_3)_\mu \ , \\ s_\mu &= \left[z_1 - \frac{1}{2}(z_2 - z_3)\right]_\mu \ , \\ &= \frac{1}{2}(2z_1 - z_2 - z_3)_\mu \ . \end{aligned} \tag{5}$$

Here $\Psi_P(r,s)$ is the three-quark wave function, while the quark positions and the proton momentum are denoted by the four-vectors z_1, z_2, z_3, and P_μ, respectively.

Example 3.9

Statements 1 and 2 are quite clear but the wave function (4) needs further explanation. First one has to take into account that the center-of-mass motion separates, i.e., the center-of-mass coordinate $Z_\mu = \frac{1}{3}(z_1+z_2+z_3)_\mu$ must not occur in (4). Then the differences between the quark coordinates are assumed to be Gaussian distributed. This can be archieved by employing Gaussian distributions for r and s. The structure of (4) can be better understood by writing a part of the exponent in its explicit form:

$$\begin{aligned}\frac{r^2}{2}+\frac{2s^2}{3} &= \frac{1}{6}\left(3z_2^2-6z_2z_3+3z_3^2+4z_1^2+z_2^2+z_3^2-4z_1z_2-4z_1z_3+2z_2z_3\right)\\ &= \frac{2}{3}\left(z_1^2+z_2^2+z_3^2-z_1z_2-z_2z_3-z_3z_2\right)\\ &= \frac{1}{3}\left[(z_1-z_2)^2+(z_1-z_3)^2+(z_2-z_3)^2\right] . \end{aligned} \tag{6}$$

We could think of employing spatial Gaussians only and plane waves with respect to the time coordinate. This represents the completely noninteracting case and ensures energy conservation at any time t. Consequently each quark carries exactly one third of the total energy. In an interacting theory the energy of every parton is a function of time and varies in general around a mean value. One can simulate this effect by using another Gaussian like

$$\exp\left(-\frac{\alpha}{4}\left(r_0^2+\boldsymbol{r}^2\right)\right) , \tag{7}$$

which is clearly a useful assumption. Note the bold letter $\boldsymbol{r}$ in the exponential, which indicates the three-vector $\boldsymbol{r}$. Expression (7) ensures that the three quarks populate neighboring positions at neighboring times and therefore describe a bound state. Unfortunately a function that is localized in space and time is not Lorentz invariant. A Lorentz-invariant wave function would be of the form $\exp(\alpha r^2/4) = \exp(\alpha(r_0^2-\vec{r}^2)/4)$. In order to restore Lorentz invariance in (7) we now use the fact that the proton momentum in the rest system is simply

$$P_\mu = (M,0,0,0) . \tag{8}$$

We therefore obtain in the rest system

$$\exp\left(-\frac{\alpha}{4}\left(r_0^2+\boldsymbol{r}^2\right)\right) = \exp\left(\frac{\alpha}{4}\left[r^2-\frac{2}{M^2}(P\cdot r)^2\right]\right) . \tag{9}$$

The same holds for the s-dependent part of the wave function (4). The total wave function has been normalized in such a way that

$$\int \mathrm{d}^4r\,\mathrm{d}^4s\,\Psi_P^2(r,s) = 1 \tag{10}$$

holds. Starting with (4) we now determine the mean square charge radius of the proton, which is given by ($P_\mu = (M,0,0,0)$):

$$\langle r^2\rangle_\mathrm{c} = \int \mathrm{d}^4r\,\mathrm{d}^4s \sum_i q_i(z_i-Z)^2\,\Psi_P^2(r,s) . \tag{11}$$

Example 3.9

With the help of the center-of-mass coordinate $\boldsymbol{Z} = \frac{1}{3}(z_1 + z_2 + z_3)$ we get

$$\begin{aligned}\sum_i q_i(z_i - \boldsymbol{Z})^2 &= -\frac{1}{3}(z_1 - \boldsymbol{Z})^2 + \frac{2}{3}(z_2 - \boldsymbol{Z})^2 + \frac{2}{3}(z_3 - \boldsymbol{Z})^2 \\ &= -\frac{1}{27}(2z_1 - z_2 - z_3)^2 + \frac{2}{27}(2z_2 - z_1 - z_3)^2 + \frac{2}{27}(2z_3 - z_1 - z_2)^2 \\ &= \frac{1}{27}\Big(-4z_1^2 - z_2^2 - z_3^2 + 4z_1 \cdot z_2 \\ &\quad + 4z_1 \cdot z_3 - 2z_2 \cdot z_3 + 2z_1^2 + 8z_2^2 + 2z_3^2 - 8z_2 \cdot z_1 - 8z_2 \cdot z_3 \\ &\quad + 4z_1 \cdot z_3 + 2z_1^2 + 2z_2^2 + 8z_3^2 - 8z_3 \cdot z_1 - 8z_3 \cdot z_2 + 4z_1 \cdot z_2 \Big) \\ &= \frac{1}{27}\left(9z_2^2 + 9z_3^2 - 18z_2 \cdot z_3\right) \\ &= \frac{1}{3}\boldsymbol{r}^2 \ . \end{aligned} \tag{12}$$

In the center-of-mass system this yields

$$\begin{aligned}\langle r^2 \rangle_{\mathrm{c}} &= \frac{1}{3}\left(\frac{\alpha^2}{3\pi^2}\right)^2 \int \mathrm{d}^4 s \exp\left[-\frac{2\alpha}{3}\left(s_0^2 + \boldsymbol{s}^2\right)\right] \int \mathrm{d}r_0\, \mathrm{d}^3 r\, \mathrm{e}^{-\frac{\alpha}{2} r_0^2}\, \mathrm{e}^{-\frac{\alpha}{2} r^2}\, r^2 \\ &= \frac{\alpha\sqrt{\alpha}}{6\pi\sqrt{3\pi}} 4\pi \int_0^\infty \mathrm{d}r\, r^4\, \mathrm{e}^{-\frac{\alpha}{2} r^2} = \frac{1}{\alpha} \ . \end{aligned} \tag{13}$$

This can be used to fix the yet undetermined parameter α in the wave function Ψ_P of (4). Thus we have to insert the value

$$\alpha = \frac{1}{\langle r^2 \rangle_{\mathrm{c}}} \approx 1.5\,\mathrm{fm}^{-2} \tag{14}$$

for α. Up to now we have described the internal structure of the proton with the ansatz (4). The free proton moves, however, with the total momentum P. Thus, from (4) we then get for the total proton wave function

$$\begin{aligned}\Psi_P(r, s, Z) = \frac{\alpha^2}{3\pi^2} \frac{\mathrm{e}^{-\mathrm{i}P \cdot Z}}{\sqrt{VT}} \exp\Bigg\{ \frac{\alpha}{2}\Bigg[\frac{1}{2}\left(r^2 - 2\left(\frac{P \cdot r}{M}\right)^2 \right) \\ + \frac{2}{3}\left(s^2 - 2\left(\frac{P \cdot s}{M}\right)^2 \right) \Bigg] \Bigg\} \end{aligned} \tag{15}$$

with the normalization

$$\int \mathrm{d}^4 r\, \mathrm{d}^4 s\, \mathrm{d}^4 Z\, \Psi_P^\dagger(r, s, Z)\, \Psi_P(r, s, Z) = 1 \ . \tag{16}$$

Here, VT denotes the integration volume for Z.

Example 3.9

In order to understand the following, note that the *one-particle Wigner function*

$$f(Z,p) := \int \frac{\mathrm{d}^4 v}{(2\pi)^4}\, \mathrm{e}^{-\mathrm{i}p_\mu v^\mu} \Psi^\dagger \left(Z_\mu - \frac{1}{2}v_\mu\right) \Psi\left(Z_\mu + \frac{1}{2}v_\mu\right) \tag{17}$$

yields the probability for finding a particle at a position Z_μ with a momentum P_μ. Here $\Psi(x)$ denotes the one-particle wave function. Now it is easy to evaluate from (15) the distribution function $f(x)$ of, for example, the first parton (at position $z_{1\mu}$ in space–time):

$$\begin{aligned} f_1(z_1, p_\mu) = N \int \mathrm{d}^4 v\, \mathrm{e}^{-\mathrm{i}p_\mu v^\mu} \int \mathrm{d}^4 z_2 \int \mathrm{d}^4 z_3 \\ \times \Psi_P^\dagger(r,s,Z)\Big|_{z_{1,\mu}\to s_{1,\mu}+\frac{1}{2}v_\mu} \Psi_P(r,s,Z)\big|_{z_{1,\mu}\to s_{1,\mu}-\frac{1}{2}v_\mu} \,. \end{aligned} \tag{18}$$

The normalization constant N is determined later. The probability of finding a parton with momentum $p = xP$ at position z_1 is

$$\begin{aligned} f_1(x) &= \int \mathrm{d}^4 z_1\, f_1(z_1, xP) \\ &= N \int \mathrm{d}^4 v\, \mathrm{e}^{-\mathrm{i}P_\mu v^\mu x} \int \mathrm{d}^4 z_1\, \mathrm{d}^4 z_2\, d^4 z_3 \\ &\quad \times \Psi_P^\dagger\left(r, s+\frac{1}{2}v, Z+\frac{1}{6}v\right) \Psi_P\left(r, s-\frac{1}{2}v, Z-\frac{1}{6}v\right) \\ &= N \int \mathrm{d}^4 v\, \mathrm{e}^{-\mathrm{i}P_\mu v^\mu\left(x-\frac{1}{3}\right)} \int \mathrm{d}^4 z_1\, \mathrm{d}^4 z_2\, \mathrm{d}^4 z_3\, \frac{\alpha^4}{9\pi^4 VT} \\ &\quad \times \exp\left\{\alpha\left[\frac{1}{2}\left(r^2 - 2\left(\frac{P\cdot r}{M}\right)^2\right)\right.\right. \\ &\quad \left.\left. + \frac{2}{3}\left(s^2 - 2\left(\frac{P\cdot s}{M}\right)^2\right) + \frac{1}{6}\left(v^2 - 2\left(\frac{P\cdot v}{M}\right)^2\right)\right]\right\} \,. \end{aligned} \tag{19}$$

The integration over z_1, z_2, and z_3 is now transformed into one over r, s, and z. The corresponding functional determinant is

$$\left|\frac{\partial(r,s,z)}{\partial(z_1,z_2,z_3)}\right| = \begin{vmatrix} 0 & 1 & -1 \\ 1 & -\frac{1}{2} & \frac{1}{2} \\ \frac{1}{3} & \frac{1}{3} & \frac{1}{3} \end{vmatrix} = -\frac{1}{3} \,. \tag{20}$$

Because we have four-dimensional integrals, the determinant occurs four times, i.e. we get the factor $(-3)^4$. By employing (16) we can reduce (19) to

$$f_1(x) = N' \int \mathrm{d}^4 v\, \exp\left(-\mathrm{i}P^\mu v_\mu\left(x - \frac{1}{3}\right)\right) \exp\left(\frac{\alpha}{6}\left[v^2 - 2\left(\frac{P\cdot v}{M}\right)^2\right]\right) , \tag{21}$$

Example 3.9

where $N' = 3^4 N$. Since this expression is Lorentz invariant, we can again evaluate it in the rest system:

$$\begin{aligned} f_1(x) &= N' \int \mathrm{d}^4 v \, \exp\left(-\mathrm{i} M v_0 \left(x - \tfrac{1}{3}\right)\right) \exp\left(-\tfrac{\alpha}{6}\left(v_0^2 + \boldsymbol{v}^2\right)\right) \\ &= N' \left(\frac{6\pi}{\alpha}\right)^{3/2} \int \mathrm{d}v_0 \, \exp\left(-\tfrac{\alpha}{6}\left[v_0 + \mathrm{i}\tfrac{3M}{\alpha}\left(x - \tfrac{1}{3}\right)\right]^2\right) \exp\left(-\tfrac{3M^2}{2\alpha}\left(x - \tfrac{1}{3}\right)^2\right) \\ &= N' \left(\frac{6\pi}{\alpha}\right)^2 \exp\left(-\tfrac{3M^2}{2\alpha}\left(x - \tfrac{1}{3}\right)^2\right) . \end{aligned} \tag{22}$$

Here the standard formula $\int_{-\infty}^{\infty} \mathrm{e}^{-x^2} \mathrm{d}x = \sqrt{\pi}$ has been employed. Now we recognize the following. For the simple quark model considered here the distribution functions are peaked around $x = 1/3$. This can immediately be understood, because it only means that each quark carries on average one third of the total momentum. Figure 3.12 shows, however, that the experimentally observed distribution functions increase monotonically for small values of x. Obviously every charged particle has less momentum than one would expect from our simple quark model. Consequently there must be more partons in a nucleon than those three, which correspond to valence quarks.

Because of the approximations leading to (22), x can become larger than 1 and even assume negative values. In these cases, however, $f(x)$ is more or less zero. These difficulties are mainly due to assumption (4), which contains arbitrarily large parton momenta (being the Fourier transform of a Gaussian distribution: indeed, (22) is again a Gaussian). Consequently $f_1(x)$ should be normalized according to

$$1 = \int_{-\infty}^{\infty} f_1(x) \mathrm{d}x = N' \left(\frac{6\pi}{\alpha}\right)^2 \sqrt{\frac{\pi 2\alpha}{3M^2}} , \tag{23}$$

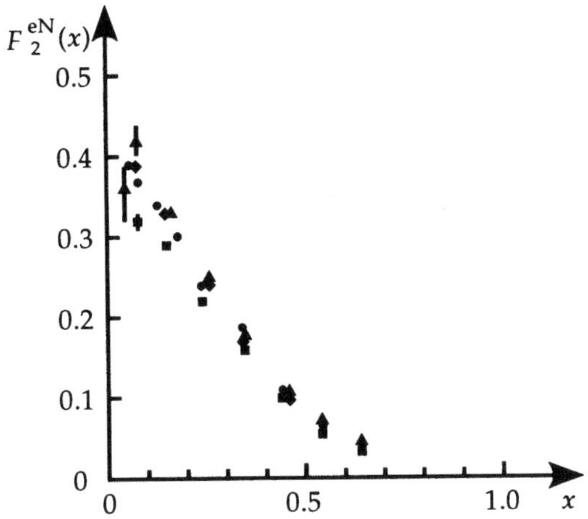

Fig. 3.12. The experimentally observed F_2 structure function of deep inelastic electron–nucleon scattering

Example 3.9

which gives

$$f_1(x) = \sqrt{\frac{3M^2}{2\pi\alpha}} \exp\left(-\frac{3M^2}{2\alpha}\left(x-\frac{1}{3}\right)^2\right) \ . \qquad (24)$$

The function (3),

$$\begin{aligned} F_2^{\mathrm{eN}}(x) &= \frac{1}{2}\left[F_2^{\mathrm{ep}}(x) + F_2^{\mathrm{en}}(x)\right] \\ &= 3 \times \frac{5}{18} \times f_1(x)x \ , \end{aligned} \qquad (25)$$

obtained in this way is depicted in Fig. 3.13. The qualitative deviation from experimental results is considerable. The failure of the model, however, yields valuable information about specific features of the quark–quark interaction, which help to obtain better results. In addition note that

$$\begin{aligned} 3\int\limits_{-\infty}^{\infty} x f_1(x)\mathrm{d}x &= 3\int\limits_{-\infty}^{\infty}\left(x+\frac{1}{3}\right)\sqrt{\frac{3M^2}{2\pi\alpha}}\exp\left(-\frac{3M^2}{2\alpha}x^2\right)\mathrm{d}x \\ &= 3\left(0+\frac{1}{3}\right) = 1 \end{aligned} \qquad (26)$$

holds. Therefore the three quarks together carry the whole proton momentum. This result is not surprising. It is inherently built into this naive model. In other words: Owing to our assumptions, this must obviously be true, and therefore (26) represents a test of our calculation.

In the next section we shall discuss a more refined model for the nucleon, i.e. the MIT bag model. Later, in Sect. 5.6, we shall see what the structure functions of this model look like. In fact they do not agree much better with the data than

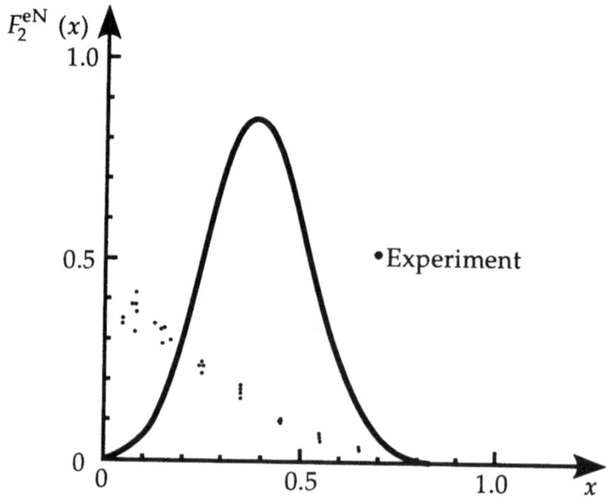

Fig. 3.13. Comparison of the experimentally observed F_2 structure function of deep inelastic electron–nucleon scattering with our simple model calculation

our present results. We shall then also see how by adding ad hoc the Q^2 evolution from QCD the confrontation of theory with experiment improves and the agreement becomes much more acceptable.

3.3 The MIT Bag Model

Before turning in the following chapters to today's prevailing theory of quark–quark interaction, we shall first discuss a specific 'bag model', the MIT bag model, in more detail.[9] Since no free quarks have been observed experimentally, one imagines that the quarks are tightly confined inside the hadrons. Inside of this confinement volume they behave mainly as free particles. All bag models must be regarded as pure phenomenology. It is at present unclear how strong any relationships between such models and QCD are. Should the confinement problem one day be solved from the QCD equations (which we shall discuss in the next chapter), it might turn out that the model assumptions of the MIT bag are unphysical. Besides this basic problem, there are also difficulties inherent to the MIT bag model. The rigid boundary condition can lead to spurious motions, e.g., oscillations of all quarks with respect to the bag, and it is not Lorentz invariant.

These disadvantages are set off by the great advantage that nearly all interesting processes and quantities can be calculated in a bag-model framework. Sometimes there are quite far-reaching approximations involved, but in total these recipes allow for quite a good phenomenological understanding of subhadronic physics. This might be found more satisfactory than a strictly formal theory of quark–quark interaction which fails to predict many physically interesting quantities owing to mathematical difficulties (insufficiencies).

We shall now formulate the MIT bag model. Start from the fact that the quark–quark interaction makes it impossible to separate colored quarks. This is most easily implemented by specifying some surface and demanding that the color current through it vanishes. This color current is analogous to the electromagnetic current and reads

$$\hat{J}^{\alpha}_{\mu} = (\overline{\mathrm{q}}_{\mathrm{r}}, \overline{\mathrm{q}}_{\mathrm{b}}, \overline{\mathrm{q}}_{\mathrm{g}})\lambda^{\alpha}\gamma_{\mu}\begin{pmatrix} \mathrm{q}_{\mathrm{r}} \\ \mathrm{q}_{\mathrm{b}} \\ \mathrm{q}_{\mathrm{g}} \end{pmatrix} \tag{3.94}$$

with the eight color SU(3) matrices λ^a. The subscripts stand for the three colors ('red', 'blue', and 'green'). Let the chosen surface be characterized by a normal

[9] See also W. Greiner and B. Müller: *Quantum Mechanics – Symmetries*, 2nd ed., (Springer, Berlin, Heidelberg 1994), P. Hasenfratz, J. Kuti, Phys. Rep. 40 **C**, 75 (1978).

vector n_μ. Then the desired condition can be written as

$$n^\mu \cdot \hat{j}_\mu^\alpha = 0\Big|_{\text{surface}} \quad , \qquad \alpha = 1, 2, 3, \ldots, 8 \ . \tag{3.95}$$

By introducing a four-dimensional normal vector we have reached a covariant form, but this does not correspond to a general covariance of the model because we must still specify the bag surface. Indeed we restrict ourselves to purely spatial surfaces, i.e., $n_\mu \to \boldsymbol{n}$:

$$-\boldsymbol{n} \cdot \boldsymbol{J}^\alpha \Big|_{R=R(\theta,\varphi)} = 0 \ . \tag{3.96}$$

To simplify (3.96) further, color independence is demanded of the internal quark wave function, i.e., $\mathrm{q}_i(x) = \mathrm{q}(x)$, $i =$r, b, g. In this way we obtain the 'quadratic bag boundary condition'

$$\boldsymbol{n} \cdot \bar{\mathrm{q}} \boldsymbol{\gamma} \mathrm{q} \Big|_{R=R(\theta,\varphi)} = 0 \ . \tag{3.97}$$

The expression $\boldsymbol{n} \cdot \boldsymbol{\gamma}$ has the property that its square is the negative unit matrix:

$$(\boldsymbol{n} \cdot \boldsymbol{\gamma})^2 = n_i n_j \frac{1}{2}(\gamma_i \gamma_j + \gamma_j \gamma_i) = -n_i n_i \mathbb{1} = -(\boldsymbol{n})^2 \cdot \mathbb{1} = -\mathbb{1} \ . \tag{3.98}$$

Its eigenvalues are accordingly $\pm \mathrm{i}$. Now, each quark state can be expanded into the corresponding eigenvectors, and, how wonderful, (3.97) is satisfied for just these eigenstates. This can be seen in the following way: From

$$\boldsymbol{n} \cdot \boldsymbol{\gamma}\, \mathrm{q}_+ = \mathrm{i} \mathrm{q}_+ \tag{3.99}$$

it follows by Hermitian conjugation that

$$\begin{aligned} &(\mathrm{q}_+^\dagger \boldsymbol{n} \cdot \boldsymbol{\gamma}^\dagger)\gamma_0 = -\mathrm{i}\bar{\mathrm{q}}_+ \ , \\ \Rightarrow \quad &\bar{\mathrm{q}}_+ \boldsymbol{n} \cdot \boldsymbol{\gamma} = -\mathrm{i}\bar{\mathrm{q}}_+ \ . \end{aligned} \tag{3.100}$$

If one multiplies (3.99) by $\bar{\mathrm{q}}_+$ from the left and (3.100) by q_+ from the right and adds both equations, then it follows that

$$\bar{\mathrm{q}}_+ \boldsymbol{n} \cdot \boldsymbol{\gamma}\, \mathrm{q}_+ = 0 \tag{3.101}$$

and similarly for the eigenvalue $-\mathrm{i}$. By restricting ourselves to eigenvectors of $\boldsymbol{n} \cdot \boldsymbol{\gamma}$, we can thus always guarantee that (3.97) is satisfied. In other words, instead of solving (3.97), which is bilinear in the quark wave function $q(x)$, one may solve the much easier linear equations

$$\boldsymbol{n} \cdot \boldsymbol{\gamma}\, \mathrm{q} = \pm \mathrm{i} \mathrm{q} \ . \tag{3.102}$$

The disadvantage of this procedure is that other solutions of (3.97) that are not eigenvectors of $\boldsymbol{n} \cdot \boldsymbol{\gamma}$ are excluded in this way.

It will be shown in Exercise 3.10 that eigenvectors to the eigenvalue $-\mathrm{i}$ are just the antiparticle solutions to those with the eigenvalue $+\mathrm{i}$. We can therefore restrict ourselves to one sign. Normally

$$\mathrm{i}\boldsymbol{n}(\theta,\varphi)\cdot\boldsymbol{\gamma}\,\mathrm{q}(x) = -\mathrm{q}(x)\Big|_{R=R(\theta,\varphi)} \qquad (3.103)$$

is chosen as the particle solution.

In principle we can choose any bag shape, i.e., an arbitrary function $R = R(\theta,\varphi)$ and a corresponding normal vector $\boldsymbol{n}(\theta,\varphi)$. The simplest shape is naturally a sphere, i.e.,

$$\begin{aligned} &R(\theta,\varphi) = R = \text{const}\;, \quad \boldsymbol{n}(\theta,\varphi) = \boldsymbol{e}_r\;, \\ &\Rightarrow \quad -\mathrm{i}\,\boldsymbol{e}_r\cdot\boldsymbol{\gamma}\,\mathrm{q}(|\boldsymbol{x}| = R) = \mathrm{q}(|\boldsymbol{x}| = R)\;. \end{aligned} \qquad (3.104)$$

Obviously there now remains only one parameter that is not fixed by the model assumptions, the bag radius R. Since it is unsatisfactory to choose R arbitrarily for every hadron, one is lead to still another form of the boundary condition. This is obtained by demanding that the pressure of the quarks on the bag surface be constant. The model assumption leading to this is that the vacuum around the bag is a complex state exerting some kind of pressure on the bag. If this exceeds the interior pressure, the bag shrinks; otherwise it inflates further. How we can visualize this pressure as arising from the interacting fields in the vacuum is discussed in Sect. 7.2. In the following we shall derive this pressure boundary condition.

EXERCISE

3.10 Antiquark Solutions in a Bag

Problem. Show that a quark wave function obeying the equation

$$\mathrm{i}\boldsymbol{n}\cdot\boldsymbol{\gamma}(x)\mathrm{q}(x) = -\mathrm{q}(x)\Big|_{R=R(\vartheta,\phi)} \qquad (1)$$

corresponds to antiquark solutions that fulfill

$$\mathrm{i}\boldsymbol{n}\cdot\boldsymbol{\gamma}(x)\tilde{\mathrm{q}}(x) = \tilde{\mathrm{q}}(x)\;. \qquad (2)$$

Solution. We have to recall that antiparticle wave functions are $\hat{C}\hat{P}\hat{T}$ transforms of the corresponding particle solutions:

$$\tilde{\mathrm{q}}(x) = \hat{C}\hat{P}\hat{T}\mathrm{q}(x)\;. \qquad (3)$$

Indeed $\hat{C}\hat{P}\hat{T}$ transforms a spinor $\Psi(x)$ into a spinor $\Psi(-x)$ up to a phase factor, i.e. a particle moving forward in space and time is transformed into one moving backward in space and time. This corresponds to the Feynman–Stueckelberg interpretation of antiparticles.[10]

[10] See also W. Greiner and J. Reinhardt: *Field Quantization* (Springer, Berlin, Heidelberg 1996) and W. Greiner: *Relativistic Quantum Mechanics – Wave Equations*, 3rd ed. (Springer, Berlin, Heidelberg 2000).

Exercise 3.10

In the standard representation

$$\hat{C} = \mathrm{i}\gamma_2 \hat{K}, \qquad \hat{P} = \gamma_0, \qquad \hat{T} = \mathrm{i}\gamma_1\gamma_3 \hat{K},$$

$\hat{K}$ being complex comjugation, and therefore

$$\hat{C}\hat{P}\hat{T} \ldots = \gamma_0 \mathrm{i}\gamma_2 \left(\mathrm{i}\gamma_1\gamma_3(\ldots)^*\right)^* . \tag{4}$$

Applying (4) to (1) we obtain

$$\begin{aligned}
\tilde{\mathrm{q}}(x) &= \hat{C}\hat{P}\hat{T}\Big[-\mathrm{i}\boldsymbol{n}\cdot\boldsymbol{\gamma}\mathrm{q}(x)\Big] \\
&= \gamma_0\mathrm{i}\gamma_2\Big[\mathrm{i}\gamma_1\gamma_3(\mathrm{i}\boldsymbol{n}\cdot\boldsymbol{\gamma}^*)\big(\mathrm{q}(x)\big)^*\Big]^* \\
&= \gamma_0\mathrm{i}\gamma_2\Big[(-\mathrm{i}\boldsymbol{n}\cdot\boldsymbol{\gamma})\mathrm{i}\gamma_1\gamma_3\big(\mathrm{q}(x)\big)^*\Big]^* \\
&= \gamma_0\mathrm{i}\gamma_2(\mathrm{i}\boldsymbol{n}\cdot\boldsymbol{\gamma}^*)\Big[\mathrm{i}\gamma_1\gamma_3\big(\mathrm{q}(x)\big)^*\Big]^* \\
&= \mathrm{i}\boldsymbol{n}\cdot\boldsymbol{\gamma}\gamma_0\mathrm{i}\gamma_2\Big[\mathrm{i}\gamma_1\gamma_3\big(\mathrm{q}(x)\big)^*\Big]^* \\
&= \mathrm{i}\boldsymbol{n}\cdot\boldsymbol{\gamma}\tilde{\mathrm{q}}(x) \ .
\end{aligned} \tag{5}$$

Here we have repeatedly made use of the fact that γ_1 and γ_3 are real and that γ_2 is purely imaginary.

The antiparticle solutions therefore fulfill the modified boundary condition (2), i.e. we have only to determine the solutions of (1) in order to obtain the quark spectrum in the MIT bag.

To calculate the bag pressure we start from the energy momentum tensor for Dirac particles

$$T_{\mu\nu} = \sum_{\mathrm{q}} \frac{\mathrm{i}}{2}\left(\bar{\mathrm{q}}\gamma_\mu \frac{\partial}{\partial x^\nu}\mathrm{q} - \bar{\mathrm{q}}\frac{\overleftarrow{\partial}}{\partial x^\nu}\gamma_\mu \mathrm{q}\right) . \tag{3.105}$$

Its canonical form reads[11]

$$T_{\mu\nu} = \sum_{\mathrm{q}} \bar{\mathrm{q}}\mathrm{i}\gamma_\mu \frac{\partial}{\partial x^\nu}\mathrm{q} - g_{\mu\nu}\left(\bar{\mathrm{q}}\mathrm{i}\gamma_\lambda \frac{\partial}{\partial x_\lambda}\mathrm{q}\right) .$$

The last term vanishes because $(\not{p} - m)q = 0$. Writing the first term as $\frac{\mathrm{i}}{2}(\bar{\mathrm{q}}\boldsymbol{\partial}_\nu\gamma_\mu\mathrm{q} + \bar{\mathrm{q}}\boldsymbol{\partial}_\nu\gamma_\mu\mathrm{q})$ and integrating the second expression by parts leads to (3.105).

[11] W. Greiner and J. Reinhardt: *Field Quantization* (Springer, Berlin, Heidelberg 1996).

Using the boundary condition (3.103), which in its most general form is

$$\begin{aligned} n_\mu \gamma^\mu \mathrm{q} &= -\mathrm{i}\mathrm{q}\Big|_{\text{surface}} \,, \\ \bar{\mathrm{q}} n_\mu \gamma^\mu &= \mathrm{i}\bar{\mathrm{q}}\Big|_{\text{surface}} \,, \end{aligned} \tag{3.106}$$

it follows that

$$\begin{aligned} n^\mu T_{\mu\nu}\Big|_{\text{surface}} &= -\sum_{\mathrm{q}} \frac{1}{2}\left(\bar{\mathrm{q}}\frac{\partial}{\partial x^\nu}\mathrm{q} + \bar{\mathrm{q}}\frac{\overleftarrow{\partial}}{\partial x^\nu}\mathrm{q}\right)\Bigg|_{\text{surface}} \\ &= -\frac{1}{2}\frac{\partial}{\partial x^\nu}\sum_{\mathrm{q}} \bar{\mathrm{q}}\mathrm{q}\Big|_{\text{surface}} \,. \end{aligned} \tag{3.107}$$

With (3.101) it also follows from (3.99) that

$$\bar{\mathrm{q}}\mathrm{q}\Big|_{\text{surface}} = 0 \,. \tag{3.108}$$

The gradient in (3.107) therefore points in the direction of the outward normal. According to the usual definition the absolute value of $n^\mu T_{\mu\nu}$ is just the pressure, so we conclude that

$$n^\mu T_{\mu\nu}\Big|_{\text{surface}} = -n_\nu P_{\mathrm{D}} \,. \tag{3.109}$$

Since $n_\nu = (0, -\boldsymbol{n})$, the force exerted on the volume element by the pressure is directed outward. The Dirac pressure ($n^2 = -1$)

$$P_{\mathrm{D}} = n^\mu n^\nu T_{\mu\nu} = -\frac{1}{2} n^\nu \frac{\partial}{\partial x^\nu} \sum_{\mathrm{q}} \mathrm{q}\bar{\mathrm{q}}\Big|_{\text{surface}} \tag{3.110}$$

is positive. For $n^\nu = (0, \boldsymbol{n})$ this becomes with $n^\nu \partial/\partial x^\nu = (0, \boldsymbol{n})\left(\partial/\partial x^0, -\boldsymbol{\nabla}\right) = -\boldsymbol{n}\cdot\boldsymbol{\nabla}$

$$P_{\mathrm{D}} = \frac{1}{2}\boldsymbol{n}\cdot\boldsymbol{\nabla}\sum_{\mathrm{q}} \bar{\mathrm{q}}\mathrm{q}\Big|_{\text{surface}} \,. \tag{3.111}$$

According to the model, this quark pressure must equal a constant exterior pressure B,

$$B = \frac{1}{2}\boldsymbol{n}\cdot\boldsymbol{\nabla}\sum_{\mathrm{q}} \bar{\mathrm{q}}\mathrm{q}\Big|_{R=R(\theta,\varphi)} \,. \tag{3.112}$$

Equations (3.112) and (3.104), together with the convention that the quarks are to be considered free inside the bag, define the MIT bag model. By introducing the new parameter B, the radii of all hadron bags are fixed.

In the following we shall investigate *spherical bags*.[12] This is for sure the most obvious assumption for hadron ground states and has the additional advantage that the solutions can be found mostly analytically. We are thus looking for solutions of the stationary free and (for the time being) massless Dirac equation. The finite-mass case is discussed in Exercise 3.11. The Dirac equation for massless particles reads

$$\not{p}\Psi = 0 \tag{3.113}$$

with the boundary conditions

$$-\boldsymbol{e}_r\cdot\boldsymbol{\gamma}\Psi = \Psi\Big|_{|\boldsymbol{x}|=R} \quad \text{or} \quad \begin{pmatrix} 0 & -\mathrm{i}\sigma_r \\ \mathrm{i}\sigma_r & 0 \end{pmatrix}\Psi = \Psi\Big|_{|\boldsymbol{x}|=R} \tag{3.114}$$

and

$$-\frac{1}{2}\frac{\partial}{\partial r}\sum_{\mathrm{q}}\bar{\mathrm{q}}\mathrm{q}\Big|_{|\boldsymbol{x}|=R} = B\ . \tag{3.115}$$

Owing to the spherical symmetry of the problem, an expansion into spherical spinors suggests itself, for example, as in the solution of the hydrogen problem[13]

$$\Psi = \begin{pmatrix} g(r)\,\chi^\mu_\kappa(\theta,\varphi) \\ -\mathrm{i}f(r)\,\chi^\mu_{-\kappa}(\theta,\varphi) \end{pmatrix}\mathrm{e}^{-\mathrm{i}Et}\ . \tag{3.116}$$

The operator $\boldsymbol{\alpha}\cdot\boldsymbol{p}$ is then

$$\boldsymbol{\alpha}\cdot\boldsymbol{p} = -\mathrm{i}\alpha_r\frac{\partial}{\partial r} + \mathrm{i}\frac{\alpha_r}{r}(\beta\hat{K}-1) \tag{3.117}$$

with

$$\hat{K}\chi^\mu_\kappa = -\kappa\chi^\mu_\kappa\ . \tag{3.107a}$$

The operator $\hat{K} = \beta(\boldsymbol{\Sigma}\cdot\boldsymbol{L}+1)$ has eigenvalues $-\kappa$ and κ, with κ being defined by

$$\begin{aligned} \kappa &= -j-\frac{1}{2} \quad \text{for} \quad \ell = j-\frac{1}{2}\ , \\ \kappa &= j+\frac{1}{2} \quad \text{for} \quad \ell = j+\frac{1}{2}\ . \end{aligned} \tag{3.118}$$

With $\alpha_r = \begin{pmatrix} 0 & \sigma_r \\ \sigma_r & 0 \end{pmatrix}$ we obtain from (3.113)

$$-\mathrm{i}\begin{pmatrix} 0 & \sigma_r \\ \sigma_r & 0 \end{pmatrix}\left(\frac{\partial}{\partial r}+\frac{1}{r}-\frac{\beta}{r}\hat{K}\right)\Psi = E\Psi\ . \tag{3.119}$$

[12] Deformed bags are discussed in D. Vasak, R. Shanker, B. Müller and W. Greiner: J. Phys. G **9**, 511 (1983) in connection with the study of fissioning bags as a form of overcritical quark–antiquark production.

[13] See e.g. W. Greiner: *Relativistic Quantum Mechanics – Wave Equations*, 3rd ed. (Springer, Berlin, Heidelberg 2000).

Ansatz (3.116) then yields by virtue of

$$\sigma_r \, \chi_\kappa^\mu = -\chi_{-\kappa}^\mu \tag{3.120}$$

the two coupled equations

$$\left(\frac{\mathrm{d}}{\mathrm{d}r} + \frac{1}{r} - \frac{\kappa}{r}\right) f(r) = E\, g(r) \tag{3.121a}$$

and

$$\left(\frac{\mathrm{d}}{\mathrm{d}r} + \frac{1}{r} + \frac{\kappa}{r}\right) g(r) = -E\, f(r) \tag{3.121b}$$

or

$$\left(\frac{\mathrm{d}}{\mathrm{d}(Er)} + \frac{1-\kappa}{Er}\right) f = g \tag{3.122a}$$

and

$$\left(\frac{\mathrm{d}}{\mathrm{d}(Er)} + \frac{1+\kappa}{Er}\right) g = -f \ . \tag{3.122b}$$

We replace $Er =: z$, take the derivative of (3.122a) with respect to z and eliminate $\frac{\mathrm{d}}{\mathrm{d}z} g$ via (3.122b):

$$\left(\frac{\mathrm{d}^2}{\mathrm{d}z^2} - \frac{1-\kappa}{z^2} + \frac{1-\kappa}{z}\frac{\mathrm{d}}{\mathrm{d}z}\right) f = \frac{\mathrm{d}}{\mathrm{d}z} g = -\frac{1+\kappa}{z} g - f \ . \tag{3.121}$$

From (3.122a) it follows that

$$\frac{1+\kappa}{z} g = \left[\frac{1+\kappa}{z}\frac{\mathrm{d}}{\mathrm{d}z} + \frac{1-\kappa^2}{z^2}\right] f \tag{3.122}$$

and, by inserting this into (3.121), one arrives at

$$\left[\frac{\mathrm{d}^2}{\mathrm{d}z^2} + \frac{2}{z}\frac{\mathrm{d}}{\mathrm{d}z} + \frac{\kappa(1-\kappa)}{z^2} + 1\right] f = 0 \ . \tag{3.123}$$

This is exactly the differential equation satisfied by the Bessel function. As f and g may not diverge more strongly than $1/r$ at the origin, we must choose the j_n functions:

$$f(r) = \mathrm{const} \times j_\ell(Er) \ , \quad \ell = \begin{cases} -\kappa & \text{for } \kappa < 0 \\ \kappa - 1 & \text{for } \kappa > 0 \end{cases} . \tag{3.124}$$

Analogously it follows that

$$g(r) = \mathrm{const} \times j_\ell(Er) \ , \quad \ell = \begin{cases} -\kappa - 1 & \text{for } \kappa < 0 \\ \kappa & \text{for } \kappa > 0 \end{cases} . \tag{3.125}$$

The factor from (3.125) can be absorbed into overall normalization. It remains to determine the relative factor A between the f and g functions from (3.122):

$$f(r) = j_{\bar{\ell}}(Er) \ , \quad g(r) = A\, j_\ell(Er) \ , \tag{3.126}$$

$$\Rightarrow \quad \left(\frac{\mathrm{d}}{\mathrm{d}z} + \frac{1-\kappa}{z}\right) j_{\bar{\ell}}(z) = A\, j_\ell(z) \ , \tag{3.127a}$$

$$\left(\frac{\mathrm{d}}{\mathrm{d}z} + \frac{1+\kappa}{z}\right) A\, j_\ell(z) = -j_{\bar{\ell}}(z) \ . \tag{3.127b}$$

For $\kappa < 0$, (3.127) gives for $A = 1$ only the usual recursion relations between spherical Bessel functions of order ℓ and $\bar{\ell} = \ell - \mathrm{sgn}(\kappa)$, while for $\kappa > 0$, (3.127) leads to $A = -1$.

The wave function of a massless quark in an MIT bag is thus

$$\Psi = N \begin{pmatrix} j_\ell(Er)\, \chi_\kappa^\mu(\theta,\phi) \\ \mathrm{i}\, \mathrm{sgn}(\kappa)\, j_{\bar{\ell}}(Er)\, \chi_{-\kappa}^\mu(\theta,\phi) \end{pmatrix} \mathrm{e}^{-\mathrm{i}Et} \ . \tag{3.128}$$

The normalization will be determined later. First we check whether a wave function of the structure (3.128) can satisfy the boundary conditions at all. We start with the linear boundary condition (3.114). Using (3.128) and (3.120) it reduces to

$$\begin{pmatrix} -\,\mathrm{sgn}(\kappa)\, j_{\bar{\ell}}(ER)\, \chi_\kappa^\mu(\theta,\phi) \\ -\mathrm{i}\, j_\ell(ER)\, \chi_{-\kappa}^\mu(\theta,\phi) \end{pmatrix} = \begin{pmatrix} j_\ell(ER)\, \chi_\kappa^\mu(\theta,\phi) \\ \mathrm{i}\, \mathrm{sgn}(\kappa)\, j_{\bar{\ell}}(ER)\, \chi_{-\kappa}^\mu(\theta,\phi) \end{pmatrix} \tag{3.129}$$

or

$$j_\ell(ER) = -\,\mathrm{sgn}(\kappa)\, j_{\bar{\ell}}(ER) \ . \tag{3.130}$$

Thus the linear MIT boundary condition can indeed be satisfied and yields an eigenvalue equation for ER. For $\kappa = -1$ we obtain, for example, the solutions

$$ER|_{\kappa=-1} \approx 2.043, \ \ 5.396, \ \ 8.578, \ \ldots \ . \tag{3.131}$$

Some interesting properties of the quark spectrum can be read from the asymptotic expansions of the Bessel function.

(1) For a given κ, e.g., $\kappa = -1$, the levels become equidistant for large energies. This follows immediately from

$$j_l(z) \simeq \frac{1}{z} \cos\left(z - \frac{l+1}{2}\pi\right) \quad \text{for } z \to \infty \ . \tag{3.132}$$

For $\kappa = -1$, (3.122) therefore asymptotically becomes

$$\cos\left(ER - \frac{\pi}{2}\right) = \cos(ER - \pi) \quad \text{or} \quad \tan(ER) = -1 \ , \tag{3.133}$$

which is obviously periodic in ER.

(2) The smaller $|\kappa|$ becomes, the higher the lowest eigenvalue will be. For very large values of l,

$$j_l(z) \simeq \frac{1}{2\sqrt{zl}}\left(\frac{ez}{2l}\right)^l \quad , \quad \ell \to \infty \ . \tag{3.134}$$

Condition (3.122) can only be satisfied if

$$\begin{aligned} 1 &= \left|\frac{j_\ell(z)}{j_{\ell-1}(z)}\right| \frac{\sqrt{l-1}}{\sqrt{l}} \frac{(l-1)^{l-1}}{l^l} \left(\frac{l\cdot z}{2}\right)^l \left(\frac{2}{l\cdot z}\right)^{l-1} \\ &= \frac{\sqrt{l-1}}{\sqrt{l}} \frac{(l-1)^{l-1}}{l^l} \left(\frac{l\cdot z}{2}\right) \simeq \frac{z(\ell-1)^{\ell-1}\sqrt{\ell-1}}{\ell^\ell\sqrt{\ell}} < \frac{z}{\ell} \\ \Rightarrow \quad & z > \ell > |\kappa| - 1 \ . \end{aligned} \tag{3.135}$$

In the last but one step, $l \approx 2$ and $(l-1) = l$ for large l has been used. The smallest eigenvalue is thus always larger than $|\kappa| - 1$. These two properties can also be seen in Fig. 3.14, which shows the lowest-lying energy states.

What about the quadratic boundary condition (3.115); can it also be satisfied? Inserting (3.128) into (3.115), we obtain

$$\begin{aligned} B = -\frac{1}{2}\frac{\mathrm{d}}{\mathrm{d}R}\sum_{\mathrm{q}} N_{\mathrm{q}}^2 \Big[& j_{\ell_{\mathrm{q}}}^2(E_{\mathrm{q}}R)\, \chi_{\kappa_{\mathrm{q}}}^{\mu_{\mathrm{q}}}(\theta,\phi)^+\, \chi_{\kappa_{\mathrm{q}}}^{\mu_{\mathrm{q}}}(\theta,\phi) \\ & - j_{\bar{\ell}_{\mathrm{q}}}^2(E_{\mathrm{q}}R)\, \chi_{-\kappa_{\mathrm{q}}}^{\mu_{\mathrm{q}}}(\theta,\phi)^+\, \chi_{-\kappa_{\mathrm{q}}}^{\mu_{\mathrm{q}}}(\theta,\phi)\Big] \ . \end{aligned} \tag{3.136}$$

Equation (3.136) can only be satisfied if the right-hand side no longer depends on θ and ϕ. This is only the case for $\kappa = \pm 1$, or when for higher κ states the q sum runs over all μ quantum numbers (shells). As we are primarily interested in $\mathrm{q}\bar{\mathrm{q}}$ and qqq systems (mesons and baryons), we can neglect this last possibility. The quadratic boundary condition can therefore only be satisfied for $|\kappa| = 1$. For this case, $|\kappa| = 1$, it holds that

$$\chi_\kappa^{\mu+}\, \chi_\kappa^\mu = \chi_{-\kappa}^{\mu+}\, \chi_{-\kappa}^\mu = \frac{1}{4\pi} \quad , \kappa = 1 \ , \tag{3.137}$$

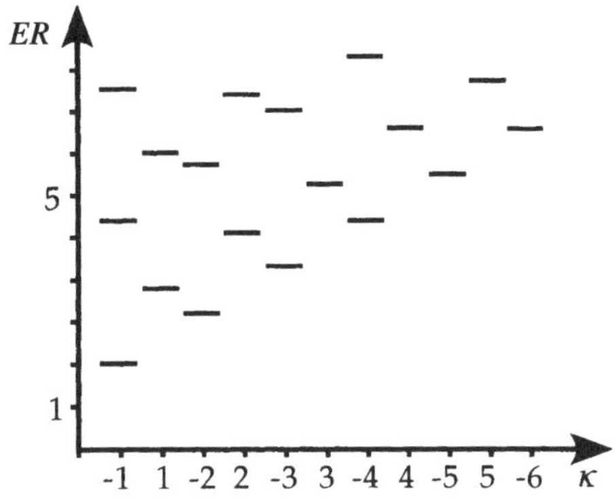

Fig. 3.14. The lowest-energy eigenvalues of the quark states in the MIT bag. R denotes the bag radius

and (3.136) becomes

$$B = -\frac{1}{4\pi}\sum_{\mathrm{q}} N_{\mathrm{q}}^2 \Bigg(j_{\ell_{\mathrm{q}}}(E_{\mathrm{q}}R) E_{\mathrm{q}} \frac{\mathrm{d}}{\mathrm{d}(E_{\mathrm{q}}R)} j_{\ell_{\mathrm{q}}}(E_{\mathrm{q}}R) \\ - j_{\bar{\ell}_{\mathrm{q}}}(E_{\mathrm{q}}R) E_{\mathrm{q}} \frac{\mathrm{d}}{\mathrm{d}(E_{\mathrm{q}}R)} j_{\bar{\ell}_{\mathrm{q}}}(E_{\mathrm{q}}R) \Bigg) . \tag{3.138}$$

When all quarks are in the lowest state with $\kappa = -1$, it follows from (3.130) and from the following properties of the $j's$, i.e.

$$\begin{aligned} \frac{\mathrm{d}}{\mathrm{d}(z)} j_l(z) &= \frac{1}{2l+1}\left[l j_{l-1}(z) - (l+1) j_{l+1}(z) \right] , \\ j_{l+1}(z) + j_{l-1}(z) &= \frac{2l+1}{z} j_l(z) , \end{aligned} \tag{3.139}$$

that

$$\frac{\mathrm{d}}{\mathrm{d}(ER)} j_{\ell_{\mathrm{q}}}(ER) = \frac{\mathrm{d}}{\mathrm{d}(ER)} j_0(ER) = -j_1(ER) = -j_0(ER) , \tag{3.140a}$$

$$\begin{aligned} \frac{\mathrm{d}}{\mathrm{d}(ER)} j_{\bar{\ell}_{\mathrm{q}}}(ER) &= \frac{\mathrm{d}}{\mathrm{d}(ER)} j_1(ER) = j_0(ER) - \frac{2}{3} j_2(ER) - \frac{2}{3} j_0(ER) \\ &= j_0(ER) - \frac{2}{ER} j_1(ER) \\ &= j_0(ER) - \frac{2}{ER} j_0(ER) , \end{aligned} \tag{3.140b}$$

and therefore

$$B = \frac{1}{4\pi} N_{\kappa=-1}^2 N_{\mathrm{q}} 2E\, j_0^2(ER) \left(1 - \frac{1}{ER} \right) , \tag{3.141}$$

where N_{q} indicates the number of quarks in the bag.

In Exercise 3.11, Equation (3), we put $m = 0$ and $\kappa = -1$ and obtain for the normalization constant

$$N_{\kappa=-1}^2 = \frac{ER}{2R^3(ER-1)\, j_0^2(ER)} , \tag{3.142}$$

and consequently

$$B = \frac{1}{4\pi} N_{\mathrm{q}} \frac{E}{R^3} \quad \text{or} \quad R^4 = \frac{N_{\mathrm{q}} ER}{4\pi B} . \tag{3.143}$$

Equation (3.136) thus yields for any quark content (for full quark shells only!) the corresponding bag radius as a function of the bag pressure constant B.

EXERCISE

3.11 The Bag Wave Function for Massive Quarks

Problem. Show that the bag wave function of massive quarks is

$$\Psi = N\begin{pmatrix} j_{\ell_\kappa}(pr)\,\chi^\mu_\kappa \\ \mathrm{i}\frac{p}{E+m}\,\mathrm{sgn}(\kappa)\; j_{\bar{\ell}_\kappa}(pr)\,\chi^\mu_{-\kappa} \end{pmatrix} \mathrm{e}^{-\mathrm{i}Et} \tag{1}$$

with

$$E = \sqrt{p^2+m^2}\;,$$

$$\ell_\kappa = \begin{cases} -\kappa-1 & \text{for } \kappa<0 \\ \kappa & \text{for } \kappa>0 \end{cases}\;,$$

$$\bar{\ell}_\kappa = \begin{cases} -\kappa & \text{for } \kappa<0 \\ \kappa-1 & \text{for } \kappa>0 \end{cases}\;.$$

Show, furthermore, that the normalization condition

$$\int\limits_{\text{bag}} \mathrm{d}^3r\; \Psi^\dagger\Psi = 1 \tag{2}$$

leads to

$$N = \frac{p}{\sqrt{2E(ER+\kappa)+m}}\,\frac{1}{|j_{\ell_\kappa}(pR)|R}\;, \tag{3}$$

R being the bag radius.

Solution. Adding the mass term to (3.113) yields

$$(\not{p}-m)\Psi = 0\;. \tag{4}$$

Correspondingly (3.119) becomes

$$\left[-\mathrm{i}\begin{pmatrix} 0 & \sigma_r \\ \sigma_r & 0 \end{pmatrix}\left(\frac{\partial}{\partial r}+\frac{1}{r}-\frac{\beta}{r}\hat{K}\right)+\beta m\right]\Psi = E\,\Psi \tag{5}$$

and (3.121) then becomes

$$\left(\frac{\mathrm{d}}{\mathrm{d}r}+\frac{1}{r}-\frac{\kappa}{r}\right)f(r)-(E-m)g(r) = 0 \tag{6a}$$

and

$$\left(\frac{\mathrm{d}}{\mathrm{d}r}+\frac{1}{r}+\frac{\kappa}{r}\right)g(r)+(E+m)f(r) = 0\;. \tag{6b}$$

Exercise 3.11

Now we insert $g(r) = j_{\ell_\kappa}(pr)$ and $f(r) = -[p/(E+m)]\,\mathrm{sgn}(\kappa)\, j_{\bar{\ell}_\kappa}(pr)$ and evaluate the left-hand sides of (6a) and (6b), respectively:

$$\frac{-p^2}{E+m}\,\mathrm{sgn}(\kappa)\left(\frac{\mathrm{d}}{\mathrm{d}(pr)} + \frac{1-\kappa}{pr}\right) j_{\bar{\ell}_\kappa}(pr) - (E-m)\, j_{\ell_\kappa}(pr) = 0 \tag{7a}$$

and

$$p\left(\frac{\mathrm{d}}{\mathrm{d}(pr)} + \frac{1+\kappa}{pr}\right) j_{\ell_\kappa}(pr) - p\ \mathrm{sgn}(\kappa)\ j_{\bar{\ell}_\kappa}(pr) = 0\ . \tag{7b}$$

With the help of recursion relations (3.127) ($c = -\,\mathrm{sgn}(\kappa)$) we then obtain

$$\frac{p^2}{E+m}\, j_{\ell_\kappa}(pr) - \frac{p^2}{E+m}\, j_{\ell_\kappa}(pr) = 0 \tag{8a}$$

and

$$p\,\mathrm{sgn}(\kappa)\ j_{\bar{\ell}_\kappa}(pr) - p\,\mathrm{sgn}(\kappa)\ j_{\bar{\ell}_\kappa}(pr) = 0 \quad . \tag{8b}$$

This yields the wave function of a massive quark in the MIT bag

$$\Psi = N\begin{pmatrix} j_{l\kappa}(pr)\chi_\kappa^\mu \\ i\frac{p}{E+m}\,\mathrm{sgn}(\kappa) j_{\bar{l}\kappa}(pr)\chi_{-\kappa}^\mu \end{pmatrix} e^{-iEt}$$

which transforms into (3.128) for $m \to 0$. Accordingly we obtain from (3.129) and (3.130) by substituting $\mathrm{sgn}(\kappa) \to [p/(E+m)]\,\mathrm{sgn}(\kappa)$ the modified linear boundary condition

$$j_{l\kappa}(pr) = -\,\mathrm{sgn}(\kappa)\frac{p}{E+m} j_{\bar{l}\kappa}(pr)\ . \tag{9}$$

By means of (9) we then evaluate the normalization integral:

$$\begin{aligned} 1 &= \int_{\text{bag}} \mathrm{d}^3 r\ \Psi^\dagger \Psi \\ &= \int_0^R \mathrm{d}r\, r^2 \left[j_{\ell_\kappa}^2(pr) + \frac{p^2}{(E+m)^2}\ j_{\bar{\ell}_\kappa}^2(pr)\right] N^2\ . \end{aligned} \tag{10}$$

Employing the differentiation and recursion formulas of the Bessel functions (see (3.139))

$$\begin{aligned} j_n'(z) &= \frac{n}{z} j_n(z) - j_{n+1}(z)\ , \\ j_{n-1}'(z) &= \frac{n-1}{z} j_{n-1}(z) - j_n(z)\ , \\ j_{n+1}'(z) &= -\frac{n+2}{z} j_{n+1}(z) + j_n(z)\ , \\ j_{n+1}(z) + j_{n-1}(z) &= \frac{2n+1}{z} j_n(z) \end{aligned} \tag{12}$$

we get *Exercise 3.11*

$$
\begin{aligned}
&\frac{\mathrm{d}}{\mathrm{d}z}\left[z^3\Big(j_n^2(z)-j_{n-1}(z)\;j_{n+1}(z)\Big)\right]\\
&\quad =3z^2\Big(j_n^2(z)-j_{n-1}(z)\;j_{n+1}(z)\Big)+z^3\Bigg[2\;j_n(z)\Big(-j_{n+1}(z)+\frac{n}{z}j_n(z)\Big)\\
&\qquad -j_{n-1}(z)\Big(-\frac{n+2}{z}j_{n+1}(z)+j_n(z)\Big)\\
&\qquad -\Big(\frac{n-1}{z}j_{n-1}(z)-j_n(z)\Big)j_{n+1}(z)\Bigg]\\
&\quad =3z^2\Big(j_n^2(z)-j_{n-1}(z)\;j_{n+1}(z)\Big)\\
&\qquad +z^3\left[\frac{2n}{z}j_n^2(z)-j_n(z)\Big(j_{n+1}(z)+j_{n-1}(z)\Big)+\frac{3}{z}j_{n-1}(z)\;j_{n+1}(z)\right]\\
&\quad =3z^2\;j_n^2(z)+z^3\left[\frac{2n}{z}j_n^2(z)-\frac{2n+1}{z}j_n^2(z)\right]\\
&\quad =2z^2\;j_n^2(z)\ .
\end{aligned}
\tag{13}
$$

Terms of the form $z^2 j_n^2(z)$ appear in the integral (10). Therefore, relation (13) allows a direct evaluation of that integral:

$$
\begin{aligned}
1=N^2\frac{R^3}{2}\Bigg[&j_{\ell_\kappa}^2(pR)-j_{\ell_\kappa-1}(pR)\;j_{\ell_\kappa+1}(pR)\\
&+\frac{p^2}{(E+m)^2}\Big(j_{\bar{\ell}_\kappa}^2(pR)-j_{\bar{\ell}_\kappa-1}(pR)\;j_{\bar{\ell}_\kappa+1}(pR)\Big)\Bigg]\ .
\end{aligned}
\tag{14}
$$

Equation (14) can be simplified using (9). For $\kappa<0$ we have (see (12))

$$
\begin{aligned}
\bar{\ell}_\kappa&=\ell_\kappa+1\ ,\quad \ell_\kappa=-\kappa-1\ ,\\
j_{\ell_\kappa+1}(pR)&=\frac{E+m}{p}j_{\ell_\kappa}(pR)\ ,\\
j_{\ell_\kappa-1}(pR)&=\frac{2\ell_\kappa+1}{pR}j_{\ell_\kappa}(pR)-j_{\ell_\kappa+1}(pR)\\
&=\left(\frac{2\ell_\kappa+1}{pR}-\frac{E+m}{p}\right)j_{\ell_\kappa}(pR)\ ,\\
j_{\ell_\kappa+2}(pR)&=\frac{2\ell_\kappa+3}{pR}j_{\ell_\kappa+1}(pR)-j_{\ell_\kappa}(pR)\\
&=\left(\frac{2\ell_\kappa+3}{pR}\frac{E+m}{p}-1\right)j_{\ell_\kappa}(pR)\ .
\end{aligned}
\tag{15}
$$

Exercise 3.11

Equation (14) therefore becomes

$$\begin{aligned}
1 &= N^2 \frac{R^3}{2} j^2_{\ell_\kappa}(pR)\Bigg[2-\left(\frac{2\ell_\kappa+1}{pR}-\frac{E+m}{p}\right)\frac{E+m}{p} \\
&\quad + \frac{p^2}{(E+m)^2}\left(1-\frac{2\ell_\kappa+3}{pR}\frac{E+m}{p}\right)\Bigg] \\
&= N^2 \frac{R^3}{2} j^2_{\ell_\kappa}(pR)\Bigg[2-2\ell_\kappa\left(\frac{E+m}{p^2R}+\frac{1}{(E+m)R}\right) \\
&\quad -\frac{E+m}{p^2R}-\frac{3}{(E+m)R}+\frac{E+m}{E-m}+\frac{E-m}{E+m}\Bigg] \\
&= N^2 \frac{R^3}{2} j^2_{\ell_\kappa}(pR)\left(-2\ell_\kappa\frac{2E}{p^2R}-\frac{4E-2m}{p^2R}+\frac{4E^2}{p^2}\right) \\
&= N^2 \frac{R^2}{p^2} j^2_{\ell_\kappa}(pR)\left(2\kappa E+m+2E^2R\right) \ .
\end{aligned} \tag{16}$$

Hence

$$N = \frac{p}{R|j_{\ell_\kappa}(pr)|\sqrt{2\kappa E+m+2E^2R}} \ . \tag{17}$$

For $\kappa > 0$ an analogous calculation yields

$$\begin{aligned}
\bar{\ell}_\kappa &= \ell_\kappa - 1 \ , \quad \ell_\kappa = \kappa \ , \\
j_{\ell_\kappa-1}(pR) &= -\frac{E+m}{p} j_{\ell_\kappa}(pR) \ , \\
j_{\ell_\kappa+1}(pR) &= \frac{2\ell_\kappa+1}{pR} j_{\ell_\kappa}(pR) - j_{\ell_\kappa-1}(pR) \\
&= \left(\frac{2\ell_\kappa+1}{pR}+\frac{E+m}{p}\right) j_{\ell_\kappa}(pR) \quad , \\
j_{\ell_\kappa-2}(pR) &= \frac{2\ell_\kappa-1}{pR} j_{\ell_\kappa-1}(pR) - j_{\ell_\kappa}(pR) \\
&= -\left(\frac{E+m}{p}\frac{2\ell_\kappa-1}{pR}+1\right) j_{\ell_\kappa}(pR) \ .
\end{aligned} \tag{18}$$

Equation (14) now has the form

$$\begin{aligned}
1 &= N^2 \frac{R^3}{2} j^2_{\ell_\kappa}(pr)\Bigg[2+\left(\frac{2\ell_\kappa+1}{pR}+\frac{E+m}{p}\right)\frac{E+m}{p} \\
&\quad + \frac{p^2}{(E+m)^2}\left(\frac{E+m}{p}\frac{2\ell_\kappa-1}{pR}+1\right)\Bigg] \\
&= N^2 \frac{R^3}{2} j^2_{\ell_\kappa}(pr)\left(2\ell_\kappa\frac{2E}{p^2R}+\frac{2m}{p^2R}+\frac{4E^2}{p^2}\right) \\
&= N^2 \frac{R^2}{p^2} j^2_{\ell_\kappa}(pr)\left(2\kappa E+m+2E^2R\right) \ .
\end{aligned} \tag{19}$$

Exercise 3.11

Therefore N is given by (17) also for $\kappa > 0$. The corresponding normalization factors for massless quarks are obtained by setting $p = E$ and $m = 0$, i.e. $N_\kappa(m=0) = \frac{E}{R|j_{l\kappa}(ER)|}\frac{1}{\sqrt{2E(\kappa+ER)}}$ (see also (3.142)).

Even in the simplest case, however, (3.143) leads to difficulties. From this equation a nucleon (three almost massless quarks in the ground state) and a pion (one quark and one antiquark) should have almost the same radius:

$$\frac{R_\mathrm{N}}{R_\pi} = \left(\frac{3}{2}\right)^{1/4} = 1.107 \quad . \tag{3.144}$$

This can be translated by (3.143) and (3.130) into a ratio of masses

$$\frac{M_\mathrm{N}}{M_\pi} = \frac{3\cdot 2.043/R_\mathrm{N}}{2\cdot 2.043 R_\pi} = \frac{3}{2}\left(\frac{2}{3}\right)^{(1/4)} = 1.36 \ .$$

This result does not agree with physical reality. The prediction for the pion mass is correspondingly much too large. In order to obtain a more realistic model we therefore have to go beyond the simple assumptions made so far. Before turning to applications of the MIT bag model, we want to discuss the necessary additional assumptions. The main problem is the nonphysical equation (3.133). It can be changed by introducing additional pressure terms with different R dependence. Let us first consider the total energy, where the effect of these new terms can be seen most clearly. So far the total energy consists of two terms: the volume energy due to the external pressure B and the single-particle energies, i.e.

$$E = B\,\frac{4\pi}{3}R^3 + \sum_\mathrm{q}\frac{\omega_\mathrm{q}}{R} \ , \qquad \omega_\mathrm{q} = E_\mathrm{q}\,R = \mathrm{const} \ . \tag{3.145}$$

For $\sum_\mathrm{q}\omega_\mathrm{q} \longrightarrow N_\mathrm{q}\omega_\mathrm{q}$ we immediately recognize that a variation with respect to R leads to (3.143). The first term added is

$$E_0 = -\frac{Z}{R} \ , \qquad Z = \mathrm{const} \ . \tag{3.146}$$

Equation (3.143) then assumes the form

$$R^4 = \frac{N_\mathrm{q}\omega_\mathrm{q} - Z}{4\pi B} \ . \tag{3.147}$$

A sufficiently large value of Z therefore allows us to adjust the mass and radius difference between mesons and baryons. Physically this new term is interpreted as the *Casimir energy*.[14] The bag yields a lower bound for the zero-point energy of the gluon field. The smaller the bag, the bigger the zero-point energy

[14] See G. Plunien, B. Müller, and W. Greiner: Phys. Rep. **134**, 87 (1986), see also W. Greiner: *Quantum Mechanics – Special Chapters* (Springer, Berlin, Heidelberg 1998).

of the vacuum oscillations. It is therefore clear that E_0 vanishes for $R \to \infty$ and diverges for $R \to 0$. For very small values of R, however, the bag boundary condition no longer makes any sense, because it is only an effective description of very complicated microscopic processes. Hence R must not assume values smaller than the typical length scale of the reactions considered. If the typical energies become larger than $\sim 1\,\mathrm{GeV}^2$ we have to apply perturbative QCD, which models the interaction between quarks and gluons. Individual quark–gluon interactions cannot be described by simple boundary conditions. Therefore $R < 1/Q$ should not become smaller than 0.2 fm. Strictly speaking, we know neither the explicit functional form nor the sign of E_0. The results of simple model calculations do not justify (3.136) and are, in addition, completely unreliable, since the self-interaction of the gluons (i.e., the non-Abelian structure of QCD, see Chap. 4) has only been treated in a rough approximation. Equation (3.136) should therefore be considered a phenomenological correction term with unknown physical origin.

The next problem is to describe the *mass splitting* within the baryon multiplet. Since all strange particles are considerable heavier than those with strangeness $S=0$, the introduction of quark masses seems to be a reasonable first step. Therefore in the following we make use of the massive eigenfunctions and the corresponding *energy eigenvalues* derived in Exercise 3.11. The masses of u, d, and s quarks are treated as free parameters. This generalization, however, is not sufficient to describe both the relatively small splitting between Σ and N and the huge splitting between K and π. A further correction is necessary that assumes different values for mesons and baryons. We therefore introduce the following interaction between the quarks (see Example 3.12):

$$E_{\mathrm{q}G} = \alpha_{\mathrm{c}}\, N \cdot \sum_{i<j} (\boldsymbol{\sigma}_i \cdot \boldsymbol{\sigma}_j)\, \frac{\mu_i \mu_j}{R^3} \left(1 + 2 \int_0^R \frac{\mathrm{d}r}{r^4}\, \mu_i \mu_j \right) \tag{3.148}$$

with

$$N = \begin{cases} 2 & \text{for a baryon} \\ 4 & \text{for a meson} \end{cases} \quad .$$

μ_i and α_{c} denote the magnetic moment of the quark with index i and the coupling constant of the color interaction, respectively.

Equation (3.148) is obtained by taking the color magnetic interaction between the quarks into account. For a more detailed treatment of these matters we refer to Chap. 4, where the QCD equations are discussed. At this point it need only be mentioned that the derivation of (3.148) is not consistent. Again the gluonic self-interaction has been neglected. The electric and magnetic parts of the remaining interaction (which then look exactly like the electromagnetic interaction) are treated in a different way. In order to justify (3.148) we must therefore postulate that all contributions except the one-gluon exchange are described by the bag boundary condition. Since α_{c} is treated more or less as a free parameter (within certain limits), (3.148) can also be interpreted as a phenomenological

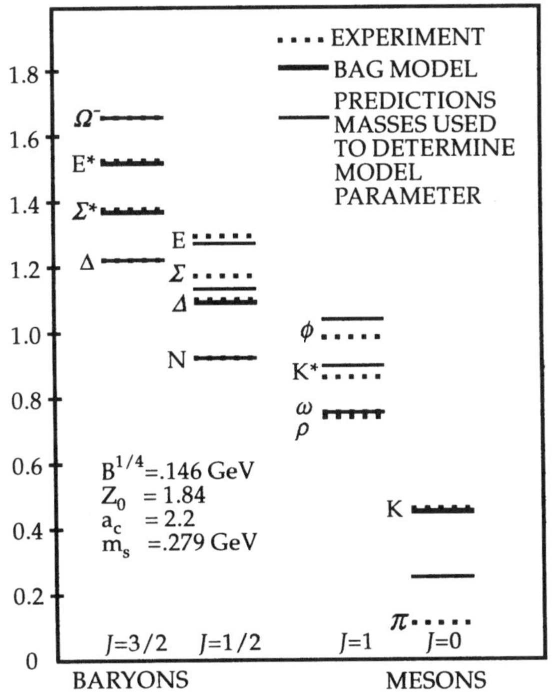

Fig. 3.15. The MIT-bag-model fit for the lightest mesons and baryons (from P. Hasenfratz and J. Kuti: The Quark Bag Model, Phys. Rep. **40**, 75 (1978))

correction. Summarizing the above arguments we find that the total energy assumes the form

$$E = \frac{4\pi}{3} R^3 B + \frac{\sum_q \omega_q - Z}{R} + E_{qG} \ . \tag{3.149}$$

The bag radius is determined by finding the minimum of $E(R)$:

$$\frac{\partial E}{\partial R} = 0 \ , \quad \frac{\partial^2 E}{\partial R^2} > 0 \Rightarrow R \ , \tag{3.150}$$

and all features of the specific bags can be obtained by using the wave functions introduced in Exercise 3.11.

A total of about 25 to 30 experimentally observed values for masses, magnetic moments, averaged charge radii, axial coupling constants, and so on is available for fitting the six parameters B, Z, α_c, m_s, m_u, and m_d. The agreement achieved in such a fit is in general better than 30%.

Let us start with the masses. Figure 3.15 depicts both the theoretical and the experimental values for the following set of parameters:[15]

$$\begin{aligned} &B = (146\,\mathrm{MeV})^4 \ , \quad Z = 1.84 \ , \quad \alpha_c = 2.2 \ , \\ &m_u = 0\,\mathrm{MeV} \ , \quad m_d = 0\,\mathrm{MeV} \ , \quad m_s = 279\,\mathrm{MeV} \ . \end{aligned} \tag{3.151}$$

[15] See T. De Grand, R.L. Jaffe, K. Johnson, J. Kiskis: Phys. Rev. D **12**, 2060 (1975).

Clearly all the masses except that of the pion are quite well described. In general the modified MIT model just described provides a satisfactory description of the light hadrons except for the pion. One possible way out of this dilemma is a combination of the bag model with a specific treatment of the pions. But we are not going to discuss these so-called *hybrid bag models*.

Instead we evaluate the averaged charge radius and the magnetic moment of the proton within the context of the MIT bag model. The squared charge radius is defined as

$$\langle r^2\rangle_{\mathrm{ch}} = \frac{1}{e}\int \mathrm{d}^3r_1 \int \mathrm{d}^3r_2 \int \mathrm{d}^3r_3\, \Psi_{\mathrm{p}}^\dagger \left(\hat{Q}\hat{r}^2\right)\Psi_{\mathrm{p}} \ . \tag{3.152}$$

Here Ψ_{p} denotes the proton wave function, which can be decomposed into quark wave functions as follows[16]:

$$\begin{aligned}\Psi_{\mathrm{p}}\left(m_s = \frac{1}{2}\right) = \frac{1}{3\sqrt{2}}\Big(&2u^\uparrow(1)u^\uparrow(2)d^\downarrow(3) - u^\uparrow(1)u^\downarrow(2)d^\uparrow(3)\\ &- u^\downarrow(1)u^\uparrow(2)d^\uparrow(3) - u^\uparrow(1)d^\uparrow(2)u^\downarrow(3) + u^\uparrow(1)2d^\downarrow(2)u^\uparrow(3)\\ &- u^\downarrow(1)d^\uparrow(2)u^\uparrow(3) - d^\uparrow(1)u^\uparrow(2)d^\downarrow(3)\\ &- d^\uparrow(1)u^\downarrow(2)d^\uparrow(3) + 2d^\downarrow(1)u^\uparrow(2)d^\uparrow(3)\Big) \ .\end{aligned} \tag{3.153}$$

$u^\uparrow(1)$ denotes the wave function of a u quark with spin projection $m_s = +1/2$ attached to quark number one. Inserting (3.153) into (3.152) and employing the *quark wave functions* (3.128) (owing to (3.151), the up and down quarks are assumed to be massless) leads to (see Exercise 3.13)

$$\langle r^2\rangle_{\mathrm{ch}} = 0.53\ R^2 \ . \tag{3.154}$$

The bag radius R follows by minimizing (3.149), yielding $R = 1\mathrm{fm}$ (the parameters have been correspondingly adjusted). Hence

$$\sqrt{\langle r^2\rangle_{\mathrm{ch}}} = 0.73\ \mathrm{fm} \ , \tag{3.155}$$

which agrees with the experimental value to within 20%:

$$\sqrt{\langle r^2\rangle}(\mathrm{exp}) = 0.88 \pm 0.03\ \mathrm{fm} \ . \tag{3.156}$$

The magnetic moment is obtained by evaluating the expectation value of the corresponding operator:

$$\hat{\boldsymbol{\mu}} = \frac{\hat{Q}}{2}\hat{\boldsymbol{r}}\times\hat{\boldsymbol{\alpha}} \ , \tag{3.157}$$

$$\mu_{\mathrm{p}} = \int \mathrm{d}^3r_1 \int \mathrm{d}^3r_2 \int \mathrm{d}^3r_3\, \Psi_{\mathrm{p}}^\dagger \sum_i \left(\frac{\hat{Q}_i}{2}\,\hat{\boldsymbol{r}}_i\times\hat{\boldsymbol{\alpha}}_i\right)\Psi_{\mathrm{p}} \ . \tag{3.158}$$

[16] See also W. Greiner and B. Müller: *Quantum Mechanics – Symmetries*, 2nd ed., (Springer, Berlin, Heidelberg 1994), P. Hasenfratz, J. Kuti, Phys. Rep. 40 **C**, 75 (1978).

We again insert the wave functions (3.153) and (3.128). A detailed calculation, which will be performed in Exercise 3.14, yields

$$\frac{|\mu_{\rm p}|}{\mu_{\rm K}} = 1.9 \ , \tag{3.159}$$

where $\mu_0 = e/2m_{\rm p}$ denotes the nuclear magneton and $m_{\rm p}$ is the proton mass.

EXAMPLE

3.12 Gluonic Corrections to the MIT Bag Model

Although the general structure of hadrons, i.e. the masses of baryons, are rather well described in the framework of the simple bag model of noninteracting quarks, the mass splitting in the baryon multiplett cannot be accounted for properly.

We will study the quark interaction energy due to their coupling to colored gluons. The interaction will be calculated only to lowest order in $\alpha_s = g^2/4\pi$. The appropriate diagrams are given in Fig. 3.16.

The gluon interaction will lift the spin degeneracies of the simple model, splitting the nucleon from the Δ resonance, and the ϱ from the π.

To lowest order in α_s the non-Abelian gluon self-coupling does not contribute and the gluons act as eight independent Abelian fields.

Calculating the color field, therefore, is equivalent to finding the solution of the classical Maxwell equations under the bag boundary conditions

$$\boldsymbol{n} \cdot \sum_i \boldsymbol{E}_i^A = 0 \ , \tag{1a}$$

$$\boldsymbol{n} \times \sum_i \boldsymbol{B}_i^A = 0 \ , \tag{1b}$$

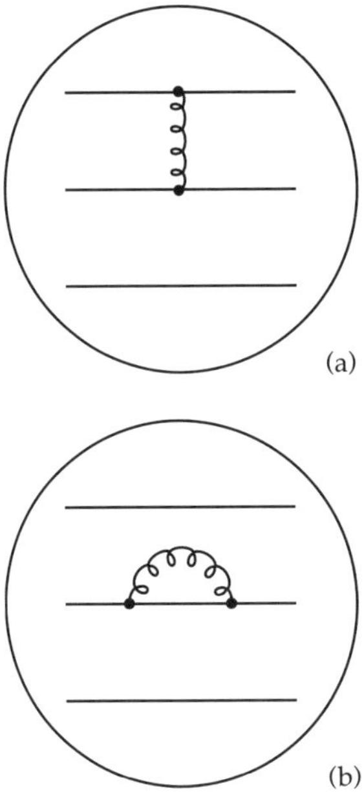

Fig. 3.16a,b. Lowest-order gluon interaction diagram for a baryon. Mesons are similar. (a) Gluon exchange; (b) gluon self energy

where the index i refers to the number of quarks in the bag. The index A denotes color and runs from 1 to 8. $\boldsymbol{E}^A$ and $\boldsymbol{B}^A$ are gluon electric and magnetic field components. These boundary conditions are necessary to confine the gluons to the bag. The electrostatic interaction energy of a static charge distribution i is

$$\Delta E_E = \frac{1}{2} \sum_{i,j} \int \mathrm{d}^3 r \boldsymbol{E}_i^A \cdot \boldsymbol{E}_j^A \ . \tag{2}$$

Similarly, the magnetostatic interaction energy is

$$\Delta E_\mu = -\frac{1}{2} \sum_{i,j} \int \mathrm{d}^3 r \boldsymbol{B}_i^A \cdot \boldsymbol{B}_j^A \ . \tag{3}$$

$\boldsymbol{E}^A$ and $\boldsymbol{B}^A$ are determined from the quark charge and current distributions by Maxwell's equation and the boundary conditions (1a) and (1b). The terms with

Example 3.12

$i = j$ are due to the self-energy diagram in Fig. 3.16b. These diagrams contribute to a renormalization of the quark mass. A proper treatment would separate this contribution of renormalization from the rest of Fig. 3.16b, since its effect is already included in the phenomenological quark mass one uses.

To avoid double counting, one usually neglects the self-energy contribution for the calculation of the magnetic energies, whereas it is taken into account in the calculation of the electric energies.

Equation (3) therefore may be rewritten as

$$\Delta E_\mu = -g^2 \sum_{i,j} \int d^3r \, \boldsymbol{B}_i^A \cdot \boldsymbol{B}_j^A \ . \tag{4}$$

The color magnetic field must satisfy

$$\begin{aligned} &\nabla \times \boldsymbol{B}_i^A = \boldsymbol{j}_i^A, \quad r < R \ , \\ &\nabla \cdot \boldsymbol{B}_i^A = 0, \quad r < R \ , \\ &\boldsymbol{n} \times \sum_i \boldsymbol{B}_i = 0, \quad r = R, \end{aligned} \tag{5}$$

where $\boldsymbol{j}_i^A$ is the color current of quark i:

$$\begin{aligned} \boldsymbol{j}_i^A &= q_i^\dagger \boldsymbol{\alpha} \frac{\lambda^A}{2} q_i \\ &= -\frac{3}{8\pi} \boldsymbol{e}_r \times \boldsymbol{\sigma}_i \lambda_i^A \frac{\mu_i'(r)}{r^3} \ . \end{aligned} \tag{6}$$

Here $\mu_i'(r)$ is the scalar magnetization density of a quark of mass m_i in the lowest cavity eigenstate:

$$\mu_i(m_i, R) = \left(\frac{1}{2} \int_{V_B} q_i^\dagger(\boldsymbol{r})(\boldsymbol{r} \times \boldsymbol{\alpha}) q_i(\boldsymbol{r}) \right)_z d^3r = \int_0^R d r \mu_i'(r) \ . \tag{7}$$

The integral over $\mu_i'(r)$ yields the magnetic moment of quark i.

Equation (5) may be integrated to determine $\boldsymbol{B}_i^A(\boldsymbol{r})$:

$$\begin{aligned} \boldsymbol{B}_i^A(\boldsymbol{r}) = &\frac{1}{8\pi} \lambda_i^A \boldsymbol{\sigma}_i \left[2M_i(r) + \mu(m_i, R)/R^3 - \mu(m_i, r)/r^3 \right] \\ &+ \frac{3}{8\pi} \lambda_i^A \boldsymbol{e}_r (\boldsymbol{e}_r \cdot \boldsymbol{\sigma}_i) \mu(m_i, r)/r^3 \end{aligned} \tag{8}$$

with

$$M_i(r) = \int_r^R \mu_i'(r') d r' \ . \tag{9}$$

Example 3.12

Inserting (8) into (4) yields

$$\Delta E_\mu = -\frac{3}{4}\alpha_s \sum_{i,j}(\sigma_i \lambda_i^A)(\sigma_j \lambda_j^A)\mu(m_i, R)\mu(m_j, R)/R^3 \times \left(1 + 2\int_0^R \mathrm{d}r \mu(m_i, R)\mu(m_j, R)/r^4\right) . \quad (10)$$

The colour and spin dependences of (10) may be simplified considerably. For a colour-singlet meson

$$\left(\lambda_1^A \lambda_2^A\right)|M\rangle = 0 \quad (11)$$

we find, after squaring,

$$\sum_A (\lambda_i^A)^2 = 16/3 \quad (12)$$

and therefore $\sum_A \lambda_1^A \lambda_2^A = -16/3$.

Likewise for baryons

$$\sum_{i=1}^{3} \lambda_i^A |B\rangle = 0 \quad (13)$$

and accordingly

$$\sum_A \lambda_i^A \lambda_j^A = -8/3, \quad i \neq j . \quad (14)$$

The final expression for the magnetic interaction energy is

$$\Delta E_\mu = N\alpha_s \sum_{i=j}(\sigma_i \cdot \sigma_j)\mu(m_i, R)\mu(m_j, R)/R^3 \times \left(1 + 2\int_0^R \mathrm{d}r \mu(m_i, R)\mu(m_j, R)/r^4\right) , \quad (15)$$

where $N = 2$ for a baryon and $N = 4$ for a meson. For a nucleon (three massless $1s_{1/2}$ quarks) one gets

$$\Delta E_\mu^N \approx -3.5\frac{\alpha_s}{R} . \quad (16)$$

The gluon electrostatic energy can be calculated along the same line. The color electric field for a single quark must satisfy

$$\nabla \cdot \boldsymbol{E}_i^A = j_i^{0A}, \quad r < R \quad (17a)$$

$$\nabla \times \boldsymbol{E}_i^A = 0, \quad r < R \quad (17b)$$

$$\sum_i \boldsymbol{n} \cdot \boldsymbol{E}_i^A = 0, \quad r = R , \quad (17c)$$

Example 3.12

where $j_i^{0A}(\boldsymbol{r})$ is a single quarks' color charge density

$$\begin{aligned} j_i^{0A} &= q_i^{\dagger}\frac{\lambda^A}{2}q_i \\ &= \frac{\lambda^A}{8\pi r^2}\varrho_i'(r) \ . \end{aligned} \tag{18}$$

Here $\varrho_i'(r)$ is the charge density of a quark of mass m_i in the lowest cavity eigenmode and satisfies

$$\int\limits_0^R \mathrm{d}r'\varrho_i'(r') = 1 \ . \tag{19}$$

The color electric field is obtained from Gauss' law:

$$\boldsymbol{E}_i^A = \frac{\lambda^A}{8\pi r^2}\boldsymbol{e}_r\varrho_i(r) \ , \tag{20}$$

where $\varrho_i(r)$ is the integral over $\varrho_i'(r)$ out to a given radius r.

Now if all of the quarks in a given hadron have the same mass, then $\varrho_i(r)$ is independent of the index i and the total color electric field is given by

$$\boldsymbol{E}^A = \frac{\boldsymbol{e}_r}{8\pi r^2}\varrho(r)\sum_i \lambda_i^A \ . \tag{21}$$

For a color-singlet hadron, $\sum\lambda^A|H\rangle = 0$. Therefore $\boldsymbol{E}^A = 0$. From (2) we see that $\Delta E_E = 0$. Notice that it was essential to include the static self-interaction $(\boldsymbol{E}_i^A \cdot \boldsymbol{E}_i^A)$ to obtain this correlation. In this approximation the energy shift in the baryon multiplett due to gluonic interaction is entirely attributed to the magnetic interaction.

EXERCISE

3.13 The Mean Charge Radius of the Proton

Problem. Evaluate the right-hand side of (3.142):

$$\langle r^2\rangle_{\mathrm{ch}} = \frac{1}{e}\int \mathrm{d}^3r_1 \int \mathrm{d}^3r_2 \int \mathrm{d}^3r_3 \ \Psi_\mathrm{p}^\dagger(\hat{Q}r^2)\Psi_\mathrm{p} \ . \tag{1}$$

Solution. Since r^2 does not change the quantum number m_s, all spin orientations yield the same value. Therefore we can restrict our calculation to one specific case,

$$\begin{aligned} \Psi_\mathrm{p}\left(m_s = \frac{1}{2}\right) \to \frac{1}{\sqrt{6}}\Big(&2u^\uparrow(1)u^\uparrow(2)d^\downarrow(3) \\ &-u^\uparrow(1)d^\uparrow(2)u^\downarrow(3) - d^\uparrow(1)u^\uparrow(2)u^\downarrow(3)\Big) \ . \end{aligned} \tag{2}$$

Exercise 3.13

The charge operator $\hat{Q}r^2$ acts on one quark at a time. For the proton it can be replaced by

$$\hat{Q}r^2 \to Q_1 r_1^2 + Q_2 r_2^2 + Q_3 r_3^2 \ . \tag{3}$$

Equation (1) then becomes

$$\begin{aligned}\langle r^2\rangle_{\text{ch}} = \frac{1}{e}\int d^3r_1 \int d^3r_2 \int d^3r_3 \frac{1}{6}\Big(&2u^{\uparrow}(1)u^{\uparrow}(2)d^{\downarrow}(3)\\ &-u^{\uparrow}(1)d^{\uparrow}(2)u^{\downarrow}(3) - d^{\uparrow}(1)u^{\uparrow}(2)u^{\downarrow}(3)\Big)^{\dagger}\\ &\times (Q_1r_1^2+Q_2r_2^2+Q_3r_3^2)\Big(2u^{\uparrow}(1)u^{\uparrow}(2)d^{\downarrow}(3)\\ &-u^{\uparrow}(1)d^{\uparrow}(2)u^{\downarrow}(3) - d^{\uparrow}(1)u^{\uparrow}(2)u^{\downarrow}(3)\Big) \ .\end{aligned} \tag{4}$$

All quarks are assumed to be massless and in the same state (namely the 1s state), i.e., we do not distinguish between up and down quarks. Therefore r_2^2 and r_3^2 can be replaced in the integrand by r_1^2:

$$Q_1r_1^2+Q_2r_2^2+Q_3r_3^2 \to (Q_1+Q_2+Q_3)r_1^2 = er_1^2 \ . \tag{5}$$

Now the integrals over $\boldsymbol{r}_2$ and $\boldsymbol{r}_3$ can easily be evaluated by using the orthogonality relations

$$\int \mathrm{d}^3r_2\, q^{s\dagger}(2)\, q'^{s'}(2) = \delta_{qq'}\,\delta_{ss'} \ , \qquad \left\{\begin{array}{l} q,\ q' = u,\ d \\ s,\ s' = \uparrow,\ \downarrow \end{array}\right. \ . \tag{6}$$

$$\langle r^2\rangle_{\text{ch}} = \int \mathrm{d}^3r_1\, r_1^2\, \frac{1}{6}\Big(4u^{\uparrow\dagger}(1)u^{\uparrow}(1) + u^{\uparrow\dagger}(1)u^{\uparrow}(1) + d^{\uparrow\dagger}(1)d^{\uparrow}(1)\Big) \ . \tag{7}$$

Owing to the assumption made above, the wave functions of u and d yield the same contributions, and (7) simplifies to

$$\langle r^2\rangle_{\text{ch}} = \int d\Omega_1 \int dr_1\, r_1^4\, u^{\uparrow\dagger}(1)u^{\uparrow}(1) \ . \tag{8}$$

Now we insert the explicit form of the wave function (3.128) and skip the index 1 in the remaining calculation:

$$\langle r^2\rangle_{\text{ch}} = N^2 \int \mathrm{d}r \int \mathrm{d}\Omega r^4 \left[j_0^2(Er)\, \chi_1^{\frac{1}{2}\dagger}(\Omega)\, \chi_1^{\frac{1}{2}}(\Omega) + j_1^2(Er)\, \chi_{-1}^{\frac{1}{2}\dagger}(\Omega)\, \chi_{-1}^{\frac{1}{2}}(\Omega)\right] \ . \tag{9}$$

Because of the orthogonality of the spherical spinors and because of (3.142), (9) assumes the form

$$\langle r^2\rangle_{\text{ch}} = \frac{ER}{2R^3(ER-1)j_0^2(ER)} \int_0^R dr\, r^4 \left(j_0^2(Er) + j_1^2(Er)\right) \ . \tag{10}$$

Exercise 3.13

The remaining integral can again be evaluated by means of the recursion relations or simply by inserting the explicit expressions

$$
\begin{aligned}
j_0(r) &= \frac{\sin z}{z} , \\
j_1(z) &= \frac{\sin z}{z^2} - \frac{\cos z}{z} .
\end{aligned} \tag{11}
$$

Hence

$$
\begin{aligned}
\langle r^2 \rangle_{\text{ch}} &= \frac{ER(ER)^2}{2R^3(ER-1)\sin^2(ER)} \\
&\quad \times \int_0^R \mathrm{d}r\, r^4 \left[\frac{\sin^2(Er)}{E^4 r^4} - 2\frac{\sin(Er)\cos(Er)}{E^3 r^3} + \frac{1}{E^2 r^2} \right] \\
&= \frac{1}{2E(ER-1)\sin^2(ER)} \\
&\quad \times \left[-r\sin^2(Er) + r - \frac{1}{E}\sin(Er)\cos(Er) + \frac{1}{3} r^3 E^2 \right]_0^R .
\end{aligned} \tag{12}
$$

The boundary condition

$$
\begin{aligned}
& j_0(ER) = j_1(ER) \\
\Rightarrow \quad & ER\cos(ER) = (1-ER)\sin(ER)
\end{aligned} \tag{13}
$$

simplifies (12) to give

$$
\begin{aligned}
\langle r^2 \rangle_{\text{ch}} &= \frac{1}{2E(ER-1)\sin^2(ER)} \\
&\quad \times \left[\frac{1}{E}(1-ER)\sin(ER)\,\cos(ER) - \frac{1}{E}\sin(ER)\cos(ER) + \frac{1}{3}R^3 E^2 \right] \\
&= \frac{-ER\sin(ER)\cos(ER) + \frac{1}{3}R^3 E^3}{2E^2(ER-1)\sin^2(ER)} \\
&= 0.53\, R^2 .
\end{aligned} \tag{14}
$$

In the last step we have inserted the numerical value for ER, which, according to (3.131) is equal to 2.0428.

EXERCISE

3.14 The Magnetic Moment of the Proton

Problem. Evaluate the right-hand side of (3.158) with the help of the wave functions (3.153) and (3.128).

Solution. The static magnetic moment is defined as

$$\boldsymbol{m} = \frac{1}{2}\int \boldsymbol{r}\times\boldsymbol{j}(r)\mathrm{d}^3r \ . \tag{1}$$

Using the definition of the electromagnetic current of quark i,

$$j_i^\mu = Q_i\bar{q}_i\gamma^\mu q_i \ , \tag{2}$$

we get for the spatial component

$$\frac{1}{2}\boldsymbol{r}\times\boldsymbol{j}_i = \frac{Q_i}{2}\bar{q}_i(\boldsymbol{r}\times\boldsymbol{\gamma})q_i = \frac{Q_i}{2}q_i^\dagger(\boldsymbol{r}\times\boldsymbol{\alpha})q_i \ , \tag{3}$$

where $\gamma^0\boldsymbol{\gamma} = \boldsymbol{\alpha}$, and therefore the operator of the magnetic moment, is

$$\frac{Q_i}{2}(\boldsymbol{r}\times\boldsymbol{\alpha}) \ . \tag{4}$$

Since the SU(6) wave function of the proton is symmetric under permutations of the indices 1, 2, and 3, we have

$$\begin{aligned}
&\int \Psi_\mathrm{p}^\dagger\left(\frac{\hat{Q}_1}{2}\hat{\boldsymbol{r}}_1\times\hat{\boldsymbol{\alpha}}_1\right)\Psi_\mathrm{p}\,\mathrm{d}^3r_1\,\mathrm{d}^3r_2\,\mathrm{d}^3r_3\\
&\quad=\int \Psi_\mathrm{p}^\dagger\left(\frac{\hat{Q}_2}{2}\hat{\boldsymbol{r}}_2\times\hat{\boldsymbol{\alpha}}_2\right)\Psi_\mathrm{p}\,\mathrm{d}^3r_1\,\mathrm{d}^3r_2\,\mathrm{d}^3r_3\\
&\quad=\int \Psi_\mathrm{p}^\dagger\left(\frac{\hat{Q}_3}{2}\hat{\boldsymbol{r}}_3\times\hat{\boldsymbol{\alpha}}_3\right)\Psi_\mathrm{p}\,\mathrm{d}^3r_1\,\mathrm{d}^3r_2\,\mathrm{d}^3r_3 \ .
\end{aligned} \tag{5}$$

Therefore (3.158) can in a first step be simplified to

$$\begin{aligned}
\mu_\mathrm{p} &= 3\int \mathrm{d}^3r_1\,\mathrm{d}^3r_2\,\mathrm{d}^3r_3\,\Psi_\mathrm{p}^\dagger\left(\frac{\hat{Q}_1}{2}\hat{\boldsymbol{r}}_1\times\hat{\boldsymbol{\alpha}}_1\right)\Psi_\mathrm{p}\\
&= \frac{1}{6}\int \mathrm{d}^3r_1\Bigg[10u^\uparrow(1)^\dagger\left(\frac{\hat{Q}_1}{2}\hat{\boldsymbol{r}}_1\times\hat{\boldsymbol{\alpha}}_1\right)u^\uparrow(1)+2u^\downarrow(1)^\dagger\left(\frac{\hat{Q}_1}{2}\hat{\boldsymbol{r}}_1\times\hat{\boldsymbol{\alpha}}_1\right)u^\downarrow(1)\\
&\quad+4d^\downarrow(1)^\dagger\left(\frac{\hat{Q}_1}{2}\hat{\boldsymbol{r}}_1\times\hat{\boldsymbol{\alpha}}_1\right)d^\downarrow(1)+2d^\uparrow(1)^\dagger\left(\frac{\hat{Q}_1}{2}\hat{\boldsymbol{r}}_1\times\hat{\boldsymbol{\alpha}}_1\right)d^\uparrow(1)\Bigg] \ .
\end{aligned} \tag{6}$$

Exercise 3.14

Here we have drawn on the orthogonality of the quark wave function

$$\int d^3r\, q^{s\dagger} q'^{s'} = \delta_{qq'}\,\delta_{ss'}\ , \quad q = u,\ d\ , \quad s = \uparrow, \downarrow\ . \tag{7}$$

Next we insert the quark charges and make use of the fact that up and down quarks are described by the same spatial wave function. In addition we omit the index 1:

$$\begin{aligned}\boldsymbol{\mu}_p &= \frac{e}{6}\int d^3r \left[\frac{10}{3}u^{\uparrow\dagger}(\hat{\boldsymbol{r}}\times\hat{\boldsymbol{\alpha}})u^{\uparrow} + \frac{2}{3}u^{\downarrow\dagger}(\hat{\boldsymbol{r}}\times\hat{\boldsymbol{\alpha}})u^{\downarrow}\right.\\ &\qquad \left. - \frac{2}{3}d^{\downarrow\dagger}(\hat{\boldsymbol{r}}\times\hat{\boldsymbol{\alpha}})d^{\downarrow} - \frac{1}{3}d^{\uparrow\dagger}(\hat{\boldsymbol{r}}\times\hat{\boldsymbol{\alpha}})d^{\uparrow}\right]\\ &= \frac{e}{2}\int d^3r u^{\uparrow\dagger}(\hat{\boldsymbol{r}}\times\hat{\boldsymbol{\alpha}})u^{\uparrow}\ . \end{aligned} \tag{8}$$

In order to evaluate this expression we insert into it (3.128) in the form

$$u^{\uparrow} = N\begin{pmatrix} j_0(Er)\,\chi^{\frac{1}{2}}_{-1} \\ \mathrm{i}\, j_1(Er)\,\hat{\sigma}_r\,\chi^{\frac{1}{2}}_{-1}\end{pmatrix} \tag{9}$$

and obtain

$$\begin{aligned}\boldsymbol{\mu}_p &= \frac{e}{2}N^2\int_0^R dr\, r^2\int d\Omega\left(j_0(Er)\chi^{\frac{1}{2}\dagger}_{-1},\ -\mathrm{i}\, j_1(Er)\chi^{\frac{1}{2}\dagger}_{-1}\hat{\sigma}_r\right)\\ &\quad\times\begin{pmatrix} 0 & \hat{\boldsymbol{r}}\times\hat{\sigma} \\ \hat{\boldsymbol{r}}\times\hat{\sigma} & 0\end{pmatrix}\begin{pmatrix} j_0(Er)\,\chi^{\frac{1}{2}}_{-1} \\ \mathrm{i}\, j_1(Er)\,\hat{\sigma}_r\,\chi^{\frac{1}{2}}_{-1}\end{pmatrix}\\ &= \frac{e}{2}N^2\int_0^R dr\, r^2\int d\Omega\, \mathrm{i}\, j_0(Er)\, j_1(Er)\chi^{\frac{1}{2}\dagger}_{-1}\left[\left(\hat{\boldsymbol{r}}\times\hat{\sigma}\right)\hat{\sigma}_r - \hat{\sigma}_r\left(\hat{\boldsymbol{r}}\times\hat{\sigma}\right)\right]\chi^{\frac{1}{2}}_{-1}. \end{aligned} \tag{10}$$

By means of the commutation relations of the σ matrices, the term inside the brackets simplifies to

$$\begin{aligned}\left[\varepsilon_{ijk}r_j\hat{\sigma}_k,\ \hat{\sigma}_\ell r_\ell\right] &= \varepsilon_{ijk}\left[\hat{\sigma}_k,\ \hat{\sigma}_\ell\right]r_j r_\ell\\ &= \varepsilon_{ijk}2\mathrm{i}\varepsilon_{k\ell m}\hat{\sigma}_m r_j r_\ell\\ &= 2\mathrm{i}\left(\delta_{\ell i}\delta_{mj} - \delta_{\ell j}\delta_{im}\right)r_j r_\ell\hat{\sigma}_m\\ &= 2\mathrm{i}r_i\hat{\boldsymbol{\sigma}}\cdot\boldsymbol{r} - 2\mathrm{i}r^2\hat{\sigma}_i\ . \end{aligned} \tag{11}$$

Exercise 3.14

Hence

$$\boldsymbol{\mu}_{\rm p} = \frac{e}{2}N^2 \int_0^R \mathrm{d}r\, r^2 \int \mathrm{d}\Omega\, \mathrm{i}\, j_0(Er)\, j_1(Er) \frac{1}{4\pi} (\,1\;\;0\,) \left[2\mathrm{i}\boldsymbol{r}\sigma_r - 2\mathrm{i}r\boldsymbol{\sigma}\right] \begin{pmatrix} 1 \\ 0 \end{pmatrix}$$
$$= \frac{eN^2}{4\pi} \int_0^R \mathrm{d}r\, r^3\, j_0(Er)\, j_1(Er) \int \mathrm{d}\Omega \left[\boldsymbol{e}_3 - \cos\theta \begin{pmatrix} \sin\theta\cos\phi \\ \sin\theta\sin\phi \\ \cos\theta \end{pmatrix} \right] , \tag{12}$$

where we have used

$$(\sigma_j)_{11} = \delta_{j3} \tag{13}$$

and

$$\left(\boldsymbol{\sigma}\cdot\frac{\boldsymbol{r}}{r}\right)_{11} = \frac{r_3}{r} = \cos\theta \; . \tag{14}$$

Now we can easily perform the ϕ, θ, and r integrations:

$$\boldsymbol{\mu}_{\rm p} = \frac{eN^2}{2}\boldsymbol{e}_3 \int_0^R \mathrm{d}r\, r^3\, j_0(Er)\, j_1(Er) \int_{-1}^{1} \mathrm{d}\cos\theta \left(1-\cos^2\theta\right)$$
$$= \boldsymbol{e}_3 \frac{2}{3} N^2 e \int_0^R \mathrm{d}r\, r^3 \frac{\sin(Er)}{Er} \left(\frac{\sin(Er)}{E^2r^2} - \frac{\cos(Er)}{Er} \right)$$
$$= \boldsymbol{e}_3 \frac{2}{3}\frac{N^2}{E^3} e \left(-\frac{r}{2}\sin^2(Er) - \frac{3}{4E}\sin(Er)\cos(Er) + \frac{3}{4}r \right)_0^R$$
$$= \boldsymbol{e}_3 \frac{2}{3}\frac{N^2}{E^4} e \left(-\frac{1}{2}\omega\sin^2\omega - \frac{3}{4}\sin\omega\cos\omega + \frac{3}{4}\omega \right) \; . \tag{15}$$

Here $\omega = ER$. With the boundary condition

$$\omega\cos\omega = (1-\omega)\sin\omega \tag{16}$$

the expression inside the brackets can further be simplified:

$$-\frac{3}{4}\left(\omega\sin^2\omega + \sin\omega\cos\omega - \omega\right) + \frac{1}{4}\omega\sin^2\omega$$
$$= -\frac{3}{4}\left(\sin^2\omega - \omega\cos\omega\sin\omega + \sin\omega\cos\omega - \omega\right) + \frac{1}{4}\omega\sin^2\omega$$
$$= -\frac{3}{4}\left(\sin^2\omega + \omega\cos^2\omega - \omega\right) + \frac{1}{4}\omega\sin^2\omega$$
$$= -\frac{3}{4}(1-\omega)\sin^2\omega + \frac{1}{4}\omega\sin^2\omega$$
$$= \frac{1}{4}(4\omega-3)\sin^2\omega \; . \tag{17}$$

Exercise 3.14

Finally we insert N^2 from (3.142)

$$\begin{aligned}\boldsymbol{\mu}_{\mathrm{p}} &= \boldsymbol{e}_3 \frac{\omega e}{3R^3(\omega-1)} \frac{\omega^2}{\sin^2\omega} \frac{1}{E^4} \frac{1}{4}(4\omega-3)\sin^2\omega \\ &= \boldsymbol{e}_3 \frac{eR}{\omega(\omega-1)} \frac{4\omega-3}{12} = 0.203\, eR\boldsymbol{e}_3 \end{aligned} \tag{18}$$

with $\omega = 2.04$.

Clearly $\boldsymbol{\mu}_{\mathrm{p}}$ has to be parallel to the z direction, because in the spherical bag model the quantization axis of the angular momentum is the only direction of special importance.

Since $\boldsymbol{\mu}_{\mathrm{p}}$ defines a direction, this special direction can be chosen to be identical with the quantization axis mentioned. In units of the nuclear magneton $\mu_0 = e/2m_{\mathrm{p}}$ the absolute value of the magnetic moment of the proton is (with $R = 1$ fm)

$$\begin{aligned}\frac{|\boldsymbol{\mu}_{\mathrm{p}}|}{\mu_0} &= 0.203 \times 1\,\mathrm{fm} \times 2 \times 938.28\,\mathrm{MeV} = 1.9\ , \\ \mu_0 &= \frac{e}{2m_{\mathrm{p}}}\ . \end{aligned} \tag{19}$$

The value (3.159) differs considerably from the experimental observation

$$\left|\boldsymbol{\mu}_{\mathrm{p}}\right| (\mathrm{exp}) = 2.79\, \mu_0\ . \tag{3.160}$$

The magnetic moments predicted for the other hadrons are too small as well. The ratios of these moments, however, are quite well described.

Table 3.3 compares the MIT model results for the charge radii and the magnetic moments with the corresponding experimental data.

There are other interesting quantities that can be evaluated in the context of the MIT bag model. One can usually achieve an agreement similar to that for the masses, charge radii, and magnetic moments. Later we shall discuss at a much higher technical level the predictions made by the bag model for the proton structure functions. Let us now summarize the advantages and disadvantages of the MIT bag model.

Advantages: The MIT bag model provides a phenomenological description of hadrons that is quantitatively semicorrect and therefore allows for rough estimates for all quantities not yet experimentally observed. One can furthermore estimate what phenomenological consequences follow from additional quark interactions. For example, the bag model allows us to connect the known weak interaction of hadrons to the weak interaction of quarks.

Table 3.3. Comparison of charge radii and magnetic moments with the corresponding results of the MIT bag model

Particles	$\langle r^2\rangle_{ch}$ (exp.)	$\langle r^2\rangle_{ch}$ (theor.)
p	0.88 ± 0.03 fm	0.73 fm
n	-0.12 ± 0.01 fm	0
π	0.78 ± 0.10 fm	0.5 fm

Particles	$\frac{\mu}{\mu_0}$ (exp.)	$\frac{\mu}{\mu_0}$ (theor.)	$\frac{\mu}{\mu_p}$ (exp.)	$\frac{\mu}{\mu_p}$ (theor.)
p	2.793	1.90	–	–
n	−1.913	−1.27	−0.68	2/3
Λ	0.613 ± 0.004	−0.48	-0.219 ± 0.001	−0.26
Σ^+	2.38 ± 0.02	1.84	0.85 ± 0.01	0.97
Σ^0	–	0.59	–	0.31
Σ^-	-1.14 ± 0.05	−0.68	-0.36 ± 0.02	−0.36
Ξ^0	-1.25 ± 0.014	−1.06	-0.448 ± 0.005	−0.56
Ξ^-	-0.69 ± 0.04	−0.44	-0.25 ± 0.01	−0.23

Disadvantages: The bag model is conceptually unsatisfying. The physical meaning of the specific correction terms is unclear and the model is neither renormalizable nor Lorentz invariant. In addition one obtains similarly good results with completely different approaches (see Sect. 3.1). The basic physical assumptions of these models, i.e., for the MIT bag the rigid boundary condition and the free motion of the quarks inside the bag, differ drastically. Therefore at least some of them should be wrong.

4. Gauge Theories and Quantum-Chromodynamics

Nowadays the common model of quark–quark interactions is an SU(3) gauge theory in a degree of freedom, arbitrarily called “color”. There are meanwhile many experimental facts supporting this model, as we shall discuss later in more detail. It seems by now also to be proven that QCD (quantum chromodynamics) is able to correctly describe the most pronounced feature of quark–quark interactions, i.e., confinement, and that it will generate, for example, the correct hadron masses. The results obtained so far are all compatible with the phenomenological properties, but in many cases the accuracy of calculations is still rather low. Typical uncertainties for hadron masses, for example, are 10%. In fact, the general acceptance of QCD is based not only on its own achievements, but also to a large extent on the outstanding success of the gauge theory of weak and electromagnetic interactions. For this reason we begin with a brief overview of the Glashow–Salam–Weinberg model. (The Glashow–Salam–Weinberg model combined with QCD is usually called the *standard model*). The typical features of gauge theories are discussed in the context of this specific model.

4.1 The Standard Model: A Typical Gauge Theory

The ideas of local internal symmetries and gauge transformations were introduced quite a long time ago. But the actual reason for developing the modern schemes for general non-Abelian gauge theories was the lack of a renormalizable field theory for massive spin-1 particles.

Spin-0 and spin-$\frac{1}{2}$ particles are described by the well-known method of field quantization, employing the Klein–Gordon and the Dirac Lagrangians, respectively. This procedure also applies to massless spin-1 particles, despite some complications owing to the four-potentials A_μ not being physical observables. The theory obtained in this way proved to be excellent, in particular the part known as quantum electrodynamics. However, a consistent field theory for massive vector particles, especially a field theory of weak interactions, could not be formulated. This problem even led some physicists to doubt the whole concept of local field theories. On the other hand, having overcome that difficulty, quantum field theory is today considered the correct theory of elementary particles (the validity of a possible alternative, the so-called “string models”, is still heavily disputed).

We shall not analyze the properties of gauge theories in this volume. A deeper understanding requires a lot of formal knowledge, which we shall present in another book. At this point we restrict ourselves to describing the general structure of gauge theories and we simply state that all theories built on these principles are renormalizable.

The standard model contains the well-known electromagnetic interaction. Here the photon is described by the free Lagrangian

$$L_0 = -\frac{1}{4} F_{\mu\nu} F^{\mu\nu} \ , \qquad F_{\mu\nu} = \partial_\mu A_\nu - \partial_\nu A_\mu \ , \tag{4.1}$$

and its coupling to, for example, the electron via the interaction term

$$L_{\text{int}} = \overline{\Psi}(\hat{p}_\mu - eA_\mu)\gamma^\mu \Psi \ . \tag{4.2}$$

These equations already include the most basic assumption of quantum field theory: elementary particles have to be described by local fields and point interactions. As a matter of fact, only the four-potentials A_μ and not the "physical" fields $\boldsymbol{E}$ and $\boldsymbol{B}$ can fulfill these requirements. If $\boldsymbol{E}$ and $\boldsymbol{B}$ are regarded as elementary, then the electromagnetic interaction has to be nonlocal. The best proof of this is the Aharonov–Bohm effect.[1] The phase of a charged particle's wave function is influenced by magnetic fields located in an area where the wave function itself is zero.

As is well known, Lagrangians (4.1) and (4.2) are invariant under the transformation

$$\begin{aligned} A_\nu(x) &\to A'_\nu(x) = A_\nu(x) + \partial_\nu \theta(x) \ , \\ \Psi(x) &\to \Psi'(x) = \mathrm{e}^{\mathrm{i}e\theta(x)}\Psi(x) \ . \end{aligned} \tag{4.3}$$

Equation (4.3) expresses the fact that not all of the four fields A_μ correspond to physical degrees of freedom. Physical quantities must not depend on the arbitrary "gauge angle" $\theta(x)$. This requirement leads to relations between the propagator and the vertex function, for example, which is important to prove the renormalizability of QED (the so-called Ward identity).[2]

From the gauge symmetry of QED we are led directly to that of a general gauge group by replacing the complex-valued functions A_ν by matrix functions. If $\hat{\lambda}^j$, $j = 1, 2, \dots, N$ is a basis in the chosen matrix space, we correspondingly have

$$\begin{aligned} A_\nu(x) &\to \sum_j \frac{\hat{\lambda}^j}{2} A^j_\nu(x) =: \hat{A}_\nu(x) \ , \\ \theta(x) &\to \sum_j \frac{\hat{\lambda}^j}{2} \theta^j(x) =: \hat{\theta}(x) \ . \end{aligned} \tag{4.4}$$

---

[1] See W. Greiner: *Quantum Mechanics – Special Chapters* (Springer, Berlin, Heidelberg 1998).

[2] See W. Greiner and J. Reinhardt: *Quantum Electrodynamics*, 2nd. ed. (Springer, Berlin, Heidelberg 1994).

The specific choice of matrices determines the underlying symmetry group. If one chooses, for example, traceless Hermitian 3×3 matrices, then the transformations

$$\exp\left(-\mathrm{i}g\sum_j \frac{\hat{\lambda}^j}{2}\theta^j(x)\right) \tag{4.5}$$

are nothing other than a three-dimensional representation of the SU(3) transformations, where we replaced the electron coupling constant $-e$ by a generalised constant g. On the other hand, it is clear that any other representation of SU(3) instead of the chosen matrices would also have been possible. Since the $\hat{\lambda}$ matrices in general do not commute, the equations have to be slightly changed in the case of (4.4) (in addition, in (4.1)–(4.3) we replace e by $-g$):

$$\begin{aligned} L_0 &= -\frac{1}{4}\hat{F}^a_{\mu\nu}\hat{F}^{a\,\mu\nu} = -\frac{1}{2}\,\mathrm{tr}\left(\hat{F}_{\mu\nu}\hat{F}^{\mu\nu}\right) \ , \\ \hat{F}_{\mu\nu} &= \partial_\mu \hat{A}_\nu - \partial_\nu \hat{A}_\mu - \mathrm{i}g\left[\hat{A}_\mu, \hat{A}_\nu\right] \ , \end{aligned} \tag{4.6}$$

$$L_{\mathrm{int}} = \overline{\Psi}\left(\hat{p}_\mu + g\hat{A}_\mu\right)\gamma^\mu\Psi \ , \tag{4.7}$$

$$\begin{aligned} \hat{A}_\nu(x) &\to \hat{A}'_\nu(x) = \mathrm{e}^{-\mathrm{i}g\hat{\theta}(x)}\left(\hat{A}_\nu(x) + \frac{\mathrm{i}}{g}\partial_\nu\right)\mathrm{e}^{+\mathrm{i}g\hat{\theta}(x)} \ , \\ \Psi(x) &\to \Psi'(x) = \mathrm{e}^{-\mathrm{i}g\hat{\theta}(x)}\Psi(x) \ . \end{aligned} \tag{4.8}$$

Now $\Psi(x)$ must be defined as a vector corresponding to the choice of $\hat{A}_\nu(x)$. (In principle Ψ could also be a matrix; then the trace of L_{int} would have to be evaluated. Such cases occur in models trying to unify the electroweak and strong interaction with a single symmetry group.) The main content of gauge field theories is contained in the quantized equations (4.6) and (4.7).

With these few remarks we have already formulated the basic ideas of gauge theories. But, like many others, this particular physical concept can be further investigated in two different directions. The first is to study the consequences of (4.6) and (4.7) in more and more detail. The second deals with the basics of these equations and tries to discover deeper-lying foundations. Persuing the latters we could, for example, give a general geometric formulation of the principles of gauge theories. Being geometric, this formulation shows the same structure as the basic equations in the theory of general relativity. Such a correspondence gives rise to the hope that the unification of the theory of gravitation and quantum theory, which deals with all other kinds of interaction, is no longer completely unlikely. However, it is still not clear whether this correspondence expresses a real similarity of both theories or whether it just follows from the generality of geometric considerations. Some elements of the geometric formulation of gauge theories are discussed in Example 4.1. Here we will discuss only the phenomenological consequences of (4.6) and (4.7).

EXAMPLE

4.1 The Geometric Formulation of Gauge Symmetries

For the sake of simplicity we consider a specific example, namely a set of spinor fields $\Psi_i(x)$, $i = 1, 2, \ldots, N$, combined as a vector

$$\Psi = \begin{pmatrix} \Psi_1(x) \\ \Psi_2(x) \\ \vdots \\ \Psi_N(x) \end{pmatrix} . \tag{1}$$

Furthermore we assume a local symmetry such that

$$\Psi'(x) = \exp\left(+\mathrm{i}g\hat{\theta}(x)\right)\Psi(x) \tag{2}$$

is equivalent to $\Psi(x)$. This situation is analogous to an arbitrary, e.g., contravariant, vector and its Lorentz transform:

$$V^\mu(x) \ , \quad V'^\mu(x) = \Lambda^\mu{}_\nu(x)V^\nu(x) \ . \tag{3}$$

The Lorentz transformation $\Lambda(x)$ corresponds in this case to the gauge transformation $\exp\left(-\mathrm{i}g\hat{\theta}(x)\right)$. A well-known technical problem in the theory of relativity is the definition of a vector (in general a tensor) derivative. The evaluation of a quotient of differences

$$\frac{V^\mu(x+h) - V^\mu(x)}{h} \tag{4}$$

yields additional terms owing to the dependence of the metric tensor $g_{\mu\nu}(x)$ on the position vector x.

This problem leads to the definition of the *covariant derivative*

$$V^\mu_\nu := \partial_\nu V^\mu + \Gamma^\mu{}_{\alpha\nu}V^\alpha \tag{5}$$

with so-called *Christoffel symbols* $\Gamma^\mu{}_{\alpha\nu}$,

$$\Gamma^\mu{}_{\alpha\nu} = -\frac{1}{2}g^{\mu\sigma}\left(\partial_\nu g_{\alpha\sigma} + \partial_\alpha g_{\sigma\nu} - \partial_\sigma g_{\nu\alpha}\right) \ , \tag{6}$$

representing the position dependence of the metric, that is, the local coordinate systems. The covariant derivative (5) constructed in this way is invariant under Lorentz transformations.

In a completely analogous manner we can write (5) for the spinor fields Ψ:

$$\partial_\mu\Psi(x) \to \left(\partial_\mu + \hat{\Gamma}_\mu(x)\right)\Psi(x) =: D_\mu\Psi(x) \ , \tag{7}$$

where $\hat{\Gamma}_\mu(x)$ is a position-dependent matrix with respect to the vectors Ψ. $\hat{\Gamma}_\mu(x)$ is determined by the requirement that the covariant derivative $D_\mu\Psi(x)$ has to

Example 4.1

be invariant under gauge transformations $\exp\left(-\mathrm{i}g\hat{\theta}\right)$ that correspond to the Lorentz transformations. Writing $-\mathrm{i}g\hat{A}_\mu$ instead of $\hat{\Gamma}_\mu$ we have

$$\mathrm{i}D_\mu\Psi(x) = \left(\hat{p}_\mu + g\hat{A}_\mu\right)\Psi(x) \ , \tag{8}$$

$$\Psi(x) \to \Psi'(x) = \mathrm{e}^{\mathrm{i}g\hat{\theta}(x)}\Psi(x) \ , \tag{9}$$

$$\hat{A}_\mu \to \hat{A}'_\mu = \mathrm{e}^{\mathrm{i}g\hat{\theta}(x)}\left(\hat{A}_\mu(x) - \frac{\mathrm{i}}{g}\partial_\mu\right)\mathrm{e}^{-\mathrm{i}g\hat{\theta}(x)} \ . \tag{10}$$

Interpreted geometrically, (5) is an effect of parallel transport. To evaluate (4), the vector V^μ has to be transported from the position $x+h$ to the point x. But in a position-dependent coordinate system this transport changes the vector's coordinate representation (think, for example, of the coordinates r, θ, and φ of a fixed vector which moves on the surface of a sphere). $\Gamma^\mu_{\alpha\nu}V^\alpha$ describes this change, which has to be added to $\partial_\nu V^\mu$ in order to reveal the physical, coordinate-independent change of the vector V^μ. Correspondingly, $\hat{A}'_\mu - \hat{A}_\mu$ represents the change of Ψ due to the position dependence of the gauge (see Fig. 4.1). Since the covariant derivatives are gauge independent and invariant under Lorentz transformations, this is also valid for their commutator:

$$\begin{aligned}\left[\hat{D}_\mu, \hat{D}_\nu\right] &= \left[(\partial_\mu - \mathrm{i}g\hat{A}_\mu), (\partial_\nu - \mathrm{i}g\hat{A}_\nu)\right] \\ &= -\mathrm{i}g\partial_\mu\hat{A}_\nu + \mathrm{i}g\partial_\nu\hat{A}_\mu - g^2\left[\hat{A}_\mu, \hat{A}_\nu\right] \\ &= -\mathrm{i}g\left(\partial_\mu\hat{A}_\nu - \partial_\nu\hat{A}_\mu - \mathrm{i}g\left[\hat{A}_\mu, \hat{A}_\nu\right]\right) \\ &= -\mathrm{i}g\hat{F}_{\mu\nu} \ . \end{aligned} \tag{11}$$

The $\hat{F}_{\mu\nu}$ represent the physical part of the gauge fields, i.e., the part that cannot be changed by simply choosing another gauge. Correspondingly, in the theory of relativity we obtain the curvature tensor $R^\mu{}_{\nu\sigma\varrho}$, describing the physical part of space curvature, i.e., the part that is not due only to the chosen coordinate system. Also $R^\mu{}_{\nu\sigma\varrho}$ is simply the commutator of the covariant derivatives (5), with additional terms occurring for the covariant indices. For example, for a tensor it holds that

$$T^\beta{}_{\alpha;\tau} = T^\beta{}_{\alpha,\tau} + \Gamma^\beta_{\mu\tau}T^\mu{}_\alpha - \Gamma^\mu_{\alpha\tau}T^\beta{}_\mu \ . \tag{12}$$

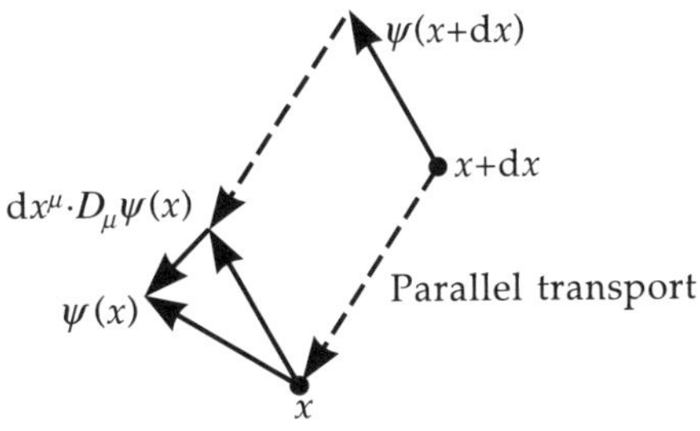

Fig. 4.1. The interpretation of $\hat{A}$ as a description of parallel transport

Example 4.1

Because of the outlined similarity one could also focus on gauge symmetries and interpret the general theory of relativity as a special gauge theory. This kind of consideration would shift the whole discussion to the formal definition of a gauge group. Terms like parallel transport or curvature would not play any role. Covariant derivatives and field tensors $\hat{F}_{\mu\nu}$ would be defined by their invariance properties. Only at the very end of such a treatment would one try to connect the obtained structures with physics. Consequently one is free to choose between understanding general relativity as a gauge theory and interpreting gauge theories geometrically.

To formulate the standard model we need an additional concept, namely the idea of spontaneous symmetry breaking. Again the basic idea can be explained quite easily. If scalar fields are considered, interaction terms of the form ϕ^4 and ϕ^2 can be renormalized. Therefore the Lagrangian

$$L = -\Phi^+ \Box \Phi - \frac{\lambda}{4}\left(\Phi^+\Phi - v^2\right)^2 \tag{4.9}$$

sets up a well-defined field theory. (Such scalar quantum field theories are in fact somewhat problematic. It is possible that consistent renormalization requires the coupling constant to be zero. This possibility is called "triviality" and implies that (4.9) makes sense only if the scalar Higgs particle, for example, has some internal structure on the 1–10 TeV scale. For its phenomenological properties at the energies we consider this is irrelevant.) Within the meaning of the Lagrange formalism the second term has to be interpreted as a contribution to the potential energy:

$$V = \frac{\lambda}{4}\left(\Phi^+\Phi - v^2\right)^2 \ . \tag{4.10}$$

Clearly the energy is minimal for $|\phi| = v = \text{const}$. Hence this is the physical ground state. On the other hand, a problem occurs if the theory is quantized in the standard way by means of commutator relations between field operators. Since $\langle 0|\hat{\phi}|0\rangle \equiv 0$, the state with $v = |\phi| = \langle \text{vac}|\hat{\phi}|\text{vac}\rangle$ cannot in principle be constructed with one-particle excitations on the usual ground state. Therefore the physical vacuum state $|\text{vac}\rangle$ is not identical with the field theoretical vacuum state, $|\text{vac}\rangle \neq |0\rangle$. The physical excitations are only obtained if fluctuations around $\phi = v$ are considered and if $\hat{\phi}(x)$ is quantized in the usual manner:

$$\Phi(x) \to v + \hat{\Phi}(x) \ , \quad \langle \text{vac}|\hat{\Phi}|\text{vac}\rangle = 0 \ . \tag{4.11}$$

There is no deeper physical reason for the fact that the excitations around $\phi = 0$ and $\phi = v$ cannot be expressed by each other. It only means that the canonical quantization scheme is not complete. In fact it is possible to choose another quantization method without this disadvantage, e.g. quantization with functional integrals. We cite here without proof the following essential statement:

If a (gauge) theory is renormalizable for $\Phi(x) = v + \lambda(x)$, i.e., for quantization around $\Phi(x) = v$, then it is also renormalizable for quantization around $\Phi(x) = 0$ and vice versa.

This makes the following trick possible. All particle masses M_i are replaced by

$$M_i \to \frac{\Phi(x)}{v} M_i \ . \tag{4.12}$$

For physical states one has to set $\phi(x) = v + \chi(x)$. This yields just M_i plus an additional interaction term, whose meaning will be explained later. So long as this term does not cause significant physical effects, the original theory has practically not been changed. In this case the theory is then exactly renormalizable if it is also renormalizable for a quantization around $\phi(x) = 0$, i.e., for vanishing mass terms. As it is possible to prove that certain classes of massless gauge theories are renormalizable, this is also true for the resulting theories with masses generated by (4.12), i.e., by spontaneous symmetry breaking. The crucial point is that spontaneous symmetry breaking preserves gauge invariance, which is essential for the proof of renormalizability.

In general, $\phi(x)$ is not a scalar with respect to the current gauge symmetry (i.e. not a singlet) but rather a multiplet. The vacuum expectation value v then defines a particular direction within this multiplet. Because of this, one speaks of *spontaneous symmetry breaking*. This expression is quite misleading since the symmetry is really preserved. The choice of v should instead be compared with the choice of a specific direction of quantization in a spherically symmetric problem. Now we want to discuss spontaneous symmetry breaking for the special case of the standard model.

The standard model describes the electromagnetic and weak interactions. The latter was first understood as a four-fermion interaction between, for example, a neutron, proton, electron, and antineutrino. But it has been clear for a long time that this can be an effective description only for a microscopic process, e.g., the exchange of charged so-called W bosons (see Fig. 4.2). Since the W fields are vectors and mediate an interaction (just like the photon), it is natural to regard them as gauge fields of a gauge group. Choosing possible candidates for this gauge group does not cause any difficulties, if the following points are considered:

(1) The leptons exclusively couple to the corresponding neutrinos, for example the electron only to the electron neutrino. Therefore we have to choose

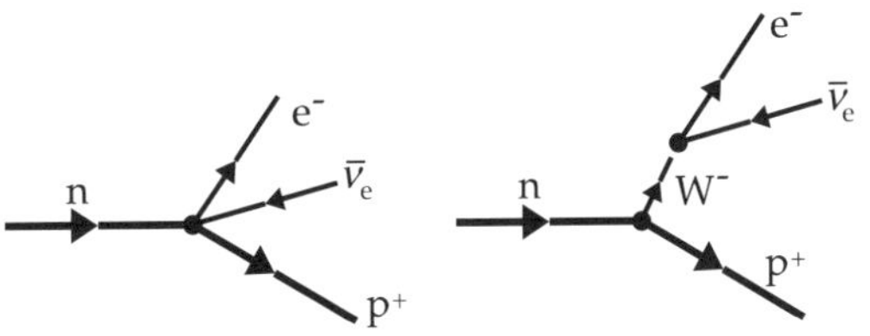

Fig. 4.2. The Fermi interaction and its interpretation as the exchange of charged W bosons

a group with an irreducible two-dimensional representation if we want to avoid additional artificial assumptions.

(2) Except for W^+ and W^- only a minimum number of additional gauge fields should be postulated. Hence the number of generators should be slightly bigger than two.

These requirements lead directly to the group SU(2). But the SU(2) contains three generators. One could therefore try to identify the additional third gauge field with the photon. However, it is not possible to carry out this idea consistently. The construction principles which have been outlined so far and which will be applied in the following example demand a new neutral particle, the Z^0, to be postulated. This involves the existence of a new interaction: so-called neutral currents. The experimental verification of this interaction and of the W and Z particles is one of the greatest triumphs of particle physics.

Hence SU(2) × U(1) is the gauge group of the GSW model. The most frequently used representation of the matrices λ_i is

$$\lambda_1 = \begin{pmatrix} 0 & 1 \\ 1 & 0 \end{pmatrix} , \quad \lambda_2 = \begin{pmatrix} 0 & -\mathrm{i} \\ \mathrm{i} & 0 \end{pmatrix} , \quad \lambda_3 = \begin{pmatrix} 1 & 0 \\ 0 & -1 \end{pmatrix} , \quad \lambda_4 = \begin{pmatrix} 1 & 0 \\ 0 & 1 \end{pmatrix} , \tag{4.13}$$

i.e., nothing more than the Pauli matrices for SU(2) and the unit matrix for U(1) (remember that the generators act on the doublet states). If we denote the corresponding gauge fields by W_1^μ, W_2^μ, W_3^μ, and B^μ, then $\hat{F}_{\mu\nu}$ from (4.6) is $(i, j, k = 1, 2, 3)$

$$\hat{F}_{\mu\nu} = \partial_\mu \left(\frac{\lambda^j}{2} W_\nu^j + \frac{\mathbb{1}}{2} \cdot B_\nu \right) - \partial_\nu \left(\frac{\lambda^j}{2} W_\mu^j + \frac{\mathbb{1}}{2} \cdot B_\mu \right) + g \varepsilon_{ijk} \lambda^k W_\mu^i W_\nu^j , \tag{4.14}$$

since $\mathbb{1}$ commutes with all matrices. Because

$$\mathrm{tr}\{\lambda_i \lambda_j\} = 2\delta_{ij} , \quad \mathrm{tr}\{\lambda_i\} = 0 , \tag{4.15}$$

L_0 is divided into two contributions, where we replace W_μ^i and B_μ by $\frac{1}{2} W_\mu^j$ and $\frac{1}{2} B_\mu$, respectively, in order to obtain the usual factor (usually $t_i = \frac{1}{2}\lambda_i$ rather than λ_i are chosen for the matrices):

$$L_0 = -\frac{1}{4} W_{\mu\nu}^j W^{j\mu\nu} - \frac{1}{4} B_{\mu\nu} B^{\mu\nu} , \tag{4.16}$$

$$W_{\mu\nu}^j = \partial_\mu W_\nu^j - \partial_\nu W_\mu^j - g \varepsilon_{ijk} W_\mu^i W_\nu^k , \tag{4.17}$$

$$B_{\mu\nu} = \partial_\mu B_\nu - \partial_\nu B_\mu . \tag{4.18}$$

Also (4.7) can be written down directly for the gauge group SU(2) × U(1). The only real novelty is the appearance of two coupling constants g and g' for the two gauge groups SU(2) and U(1):

$$L_{\mathrm{int}} = \overline{\Psi} \left(p_\mu + g\, W_\mu^j \frac{\lambda^j}{2} + g' B_\mu \right) \gamma^\mu \Psi . \tag{4.19}$$

Clearly the total Lagrangian $L_0 + L_{\mathrm{int}}$ is then invariant under the transformations

$$\frac{\lambda^j}{2} W^j_\mu \to \exp\left(\mathrm{i}g\theta^j \frac{\lambda^j}{2}\right)\left(\frac{\lambda^j}{2} W^j_\mu + \frac{\mathrm{i}}{g}\partial_\mu\right)\exp\left(-\mathrm{i}g\theta^j \frac{\lambda^j}{2}\right) , \tag{4.20}$$

$$\begin{aligned}\frac{\mathbb{1}}{2} B_\mu &\to \exp\left(\mathrm{i}g' \frac{\mathbb{1}}{2}\theta'(x)\right)\left(\frac{\mathbb{1}}{2} B_\mu + \frac{\mathrm{i}}{g'}\partial_\mu\right)\exp\left(-\mathrm{i}g' \frac{\mathbb{1}}{2}\theta'(x)\right)\\ &= B_\mu + \partial_\mu \theta'(x) ,\end{aligned} \tag{4.21}$$

$$\Psi(x) \to \exp\left(\mathrm{i}g\theta^j(x)\frac{\lambda^j}{2}\right)\exp\left(\mathrm{i}g'\frac{\mathbb{1}}{2}\theta'(x)\right)\Psi(x) . \tag{4.22}$$

Since the U(1) transformations commute, (4.21) has a very simple form (just as in electrodynamics). Because of this simple feature one can allow different fermions to transform differently under U(1). If we replace (4.19) and (4.22) by

$$L_{\mathrm{int}} = \sum_j \overline{\Psi}_{\mathrm{f}} \left(p_\mu + g W^j_\mu \frac{\lambda^j}{2} + g' \frac{y_f}{2} B_\mu\right)\gamma^\mu \Psi_{\mathrm{f}} , \tag{4.23}$$

$$\Psi(x) \to \exp\left(\mathrm{i}g\theta^j(x)\frac{\lambda^j}{2}\right)\exp\left(\mathrm{i}g'\frac{y_f}{2}\theta'(x)\right)\Psi_{\mathrm{f}} \tag{4.24}$$

with arbitrary numbers y_f, then the $SU(2) \times U(1)$ symmetry still remains valid.

As mentioned earlier, the spinor doublet is identified with, for example, the electron and the electron–neutrino field; usually ν_{e} is written as the upper component:

$$\Psi_{\mathrm{f}} = \begin{pmatrix}\nu_{\mathrm{e}}\\ \mathrm{e}\end{pmatrix}_{\mathrm{L}} . \tag{4.25}$$

Here we also took the experimenal observation into account that only left-handed leptons and quarks interact weakly:

$$e_{\mathrm{L}} = \frac{1-\gamma_5}{2} e . \tag{4.26}$$

Equation (4.25) then yields the structure of the remaining fermion doublets:

$$\Psi_{\mathrm{f}} : \begin{pmatrix}\nu_\mu\\ \mu\end{pmatrix}_{\mathrm{L}}, \begin{pmatrix}\nu_\tau\\ \tau\end{pmatrix}_{\mathrm{L}}, \begin{pmatrix}\mathrm{u}\\ \mathrm{d}'\end{pmatrix}_{\mathrm{L}}, \begin{pmatrix}\mathrm{c}\\ \mathrm{s}'\end{pmatrix}_{\mathrm{L}}, \begin{pmatrix}\mathrm{t}\\ \mathrm{b}'\end{pmatrix}_{\mathrm{L}}, \dots . \tag{4.27}$$

Here the quark fields d′, s′, and b′ are orthogonal superpositions of the mass eigenstates d, s, and b. The fact that the mass eigenstates differ from the eigenstates with respect to the weak interaction is one of the most fascinating features of the weak interaction. It is related to exotic effects such as kaon oscillations and could be responsible for CP violation.[3]

[3] See W. Greiner and B. Müller: *Gauge Theory of Weak Interactions*, 3rd. ed. (Springer, Berlin, Heidelberg 2000).

Lastly we still have to define the Higgs sector. For Higgs particles we choose also the lowest SU(2) representation, namely the doublet:

$$\phi = \begin{pmatrix} \phi_1 + \mathrm{i}\phi_2 \\ \phi_3 + \mathrm{i}\phi_4 \end{pmatrix} , \quad \phi_1, \phi_2, \phi_3, \text{ and } \phi_4 \text{ are real.} \tag{4.28}$$

To maintain the gauge invariance of the theory, for the ϕ field also all derivatives have to be replaced by covariant derivatives (see Example 4.1). This yields the Higgs Lagrangian

$$L_{\mathrm{H}} = \left| \left(\partial_\mu - \mathrm{i}g \frac{\lambda^j}{2} W_\mu^j - \mathrm{i}g' \frac{y_{\mathrm{H}}}{2} B_\mu \right) \phi \right|^2 - \frac{\lambda}{4} \left(\phi^+ \phi - v^2 \right)^2 . \tag{4.29}$$

If now ϕ in L_{H} is replaced by means of (4.11) by

$$\phi \to \begin{pmatrix} 0 \\ v \end{pmatrix} + \chi , \tag{4.30}$$

for example, then (4.29) generates the gauge-field mass terms

$$\begin{aligned} L_{\mathrm{H}} &\to \frac{1}{4} \left| \begin{pmatrix} g(W_\mu^1 - \mathrm{i}W_\mu^2) \\ g' y_{\mathrm{H}} B_\mu - g W_\mu^3 \end{pmatrix} \right|^2 v^2 \\ &= \frac{g^2 v^2}{4} (W_1^2 + W_2^2) + \frac{v^2}{4} \left(g' y_{\mathrm{H}} B_\mu - g W_\mu^3 \right)^2 . \end{aligned} \tag{4.31}$$

Now we expand B_μ and W_μ^3 into two new fields A_μ and Z_μ,

$$\begin{aligned} B_\mu &= \frac{1}{\sqrt{g^2 + g'^2 y_{\mathrm{H}}^2}} (g' y_{\mathrm{H}} Z_\mu - g A_\mu) , \\ W_\mu^3 &= \frac{1}{\sqrt{g^2 + g'^2 y_{\mathrm{H}}^2}} (-g Z_\mu - g' y_{\mathrm{H}} A_\mu) , \end{aligned} \tag{4.32}$$

and obtain

$$L_{\mathrm{H}} \to \frac{g^2 v^2}{4} (W_1^2 + W_2^2) + \frac{v^2}{4} (g^2 + g'^2 y_{\mathrm{H}}^2) Z^2 . \tag{4.33}$$

The A_μ field remains massless and is consequently identified with the photon. One can also absorb y_{H} in the definition of g' and with these identifications all other y values follow. y, the so-called "*weak hypercharge*", has to be chosen in such a way that all particles get their correct electric charges, utilizing the well-known Gell-Mann–Nishijima formula

$$y = 2(Q - t_3) \quad \text{or} \quad Q = \frac{y}{2} + t_3 , \tag{4.34}$$

where Q and t_3 denote the electric charge and the weak isospin of each particle, respectively. Finally it is common practice also to generate the fermion masses by spontaneous symmetry breaking. This can be achieved by replacing all mass terms by corresponding couplings to the Higgs field:

$$m_e\,\bar{e}e \to \frac{m_e}{v}\begin{pmatrix}\nu_e\\ e\end{pmatrix}_L^+ \gamma_0\,\phi\, e_R + \bar{e}_R\,\phi^+\begin{pmatrix}\nu_e\\ e\end{pmatrix}_L \,. \tag{4.35}$$

With these remarks we conclude our brief sketch of the standard model, which has been experimentally verified to a remarkable degree. For an extensive discussion we refer to *Relativistic Quantum Mechanics*[4]. At this point we turn to the question of what an analogous theory for quark–quark interactions has to look like.

4.2 The Gauge Theory of Quark–Quark Interactions

In order to construct a gauge theory of quark–quark interactions, one has first of all to determine the number of "charge states", i.e., the number of different kinds of quarks with respect to the new interaction. Following our discussion in Sect. 1.1 the minimum number of quarks needed is three. Now we want to give several arguments indicating that there are exactly three quark states for every quark flavor, i.e., three colors ($N_c = 3$).

The results of high-energy e^+e^- reactions provide the first argument. Here hadrons are created by pair annihilation followed by the creation of quark pairs. Suppose that the interaction in the final channel does not influence the cross section (this assumption is confirmed by deep inelastic scattering experiments (see Chap. 3) and follows from asymptotic freedom). Then the cross section is proportional to the square of the amplitude.

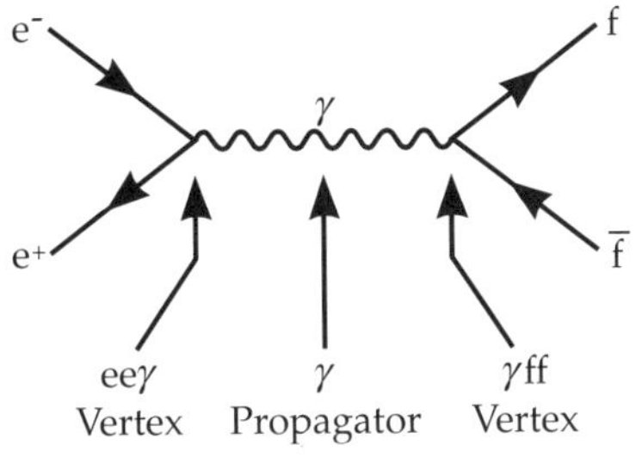

Fig. 4.3. The elementary interaction graph for high energy electron–positron annihilation to fermion plus antifermion: $e^+ + e^- \to f + \bar{f}$

For a given choice of $f\,\bar{f}$, the amplitude for the transition is found by multiplying three factors, they are: $-\sqrt{\alpha}$ for the eeγ vertex, $1/W$ for the γ propagator, and $Q_f\sqrt{\alpha}$ for the γff vertex, where α is the fine-structure constant, Q_f is the charge on f in units of the positron charge, and W is the total centre-of-mass energy. Then the total amplitude is proportional to their product $-Q_f\alpha/W$, and the

[4] W. Greiner: *Relativistic Quantum Mechanics – Wave Equations*, 3rd ed. (Springer, Berlin, Heidelberg 2000).

total cross section, σ, is proportional to $Q_f^2\alpha^2/W^2$. The missing factors in the cross section are $4\pi/3(hc)^2 N_c$ (N_c is the color factor), so that

$$\sigma(e^+e^- \to \mu^+\mu^-) = \frac{4\pi}{3}\frac{\alpha^2}{W^2} , \quad Q_f = -1,\ N_c = 1 , \tag{4.36}$$

and

$$\sigma(e^+e^- \to u\bar{u}) = \frac{4\pi}{3}\left(\frac{2}{3}\right)^2 N_c \frac{\alpha^2}{W^2} , \quad Q_f = +2/3,\ N_c = 3 , \tag{4.37}$$

and

$$\sigma(e^+e^- \to \text{all } q\bar{q}) = \frac{4\pi}{3}\left(\sum_q Q_q^2\right) N_c \frac{\alpha^2}{W^2} , \quad N_c = 3 . \tag{4.38}$$

Here we have put again $h = c = 1$, so that the $(hc)^2$ factors appearing in these formulas also equal 1. If ff is a $q\bar{q}$ pair, then it is assumed that there is a probability of 1 for becoming hadrons after the production of the $q\bar{q}$ pair. Then

$$R = \frac{\sigma(e^+e^- \to \text{hadrons})}{\sigma(e^+e^- \to \mu^+\mu^-)} = \left(\sum_q Q_q^2\right) N_c . \tag{4.39}$$

Note that the cross sections are scale invariant and that these formulae apply only if $W \gg 2m_f$.

Obviously the ratio of the cross sections is simply given by the charges of the particles (provided the particle masses are negligible compared with the energy of the e^+e^- pair). In detail one obtains:

$$R = \frac{\sigma(e^+e^- \to \bar{q}q)}{\sigma(e^+e^- \to \mu^+\mu^-)} = \left[\left(\frac{2}{3}\right)^2 + \left(\frac{-1}{3}\right)^2 + \left(\frac{-1}{3}\right)^2\right] N_c = \frac{2}{3} N_c$$

$$\text{for} \quad E_{cm} > 2\,\text{GeV} \gg 2m_u ,\quad 2m_d ,\quad 2m_s ,$$

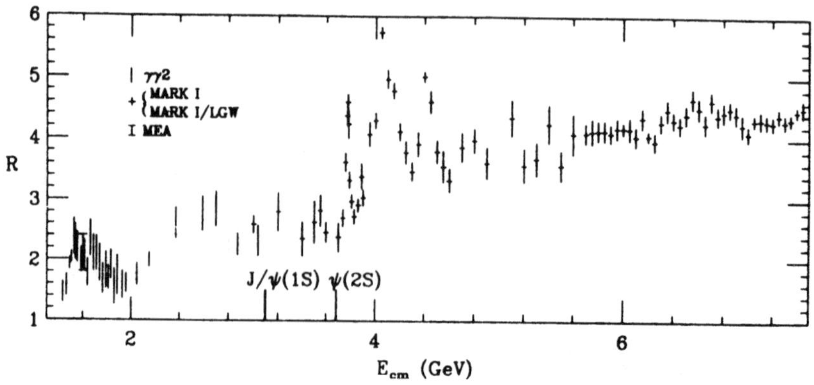

Fig. 4.4a. The ratio of the cross sections for hadronic and muonic reactions in e^+e^- pair annihilation, $R = \sigma(e^+e^- \to \text{hadrons})/\sigma(e^+e^- \to \mu^+\mu^-)$. (From *Review of Particle Properties:* Phys. Rev. D **45** (1992)). A comparison with theory including radiative corrections is presented in Fig. 7.15

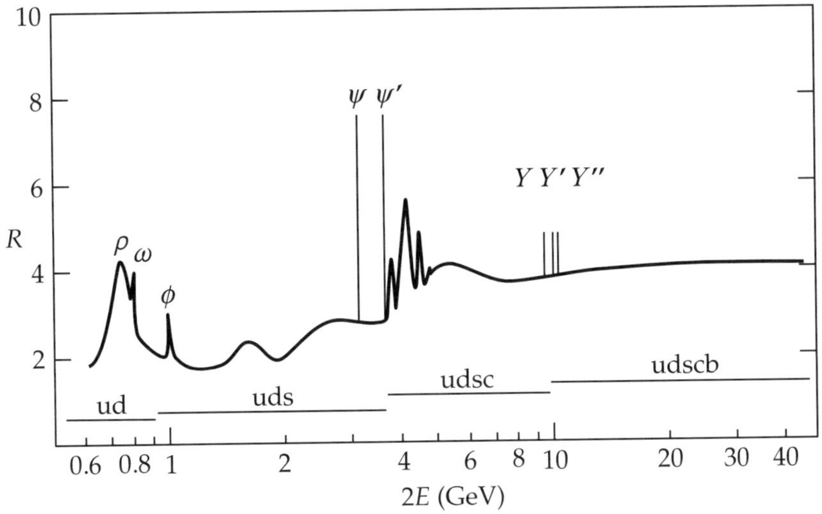

Fig. 4.4b. The ratio $R = \sigma(e^+e^- \to \text{hadrons}) / \sigma(e^+e^- \to \mu^+\mu^-)$ as a function of the total center-of-mass energy $E_{cm} = 2E$ (each beam has energy E). The curve is drawn through many measured points (see Fig. 4.4a) and represents only a rough average. However, it does reproduce the major features. The peaks are labelled by the conventional symbol representing the various vector mesons. The narrow peaks due to the Ψ and Υ families are inidcated by vertival lines: the actual values of the peak cross-section are not represented. The above $B\bar{B}$-threshold Υ states are ommited. The horizontal lines marked with quark flavours are the values expected away from resonances and without the colour factor N_c

$$R = \left[\left(\frac{2}{3}\right)^2 + \left(\frac{-1}{3}\right)^2 + \left(\frac{-1}{3}\right)^2 + \left(\frac{2}{3}\right)^2\right] N_c = \frac{10}{9} N_c$$

$$\text{for } E_{cm} > 3\ \text{GeV} > 2m_c\ , \tag{4.40}$$

$$R = \left[\left(\frac{2}{3}\right)^2 + \left(\frac{-1}{3}\right)^2 + \left(\frac{-1}{3}\right)^2 + \left(\frac{2}{3}\right)^2 + \left(\frac{-1}{3}\right)^2\right] N_c = \frac{11}{9} N_c$$

$$\text{for } E_{cm} > 10\ \text{GeV} > 2m_b\ .$$

Here N_c denotes the number of colors. If we compare this prediction to the measured values (Fig. 4.4), it becomes clear that the assumptions made above lead to the value $N_c = 3$. The second argument is quite similar. We consider the two-photon decay of the π^0 meson shown in Fig. 4.5. Owing to the relatively small pion mass only the u and d quarks contribute to this graph. The coupling to these quarks is well understood from studies of the pion–baryon interaction, and the decay width

$$\Gamma = 7.63\ \text{eV} \left(\frac{N_c}{3}\right)^2 \tag{4.41}$$

is obtained. The experimental value

$$\Gamma_{exp} = 7.57 \pm 0.32\ \text{eV} \tag{4.42}$$

obviously again favors $N_c = 3$. As some theoretical assumptions enter the derivation of (4.37), this argument is somewhat less direct than the first one.

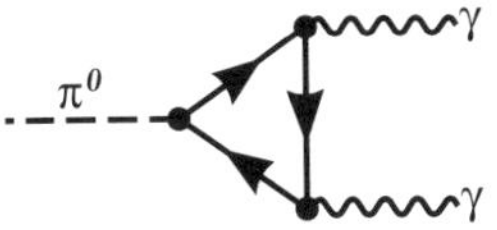

Fig. 4.5. Graph for the $\pi^0 \to 2\gamma$ decay

The last argument we want to discuss is of a purely theoretical nature. From a strictly empirical point of view it is therefore the least well founded. From the

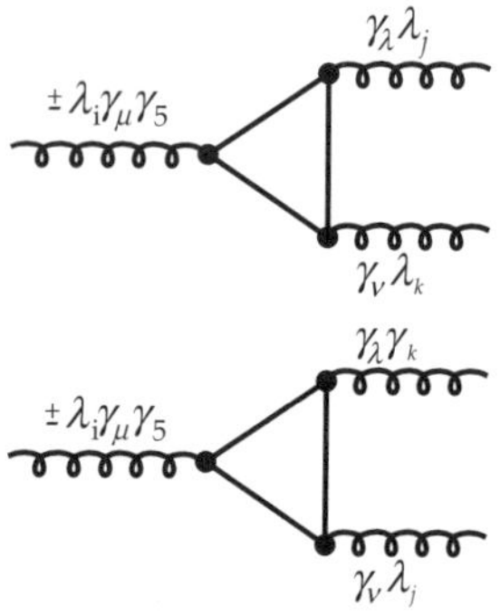

Fig. 4.6. Potentially divergent graphs of the standard model. The plus holds for right-handed fermions and the minus sign for left-handed fermions. The upper and lower diagrams differ in the exchange of the outgoing gluons.

theoretical point of view it is, on the contrary, the most fascinating, since it states that the standard model is only internally consistent for three colors. This implies that there must be a fundamental symmetry linking the electroweak sector of the standard model to the QCD sector and thus motivates the search for a Grand Unified Theory.

In the last section we claimed that all gauge theories with or without spontaneous symmetry breaking are renormalizable. However, this statement is not valid in full generality, since there are graphs in the standard model, for example, that are divergent and not subject to the general proof of renormalizability. They can all be traced to the *triangle anomaly*, which is discussed in detail in *Gauge Theory of Weak Interactions*.[5] In essence the anomaly can be reduced to the time-ordered product of one axial vector current and two vector currents:

$$T^{ijk}_{\mu\nu\lambda}(k_1,k_2,q) = \mathrm{i}\int \mathrm{d}x_1\mathrm{d}x_2\ \exp(\mathrm{i}k_1\cdot x_1 + \mathrm{i}k_2\cdot x_2) \\ \langle 0 \mid A^i_\mu(x_1)V^k_\nu(x_2)V^j_\lambda(0)\mid 0\rangle \ , \tag{4.43}$$

as depicted in Fig. 4.6. The currents are

$$\begin{aligned} A^i_\mu(x) &= \bar{\psi}(x)\frac{\hat{\lambda}^i}{2}\gamma_\mu\gamma_5\psi(x) \ , \\ V^i_\mu(x) &= \bar{\psi}(x)\frac{\hat{\lambda}^i}{2}\gamma_\mu\psi(x) \ . \end{aligned} \tag{4.44}$$

In the standard model there exist chiral fermions that exhibit vector (γ_μ) as well as axial vector ($\gamma_\mu\gamma_5$) character, which can be seen by considering the weak current of the GSW theory

$$J^i_\mu \sim \bar{\psi}(x)\frac{\hat{\lambda}^i}{2}\,(1\pm\gamma_5)\,\gamma_\mu\psi(x) \ . \tag{4.45}$$

Since for right- and left-handed fermions γ_5 enters with different sign they contribute with opposite sign to the anomalous triangle diagram. Left-handed fermions are members of the doublet representation, $\hat{\lambda}^i$, thus denoting the $\mathrm{SU_L}(2)\times\mathrm{U_Y}(1)$ 2×2 matrices, being essentially the three Pauli matrices $\hat{\tau}_i$ plus the unit matrix. The latter is proportional to the hypercharge. Right-handed fermions are members of the singlet representation. Thus in the right-handed case the $\hat{\lambda}^i$ reduce to a single number, being the hypercharge of the right-handed fermion. We shall now demonstrate that the quark and lepton contributions to the triangle diagram cancel exactly. When calculating the diagram we have to take the trace of the $\hat{\lambda}$ matrices and take into account the exchange diagram. The result is proportional to

$$\mathrm{tr}\left(\hat{\lambda}_i\hat{\lambda}_j\hat{\lambda}_k\right) + \mathrm{tr}\left(\hat{\lambda}_i\hat{\lambda}_k\hat{\lambda}_j\right) = \mathrm{tr}\left(\hat{\lambda}_i\{\hat{\lambda}_j,\hat{\lambda}_k\}\right) \ . \tag{4.46}$$

[5] W. Greiner and B. Müller: *Gauge Theory of Weak Interactions*, 3rd ed. (Springer, Berlin, Heidelberg 2000).

Let us first investigate the doublet contributions to the left-handed fermions. For pure $SU_L(2)$ the corresponding coeficient vanishes:

$$\mathrm{tr}\left(\hat{\tau}_i\{\hat{\tau}_j,\hat{\tau}_k\}\right) = 2\delta_{jk}\,\mathrm{tr}\left(\hat{\tau}_i\right) = 0\ . \tag{4.47}$$

The same is the case if two of the $\hat{\lambda}$ matrices are the hypercharge:

$$\mathrm{tr}\left(\hat{\tau}_i\{\hat{Y},\hat{Y}\}\right) = 2\hat{Y}^2\,\mathrm{tr}\left(\hat{\tau}_i\right) = 0\ . \tag{4.48}$$

Thus we have to consider only those cases in which one or all of the $\hat{\lambda}$ matrices are the hypercharge $\hat{Y}$. We have

$$\mathrm{tr}\left(\hat{Y}\{\hat{\tau}_j,\hat{\tau}_k\}\right) = 2\delta_{jk}\,\mathrm{tr}\left(\hat{Y}\right)\ , \tag{4.49}$$

and thus the anomaly contribution is proportional to Y. Summing over all doublets of one generation we arrive at

$$\sum_{\text{doublets}} Y = \sum_{\text{lh. quarks}} Y + \sum_{\text{lh. leptons}} Y\ , \tag{4.50}$$

which vanishes by explicit calculation for the fermion assignments in each generation, as can be read off from Table 4.1. Indeed one has

$$\begin{aligned}\sum_{\text{lh. quarks}} Y &= 3\left(\frac{1}{3}\times 2\right) = 2\ ,\\ \sum_{\text{lh. leptons}} Y &= -1\times 2 = -2\ .\end{aligned} \tag{4.51}$$

The factor 3 in the sum over left-handed quarks is due to the color degree of freedom. For the case when all $\hat{\lambda}$ matrices are the hypercharge we find that the potentially anomalous contribution has a coefficient given by

$$\mathrm{tr}\left(\hat{Y}\hat{Y}\hat{Y}\right) \sim Y^3\ . \tag{4.52}$$

	T	T_3	y	Q
ν_e	$+\frac{1}{2}$	$+\frac{1}{2}$	-1	$+0$
e_L	$+\frac{1}{2}$	$-\frac{1}{2}$	-1	-1
e_R	$+0$	$+0$	-2	-1
u_L	$+\frac{1}{2}$	$+\frac{1}{2}$	$+\frac{1}{3}$	$+\frac{2}{3}$
$(d_C)_L$	$+\frac{1}{2}$	$-\frac{1}{2}$	$+\frac{1}{3}$	$-\frac{1}{3}$
u_R	$+0$	$+0$	$+\frac{4}{3}$	$+\frac{2}{3}$
$(d_C)_R$	$+0$	$+0$	$-\frac{2}{3}$	$-\frac{1}{3}$

Table 4.1. The quantum numbers of the particles of the first fermion family of the standard model

Now we have to consider also contributions of the right-handed singlets which enter with opposite sign:

$$\sum_{\text{lh. quarks}} Y^3 - \sum_{\text{rh. quarks}} Y^3 = 3\left(\frac{1}{3}\right)^3 \times 2 - \left[\left(\frac{4}{3}\right)^3 + 3\left(\frac{-2}{3}\right)^3\right] = -6 ,$$
$$\sum_{\text{lh. leptons}} Y^3 - \sum_{\text{rh. leptons}} Y^3 = (-1)^3 \times 2 - (-2)^3 = 6 . \tag{4.53}$$

If there are exactly three times as many quarks as leptons their contributions cancel exactly, i.e. the large loop momenta do not contribute to the graphs in Fig. 4.6, which then are no longer divergent. Assuming in addition that there are just as many kinds of leptons as quark flavors, we conclude that there must exist three colors.

This argument might have a problem, since the graphs of Fig. 4.6 might be renormalizable in the context of an improved field theory or in a theory going far beyond field theory. We know, however, that they cannot be renormalized on the basis of the present understanding and with any technique developed so far. The role of the anomaly is so fascinating just because it seems to point to something beyond the standard model. It has several very interesting consequences; for example, owing to the anomaly, the standard model should violate separate lepton and baryon number conservation, conserving only B–L, which again shows that it is intimately connected with any unifying model.

Altogether the arguments discussed are sufficiently convincing to demand that every theory of quark–quark interactions has to start with three color states. In analogy to the standard model it is quite natural to regard them as the fundamental representation of a SU(3) color group. It is actually relatively easy to see that SU(3) is the only possible compact semi-simple Lie group. These Lie groups are completely classified and one can look up their irreducible representations in, for example, R. Slansky, Phys. Rep. **79**, p. 1–128. SU(3) is the only group which has complex irreducible triplets. The only alternative to standard QCD are, therefore, groups that are spontaneously broken to a SU(3) symmetry group plus some high mass residues of the original, larger group. Consequently the basic equations are just (4.6) and (4.7), where we have changed the sign of g and $\hat{\theta}(x)$ in order to agree with common conventions:

$$\hat{A}_\nu(x) = \sum_{a=1}^{8} \frac{1}{2}\hat{\lambda}_a \, A_\nu^a(x) \qquad (\hat{\lambda}_a = \text{SU(3) matrices}) ,$$
$$\hat{\theta}(x) = \sum_{a=1}^{8} \frac{1}{2}\hat{\lambda}_a \theta^a(x) , \tag{4.54}$$

and

$$\hat{A}_\nu(x) \to \hat{A}'_\nu(x) = \mathrm{e}^{\mathrm{i}g\bar{\theta}(x)} \left(\bar{A}_\nu(x) + \frac{\mathrm{i}}{g}\partial_\nu\right) \mathrm{e}^{-\mathrm{i}g\bar{\theta}(x)} ,$$
$$\Psi(x) \to \Psi'(x) = \mathrm{e}^{\mathrm{i}g\hat{\theta}(x)}\Psi(x) . \tag{4.55}$$

The quark wave function is then a color-SU(3) triplet

$$\Psi(x) = \begin{pmatrix} \Psi_{\mathrm{r}}(x) \\ \Psi_{\mathrm{b}}(x) \\ \Psi_{\mathrm{g}}(x) \end{pmatrix} \tag{4.56}$$

and the Lagrangian is

$$L = -\frac{1}{2}\,\mathrm{tr}(\hat{F}_{\mu\nu}\hat{F}^{\mu\nu}) + \bar{\Psi}(p_\mu + g\hat{A}_\mu)\gamma^\mu\Psi \tag{4.57}$$

with the tensor of the gluon field strength

$$\begin{aligned}
\hat{F}_{\mu\nu} &= \partial_\mu \hat{A}_\nu - \partial_\nu \hat{A}_\mu - \mathrm{i}g\left[\hat{A}_\mu, \hat{A}_\nu\right] \\
&= \partial_\mu A^a_\nu \frac{\lambda^a}{2} - \partial_\nu A^a_\mu \frac{\lambda^a}{2} - \mathrm{i}g A^a_\mu\, A^b_\nu \underbrace{\left[\frac{\lambda^a}{2}, \frac{\lambda^b}{2}\right]}_{\mathrm{i}\,f^{abc}\frac{\lambda^c}{2}} \\
&= \left(\partial_\mu A^a_\nu - \partial_\nu A^a_\mu + g f^{abc} A^b_\mu A^c_\nu\right)\frac{\lambda^a}{2} \\
&\equiv F^a_{\mu\nu}\frac{\lambda^a}{2}\;.
\end{aligned} \tag{4.58}$$

Equations (4.54) to (4.58) completely define the theory. They lead to the Feynman rules and allow the evaluation of arbitrary graphs. This procedure has been very successful for all field theories discussed so far, but in the case of QCD it has to be applied with care. Here the essential question is under what circumstances does such a perturbative expansion in the coupling constant, as Feynman rules imply, converge. One finds that the smallness of the coupling constant is an insufficient criterion. It is still not completely understood, for example, whether the perturbative treatment of QED really converges or whether this is only an asymptotic expansion. In the latter case the evaluation of additional classes of graphs would yield improvements only to a certain order in α. The consideration of higher-order terms then would again worsen the agreement with experimental observations. Unfortunately the QCD coupling constant has to be chosen in such a way that the convergence of the resulting perturbation series is not clear. To be more precise the convergence depends on the value of the momenta that occur. In the case of large momenta, for example, in deep inelastic electron–nucleon scattering reactions, we shall show that perturbative calculations agree very well with experimental results. But for small momenta no convergence can be observed; see the discussion of the running coupling constant in Example 4.4. This fact, however, fits quite well into the general picture, since quark confinement, for instance, is supposed to be a nonperturbative effect. To what extent is this behavior characteristic for the SU(3) theory? In fact only the way in which g decreases with increasing momentum is typical for the SU(3) theory. An SU(4) or SU(2) theory would lead to a stronger or weaker dependence, respectively. For a U(1) theory, e.g., QED, even the opposite behavior occurs. Here perturbation theory converges very well at small momenta and breaks down at unphysically high scales (of the order of 10^{20} GeV).

General features are thus not necessarily characteristic for the chosen theory. Only a thorough and detailed analysis of many processes with different momentum transfers can prove the validity of the QCD equations (4.49)–(4.53). The completion of such an analysis is still out of sight, but the evidence in favor of an SU(3) color gauge theory is meanwhile so strong that at present hardly anybody seriously doubts its validity.

Before we turn in the next chapter to perturbative QCD for large momentum transfers, Example 4.2 introduces the Feynman rules for QCD. Finally, Example 4.3 gives more details on the dependence of the coupling constant on momentum transfer.

EXAMPLE

4.2 The Feynman Rules for QCD

First we discuss the vertices, starting with the *quark–gluon vertex*, which can be deduced from the following part of the Lagrangian:

$$L_{\text{int}} = \bar{\Psi} g \frac{\lambda^a}{2} \gamma_\mu A^{a\mu} \Psi \ . \tag{1}$$

Throughout this problem latin letters $a, b, c, \dots$ denote the SU(3) index. One has to sum over indices occurring twice ($a = 1, \dots, 8$).

In general, Feynman rules are obtained by varying the corresponding action integral in momentum space. Here

$$\frac{\delta^3}{\delta\Psi^i_\gamma(p)\delta\bar{\Psi}^j_\beta(p')\delta A^{b\nu}(k)} \left[\int \bar{\Psi}(p_1) g \frac{\lambda^a}{2} \gamma_\mu A^{a\mu}(p_2)\Psi(p_3) \right.$$
$$\left. \times (2\pi)^4 \delta^4(p_1 + p_2 - p_3)\, \mathrm{d}^4 p_1\, \mathrm{d}^4 p_2\, \mathrm{d}^4 p_3 \right] \tag{2}$$

has to be evaluted. Using

$$\frac{\delta\Psi(p_1)^i_\gamma}{\delta\Psi(p')^j_\beta} = \delta^4(p_1 - p')\delta_{ij}\delta_{\gamma\beta} \ , \quad \text{etc.} \tag{3}$$

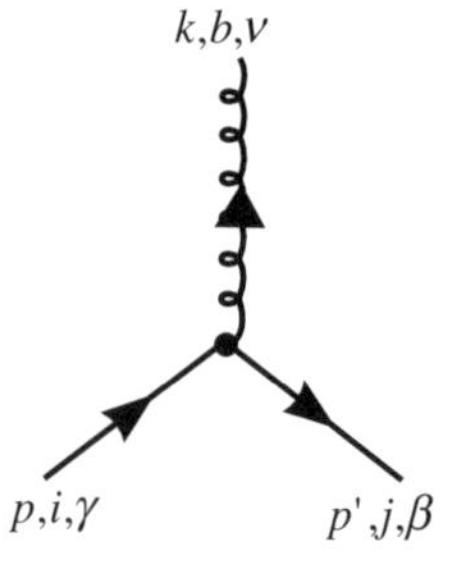

we can simplify the quark–gluon vertex to

$$g \left(\frac{\lambda^b}{2}\right)_{ji} (\gamma_\nu)_{\beta\gamma} (2\pi)^4\, \delta^4(p' + k - p) \ . \tag{4}$$

Since the gauge field is non-Abelian, the gluon Lagrangian also contains second and third powers of the field $\hat{A}_\mu$, which lead to gluon–gluon interactions.

The *three-gluon vertex* is evaluted in an analogous way. From the Lagrangian we obtain, with

$$\mathrm{tr}\left\{\lambda^a \left[\lambda^b, \lambda^c\right]\right\} = 2\mathrm{i}\, f_{bcd}\, \mathrm{tr}\left\{\lambda^a \lambda^d\right\} = 4\mathrm{i}\, f_{abc} \ , \tag{5}$$

Example 4.2

the following three-gluon terms:

$$
\begin{aligned}
-\frac{1}{2}\mathrm{tr}\left\{\hat{F}_{\mu\nu}\hat{F}^{\mu\nu}\right\}
&= -\frac{1}{2}\mathrm{tr}\left\{F^a_{\mu\nu}\,F^{\mu\nu b}\frac{\lambda^a}{2}\frac{\lambda^b}{2}\right\}\\
&= -\frac{1}{8}F^a_{\mu\nu}\,F^{\mu\nu b}\mathrm{tr}\left\{\lambda^a\lambda^b\right\}\\
&= -\frac{1}{4}F^a_{\mu\nu}\,F^{\mu\nu a}\\
&= -\frac{1}{4}\left\{\left(\partial_\mu A^a_\nu-\partial_\nu A^a_\mu\right)g f^{abc}A^{b\mu}A^{c\nu}\right.\\
&\qquad\left.+g f^{abc}A^b_\mu A^c_\nu\left(\partial^\mu A^{\nu a}-\partial^\nu A^{\mu a}\right)\right\}+\dots\\
&= -\frac{g}{2}\left\{\left(\partial_\mu A^a_\nu-\partial_\nu A^a_\mu\right)f^{abc}A^{b\mu}A^{c\nu}\right\}+\dots\\
&\qquad -g\partial_\mu A^a_\nu\,f^{abc}A^{b\mu}A^{c\nu}\;. \qquad (6)
\end{aligned}
$$

In the last step $\left(\partial_\mu A^a_\nu-\partial_\nu A^a_\mu\right)=-\left(\partial_\nu A^a_\mu-\partial_\mu A^a_\nu\right)$ and $f^{abc}=-f^{acb}$ have been used. The dots ... indicate noncubic terms.

Again this is transformed into momentum space and varied with respect to the fields:

$$
\begin{aligned}
&\frac{\delta^3}{\delta A^r_\varrho(k_1)\delta A^s_\sigma(k_2)\delta A^t_\tau(k_3)}\\
&\quad\times\left\{-\int\frac{g}{2}\left[\mathrm{i}p_{1,\mu}A^a_\nu(p_1)-\mathrm{i}p_{1,\nu}A^a_\mu(p_1)\right]f_{abc}A^{b\mu}(p_2)\right.\\
&\quad\left.\times A^{c\nu}(p_3)(2\pi)^4\delta^4(p_1+p_2+p_3)\,\mathrm{d}^4p_1\,\mathrm{d}^4p_2\,\mathrm{d}^4p_3\right\}\\
&=\frac{\delta^2}{\delta A^r_\varrho(k_1)\delta A^s_\sigma(k_2)}\left\{-\mathrm{i}\frac{g}{2}\int\left[k_{3,\mu}f_{tbc}A^{b\mu}(p_2)A^{c\tau}(p_3)\right.\right.\\
&\quad\left.(2\pi)^4\delta^4(k_3+p_2+p_3)\,\mathrm{d}^4p_2\,\mathrm{d}^4p_3\right]\\
&\quad+\mathrm{i}\frac{g}{2}\int\left[k_{3,\nu}f_{tbc}A^{b\tau}(p_2)A^{c\nu}(p_3)\right.\\
&\quad\left.(2\pi)^4\delta^4(k_3+p_2+p_3)\,\mathrm{d}^4p_2\,\mathrm{d}^4p_3\right]\\
&\quad-\mathrm{i}\frac{g}{2}\int\left[p_{1,\mu}A^a_\nu(p_1)-p_{1,\nu}A^a_\mu(p_1)\right]f_{abc}\delta^b_t g^{\mu\tau}A^{b,\mu}(p_3)\\
&\quad(2\pi)^4\delta^4(k_3+p_1+p_3)\,\mathrm{d}^4p_2\,\mathrm{d}^4p_3\\
&\quad-\mathrm{i}\frac{g}{2}\int\left[p_{1,\mu}A^a_\nu(p_1)-p_{1,\nu}A^a_\mu(p_1)\right]f_{abc}\delta^c_t g^{\nu\tau}A^{c,\nu}(p_2)\\
&\quad\left.(2\pi)^4\delta^4(k_3+p_1+p_2)\,\mathrm{d}^4p_1\,\mathrm{d}^4p_2\right\}
\end{aligned}
$$

Example 4.2

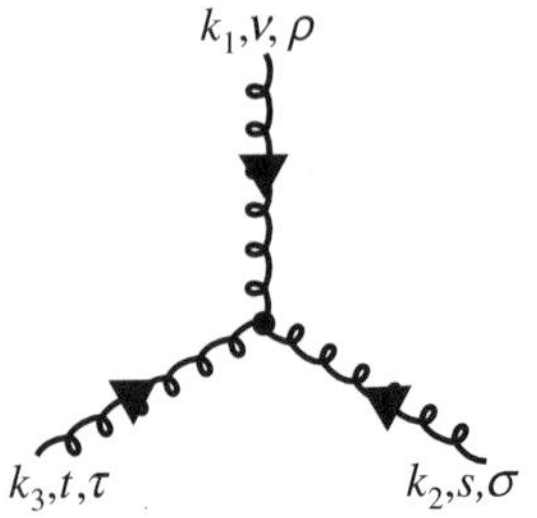

$$\begin{aligned}
&= \frac{\delta^2}{\delta A^r_\varrho(k_1)\delta A^s_\sigma(k_2)} \left\{ -\mathrm{i}\frac{g}{2}\int \Big[k_{3,\mu} f_{tbc} A^{b\mu}(p_2) A^{c\tau}(p_1) \right.\\
&\quad - k_{3,\nu} f_{tbc} A^{b\tau}(p_1) A^{c\nu}(p_2) + 2 f_{tca} A^{c\nu}(p_1)\big(p_2{}^\tau A^a_\nu(p_2)\\
&\quad \left. - p_{2,\nu} A^{a\tau}(p_2)\big)\Big](2\pi)^4\delta^4(p_1+p_2+k_3)\, \mathrm{d}^4p_1\, \mathrm{d}^4p_2 \right\}\\
&= \frac{\delta}{\delta A^r_\varrho(k_1)}(-\mathrm{i}g)\int \Big[k_{3,\mu} f_{tbs} A^{b\mu}(p) g^{\sigma\tau} + k_3{}^\sigma f_{tsc} A^{c\tau}(p)\\
&\quad + f_{tsa}\big(p^\tau A^{a\sigma}(p) - p^\sigma A^{a\tau}(p)\big) + k_2{}^\tau f_{tcs} A^{c\sigma}(p)\\
&\quad - k_{2,\nu} f_{tcs} A^{c\nu}(p) g^{\tau\sigma}\Big](2\pi)^4 \mathrm{d}^4(p+k_2+k_3)\mathrm{d}^4p\\
&= -\mathrm{i}g\Big[k_3{}^\varrho f_{trs} g^{\sigma\tau} + k_3{}^\sigma f_{tsr} g^{\varrho\tau} + f_{tsr}\big(k_1{}^\tau g^{\sigma\varrho} - k_1{}^\sigma g^{\tau\varrho}\big)\\
&\quad + f_{trs}\big(k_2{}^\tau g^{\sigma\varrho} - k_2{}^\varrho g^{\tau\sigma}\big)\Big](2\pi)^4\delta^4(k_1+k_2+k_3)\\
&= \mathrm{i}g f_{rst}\Big[\big(k_1{}^\tau - k_2{}^\tau\big) g^{\varrho\sigma} + \big(k_2{}^\varrho - k_3{}^\varrho\big) g^{\sigma\tau}\\
&\quad + \big(k_3{}^\sigma - k_1{}^\sigma\big) g^{\tau\varrho}\Big](2\pi)^4\delta^4(k_1+k_2+k_3)\ .
\end{aligned} \tag{7}$$

Also the *four-gluon vertex* can be determined in the same manner. Using (4.53), the relevant part of the Lagrangian is

$$\begin{aligned}
&-\frac{1}{2}\mathrm{tr}\left\{\hat{F}_{\mu\nu}\hat{F}^{\mu\nu}\right\}\\
&\to -\frac{1}{2}A^e_\mu A^f_\nu A^{r\mu} A^{s\nu}\frac{-g^2}{16}\mathrm{tr}\left\{\left[\lambda^e,\lambda^f\right]\left[\lambda^r,\lambda^s\right]\right\}\\
&= -\frac{1}{2}A^e_\mu A^f_\nu A^{r\mu} A^{s\nu}\frac{g^2}{4}\mathrm{tr}\left\{\lambda^g\lambda^t\right\} f_{efg} f_{rst}\\
&= -\frac{g^2}{4} f_{efg} f_{rsg} A^e_\mu A^f_\nu A^{r\mu} A^{s\nu}\ .
\end{aligned} \tag{8}$$

Here the well-known relations $\left[\lambda^e,\lambda^f\right]_- = \mathrm{i}2 f_{efg}\lambda^g$ and $\mathrm{tr}\left\{\lambda^g\lambda^t\right\} = 2\delta^{gt}$ have been applied. Performing the standard transformation to momentum space simply yields all possible permutations of the four index pairs (a,α), (b,β), (c,γ), and (d,δ), i.e., a total of $4! = 24$ terms:

$$\begin{aligned}
&-\frac{1}{4}g^2 f_{efg} f_{rsg}\Big[\delta_{ae}\delta_{bf}\delta_{cr}\delta_{ds} g^{\alpha\gamma} g^{\beta\delta} + \delta_{ae}\delta_{bf}\delta_{cs}\delta_{dr} g^{\alpha\delta} g^{\beta\gamma}\\
&\quad + \delta_{ae}\delta_{br}\delta_{cf}\delta_{ds} g^{\alpha\beta} g^{\gamma\delta} + \delta_{ae}\delta_{br}\delta_{cs}\delta_{df} g^{\alpha\beta} g^{\gamma\delta}\\
&\quad + \delta_{ae}\delta_{bs}\delta_{cr}\delta_{df} g^{\alpha\gamma} g^{\beta\delta} + \delta_{ae}\delta_{bs}\delta_{cf}\delta_{dr} g^{\alpha\delta} g^{\beta\gamma}\\
&\quad + (e\to r,\, f\to s) + (e\to f,\, r\to s) + (e\to s,\, f\to r)\Big]\\
&= -\frac{g^2}{4}\Big[f_{gab} f_{gcd}\left(g^{\alpha\gamma} g^{\beta\delta} - g^{\alpha\delta} g^{\beta\gamma}\right)\\
&\quad + f_{gac} f_{gdb}\left(g^{\alpha\delta} g^{\beta\gamma} - g^{\alpha\beta} g^{\gamma\delta}\right) + f_{gad} f_{gbc}\left(g^{\alpha\beta} g^{\gamma\delta} - g^{\alpha\gamma} g^{\beta\delta}\right)\Big]4\ .
\end{aligned} \tag{9}$$

Example 4.2

Because of the permutation relations concerning $f_{efg}f_{rsg}$, the three terms denoted by $(e \to r,\ f \to s)$, etc. contribute the same as the first one. The complete vertex factor is then

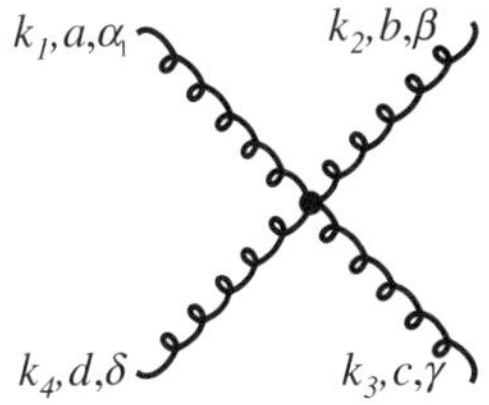

$$\begin{aligned}
&-g^2\left[f_{eab}f_{ecd}\left(g^{\alpha\gamma}g^{\beta\delta}-g^{\alpha\delta}g^{\beta\gamma}\right)\right.\\
&+f_{eac}f_{edb}\left(g^{\alpha\delta}g^{\beta\gamma}-g^{\alpha\beta}g^{\gamma\delta}\right)\\
&\left.+f_{ead}f_{ebc}\left(g^{\alpha\beta}g^{\gamma\delta}-g^{\alpha\gamma}g^{\beta\delta}\right)\right]\\
&\times(2\pi)^4\,\delta^4(k_1+k_2+k_3+k_4)\ . \qquad (10)
\end{aligned}$$

In principle the propagators are identical with those in QED. Hence the quark propagator is

$$\frac{1}{p\cdot\gamma-m+\mathrm{i}\varepsilon} \qquad (11)$$

and the gluon propagator could be chosen to be

$$-\frac{g_{\mu\nu}\delta_{ab}}{k^2+\mathrm{i}\varepsilon}\ . \qquad (12)$$

In QED the rule holds that outer lines only propagate real physical degrees of freedom, in particular transverse photons. On the other hand, inner lines (those occurring in loops) represent all degrees of freedom. The latter is possible because the nonphysical components do not contribute to closed loops. However, in QCD additional vertices occur resulting in additional loops that no longer exhibit this feature, e.g.

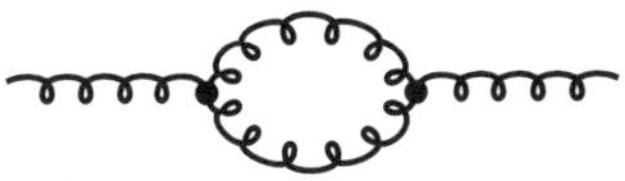

Therefore one has for every inner line to subtract the nonphysical gluonic contributions. This can be achieved by introducing an artificial particle without any physical meaning. The couplings and the propagator of this particle are chosen in such a way that graphs containing this particle cancel the nonphysical gluonic contributions. The corresponding graph is

Here the dashed line denotes the new particle just mentioned. Because of their nonphysical nature these particles are called "ghosts" or "ghost fields". We shall not discuss here how these fields are exactly defined and how their couplings are determined. We only list the corresponding Feynman rules. Example 4.3 may help us to understand the ghost idea somewhat better.

Example 4.2

One should be aware of the "i" in the expression for the vertex. This different phase compared with (7) shows that certain cancellations of ghost and gluon loops are possible. In that way, if the expression (21) below is implemented twice in the above ghost-loop diagram, additional "i" results in a minus-sign, so that the ghost-loop amplitude

partly cancels the gluon-loop amplitude

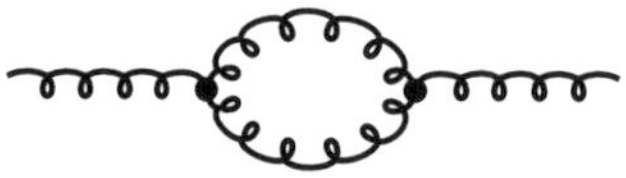

The Feynman rules discussed above are summarized in (15)–(21). In order to indicate the different possible gauges, the gluon propagator is given in its most general form, assuming $\partial_\mu A^\mu = 0$ (transverse gauge). From QED we already know the Landau gauge ($\lambda = 1$) and the transverse propagator ($\lambda \to 0$). In fact an arbitrary mixing of these two cases can also be chosen. Gauge invariance implies that in the evaluation of a physical process all λ-dependent terms cancel each other or are equal to zero. Therefore λ can be arbitrarily chosen, i.e. different from $\lambda = 1,\ \lambda \to 0$:

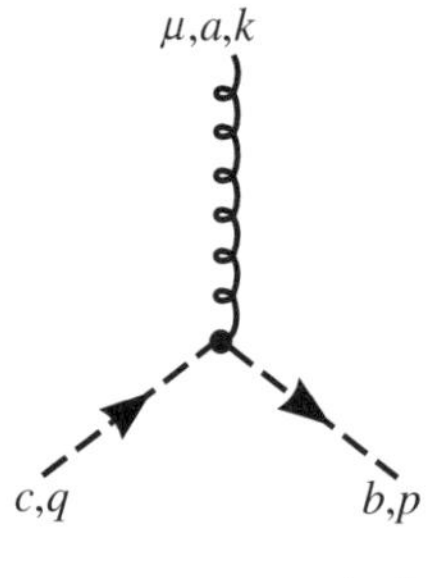

$$-\mathrm{i}g f_{abc} p_\mu (2\pi)^4\, \delta^4(p-k-q) \tag{13}$$

and

$$\frac{\delta_{ij}}{p^2 + \mathrm{i}\varepsilon} \tag{14}$$

Note that all Feynman graphs are frequently multiplied by an additional factor i because the starting point of a perturbative expansion is really i $\int L(x)\, \mathrm{d}^4x$ and not $\int L(x)\, \mathrm{d}^4x$. However, this leads only to an unimportant overall phase.

Example 4.2

$$-\delta_{ab}\left[\frac{g_{\mu\nu}}{k^2+\mathrm{i}\varepsilon}-(1-\lambda)\frac{k_\mu k_\nu}{(k^2+\mathrm{i}\varepsilon)^2}\right] , \tag{15}$$

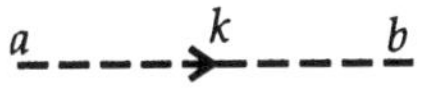

$$\frac{\delta_{ab}}{k^2+\mathrm{i}\varepsilon} , \tag{16}$$

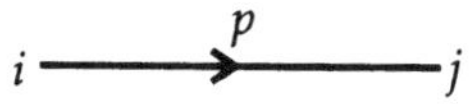

$$\frac{\delta_{ij}}{\not p-m+\mathrm{i}\varepsilon} , \tag{17}$$

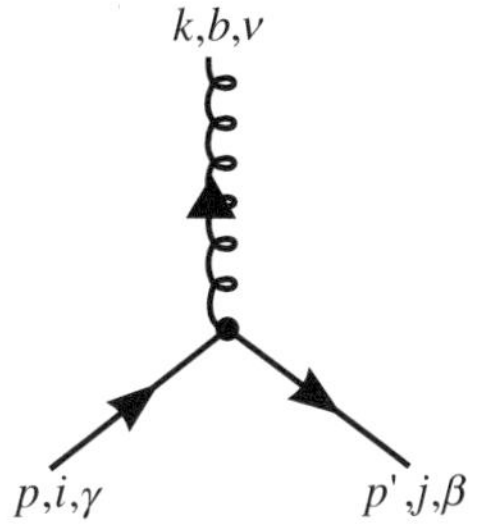

$$g\left(\frac{1}{2}\lambda^b\right)_{ji}(\gamma_\nu)_{\beta\gamma}(2\pi)^4\,\delta^4(p'+k-p) , \tag{18}$$

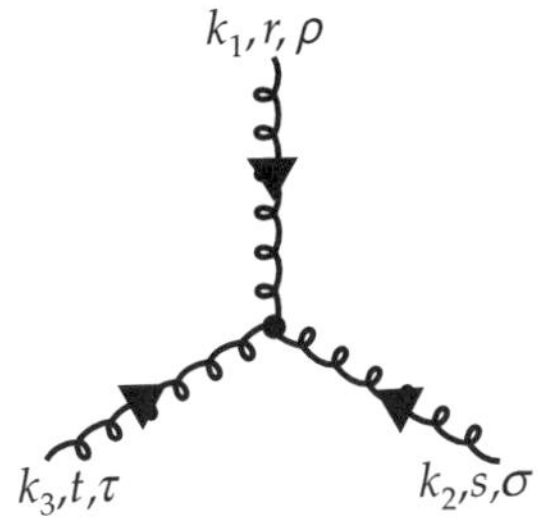

$$\mathrm{i}g f_{rst}\Big[\left(k_1^\tau-k_2^\tau\right)g^{\varrho\sigma}+\left(k_2^\varrho-k_3^\varrho\right)g^{\sigma\tau} + \left(k_3^\sigma-k_1^\sigma\right)g^{\tau\varrho}\Big](2\pi)^4\,\delta^4(k_1+k_2+k_3) , \tag{19}$$

Example 4.2

$$-g^2\Big[f_{eab}f_{ecd}\big(g^{\alpha\gamma}g^{\beta\delta}-g^{\alpha\delta}g^{\beta\gamma}\big)+f_{eac}f_{edb}\big(g^{\alpha\delta}g^{\beta\gamma}-g^{\alpha\beta}g^{\gamma\delta}\big)$$
$$+f_{ead}f_{ebc}\big(g^{\alpha\beta}g^{\gamma\delta}-g^{\alpha\gamma}g^{\beta\delta}\big)\Big](2\pi)^4\,\delta^4(k_1+k_2+k_3+k_4)\ , \tag{20}$$

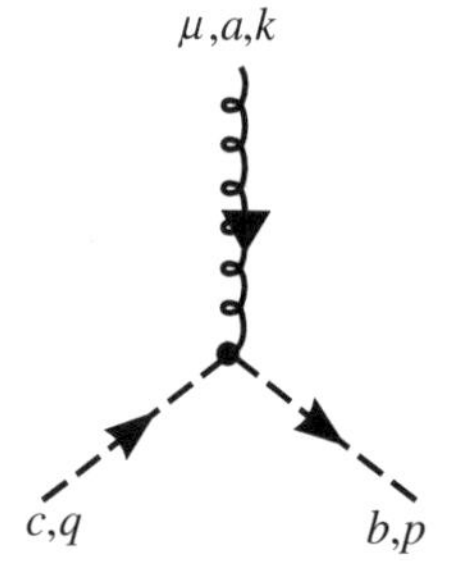

$$-\mathrm{i}g f_{abc}\,p_\mu(2\pi)^4\,\delta^4(p-k-q)\ . \tag{21}$$

EXAMPLE

4.3 Fadeev–Popov Ghost Fields

To define the gluon propagator (see (15) of the last Example) properly one has to specify a gauge. Otherwise the equation of motion (Maxwell equation) of the free gluon field cannot be inverted to give the Green function of the field equation. This problem is already familiar from Abelian theories as QED. If we write the Lagrangian of the free photon field as

$$\begin{aligned}\mathcal{L}&=-\frac{1}{4}F^2\\&=-\frac{1}{2}\partial_\mu A_\nu(\partial^\mu A^\nu-\partial^\nu A^\mu)\\&=\frac{1}{2}A_\nu\left(g^{\mu\nu}\partial^2-\partial^\mu\partial^\nu\right)A_\mu\ ,\end{aligned} \tag{1}$$

Example 4.3

where a surface term is discarded after partial integration, the photon propagator $D_{\mu\nu}$ is the inverse of the sandwiched operator in (1)

$$\left(g^{\mu\nu}\partial^2 - \partial^\mu\partial^\nu\right) D_{\nu\lambda}(x-y) = \delta^\mu_\lambda\,\delta(x-y)\ . \tag{2}$$

If we multiply (2) by the operator ∂_μ, giving

$$(0\cdot\partial^\nu)D_{\nu\lambda}(x-y) = \partial_\lambda\delta(x-y)\ , \tag{3}$$

we see that $D_{\mu\nu}$ is infinite. This is because the operator $\left(g^{\mu\nu}\partial^2 - \partial^\mu\partial^\nu\right)$ has no inverse. This is the projection operator onto tranverse modes. It is a general feature of projection operators that they do not have an inverse, e.g. applied to an arbitrary function $\partial_\mu\Lambda$ it gives

$$\left(g^{\mu\nu}\partial^2 - \partial^\mu\partial^\nu\right)\partial_\mu\Lambda = (\partial^\nu\partial^2 - \partial^2\partial^\nu)\Lambda = 0\ . \tag{4}$$

Indeed the operator has a zero eigenvalue and therefore cannot be inverted. The physical reason behind this is that one has to make sure to propagate only physical degrees of freedom. All fields that are related only by gauge transformations $A_\mu \to A_\mu + \partial_\mu\Lambda$ are propagated as well, which clearly gives an infinite contribution. This can be cured by fixing a particular gauge. To be definite we impose the Lorentz condition

$$\partial_\mu A^\mu = 0\ , \tag{5}$$

which can be included in the Lagrangian in the general form

$$\begin{aligned}\mathcal{L}_{\text{fix}} &= -\frac{1}{2\lambda}(\partial_\mu A^\mu)^2 \\ &= \frac{1}{2\lambda}A^\mu g_{\mu\nu}\partial^2 A^\nu\ ,\end{aligned} \tag{6}$$

where λ is some arbitrary gauge parameter. Again a surface term has been neglected. If we include the gauge fixing in (1), the arising operator is well defined,

$$\frac{1}{2}A^\mu\left(g_{\mu\nu}\partial^2 - (1-\lambda^{-1})\partial_\mu\partial_\nu\right)A^\nu\ , \tag{7}$$

and can be inverted to give the photon propagator in momentum space,

$$D_{\mu\nu}(k) = -\frac{1}{k^2}\left(g_{\mu\nu} - (1-\lambda)\frac{k_\mu k\nu}{k^2}\right)\ , \tag{8}$$

where $\lambda \to 1$ corresponds to the Feynman and $\lambda \to 0$ to the Landau gauge. However, in QED such a gauge-fixing term does not affect the general physics of the theory because the field $\chi = \partial_\mu A^\mu$, i.e. the longitudinal part of the photon field does not interact with physical degrees of freedom. According to the Abelian Maxwell equation it obeys a free field equation $\partial^2\chi = 0$ and therefore

Example 4.3

does not mix with the transverse part of the photon field. The corresponding situation for non-Abelian theories is much more complicated. For instance in QCD the gauge-fixing field $\chi^a = \partial^\mu A^a_\mu$ obeys the non-Abelian extension of the wave equation

$$\partial^2 \chi^a + g f^{abc} A^{b\,\mu} \partial_\mu \chi^c = 0 \ . \tag{9}$$

Therefore the unphysical χ particles, i.e. the longitudinal part of the gluon field, can interact with the transverse (physical) components of A^a_μ. Those unphysical longitudinal components contribute to gluon loops and therefore have to be substracted. This substraction can be done by the introduction of unphysical ghost fields η^a that exactly cancel the χ^a fields. The introduction of the η^a fields is most conviently done in the framework of path integral quantization[6] and was first done by De Witt, Fadeev, and Popov. In this framework it can be shown that the so-called Fadeev–Popov ghost fields obey the same equation (9) as the scalar χ^a fields, although they have to be quantized as fermion fields. Then, according to Fermi–Dirac statistics, each closed loop of ghost fields has to be accompanied by a factor -1. For each gluon loop one has to include one ghost loop which cancels exactly the longitudinal part of the gluons. The complete gauge-fixing term then can be written as

$$\mathcal{L} = \mathcal{L}_{\text{fix}} + \mathcal{L}_{\text{F.P.}} \tag{10}$$

with

$$\begin{aligned} \mathcal{L}_{\text{F.P.}} &= \partial_\mu \eta^{a\,\dagger} \left(\partial^\mu \eta^a + g f^{abc} A^{b,\mu} \eta^c \right) \\ &= -\eta^{a\,\dagger} \partial^\mu D^{ac}_\mu \eta^c \ , \end{aligned} \tag{11}$$

where again a surface term is neglected. From this Lagrangian one can derive the Feynman rules for the ghost propagators and ghost vertices.

EXAMPLE

4.4 The Running Coupling Constant

We discussed above that in general a perturbation series converges for a given coupling constant only at some momentum transfers. This behavior is easily understood, since the renormalized coupling constant is defined for a specific momentum transfer. In QED, for example, one can choose the incoming and outgoing photons to be on the mass shell, and then

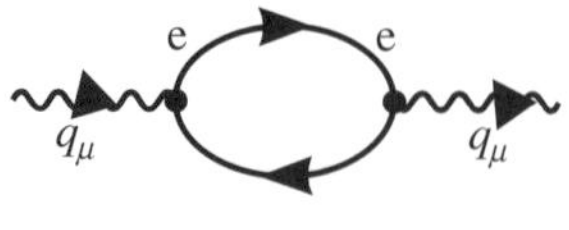

[6] See, e.g., W. Greiner and J. Reinhard: *Field Quantization* (Springer, Berlin, Heidelberg, 1996).

Example 4.4

$$e_{\mathrm{R}} = e\left[1 - \frac{\alpha}{2\pi}\log\left(\frac{\Lambda^2}{m^2}\right)\right]^{\frac{1}{2}} \qquad \text{for} \quad q^2 = 0 \ . \tag{1}$$

For other values of q^2 this graph then yields a finite, q^2-dependent renormalization, i.e., a finite additional contribution to every electromagnetic process. These corrections can be given analytically and taken into account by a redefinition of the electric charge. We obtain in a massless QED [7]

$$e_{\mathrm{R}}^2(q^2) = \frac{e_{\mathrm{R}}^2(q^2 = -M^2)}{1 - Ce_{\mathrm{R}}^2(q^2 = -M^2)\ln\left(q^2/-M^2\right)} \ . \tag{2}$$

The value of the constant C depends on the number of fermions considered by the theory and on the fermion charges. From (2) it also follows that $q^2 = 0$ is a special value, quite unsuitable for renormalization (here e_{R} is equal to zero). Therefore $q^2 = -M^2$ is usually chosen as the renormalization point. Equation (2) denotes the additional contributions to all massless QED graphs caused by vacuum polarization. $e_{\mathrm{R}}^2(q^2)$ is valid only for a certain q^2 range: if $q^2 < 0$, $Ce_{\mathrm{R}}^2(-M^2)\ln\left(q^2/-M^2\right) < 1$. In this range it can be interpreted as a physical charge. For $q^2 > 0$, $e_{\mathrm{R}}^2(q^2)$ becomes formally complex and such an interpretation is no longer possible. In the latter case $e_{\mathrm{R}}^2(q^2)$ reflects the fact that the following graphs lead to different phases.

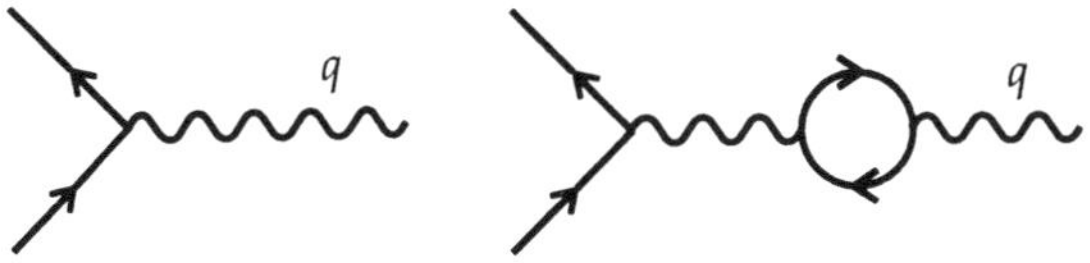

(3)

The divergence of this "charge" for

$$Ce_{\mathrm{R}}^2(-M^2)\ln\left(\frac{q^2}{-M^2}\right) \to 1$$

indicates the breakdown of perturbation theory. For values of $-q^2$ this large the contributions of two-loop corrections become as large as those of one-loop graphs in (1), and so on. If the higher corrections are also taken into account, we obtain instead of (2) the nonconvergent expression

$$e^2(q^2) = \frac{e^2(-M^2)}{1 - Ce^2(-M^2)\ln\left(\frac{q^2}{-M^2}\right) + C'e^4(-M^2)\ln^2\left(\frac{q^2}{-M^2}\right) + \dots} \ . \tag{4}$$

Since QCD is a scale-free theory, it is possible to define a running coupling constant for it, which takes the finite corrections due to the graphs shown in Fig. 4.7 into account. We get

$$\alpha_{\mathrm{s}}(q^2) = \frac{\alpha_{\mathrm{s}}(-M^2)}{1 + \frac{11 - \frac{2}{3}N_{\mathrm{f}}(q^2)}{4\pi}\alpha_{\mathrm{s}}(-M^2)\ln\left(\frac{q^2}{-M^2}\right)} \ , \tag{5}$$

[7] See W. Greiner and J. Reinhardt: *Quantum Electrodynamics*, 2nd. ed. (Springer, Berlin, Heidelberg 1994)

Fig. 4.7. QCD graphs taken into account by $\alpha_s(q^2)$

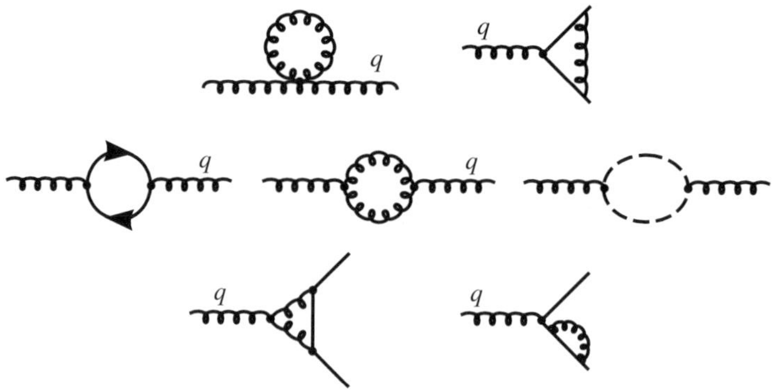

where N_f denotes the number of quark flavors with a mass much smaller than $\frac{1}{2}\sqrt{-q^2}$. We shall derive (5) in Sect. 4.4 after having introduced the necessary techniques in Sect. 4.3. Here we only want to discuss its phenomenological meaning. Equation (5) includes two parameters M^2 and $\alpha_s(M^2)$, which, however, are not independent of each other. It is possible to introduce a quantity $\Lambda = \Lambda(M)$ such that

$$\frac{11 - \frac{2}{3}N_f(q^2)}{4\pi}\alpha_s(-M^2)\ln(\Lambda^2/M^2) = -1 \ , \tag{6}$$

leading to

$$\alpha_s(q^2) = \frac{4\pi}{(11 - \frac{2}{3}N_f(q^2))\ln(-q^2/\Lambda^2)} \ . \tag{7}$$

Hence the running coupling constant is fixed for all momentum transfers by one parameter. Λ can be determined by investigating highly energetic e^+e^- pair annihilation and many other processes. The "world average" of these results is

$$\alpha_s\big(-(34\,\text{GeV})^2\big) \approx 0.14 \pm 0.02 \ . \tag{8}$$

Clearly the q^2 dependence of the strong coupling constant is determined by N_f, i.e., the number of quarks with $M^2 < |q^2|$. In Fig. 4.8, (7) is plotted for $N_f = 6$

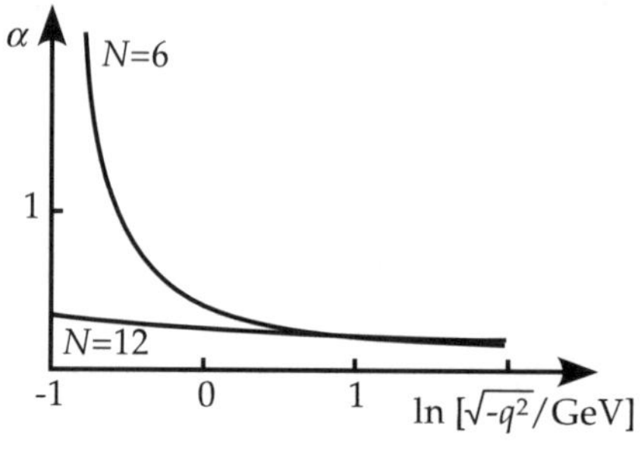

Fig. 4.8. Running coupling constants for the unphysical case of 6 and 12 different massless quarks. Both curves assume $\alpha(-100\,\text{GeV}^2) = 0.2$

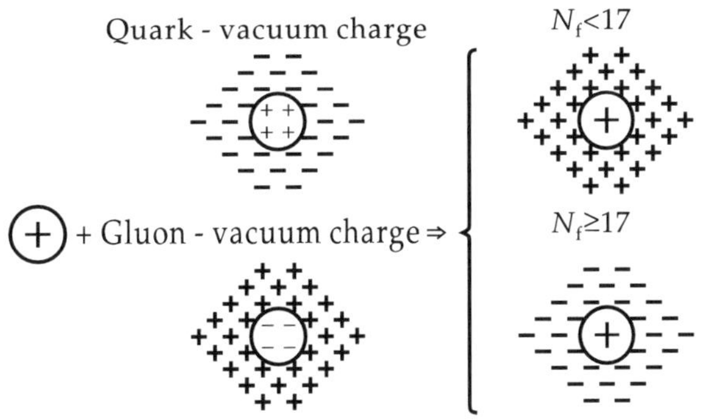

Fig. 4.9. Charge screening in QCD. N_f denotes the number of (massless) quarks

and $N_f = 12$. Λ has been fixed to obey (8). The plot for $N_f = 6$ increases rapidly for small values of $\sqrt{-q^2}$, i.e., the interaction can certainly not be described by perturbation theory below 500 MeV energy transfer. This conclusion is not valid for $N_f = 12$. Since at these energies the real physical hadrons have a strongly correlated structure rather than appearing as a group of free quarks, it follows immediately that the number of "light" quarks cannot be much greater than six. In fact only two "light"-quark doublets have been discovered. This example yields

$$\alpha_s\left(-(100\,\mathrm{GeV})^2\right) = 0.2 \quad \Rightarrow \quad \Lambda = 112\,\mathrm{MeV} \quad \text{for} \quad N_f = 6 \ . \tag{9}$$

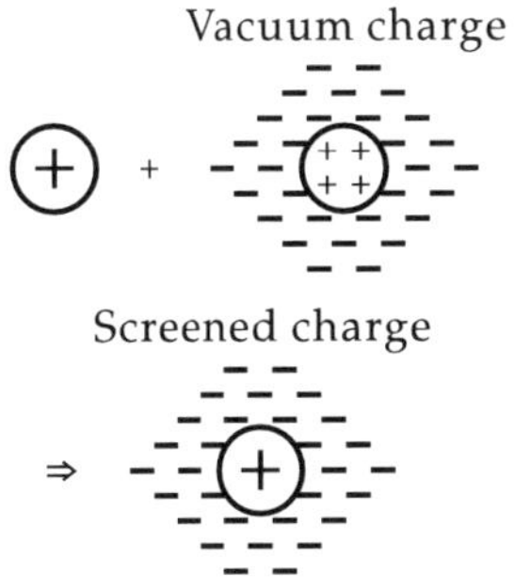

Fig. 4.10. Charge screening in QED

The currently discussed values for the scale parameter of QCD, Λ_{QCD}, range from 100 MeV to 300 MeV. It should be mentioned that Fig. 4.8 contains an invalid simplification. In contradiction to this figure, N_f decreases with decreasing $\sqrt{-q^2}$, because an increasing number of quarks must be considered massive.

Equation (7) can be interpreted as antiscreening of the charge unlike the usual screening in QED. The vacuum polarization graph in (1) leads in the case of QED to a polarization of the vacuum, which in the context of field theory is a medium with well-defined, nontrivial features.[8] This behavior is depicted in Fig. 4.10 for an extended charge. In QCD the corresponding graph with a virtual quark loop shows the same effect. Here, however, there is also a gluon contribution. Since the gluons carry a charge, their virtual excitations can also be polarized. Therefore the analogous graph in Fig. 4.9 is somewhat more complicated. In particular, the total charge distribution depends on the dominance of the quark or gluon part. Although the latter is defined by the gauge group, the first part increases with every additional quark until it dominates the theory and QCD shows the same behavior as QED. From this argument it also follows that the antiscreening in SU(N) gauge theories in principle increases with N. Even the explicit form of that dependence can be understood. The gluon contribution is simply proportional to the number of possible permutations for the loop gluons in Fig. 4.11.

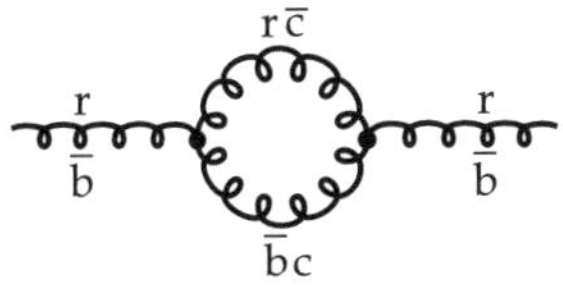

Fig. 4.11. Gluon vacuum polarization for QCD

[8] See W. Greiner and J. Reinhardt: *Quantum Electrodynamics*, 2nd. ed. (Springer, Berlin, Heidelberg 1994)

Example 4.4

For QCD, i.e. SU(3), c can assume three values; for SU(N) there are correspondingly N values. Therefore (7) becomes, in the case of an SU(N) gauge theory,

$$\alpha_N(q^2) = \frac{4\pi}{\left(11 \times \frac{1}{3}N - \frac{2}{3}N_{\mathrm{f}}\right) \ln\left(-q^2/\Lambda_N^2\right)} \ . \tag{10}$$

4.3 Dimensional Regularization

Currently so-called *dimensional regularization* is considered to be the standard procedure for regularizing quantum field theories. Only in special cases, where some specific disadvantages of this method show up, are other procedures, such as Pauli–Villars regularization or the momentum-cutoff method, employed. For lattice gauge theories, however, regularization is automatically provided by the lattice constant, which necessarily yields a rather special regularization scheme. In the following we discuss only dimensional regularization.

The basic observation that motivated dimensional regularization is that only logarithmic divergences in standard quantum field theories are encountered and these vanish for any dimensionality smaller than 4. Thus if it is possible to define a generalized integration for noninteger dimensions "d" the usual divergences will show up as poles in "$d-4$" which should be relatively easy to isolate. To realize this idea it is advisable first to simplify the divergences that occur by a trick known as "*Wick rotation*", namely by continuing it to imaginary energies. As mentioned above, only logarithmic divergences are encountered, so a typical divergent integral has the form

$$\mathrm{i}\int \mathrm{d}^4k \frac{1}{(k^2 - m_1^2 + \mathrm{i}\varepsilon)} \frac{1}{\left[(p+k)^2 - m_2^2 + \mathrm{i}\varepsilon\right]} \ . \tag{4.59}$$

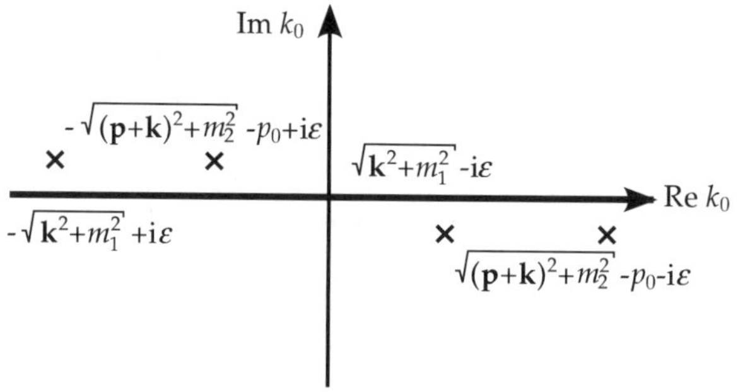

Fig. 4.12. Poles of the integrand in (4.54) for $p_0 < \sqrt{(\boldsymbol{p}+\boldsymbol{k})^2 + m_2^2}$

One awkward property of this integral is that its behavior for large k_μ is not uniform, since $k^2 = (k^0)^2 - (\boldsymbol{k})^2$ can stay small even when k^0 and $|\boldsymbol{k}|$ both become large. This problem can be circumvented by continuing k^0 to $\mathrm{i}k_E^0$ $(\boldsymbol{k} \to \boldsymbol{k}_E)$, implying that

$$k^2 \to -(k_E^0)^2 - (\boldsymbol{k}_E)^2 = -k_E^2 \ ,$$

with $k_E^2 = (k_E^0)^2 + (\boldsymbol{k}_E)^2$. $(k_E)_\mu$ is obviously a Cartesian vector that becomes uniformly large if $(k_E)_\mu \to \infty$, so it is much easier to analyze the ultraviolet divergences for $(k_E)_\mu$ than for k_μ.

To perform the Wick rotation one has to deform the integration path of k^0 in the complex plane. Figures 4.12 and 4.13 show the positions of the poles. We now use

$$\int_{-\infty}^{\infty} \mathrm{d}k^0 \cdots + \int_{+\mathrm{i}\infty}^{-\mathrm{i}\infty} \mathrm{d}k^0 \cdots + \int_{\mathcal{C}_1} \mathrm{d}k^0 \cdots + \int_{\mathcal{C}_2} \mathrm{d}k^0 \ldots$$

$$= \begin{cases} 0 & \text{for Fig. 4.12} \\ \text{residue for } k_0 = \sqrt{(\boldsymbol{p}+\boldsymbol{k})^2 - m_2^2} - p_0; & \text{for Fig. 4.13} \end{cases} \ . \tag{4.60}$$

The ε prescription in (4.59) guarantees that the integrals over the arcs at infinity, $\int_{\mathcal{C}_1} \mathrm{d}k^0$ and $\int_{\mathcal{C}_2} \mathrm{d}k^0$, vanish. This can easily be seen if we use

$$\frac{1}{(k^2 - m_1^2 + \mathrm{i}\varepsilon)} \frac{1}{(p+k)^2 - m_2^2 + \mathrm{i}\varepsilon}$$
$$= -\mathrm{i} \int_0^\infty \mathrm{d}\alpha\, \mathrm{e}^{\mathrm{i}\alpha(k^2 - m_1^2 + \mathrm{i}\varepsilon)} (-\mathrm{i}) \int_0^\infty \mathrm{d}\beta\, \mathrm{e}^{\mathrm{i}\beta[(p+k)^2 - m_2^2 + \mathrm{i}\varepsilon]} \ . \tag{4.61}$$

From $k^0 = R\,\mathrm{e}^{\mathrm{i}\varphi}$, $0 \le \varphi \le \frac{\pi}{2}$, and $\pi \le \varphi \le \frac{3}{2}\pi$ with $R \to \infty$, we get

$$\mathrm{Im}(k^2) = R^2 \sin(2\varphi) + O(R) > 0 \ , \tag{4.62}$$

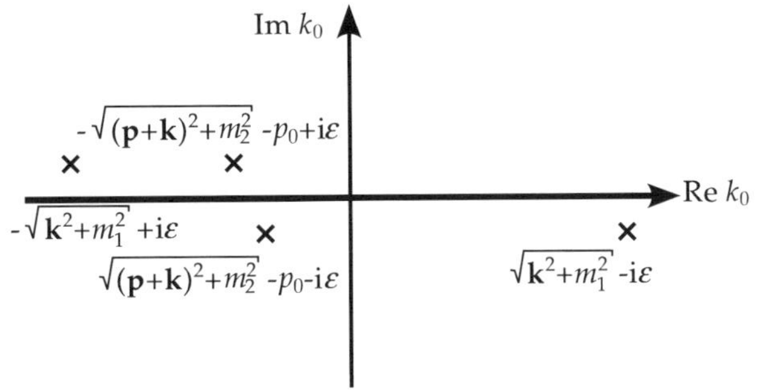

Fig. 4.13. Poles of the integrand in (4.54) for $p_0 > \sqrt{(\boldsymbol{p}+\boldsymbol{k})^2 + m_2^2}$

such that all terms vanish exponentially. The result of the Wick rotation is therefore

$$
\begin{aligned}
&\mathrm{i}\int\limits_{-\infty}^{\infty}\mathrm{d}^4k\,\frac{1}{k^2-m_1^2+\mathrm{i}\varepsilon}\,\frac{1}{(p+k)^2-m_2^2+\mathrm{i}\varepsilon}\\
&\quad=-\int\limits_{-\infty}^{\infty}\mathrm{d}^4k_E\,\frac{1}{k_E^2+m_1^2-\mathrm{i}\varepsilon}\,\frac{1}{(p_E+k_E)^2+m_2^2-\mathrm{i}\varepsilon}\\
&\qquad+\text{ possibly a finite residuum }.
\end{aligned}
\tag{4.63}
$$

The residuum that possibly appears is finite. As we intend to isolate and substract the divergent part of the integral (4.59) we do not have to bother about this finite contribution. *We define the renormalized integral as* (4.59) *minus the divergent part of* (4.53). This definition is not affected by the finite residuum.

Next we proceed to define an abstract mathematical operation which we want to call the d-dimensional integral:

$$
\int \mathrm{d}^d k_E\, f(k_E)\ .
$$

First we impose the following conditions:

(1) Linearity

$$
\begin{aligned}
&\int \mathrm{d}^d k_E[a\, f(k_{E\mu})+bg(k_{E\mu})]\\
&\quad= a\int \mathrm{d}^d k_E\, f(k_{E\mu})+b\int \mathrm{d}^d k_E\, g(k_{E\mu})\ .
\end{aligned}
\tag{4.64}
$$

(2) Invariance of the integral under finite shifts (p_μ finite)

$$
\int \mathrm{d}^d k_E f(k_{E\mu}) = \int \mathrm{d}^d k_E f(k_{E\mu}+p_{E\mu})\ . \tag{4.65}
$$

(3) A scaling property

$$
\int \mathrm{d}^d k_E f(\lambda k_{E\mu}) = \lambda^{-d}\int \mathrm{d}^d k_E f(k_{E\mu})\ . \tag{4.66}
$$

From these definitions it is not clear what is meant by a Lorentz-vector k_μ in d dimensions. The point is that it is always sufficient to treat scalar integrals. Any vectorlike integral is completely specified by its values if contracted with few linearly independent vectors and those scalar integrals can be analytically continued. Bearing this in mind it is, however, helpful to use vectors like k_μ as a shorthand for a collection of suitably defined scalar products.

The d-dimensional integral is now defined by its action on a set of basis functions

$$\lambda(A, P_\mu) = \mathrm{e}^{-A(k+P)^2} \ . \tag{4.67}$$

Properties (4.64) and (4.65) guarantee that all functions decomposed in this basis can be integrated once the integral

$$\int \mathrm{d}^d k_E \, \mathrm{e}^{-Ak_E^2} \tag{4.68}$$

is known. Here the scaling property (4.66) becomes important, since it reduces any such integral to a single one:

$$\int \mathrm{d}^d k_E \, \mathrm{e}^{-Ak_E^2} = A^{-\frac{d}{2}} \int \mathrm{d}^d k_E \, \mathrm{e}^{-k_E^2} \ . \tag{4.69}$$

As we want the d-dimensional integral to coincide with the usual one for integer values of d we impose the condition

$$\int \mathrm{d}^d k_E \, \mathrm{e}^{-k_E^2} = \pi^{\frac{d}{2}} \ . \tag{4.70}$$

This relation is verified for integer d in Exercise 4.5. We note for later use that in Exercise 4.5, (8), we also derive

$$\int_{-\infty}^{\infty} \mathrm{d}^d k \, \mathrm{e}^{-k^2} = \frac{1}{(d-2)!!} \left\{ \begin{array}{ll} (2\pi)^{\frac{d}{2}} & d \text{ even} \\ 2\,(2\pi)^{\frac{d-1}{2}} & d \text{ odd} \end{array} \right\} \int_0^{\infty} \mathrm{d}k \, k^{d-1} \mathrm{e}^{-k^2} \ . \tag{4.71}$$

Now, for any function $f(k)$ which can be expanded into Gaussians, i.e.

$$f(k) = \sum_n c_n \, \mathrm{e}^{-A_n(k+p_n)^2} \ , \tag{4.72}$$

one obtains

$$\int_{-\infty}^{\infty} \mathrm{d}^d k_E \, f(k_E^2) = \frac{1}{(d-2)!!} \left\{ \begin{array}{ll} (2\pi)^{\frac{d}{2}} & d \text{ even} \\ 2\,(2\pi)^{\frac{d-1}{2}} & d \text{ odd} \end{array} \right\} \times \int_0^{\infty} \mathrm{d}k_E \, k_E^{d-1} f(k_E^2) \ . \tag{4.71b}$$

With $\Gamma\left(\frac{1}{2}\right) = \sqrt{\pi}$ and the unifying relation

$$\frac{2(\pi)^{\frac{d}{2}}}{\Gamma\left(\frac{d}{2}\right)} = \left\{ \begin{array}{lll} d \text{ even}: \ 2\pi^{\frac{d}{2}} \times \frac{2}{d-2} \times \frac{2}{d-4} \times \cdots \times \frac{2}{2} & = & \frac{(2\pi)^{\frac{d}{2}}}{(d-2)!!} \\ d \text{ odd}: \ 2\pi^{\frac{d}{2}} \times \frac{2}{d-2} \times \frac{2}{d-4} \times \cdots \times \frac{2}{1} \times \frac{1}{\sqrt{\pi}} & = & \frac{2(2\pi)^{\frac{d-1}{2}}}{(d-2)!!} \end{array} \right. \tag{4.73}$$

(4.71b) can be written as

$$\int_{-\infty}^{\infty} \mathrm{d}^d k_E\, f(k_E^2) = \frac{2(\pi)^{\frac{d}{2}}}{\Gamma\left(\frac{d}{2}\right)} \int_0^{\infty} \mathrm{d}k_E\, k_E^{d-1} f(k_E^2) \ . \tag{4.71c}$$

This equation is most important and will be used constantly during dimensional regularization.

EXERCISE

4.5 The d-Dimensional Gaussian Integral

Problem. Derive (4.65) for integer d.

Solution. Omitting the subscript "E" in this exercise, we define for integer d and Euclidian k_μ, $\mu = 0, 1, 2, \dots, d-1$,

$$\begin{aligned}
k_0 &= k \cos\vartheta_1 \ , \\
k_1 &= k \sin\vartheta_1 \cos\vartheta_2 \ , \\
k_2 &= k \sin\vartheta_1 \sin\vartheta_2 \cos\vartheta_3 \ , \\
&\vdots \\
k_{d-2} &= k \sin\vartheta_1 \sin\vartheta_2 \dots \sin\vartheta_{d-2} \cos\vartheta_{d-1} \ , \\
k_{d-1} &= k \sin\vartheta_1 \sin\vartheta_2 \dots \sin\vartheta_{d-2} \sin\vartheta_{d-1}
\end{aligned} \tag{1}$$

and calculate the Jacobian row by row

$$\begin{aligned}
J &= \frac{\partial(k_0 k_1 k_2 \dots k_{d-1})}{\partial(k, \vartheta_1, \vartheta_2 \dots \vartheta_{d-1})} \\
&= \begin{vmatrix}
\cos\vartheta_1 & -k\sin\vartheta_1 & 0 & \dots \\
\sin\vartheta_1\cos\vartheta_2 & k\cos\vartheta_1\cos\vartheta_2 & -k\sin\vartheta_1\sin\vartheta_2 & \dots \\
\sin\vartheta_1\sin\vartheta_2\cos\vartheta_3 & k\cos\vartheta_1\sin\vartheta_2\cos\vartheta_3 & k\sin\vartheta_1\cos\vartheta_2\cos\vartheta_3 & \dots \\
\vdots & \vdots & \vdots & \ddots
\end{vmatrix} \\
&= k^{d-1}\left\{\cos\vartheta_1\cos\vartheta_1(\sin\vartheta_1)^{d-2} + \sin\vartheta_1\sin\vartheta_1(\sin\vartheta_1)^{d-2}\right\} \\
&\quad\times \begin{vmatrix}
\cos\vartheta_2 & -\sin\vartheta_2 & \dots \\
\sin\vartheta_2\cos\vartheta_3 & \cos\vartheta_2\cos\vartheta_3 & \dots \\
\vdots & \vdots & \ddots
\end{vmatrix} = \dots \\
&= k^{d-1}(\sin\vartheta_1)^{d-2}(\sin\vartheta_2)^{d-3}\dots(\sin\vartheta_{d-3})^2 \\
&\quad\times \begin{vmatrix}
\cos\vartheta_{d-2} & -\sin\vartheta_{d-2} & 0 \\
\sin\vartheta_{d-2}\cos\vartheta_{d-1} & \cos\vartheta_{d-2}\cos\vartheta_{d-1} & -\sin\vartheta_{d-2}\sin\vartheta_{d-1} \\
\sin\vartheta_{d-2}\sin\vartheta_{d-1} & \cos\vartheta_{d-2}\sin\vartheta_{d-1} & \sin\vartheta_{d-2}\cos\vartheta_{d-1}
\end{vmatrix}
\end{aligned}$$

Exercise 4.5

$$
\begin{aligned}
&= k^{d-1}(\sin\vartheta_1)^{d-2}(\sin\vartheta_2)^{d-3}\dots(\sin\vartheta_{d-2}) \\
&\quad\times\begin{vmatrix} \cos\vartheta_{d-1} & -\sin\vartheta_{d-1} \\ \sin\vartheta_{d-1} & \cos\vartheta_{d-1} \end{vmatrix} \\
&= k^{d-1}(\sin\vartheta_1)^{d-2}(\sin\vartheta_2)^{d-3}\dots(\sin\vartheta_{d-2}) \ . \qquad (2)
\end{aligned}
$$

The Gaussian integral thus becomes

$$
\begin{aligned}
&\int\limits_{-\infty}^{\infty} \mathrm{d}^d k \, \mathrm{e}^{-k^2} \qquad (3) \\
&= \int\limits_0^{\infty} \mathrm{d}k \, k^{d-1} \, \mathrm{e}^{-k^2} \int\limits_0^{\pi} (\sin\vartheta_1)^{d-2} \, \mathrm{d}\vartheta_1 \dots \int\limits_0^{\pi} \sin\vartheta_{d-2} \, \mathrm{d}\vartheta_{d-2} \int\limits_0^{\pi} \mathrm{d}\vartheta_{d-1} \ .
\end{aligned}
$$

We use partial integration to get

$$
\begin{aligned}
&\int\limits_0^{\pi} (\sin\vartheta)^{n-1} \sin\vartheta \, \mathrm{d}\vartheta \\
&= \Big[(\sin\vartheta)^{n-1}(-\cos\vartheta)\Big]_0^{\pi} - (n-1)\int\limits_0^{\pi} (\sin\vartheta)^{n-2}\cos\vartheta(-\cos\vartheta)\mathrm{d}\vartheta \\
&= (n-1)\int\limits_0^{\pi} \Big[(\sin\vartheta)^{n-2} - (\sin\vartheta)^{n}\Big]\mathrm{d}\vartheta \qquad (4)
\end{aligned}
$$

or

$$
\int\limits_0^{\pi} (\sin\vartheta)^n \mathrm{d}\vartheta = \frac{n-1}{n}\int\limits_0^{\pi} (\sin\vartheta)^{n-2}\mathrm{d}\vartheta \ .
$$

For even $n \geq 2$ we thus get

$$
\int\limits_0^{\pi} (\sin\vartheta)^n \mathrm{d}\vartheta = \frac{(n-1)!!}{n!!}\int\limits_0^{\pi} \mathrm{d}\vartheta = \frac{(n-1)!!}{n!!}\pi \ , \qquad (5)
$$

and for $n = 0$ we have

$$
\int\limits_0^{\pi} \mathrm{d}\vartheta = \pi \ . \qquad (6)
$$

For odd n we find

$$
\begin{aligned}
\int\limits_0^{\pi} (\sin\vartheta)^n \mathrm{d}\vartheta &= \frac{(n-1)!!}{n!!}\int\limits_0^{\pi} \sin\vartheta \, \mathrm{d}\vartheta \\
&= \frac{(n-1)!!}{n!!} \times 2 \ . \qquad (7)
\end{aligned}
$$

Exercise 4.5

Inserting all these equations into (3) we get

$$\int\limits_{-\infty}^{\infty} \mathrm{d}^d k\, \mathrm{e}^{-k^2} = \int\limits_0^{\infty} \mathrm{d}k\, k^{d-1} \mathrm{e}^{-k^2} I_d \frac{1}{(d-2)!!} \;,$$

$$d \text{ even} \Rightarrow I_d = (2\pi)^{\frac{d}{2}} \;,$$

$$d \text{ odd} \Rightarrow I_d = 2 \times (2\pi)^{\frac{d-1}{2}} \;. \tag{8}$$

On the other hand, partial integration gives

$$\begin{aligned}\int\limits_0^{\infty} \mathrm{d}k\, k^{d-1}\, \mathrm{e}^{-k^2} &= \frac{1}{2} \int\limits_0^{\infty} \mathrm{d}k\, k^{d-2} 2k\, \mathrm{e}^{-k^2} \\ &= \frac{d-2}{2} \int\limits_0^{\infty} \mathrm{d}k\, k^{d-3}\, \mathrm{e}^{-k^2} \\ &= (d-2)!!K_d \;,\end{aligned} \tag{9}$$

$$d \text{ even} \Rightarrow K_d = \left(\frac{1}{2}\right)^{\frac{d-2}{2}} \int\limits_0^{\infty} \mathrm{d}k\, k\mathrm{e}^{-k^2} = \left(\frac{1}{2}\right)^{\frac{d}{2}} \;,$$

$$d \text{ odd} \Rightarrow K_d = \left(\frac{1}{2}\right)^{\frac{d-1}{2}} \int\limits_0^{\infty} \mathrm{d}k\, \mathrm{e}^{-k^2} = \left(\frac{1}{2}\right)^{\frac{d-1}{2}} \frac{\sqrt{\pi}}{2} \;. \tag{10}$$

All together this gives

$$\int\limits_{-\infty}^{\infty} \mathrm{d}^d k\, \mathrm{e}^{-k^2} = \frac{1}{(d-2)!!} (d-2)!!\; I_d K_d = \pi^{\frac{d}{2}} \;, \tag{11}$$

which completes our proof.

With (4.67) all integrals can be expanded in a sum of Gaussians,

$$f(k) = \sum_n c_n\, \mathrm{e}^{-A_n (k_E + p_{nE})^2} \;, \tag{4.74}$$

and $f(k)$ can now be integrated d-dimensionally, i.e.

$$\int \mathrm{d}^d k_E\; f(k_E) = \sum_n c_n\, A_n^{-\frac{d}{2}}\, \pi^{\frac{d}{2}} \;. \tag{4.71d}$$

However, it is not clear whether this sum converges. If it does for some integer d, i.e., if the original integrand converges, then the expression (4.71d) also converges. This is not true for divergent integrals. In this case the d-dimensional integral may still be finite. Thus all finite parts of the momentum integrals are reproduced. The infinite parts, associated, for example, with integrands like $1/(k_E^2)^\alpha$ or $(k_E^2)^\alpha$, which we want to subtract, will lead to infinite nonconvergent sums over Gaussians, the resulting term on the right-hand side of (4.71b). These will be subtracted during renormalization. We will see below that these infinite parts are different for d-dimensional integrals than for normal ones. Actually many divergences of usual integrals vanish automatically for d-dimensional integrals, and the important points to keep in mind throughout this discussion are:

d-dimensional convergent integrals for $d \to n$ (where n is an integer) coincide with the normal convergent n-dimensional integrals.

n-dimensional divergent integrals are in general different from the corresponding $d \to n$ limit of the d-dimensional integral. The latter is in general less divergent.

For example, for dimensional reasons the integral

$$\int \mathrm{d}^d k \, \frac{1}{(k^2+m^2)^\mu} \tag{4.75}$$

behaves as $(m^2)^{d/2-\mu}$. Thus for $d=4$ we will encounter a logarithmic divergence for $\mu=2$, and a powerlike divergence for $\mu<2$ which has to be regularized by a cut-off in conventional n-dimensional integration. However, we will show (see (4.81) below) that in dimensional regularization

$$\int \mathrm{d}^d k \, \frac{1}{(k^2+m^2)^\mu} \sim \Gamma(\mu-d/2)\,(m^2)^{d/2-\mu} \tag{4.76}$$

does pick up only logarithmic divergences even for $\mu<2$. For instance, for $\mu=1$ and $d=4+2\varepsilon$ one gets

$$\begin{aligned}\Gamma(1-2-\varepsilon) = \Gamma(-1-\varepsilon) = \frac{-\Gamma(-\varepsilon)}{1+\varepsilon} &= +(1-\varepsilon)\left(\frac{1}{\varepsilon}+\gamma_\mathrm{E}\right) \\ &= \frac{1}{\varepsilon}+\gamma_\mathrm{E}-1+\mathcal{O}(\varepsilon)\end{aligned} \tag{4.77}$$

which clearly is only a (dimensionless) logarithmic divergence.

It is crucial to understand that our definition of a d-dimensional integral does not always reduce to a normal integral in integer d dimensions. Instead it has peculiar properties, for example,

$$\int \mathrm{d}^d k_E \, \left(k_E^2\right)^\nu = 0 \qquad \text{for any } \nu \;, \tag{4.78}$$

which follows directly from the scaling axiom (4.66). Indeed we have

$$\int_{-\infty}^{\infty} \mathrm{d}^d k_E \left(\lambda^2 k_E^2\right)^{\nu} = \lambda^{2\nu} \int_{-\infty}^{\infty} \mathrm{d}^d k_E \left(k_E^2\right)^{\nu} = \lambda^{-d} \int_{-\infty}^{\infty} \mathrm{d}^d k_E \left(k_E^2\right)^{\nu} . \quad (4.79)$$

Because this holds for any λ, ν, d, one is directly led to (4.78). Actually $\nu = -d/2$ is the only power for which (4.78) does not imply the vanishing of this integral, but since we make critical use of the fact that our renormalized integrals are analytic functions of all variables involved, like ν in (4.78), we have also to define

$$\int \mathrm{d}^d k_E \left(k_E^2\right)^{-\frac{d}{2}} = 0 . \quad (4.80)$$

This derivation of (4.78) is actually not without problems, because it assumes that $\int \mathrm{d}^d k_E \,(k_E^2)^{\nu}$ exists, which is only true after a specific definition such as $\lim_{m^2 \to 0} \int \mathrm{d}^d k_E \,(k_E^2 + m^2)^{\nu}$. To be on the safe side it is better to regard (4.78) as a definition that is made plausible by (4.65). A peculiar feature of (4.78) seems to be that it leads to finite results even for theories for which arbitrarily high divergences occur. While this is formally true it is of no importance since such theories cannot be renormalized. They would require an infinite number of renormalization constants and counterterms and thus cannot be formulated consistently. As an example let us mention a scalar field theory with a ϕ^6 coupling. This coupling leads to divergent graphs like

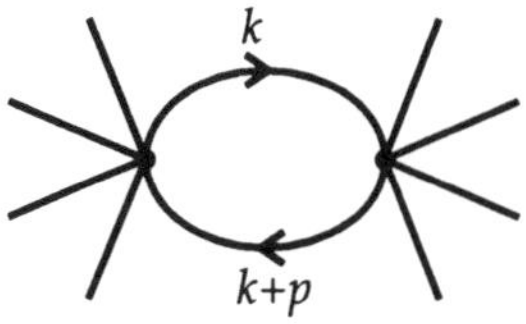

which is divergent $\left(\sim \int \mathrm{d}^4 k \left[1/k^2 (k+p)^2\right]\right)$ and must be renormalized, leading to a counter term of the form

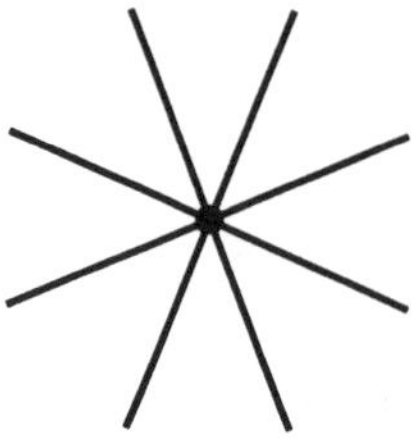

which is a ϕ^8 interaction. Continuing this argument we find that the renormalized theory has infinitely many interaction terms of the general form $\sum_{n=3}^{\infty} c_n \phi^{2n}$, and no consistent theory can be formulated. Note that for a ϕ^4 interaction the counterterm is again ϕ^4. This is the reason why scalar theories up to ϕ^4 terms in the

Lagrangian are renormalizable. In fact only those theories can be renormalized that have at most logarithmic divergences, and thus integrals like

$$\int \mathrm{d}^4k \left(k^2\right)^3$$

cannot occur. For renormalizable theories the differences between n-dimensional integrals and d-dimensional integrals (with the limit $d \to n$) is therefore irrelevant.

Equation (4.78) is essential for deriving the following basic formula:

$$I_{\nu,\mu} = \int \mathrm{d}^d k_E \, \frac{\left(k_E^2\right)^\nu}{\left(k_E^2+M^2\right)^\mu} = \pi^{\frac{d}{2}} \left(M^2\right)^{\nu-\mu+\frac{d}{2}} \frac{\Gamma\left(\nu+\frac{d}{2}\right)\Gamma\left(\mu-\nu-\frac{d}{2}\right)}{\Gamma\left(\frac{d}{2}\right)\Gamma(\mu)} \,, \tag{4.81}$$

which we shall prove next. Using (4.78) we can rewrite $I_{\nu,\mu}$ in the following manner:

$$\begin{aligned} I_{\nu,\mu} &= \int \mathrm{d}^d k_E \, \frac{\left(k_E^2\right)^\nu}{\left(k_E^2+M^2\right)^\mu} = M^{2\nu-2\mu+d} \int \mathrm{d}^d k_E \, \frac{\left(k_E^2\right)^\nu}{\left(k_E^2+1\right)^\mu} \\ &= M^{2\nu-2\mu+d} \int \mathrm{d}^d k_E \left[\frac{\left(k_E^2\right)^\nu}{\left(k_E^2+1\right)^\mu} - \sum_n \left(k_E^2\right)^{\nu_n} c_n \right] \end{aligned} \tag{4.82}$$

with arbitrary constant coefficients ν_n and c_n. These coefficients are now chosen such that they cancel all potential divergences, i.e., all high powers of k^2. The integral is then convergent and we simply apply (4.71c):

$$I_{\nu,\mu} = M^{2\nu-2\mu+d} \frac{2\pi^{\frac{d}{2}}}{\Gamma\left(\frac{d}{2}\right)} \int_0^\infty \frac{\mathrm{d}k_E^2}{2} \left(k_E^2\right)^{\frac{d}{2}-1} \left[\frac{\left(k_E^2\right)^\nu}{\left(k_E^2+1\right)^\mu} - \sum_n \left(k_E^2\right)^{\nu_n} c_n \right] . \tag{4.83}$$

The trick is now to derive by partial integration a recurrence relation for the $I_{\nu,\mu}$:

$$\begin{aligned} I_{\nu,\mu} = {} & \frac{\left(\nu+\frac{d}{2}-1\right)}{\mu-1} I_{\nu-1,\mu-1} + M^{2\nu-2\mu+d} \frac{2\pi^{\frac{d}{2}}}{\Gamma\left(\frac{d}{2}\right)} \frac{1}{2} \\ & \times \left\{ \left(k_E^2\right)^{\nu+\frac{d}{2}-1} \left[\frac{\left(k_E^2+1\right)^{1-\mu}}{1-\mu} - \sum_n \left(k_E^2\right)^{\nu_n+1-\nu} \frac{c_n}{\nu_n+1} \right] \right\}_0^\infty . \end{aligned} \tag{4.84}$$

The second term can obviously be absorbed into the c_ns. In other words, c_n can be chosen such that it vanishes. This leaves us with the relation

$$I_{\nu,\mu} = \frac{\nu+\frac{d}{2}-1}{\mu-1} I_{\nu-1,\mu-1} \qquad \text{for all real } \nu, \mu \ . \tag{4.85}$$

In view of the relation $\Gamma(z+1) = z\Gamma(z)$, which uniquely defines the gamma function, (4.85) suggests (d is fixed) the following ansatz for $I_{\nu,\mu}$:

$$I_{\nu,\mu} = \frac{\Gamma\left(\nu+\frac{d}{2}\right)}{\Gamma(\mu)} K(\nu-\mu) \tag{4.86}$$

with an arbitrary function $K(\nu-\mu)$. To fix $K(\nu-\mu)$ we use the fact that

$$\frac{\left(k_E^2\right)^\nu}{\left(k_E^2+1\right)^\mu} = \frac{\left(k_E^2\right)^\nu}{\left(k_E^2+1\right)^\mu}\left[\left(k_E^2+1\right)-k_E^2\right] , \tag{4.87}$$

$$I_{\nu,\mu} = I_{\nu,\mu-1} - I_{\nu+1,\mu} . \tag{4.88}$$

Inserting (4.86) into (4.88) gives

$$\begin{aligned}
\frac{\Gamma\left(\nu+\frac{d}{2}\right)}{\Gamma(\mu)} K(\nu-\mu) &= \left[\frac{\Gamma\left(\nu+\frac{d}{2}\right)}{\Gamma(\mu-1)} - \frac{\Gamma\left(\nu+\frac{d}{2}+1\right)}{\Gamma(\mu)}\right] K(\nu-\mu+1) \\
\Rightarrow K(\nu-\mu) &= \left[(\mu-1)-\left(\nu+\frac{d}{2}\right)\right] K(\nu-\mu+1) \\
&= \left(\mu-\nu-1-\frac{d}{2}\right) K(\nu-\mu+1)
\end{aligned} \tag{4.89}$$

$$\Rightarrow K(\nu-\mu) = \Gamma\left(\mu-\nu-\frac{d}{2}\right) \text{const} , \tag{4.90}$$

where we have again used that the relation $\Gamma(x+1) = x\Gamma(x)$ defines the gamma function uniquely, up to a constant. The remaining constant in

$$\int \mathrm{d}^d k_E \frac{\left(k_E^2\right)^\nu}{\left(k_E^2+M^2\right)^\mu} = \frac{2\pi^{d/2}}{\Gamma\left(\frac{d}{2}\right)} M^{2\nu-2\mu+d} \frac{\Gamma\left(\mu-\nu-\frac{d}{2}\right)\Gamma\left(\nu+\frac{d}{2}\right)}{\Gamma(\mu)} \text{const} \tag{4.91}$$

can easily be fixed. We choose $\nu = 1-d/2$ and $\mu = 2$. With (4.71c) we get, on the one hand,

$$\begin{aligned}
\int_0^\infty \mathrm{d}^d k_E \frac{\left(k_E^2\right)^{1-d/2}}{\left(k_E^2+M^2\right)^2} &= \frac{2\pi^{d/2}}{\Gamma\left(\frac{d}{2}\right)} \int_0^\infty \frac{\mathrm{d}k_E^2}{2} \frac{1}{\left(k_E^2+M^2\right)^2} \\
&= \frac{2\pi^{d/2}}{\Gamma\left(\frac{d}{2}\right)} \frac{1}{2} \left[\frac{-1}{k_E^2+M^2}\right]_0^\infty = \frac{\pi^{d/2}}{\Gamma\left(\frac{d}{2}\right)} \frac{1}{M^2}
\end{aligned} \tag{4.92}$$

and from (4.91), on the other hand,

$$\frac{2\pi^{d/2}}{\Gamma\left(\frac{d}{2}\right)} M^{2-d-4+d} \frac{\Gamma(2-1)\,\Gamma(1)}{\Gamma(2)} \text{const} = \frac{2\pi^{d/2}}{\Gamma\left(\frac{d}{2}\right)} \frac{1}{M^2} \text{const} . \tag{4.93}$$

Comparing both expressions we conclude that const $= 1/2$, or

$$\int \mathrm{d}^d k_E \frac{\left(k_E^2\right)^\nu}{\left(k_E^2+M^2\right)^\mu} = \frac{\pi^{d/2}}{\Gamma\left(\frac{d}{2}\right)} M^{2\nu-2\mu+d} \frac{\Gamma\left(\mu-\nu-\frac{d}{2}\right)\Gamma\left(\nu+\frac{d}{2}\right)}{\Gamma(\mu)} \ . \tag{4.94}$$

This is an important formula, since any integral appearing in the calculations of Feynman graphs can be brought into this form. The steps are:

(1) Introduce Feynman parameters according to

$$\begin{aligned} &\frac{1}{a_1^{\alpha_1} a_2^{\alpha_2} \dots a_m^{\alpha_m}} \\ &= \frac{\Gamma(\alpha_1+\alpha_2 \dots \alpha_m)}{\Gamma(\alpha_1)\Gamma(\alpha_2)\dots\Gamma(\alpha_m)} \int\limits_0^1 \mathrm{d}x_1 \int\limits_0^{x_1} \mathrm{d}x_2 \dots \int\limits_0^{x_{m-2}} \mathrm{d}x_{m-1} \\ &\quad \times \frac{x_{m-1}^{\alpha_1-1}(x_{m-2}-x_{m-1})^{\alpha_2-1}\dots(1-x_1)^{\alpha_m-1}}{[a_1 x_{m-1}+a_2(x_{m-2}-x_{m-1})+\dots+a_m(1-x_1)]^{\alpha_1+\alpha_2\dots\alpha_m}} \ . \end{aligned} \tag{4.95}$$

This relation will be derived in Exercise 4.7.

(2) Shift the momentum variable of the loop k_μ such that the linear term in the denominator vanishes:

$$\int \mathrm{d}^d k_E \frac{f(k_{E\mu})}{\left(k_E^2+m^2\right)^\beta} \ . \tag{4.96}$$

(3) Use the fact that all k_μ have to be contracted with some other k_λ according to $k_\mu k_\lambda \to g_{\mu\lambda} k^2/d$. Otherwise the term vanishes if individual components k_μ are inverted, $k_\mu \to -k_\mu$. The resulting expressions are sums of terms of the form (4.94).

For convenience we give a list of expressions which shows the results from different terms following steps 2 and 3. The integrals on the left-hand sides are in Minkowski space; therefore a factor "i" appears on the right, owing to the Wick rotation.

$$\int \mathrm{d}^d k \frac{1}{\left(k^2+2k\cdot p+m^2\right)^\alpha} = \frac{\mathrm{i}\pi^{d/2}}{\left(m^2-p^2\right)^{\alpha-d/2}} \frac{\Gamma(\alpha-d/2)}{\Gamma(\alpha)} \ . \tag{4.97}$$

$$\int \mathrm{d}^d k \frac{k_\mu}{\left(k^2+2k\cdot p+m^2\right)^\alpha} = \frac{\mathrm{i}\pi^{d/2}}{\left(m^2-p^2\right)^{\alpha-d/2}} \frac{\Gamma(\alpha-d/2)}{\Gamma(\alpha)}(-p_\mu) \ . \tag{4.98}$$

$$\begin{aligned} \int \mathrm{d}^d k \frac{k^2}{\left(k^2+2k\cdot p+m^2\right)^\alpha} &= \frac{\mathrm{i}\pi^{d/2}}{\left(m^2-p^2\right)^{\alpha-d/2}} \frac{1}{\Gamma(\alpha)}\left[\Gamma\left(\alpha-\frac{d}{2}\right)p^2\right. \\ &\quad \left. +\Gamma\left(\alpha-1-\frac{d}{2}\right)\frac{d}{2}\left(m^2-p^2\right)\right] \ . \end{aligned} \tag{4.99}$$

$$\int \mathrm{d}^d k \frac{k_\mu k_\nu}{\left(k^2+2k\cdot p+m^2\right)^\alpha} = \frac{\mathrm{i}\pi^{d/2}}{\left(m^2-p^2\right)^{\alpha-d/2}} \frac{1}{\Gamma(\alpha)}\left[\Gamma\left(\alpha-\frac{d}{2}\right) p_\mu p_\nu + \Gamma\left(\alpha-1-\frac{d}{2}\right)\frac{1}{2} g_{\mu\nu}\left(m^2-p^2\right)\right] . \tag{4.100}$$

$$\int \mathrm{d}^d k \frac{k_\mu k_\nu k_\lambda}{\left(k^2+2k\cdot p+m^2\right)^\alpha} = \frac{\mathrm{i}\pi^{d/2}}{\left(m^2-p^2\right)^{\alpha-d/2}} \frac{1}{\Gamma(\alpha)}\left[-\Gamma\left(\alpha-\frac{d}{2}\right) p_\mu p_\nu p_\lambda - \Gamma\left(\alpha-1-\frac{d}{2}\right)\left(m^2-p^2\right) \times \frac{1}{2}\left(g_{\mu\nu}p_\lambda + g_{\mu\lambda}p_\nu + g_{\nu\lambda}p_\mu\right)\right] . \tag{4.101}$$

To check that (4.94) also reproduces (4.80) we use the following property of the Γ function:

$$\begin{aligned} &\Gamma(\beta+1) = \beta\Gamma(\beta) \\ \Rightarrow &\lim_{\beta\to 0}\Gamma(\beta) = \lim_{\beta\to 0}\frac{1}{\beta}\Gamma(\beta+1) \\ &= \lim_{\beta\to 0}\frac{1}{\beta}\left[1+\beta\Gamma'(1)+O\left(\beta^2\right)\right] = \frac{1}{\beta}+\Gamma'(1) . \end{aligned} \tag{4.102}$$

This implies that

$$\begin{aligned} &\lim_{\mu\to 0}\int \mathrm{d}^d k_E \frac{\left(k_E^2\right)^\nu}{\left(k_E^2+M^2\right)^\mu} \\ &= \lim_{\mu\to 0}\frac{\pi^{d/2}}{\Gamma\left(\frac{d}{2}\right)} M^{2\nu+d}\,\Gamma\left(-\nu-\frac{d}{2}\right)\Gamma\left(\nu+\frac{d}{2}\right)\mu = 0 . \end{aligned} \tag{4.103}$$

Equation (4.102) also allows us to write down in a uniform manner all the divergences that can occur on the right-hand side of (4.94). Such divergences occur if $\mu-\nu-d/2$ or $\nu+d/2$ are negative integers or zero. The Γ function has no zeros on the positive real axes, so that $1/\Gamma(\mu)$ does not create any divergences. With

$$\lim_{\varepsilon\to 0}\Gamma(-n+\varepsilon) = \lim_{\varepsilon\to 0}\frac{\Gamma(1-n+\varepsilon)}{-n+\varepsilon} = \lim_{\varepsilon\to 0}\frac{\Gamma(2-n+\varepsilon)}{(-n+\varepsilon)(1-n+\varepsilon)} = \ldots \tag{4.104a}$$

all divergences can be related to $\lim_{\varepsilon\to 0}\Gamma(\varepsilon)$ from (4.94). Let us discuss as an example

$$\begin{aligned} \lim_{\varepsilon\to 0}\Gamma(-3+\varepsilon) &= \lim_{\varepsilon\to 0}\frac{\Gamma(\varepsilon)}{(-3+\varepsilon)(-2+\varepsilon)(-1+\varepsilon)} \\ &= \lim_{\varepsilon\to 0}\left(-\frac{1}{6}\right)\left(1+\frac{\varepsilon}{3}+\frac{\varepsilon}{2}+\varepsilon\right)\left(\frac{1}{\varepsilon}+\Gamma'(1)\right) \end{aligned}$$

$$
\begin{aligned}
&= \lim_{\varepsilon\to 0}\left(\frac{-1}{6\varepsilon}\right) - \frac{1}{6}\left(\Gamma'(1) + \frac{1}{3} + \frac{1}{2} + 1\right) \\
&= -\frac{\Gamma'(1)+\frac{11}{6}}{6} - \frac{1}{6}\lim_{\varepsilon\to 0}\frac{1}{\varepsilon} \ . \qquad (4.104\text{b})
\end{aligned}
$$

All possible divergences (which can be only logarithmic for renormalizable theories) are therefore proportional to $1/\varepsilon$, and (4.94) and (4.104a,b) already determine dimensional regularization.

Dimensional regularization, i.e., the separation of the divergent and finite parts, is realized by identifying the $1/\varepsilon$ terms with the divergent part and the rest, which is the finite one. The $1/\varepsilon$ terms are then absorbed into the renormalization constants, such as mass, charge, ..., during renormalization.

To end this section, let us mention a second valuable formula for dimensional regularization ($d = 4+2\varepsilon$):

$$
\int \mathrm{d}^d x \frac{\mathrm{e}^{\mathrm{i}\boldsymbol{p}\cdot\boldsymbol{x}}}{(-x^2)^\nu} = -\mathrm{i}\pi^2 \frac{\Gamma(2-\nu+\varepsilon)}{\Gamma(\nu)} \left(\frac{-4\pi^2\mu^2}{p^2}\right)^{\varepsilon} \left(\frac{-p^2}{4}\right)^{\nu-2} , \qquad (4.105)
$$

which allows us to perform d-dimensional Fourier transformations. It is derived in Exercise 4.6.

EXERCISE

4.6 The d-Dimensional Fourier Transform

Problem. Derive the equation

$$
\int \mathrm{d}^d x \, \frac{\mathrm{e}^{\mathrm{i}\boldsymbol{p}\cdot\boldsymbol{x}}}{(-x^2)^\nu} = -\mathrm{i}\pi^2 \frac{\Gamma(2-\nu+\varepsilon)}{\Gamma(\nu)} \left(\frac{-4\pi^2\mu^2}{p^2}\right)^{\varepsilon} \left(\frac{-p^2}{4}\right)^{\nu-2} . \qquad (1)
$$

Solution. In three dimensions plane waves can be expanded into Bessel functions and Legendre polynomials according to

$$
\mathrm{e}^{\mathrm{i}\boldsymbol{p}\cdot\boldsymbol{x}} = \mathrm{e}^{\mathrm{i}|\boldsymbol{p}|\,|\boldsymbol{x}|\,\cos\theta} = \dots \ . \qquad (2)
$$

To prove (1) we need the generalization of this expansion to arbitrary dimensions. The exponential itself looks the same for any Euclidian dimension, namely

$$
\mathrm{e}^{\mathrm{i}\boldsymbol{p}\cdot\boldsymbol{x}} = \mathrm{e}^{\mathrm{i}|\boldsymbol{p}|\,|\boldsymbol{x}|\,\cos\theta} \ . \qquad (3)
$$

Therefore the dimensionality enters only in the orthogonality property. The functions of θ that we call $C_i(\theta)$ shall be orthogonal with the weight $(\sin\theta)^{d-2}$,

Exercise 4.6

because the d-dimensional volume element is proportional to $(\sin\vartheta)^{d-2}$ (see Exercise 4.5, (2)):

$$\begin{aligned}\int \mathrm{d}^d\Omega\, C_i(\cos\theta)\, C_j(\cos\theta) &\sim \int_0^\pi \mathrm{d}\theta (\sin\theta)^{d-2}\, C_i(\cos\theta)\, C_j(\cos\theta)\\ &= \int_{-1}^{1} \mathrm{d}\cos\theta\, (\sin^2\theta)^{(d-3)/2}\, C_i(\cos\theta)\, C_j(\cos\theta)\\ &= \int_{-1}^{1} \mathrm{d}x\, (1-x^2)^{(d-3)/2}\, C_i(x)\, C_j(x)\ . \end{aligned}\tag{4}$$

Orthogonal polynomials with the weight $(1-x^2)^{d-1/2}$ are the "Gegenbauer polynomials" $C_n^{(\alpha)}(x)$.[9] The important properties for us are

$$\mathrm{e}^{\mathrm{i}px\cos\theta} = \Gamma(\nu)\left(\frac{px}{2}\right)^{-\nu}\sum_{k=0}^{\infty}(\nu+k)\, i^k J_{\nu+k}(px)\, C_k^{(\nu)}(\cos\theta)\tag{5}$$

with arbitrary ν,

$$\int_{-1}^{1} \mathrm{d}x\, (1-x^2)^{\alpha-1/2}\, C_n^{(\alpha)}(x) C_{n'}^{(\alpha)}(x) = \delta_{nn'}\frac{\pi 2^{1-2\alpha}\Gamma(n+2\alpha)}{n!\,(n+\alpha)\,[\Gamma(\alpha)]^2}\quad \left(\alpha>-\frac{1}{2}\right)\ ,\tag{6}$$

and

$$C_0^{(\alpha)}(x) = 1\ .\tag{7}$$

From (6) and (7) we find that

$$\begin{aligned}\int_{-1}^{1} \mathrm{d}x\, (1-x^2)^{\alpha-1/2} C_n^{(\alpha)}(x) &= \int_{-1}^{1} \mathrm{d}x\, (1-x^2)^{\alpha-1/2} C_n^{(\alpha)}(x) C_0^{(\alpha)}(x)\\ &= \delta_{n0}\frac{\pi 2^{1-2\alpha}\Gamma(2\alpha)}{\alpha[\Gamma(\alpha)]^2}\\ &= \delta_{n0}\int_{-1}^{1} \mathrm{d}x\, (1-x^2)^{\alpha-1/2}\ . \end{aligned}\tag{8}$$

[9] Their properties can be found, for example, in M. Abramowitz and A. Stegun: *Handbook of Mathematical Functions*, Chap. 22. (Dover, 1972).

Exercise 4.6

Comparing (8) with the d-dimensional integral (4) we find that α has to be chosen as $\alpha = \frac{d}{2} - 1$. Using (6), relation (4) can now be denoted as

$$\int \mathrm{d}^d\Omega\, C_i^{d/2-1}(\cos\theta)\, C_j^{d/2-1}(\cos\theta) = \delta_{ij} \frac{\pi 2^{3-d}\,\Gamma(i-2+d)}{i!\,\left(i-1+\frac{d}{2}\right)\,\left[\Gamma\left(\frac{d}{2}-1\right)\right]^2} \ ,$$

or

$$\int \mathrm{d}^d\Omega\, C_i^{\alpha}(\cos\theta)\, C_j^{\alpha}(\cos\theta) = \delta_{ij} \frac{\pi 2^{1-2\alpha}\,\Gamma(i+2\alpha)}{i!\,(i+\alpha)\,[\Gamma(\alpha)]^2} \quad \left(\alpha = \frac{d}{2} - 1\right) \ , \tag{9}$$

and, furthermore,

$$\int \mathrm{d}^d\Omega = \int\limits_{-1}^{1} \mathrm{d}x\,(1-x^2)^{(d-3)/2} = \frac{\pi 2^{3-d}\,\Gamma(d-2)}{\left(\frac{d}{2}-1\right)\,\left[\Gamma\left(\frac{d}{2}-1\right)\right]^2} \ . \tag{10}$$

The angular integral is now easy to perform. We first substitute $x^0 \to -\mathrm{i}x_{\mathrm{E}}^0$ to go to Euclidian coordinates. We also substitute $p^0 \to \mathrm{i}\,p_{\mathrm{E}}^0$. Note that always the time components only are Wick-rotated!

$$\begin{aligned}
&\int \mathrm{d}^d x\, \frac{\mathrm{e}^{\mathrm{i}\boldsymbol{p}\cdot\boldsymbol{x}}}{(-x^2)^{\nu}} \\
&= -\mathrm{i} \int \mathrm{d}^d x_{\mathrm{E}}\, \frac{\mathrm{e}^{\mathrm{i}\,p_{\mathrm{E}}x_{\mathrm{E}}\cos\theta}}{(x_{\mathrm{E}}^2)^{\nu}} \\
&= -\mathrm{i} \int\limits_0^{\infty} \mathrm{d}x_{\mathrm{E}}\, \frac{x_{\mathrm{E}}^{d-1}}{(x_{\mathrm{E}}^2)^{\nu}}\, \Gamma(\alpha) \left(\frac{p_{\mathrm{E}}x_{\mathrm{E}}}{2}\right)^{-\alpha} \\
&\quad \times \sum_{k=0}^{\infty} (\alpha+k)\mathrm{i}^k\, J_{\alpha+k}(p_{\mathrm{E}}x_{\mathrm{E}}) \int \mathrm{d}^d\Omega\, C_k^{(\alpha)}(\cos\theta) \\
&= -\mathrm{i} \int\limits_0^{\infty} \mathrm{d}x_{\mathrm{E}}\, (x_{\mathrm{E}})^{d-1-2\nu}\, \Gamma(\alpha) \left(\frac{p_{\mathrm{E}}x_{\mathrm{E}}}{2}\right)^{-\alpha} \alpha J_{\alpha}(p_{\mathrm{E}}x_{\mathrm{E}}) \int \mathrm{d}^d\Omega \\
&= -\mathrm{i} \frac{2\pi^{d/2}}{\Gamma(d/2)}\, \Gamma(d/2-1) \left(\frac{p_{\mathrm{E}}}{2}\right)^{1-d/2} \left(\frac{d}{2}-1\right) \\
&\quad \times \int\limits_0^{\infty} \mathrm{d}x_{\mathrm{E}}\, (x_{\mathrm{E}})^{d-1-2\nu-d/2+1}\, J_{d/2-1}(p_{\mathrm{E}}x_{\mathrm{E}}) \\
&= -\mathrm{i}2\pi^{d/2} \left(\frac{p_{\mathrm{E}}}{2}\right)^{1-d/2} \int\limits_0^{\infty} \mathrm{d}x_{\mathrm{E}}\, x_{\mathrm{E}}^{d/2-2\nu}\, J_{d/2-1}(p_{\mathrm{E}}x_{\mathrm{E}}) \\
&= -\mathrm{i}2\pi^{d/2} \left(\frac{p_{\mathrm{E}}}{2}\right)^{1-d/2} (p_{\mathrm{E}})^{-1-d/2+2\nu} \int\limits_0^{\infty} \mathrm{d}y\, y^{d/2-2\nu}\, J_{d/2-1}(y) \ .
\end{aligned} \tag{11}$$

Exercise 4.6

The remaining integral can be found in appropriate integral tables. [10]

$$\int_0^\infty \mathrm{d}y\, y^{d/2-2\nu} J_{d/2-1}(y) = 2^{d/2-2\nu} \frac{\Gamma\left(\frac{d}{4}+\frac{d}{4}-\nu\right)}{\Gamma\left(\frac{d}{4}-\frac{d}{4}+\nu\right)} \,. \tag{12}$$

Putting all this together we get

$$\int \mathrm{d}^d x \, \frac{\mathrm{e}^{\mathrm{i}p\cdot x}}{(-x^2)^\nu} = -\mathrm{i}\pi^{d/2} (p_\mathrm{E})^{2\nu-d}\, 2^{d/2}\, 2^{d/2-2\nu} \frac{\Gamma(d/2-\nu)}{\Gamma(\nu)} \,. \tag{13}$$

Finally we insert again $d = 4+2\varepsilon$ to obtain the result

$$\int \mathrm{d}^d x \, \frac{\mathrm{e}^{\mathrm{i}p\cdot x}}{(-x^2)^\nu} = -\mathrm{i}\pi^2 \left(\frac{p_E^2}{4}\right)^{\nu-2} \left(\frac{4\pi}{p_E^2}\right)^{\varepsilon} \frac{\Gamma(2-\nu+\varepsilon)}{\Gamma(\nu)} \,, \tag{14}$$

which completes our proof. Note that $p_E^2 = -p^2$!

EXERCISE

4.7 Feynman Parametrization

Problem. Prove the most general form of the Feynman parametrization,

$$\prod_{i=1}^{n} \frac{1}{a_i^{A_i}} = \frac{\Gamma(A)}{\prod_{i=1}^{n} \Gamma(A_i)} \int_0^1 \frac{\prod_{i=1}^{n} \mathrm{d}x_i\, x_i^{A_i-1}}{\left(\sum_{i=1}^{n} a_i x_i\right)^A} \,\delta\left(1-\sum_{i=1}^{n} x_i\right) \,, \tag{1}$$

with $A = \sum_{i=1}^{n} A_i$ by means of mathematical induction. The a_i $(i = 1, 2, \ldots, n)$ are arbitrary complex numbers.

Solution. To prove (1) we start with the simple formula

$$\frac{1}{a\cdot b} = \int_0^1 \frac{\mathrm{d}x}{(ax-b(1-x))^2} \,, \tag{2}$$

which is obtained by observing that

$$\frac{1}{a\cdot b} = \frac{1}{b-a}\left(\frac{1}{a}-\frac{1}{b}\right) = \frac{1}{b-a}\int_a^b \frac{\mathrm{d}z}{z^2} \,. \tag{3}$$

[10] See, for example, I. Gradshtein and I. Ryshik: *Tables of Series, Products and Integrals*, No. 6.151.14 (Harri Deutsch, Frankfurt am Main 1981).

Exercise 4.7

Substituting $z = ax + b(1-x)$ one obtains (2). Equation (2) can be further generalized to arbitrary powers by taking derivatives

$$\frac{\partial^A}{\partial a^A}\frac{\partial^B}{\partial b^B} \tag{4}$$

on both sides so that

$$\begin{aligned}\frac{1}{a^A \cdot b^B} &= \frac{\Gamma(A+B)}{\Gamma(A)\Gamma(B)}\int\limits_0^1 \mathrm{d}x\,\frac{x^{A-1}(a-x)^{B-1}}{(ax+b(1-x))^{A+B}}\\ &= \frac{\Gamma(A+B)}{\Gamma(A)\Gamma(B)}\int\limits_0^1 \mathrm{d}x\,\mathrm{d}y\,\frac{x^{A-1}y^{B-1}}{(ax+by)^{A+B}}\,\delta(1-x-y)\ .\end{aligned} \tag{5}$$

Thus we have proven (1) for $n = 2$. Now we assume that the formula (1) holds for n and prove that it is also valid for $n+1$. Let us start with the expression

$$\frac{1}{a_{n+1}^{A_{n+1}}}\prod_{i=1}^{n}\frac{1}{a_i^{A_i}} = \frac{\Gamma(A)}{\prod_{i=1}^{n}\Gamma(A_i)}\int\limits_0^1 \frac{\prod_{i=1}^{n}\mathrm{d}x_i\,x_i^{A_i-1}}{\left(\sum_{i=1}^{n}a_i x_i\right)^A}\,\delta\left(1-\sum_{i=1}^{n}x_i\right)\frac{1}{a_{n+1}^{A_{n+1}}}\ , \tag{6}$$

where we multiplied (1) on both sides by $1/a_{n+1}^{A_{n+1}}$. Applying (5) we can rewrite (6) to obtain

$$\begin{aligned}\prod_{i=1}^{n}\frac{1}{a_i^{A_i}} &= \frac{\Gamma(A)}{\prod_{i=1}^{n}\Gamma(A_i)}\int\limits_0^1\prod_{i=1}^{n}\mathrm{d}x_i x_i^{A_i-1}\,\delta\left(1-\sum_{i=1}^{n}x_i\right)\frac{\Gamma(A+A_{n+1})}{\Gamma(A)\Gamma(A_{n+1})}\\ &\quad\times\int\limits_0^1 \mathrm{d}x_{n+1}\mathrm{d}y\frac{y^{A-1}x_{n+1}^{A_{n+1}-1}}{\left(y\sum_{i=1}^{n}a_i x_i + a_{n+1}x_{n+1}\right)^{A+A_{n+1}}}\,\delta\left(1-y-x_{n+1}\right)\ .\end{aligned} \tag{7}$$

Now we change variables $\tilde{x}_i = yx_i$ $(i = 1, 2, \ldots n)$ in the above equation and perform the y integration. After replacing $\tilde{x}_i$ by x_i we obtain

$$\prod_{i=1}^{n+1}\frac{1}{a_i^{A_i}} = \frac{\Gamma(A+A_{n+1})}{\prod_{i=1}^{n+1}\Gamma(A_i)}\int\limits_0^1\frac{\prod_{i=1}^{n+1}\mathrm{d}x_i x_i^{A_i-1}}{\left(\sum_{i=1}^{n+1}a_i x_i\right)^{A+A_{n+1}}}\,\delta\left(1-\sum_{i=1}^{n+1}x_i\right)\ . \tag{8}$$

This completes the proof of (1).

4.4 The Renormalized Coupling Constant of QCD

We shall now use dimensional regularization to calculate the renormalized QCD coupling constant to lowest order. The divergent graphs contributing to the renormalization of g_s belong to three classes, shown in Figs. 4.14, 4.15, and 4.16. Let us work them out one by one.

Graph (a1). We start with vacuum polarization, in particular with the fermion graph (a1) in Fig. 4.14. This graph is depicted in greater detail in Fig. 4.17. One may wonder why the quantum numbers of the outgoing gluon k', a', μ' are not identical with those of the ingoing gluon. Of course, this is the most general case. It will turn out that indeed $a = a'$ (see (4.119)) and $k = k'$ (momentum conservation), but $\mu \neq \mu'$. Furthermore, in (4.133) and below, we shall see that only the sum of all four vacuum polarization graphs $(a1)$–$(a4)$ is

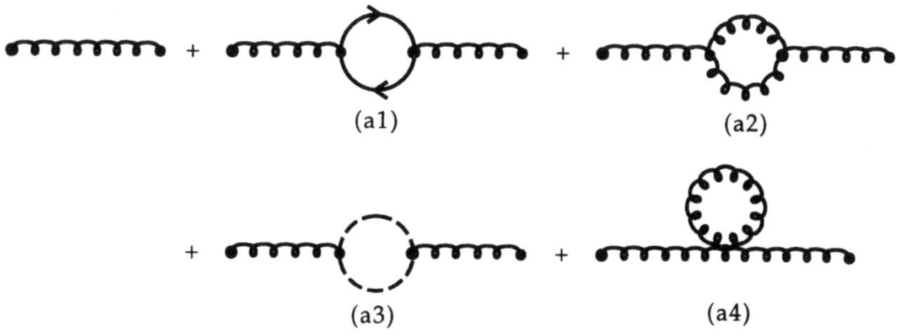

Fig. 4.14. The vacuum polarization graphs of QCD

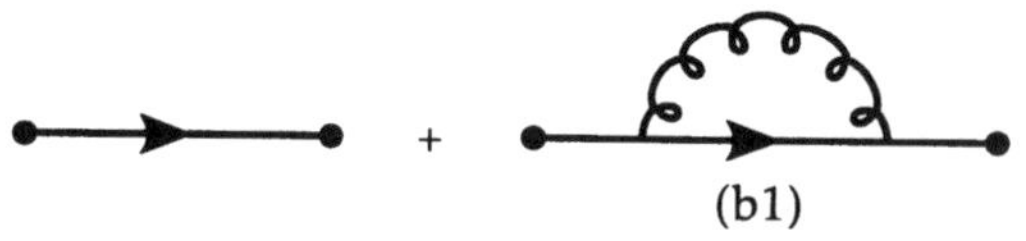

Fig. 4.15. The self energy graphs of QCD

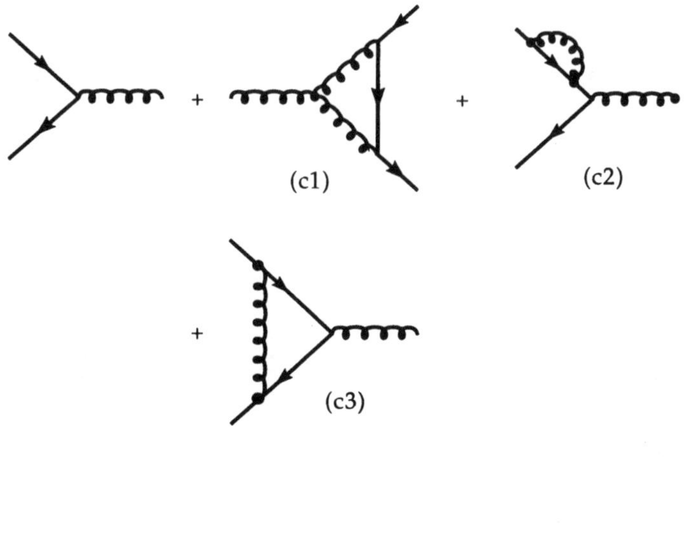

Fig. 4.16. The vertex correction graphs of QCD

proportional to $\frac{1}{k^2}(k_\mu k_{\mu'} - g_{\mu\mu'}k^2)$, i.e. the projector which ensures gauge invariance. Using the QCD Feynman rules it is easy to write down the polarization tensor:

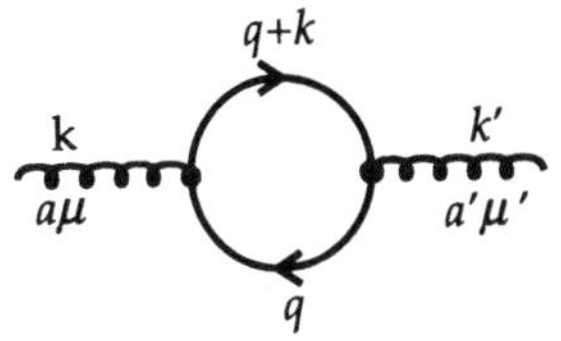

Fig. 4.17. The variables chosen for the quark loop contribution to vacuum polarization

$$\Pi^{a\,a'\,(\mathrm{a1})}_{\mu\mu'}(k) = \sum_{i=1}^{N_\mathrm{F}} g^2 \int \frac{\mathrm{d}^4q}{(2\pi)^4} \times \mathrm{tr}\left\{\frac{\lambda^a}{2}\frac{\lambda^{a'}}{2}\frac{\not{q}+\not{k}+m}{(q+k)^2-m^2+\mathrm{i}\eta}\gamma_\mu\frac{\not{q}+m}{q^2-m^2+\mathrm{i}\eta}\gamma_{\mu'}\right\} . \tag{4.106}$$

N_F is the number of quark flavors in the theory, or more precisely the number of quarks with $m^2 \ll |k^2|$. The whole calculation can be carried through for arbitrary m^2, but one finds that the contribution is suppressed for $m^2 > |k^2|$. N_F therefore counts only the light flavors and for simplicity we can set $m = 0$ in what follows whenever this does not lead to infrared divergencies. The trace over the color indices gives simply

$$\mathrm{tr}\left\{\frac{\lambda^a}{2}\frac{\lambda^{a'}}{2}\right\} = \delta_{a\,a'}\frac{1}{2} . \tag{4.107}$$

As usual we introduce Feynman parameters

$$\begin{aligned}&\frac{1}{(q+k)^2-m^2+\mathrm{i}\eta}\,\frac{1}{q^2-m^2+\mathrm{i}\eta}\\ &\quad= \int_0^1 \mathrm{d}z\,\frac{1}{[q^2+2kqz+(k^2-m^2)z-m^2(1-z)+\mathrm{i}\eta]^2}\\ &\quad= \int_0^1 \mathrm{d}z\,\frac{1}{[(q+kz)^2+k^2z(1-z)-m^2+\mathrm{i}\eta]^2} .\end{aligned} \tag{4.108}$$

To get rid of the linear term in the denominator we substitute $q_\mu \to q'_\mu = q_\mu - k_\mu z$:

$$\Pi^{(\mathrm{a1})}_{\mu\mu'}(k) = g^2\frac{N_\mathrm{F}}{2}\int_0^1 \mathrm{d}z \int_0^\infty \frac{\mathrm{d}^4q}{(2\pi)^4}\,\frac{\mathrm{tr}\left\{[\not{q}+\not{k}(1-z)]\gamma_\mu[\not{q}-\not{k}z]\gamma_{\mu'}\right\}}{[q^2+k^2z(1-z)-m^2+\mathrm{i}\eta]^2} , \tag{4.109}$$

where we have dropped the color indices, i.e. $\Pi^{ab}_{\mu\nu} = \delta^{ab}\Pi_{\mu\nu}$. Here the term proportional to m^2 in the nominator has already been dropped. Next we take the trace and neglect all terms proportional to odd powers of q. They are zero as can be seen by substituting q_ν for $-q_\nu$. For the same reason in intergrations over even powers of q only diagonal terms contribute:

$$\int \mathrm{d}^4q\; q_\mu q_\nu\; f(q^2) = g_{\mu\nu} \times \text{const} = \frac{1}{4}\int \mathrm{d}^4q\; g_{\mu\nu}\, q^2 f(q^2) . \tag{4.110}$$

The left-hand side is a Lorentz tensor. After integrating over q the only possible Lorentz tensor is $g_{\mu\nu}$. By contracting both sides with $g^{\mu\nu}$ we see that the factor $1/4$ is necessary to have the correct normalization ($\mathrm{tr}\, g = 4$). Therefore (4.109) becomes

$$\begin{aligned}
\Pi^{(\mathrm{a1})}_{\mu\mu'}(k) &= g^2 \frac{N_\mathrm{F}}{2} 4 \int_0^1 \mathrm{d}z \\
&\quad \times \int_0^\infty \frac{\mathrm{d}^4 q}{(2\pi)^4} \frac{2q_\mu q_{\mu'} - q^2 g_{\mu\mu'} - z(1-z)(2k_\mu k_{\mu'} - k^2 g_{\mu\mu'})}{[q^2 + k^2 z(1-z) - m^2 + \mathrm{i}\eta]^2} \\
&= 2g^2 N_\mathrm{F} \int_0^1 \mathrm{d}z \\
&\quad \times \int_0^\infty \frac{\mathrm{d}^4 q}{(2\pi)^4} \frac{-\frac{1}{2} q^2 g_{\mu\mu'} - z(1-z)(2k_\mu k_{\mu'} - k^2 g_{\mu\mu'})}{[q^2 + k^2 z(1-z) - m^2 + \mathrm{i}\eta]^2} \, .
\end{aligned} \tag{4.111}$$

Now only q^2 appears in the integrand, so it can safely be continued to Euclidian space, $q_\mu \to \tilde{q}_\mu = (\mathrm{i}q^0, \boldsymbol{q})$, and translated into a d-dimensional integral:

$$\Pi^{(\mathrm{a1})}_{\mu\mu'}(k) \Rightarrow 2g^2 N_\mathrm{F} \int_0^1 \mathrm{d}z \int_0^\infty \mathrm{i} \frac{\mathrm{d}^d \tilde{q}}{(2\pi)^4} \mu^{4-d} \frac{\frac{1}{2}\tilde{q}^2 g_{\mu\mu'} - z(1-z)(2k_\mu k_{\mu'} - k^2 g_{\mu\mu'})}{[\tilde{q}^2 - k^2 z(1-z) + m^2 - \mathrm{i}\eta]^2} \, . \tag{4.112}$$

We have introduced a dimensional quantity μ to keep the total dimension of the expression unchanged. The integration is four-dimensional, i.e. $\int \mathrm{d}^4 \ldots$. If this is changed into $\int \mathrm{d}^d \ldots$ with the same integrand, a factor μ^{4-d} ensures the same dimension of the whole expression. With (4.81) it follows that

$$\begin{aligned}
\Pi^{(\mathrm{a1})}_{\mu\mu'}(k) &= \frac{2\mathrm{i}g^2 N_\mathrm{F}}{(2\pi)^4} \int_0^1 \mathrm{d}z \frac{\pi^{\frac{d}{2}} \mu^{4-d}}{\Gamma\left(\frac{d}{2}\right)} \left[m^2 - k^2 z(1-z) - \mathrm{i}\eta\right]^{\frac{d}{2}-2} \\
&\quad \times \frac{\Gamma\left(2-\frac{d}{2}\right)\Gamma\left(\frac{d}{2}\right)}{\Gamma(2)} [-z(1-z)] \left(2k_\mu k_{\mu'} - k^2 g_{\mu\mu'}\right) \\
&\quad + \frac{2\mathrm{i}g^2 N_\mathrm{F}}{(2\pi)^4} \frac{g_{\mu\mu'}}{2} \int_0^1 \mathrm{d}z \frac{\pi^{\frac{d}{2}} \mu^{4-d}}{\Gamma\left(\frac{d}{2}\right)} \left[m^2 - k^2 z(1-z) - \mathrm{i}\eta\right]^{\frac{d}{2}-1} \\
&\quad \times \frac{\Gamma\left(1-\frac{d}{2}\right)\Gamma\left(1+\frac{d}{2}\right)}{\Gamma(2)} \, .
\end{aligned} \tag{4.113}$$

The terms on the right-hand side are divergent. For $d = 4 - 2\varepsilon$ the first is proportional to

$$\Gamma\left(2 - \frac{d}{2}\right) = \Gamma(\varepsilon) = \frac{1}{\varepsilon} + \Gamma'(1) + O(\varepsilon) \ , \tag{4.114}$$

while the second term is proportional to

$$\begin{aligned}\frac{\Gamma\left(1-\frac{d}{2}\right)\Gamma\left(1+\frac{d}{2}\right)}{\Gamma\left(\frac{d}{2}\right)} &= \frac{4-2\varepsilon}{2}\Gamma(-1+\varepsilon) \\ &= (2-\varepsilon)\frac{\Gamma(\varepsilon)}{(-1+\varepsilon)} \\ &= (2-\varepsilon)(-1-\varepsilon)\left(\frac{1}{\varepsilon}+\Gamma'(1)\right)+O(\varepsilon) \\ &= -\frac{2}{\varepsilon} - 1 - 2\Gamma'(1) + O(\varepsilon) \ . \end{aligned}\tag{4.115}$$

We thus have

$$\begin{aligned}\Pi^{(\mathrm{a1})}_{\mu\mu'}(k) = {} & \frac{2\mathrm{i}g^2 N_\mathrm{F}}{(2\pi)^4}\,\mu^{4-d}\,\pi^{\frac{d}{2}}\left(\frac{1}{\varepsilon}+\Gamma'(1)\right) \\ & \times \int_0^1 \mathrm{d}z\,\frac{-z(1-z)}{[m^2-k^2z(1-z)]^\varepsilon}\left(2k_\mu k_{\mu'} - k^2 g_{\mu\mu'}\right) \\ & - \frac{2\mathrm{i}g^2 N_\mathrm{F}}{(2\pi)^4}\,\mu^{4-d}\,\pi^{\frac{d}{2}}\left(\frac{2}{\varepsilon}+1+2\Gamma'(1)\right) \\ & \times \int_0^1 \mathrm{d}z\,[m^2-k^2z(1-z)]^{1-\varepsilon}\frac{1}{2}g_{\mu\mu'} \ . \end{aligned}\tag{4.116}$$

It can now be seen that m^2 can safely be neglected. For this purpose the integrand is expanded in m^2/k^2:

$$\begin{aligned}\left[m^2-k^2z(1-z)\right]^{-\varepsilon} &= (-k^2)^{-\varepsilon}(z(1-z))^{-\varepsilon}\left(1-\frac{m^2}{k^2z(1-z)}\right)^{-\varepsilon} \\ &= (-k^2)^{-\varepsilon}(z(1-z))^{-\varepsilon} \\ &\quad\times\left(1+\varepsilon\frac{m^2}{k^2}\frac{1}{z(1-z)}+\mathcal{O}\left(\frac{m^4}{k^4}\right)\right) \ . \end{aligned}\tag{4.117}$$

The leading z integral can be performed easily,

$$\begin{aligned}\int_0^1 \mathrm{d}z\,\left[z(1-z)\right]^{1-\varepsilon} &= \frac{\Gamma(2-\varepsilon)\Gamma(2-\varepsilon)}{\Gamma(4-\varepsilon)} = \frac{\left[1-\varepsilon\Gamma'(2)\right]^2}{3!-2\varepsilon\Gamma'(4)} \\ &= \frac{1}{3!} - \frac{\varepsilon}{3}\Gamma'(2) + \frac{\varepsilon}{3}\Gamma'(4) \ , \end{aligned}\tag{4.118}$$

and turns out to be finite for $\varepsilon \to 0$. Thus, indeed, we can safely neglect the quark mass since our expression is infrared safe. We get

$$\Pi^{(a1)}_{\mu\mu'}(k) = \frac{2ig^2 N_F}{16\pi^2}\,\pi^\varepsilon\,\mu^{2\varepsilon}\left(-2k_\mu k_{\mu'} + 2k^2 g_{\mu\mu'}\right)\left(-k^2\right)^{-\varepsilon}\frac{1}{3!}\left(\frac{1}{\varepsilon}+\text{const}\right)$$
$$= ig^2\frac{N_F}{16\pi^2}\,\frac{2}{3}\left(\frac{\pi\mu^2}{-k^2}\right)^\varepsilon\left(\frac{1}{\varepsilon}+\text{const}\right)\left(g_{\mu\mu'}k^2 - k_\mu k_{\mu'}\right)+\dots\,. \tag{4.119}$$

Finally, we expand $y^\varepsilon = 1+\varepsilon\ln(y)$:

$$\Pi^{(a1)}_{\mu\mu'}(k) = g^2 i\frac{N_F}{16\pi^2}\,\frac{2}{3}\left[\frac{1}{\varepsilon}+\ln\left(\frac{\pi\mu^2}{-k^2}\right)+\text{const}\right]\left(g_{\mu\mu'}k^2 - k_\mu k_{\mu'}\right)\,. \tag{4.120}$$

Note that every $1/\varepsilon$ term is always accompanied by a term $\ln(-k^2/\mu^2)$.

With $\ln(-k^2/\pi\mu^2) = \ln(-k^2/\mu^2) - \ln(\pi)$ we can finally write

$$\Pi^{(a1)}_{\mu\mu'}(k) = g^2 i\,\frac{N_F}{16\pi^2}\,\frac{2}{3}\left(\frac{1}{\varepsilon}-\ln\left(\frac{-k^2}{\mu^2}\right)+\text{const}\right)\left(g_{\mu\mu'}k^2 - k_\mu k_{\mu'}\right)\,. \tag{4.121}$$

This completes the calculation of the quark loop graph of Fig. 4.17.

Graph (a2). Next we calculate the gluon loop with two 3-gluon vertices, applying (19) of Example 4.2 twice:

$$\Pi^{aa'\,(a2)}_{\mu\mu'}(k) = g^2\frac{1}{2}\int\frac{d^4k}{(2\pi)^4}\Big\{f^{abc}[g_{\mu\lambda}(k-q)_\nu + g_{\nu\mu}(-q-2k)_\lambda$$
$$+ g_{\lambda\nu}(q+q+k)_\mu]$$
$$\times f^{a'bc}\left[g_{\mu'}{}^{\lambda}(-k+q)^\nu + g^\nu{}_{\mu'}(q+2k)^\lambda + g^{\nu\lambda}(-2q-k)_{\mu'}\right]\Big\}$$
$$\times\frac{1}{(q+k)^2+i\eta}\,\frac{1}{q^2+i\eta}\,. \tag{4.122}$$

The factor $1/2$ in front follows from combinatorics. Such combinatoric factors are fairly easy to derive, as we shall now demonstrate for all the graphs relevant for our calculation.

Let us start with the gluon loop. The 3-gluon vertex contains 6 terms, corresponding to the 3! orientations of the 3-gluon vertex. The total symmetry factor is therefore $(3!)^2\cdot\frac{1}{2!}$ (see Fig. 4.19). The factor $\frac{1}{2!}$ stems from the general pertubation theory series, which reads

$$\sum_{n=0}^{\infty}\frac{1}{n!}(-i)^n\int d^4x_1\cdots d^4x_n T\left(\hat{H}_i(t_1)\hat{H}_i(t_2)\cdots\hat{H}_i(t_n)\right)\,,$$

where $\hat{H}_i$ is the interaction Hamilton density. The second-order term acquires the factor $\frac{1}{2!}$. For more details see W. Greiner and J. Reinhardt.[11]

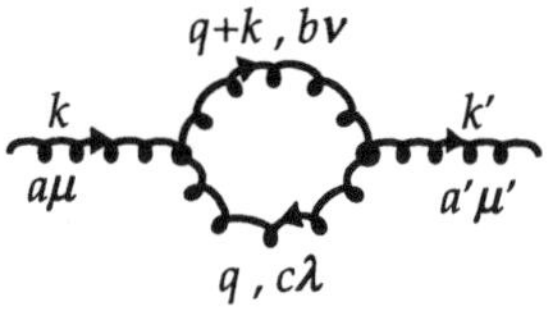

Fig. 4.18. The chosen variables from the first gluon loop contribution to vacuum polarization. For the distinction of the quantum numbers k', a', μ' of the outgoing gluon from those of the incoming gluon, see discussion at the beginning of this paragraph

[11] W. Greiner and J. Reinhardt: *Field Quantization* (Springer, Berlin, Heidelberg, New York 1995).

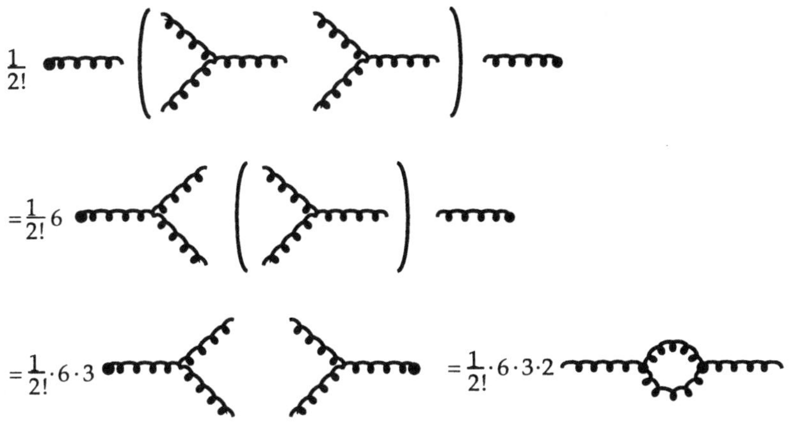

Fig. 4.19. The symmetry factor for the graph in Fig. 4.18. We get from the first to the second line a factor of $6 = 3 \cdot 2$ since we have three possibilities to connect the left external gluon line with one leg of the three gluon vertices. From the second to the third line we aquire a factor of 3 due to the same reasoning. The connection of the two remaining three gluon vertices again is possible in two different ways. The factor $1/2!$ is due to the fact that we are considering second order in perturbation theory

As the two factors $6^2 = (3!)^2$ are absorbed in the 2×6 terms of the two 3-gluon vertices we are left with a factor $1/2$. The corresponding factors for the other graphs are derived in Fig. 4.20.

Following this scheme it is easy to obtain (see Figs. 4.20–4.24) any symmetry factor of interest. The whole expression for $\Pi^{aa'\,(\mathrm{a2})}_{\mu\mu'}$ is obviously very similar to (4.106) except for the different nominator, which we simplify first using

$$f^{abc} f^{a'bc} = \delta_{a\,a'} C_2 = \delta_{a\,a'} N \quad \text{for} \quad \mathrm{SU}(N) \ . \tag{4.123}$$

To obtain the symmetry factor for the second graph in Fig. 4.16 a general remark will be helpful. To obtain symmetry factors for n-th order pertubation theory we always have to deal with contributions like the one depicted below. Very schematically we displayed the n-th order contribution that can occur in QCD due to the 4 possible interaction vertices.

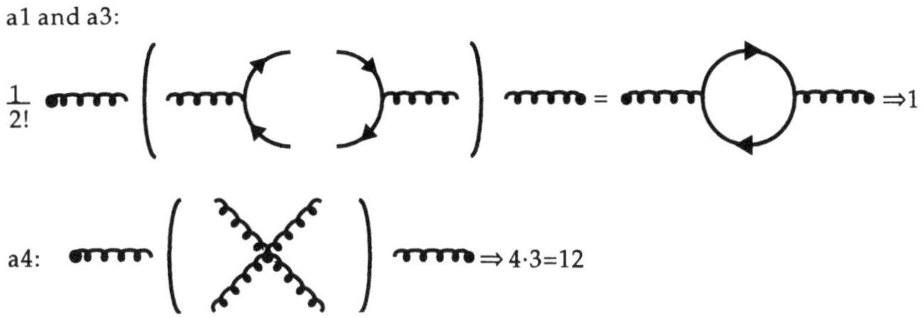

Fig. 4.20. The symmetry factor for the other graphs in Fig. 4.14

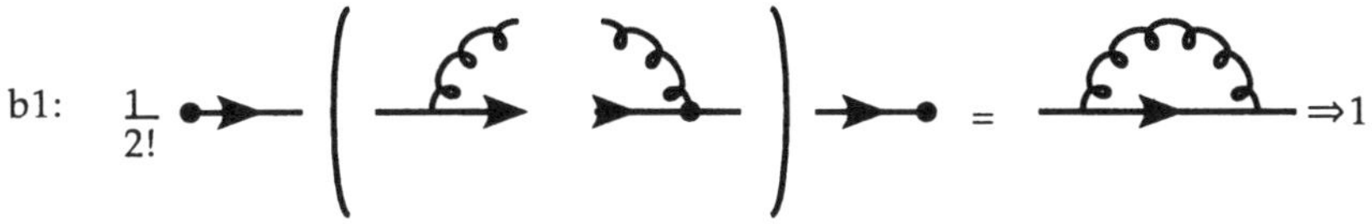

Fig. 4.21. The symmetry factor for the second graph in Fig. 4.15

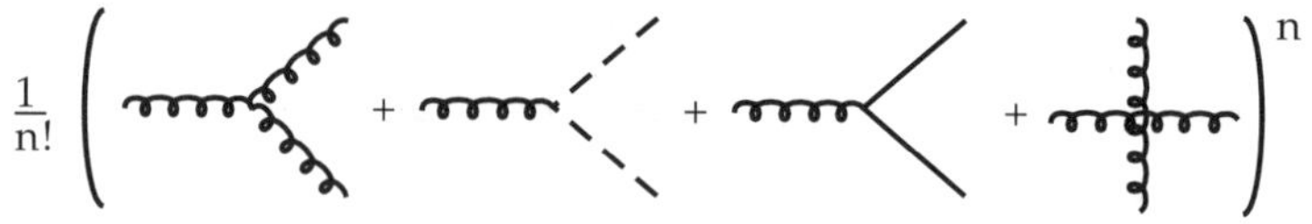

Since we are interested in an interaction which includes 2 quark–quark–gluon vertices and 1 3-gluon vertex we can restrict ourselves to

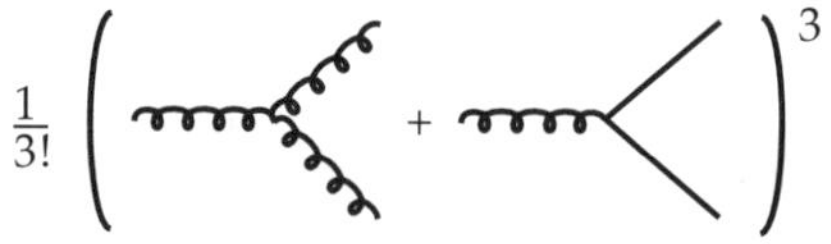

Now we simply use the binomial expansion to glue the vertices together in such a manner that we get the diagram of interest:

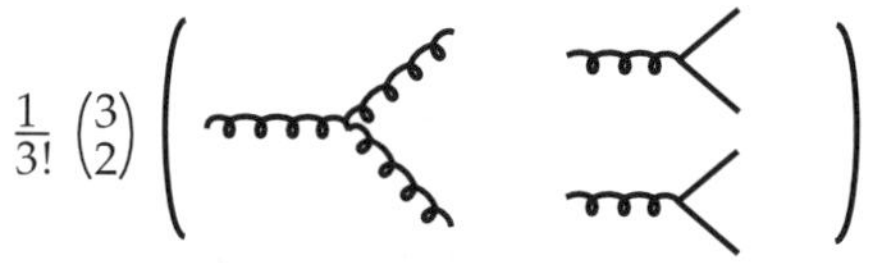

Inserting the external lines, we thus proceed to Fig. 4.22.
We get back to the calculation of the graph (a2) in (4.122):

$$
\begin{aligned}
&\{\dots\}\\
&= 3\delta_{a\,a'}\Big[-g_{\mu\mu'}(k-q)^2+(k-q)_{\mu'}(q+2k)_{\mu}-(k-q)_{\mu}(k+2q)_{\mu'}\\
&\quad -(q-k)_{\mu}(2k+q)_{\mu'}-g_{\mu\mu'}(q+2k)^2+(2q+k)_{\mu'}(q+2k)_{\mu}\\
&\quad +(q-k)_{\mu'}(2q+k)_{\mu}+(q+2k)_{\mu'}(2q+k)_{\mu}\\
&\quad -4(2q+k)_{\mu}(2q+k)_{\mu'}\Big]\\
&= 3\delta_{a\,a'}\Big[-g_{\mu\mu'}\left(k^2-2k\cdot q+q^2+q^2+4k\cdot q+4k^2\right)\\
&\quad +k_{\mu}k_{\mu'}(2-1+2+2-1+2-4)\\
&\quad +k_{\mu}q_{\mu'}(-2-2+1+4+1+1-8)\\
&\quad +q_{\mu}k_{\mu'}(1+1-2+1-2+4-8)\\
&\quad +q_{\mu}q_{\mu'}(-1+2-1+2+2+2-16)\Big]\\
&= 3g^2\delta_{a\,a'}\Big[-g_{\mu\mu'}\left(2q^2+2k\cdot q+5k^2\right)+2k_{\mu}k_{\mu'}\\
&\quad -5k_{\mu}q_{\mu'}-5k_{\mu'}q_{\mu}-10q_{\mu}q_{\mu'}\Big]\,.
\end{aligned}
\tag{4.124}
$$

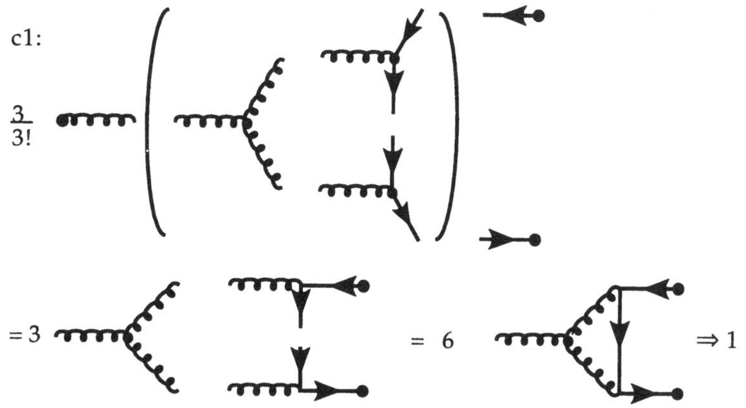

Fig. 4.22. The symmetry factor for the second graph in Fig. 4.16 (3rd-order perturbation theory implies $1/3!$ binomial for the combination of one 3-gluon vertex and two quark–gluon vertices: $\binom{3}{2} = \frac{3!}{2!\,1!} = 3$)

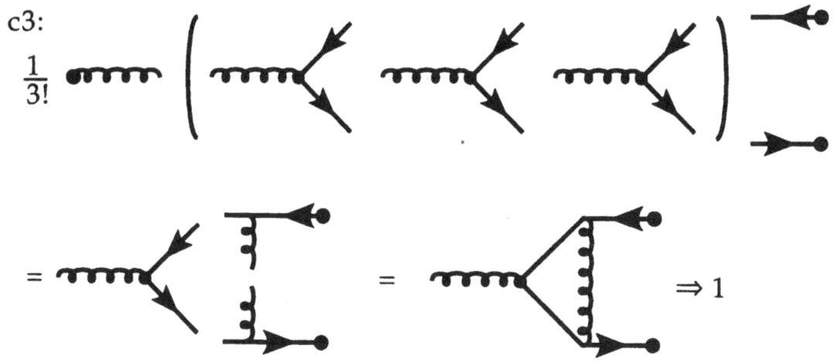

Fig. 4.23. The symmetry factor for the fourth graph in Fig. 4.16

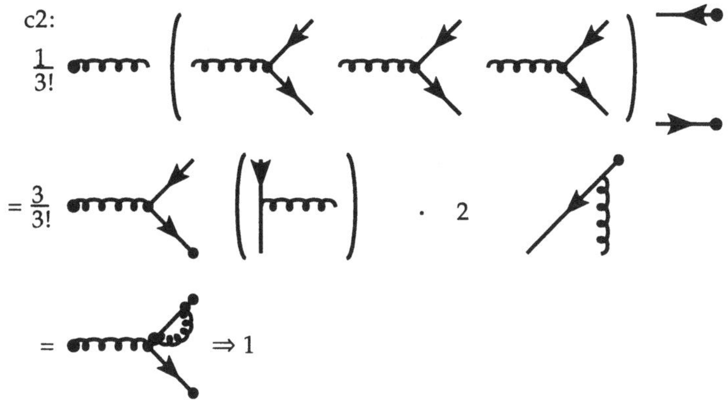

Fig. 4.24. The symmetry factor for the third graph in Fig. 4.16

We again substitute

$$q_\mu \to q_\mu - k_\mu z \tag{4.125}$$

and keep only the terms with even powers in q_μ:

$$\{\dots\} \Rightarrow 3\delta_{a\,a'}\Big[-g_{\mu\mu'}\left(2q^2 + 2k^2z^2 - 2k^2z + 5k^2\right) \\ +2k_\mu k_{\mu'} + 5k_\mu k_{\mu'}z + 5k_\mu k_{\mu'}z - 10k_\mu k_{\mu'}z^2 - 10q_\mu q_{\mu'}\Big]$$

$$\Rightarrow 3\delta_{a\,a'}\Big[-g_{\mu\mu'}\left(-2k^2z(1-z)+5k^2\right)$$
$$+2k_\mu k_{\mu'}+10k_\mu k_{\mu'}z(1-z)-2g_{\mu\mu'}q^2-\frac{5}{2}g_{\mu\mu'}q^2\Big]$$
$$=: 3\delta_{a\,a'}\left(A_{\mu\mu'}+B_{\mu\mu'}q^2\right)\ , \tag{4.126}$$

defining $A_{\mu\mu'}$ and $B_{\mu\mu'}$ to abbreviate the corresponding terms. Repeating the Feynman trick (4.108) for the propagator product $((q+k)^2+\mathrm{i}\eta)^{-1}\cdot(q^2+\mathrm{i}\eta)^{-1}$ and introducing again Euclidean coordinates $\tilde{q}_\mu=(\mathrm{i}q_0,\boldsymbol{q})$, the analogue to (4.112) now becomes

$$\Pi^{aa'\,(\mathrm{a2})}_{\mu\mu'}(k)=\frac{3}{2}g^2\delta_{a\,a'}\mathrm{i}\int_0^1\mathrm{d}z\int_0^\infty\frac{\mathrm{d}^d\tilde{q}}{(2\pi)^4}\,\mu^{4-d}\,\frac{-B_{\mu\mu'}\tilde{q}^2+A_{\mu\mu'}}{[\tilde{q}^2-k^2z(1-z)-\mathrm{i}\eta]^2}$$
$$=\frac{3}{2}g^2\delta_{a\,a'}\mathrm{i}\int_0^1\mathrm{d}z\,\frac{\mu^{4-d}\pi^{\frac{d}{2}}}{(2\pi)^4}\,\frac{1}{\varepsilon}\left[A_{\mu\mu'}+2B_{\mu\mu'}k^2z(1-z)\right]$$
$$\times\left(-k^2z(1-z)\right)^{-\varepsilon}+O(\varepsilon)\ . \tag{4.127}$$

In the last step we have directly inserted the analogue expression from (4.119). The z integral now becomes

$$g_{\mu\mu'}\left(2k^2+\frac{9}{2}k^22\right)\int_0^1\mathrm{d}z\ [z(1-z)]^{1-\varepsilon}-5g_{\mu\mu'}k^2\int_0^1\mathrm{d}z\ [z(1-z)]^{-\varepsilon}$$
$$+2k_\mu k_{\mu'}\int_0^1\mathrm{d}z\ [z(1-z)]^{-\varepsilon}+10k_\mu k_{\mu'}\int_0^1\mathrm{d}z\ [z(1-z)]^{1-\varepsilon}$$
$$=\frac{11}{6}g_{\mu\mu'}k^2-5g_{\mu\mu'}k^2+2k_\mu k_{\mu'}+10k_\mu k_{\mu'}\frac{1}{6}+O(\varepsilon)$$
$$=-\frac{19}{6}g_{\mu\mu'}k^2+\frac{22}{6}k_\mu k_{\mu'}+O(\varepsilon)\ , \tag{4.128}$$

where we have used (4.118) and $\int_0^1\mathrm{d}z\,[z(1-z)]^{-\varepsilon}=\Gamma(1-\varepsilon)\Gamma(1-\varepsilon)/\Gamma(2-2\varepsilon)$. This integral is a special case of the more general one[12] $B(x,y)=\int_0^1\mathrm{d}z z^{x-1}(1-z)^{y-1}=\Gamma(x)\Gamma(y)/\Gamma(x+y)$. Thus, all together we get

$$\Pi^{aa'\,(\mathrm{a2})}_{\mu\mu'}(k)=\frac{3\mathrm{i}g^2\delta_{a\,a'}}{(16\pi)^2}\frac{1}{12}\left[\frac{1}{\varepsilon}-\ln\left(\frac{-k^2}{\mu^2}\right)+\mathrm{const}\right]$$
$$\left(-19g_{\mu\mu'}k^2+22k_\mu k_{\mu'}\right)\ . \tag{4.129}$$

[12] See A.J.G. Hey and R.L. Kelly: Phys. Rep. **96**, 72 (1983).

This expression is not gauge invariant ($k^\mu \Pi^{(a2)}_{\mu\mu'} \neq 0$), which is quite reasonable, since, as explained in Examples 4.2 and 4.3, gauge invariance is only restored by including the ghost terms. More precisely the ghost fields are constructed such that they cancel the contribution from the unphysical degrees of freedom of the gluon field. These unphysical degrees of freedom violate the gauge symmetry.

Graph (a3). To see how this happens we next calculate the ghost contribution,

$$\Pi^{aa'\,(a3)}_{\mu\mu'}(k) = -g^2 f^{abc} f^{a'cb} \int \frac{d^4q}{(2\pi)^4} \frac{q_{\mu'}(k+q)_\mu}{[(q+k)^2+i\eta](q^2+i\eta)} \ . \tag{4.130}$$

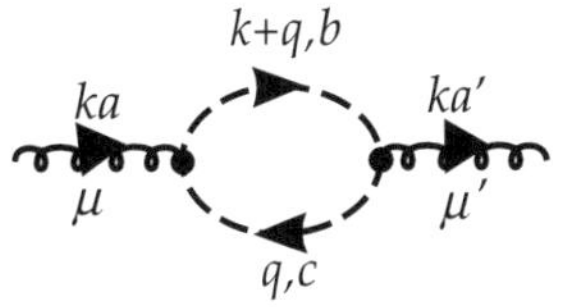

Fig. 4.25. The chosen variables for the ghost loop contribution to vacuum polarization. b and c describe the colour of the ghosts

Note that the color indices at f^{abc} and $f^{a'cb}$ are constructed according to the order: gluon, outgoing ghost, incoming ghost (see Example 4.2, (21)). The difference in sign is due to the fact that the ghost fields anticommute. Using (4.123) and repeating the introduction of a Feynman parameter (see (4.108)) and shifting according to $q_\mu \to q_\mu - k_\mu z$, we obtain

$$\Pi^{aa'\,(a3)}_{\mu\mu'}(k) = 3g^2\delta_{a\,a'} \int_0^1 dz \int \frac{d^4q}{(2\pi)^4} \frac{\frac{1}{4}g_{\mu\mu'}q^2 - k_\mu k_{\mu'} z(1-z)}{\left(q^2+k^2z(1-z)+i\eta\right)^2} \ , \tag{4.131}$$

where we also made use of $q_{\mu'}q_\mu = \frac{1}{4}g_{\mu\mu'}q^2$ according to (4.110) and dropped all terms linear in q in the nominator. After introducing Euclidean coordinates and translating (4.130) into a d-dimensional integral (as was done in the step of passing from (4.111) to (4.112)), (4.130) becomes

$$\begin{aligned}\Pi^{aa'\,(a3)}_{\mu\mu'}(k) &= \frac{3g^2\delta_{a\,a'}i}{16\pi^2}\left(\frac{\pi\mu^2}{-k^2}\right)^\varepsilon\left(\frac{1}{\varepsilon}+\text{const}\right)\\ &\quad \int_0^1 dz \left(-\frac{2}{4}g_{\mu\mu'}k^2 - k_\mu k_{\mu'}\right)(1-z)z\\ &= \frac{3g^2\delta_{a\,a'}i}{16\pi^2}\left[\frac{1}{\varepsilon} - \ln\left(\frac{-k^2}{\mu^2}\right)+\text{const}\right]\\ &\quad \frac{1}{12}\left(-g_{\mu\mu'}k^2 - 2k_\mu k_{\mu'}\right) \ .\end{aligned} \tag{4.132}$$

Together this gives

$$\begin{aligned}&\Pi^{aa'\,(a2)}_{\mu\mu'}(k) + \Pi^{aa'\,(a3)}_{\mu\mu'}(k)\\ &= \frac{3g^2\delta_{a\,a'}i}{16\pi^2}\left[\frac{1}{\varepsilon} - \ln\left(\frac{-k^2}{\mu^2}\right)+\text{const}\right]\frac{20}{12}\left(k_\mu k_{\mu'} - g_{\mu\mu'}k^2\right) \ .\end{aligned} \tag{4.133}$$

The sum of the gluon-plus-ghost vacuum graph is thus again gauge invariant. This illustrates nicely the role of the ghost fields, namely to subtract out the unphysical components of the A_μ field, which are fixed by the chosen gauge. The graph (a4) of Fig. 4.14 gives no contribution. This is best seen from the defining equation (4.78) for dimensional regularization:

$$\int \mathrm{d}^d q \, \frac{1}{q^2} = 0 \ .$$

The bubble is proportional to such a term because it does not depend on the external momentum. The sum of all graphs shown in Fig. 4.14 is needed in the renormalization procedure following Fig. 4.29. We denote it here as

$$\Pi^{(a1)}_{\mu\nu}(k) + \Pi^{(a2)}_{\mu\nu}(k) + \Pi^{(a3)}_{\mu\nu}(k) = \left(k_\mu k_\nu - k^2 \, g_{\mu\nu}\right) Z^a \ ,$$

where the divergent factor

$$Z^a = Z^{(a1)} + Z^{(a2)} + Z^{(a3)} = \frac{g^2}{16\pi^2}\left[\frac{1}{\varepsilon} - \ln\left(\frac{-k^2}{\mu^2}\right)\right]\left[\frac{20}{12}\, C_2 - \frac{2}{3}\, N_f\right]$$

will be needed in the renormalization procedure following Fig. 4.29 below. The factor C_2 will be discussed below, after (4.137).

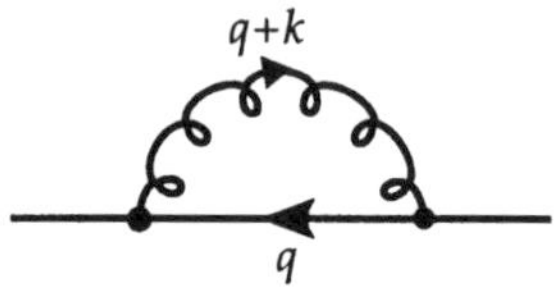

Fig. 4.26. The chosen variables for the self-energy graph

Graph (b1). Next we calculate the self-energy graph

$$\Sigma^{\mathrm{b}}(k) = g^2 \int \frac{\mathrm{d}^4 q}{(2\pi)^4} \gamma_\mu \frac{\not{q}}{q^2 + \mathrm{i}\eta} \gamma^\mu \frac{1}{(q+k)^2 + \mathrm{i}\eta} \mathrm{tr}\left(\frac{\lambda_a}{2}\frac{\lambda_a}{2}\right)\frac{1}{3} \ . \tag{4.134}$$

In this case we have to average over three quark colors, which leads to the factor 1/3. Making use of $\gamma_\mu \not{q} \gamma^\mu = \gamma_\mu q_\nu \gamma^\nu \gamma^\mu = q_\nu \gamma_\mu (2g^{\mu\nu} - \gamma^\mu \gamma^\nu) = 2\not{q} - \gamma_\mu \gamma^\mu \not{q} = 2\not{q} - 4\not{q} = -2\not{q}$, we have

$$\Sigma^{\mathrm{b}}(k) = \frac{-2}{3} 4 g^2 \int \frac{\mathrm{d}^4 q}{(2\pi)^4} \frac{\not{q}}{[(q+k)^2 + \mathrm{i}\eta][q^2 + \mathrm{i}\eta]} \ . \tag{4.135}$$

Following the same steps as before gives

$$\begin{aligned} \Sigma^{\mathrm{b}}(k) &= -\frac{8}{3} g^2 \frac{\mathrm{i}}{16\pi^2}\left(\frac{1}{\varepsilon} - \ln\frac{-k^2}{\mu^2} + \mathrm{const}\right)(-\not{k}) \int_0^1 \mathrm{d}z\, z \\ &= \frac{4}{3} g^2 \frac{\mathrm{i}}{16\pi^2}\left(\frac{1}{\varepsilon} - \ln\frac{-k^2}{\mu^2} + \mathrm{const}\right)\not{k} \\ &\equiv \mathrm{i}\not{k} Z^{\mathrm{b}} \ . \end{aligned} \tag{4.136}$$

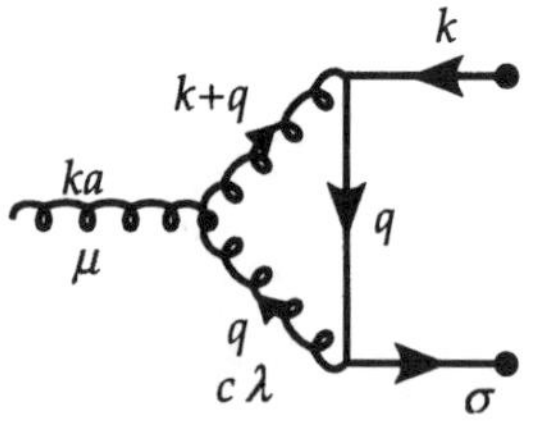

Fig. 4.27. The chosen variables for the first vertex correction graph. The assignment of momenta k and 0 for the outgoing quark lines corresponds to the special Lorentz system as explained in the text

The latter expression defines the divergent factor Z^{b}, which will be important in the renormalization procedure below (see Fig. 4.29 and (4.140)).

Graph (c1). We turn now to the vertex corrections Γ^a_μ; see Fig. 4.27 and (19) of Example 4.2. Γ^a_μ is defined as $\Lambda^a_\mu = \lambda^a/2\gamma_\mu + \Gamma^a_\mu$ (Γ^a_μ corrects the bare vertex $\lambda^a/2\gamma_\mu$):

$$\begin{aligned} \Gamma^{a\,(\mathrm{c1})}_\mu = -\mathrm{i}g^3 \int \frac{\mathrm{d}^4 q}{(2\pi)^4} f^{abc}\left[g_{\mu\lambda}(k-q)_\nu + g_{\nu\mu}(-q-2k)_\lambda + g_{\lambda\nu}(2q+k)_\mu\right] \\ \times \frac{\lambda^b}{2}\frac{\lambda^c}{2} \frac{\gamma^\nu \not{q} \gamma^\lambda}{(q^2 + \mathrm{i}\eta)^2[(q+k)^2 + \mathrm{i}\eta]} \ . \end{aligned} \tag{4.137}$$

Here the quark propagator $\frac{1}{\not{q}+m}$ was taken in the limit $m \to 0$, i.e. $\frac{1}{\not{q}+m} \to \frac{\not{q}}{\not{q}^2+\mathrm{i}\eta}$. This, together with the one gluon propagator $\frac{1}{\not{q}^2+\mathrm{i}\eta}$, yields the above result.

To make things easy we choose a very special kinematics: we choose the outgoing quark to have zero momentum. Or, in other words, we calculate in a specific Lorentz frame where this is the case, relying on the fact that the divergencies and thus the renormalization constants for coupling constants, masses, and wave functions are Lorentz invariant. The divergent term is not affected by this; only the finite ones are, and this makes things much easier. Using $-\mathrm{i} f^{abc}\frac{\lambda^b}{2}\frac{\lambda^c}{2} = -\frac{\mathrm{i}}{2} f^{abc}\left[\frac{\lambda^b}{2}, \frac{\lambda^c}{2}\right] = \frac{1}{2} f^{abc} f^{dbc}\frac{\lambda^d}{2} = N\frac{\delta_{ad}}{2}\frac{\lambda^d}{2} \equiv \frac{C_2}{2}\frac{\lambda^a}{2}$, where we have utilized (4.123), we get

$$
\begin{aligned}
\Gamma_\mu^{a\,(\mathrm{c1})} &= \frac{1}{2}C_2\frac{\lambda^a}{2}g^3\int\frac{\mathrm{d}^4q}{(2\pi)^4}\int_0^1 \mathrm{d}z\,\frac{\Gamma(3)}{\Gamma(2)\Gamma(1)}(1-z) \\
&\quad\times\frac{(\not{k}\not{q}-q^2)\gamma_\mu-\gamma_\mu(q^2+2\not{k}\not{q})-2\not{q}(2q+k)_\mu}{(q^2+2qkz+k^2z+\mathrm{i}\eta)^3} \\
&= \frac{1}{2}C_2\frac{\lambda^a}{2}2g^3\int_0^1 \mathrm{d}z\,(1-z)\int\frac{\mathrm{d}^4q}{(2\pi)^4}\frac{-2q^2\gamma_\mu-q^2\gamma_\mu+\dots}{(q^2+k^2z(1-z)+\mathrm{i}\eta)^3} \\
&= \mathrm{i}C_2\frac{\lambda^a}{2}g^3(+3)\int_0^1 \mathrm{d}z\,(1-z)\frac{\mu^{4-d}}{16\pi^2}\gamma_\mu\pi^\varepsilon\left(-k^2z(1-z)\right)^{-\varepsilon} \\
&\quad\times\frac{\Gamma(3)\Gamma(\varepsilon)}{\Gamma(2)\Gamma(3)}+\dots \\
&= \mathrm{i}C_2\frac{\lambda^a}{2}3g^3\frac{1}{2}\frac{\gamma_\mu}{16\pi^2}\left(\frac{1}{\varepsilon}-\ln\frac{-k^2}{\mu^2}+\dots\right)\,.
\end{aligned} \tag{4.138}
$$

Graph (c3). Adopting again the same kinematics as for the calculation of $\Gamma_\mu^{a\,(\mathrm{c1})}$ (see Fig. 4.27), the second vertex graph Fig. 4.28 is

$$
\begin{aligned}
\Gamma_\mu^{a\,(\mathrm{c3})} &= g^3\int\frac{\mathrm{d}^4q}{(2\pi)^4}\frac{\lambda^b}{2}\gamma^\lambda\frac{\lambda^a}{2}(\not{q}+\not{k})\gamma_\mu\not{q}\frac{\lambda^b}{2}\gamma_\lambda\frac{1}{(q^2+\mathrm{i}\eta)^2\left((q+k)^2+\mathrm{i}\eta\right)} \\
&= g^3\left(\frac{4}{3}-\frac{1}{2}C_2\right)\frac{\lambda^a}{2}(-2)\int\frac{\mathrm{d}^4q}{(2\pi)^4}\frac{\frac{1}{4}(-2)\gamma_\mu q^2}{(q^2+\mathrm{i}\eta)^2\left((q+k)^2+\mathrm{i}\eta\right)} \\
&= g^3\left(\frac{4}{3}-\frac{C_2}{2}\right)\frac{\mathrm{i}\lambda^a}{2}\frac{\gamma_\mu}{16\pi^2}\left(\frac{1}{\varepsilon}-\ln\frac{-k^2}{\mu^2}+\dots\right)\,.
\end{aligned}
$$

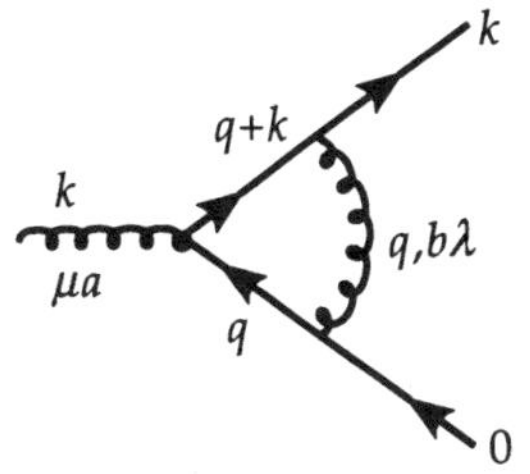

Fig. 4.28. The chosen variables for the second vertex correction graph

All third-order vertex correction graphs (C1) and (C2) together add up to

$$
\Gamma_\mu^{a\,(\mathrm{c})} = g^3\mathrm{i}\frac{\lambda^a}{2}\frac{\gamma_\mu}{16\pi^2}\left(\frac{1}{\varepsilon}-\ln\frac{-k^2}{\mu^2}+\dots\right)\left(-\frac{8}{3}+\frac{4}{3}+\frac{3}{2}C_2-\frac{C_2}{2}\right)
$$

$$
\begin{aligned}
&= g^3 \mathrm{i} \frac{\lambda^a}{2} \frac{\gamma_\mu}{16\pi^2} \left(\frac{1}{\varepsilon} - \ln \frac{-k^2}{\mu^2} + \dots \right) \left(C_2 - \frac{4}{3} \right) \\
&\equiv \mathrm{i} g \frac{\lambda^a}{2} \gamma_\mu Z^{\mathrm{c}} \ .
\end{aligned} \tag{4.139}
$$

The divergent factor Z^{c} defined in the last step will become important in the renormalization procedure below.

Now we can renormalize the coupling constant. The way in which the various contributions combine is shown in Fig. 4.29. The renormalization procedure is in fact rather easy. The square root of the vacuum polarization factor and the self-energy factor multiply each of the respective lines. Thus the quark–gluon vertex acquires the factors $(1+Z^{\mathrm{b}})^{\frac{1}{2}}\,(1+Z^{\mathrm{b}})^{\frac{1}{2}}\,(1+Z^{\mathrm{c}})^{1}\,(1+Z^{\mathrm{a}})^{\frac{1}{2}}$. Note, that this convention differs from the one used in QED,[13] where the renormalization constants $\tilde{Z}$ are related to the one used here by $\tilde{Z} = 1 + Z$. Recall that we defined the renormalization constants as (see (4.113), (4.116), (4.120))

$$
\Gamma_\mu^{a(c)} = \mathrm{i} g \frac{\lambda^a}{2} \gamma_\mu\, Z^{(c)} \ , \quad \Sigma^{(b)} = \mathrm{i} \gamma_\mu k^\mu\, Z^b \ , \quad \Pi^{(a)}_{\mu\mu'} \mathrm{i}(k_\mu k_{\mu'} - g_{\mu\mu'})\, Z^a \ . \tag{4.140}
$$

The renormalized quark–gluon vertex requires the full vertex correction $Z^{(c)}$, half of the correction to the gluon propagator and two times half of the corrections from the self-energy. Graphically we can depict that as follows:

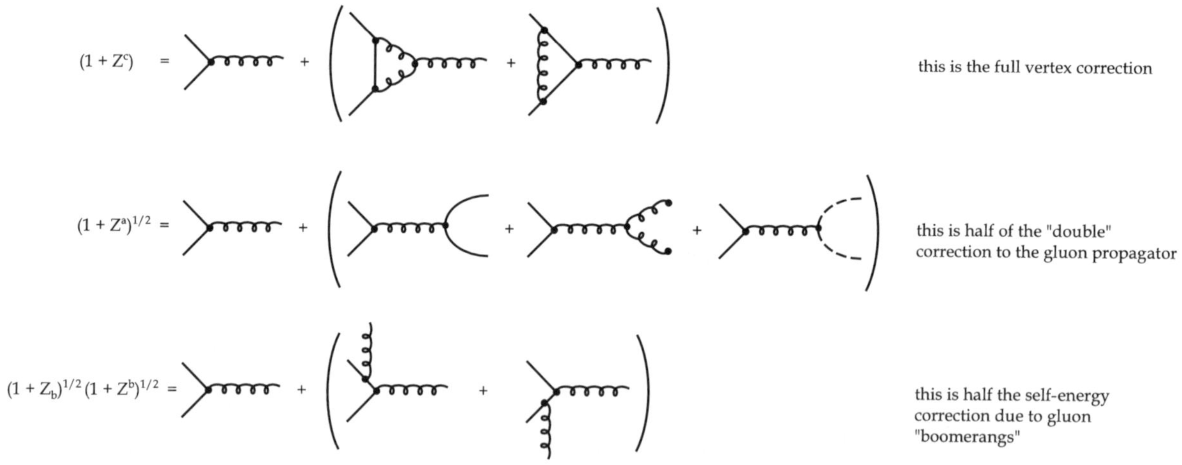

The $\sqrt{(1+Z)} \cdot \sqrt{(1+Z)}$ partiton of the Z factors over two vertices is illustrated in Fig. 4.29 for quark–quark scattering with gluon exchange.

The full quark–quark scattering amplitude with its various contributions up to $O(g_0^6)$ is shown in Fig. 4.30.

---

[13] See W. Greiner and J. Reinhardt: *Quantum Electrodynamics*, 2nd. ed. (Springer, Berlin, Heidelberg 1994).

Adding all contributions, the total vertex is given by

$$
\begin{aligned}
\Gamma_\mu^{a(\text{total})} &= -\frac{\lambda^a}{2} g_0 \gamma_\mu \left(1+Z^{\mathrm{b}}\right)^{\frac{1}{2}} \left(1+Z^{\mathrm{b}}\right)^{\frac{1}{2}} \left(1+Z^{\mathrm{c}}\right)^{1} \left(1+Z^{\mathrm{a}}\right)^{\frac{1}{2}} \qquad (4.141)\\
&= -\mathrm{i}\frac{\lambda^a}{2} g_0 \gamma_\mu \left\{1 - \frac{g^2}{16\pi^2}\left[\frac{1}{2}\left(\frac{20}{12}C_2 - \frac{2}{3}N_{\mathrm{F}}\right)\right.\right.\\
&\quad \left.\left. -\frac{1}{2}\times 2\times\frac{4}{3} + \frac{4}{3} - \frac{C_2}{2} + \frac{3}{2}C_2\right]\left[\frac{1}{\varepsilon} - \ln\left(\frac{-k^2}{\mu^2}\right) + \dots\right]\right\}\\
&= -\mathrm{i}\frac{\lambda^a}{2} g_0 \gamma_\mu \left\{1 - \frac{g^2}{16\pi^2}\left(\frac{22}{12}C_2 - \frac{1}{3}N_{\mathrm{F}}\right)\left[\frac{1}{\varepsilon} - \ln\left(\frac{-k^2}{\mu^2}\right) + \dots\right]\right\} .
\end{aligned}
$$

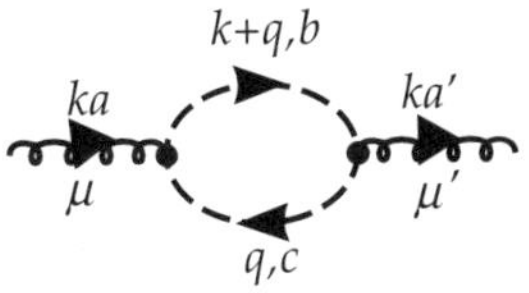

Fig. 4.29. The combination of radiative corrections contributing to the renormalization of the coupling constant g

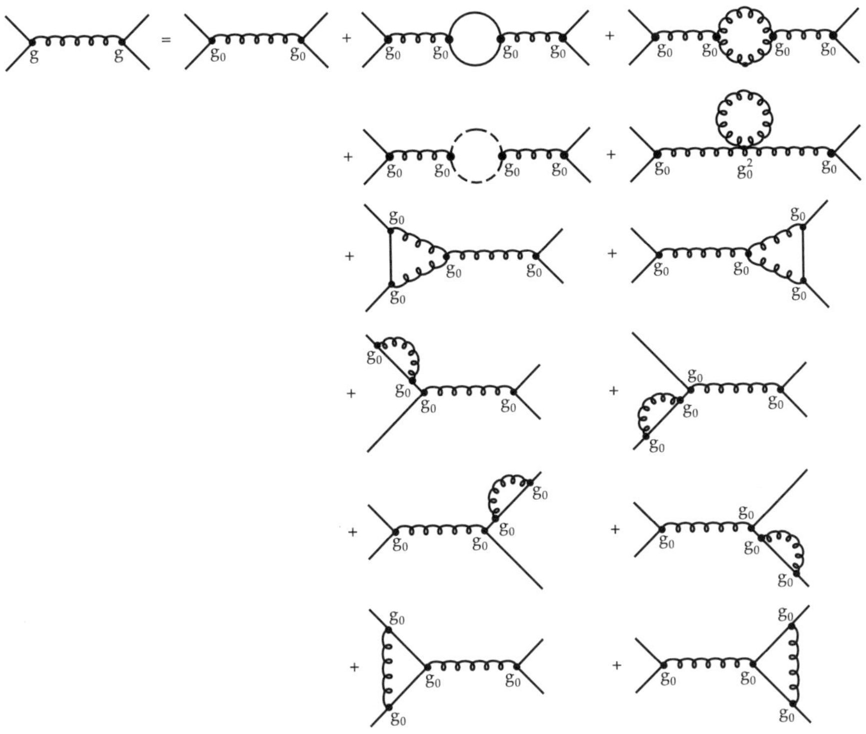

+ higher orders $\mathcal{O}(g_0^6)$ ...

Fig. 4.30. Order g_0^6 corrections to the quark–quark scattering amplitude

Here $(1+Z^{\mathrm{a}})^{\frac{1}{2}} \approx (1+(1/2)Z^{\mathrm{a}})$ has been approximated. The renormalized coupling constant is obtained from (4.141) by subtracting the value of the correction at some renormalization scale M^2:

$$
\begin{aligned}
g_{\mathrm{R}} &= g_0 - \frac{g_0^3}{16\pi^2}\left(\frac{11}{6}C_2 - \frac{1}{3}N_{\mathrm{F}}\right)\left[\frac{1}{\varepsilon} - \ln\left(\frac{-k^2}{\mu^2}\right) - \frac{1}{\varepsilon} + \ln\left(\frac{-M^2}{\mu^2}\right)\right]\\
&= g_0 + \frac{g_0^3}{16\pi^2}\left(\frac{11}{6}C_2 - \frac{1}{3}N_{\mathrm{F}}\right)\ln\left(\frac{-k^2}{M^2}\right) . \qquad (4.142)
\end{aligned}
$$

An important quantity in QCD is the so-called β *function*, which will be discussed below (see, e.g. Exercise 5.11) The β function is defined as $\beta = M\,(\partial g_R/\partial M)|_{-k^2=M^2}$. It describes the dependence of the effective renormalized coupling on the renormalization scale M^2. It can now be easily calculated:

$$\beta = M\frac{\partial g_R}{\partial M} = -\frac{g_0^3}{16\pi^2}\left(\frac{11}{6}C_2 - \frac{1}{3}N_\mathrm{F}\right)\left(\frac{2M}{M^2}\right)M$$
$$= -\frac{g_0^3}{16\pi^2}\left(\frac{11}{3}C_2 - \frac{2}{3}N_\mathrm{F}\right) + O(g_0^5)\ . \tag{4.143}$$

This is the first term in the perturbative expansion for the β function. For QCD we have to insert $C_2 = 3$. Let us reflect on the β-function issue in order to understand the underlying ideas. Iteration and summation of the one-loop corrections in terms of geometrical series leads to

$$g^2(-k^2) = \frac{g_0^2}{1 + \frac{g_0^2}{16\pi^2}\left(\frac{11}{3}C_2 - \frac{2N_\mathrm{F}}{3}\right)\log\left(\frac{-k^2}{M^2}\right)} \tag{4.144}$$

where g_0 is the coupling constant defined at the scale M^2. This is the famous one-loop running coupling constant of QCD. Instead of doing this iteration and resumming the one-loop corrections, one may follow a more formal approach, as we will show now. As mentioned above, in QCD the so-called β function allows a more general definition of the running coupling constant than just the resummation method, based on perturbation theory. The β function is defined as solution of the following differential equation

$$M\left(\frac{\partial g}{\partial M}\right)_{-k^2=M^2} \equiv \beta(g) = -b_0\, g^3 + \mathcal{O}(g^5) \tag{4.145}$$

and is determined order by order in perturbation theory. Above, in (4.143) and (4.144), we have determined the coefficient b_0 entering on the right-hand side of (4.146) as

$$b_0 = \frac{1}{16\pi^2}\left(\frac{11}{3}C_2 - \frac{2N_\mathrm{F}}{3}\right)\ . \tag{4.146}$$

It is important to notice that (4.145) constitutes a differential equation for the coupling constant g in its dependendce on the renormalization scale M, while (4.146) is based on a resummation of one-loop diagrams. There will be situations in which the function $\beta(g)$ can be determined nonperturbatively. Then (4.145) leads to more or less non perturbative expressions for the running coupling constant. This remark may demonstrate the advantage of the β-function method for the calculation of $g(-k^2)$. We follow this idea and solve the differential equation (4.145) to get an explicit solution for $g(-k^2)$ in the one-loop approximation. Separation of the variables g and M yields

$$\int\limits_{g(M^2)}^{g(-k^2)} \frac{\mathrm{d}g}{g^3} = -b_0 \int\limits_{M^2}^{(-k^2)^2} \frac{\mathrm{d}M}{M} \tag{4.147}$$

and therefore

$$\frac{1}{g^2(M^2)} - \frac{1}{g^2(-k^2)} = -b_0 \ln\left(\frac{-k^2}{M^2}\right) . \tag{4.148}$$

Solving for $g^2(-k^2) = 4\pi\alpha_s(-k^2)$ yields

$$\alpha_s(-k^2) = \frac{\alpha_s(M^2)}{1 + \frac{\alpha_s}{4\pi}\left(11 - \frac{2N_\mathrm{F}}{3}\right)\ln\left(\frac{-k^2}{M^2}\right)} , \tag{4.149}$$

which is the running coupling constant up to one-loop order (note that $C_2 = 3$ in QCD).

It is instructive to compare our results (4.142) with those obtained for a different gauge, namely the Landau gauge, in which the gluon propagator is

$$\frac{-g_{\mu\nu} + \frac{k_\mu k_\nu}{k^2}}{k^2 + \mathrm{i}\eta} . \tag{4.150}$$

In this gauge, instead of (4.133) one obtains

$$\begin{aligned} \Pi^{(\mathrm{a}_2)}_{\mu\mu'} + \Pi^{(\mathrm{a}_3)}_{\mu\mu'}\big|_{\mathrm{Landau}} &= \mathrm{i}\frac{3\delta_{aa'}}{16\pi^2}\left[\frac{1}{\varepsilon} - \ln\left(\frac{-k^2}{\mu^2}\right) + \dots\right] \\ &\quad \times \frac{26}{12}(k_\mu k_{\mu'} - g_{\mu\mu'}k^2) , \end{aligned} \tag{4.151}$$

instead of (4.136)

$$\Sigma^{\mathrm{b}}(k) = 0 , \tag{4.152}$$

and instead of (4.141)

$$\Gamma^{a\,(\mathrm{c})}_{\mu} = \mathrm{i}\frac{\lambda^a}{2}\frac{\gamma_\mu}{16\pi^2}\left[\frac{1}{\varepsilon} - \ln\left(\frac{-k^2}{\mu^2}\right) + \dots\right]\frac{3}{4}C_2 . \tag{4.153}$$

Together this gives

$$\begin{aligned} g_\mathrm{R} &= g_0 - \frac{g_0^3}{16\pi^2}\left(\frac{13}{12}C_2 + \frac{3}{4}C_2 - \frac{1}{3}N_\mathrm{F}\right)\left[\frac{1}{\varepsilon} - \ln\left(\frac{-k^2}{\mu^2}\right) + \dots\right] \\ &= g_0 - \frac{g_0^3}{16\pi^2}\left(\frac{11}{6}C_2 - \frac{1}{3}N_\mathrm{F}\right)\left[\frac{1}{\varepsilon} - \ln\left(\frac{-k^2}{\mu^2}\right) + \dots\right] . \end{aligned} \tag{4.154}$$

Thus the β function as an observable is gauge independent, as it must be, but the renormalization of the wave function owing to the self-energy graph and the vacuum polarization graphs is gauge dependent. For completeness let us also cite the result for the β function to third order:[14]

$$\beta(g) = -\frac{g_0^3}{(4\pi)^2}\left(\frac{11}{3}C_2 - \frac{3}{2}N_F\right) - \frac{g_0^5}{(4\pi)^4}\left(\frac{34}{3}C_2^2 - 2C_1N_F - \frac{10}{3}C_2N_F\right)$$
$$- \frac{g_0^7}{(4\pi)^6}\left(\frac{2857}{54}C_2^3 + C_1^2N_F - \frac{205}{18}C_1C_2N_F\right.$$
$$\left. - \frac{1415}{54}C_2^2N_F + \frac{11}{9}C_1N_F^2 + \frac{79}{54}C_2N_F^2\right) \tag{4.155}$$

with $C_2 = N$ (see (4.124)) for SU(N) and $C_1 = \mathrm{tr}\frac{\lambda^a}{2}\frac{\lambda^a}{2}/N = 4/3$ for quarks in the fundamental representation.

For QCD this simplifies to

$$\beta(g) = -\frac{g^3}{(4\pi)^2}\left(11 - \frac{2}{3}N_F\right) - \frac{g^5}{(4\pi)^4}\left(102 - \frac{38}{3}N_F\right)$$
$$- \frac{g^7}{(4\pi)^6}\left(\frac{2857}{2} - \frac{5033}{18}N_F + \frac{325}{54}N_F^2\right) . \tag{4.156}$$

Up to second order this can be translated into an explicit solution for $\alpha_s = g_R^2/4\pi$

$$\alpha_s(\mu) = \frac{12\pi}{(33-2N_F)\ln\left(\frac{\mu^2}{\Lambda^2}\right)}\left[1 - \frac{6(153-19N_F)}{(33-2N_F)^2}\frac{\ln\left[\ln\left(\frac{\mu^2}{\Lambda^2}\right)\right]}{\ln\left(\frac{\mu^2}{\Lambda^2}\right)}\right] , \tag{4.157}$$

where we have introduced the famous Λ_{QCD} parameter (see Example 4.3). It is explained there that $\Lambda_{QCD} = \Lambda_{QCD}\left(M^2\right)$, i.e., it is a function of the renormalization scale M^2. To third order, such an explicit expansion cannot be given. This is because one actually expands simultaneously in $\ln\mu^2/\Lambda^2$ and $\ln\left[\ln\left(\mu^2/\Lambda^2\right)\right]$. Such a double expansion becomes ambiguous at higher orders.

Let us finally note that the higher terms of the β function, i.e., the cofficient of g_0^3 and g_0^5 depend on the regularization scheme used, leading to different Λ_{QCD} parameters for different renormalization schemes, which are correlated with different expressions in the large bracket on the right-hand side of (4.157). This will be explicitly discussed for the lattice gauge regularization in Sect. 7.1.

[14] See O.V. Tarasov, A.A. Vladimirov, and A.Yu. Zharkov: Phys. Lett. B **93**, 429 (1980); S.A. Larin and J.A.M. Vermaseren: Phys. Lett. B **303**, 334 (1993).

4.5 Extended Example: Anomalies in Gauge Theories

In this extended example we will discuss the origin of quantum anomalies. Let us first explain what the term quantum anomalies in field theory means. Consider the action of a classical field theory which is invariant under certain symmetry transformations. If this invariance cannot be preserved at the quantum level, i.e., when quantum corrections are taken into account, such a phenomenon is called a quantum anomaly.

We will concentrate on the physical meaning of the phenomenon and demonstrate the existence of two important anomalies in field theories: the chiral and the scale anomaly.

4.5.1 The Schwinger Model on the Circle

Before discussing the scale anomaly in QCD and its connection to the QCD β function we analyse the chiral anomaly in one of the simplest gauge field models, namely the Schwinger model on a circle. The Schwinger model is two-dimensional QED with massless Dirac fermions. The Langrange density for this model is

$$\mathcal{L} = -\frac{1}{4e_0^2}\, F^2 + \bar{\Psi}\mathrm{i}\gamma \cdot D\Psi \ , \tag{4.158}$$

where e_0 is the bare gauge coupling constant having mass dimension $D = 2$, F is the gauge field strength tensor with components

$$F_{\mu\nu} = \partial_\mu A_\nu - \partial_\nu A_\mu \ , \tag{4.159}$$

D denotes the covariant derivative with components

$$D_\mu = \partial_\mu + \mathrm{i}e_0 A_\mu \tag{4.160}$$

and Ψ is the two-component spinor field. The gamma matrices can be chosen as Pauli matrices in the following way:

$$\gamma^0 = \sigma_2 = \begin{pmatrix} 0 & -\mathrm{i} \\ \mathrm{i} & 0 \end{pmatrix}, \quad \gamma^1 = \mathrm{i}\sigma_1 = \begin{pmatrix} 0 & \mathrm{i} \\ \mathrm{i} & 0 \end{pmatrix}, \quad \gamma^5 = \sigma_3 = \begin{pmatrix} 1 & 0 \\ 0 & -1 \end{pmatrix} . \tag{4.161}$$

Note that $\gamma^5 = \gamma^0\gamma^1$. The Pauli matrices are in fact two dimensional representations of the gamma matrices since

$$\gamma^\mu\gamma^\nu + \gamma^\nu\gamma^\mu = 2\, g^{\mu\nu} \text{ with } \ g^{\mu\nu} = \gamma^{5\,\mu\nu} \ . \tag{4.162}$$

With the two dimensional antisymmetric tensor

$$\epsilon = \begin{pmatrix} 0 & 1 \\ -1 & 0 \end{pmatrix} \tag{4.163}$$

one also verifies explicitely the relations

$$\gamma^\mu \gamma^\nu = \epsilon^{\mu\nu} \gamma^5 + g^{\mu\nu} \ , \tag{4.164}$$

which will be useful later on.

The spinor $\Psi_L \equiv (\psi_1, 0)$ will be called left-handed ($\gamma^5 \Psi_L = +\Psi_L$), while the spinor $\Psi_R \equiv (0, \psi_2)$ with $\gamma^5 \Psi_R = -\Psi_R$ will be called right-handed.

In spite of the considerable simplifications compared to the four-dimensional QED, the dynamics of the model (4.158) is still too complicated for our purpose. In order to simplify the situation further let us consider the system described by (4.158) in a finite spatial domain of length L. We impose periodic boundary conditions on the gauge field and (just for convenience) antiperiodic ones for the massless Dirac fermions. Thus

$$A\left(t, -\frac{1}{2}L\right) = A\left(t, +\frac{1}{2}L\right) \ , \quad \Psi\left(t, -\frac{1}{2}L\right) = -\Psi\left(t, +\frac{1}{2}L\right) \ . \tag{4.165}$$

These conditions impose that the gauge field A and the Dirac fermions Ψ can be expanded in Fourier modes, i.e $\exp(\mathrm{i}kx\, 2\pi/L)$ for the bosons and $\exp[\mathrm{i}(k+1/2)x\, 2\pi/L)$ for the fermions.

Now let us recall that the Lagrange density (4.158) is invariant under the local gauge transformations

$$\Psi \longrightarrow \mathrm{e}^{\mathrm{i}\alpha(t,x)}\, \Psi, \quad A_\mu \longrightarrow A_\mu - \partial_\mu \alpha(t, x) \ . \tag{4.166}$$

It is quite evident that all modes for the field A_1 except for the zero mode can be gauged away. Indeed, terms of the type $a(t)\exp(\mathrm{i}kx\, 2\pi/L)$ in A_1 with nonzero momentum can be gauged away with $\alpha(t, x) = -(\mathrm{i}k)^{-1} a(t)\exp(\mathrm{i}kx\, 2\pi/L)$. This choice for the gauge function is in agreement with the boundary conditions (4.165), i.e. the gauge function is periodic on the circle with radius L. As a consequence we can treat A_1 in the most general case as x-independent.

However, the possibilities provided by gauge invariance are not exhausted yet. There exist another class of legal gauge transformations which are not periodic in x:

$$\alpha(t, x) = \frac{2\pi}{L} nx \quad n = \pm 1, \pm 2, \ldots \ . \tag{4.167}$$

Since $\partial\alpha/\partial x = \text{const.}$ and $\partial\alpha/\partial t = 0$ the periodicitiy of the gauge field is not violated. For the phase factor $\exp(\mathrm{i}\alpha)$ the analogous assertion is valid- the difference of phases at the endpoints of the interval $[-L/2, L/2]$ is equal to $2\pi n$.

As a result, we arrive at the conclusion that A_1 should not be considered in the whole interval $(-\infty, +\infty)$. The points A_1, $A_1 \pm 2\pi/L$, $A_1 = \pm 4\pi/L$, etc. are equivalent with respect to the gauge transformations (4.167) and must be identified. In other words, the variable A_1 should be considered only in the interval $[0, 2\pi/L]$. Beyond this interval we find gauge copies of this interval. In the commonly accepted terminology we may say that A_1 lives on the circle with length $2\pi/L$.

So far we considered the $\mu = 1$ component of the gauge field and this is sufficient if we consider the gauge field as "external". The $\mu = 0$ component of the gauge field is then responsible for the Coulomb interaction between the fermions. Since in two dimensions the Coulomb interaction grows linearly with the distance between two fermions (confinement) the corresponding effect is of the order $e_0 L$ at maximum. Now, if L is small, $e_o L \ll 1$, the Coulomb interaction never becomes strong and we can neglect it in first approximation.

Let us turn to *global symmetries* of the model. The Lagrange density with finite L is invariant under multiplication of the fermion field by a constant phase. According to the famous theorem of E. Noether one easily derives from this a conserved current which turns out to be identical with the usual electromagnetic current:

$$J_\mu = \bar{\Psi}\gamma_\mu \Psi \ , \quad \dot{Q}(t) = 0 \ , \quad Q(t) = \int \mathrm{d}x \ j_0(t,x) \ . \tag{4.168}$$

The Lagrange density (4.158) admits another global symmetrie:

$$\Psi \longrightarrow \mathrm{e}^{\mathrm{i}\alpha\gamma^5} \Psi \ , \tag{4.169}$$

which is called the *global axial transformation*. Again the theorem by Noether ensures the existence of a conserved current $J_{5\,\mu}$, the so called *axial current*, which is conserved at the classical level just in the same way as the electromagnetic current. Note that the axial transformation multiplies the left- and right-handed fermions by opposite phases. If the axial charge of the left-handed fermion is $Q_5 = +1$, for the right-handed fermion $Q_5 = -1$.

The conservation of Q and Q_5 is equivalent to the conservation of the number of the left-handed and right-handed fermions seperately. As we will see, in the quantized theory only the sum of chiral charges is conserved, so only one out of the two global symmetries of the classical theory survives the quantization of the theory.

4.5.2 Dirac Sea

Let us now give a heuristic derivation of the chiral anomaly in the Schwinger model on the circle before deducing it in a more rigorous manner.

In the two-dimensional electrodynamics the Dirac equation determining the energy levels of the massless fermions is given by

$$\left[\mathrm{i}\frac{\partial}{\partial t} + \sigma_3\left(\mathrm{i}\frac{\partial}{\partial x} - A_1\right)\right] \Psi = 0 \ . \tag{4.170}$$

For the kth stationary state $\Psi \sim \exp(-\mathrm{i}E_k t)\ \Psi_k(x)$ and the energy of the kth state is

$$E_k \Psi_k(x) = -\sigma_3 \left(\mathrm{i}\frac{\partial}{\partial x} - A_1\right) \Psi_k(x) \ . \tag{4.171}$$

Furthermore, the eigenfunctions are proportional to

$$\Psi_k(x) \sim \exp\left[\mathrm{i}\left(k+\frac{1}{2}\right)\frac{2\pi}{L}x\right] , \quad k=0, \pm 1, \pm 2, \dots \ . \tag{4.172}$$

As a result, we conclude that the energy of the kth level for the left-handed fermions is

$$E_{k\,L} = \left(k+\frac{1}{2}\right)\frac{2\pi}{L} + A_1 \ , \tag{4.173}$$

while for the right-handed fermions

$$E_{k\,R} = -\left(k+\frac{1}{2}\right)\frac{2\pi}{L} - A_1 \ . \tag{4.174}$$

At $A_1 = 0$ the energy levels are for the left- and right-handed fermions are degenerate. If A_1 increases, the degeneracy is lifted and the level are split. At the point $A_1 = 2\pi/L$ the structure of the energy levels is precisely the same as for $A_1 = 0$ – the degeneracy takes place again. This is the remnant of the gauge invariance of the original theory. It is important to note that the identity of the points $A_1 = 0$ and $A_1 = 2\pi/L$ is achieved in a non-trivial way. Since in passing from $A_1 = 0$ to $A_1 = 2\pi/L$ a restructuring of the fermion levels take place. All left-handed levels are shifted upwards by one interval while all right-handed levels are shifted downwards by the same one interval. This phenomenon lies on the basis of the chiral anomaly in the model at hand, as will become clear shortly.

Let us now proceed to field theory. The first task is the construction of the ground state. To this end, following the well-known Dirac-prescription we fill up all levels lying in the Dirac sea, leaving all positive-energy levels empty. We will use $\left|1_{L,R}, k\right\rangle$ for the filled and $\left|0_{L,R}, k\right\rangle$ for the empty energy levels with a given k. The subscript L (R) indicates that we deal with left-handed (right-handed) fermions.

At first, the value of A_1 is fixed in the vicintiy of zero. Then, the fermion ground state, as seen from the figure, reduces to

$$\begin{aligned} |\Psi_{\text{ferm. vac.}}\rangle = &\left(\bigotimes_{k=-1,-2,\dots} |1_L, k\rangle\right) \otimes \left(\bigotimes_{k=1,2,\dots} |0_L, k\rangle\right) \otimes \\ &\left(\bigotimes_{k=0,1,2,\dots} |1_R, k\rangle\right) \otimes \left(\bigotimes_{k=-1,-2,\dots} |0_R, k\rangle\right) . \end{aligned}$$

The Dirac sea, or all negative-energy levels are completely filled. Now let A_1 increase (adiabatically) from 0 to $2\pi/L$. At $A_1 = 2\pi/L$ the fermion ground state (4.175) describes that state which, from the point of view of the normally filled Dirac sea, contains one left-handed particle and one right-handed hole.

The interesting question is whether the quantum numbers of the fermions change in the transition process from $A_1 = 0$ to $A_1 = 2\pi/L$. Naively we would say that the appearance of the particle and the hole does not change the electric charge. In other words, the electromagnetic current is conserved. On the other hand, the axial charges of the left handed particle and the right-handed hole are the same ($Q_5 = 1$) and, hence, in the transition at hand

$$\Delta Q_5 = 2 \ . \tag{4.175}$$

Equation (4.175) can be rewritten as $\Delta Q_5 = (L/\pi)\, \Delta A_1$. Dividing by the transition time we get

$$\dot{Q}_5 = \frac{L}{\pi}\, \dot{A}_1 \ , \tag{4.176}$$

which implies, in turn, that the conserved charge is given by

$$\tilde{Q}_5 = \int \mathrm{d}x \, \left(J_{5\,\mu} - \frac{1}{\pi} A_1 \right) \ . \tag{4.177}$$

The conserved current corresponding to this charge is

$$\tilde{J}^{5\,\mu} = J^{5\,\mu} - \frac{1}{\pi}\epsilon^{\mu\nu} A_\nu \ , \quad \partial \cdot \tilde{J}^5 = 0 \ , \quad \partial \cdot J^5 = \frac{1}{\pi}\epsilon^{\mu\nu}\partial_\mu A_\nu \ , \tag{4.178}$$

where $\epsilon^{01} = -\epsilon^{10} = -\epsilon_{01} = 1$. The third equality in (4.178) is the famous *axial anomaly* in the Schwinger model. We succeeded in deriving it by "hand-waving" arguments by inspecting the picture of motion of the fermion levels in the external field $A_1(t)$. It turns out that in this language the axial (or chiral) anomaly presents a widely known phenomenon: the crossing of the zero point in the energy scale by a group of levels. The presence of the infinite number of levels and the Dirac picture, according to which the emergence of a filled level from the sea means the appearance of a particle while the submergence of an empty level into the sea is equivalent to the production of a hole, are the most essential elements of the whole construction. With a finite number of levels there can be no anomaly.

4.5.3 Ultraviolet Regularization

In spite of the transparent character of this heuristic derivation almost all of the "evident" arguments above can be questioned by the careful reader. Indeed, why is the fermion ground state (4.175) the appropriate choice? In what sense is the energy of this state minimal, taking into account the fact that, accoording to (4.173) and (4.174)

$$E \sim -\sum_{k=0}^{\infty} \left(k + \frac{1}{2} \right) \frac{2\pi}{L} \ , \tag{4.179}$$

and the series is divergent?

The ground state (4.175) describing the fermion sector at $A_1 = 0$ contains, in particular, the direct product of an infinite large number of filled states with negative energy. It is clear that the infinite product is ill-defined, and one cannot do without regularization in calculating physical quantities. The contribution from large momenta should be somehow cut off.

In order to preserve the gauge invariance, it is possible and convenient to use the regularization called the *Schwinger- or ε-splitting*. The regularization will provide a more solid basis to the heuristic derivation presented above. Instead of the original currents

$$J_\mu(t,x) = \bar{\Psi}(t,x)\,\gamma_\mu\,\Psi(t,x)\ , \quad J_{5\,\mu}(t,x) = \bar{\Psi}(t,x)\,\gamma_\mu\gamma_5\,\Psi(t,x)\ , \tag{4.180}$$

we introduce the regularized objects

$$\begin{aligned} J_\mu^{\text{reg}}(t,x) &= \bar{\Psi}(t,x+\varepsilon)\,\gamma_\mu\,\Psi(t,x)\exp\left(-\mathrm{i}\int\limits_x^{x+\varepsilon} dx\,A_1\right)\ , \\ J_{5\,\mu}^{\text{reg}}(t,x) &= \bar{\Psi}(t,x+\varepsilon)\,\gamma_\mu\gamma_5\,\Psi(t,x)\exp\left(-\mathrm{i}\int\limits_x^{x+\varepsilon} dx\,A_1\right)\ . \end{aligned} \tag{4.181}$$

In the calculation of physical quantities the limit $\varepsilon \longrightarrow 0$ is always implied. At the intermediate stages, however, all computations are performed with fixed ε. The exponential factor in (4.181) ensures gauge invariance of the non-local ("split") currents. Without this exponential any composite operator like the electromagnetic current or the axial current transforms under the local phase transformations (4.165) like

$$\begin{aligned} &\Psi_\alpha^\dagger(t,x+\varepsilon)\,\Psi_\beta(t,x) \\ &\qquad \longrightarrow \exp\left[-\mathrm{i}\alpha(x+\varepsilon)+\mathrm{i}\alpha(x)\right]\,\Psi_\alpha^\dagger(t,x+\varepsilon)\,\Psi_\beta(t,x)\ . \end{aligned} \tag{4.182}$$

The gauge transformation of A_1 ($A_1 \longrightarrow A_1 - \partial\alpha(x)/\partial x$) compensates for the phase factor in (4.182).

Now, there is no difficulty to calculate the electric and axial charges of the state (4.175). If

$$Q(t) = \int \mathrm{d}x\, J_0^{\text{reg}}(t,x)\ , \quad Q_5(t) = \int \mathrm{d}x\, J_{5\,0}^{\text{reg}}(t,x)\ , \tag{4.183}$$

then we get for the fermion ground state

$$Q = Q_L + Q_R\ , \quad Q_5 = Q_L - Q_R\ , \tag{4.184}$$

$$\begin{aligned} Q_L &= \sum_{k=-1}^{-\infty} \exp\left(-\mathrm{i}\varepsilon\,E_{k\,L}\right) \\ &= \sum_{k=-1}^{-\infty} \exp\left\{-\mathrm{i}\varepsilon\left[\left(k+\frac{1}{2}\right)\frac{2\pi}{L}+A_1\right]\right\}\ , \end{aligned} \tag{4.185}$$

$$Q_R = \sum_{k=0}^{+\infty} \exp\left(+\mathrm{i}\varepsilon\, E_{k\,R}\right)$$
$$= \sum_{k=0}^{+\infty} \exp\left\{-\mathrm{i}\varepsilon\left[\left(k+\frac{1}{2}\right)\frac{2\pi}{L}+A_1\right]\right\} \;. \tag{4.186}$$

In the limit $\varepsilon \longrightarrow 0$ both charges, Q_L and Q_R, turn into the sum of units – each unit represents one filled level from the Dirac sea. Performing the summation

$$Q_L = e^{-\mathrm{i}\varepsilon\, A_1} 1 - \mathrm{e}^{+\mathrm{i}\varepsilon\, 2\pi/L} \;, \tag{4.187}$$

$$Q_R = e^{-\mathrm{i}\varepsilon\, A_1} 1 - \mathrm{e}^{-\mathrm{i}\varepsilon\, 2\pi/L} \tag{4.188}$$

and expanding in ε we arrive at

$$Q_L = \frac{L}{-\mathrm{i}\varepsilon 2\pi} + \frac{L}{2\pi}\, A_1 + \mathcal{O}(\varepsilon) \;, \tag{4.189}$$

$$Q_R = \frac{L}{+\mathrm{i}\varepsilon 2\pi} - \frac{L}{2\pi}\, A_1 + \mathcal{O}(\varepsilon) \tag{4.190}$$

for the fermion ground state (4.175).

Equations (4.189) show that under our choice of the fermion ground state the charge of the vacuum vanishes, $Q = Q_L + Q_R = 0$. There is no time dependence – the charge is conserved. The axial charge consists of two terms: the first represents an infinitely large constant and the second one gives a linear A_1 dependence. In the transition from $A_1 = 0$ to $A_1 = 2\pi/L$ the axial charge changes by two units.

These conclusions are not new for us. We have found just the same from the illustrative picture described above in which the electric and axial charges of the Dirac sea are determined intuitively. Now we learned how to sum up the infinite series of units, i.e. the charges of the "left-handed" and "right-handed" seas.

It is very important to note, that there is no regularization ensuring simultaneously the gauge invariance and conservation of the axial current.

Now, we leave the issue of charges and proceed to the calculation of the fermion-sea energy, the problem which could not be solved at the naive level, without regularization. Fortunately, all necessary elements are already prepared. The fermion part of the Hamiltonian

$$H = -\Psi^{\dagger}(t,x)\,\sigma_3\left(\mathrm{i}\frac{\partial}{\partial x} - A_1\right)\Psi(t,x) \tag{4.191}$$

reduces after the ε-splitting to

$$H^{\mathrm{reg}} = -\Psi^{\dagger}(t,x+\varepsilon)\,\sigma_3\left(\mathrm{i}\frac{\partial}{\partial x} - A_1\right)\Psi(t,x)\exp\left(-\mathrm{i}\int_x^{x+\varepsilon} dx\, A_1\right) \;. \tag{4.192}$$

This formula implies the following regularized expressions for the energy of the "left-handed" and "right-handed" seas:

$$E_L = \sum_{k=-1}^{-\infty} E_{k\,L} \exp(-\mathrm{i}\varepsilon\, E_{k\,L}) \ ,$$

$$E_R = \sum_{k=0}^{+\infty} E_{k\,R} \exp(+\mathrm{i}\varepsilon\, E_{k\,R}) \ , \qquad (4.193)$$

where the energies of the individual levels $E_{k\,L,R}$ are given in (4.173) and (4.174) and the summation runs over all levels with negative energy. The expressions (4.193) have the following meaning: in the limit $\varepsilon \longrightarrow 0$ they reduce to the sum of the energies of all filled fermion levels from the Dirac sea. Notice, that E_L and E_R can be obtained by differentiating the expressions (4.185) and (4.186) for $Q_{L,R}$ with respect to ε:

$$E_L = \mathrm{i}\frac{\partial}{\partial\varepsilon}\, Q_L = \mathrm{i}\frac{\partial}{\partial\varepsilon}\ e^{-\mathrm{i}\varepsilon\, A_1} 1 - \mathrm{e}^{+\mathrm{i}\varepsilon\, 2\pi/L} \ , \qquad (4.194)$$

$$E_R = -\mathrm{i}\frac{\partial}{\partial\varepsilon}\, Q_R = -\mathrm{i}\frac{\partial}{\partial\varepsilon}\ e^{-\mathrm{i}\varepsilon\, A_1} 1 - \mathrm{e}^{-\mathrm{i}\varepsilon\, 2\pi/L} \ , \qquad (4.195)$$

where we have used (4.187) and (4.188). The total energy of the sea is then

$$\begin{aligned} E_{\text{sea}} &\equiv E_L + E_R \\ &= e^{-\mathrm{i}\varepsilon\, A_1} 1 - \cos(2\pi\varepsilon/L) \left(\frac{2\pi}{L} + \mathrm{i}A_1 \sin\left(\frac{2\pi}{L}\varepsilon\right)\right) \ . \end{aligned} \qquad (4.196)$$

Expanding in ε we get

$$\begin{aligned} E_{\text{sea}} &= \frac{(1 - \mathrm{i}\varepsilon\, A_1 + \varepsilon^2\, A_1^2 + \ldots)}{1 - 1 + \frac{1}{2}\left(\frac{2\pi}{L}\,\varepsilon\right)^2 + \ldots} \left(\frac{2\pi}{L} + \mathrm{i}A_1 \sin\left(\frac{2\pi}{L}\,\varepsilon\right) + \ldots\right) \\ &= \frac{L}{2\pi}\, A_1^2 + \mathcal{O}(\varepsilon) + \text{a constant independent of } A_1 \ . \end{aligned} \qquad (4.197)$$

One remark is in order here. From the very beginning we assumed that the dynamics of A_1 is neglectable when compared to the dynamics of the fermions. Let us verify that this is indeed justified.

The effective Lagrange density determinig the quantum mechanics of A_1 is

$$\mathcal{L} = \frac{L}{2e_0^2}\, \dot{A}_1^2 - \frac{L}{2\pi}\, A_1^2 \ . \qquad (4.198)$$

This is the ordinary harmonic oscillator with the wave function

$$\frac{L}{e_0 \pi^{\frac{3}{2}}} \exp\left(-\frac{LA_1^2}{2e_0 \pi^{\frac{1}{2}}}\right) \qquad (4.199)$$

of the ground state and the level splitting

$$\omega_A = \frac{e_0}{\pi^{\frac{1}{2}}} \ . \tag{4.200}$$

The characteristic frequencies in the fermion sector are $\omega_\Psi \sim L^{-1}$. Hence, $\omega_A/\omega_\Psi \sim Le_0 \ll 1$.

4.5.4 Standard Derivation

It will be extremely useful to discuss the connection between the picture presented above and the more standard derivation of the chiral anomaly in the Schwinger model. The following discussion will serve as a bridge between the physical picture described above and the standard approach to anomalies in QCD and other gauge theories.

We would like to demonstrate that

$$\partial \cdot J^5 = \frac{1}{2\pi}\, \epsilon^{\mu\nu} F_{\mu\nu} \ , \tag{4.201}$$

by considering directly the the divergence of the axial current. Then we need to bother only about the ultraviolet regularization. In particular, the theory can be considered in the infinite space since the finiteness of L does not affect the result stemming from the short distances.

One of the convenient methods of the ultrviolet regularization is due to Pauli and Villars. In the model at hand it reduces to the following: In addition to the original massless fermions in the Lagrange density, heavy regulator fermions are introduced with mass M_0 whereas the limit $M_0 \longrightarrow \infty$ is implied. Furthermore, each loop of the regulator fermions is supplied by an extra minus sign relatively to the normal fermion loop. The role of the regulator fermions in the low-energy regime ($E \ll M_0$) is the introduction of an ultraviolett cut-off in the formally divergent integrals corresponding to the fermion loops. Such a regularization procedure automatically guarantees the gauge invariance and the electromagnetic current conservation.

In the model regularized according to Pauli and Villars the axial current has the form

$$J^{5\,\mu} = \bar{\Psi}\, \gamma^\mu \gamma^5\, \Psi + \bar{R}\, \gamma^\mu \gamma^5\, R \ , \tag{4.202}$$

where R is the regulator fermion. In calculating the divergence of the regularized current the naive equations of motions can be used. Then,

$$\begin{aligned} \partial \cdot J^5 &= \bar{\Psi}\left(\overleftarrow{\partial} \cdot \gamma\gamma^5 + \overrightarrow{\partial} \cdot \gamma\gamma^5\right)\Psi + \bar{R}\left(\overleftarrow{\partial} \cdot \gamma\gamma^5 + \overrightarrow{\partial} \cdot \gamma\gamma^5\right) R \\ &= \bar{\Psi}\left(\mathrm{i}g\gamma \cdot A_1\gamma^5 + \mathrm{i}g\gamma^5\gamma \cdot A_1\right)\Psi \\ &\quad + \bar{R}\left(\mathrm{i}g\gamma \cdot A_1\gamma^5 + \mathrm{i}g\gamma^5\gamma \cdot A_1 + 2\mathrm{i}M_0\right) R \\ &= 2\mathrm{i}M_0\, \bar{R}\, \gamma^5\, R \ . \end{aligned} \tag{4.203}$$

The divergence is non-vanishing, i.e. the axial current is not conserved! But, as was expected, $\partial_\mu J^{5\,\mu}$ contains only the regulator anomalous term.

The next step is to contract the regulator fields and calculate the relevant diagrams. In this context we may learn a new technique, the so-called *background field technique* which we shall use here in order to calculate the right hand side of (4.203). One can write in momentum representation

$$\begin{aligned} 2\mathrm{i}M_0\, \bar{R}\, \gamma^5\, R &= -2M_0\, \mathrm{tr} \frac{\gamma^5}{\gamma \cdot \mathcal{P} - M_0} \\ &\equiv -2M_0 \int \frac{\mathrm{d}^2 p}{(2\pi)^2} \mathrm{tr}_{\mathrm{L+C}} \left\langle p \left| \frac{\gamma^5}{\gamma \cdot \mathcal{P} - M_0} \right| p \right\rangle , \end{aligned} \tag{4.204}$$

where $\mathcal{P} = \mathrm{i}D$ and $\mathrm{tr}_{\mathrm{L+C}}$ means the trace with respect to spin (Lorentz) and colour indices; $|p\rangle$ denotes a state vector describing a fermion with momentum p. Note, that we have taken into account the fact that the minus sign in the fermion loop does not appear for the regulator fermions.

Moreover,

$$\begin{aligned} \frac{1}{\gamma \cdot \mathcal{P} - M_0} &= \frac{\gamma \cdot \mathcal{P} + M_0}{\gamma \cdot \mathcal{P} \gamma \cdot \mathcal{P} - M_0^2} \\ &= \frac{\gamma \cdot \mathcal{P} + M_0}{\mathcal{P}^2 + \frac{1}{2}\mathrm{i}\epsilon^{\mu\nu} F_{\mu\nu} \gamma^5 - M_0^2} . \end{aligned} \tag{4.205}$$

The second equality here stems from the definition of the gamma matrices and the following property

$$\left[\mathcal{P}_\mu, \mathcal{P}_\nu\right] = -\left[D_\mu, D_\nu\right] = \mathrm{i}\, F_{\mu\nu} . \tag{4.206}$$

In fact, using (4.162) and (4.164) one may verify

$$\begin{aligned} \gamma \cdot \mathcal{P} \gamma \cdot \mathcal{P} &= \gamma^\mu \gamma^\nu\, \mathcal{P}_\mu \mathcal{P}_\nu = \frac{1}{2} \left(\gamma^\mu \gamma^\nu \left[\mathcal{P}_\mu, \mathcal{P}_\nu\right] + 2g^{\mu\nu}\, \mathcal{P}_\mu \mathcal{P}_\nu \right) \\ &= \mathcal{P}^2 + \frac{1}{2}\mathrm{i}\gamma^\mu \gamma^\nu\, F_{\mu\nu} = \mathcal{P}^2 + \frac{1}{2}\mathrm{i}\epsilon^{\mu\nu}\gamma^5\, F_{\mu\nu} . \end{aligned} \tag{4.207}$$

Now, since $M_0 \longrightarrow \infty$ the trace in (4.204) can be expanded in inverse powers of M_0. In this way we get

$$\begin{aligned} \mathrm{tr} \frac{\gamma^5}{\gamma \cdot \mathcal{P} - M_0} &= \mathrm{tr} \Bigg[\gamma^5 \left(\gamma \cdot \mathcal{P} + M_0\right) \\ &\times \left(\frac{1}{\mathcal{P}^2 - M_0^2} + \frac{1}{\mathcal{P}^2 - M_0^2} (-)\frac{1}{2}\mathrm{i}\epsilon^{\mu\nu} F_{\mu\nu}\gamma^5\, \frac{1}{\mathcal{P}^2 - M_0^2} + \ldots \right) \Bigg] . \end{aligned} \tag{4.208}$$

The first term in the expansion vanishes after taking the trace of the gamma matrices. The third and all other terms are irrelevant because they vanish in the limit $M_0 \longrightarrow \infty$. The only relevant term is the second one, where we can substitue the operator $\mathscr{P} = \mathrm{i}D = \mathrm{i}\partial - A$ by the momentum p since the result is explicitly proportional to the back ground field. Then

$$2\mathrm{i}M_0\ \bar{R}\gamma^5 R = -2\mathrm{i}M_0^2 \int \frac{\mathrm{d}^2 p}{(2\pi)^2} \frac{1}{\left(p^2 - M_0^2\right)^2}\ \epsilon^{\mu\nu} F_{\mu\nu}\ . \tag{4.209}$$

The integration over p can be done with the help of the Feynman integral

$$I_0 \equiv \int \frac{\mathrm{d}^n p}{(2\pi)^n} \frac{1}{(p^2 - M_0^2)^\alpha} = \mathrm{i}\frac{(-\pi)^{\frac{n}{2}}}{(2\pi)^n} \frac{\Gamma\left(\alpha - \frac{n}{2}\right)}{\alpha} \frac{1}{(-M_0^2)^{\alpha - \frac{n}{2}}}\ . \tag{4.210}$$

Thus,

$$2\mathrm{i}M_0\ \bar{R}\gamma^5 R = \frac{1}{2\pi}\ \epsilon^{\mu\nu} F_{\mu\nu}\ . \tag{4.211}$$

This computation completes the standard derivation of the anomaly. One should have a very rich imagination to be able to see in these manipulaions the simple physical nature of the phenomenon which has been described above (restructuring of the Dirac sea and level crossing). Nevertheless, this is the same phenomenon viewed from a different perspective – less transparent but more economic since we get the final result very quickly using the well-developed machinery of the diagram technique.

4.5.5 Anomalies in QCD

Before proceeding to quantum anomalies in QCD let us first list the symmetries of the classical action. We will assume that the theory contains n_f massless quarks and for the moment forget the heavy quarks which are inessential in the given context. The correction due to small u-, d-, and s-quark masses can be considered seperately if necessary.

In the chiral limit, i.e. $m_q = 0$, the classical Lagrange density is

$$\mathcal{L} = \sum_q^{n_f} \left(\bar{q}^a_{L\alpha} \mathrm{i}(\gamma)^{\alpha\beta} \cdot (D)_{ab} q^b_{L\beta} + \bar{q}^a_{R\alpha} \mathrm{i}(\gamma)^{\alpha\beta} \cdot (D)_{ab} q^b_{R\beta}\right) - \frac{1}{4} G^a_{\mu\nu}\ G^{\mu\nu}_a\ ,$$
$$q_{L,R} = \frac{1}{2}(1 \pm \gamma_5)\ q\ , \tag{4.212}$$

where q is the quark field and G is the gluon field strength tensor. The action of the classical theory is invariant under the following global transformations:

(1) Rotation of fermions of different flavours,

$$\vec{q} \longrightarrow U\vec{q}\ , \tag{4.213}$$

where $\vec{q}$ denote the n_f-tupel of quark flavours and U is a three-dimensional unitary representation of an arbitrary element of the SU(3) group. Since the left-handed and right-handed quarks enter in the Lagrange density as separate terms, the Lagrange density actually possesses $SU(3)_L \times SU(3)_R$ symmetry, called the *chiral flavour invariance*.

(2) The U(1) transformations,

$$\text{vector:} \quad q_{L,R} \longrightarrow \exp(+i\alpha)\; q_{L,R} \tag{4.214}$$

$$\text{axial:} \quad q_L \longrightarrow \exp(+i\beta)\; q_L\;, \quad q_R \longrightarrow \exp(-i\beta)\; q_R\;. \tag{4.215}$$

The physical meaning of this transformations is, that in the classical theory the charge of the left-handed and right-handed fermions are conserved seperately.

(3) The scale transformations

$$A(x) \longrightarrow \lambda\; A(\lambda x)\;, \quad q(x) \longrightarrow \lambda^{\frac{3}{2}}\; q(\lambda x)\;. \tag{4.216}$$

The scale invariance stems from the fact that the classical action contains no dimensional constants.

At the quantum level the fate of the above symmetries is different. The currents generating the chiral flavour tranformations are conserved even with the ultraviolet regularization switched on. They are anomaly-free in pure QCD. The vector U(1) invariance also stays a valid anomaly-free symmetry at the quantum level. This symmetry is responsible for the fact that the quark number is constant in any QCD process. Finally, the current generating the axial U(1) transformations and the dilatation current are not conserved at the quantum level due to anomalies, as we will show below.

4.5.6 The Axial and Scale Anomalies

As explained above, the concrete form of the anomalous relations can be established without going beyond perturbation theory, provided an appropriate ultraviolet regularization is chosen. It is important to note that in the literature two languages are used for the description of one and the same anomalous relations and many authors do even not realize the distinction between the languages.

Within the first approach one establishes an operator relation, say, between the divergence of the axial current and $G^a_{\mu\nu}\tilde{G}^{\mu\nu}_a$ (see below). Both, the axial current and $G^a_{\mu\nu}\tilde{G}^{\mu\nu}_a$ are treated within this procedure as Heisenberg operators of the quantum field theory. In order to convert the operator relations into amplitudes it is necessary to make one more step: to calculate according to the general rules the matrix elements of the operators occuring in the right-hand and left-hand sides of the anomalous equality.

Within the second approach one analyzes directly the matrix elements. More exactly, one fixes usually an external gluonic field and determines the divergence of the axial current in this field. In the absence of the external field the axial current is conserved. The existence of the anomaly implies that the axial current is not conserved and that the divergence of the axial current is locally expressible in terms of the external field. The analysis of the anomaly in the Schwinger model has been carried out just in this way.

Although one and the same letters are used in both cases – perhaps, the confusion is due to this custom – it is quite evident that the Heisenberg operator at the point x and the expression for the background field at the same point are by no means identical objects. Certainly, in the leading order

$$\langle G_{\mu\nu}\tilde{G}^{\mu\nu}\rangle = (G_{\mu\nu}\tilde{G}^{\mu\nu})_{\text{ext}} , \quad \langle G_{\mu\nu}G^{\mu\nu}\rangle = (G_{\mu\nu}G^{\mu\nu})_{\text{ext}} , \tag{4.217}$$

where $\langle \dots \rangle$ denotes in the case at hand averaging over the external gluon field while the subscript "ext" marks the external field. In the next-to-leading order, however, the right-hand side of equation (4.217) acquires an α_s correction. Therefore if the anomalies are discussed beyond the leading order it is absolutely necessary to specify what particular relations are considered: the operator relations or those of the matrix elements. Only in the one-loop approximation do both versions superficially coincide. In the remainder of this chapter the term "anomaly" will mean the operator anomalous relation.

Let us begin with the axial anomaly since it is simpler in the technical sense and a close example has been analyzed already in the Schwinger model.

The current generating the axial U(1) transformation is

$$J_5^\mu = \sum_q^{n_f} \bar{q}\,\gamma_\mu\gamma_5\,q \ . \tag{4.218}$$

Differentiating and invoking the equation of motion $\gamma \cdot Dq = 0$ we get

$$\partial \cdot J_5 = \sum_q^{n_f} \left(\bar{q}\ \overleftarrow{D} \cdot \gamma\gamma_5\, q - \bar{q}\ \gamma_5\gamma \cdot \overrightarrow{D}\ q \right) = 0 \ . \tag{4.219}$$

The lesson obtained in the Schwinger model teaches us, however, that this is not the whole story and the conservation of the axial current will be destroyed after switching on the ultraviolet regluarization. In the Schwinger model on the circle we deal with the weak coupling regime and, therfore, can choose any of the alternative lines of reasoning either based on the infrared or on the ultraviolet approaches. In QCD it is rather meaningless to speak about the infrared behaviour of quarks. To make the calculation of the anomaly reliable we must invoke only the Green functions at short distances.

We use the ε-splitting for the ultraviolet regularization and write again

$$J_{5\,\mu}^{\text{reg}} = \sum_q^{n_f} \bar{q}(x+\varepsilon)\ \gamma_\mu\gamma_5 \exp\left(ig \int_{x-\varepsilon}^{x+\varepsilon} dy \cdot A(y) \right) q(x-\epsilon) \ . \tag{4.220}$$

In what follows it is convenient (of course not obligatory) to impose the so called *Fock–Schwinger gauge*

$$(x - x_0) \cdot A^a(x) = 0 \tag{4.221}$$

where x_0 is an arbitrary point in space-time playing the role of a gauge parameter. Let us notice that the condition (4.221) is invariant under scale transformations and inversion of the coordinates, but breaks the translational symmetry. The latter restores itself in gauge invariant quantitities. In other words, the parameter x_0 should cancel out in all correlation functions induced by colourless sources. This fact may in principal serve as an additional check of correctness of the actual calculation. Nevertheless, we shall not exploit this useful property and put $x_0 = 0$ in all following considerations,

$$x \cdot A^a(x) = 0 \tag{4.222}$$

getting compact formulae for the quark propagator in the background field. The most important consequence of (4.222) is that the potential is directly expressable in terms of the gluon field strength tensor. Let us derive this result in detail. We start with the identity

$$A^a_\mu(x) \equiv \partial_\mu \left(x \cdot A^a(x)\right) - x^\nu \partial_\mu A^a_\nu(x) \ , \tag{4.223}$$

where due to (4.222) the first summand on the right-hand side vanishes. The second term can be rewritten as follows

$$x_\nu G^{a\,\mu\nu}(x) - x \cdot \partial A^a_\mu(x) \ . \tag{4.224}$$

Combining the equations (4.223) and (4.224) we get

$$(1 + x \cdot \partial) \ A^a_\mu(x) = x_\nu G^{a\,\nu\mu}(x) \ . \tag{4.225}$$

With the substitution $x = \alpha y$ one immediately sees that the left-hand side of (4.225) reduces to a full derivative

$$\frac{\mathrm{d}}{\mathrm{d}\alpha} \left(\alpha A^a_\mu(\alpha y)\right) = \alpha y^\nu G^a_{\nu\mu}(\alpha y) \ . \tag{4.226}$$

Integrating over α from 0 to 1 we arrive at

$$A^a_\mu(x) = \int_0^1 \mathrm{d}\alpha \ \alpha x^\nu G^a_{\nu\mu}(\alpha x) \ , \tag{4.227}$$

which is the desired expression.

For our purpose it is much more convenient to deal with another representation of the potential because we are interested here in the expansion in local operators $\mathcal{O}(\mathcal{A})$, i.e. the gluon field strength tensor and its covariant derivatives at the origin. This is achieved by a Taylor expansion of the gluon field strength tensor in (4.227):

$$A^a_\mu(x) = \frac{1}{2 \cdot 0!} x^\nu G^a_{\nu\mu}(0) + \dots \ , \tag{4.228}$$

where the dots denote terms with derivatives of the gluon field strength tensor.

Now let us return to the ultraviolet regularized version of the axial current (4.220) and calculate its divergence

$$\partial\cdot J_5^{\rm reg} = \sum_q^{n_f} \bar{q}(x+\varepsilon)\ \overleftarrow{\partial}\cdot\gamma\gamma^5 \exp\left(ig\int_{x-\varepsilon}^{x+\varepsilon} dy\cdot A(y)\right) q(x-\varepsilon)$$
$$+\sum_q^{n_f} \bar{q}(x+\varepsilon) \exp\left(ig\int_{x-\varepsilon}^{x+\varepsilon} dy\cdot A(y)\right) \gamma\cdot\overrightarrow{\partial}\gamma^5 q(x-\varepsilon)$$
$$-\sum_q^{n_f} \bar{q}(x+\varepsilon)\ \gamma^5\gamma\cdot\partial \exp\left(ig\int_{x-\varepsilon}^{x+\varepsilon} dy\cdot A(y)\right) q(x-\varepsilon)\ . \quad (4.229)$$

Using the equations of motions for quarks in the first and second line of (4.229) and differentiating the exponential we get

$$\partial\cdot J_5^{\rm reg} = \sum_q^{n_f} \bar{q}(x+\varepsilon)\,(-ig)\,\gamma\cdot A(x+\varepsilon)\gamma_5 \exp\left(ig\int_{x-\varepsilon}^{x+\varepsilon} dy\cdot A(y)\right) q(x-\varepsilon)$$
$$+\sum_q^{n_f} \bar{q}(x+\varepsilon) \exp\left(ig\int_{x-\varepsilon}^{x+\varepsilon} dy\cdot A(y)\right) (-)\gamma_5\, ig\gamma\cdot A(x-\varepsilon)\, q(x-\varepsilon)$$
$$-\sum_q^{n_f} \bar{q}(x+\varepsilon)\ \gamma^5 ig\gamma\cdot(A(x+\varepsilon)-A(x-\varepsilon))$$
$$\times \exp\left(ig\int_{x-\varepsilon}^{x+\varepsilon} dy\cdot A(y)\right) q(x-\varepsilon)\ . \quad (4.230)$$

Let us impose the Fock–Schwinger gauge condition and use the expansion (4.228) of the vector potential in (4.230). Expanding the exponential up to order we obtain

$$\partial\cdot J_5^{\rm reg} = -2ig\sum_q^{n_f} \bar{q}(x+\varepsilon)\ \varepsilon^\nu G_{\nu\mu}(0)\gamma^\mu\gamma^5\ q(x-\varepsilon)\ . \quad (4.231)$$

Contracting the quark lines in the loop we arrive at the following expression we arrive at the following expression

$$\partial\cdot J_5^{\rm reg} = n_f {\rm tr}_{\rm L+C}\left[-2ig\ \varepsilon_\nu G^{\nu\mu}(0)\ \gamma_\mu\gamma_5\,(-i)S(x-\varepsilon,x+\varepsilon)\right]\ , \quad (4.232)$$

where $S(x,y)$ is the massless quark propagator in the background field. The expression for this propagator as a series in the background field in the Fock–

Schwinger gauge is[15]:

$$S(x,y)=S^{(0)}(x-y)+S^{(1)}(x,y)+S^{(2)}(x,y)+\dots\ , \tag{4.233}$$

whereby

$$S^{(0)}(x-y)=\frac{1}{2\pi^2}\,\gamma_\lambda\,\frac{r^\lambda}{(-r^2)^2} \tag{4.234}$$

with $r=x-y$ and the corrections to free propagation up to Order $\mathcal{O}(g^2)$ are

$$S^{(1)}(x,y)=\frac{\mathrm{i}}{4\pi^2}\,\gamma_\lambda\frac{r^\lambda}{(-r^2)^2}\,y_\nu x_\mu\, gG^{\nu\mu}(0)+\frac{1}{8\pi^2}\,\frac{r^\lambda}{-r^2}\,g\tilde{G}_{\lambda\varrho}\gamma^\varrho\,\gamma^5 \tag{4.235}$$

$$S^{(2)}(x,y)=-\frac{1}{192\pi^2}\,\gamma_\lambda\frac{r^\lambda}{(-r^2)^2}\,\left(x^2y^2-(xy)^2\right)\,g^2G^2(0)\ . \tag{4.236}$$

Notice that under the standard choice of the origin, $y=0$, the first term in $S^{(1)}$ and the whole second order contribution $S^{(2)}$ vanishes. Inserting (4.233) in (4.232) and neglecting all term vanisihing in the limit $\varepsilon\to 0$ we find

$$\begin{aligned}\partial\cdot J_5^{\text{reg}}&=-n_f\frac{1}{2}g^2\,G^a_{\nu\mu}\tilde{G}_{a\,\alpha\beta}(0)\,\frac{\varepsilon^\nu\varepsilon^\alpha}{\varepsilon^2}\,\frac{1}{8\pi^2}\,\mathrm{tr}\left(\gamma^\mu\gamma_5\gamma^\beta\gamma_5\right)\\&=n_f\frac{\alpha_s}{4\pi}\,G^a_{\mu\nu}\tilde{G}^{\mu\nu}_a(0)\ ,\end{aligned} \tag{4.237}$$

$$\tilde{G}^{a\,\mu\nu}=\frac{1}{2}\varepsilon^{\mu\nu\varrho\sigma}G^a_{\varrho\sigma}\ . \tag{4.238}$$

Notice that quarks propagate only very short distances $\varepsilon\longrightarrow 0$. The expression (4.237) gives the axial anomaly in the one-loop approximation,

$$\partial\cdot J_5=n_f\frac{\alpha_s}{4\pi}\,G^a_{\mu\nu}\tilde{G}^{\mu\nu}_a\ . \tag{4.239}$$

This result is readily reproducible within the Pauli–Villars regularization. Higher order modifications will be discussed below.

Let us proceed now with the scale anomaly. First of all it is instructive to check that the scale transformations are generated by the current

$$J_{d\,\nu}=x^\mu T_{\mu\nu}\ , \tag{4.240}$$

where $T_{\mu\nu}$ are the components of the energy-momentum tensor (symmetric and conserved),

$$\begin{aligned}T_{\mu\nu}=&-G^a_{\mu\varrho}G^{\varrho}_{\nu\,a}+\frac{1}{4}g_{\mu\nu}\,G^a_{\varrho\sigma}G^{\varrho\sigma}_a\\&+\frac{1}{4}\sum_q^{n_f}\bar{q}\left(\gamma_\mu\overleftrightarrow{D}_\nu+\gamma_\nu\overleftrightarrow{D}_\mu\right)q\ .\end{aligned} \tag{4.241}$$

[15] V.A. Novikov, M.A. Shifman, A.I. Vainshtein and V.I. Zakharov, Fortsch. Phys. **32**, 585 (1985).

The corresponding charge can be represented as

$$Q_d \equiv \int d^3x\, J_{d\,0} = t\,H + \tilde{Q}_d\,,$$
$$\tilde{Q}_d = \int d^3x\, x^k T_{k0}\,, \quad k \in \{1,2,3\}\,, \tag{4.242}$$

where H is the Hamiltonian. For $\tilde{Q}_d$ the following algebraic relations hold

$$[\tilde{Q}_d, p_k] = \mathrm{i}p_k\,,$$
$$\left[\tilde{Q}_d, \exp\left(\mathrm{i}x_k p^k\right)\right] = -\left(x_k p^k\right)\,\exp\left(\mathrm{i}x_k p^k\right)\,. \tag{4.243}$$

Let $\mathcal{O}(x)$ be an arbitrary colourless operator. Then

$$[Q_d, \mathcal{O}(x)] = t\,[H, \mathcal{O}(x)] + [\tilde{Q}_d, \mathcal{O}(x)]$$
$$= -\mathrm{i}\frac{\partial}{\partial t}\mathcal{O}(x) + [\tilde{Q}_d, \mathcal{O}(x)]\,. \tag{4.244}$$

Moreover, the commutator $[\tilde{Q}_d, \mathcal{O}(x)]$ contains two terms, the first term corresponds to a rescaling of the space coordinate and the second one to a rescaling of the operator $\mathcal{O}$. Indeed,

$$[\tilde{Q}_d, \mathcal{O}(x)] = \left[\tilde{Q}_d, \exp\left(\mathrm{i}x_k p^k\right)\mathcal{O}(t,0)\exp\left(-\mathrm{i}x_k p^k\right)\right] \tag{4.245}$$
$$= \left[\tilde{Q}_d, \exp\left(\mathrm{i}x_k p^k\right)\right]\,\mathcal{O}(t,0)\exp\left(-\mathrm{i}x_k p^k\right)$$
$$+ \exp\left(\mathrm{i}x_k p^k\right)\mathcal{O}(t,0)\,\left[\tilde{Q}_d, \exp\left(-\mathrm{i}x_k p^k\right)\right]$$
$$+ \exp\left(\mathrm{i}x_k p^k\right)\,[\tilde{Q}_d, \mathcal{O}(t,0)]\,\exp\left(-\mathrm{i}x_k p^k\right)$$
$$= -\mathrm{i}x_k\partial^k\mathcal{O}(t,x) + \exp\left(\mathrm{i}x_k p^k\right)\,[\tilde{Q}_d, \mathcal{O}(t,0)]\,\exp\left(-\mathrm{i}x_k p^k\right)\,.$$

The commutator $[\tilde{Q}_d, \mathcal{O}(t,0)]$ should be proportional to $\mathcal{O}(t,0)$, hence

$$[D, \mathcal{O}(x)] = -\mathrm{i}\,(x\cdot\partial + d)\,\mathcal{O}(x)\,, \tag{4.246}$$

where d is a dimensionless number determined by the particular form of the operator $\mathcal{O}$.

Only the general properties of quantum field theories have been used so far. The numerical value of the coefficient d depends on the specific structure of the underlying theory. In QCD d is equal to the normal dimension of the operator $\mathcal{O}$, for nstance, if $\mathcal{O} = \mathcal{L}$ where $\mathcal{L}$ is the Lagrange density of QCD, then $d = 4$. Equation (4.246) expresses in a mathematical language the change of the units of length and mass. This explains the origin of the name "scale transformations".

Notice that the restriction on colourless operators is not superfluous. For any coloured operator the commutator with Q_d, as seen by direct calculation, does not reduce to the form (4.246). One should not be puzzled by this fact. The complication which arise are due to gauge fixing, i.e. additional terms can be eliminated by an appropriate gauge transformation.

Using the classical equations of motions we get

$$\partial \cdot J_d = T^\mu_\mu = 0 \ . \tag{4.247}$$

In the classical theory the trace of the energy-momentum tensor vanishes provided that no explicit mass term occurs in the Lagrange density. At the quatum level the energy-momentum tensor is no longer traceless and its trace is given by the scale or dilatation anomaly. The fact that the scale invariance (4.216) is lost in loops is obious. Indeed, the invariance (4.216) takes place because there are no parameters with mass dimension in the classical action. Already at the one-loop level, however, such a parameter inevitably appears in the effective action, the ultraviolet cut-off M_0.

The M_0 dependence of the effective action is known beforhand. It is very convenient to use this information, seemingly it is the shortest way to calculate the trace of the energy-momentum tensor. In the one-loop approximation the effective action can be written somewhat symbolically as

$$S_{\text{eff}} = -\frac{1}{4}\int d^4x \left(\frac{1}{g_0^2} - \frac{b}{16\pi^2}\ln M_0^2 x_{\text{char}}^2\right)\left(G^a_{\mu\nu}\, G_a^{\mu\nu}\right)_{\text{ext}} + \dots \ , \tag{4.248}$$

where $b = \frac{11}{3}N_c - \frac{2}{3}n_f$ is the first coefficient in the Gell-Mann–Low function, x_{char} denotes the characteristic length scale and the dots stand for the fermion terms. We rescaled the gluon field, $gA \longrightarrow A$, so that the coupling constant figures only as an overall factor in front of G^2. The variation of the effective action under the transformation (4.216) is

$$\delta S_{\text{eff}} = -\frac{1}{32\pi^2} b\, G^2 \ln(\lambda) \ , \tag{4.249}$$

implying that

$$\partial \cdot J_d = -\frac{1}{32\pi^2} b\, G^2 \ . \tag{4.250}$$

Returning to the standard normalization og the gluon field we finally get

$$T^\mu_\mu = -\frac{b\alpha_s}{8\pi} G^2 + \mathcal{O}(\alpha_s^2) \ . \tag{4.251}$$

4.5.7 Multiloop Corrections

Although in practical applications it is quite sufficient to limit oneself to the one-loop expressions for the anomalies (4.239) and (4.251), the question for the higher order corrections still deserves a brief discussion. Until recently it was generally believed that the question was totally solved. Namely, the axial anomaly (by the Adler–Bardeen theorem) is purely one-loop while the scale anomaly contains the complete QCD β-function in the right-hand side. In other words, $-b\alpha_s/8\pi$ in equation (4.251) is substitued by $\beta(\alpha_s)/4\pi$ if higher order corrections are taken into account.

Later it become clear, however, that the situation is far from being such simple and calls for additional studies. Surprising though it is, the standard arguments was first revealed not within QCD but in a more complex model – supersymmetric Yang–Mills theory. The minimal model of such a type includes gluons and gluinos, which are Majorana fermions in the adjoint representation of the colour group. The axial current and the dilatational current in supersymmetric theories enter one supermultiplet and, consequently, the coefficients in the chiral and dilatation anomalies cannot be different – one-loop for $\partial \cdot J_5$ and multiloop for $\partial \cdot J_d$.

Being unable to discuss here the multiloop corrections in detail, we note only that the generally accepted treatment is based on the confusion just mentioned in the beginning of the last section. The standard derivation of the Adler–Bardeen theorem seems to be valid only provided that the axial anomaly is treated as an operator equality. At the same time the relation

$$T^{\mu}_{\mu} = \frac{\beta(\alpha_s)}{4\alpha_s} \left(G^a_{\mu\nu} G^{\mu\nu}_a\right)_{\text{ext}} \tag{4.252}$$

takes place only for the matrix elements.

It is quite natural to try to reduce both anomalies to a unified form, preferably to the operator form. Then $\partial \cdot J_5$ is exhausted by the one-loop approximation (4.239), at least, within a certain ultraviolet regularization. As far as the trace anomaly (scale anomaly) is concerned, in this case we do not know even the two-loop coefficient in front of the operator G^2, to say nothing about higher-order corrections.

5. Perturbative QCD I: Deep Inelastic Scattering

In the last chapter we discussed how the QCD coupling constant depends on the transferred momenta. Because of this specific dependence, QCD can only be treated perturbatively in the case of large momentum transfers. For practical purposes, however, it is indispensable to know at what momentum values the transition between perturbative and non-perturbative effects takes place. To this end we again take a look at Fig. 1.1. The quark confinement problem will be considered later together with different models for its solution. A common feature of all these models is that they postulate nonperturbative effects. Since, for example, the masses of the N resonances in Fig. 1.1 reach a value of 2.5 GeV and since on the other hand the rest masses of the constituent up and down quarks are negligible, we can conclude that for momentum transfers less than 1 GeV, QCD is certainly still in the nonperturbative region. On the other hand, Fig. 4.4 shows that even at momentum transfers of a few GeV, quarks inside nucleons behave almost like free particles. Hence the transition from the nonperturbative to the perturbative region must take place quite rapidly, i.e., between $\sqrt{Q^2} \approx 1$ GeV and $\sqrt{Q^2} \approx 3$ GeV. According to Fig. 4.8, QCD can only yield such an immediate transition if the number of quark flavors with masses less than 1–2 GeV is not much larger than six. In fact there are only three or four such quarks: up ($m_{\mathrm{u}} = 5.6 \pm 1.1$ MeV), down ($m_{\mathrm{d}} = 9.9 \pm 1.1$ MeV), strange ($m_{\mathrm{s}} = 199 \pm 33$ MeV), and charm ($m_{\mathrm{c}} = 1.35 \pm 0.05$ GeV). The bottom quark is with $m_{\mathrm{b}} \approx 5$ GeV too heavy, and so is the top quark, which was discovered at the Fermilab Tevatron collider in 1995 and has a mass $m_{\mathrm{t}} = 174.3 \pm 5.1$ GeV. As a further consequence of this sudden transition it is almost certain that for somewhat larger momentum transfers $\sqrt{Q^2}$ all processes can be evaluated by means of the usual perturbation theory, i.e., the QCD Feynman rules. It therefore seems quite obvious to calculate QCD corrections to the parton model of deep inelastic lepton–nucleon scattering. The next section treats these questions in some detail.

5.1 The Gribov–Lipatov–Altarelli–Parisi Equations

The Gribov–Lipatov–Altarelli–Parisi equations (GLAP) describe the influence of the perturbative QCD corrections on the distribution functions that enter the parton model of deep inelastic scattering processes. At this point we investigate their structure and the functions that occur only for the two correction graphs in

Fig. 5.1. In line with the parton-model assumptions, both the scattering parton and the "emitted" gluon can be treated here as free particles. Having determined the GLAP equations for the two graphs mentioned above it is quite easy to extrapolate to their general form. We shall use the notation defined in Fig. 5.2.

Obviously these graphs are similar to those for Compton scattering. Hence their contribution to the scattering tensor $W_{\mu\nu}$ can be evaluated in analogy to the corresponding QED graphs. The first step towards this end is the determination of the correct normalization factor. $W_{\mu\nu}$ is then just a factor entering the cross section (see (3.22)); the photon progagator and the normalization factor of the incoming photon have been separated. Bearing all these facts in mind one is led to the correct result, which we want, however, to derive in a slightly different way. We start with the scattering amplitude $W^N_{\mu\mu'}$ (3.33) – see also Example 3.2:

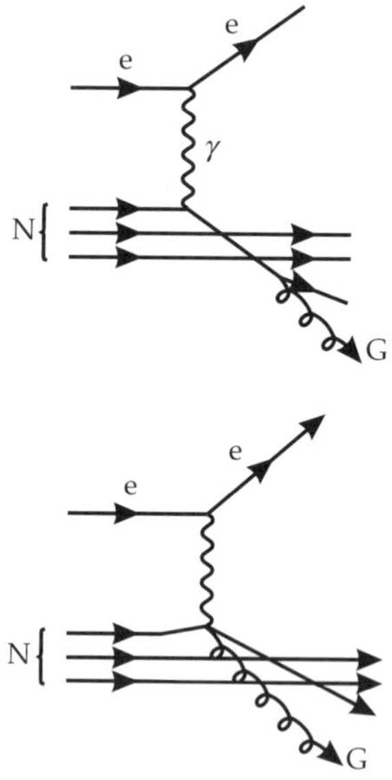

Fig. 5.1. Two correction graphs for deep-inelastic electron–nucleon scattering

$$W^N_{\mu\mu'} = \frac{1}{2\pi}\int \mathrm{d}^4x\, \mathrm{e}^{\mathrm{i}qx}\frac{1}{2}\sum_{\text{pol.}}\langle N|\hat{J}_\mu(x)\hat{J}_{\mu'}(0)|N\rangle \ . \tag{5.1}$$

Note that $\frac{1}{2}\sum_{\text{pol.}}$ stands for the averaging over the incoming nucleon spin. In order to obtain the contributions due to Fig. 5.2 we clearly have to insert

$$\hat{J}_\mu(x) = \int \hat{\bar{\Psi}}(y) g\hat{G}^a_\varrho(y)\gamma^\varrho\frac{\lambda^a}{2}S(y-x)Q_f\gamma_\mu\hat{\Psi}(x)\,\mathrm{d}^4y$$

for the transition current operator $\hat{J}_\mu(x)$ and

$$\hat{J}_\mu(x) = \int \hat{\bar{\Psi}}(x)Q_\mathrm{f}\gamma_\mu S(x-y)\frac{\lambda^a}{2}\hat{G}^a_\varrho\gamma^\varrho\hat{\Psi}(y)\,\mathrm{d}^4y \tag{5.2}$$

as the exchange term. $\hat{\Psi}$ and $\hat{G}^a_\varrho$ denote the field operators of a quark with flavor f and a gluon, respectively. Q_f is the electric charge of the quark with flavor f. The space-time coordinate x characterizes the point of interaction between photon and current. The key observation for carrying out the calculation is that we consider the quarks inside the nucleon as essentially free particles. Therefore, instead of calculating the product of current operators $\hat{J}_\mu(x)\hat{J}_\nu(0)$ in a hadronic state $\langle N|\quad|N\rangle$ we adopt free quark states $\langle\psi|\quad|\psi\rangle$. This allows to calculate the tensor $W_{\mu\nu}$ order by order perturbatively. The action of the operators $\hat{J}_\mu$ on the quark states $|\psi\rangle$ i.e. $\hat{J}_\mu|\psi\rangle$ implies that the operators $\hat{\Psi}$ can be replaced by

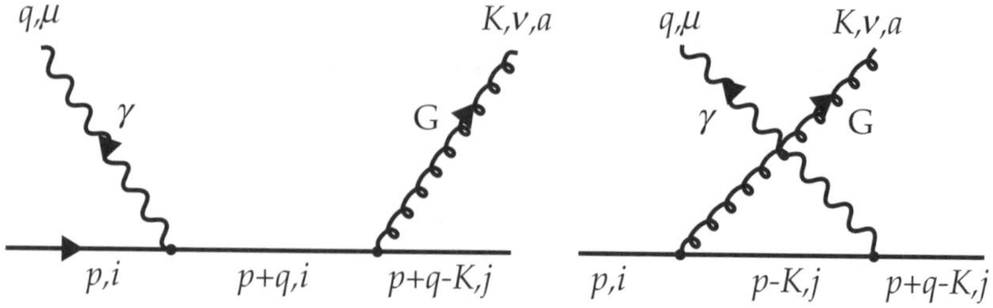

Fig. 5.2. Definition of the quantities employed. p, q and K stand for the corresponding four-momenta, while i, μ, j, ν denote the spin directions (polarizations)

the quark wave functions ψ. Hence the above equations translate into

$$W_{\mu\mu'} = \frac{1}{2\pi}\int d^4x\, e^{iqx}\frac{1}{2}\sum_{\text{pol.}}\sum_{\text{partons}} J_\mu(x)J_{\mu'}(0) \ , \tag{5.3}$$

$$J_\mu(x) = \int \overline{\Psi}(y)gG^a_\varrho(y)\gamma^\varrho\frac{\lambda^a}{2}S(y-x)Q_\mathrm{f}\gamma_\mu\Psi(x)\,d^4y$$

$$J_\mu(x) = \int \overline{\Psi}(x)Q_\mathrm{f}\gamma_\mu S(x-y)\frac{\lambda^a}{2}G^a_\varrho(y)\gamma^\varrho\Psi(y)\,d^4y \ , \tag{5.4}$$

The quark and gluon wave functions Ψ and G^a_ϱ will later on be taken as plane waves. To calculate the transition probabilities the graphs of Fig. 5.2 have to be squared. This leads us to the calculation of the graphs depicted in Fig. 5.3. Note that the intermediate propagators in Fig. 5.3 are on the mass shell because they correspond to the outgoing gluons and quarks, which are real. Transforming this into momentum space, we obtain the Feynman graphs of Fig. 5.3. We reiterate

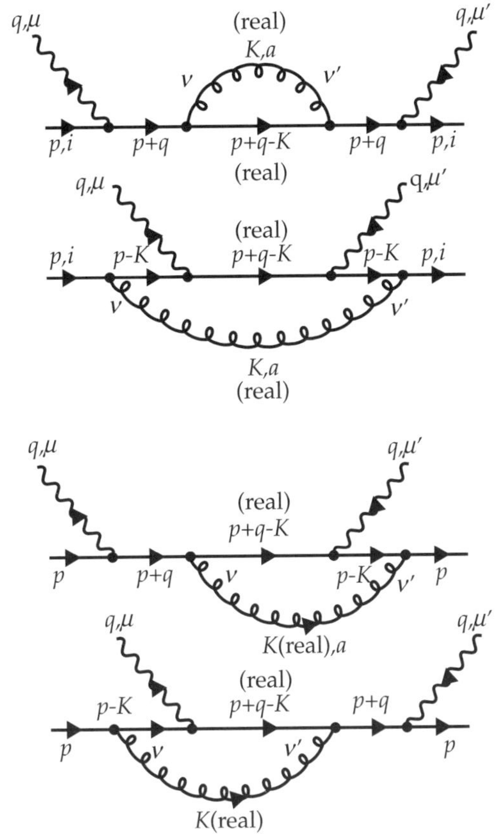

Fig. 5.3. The Feynman representation of the corrections to $W_{\mu\mu'}$ shown in Fig. 5.1. The indication (real) at the gluon and quark propagators draws attention to the fact that the intermediate quarks and gluons in this graph are actually the outgoing particles of the process considered, i.e. they are on the mass shell

that here it must be taken into account that the outgoing (supposedly massless) parton and the outgoing gluons are on the mass shell, i.e.,

$$(p+q-K)^2 = K^2 = 0 \ . \tag{5.5}$$

The usual propagators therefore have to be substituted by δ functions if these particles occur as "inner lines" in the $W_{\mu\mu'}$ graphs:

$$\begin{aligned}\frac{1}{K^2+\mathrm{i}\varepsilon} &= P(1/K^2) - \pi\mathrm{i}\delta(K^2) \to -2\pi\mathrm{i}\delta(K^2)\Theta(K_0)\\ \frac{\gamma_\mu(p^\mu+q^\mu-K^\mu)}{(p+q-K)^2+\mathrm{i}\varepsilon} &\to -2\pi\mathrm{i}\gamma_\mu(p^\mu+q^\mu-K^\mu)\delta[(p+q-K)^2]\\ &\quad\times\Theta(p_0+q_0-K_0) \ .\end{aligned} \tag{5.6}$$

The arrow $\to$ indicates that only the imaginary part of the Compton forward scattering amplitude contributes to the scattering cross section. Using the Feynman rules of Exercise 4.2 together with these modifications yields

$$\begin{aligned}W_{\mu\mu'} = \int &\left[g^2\frac{1}{3}\sum_{a,i,j}\left(\frac{\lambda^a}{2}\right)_{ji}\left(\frac{\lambda^a}{2}\right)_{ij}\right]\frac{1}{2}\left(\sum_\varepsilon \varepsilon^{*\nu}\varepsilon^{\nu'}\right)Q_\mathrm{f}^2\\ &\times\frac{1}{2}\sum_s\Big\{\bar{u}(p,s)(\gamma_{\nu'}(\not{p}-\not{K})^{-1}\gamma_{\mu'}+\gamma_{\mu'}(\not{p}+\not{q})^{-1}\gamma_{\nu'})\\ &\times(\not{p}+\not{q}-\not{K})\Theta(p_0+q_0-K_0)\delta\left[(p+q-K)^2\right]\Theta(K_0)\delta(K^2)\\ &\times\left[\gamma_\mu(\not{p}-\not{K})^{-1}\gamma_\nu+\gamma_\nu(\not{p}+\not{q})^{-1}\gamma_\mu\right]\cdot u(p,s)\Big\}\\ &\times(-4\pi^2)\frac{\mathrm{d}^4K}{(2\pi)^4}\frac{1}{2\pi} \ .\end{aligned} \tag{5.7}$$

Here we have averaged over spin s (this gives the second factor $\frac{1}{2}$: it is identical with the factor $\frac{1}{2}$ stemming from the averaging over nucleon spin in (5.1)), quark color i (this gives the factor $\frac{1}{3}$), and photon polarization ε of the initial state (this gives the first factor $\frac{1}{2}$). Note that the factor e^2 (charge squared) has been separated from $W_{\mu\nu}$. The factor $(-4\pi^2) = (-2\pi\mathrm{i})^2$ stems from the $2\pi\mathrm{i}$ factors of the gluon and quark propagators (5.6). Q_f denotes the electric charge of the flavor f, and the additional factor $1/2\pi$ is due to (5.1)! In the Feynman gauge we have

$$\sum_\varepsilon \varepsilon^{*\nu}\varepsilon^{\nu'} = -g^{\nu\nu'} \ . \tag{5.8}$$

Additional gauge terms that will appear, for example, in the Landau gauge will be proportional to q^ν or $q^{\nu'}$ and vanish if contracted with a conserved current.[1]

[1] See the discussion in W. Greiner and J. Reinhardt: *Field Quantization* (Springer Berlin, Heidelberg 1996).

We get

$$\frac{1}{3}\sum_a \mathrm{tr}\left(\frac{\lambda^a}{2}\right)^2 = \frac{1}{3}\sum_a \frac{1}{2} = \frac{4}{3} \ . \tag{5.9}$$

Utilizing the projection operator $\sum_s \overline{u}(p,s)u(p,s) = \not{p}$, which is in our normalization of the Dirac-spinors – see Sect. 2.1.4 – the correspondence to the relation $\sum_s \overline{u}(p,s)u(p,s) = \frac{\not{p}+m}{2m}$ wellknown from QED[2], we obtain

$$\begin{aligned} W_{\mu\mu'} =& \frac{4g^2}{3}\frac{Q_\mathrm{f}^2}{8\pi^3}\frac{1}{4}\int \mathrm{d}^4K \ \mathrm{tr}\left\{\not{p}\left(\gamma_\nu \frac{(\not{p}-\not{K})}{(p-K)^2}\gamma_{\mu'} + \gamma_{\mu'}\frac{(\not{p}+\not{q})}{(p+q)^2}\gamma_\nu\right)\right. \\ &\times (\not{p}+\not{q}-\not{K})\left.\left(\gamma_\mu \frac{(\not{p}-\not{K})}{(p-K)^2}\gamma^\nu + \gamma^\nu \frac{(\not{p}+\not{q})}{(p+q)^2}\gamma_\mu\right)\right\} \\ &\times \Theta(K_0)\Theta(p_0+q_0-K_0)\delta\left(K^2\right)\delta\left[(p+q-K)^2\right] \ . \end{aligned} \tag{5.10}$$

Fortunately we do not have to evaluate the trace completely, since we already know that all the information provided by $W_{\mu\mu'}$ is contained in the structure functions W_1 and W_2 (see (3.18) and (3.32)). Therefore it is sufficient to determine $p^\mu p^{\mu'} W_{\mu\mu'}$ and $W_\mu{}^\mu$. We abbreviate the trace in (5.10) by $S_{\mu\mu'}$ and delegate the evaluation of $p^\mu p^{\mu'} S_{\mu\mu'}$ and $S_\mu{}^\mu$ to Exercise 5.2. The results are

$$p^\mu p^{\mu'} S_{\mu\mu'} = 4u \ , \tag{5.11}$$

$$S_\mu{}^\mu = -8\left(\frac{s}{t}+\frac{t}{s}-\frac{2Q^2u}{st}\right) \tag{5.12}$$

with the *Mandelstam variables*

$$\begin{aligned} t &= (p-K)^2 = -2p\cdot K \ , \\ s &= (q+p)^2 = 2\nu - Q^2 \ , \\ \nu &= p\cdot q \ , \\ u &= (q-K)^2 = (q+p-K-p)^2 \\ &= -2p\cdot(q+p-K) = -2\nu - t \ . \end{aligned} \tag{5.13}$$

[2] W. Greiner and J. Reinhardt: *Quantum Electrodynamics*, 2nd ed., (Springer Berlin, Heidelberg 1996).

EXAMPLE

5.1 Photon and Gluon Polarization Vectors

Real photons and gluons have two physical degrees of freedom. Thus the simplest choice would be ($\boldsymbol{k}$ is the photon or gluon momentum) that $\boldsymbol{\varepsilon}_1$ and $\boldsymbol{\varepsilon}_2$ are defined such that

$$\begin{aligned}&(\boldsymbol{\varepsilon}_1)^2 = (\boldsymbol{\varepsilon}_2)^2 \ , \quad (\varepsilon_1)^0 = (\varepsilon_2)^0 = 0 \ , \\ &\boldsymbol{\varepsilon}_{1,2} \cdot \boldsymbol{k} = 0 \ , \quad \text{and} \quad \boldsymbol{\varepsilon}_1 \cdot \boldsymbol{\varepsilon}_2 = 0 \ . \end{aligned} \tag{1}$$

The sum over the vector particle polarization is then

$$\sum_{\varepsilon} \varepsilon^\mu \varepsilon^{*\nu} = \begin{cases} 0 & \text{if } \mu = 0 \text{ or } \nu = 0 \\ \sum_{\boldsymbol{\varepsilon}=\boldsymbol{\varepsilon}_1,\boldsymbol{\varepsilon}_2} \varepsilon^\mu \varepsilon^{*\nu} & \text{otherwise} \end{cases} \ . \tag{2}$$

This, however, is not explicitly Lorentz invariant. Therefore it is advantageous to choose different vectors. The three independent four vectors k_μ, $\varepsilon_{1\mu}$, $\varepsilon_{2\mu}$ do not completely span the four-dimensional space. For that reason we introduce an additional vector n_μ. Then the following is a rather general covariant form:

$$\Sigma^{\mu\nu} = \sum_{\varepsilon} \varepsilon^\mu \varepsilon^{*\nu} = -g^{\mu\nu} + \frac{n^\mu k^\nu + n^\nu k^\mu}{n \cdot k} - \frac{n^2 k^\mu k^\nu}{(n \cdot k)^2} \ , \tag{3}$$

that reduces to (5.6) for $n_\mu = (0, \boldsymbol{0})$. This form has the property

$$k_\mu \Sigma^{\mu\nu} = \frac{n^\nu}{n \cdot k} k^2 - \frac{n^2 k^2}{(n \cdot k)^2} k^\nu \ , \tag{4}$$

which vanishes for real photons ($k^2 = 0$!) as it should. Furthermore it fulfills

$$n_\mu \Sigma^{\mu\nu} = -n^\nu + \frac{n^2 k^\nu}{n \cdot k} + n^\nu - \frac{n^2}{n \cdot k} k^\nu = 0 \ , \tag{5}$$

implying the gauge

$$n_\mu A^\mu = 0 \ . \tag{6}$$

The specific form of n_μ for the gauge one is interested in can be read off from (6). We give two standard examples.

1. The Lorentz gauge. In this case $n_\mu = k_\mu$, so that (6) becomes $k_\mu A^\mu = 0$, which is the well-known definition of the Lorentz gauge. From (3) then follows

$$\begin{aligned}\Sigma^{\mu\nu}(1) &= -g^{\mu\nu} + \frac{2k^\mu k^\nu}{k^2} - \frac{k^\mu k^\nu}{k^2} \\ &= -\left[g^{\mu\nu} - \frac{k^\mu k^\nu}{k^2} \right] \ . \end{aligned} \tag{7}$$

2. The light-like gauge:[3]

Example 5.1

$$\Sigma^{\mu\nu}(2) = -g^{\mu\nu} + \frac{n^\mu k^\nu + n^\nu k^\mu}{n\cdot k} \ , \tag{8}$$

where n^ν is the fixed four-vector with $n^2=0$. Note, the fact that $n^2=0$ suggests the name "light-like" gauge. However, the form (3) also allows us to choose n_μ according to the problem we are treating. It might be very advantageous to choose, for example,

$$n_\mu = k_\mu + \alpha\, p_\mu \ , \tag{9}$$

where p is one of the momenta of the problem. This might be useful, for example, if $k^2=0$ and $p^2=0$, which describe outgoing partons (with negligable rest mass). Then (9) leads to

$$\begin{aligned} p_\mu \Sigma^{\mu\nu} &= -p^\nu + \frac{k\cdot p\, k^\nu + (k^\nu + \alpha\, p^\nu)\, p\cdot k}{\alpha\, p\cdot k} - \frac{2\alpha\, k\cdot p\, k\cdot p\, k^\nu}{(\alpha\, p\cdot k)^2} \\ &= -p^\nu + \frac{2k^\nu + \alpha\, p^\nu - 2k^\nu}{\alpha} = 0 \ , \end{aligned} \tag{10}$$

and also

$$k_\mu \Sigma^{\mu\nu} = 0 \ , \tag{11}$$

which allows us to set all vectors p_μ, k_μ contracted with $\Sigma^{\mu\nu}$ equal to zero.

EXERCISE

5.2 More about the Derivation of QCD Corrections to Electron–Nucleon Scattering

Problem. Evaluate $S_\mu{}^\mu$ and $p^\mu p^{\mu'} S_{\mu\mu'}$ for the trace in (5.8).

Solution. First we employ

$$\begin{aligned} \gamma_\mu \not{a} \gamma^\mu &= -2\not{a} \ , \\ \gamma_\mu \not{a}\not{b} \gamma^\mu &= 4a\cdot b \ , \\ \gamma_\mu \not{a}\not{b}\not{c} \gamma^\mu &= -2\not{c}\not{b}\not{a} \ , \end{aligned} \tag{1}$$

[3] E. Tomboulis: Phys. Rev. D **8**, 2736 (1973).

Exercise 5.2

in order to simplify $S_{\mu\mu'}$:

$$\begin{aligned}
S_{\mu\mu'} = &\frac{-2}{(p-K)^4}\mathrm{tr}\left[\not{p}(\not{p}-\not{K})\gamma_{\mu'}(\not{p}+\not{q}-\not{K})\gamma_\mu(\not{p}-\not{K})\right] \\
&+\frac{-2}{(p-K)^2(p+q)^2}\mathrm{tr}\left[\not{p}\gamma_\mu(\not{p}+\not{q})(\not{p}-\not{K})\gamma_{\mu'}(\not{p}+\not{q}-\not{K})\right] \\
&+\frac{-2}{(p-K)^2(p+q)^2}\mathrm{tr}\left[\not{p}\gamma_{\mu'}(\not{p}+\not{q})(\not{p}-\not{K})\gamma_\mu(\not{p}+\not{q}-\not{K})\right] \\
&+\frac{-2}{(p-q)^4}\mathrm{tr}\left[\not{p}\gamma_{\mu'}(\not{p}+\not{q})(\not{p}+\not{q}-\not{K})(\not{p}+\not{q})\gamma_\mu\right] \ .
\end{aligned} \tag{2}$$

Then $S_\mu{}^\mu$ is brought into the form

$$\begin{aligned}
S_\mu{}^\mu = &\frac{4}{(p-K)^4}\mathrm{tr}\left[\not{p}(\not{p}-\not{K})(\not{p}-\not{K}+\not{q})(\not{p}-\not{K})\right] \\
&-\frac{16}{(p-K)^2(p+q)^2}(p+q)\cdot(p-K)\mathrm{tr}\left[\not{p}(\not{p}-\not{K}+\not{q})\right] \\
&+\frac{4}{(p-q)^4}\mathrm{tr}\left[\not{p}(\not{p}+\not{q})(\not{p}-\not{K}+\not{q})(\not{p}+\not{q})\right] \ .
\end{aligned} \tag{3}$$

For massless quarks and real gluons, utilizing $\not{a}\not{a} = a_\mu a_\nu(\gamma^\mu\gamma^\nu + \gamma^\nu\gamma^\mu) = g^{\mu\nu}a_\mu a_\nu = a^2$, one has

$$\begin{aligned}
\not{p}^2 = p^2 = 0 \ , \\
(\not{p}-\not{K}+\not{q})^2 = (p-K+q)^2 = 0 \ , \\
(p-K)^2 = -2p\cdot K \ ,
\end{aligned}$$

because of the $\delta\left[(p+q-K)^2\right]$ function in (5.8)

$$\begin{aligned}
(\not{p}+\not{q})(\not{p}-\not{K}+\not{q}) =&(\not{p}+\not{q}-\not{K}+\not{K})(\not{p}-\not{K}+\not{q}) \\
=&(\not{p}-\not{K}+\not{q})^2+\not{K}(\not{p}-\not{K}+\not{q}) = +\not{K}(\not{p}-\not{K}+\not{q}) \ , \\
(\not{p}-\not{K})(\not{p}-\not{K}+\not{q}) =&(\not{p}-\not{K}+\not{q}-\not{q})(\not{p}-\not{K}+\not{q}) \\
=&(\not{p}-\not{K}+\not{q})^2-\not{q}(\not{p}-\not{K}+\not{q}) = -\not{q}(\not{p}-\not{K}+\not{q}) \ , \\
(p+q)^2 =&(p+q-K+K)^2 \\
=&2K\cdot(p+q-K) = 2K\cdot(p+q) \ ,
\end{aligned} \tag{4}$$

which simplifies (3) to

$$\begin{aligned}
S_\mu{}^\mu = &\frac{1}{(p\cdot K)^2}\mathrm{tr}\left[\not{p}\not{q}(\not{p}-\not{K}+\not{q})\not{K}\right] \\
&+\frac{4}{p\cdot K\,K\cdot(p+q)}(p+q)\cdot(p-K)4p\cdot(q-K) \\
&+\frac{1}{[K\cdot(p+q)]^2}\mathrm{tr}\left[\not{p}\not{K}(\not{p}-\not{K}+\not{q})\not{q}\right]
\end{aligned}$$

Exercise 5.2

$$
\begin{aligned}
&= \left(\frac{4}{(p\cdot K)^2} + \frac{4}{[K\cdot(p+q)]^2} \right) [2(p\cdot q)\,(p\cdot K)] \\
&\quad + \frac{16}{(p\cdot K)\,(K\cdot(p+q))} (p+q)\cdot(p-K)p\cdot(q-K) \ . \qquad (5)
\end{aligned}
$$

Next we introduce the Mandelstam variables

$$
\begin{aligned}
t &= (p-K)^2 = -2p\cdot K \ , \\
s &= (p+q)^2 = 2K\cdot(p+q) = 2\,p\cdot q \ , \\
u &= (q-K)^2 \\
&= (q-K+p)^2 - 2p\cdot(q-K+p) + p^2 \\
&= -2p\cdot(q-K) \ . \qquad (6)
\end{aligned}
$$

Taking into account

$$
\begin{aligned}
Q^2 \equiv -q^2 &= -(q+p-p)^2 \\
&= -2K\cdot(p+q) + 2p\cdot(q+p) \\
&= 2(q+p)\cdot(p-K) \ , \qquad (7)
\end{aligned}
$$

we finally obtain

$$
\begin{aligned}
S_\mu{}^\mu &= -8\frac{s}{t} - 8\frac{t}{s} + 16\frac{Q^2 u}{st} \\
&= -8\left(\frac{s}{t} + \frac{t}{s} - 2\frac{Q^2 u}{st} \right) \ . \qquad (8)
\end{aligned}
$$

Now we evaluate the scalar $p^\mu p^{\mu'} S_{\mu\mu'}$ in a completely analogous manner. From (2) follows

$$
p^\mu p^{\mu'} S_{\mu\mu'} = \frac{-2}{(p-K)^4} \mathrm{tr}\,[\not{p}(\not{p}-\not{K})\not{p}(\not{p}-\not{K}+\not{q})\not{p}(\not{p}-\not{K})] \ . \qquad (9)
$$

All other terms vanish because $\not{p}^2 = p^2 = 0$, and therefore (9) simplifies futher to

$$
p^\mu p^{\mu'} S_{\mu\mu'} = \frac{-2}{(p-K)^4}\ \mathrm{tr}\,[\not{p}\not{K}\not{p}(-\not{K}+\not{q})\not{p}\not{K}] \ . \qquad (10)
$$

The final simplifications are achieved by exchanging the first two factors under the trace

$$
\not{p}\not{K} = p_\mu k_\nu \gamma^\mu \gamma^\nu = p_\mu k_\nu (2g^{\mu\nu} - \gamma^\nu\gamma^\mu) = 2\,p\cdot K - \not{K}\not{p}
$$

and therefore

$$
p^\mu p^{\mu'} S_{\mu\mu'} = \frac{-4}{(p-K)^4} p\cdot K\ \mathrm{tr}\,[\not{p}(-\not{K}+\not{q})\not{p}\not{K}] + 0 \ . \qquad (11)
$$

Exercise 5.2

A similar exchange of the last two factors yields

$$\begin{aligned}
p^{\mu} p^{\mu'} S_{\mu\mu'} &= \frac{-8}{(p-K)^4}(p\cdot K)^2 \,\mathrm{tr}\,[\not{p}(\not{q}-\not{K})] \\
&= \frac{-8}{(p-K)^4}(p\cdot K)^2 \cdot 4p\cdot(q-K) \\
&= -8p\cdot(q-K) = 4u \;. \qquad (12)
\end{aligned}$$

In the last step one of the relations (4) and (6) has been used.

Since the traces in (5.11) and (5.12) are Lorentz invariant, the K integration can be performed in an arbitrary reference system. We choose the *Breit system* (see Chap. 3), i.e.,

$$p_\mu = (p, 0, 0, -p) \;, \qquad q_\mu = (0, 0, 0, \sqrt{Q^2}) \;. \tag{5.14}$$

The corresponding kinematics is depicted in Fig. 5.4. The determination of $W_{\mu\mu'}$ in (5.10) is leading us to integrals of the following form

$$\begin{aligned}
I &= \int \mathrm{d}^4K\, f(Q^2, t, \nu)\Theta(K_0)\Theta(p_0+q_0-K_0)\delta(K^2)\delta\left[(p+q-K)^2\right] \\
I &= \int \mathrm{d}K_0 K^2 \mathrm{d}K\, \mathrm{d}\Omega_K\, f(Q^2, t, \nu)\Theta(p_0+q_0-K_0) \\
&\quad \times \delta(K_0^2 - K^2)\delta\left[(p+q-K)^2\right] \;.
\end{aligned}$$

Since the process considered is clearly cylindrically symmetric, the K_0 and φ integrations can be performed immediately: Using $\delta(K_0^2 - K^2) = \delta(K_0 - K)/2K$ yields

$$I = 2\pi \int\limits_0^\infty \mathrm{d}K \int\limits_{-1}^{1} \mathrm{d}\cos\theta\, f(Q^2, t, \nu)\frac{K^2}{2K}\Theta(p_0+q_0-K)\delta\left[(p+q-K)^2\right] \;. \tag{5.15}$$

Here $f(Q^2, t, \nu)$ is any function depending on the Mandelstam variables (5.13), which will be specified later. Next we rewrite the second δ function by inserting

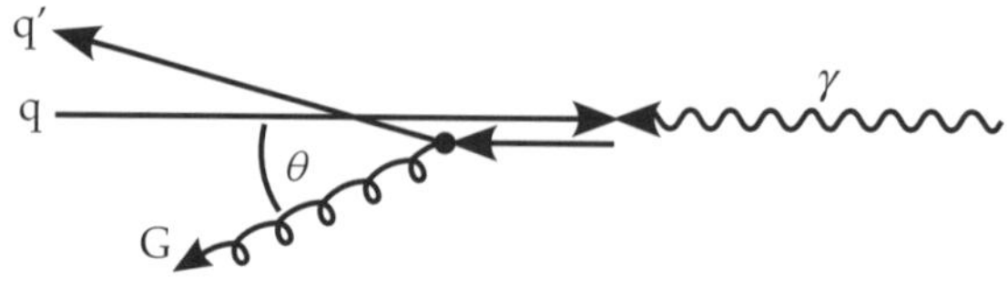

Fig. 5.4. Kinematics of the graphs depicted in Fig. 5.2 in the Breit system

the explicit form of K_μ:

$$K_\mu = (K, 0, K\sin\theta, K\cos\theta) \ , \tag{5.16}$$

$$\begin{aligned}(p_\mu + q_\mu - K_\mu)^2 &= \left(p - K, 0, -K\sin\theta, -p + \sqrt{Q^2} - K\cos\theta\right)^2 \ , \\ &= p^2 + K^2 - 2pK - \Big(K^2 + p^2 + Q^2 + 2pK\cos\theta \\ &\quad -2K\sqrt{Q^2}\cos\theta - 2p\sqrt{Q^2}\Big) \\ &= -2pK(1+\cos\theta) - Q^2 + 2\nu + 2K\sqrt{Q^2}\cos\theta \ . \end{aligned} \tag{5.17}$$

Note that in the Breit system (5.14) $\nu = p\cdot q = p\cdot\sqrt{Q^2}$, which entered (5.17). Now $\cos\theta$ is replaced by t. According to (5.13) we get with (5.14) and (5.16)

$$t = (p_\mu - K_\mu)^2 = -2pK(1+\cos\theta) \ , \tag{5.18}$$

$$(p_\mu + q_\mu - K_\mu)^2 = t - Q^2 + 2\nu - 2K\sqrt{Q^2} + 2K\frac{Q^2}{p\sqrt{Q^2}}p(1+\cos\theta) \tag{5.19}$$

$$= t(1 - Q^2/\nu) - Q^2 + 2\nu - 2K\sqrt{Q^2} \ . \tag{5.20}$$

Finally, according to (5.18), and remembering that p and K are fixed, we replace $\mathrm{d}\cos\theta$ by

$$\mathrm{d}\cos\theta \to -\frac{1}{2pK}\mathrm{d}t \tag{5.21}$$

and obtain

$$\begin{aligned} I &= 2\pi\int_0^\infty \mathrm{d}K\int \mathrm{d}t\left(\frac{-1}{2pK}\right)\frac{K}{2}\Theta(p_0 + q_0 - K) \\ &\quad\times\delta\left[t\left(1 - \frac{Q^2}{\nu}\right) - Q^2 + 2\nu - 2K\sqrt{Q^2}\right] f(Q^2, t, \nu) \\ &= -\int_{t_1}^{t_2}\frac{\pi}{4p\sqrt{Q^2}}\mathrm{d}t\ f(Q^2, t, \nu) = \frac{\pi}{4\nu}\int_{t_2}^{t_1}\mathrm{d}t\ f(Q^2, t, \nu) \ . \end{aligned} \tag{5.22}$$

The boundaries of the t integration are not yet determined. This is achieved by substituting K from (5.16) into the argument of the δ function:

$$K = \frac{-t}{2p(1+\cos\theta)} \ , \tag{5.23}$$

$$0 = t\left(1 - \frac{Q^2}{\nu}\right) - Q^2 + 2\nu + \frac{\sqrt{Q^2}t}{p(1+\cos\theta)} \ . \tag{5.24}$$

The second equation yields for t $\left(\nu = \sqrt{Q^2} \cdot p\right)$

$$t = \frac{Q^2 - 2\nu}{1 - \frac{Q^2}{\nu} + \frac{Q^2}{\nu(1+\cos\theta)}} = \frac{Q^2 - 2\nu}{1 - \frac{Q^2}{2\nu} + \frac{Q^2}{\nu}\left(\frac{1}{1+\cos\theta} - \frac{1}{2}\right)} \; . \tag{5.25}$$

Since $x = Q^2/2\nu$ is less than or equal to one (see (3.42)), the denominator is bigger than or equal to zero. The numerator is always less than or equal to zero and consequently t is a monotonically decreasing, negative function of $\cos\theta$. The boundaries t_1 and t_2 are simply

$$t_1 = t(\cos\theta = -1) = 0 \; , \quad t_2 = t(\cos\theta = 1) = -2\nu \; . \tag{5.26}$$

Now we have to investigate whether the Θ function leads to further restrictions. With (5.23), (5.25), and (5.14)

$$\begin{aligned} p_0 + q_0 - K_0 = p - K &= p - \frac{2\nu - Q^2}{2\sqrt{Q^2} + 2p(1+\cos\theta)\left(1 - \frac{\sqrt{Q^2}}{p}\right)} \\ &= p - \frac{2\nu - Q^2}{-2\sqrt{Q^2}\cos\theta + 2p(1+\cos\theta)} \\ &= p - \frac{2\nu - Q^2}{\left(4p - 2\sqrt{Q^2}\right)\cos\theta + 2p(1-\cos\theta)} \; . \end{aligned} \tag{5.27}$$

Because $x = \sqrt{Q^2}/2p \leq 1 \Rightarrow \sqrt{Q^2} \leq 2p$, the denominator is always positive. It assumes its smallest value at $\cos\theta = -1$ for $p > \sqrt{Q^2}$ and at $\cos\theta = 1$ for $p < \sqrt{Q^2}$. Let us now consider these two cases separately, i.e., for $p > \sqrt{Q^2}$

$$\frac{2\nu - Q^2}{\left(4p - 2\sqrt{Q^2}\right)\cos\theta + 2p(1-\cos\theta)} \leq \frac{2\nu - Q^2}{2\sqrt{Q^2}} = p - \frac{\sqrt{Q^2}}{2} \leq p \; , \tag{5.28}$$

because $\nu = \sqrt{Q^2}p$, and for $p < \sqrt{Q^2}$

$$\frac{2\nu - Q^2}{\left(4p - 2\sqrt{Q^2}\right)\cos\theta + 2p(1-\cos\theta)} \leq \frac{2\nu - Q^2}{4p - 2\sqrt{Q^2}} = \frac{\sqrt{Q^2}}{2} \leq p \; . \tag{5.29}$$

Hence the argument of the Θ function is positive definite:

$$p_0 + q_0 - k_0 \geq p - p = 0 \; . \tag{5.30}$$

Thus, from (5.10) and (5.22) we obtain the result

$$W_{\mu\mu'} = -\frac{4}{3}g^2 \frac{Q_f^2}{16\pi^3}\frac{\pi}{8\nu}\int\limits_{-2\nu}^{0} \mathrm{d}t \, S_{\mu\mu'} \; , \tag{5.31}$$

and from (5.11) and (5.12)

$$p^{\mu} p^{\mu'} S_{\mu\mu'} = 4u = -4(2\nu + t) \ , \tag{5.32}$$

$$S_{\mu}{}^{\mu} = -8\left(\frac{2\nu - Q^2}{t} + \frac{t}{2\nu - Q^2} + \frac{2Q^2(2\nu + t)}{t(2\nu - Q^2)}\right) \ . \tag{5.33}$$

The t integration yields a logarithmic singularity for $S_{\mu}{}^{\mu}$, which is the same kind of infrared singularity that occurs in the evaluation of the QED bremsstrahlung cross section. In fact the graphs in Fig. 5.1 can also be interpreted as bremsstrahlung processes. Before we discuss this singularity further, however, we should like to finish our calculation. First, inserting (5.11) into (5.31) is without any problem and yields

$$p^{\mu} p^{\mu'} W_{\mu\mu'} = \frac{4}{3} g^2 \frac{Q_{\mathrm{f}}^2}{16\pi^2} \frac{1}{2\nu} 2\nu^2 = \frac{4}{3} \alpha_s \frac{Q_{\mathrm{f}}^2}{8\pi} 2\nu \ , \quad \alpha_s = \frac{g^2}{4\pi} \ . \tag{5.34}$$

Second, inserting (5.12) into (5.31) leads to a logarithmically divergent integral, which we cut off at $-\lambda^2$:

$$\begin{aligned} W_{\mu}{}^{\mu} &= \frac{4}{3} \alpha_s \frac{Q_{\mathrm{f}}^2}{4\pi} \frac{1}{\nu} \left(\frac{(2\nu - Q^2)^2 + 4\nu Q^2}{(2\nu - Q^2)} \int\limits_{-2\nu}^{-\lambda^2} \frac{\mathrm{d}t}{t} + \frac{-2\nu^2}{2\nu - Q^2} + \frac{2Q^2 2\nu}{2\nu - Q^2} \right) \\ &= \frac{4}{3} \alpha_s \frac{Q_{\mathrm{f}}^2}{4\pi} \left(2\frac{1+x^2}{1-x} \ln \frac{\lambda^2 x}{Q^2} - \frac{1}{1-x} + 4\frac{x}{1-x} \right) \ . \end{aligned} \tag{5.35}$$

Here, again, $x = Q^2/2\nu$ has been introduced. Next we answer the question: How are the structure functions W_1 and W_2 or F_1 and F_2, respectively, connected with the expressions in (5.34) and (5.35)? First remember the connection between F_1 and W_1 and between F_2 and W_2, respectively. The Fs were introduced in (3.43). Adopting them, equations (3.18) and (3.43) yield the relation

$$\begin{aligned} W_{\mu\mu'}^{N} &= \left(-g_{\mu\mu'} + \frac{q_\mu q_{\mu'}}{q^2}\right) \frac{F_1}{M_{\mathrm{N}}} + \left(P_\mu - q_\mu \frac{q \cdot P}{q^2}\right) \left(P_{\mu'} - q_{\mu'} \frac{q \cdot P}{q^2}\right) \frac{F_2}{M_{\mathrm{N}} \nu} \\ &= \frac{1}{M_{\mathrm{N}}} \left[\left(-g_{\mu\mu'} + \frac{q_\mu q_{\mu'}}{q^2}\right) F_1 \right. \\ &\quad \left. + \left(P_\mu - q_\mu \frac{q \cdot P}{q^2}\right) \left(P_{\mu'} - q_{\mu'} \frac{q \cdot P}{q^2}\right) \frac{F_2}{\nu} \right] \ . \end{aligned} \tag{5.36}$$

$W_{\mu\mu'}^{N}$ is the scattering amplitude for leptons (transmitted by photons) at nucleons. Here the factor $1/M_N$ is convenient. The scattering amplitude for leptons at quarks, $W_{\mu\mu'}$, is defined without the $1/M_N$ factor. This is convenient since one deals with massless quarks. The corresponding structure functions are now denoted by $F_1^{\mathrm{Q}u}$ and $F_2^{\mathrm{Q}u}$, respectively. For quarks, (5.36) therefore becomes

$$\begin{aligned} W_{\mu\mu'} &= \left(-g_{\mu\mu'} + \frac{q_\mu q_{\mu'}}{q^2}\right) F_1^{\mathrm{Q}u} \\ &\quad + \frac{1}{\nu} \left(p_\mu - q_\mu \frac{q \cdot p}{q^2}\right) \left(p_{\mu'} - q_{\mu'} \frac{q \cdot p}{q^2}\right) F_2^{\mathrm{Q}u} \ . \end{aligned} \tag{5.37}$$

It follows with $Q^2 = -q^2$ and $p^2 = 0$ (massless quarks!) that

$$\begin{aligned} W^\mu_\mu &= (-4+1)F_1^{\mathrm{Q}u} + \left(-2\frac{(q\cdot p)^2}{q^2} + \frac{(q\cdot p)^2}{q^2}\right)\frac{1}{\nu}F_2^{\mathrm{Q}u} \\ &= -3F_1^{Qu} + \frac{\nu}{Q^2}F_2^{\mathrm{Q}u} \\ &= -3F_1^{Qu} + \frac{F_2^{\mathrm{Q}u}}{2x} \end{aligned} \tag{5.38}$$

and similarly

$$\begin{aligned} p^\mu p^{\mu'} W_{\mu\mu'} &= -\frac{\nu^2}{Q^2}F_1^{\mathrm{Q}u} + \frac{\nu^3}{Q^4}F_2^{\mathrm{Q}u} \\ &= -\frac{Q^2}{4x^2}F_1^{\mathrm{Q}u} + \frac{Q^2}{8x^3}F_2^{\mathrm{Q}u} \\ &= \frac{Q^2}{8x^3}\left(F_2^{\mathrm{Q}u} - 2xF_1^{\mathrm{Q}u}\right) . \end{aligned} \tag{5.39}$$

Obviously $p^\mu p^{\mu'} W_{\mu\mu'}$ is a measure for the violation of the Callan–Gross relation:

$$F_2^{\mathrm{Q}u} = 2xF_1^{\mathrm{Q}u} \tag{5.40}$$

by the interaction. It is now clear that instead of the two scalars W^μ_μ and $p^\mu p^{\mu'} W_{\mu\mu'}$ one may choose $F_1^{\mathrm{Q}u}$ and $F_2^{\mathrm{Q}u}$ (or any other linearly independent combination). The connection between both follows from (5.38) and (5.39) as

$$\begin{aligned} F_1^{\mathrm{Q}u}(x,Q^2) &= -\frac{1}{2}W^\mu_\mu + \frac{2x^2}{Q^2}p^\mu p^{\mu'} W_{\mu\mu'} , \\ F_2^{\mathrm{Q}u}(x,Q^2) &= -xW^\mu_\mu + 12\frac{x^3}{Q^2}p^\mu p^{\mu'} W_{\mu\mu'} . \end{aligned} \tag{5.41}$$

Using the results (5.34) and (5.35) the additional contributions to these quark structure functions due to “gluon bremsstrahlung” (see Fig. 5.1) are therefore

$$\Delta\left[F_2^{\mathrm{Q}u}(x,Q^2) - 2xF_1^{\mathrm{Q}u}(x,Q^2)\right] = \frac{4}{3}\alpha_s Q_{\mathrm{f}}^2\frac{x^2}{\pi} , \tag{5.42}$$

$$\begin{aligned} \Delta F_2^{\mathrm{Q}u}(x,Q^2) &= \frac{4}{3}\alpha_s Q_{\mathrm{f}}^2 \left\{\frac{3x^2}{2\pi} - \frac{x}{4\pi}\left[2\frac{1+x^2}{1-x}\right.\right. \\ &\quad \left.\left.\times\left(\ln\frac{\lambda^2}{Q^2} + \ln x\right) + \frac{4x-1}{1-x}\right]\right\} \\ &\approx -\frac{4}{3}\alpha_s Q_f^2 \frac{x}{2\pi}\frac{1+x^2}{1-x}\ln\lambda^2 Q^2 + \dots , \end{aligned} \tag{5.43}$$

where we have retained only the logarithmic Q^2-dependent term, which will dominate for $Q^2 \to \infty$. Now we investigate the dependence of the nucleon structure functions $F_{1,2}$ on the quark structure functions $F_1^{\mathrm{Q}u}$ and $F_2^{\mathrm{Q}u}$. To this end we have to go back to (3.70)–(3.82). In particular, according to (3.81), F_2 is given by

$$F_2(X, Q^2) = \sum_i f_i(X, Q^2) Q_i^2 \cdot X$$
$$= \sum_i \int_0^1 \mathrm{d}\xi_i \; f_i(\xi_i, Q^2) Q_i^2 \xi_i \delta(\xi_i - X) \tag{5.44}$$

with

$$p_\mu = \xi_i P_\mu \quad \text{and} \quad \int_0^1 \mathrm{d}\xi_i \; f(\xi_i) = 1 \; .$$

X denotes $Q^2/2P_\mathrm{N} \cdot q$ while x is equal to $Q^2/2p \cdot q$. The ξ_i is the fraction of the ith parton of the total momentum P_μ and $f(\xi_i)$ is the probability that a parton will have that momentum. Q_i denotes the charge of the ith parton. By definition the structure functions are factors in the scattering tensor, which turn the contributions of free pointlike particles into those of extended particles. Correspondingly one has to multiply the right-hand side of (5.44) by the factor $F_2^{\mathrm{Q}u}(x, Q^2)$ and to modify the δ function according to

$$x = \frac{Q^2}{2p \cdot q} = \frac{Q^2}{2\xi_i P_\mathrm{N} \cdot q} = \frac{X}{\xi_i} \; , \tag{5.45}$$

$$F_2(X, Q^2) = \sum_i \int \mathrm{d}\xi_i \; f_i(\xi_i) F_2^{\mathrm{Q}u,i}\left(\frac{X}{\xi_i}, Q^2\right)$$
$$= \sum_i \int_0^1 \mathrm{d}\xi_i \; f_i(\xi_i)\xi_i \int_0^1 \mathrm{d}x \, \delta(X - x\xi_i) F_2^{\mathrm{Q}u,i}(x, Q^2) \; . \tag{5.46}$$

The last step in (5.46) is easily verified. Note that we have integrated over all values of x that contribute to a given X. One might look for a missing factor Q_i^2. However, this is contained in $F_2^{\mathrm{Q}u,i}$. With the help of (5.43) we finally obtain the following contribution of the graphs in Fig. 5.1 to the nucleon structure function F_2 of deep inelastic electron–nucleon scattering:

$$\Delta F_2(X, Q_1^2, Q_2^2)_{\text{bremsstrahlung}} = \Delta F_2(X, Q_1^2) - \Delta F_2(X, Q_2^2)$$
$$= -\frac{4}{3}\alpha_s \frac{1}{2\pi} \sum_i Q_i^2 \left[\int_0^1 \mathrm{d}\xi_i \; f(\xi_i)\xi_i \int_0^1 \mathrm{d}x \frac{1}{\xi_i} \delta\left(x - \frac{X}{\xi_i}\right) \right.$$
$$\left. \times \left(\frac{1+x^2}{1-x}\right) \ln\left(Q_2^2/Q_1^2\right) \right]$$

$$= -\frac{4}{3}\alpha_s \frac{1}{2\pi} \sum_i Q_i^2 \int\limits_X^1 \mathrm{d}\xi_i \, f(\xi_i) \frac{X}{\xi_i} \frac{1+(X/\xi_i)^2}{1-(X/\xi_i)} \ln \frac{Q_2^2}{Q_1^2} \,, \quad (5.47)$$

where the lower boundary of integration is due to the fact that $\xi_i > X$ (see (5.45)). The integral is divergent, because the integrand has a singularity at $X = \xi_i$ or $x = 1$. Since $x = 1$ means elastic scattering, the gluon emitted in this reaction carries neither energy nor momentum. The occurrence of such an infrared divergence is not surprising: one encounters similar behavior in QED bremsstrahlung. There it can be shown that, up to a given order in α, the infrared divergences cancel each other if all scattering processes, including the elastic channel, are taken into account. To lowest order this is achieved by the following additional graphs:

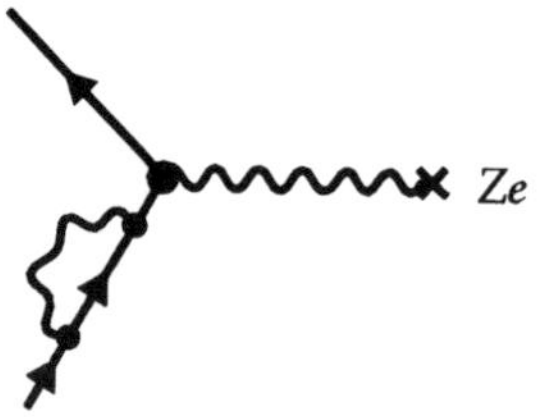

Many more graphs exist in QCD and an analogous proof is more sophisticated. The desired compensation can, to lowest order, be shown by an explicit evaluation. The result can be brought into the following form:

$$\Delta F_2(\text{bremsstrahlung} + \text{radiative corrections})$$

$$\approx -\frac{4}{3}\frac{\alpha_s}{2\pi} \sum_i Q_i^2 \ln \frac{Q_2^2}{Q_1^2} \int\limits_X^1 \mathrm{d}\xi_i \, f_i(\xi_i) \frac{X}{\xi_i} \left[\frac{1+(X/\xi_i)^2}{1-(X/\xi_i)} \right]_+ \,, \quad (5.48)$$

where $[(1+z^2)/(1-z)]_+$ is defined by

$$\int\limits_0^1 \mathrm{d}z \, F(z) \left[\frac{1+z^2}{1-z} \right]_+ = \int\limits_0^1 \mathrm{d}z \, [F(z) - F(1)] \frac{1+z^2}{1-z} \quad (5.49)$$

for every sufficiently regular function $F(z)$. Keep in mind that Q_i^2 denote the squares of the charges of the various quarks, while Q_1^2 and Q_2^2 stand for the two squared momentum transfers. The virtual diagrams which have to be taken

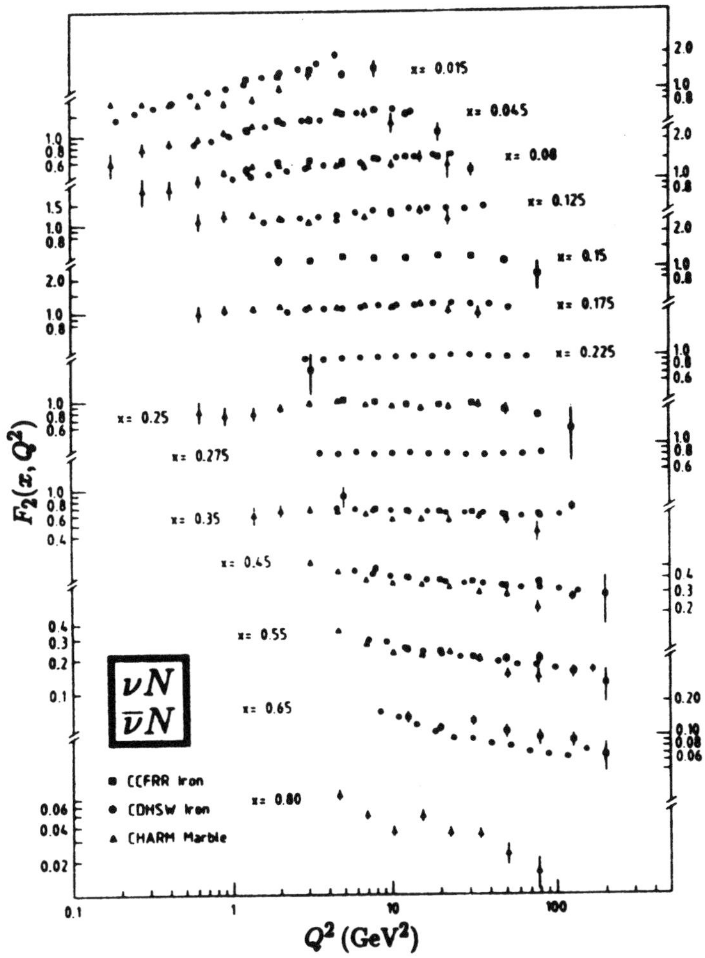

Fig. 5.5. The structure functions of neutrino–nucleon scattering. (From Review of Particle Properties, Physical Review D **45** (1992))

into account regularize the $1/(1-\frac{X}{\xi_i})$ singularity in (5.47) so that (5.48) holds. That such a cancellation has to occur can be understood by rather general arguments. The Bloch–Nordsieck theorem assures that infrared divergences cancel in inclusive cross sections.[4] (From now on we shall again write x for X.) One possible choice for $f(\xi_i)$ is the distribution function derived in the context of the simple parton model (see Fig. 3.11). In this case the result would be an expression for the scale violation of the function F_2 depending on one fit parameter α_s. From Fig. 3.5 it is clear that for $x > 0.5$ scale violations in fact occur, i.e., F_2 becomes a function of x and Q^2. In order to be more precise, F_2 decreases with increasing Q^2 for $x > 0.5$. The very same behavior can also be encountered in other scattering processes, e.g., in neutrino–nucleon scattering (Fig. 5.5). Here it is seen that for small x the Q^2 dependence is just the opposite, i.e., for

[4] F. Bloch and A. Nordsieck: Phys. Rev. **52** (1937) 54.

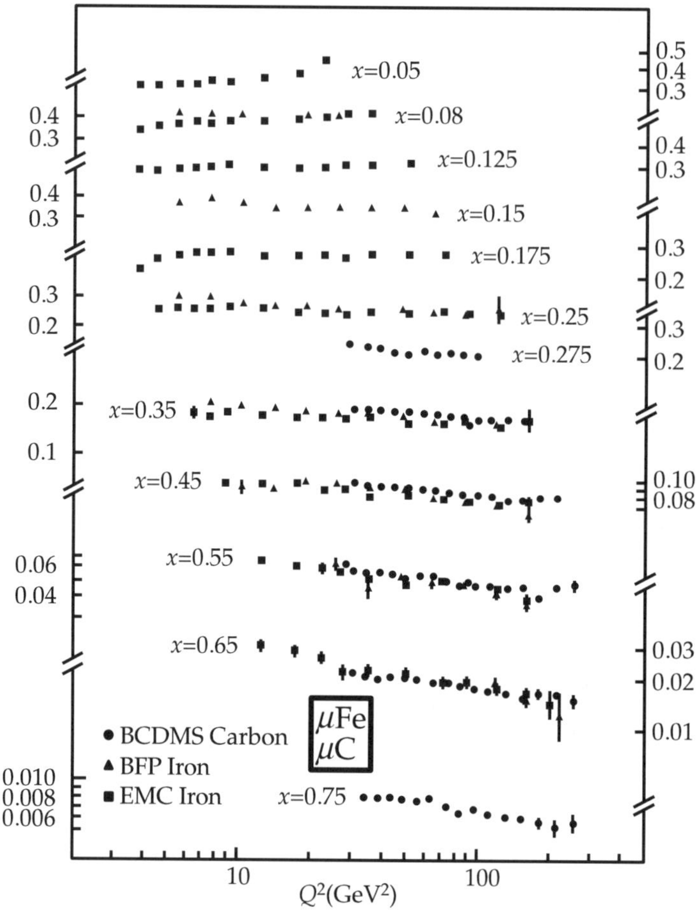

Fig. 5.6. Structure functions of muon–carbon and muon–iron scattering. (From Review of Particle Properties, Physical Review D **45** (1992))

$x < 0.1$ F_2 increases with Q^2. The data in Fig. 5.6 show the same tendency for electron–nucleon scattering and muon–nucleon scattering.

The overall behavior (see Fig. 5.7) is readily understood. For increasing momentum transfers QCD processes become more important, and on average more quarks, antiquarks, and gluons occur, between which the total momentum is distributed. Owing to (3.82), F_2 is given by

$$F_2^{\mathrm{eN}}(x, Q^2) = \sum_i f_i(x, Q^2) Q_i^2 x \ , \tag{5.50}$$

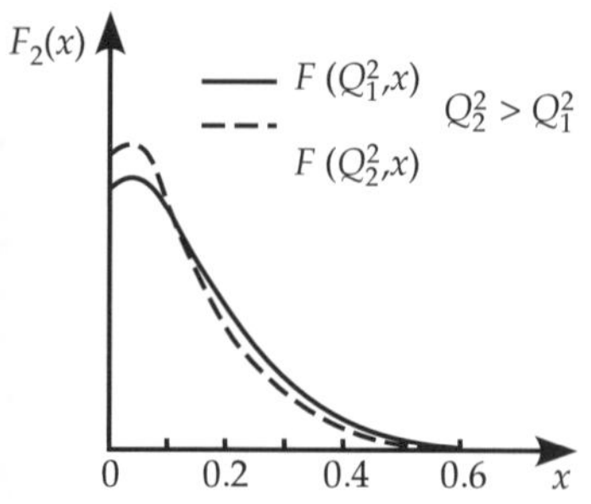

Fig. 5.7a. The schematic dependence of the structure functions on Q^2, with $Q_2^2 \gg Q_1^2$

i.e., by x times the probability that a parton carries the momentum fraction x. $F_2(x, Q^2)$ decreases for large x values with increasing Q^2. This can be recognized by inspection of Fig. 5.7a, which illustrates the interacting quarks in a nucleon. With higher momentum transfer Q^2 the number of partons in a nucleon increases. Because now the total momentum of the nucleon is distributed over more partons, the distribution function $f_i(x, Q^2)$ has to decrease, particularly for large x. Since the total momentum is constant,

$$\int \mathrm{d}x \sum_i f_i(x, Q^2)x = 1 \ , \tag{5.51}$$

an enhancement for small x values should follow. However, this can be prevented by the charge factors, for example, by Q_i^2 in (5.48). In this way the gluons contribute to the sum (5.51), but not to the F_2^{eN} structure functions. As a matter of fact, according to Chap. 3, the charged partons alone contribute only about 50% to (5.51) (see explanation to (3.88)).

Structure functions for muon–carbon and muon–iron scattering are depicted in Fig. 5.6. Again the same Q^2 dependance is observed. Furthermore, it is remarkable that in the region $0.5 < x < 0.65$ and for very small x the F_2 function for iron is considerably smaller than for μ–p scattering. Part of this discrepancy is due to the difference between neutrons and protons (the sums of the squared charges of the quarks have a ratio of 3 to 2). The remainder of the discrepancy is known as the EMC (European muon collaboration) effect, recalling the experimental collaboration that first discovered the effect.

The history of the EMC effect has been quite involved. Eventually it was found that part of the originally observed effect was due to an incorrect measurement, since data analysis for small x values is extremely difficult. However, it is now clear that the structure functions depend on the size of the nucleus. Figure 5.8 gives recent experimental results.

A number of theoretical models for the observed phenomena exist, but their physical meaning is still heavily disputed. If we disregard this problem, pertu-

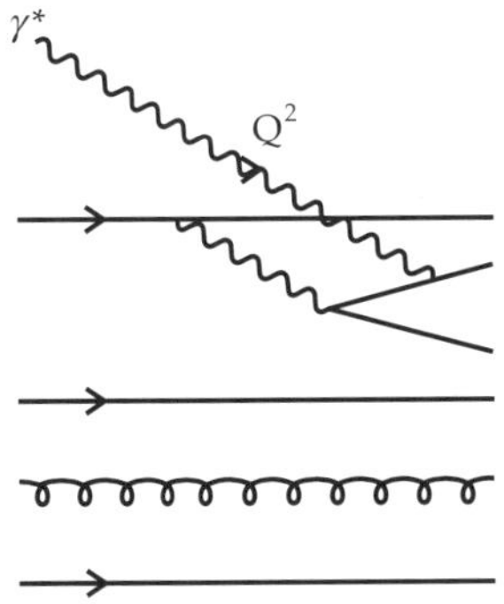

Fig. 5.7b. Contribution to the small x domain visible at higher Q^2. At higher Q^2 additional partons are visible, because more $q\bar{q}$ pairs can be excited. They contribute mostly at small x

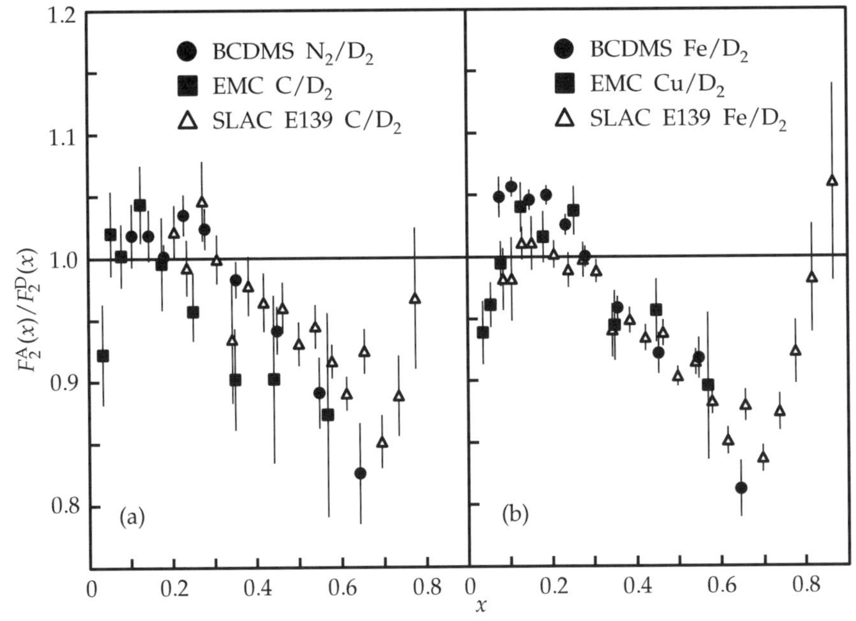

Fig. 5.8. The EMC effect: the ratio of F_2 for carbon, nitrogen, iron, and copper to F_2 for deuterium (From Review of Particle Properties, Physical Review D **45** (1992))

bative QCD yields a reasonable description of the structure functions. To this end, further corrections besides the gluon–bremsstrahlung correction have to be taken into account. They could of course be determined graph by graph, but there is a more elegant way to evaluate these corrections. Every calculation again and again employs the same techniques and the specific kinematics of the parton picture. Thus it is useful to formulate a general scheme suited for the special situation of deep inelastic scattering. In this way we are led directly to the Gribov–Lipatov–Altarelli–Parisi equations, which we derive here in an intuitive fashion. Later on we discuss a more formal derivation.

First we rewrite the result of the bremsstrahlung calculation. Inserting (5.50) into (5.48) (remember that we replaced $X \to x$ again) yields

$$\begin{aligned}\Delta F_2^{\mathrm{eN}} &= \sum_i \Delta f_i(x) Q_i^2 x \\ &\approx -\frac{4}{3}\frac{\alpha_s}{2\pi}\sum_i Q_i^2 \ln\frac{Q_2^2}{Q_1^2}\int\limits_0^1 \mathrm{d}\xi_i\, f_i(\xi_i)\frac{x}{\xi_i}\left(\left[1+\left(\frac{x}{\xi_i}\right)^2\right]\left(1-\frac{x}{\xi_i}\right)^{-1}\right)_+\end{aligned}$$

or

$$\begin{aligned}0 = \sum_i Q_i^2 \Bigg[& x\Delta f_i(x) \\ & + \frac{4}{3}\frac{\alpha_s}{2\pi}\ln\frac{Q_2^2}{Q_1^2}\int\limits_x^1 \mathrm{d}\xi_i\; f_i(\xi_i)\frac{x}{\xi_i}\left(\left[1+\left(\frac{x}{\xi_i}\right)^2\right]\left(1-\frac{x}{\xi_i}\right)^{-1}\right)_+\Bigg] .\end{aligned} \tag{5.52}$$

Since the different quarks are considered to be independent, every single term in the sum must be equal to zero:

$$\Delta f_i(x) = -\frac{4}{3}\frac{\alpha_s}{2\pi}\ln\frac{Q_2^2}{Q_1^2}\int\limits_x^1 \mathrm{d}\xi_i\,\frac{f_i(\xi_i)}{\xi_i}\left(\left[1+\left(\frac{x}{\xi_i}\right)^2\right]\left(1-\frac{x}{\xi_i}\right)^{-1}\right)_+ . \tag{5.53}$$

We define

$$P_{qq}(z) = \frac{4}{3}\left\{\frac{1+z^2}{1-z}\right\}_+ . \tag{5.54}$$

Furthermore, the index i is omitted, which, together with the replacement $\xi_i = \xi = y$, gives

$$\Delta f(x) = -\frac{\alpha_s}{2\pi}\ln\frac{Q_2^2}{Q_1^2}\int\limits_x^1 \frac{\mathrm{d}y}{y} f(y) P_{qq}\left(\frac{x}{y}\right) . \tag{5.55}$$

The omission of the quark index i indicates that, at this level, all quarks are considered equal. Now, we define the logarithmic momentum variable

$$t(Q^2) = \ln \frac{Q^2}{Q_0^2} ,$$
$$\Delta t = t(Q_1^2) - t(Q_2^2) = \ln \frac{Q_1^2}{Q_2^2} , \tag{5.56}$$

which gives

$$\frac{\Delta f(x)}{\Delta t} = \frac{\alpha_s}{2\pi} \int_x^1 \frac{\mathrm{d}y}{y} f(y) P_{qq}\left(\frac{x}{y}\right) . \tag{5.57}$$

The last equation is only valid for small Δt. Indeed, for large Δt values it must be taken into account that α_s also depends on Q^2 and also $f(y) \to f(y,t)$. The correct equation is therefore:

$$\frac{\mathrm{d}f(x,t)}{\mathrm{d}t} = \frac{\alpha_s(t)}{2\pi} \int_x^1 \frac{\mathrm{d}y}{y} f(y,t) P_{qq}\left(\frac{x}{y}\right) . \tag{5.58}$$

Equation (5.58) is already a *Gribov–Lipatov–Altarelli–Parisi (GLAP) equation*. All further QCD corrections are described by additional terms on the right-hand side.

We will discuss the physical meaning of this equation in more detail below. However, we would first like to remark that the expression $[(1+z^2)/(1-z)]_+$ can be written in different ways. Two of those used more frequently in the literature are the following:

$$P'_{qq}(z) = \frac{4}{3}\left(\frac{1+z^2}{(1-z)_+} + \frac{3}{2}\delta(1-z)\right) , \tag{5.59}$$

and

$$P''_{qq}(z) = -\frac{4}{3}(1+z) + 2\delta(1-z)$$
$$+ \lim_{\eta\to 0} \frac{8}{3}\left(\frac{1}{1-z+\eta} - \delta(1-z)\int_0^1 \mathrm{d}z' \frac{1}{1-z'+\eta}\right) . \tag{5.60}$$

The equivalence of (5.59) and (5.60) with (5.54) can be proven by applying them to an arbitrary smooth and continous function.

$$\int_0^1 P'_{qq}(z) f(z)\,\mathrm{d}z = \frac{4}{3}\int_0^1 \mathrm{d}z \left(\frac{1+z^2}{1-z} f(z) - \frac{2}{1-z} f(1)\right) + \frac{4}{3}\cdot\frac{3}{2} f(1)$$
$$= \frac{4}{3}\int_0^1 \mathrm{d}z \left(\frac{1+z^2}{1-z} f(z) - \frac{2}{1-z} f(1) + (1+z) f(1)\right)$$

$$
\begin{aligned}
&= \frac{4}{3}\int_0^1 dz \left(\frac{1+z^2}{1-z} f(z) - \frac{z^2+1}{1-z} f(1) \right) \\
&= \frac{4}{3}\int_0^1 dz \left\{ \frac{1+z^2}{1-z} \right\}_+ f(z) \ . \qquad (5.61)
\end{aligned}
$$

For $P''_{qq}(z)$ a similar calculation can be performed:

$$
\begin{aligned}
\int_0^1 P''_{qq}(z) f(z)\, dz &= -\frac{4}{3}\int_0^1 dz \left(f(z)(1+z) + \frac{8}{3}\lim_{\eta\to+0}\int_0^1 dz \frac{1}{1-z+\eta} f(z) \right) \\
&\quad + 2f(1) - \frac{8}{3} f(1) \lim_{\eta\to+0}\int_0^1 dz \frac{1}{1-z+\eta} \\
&= \lim_{\eta\to+0} \frac{8}{3}\int_0^1 dz \frac{1}{\{1-z+\eta\}_+} f(z) \\
&\quad - \frac{4}{3}\int_0^1 dz \left[f(z)(1+z) - f(1)(1+z) \right] \\
&= \frac{8}{3}\int_0^1 dz \frac{f(z)}{\{1-z\}_+} + \frac{4}{3}\int_0^1 dz \frac{z^2-1}{1-z} \left[f(z) - f(1) \right] \\
&= \frac{8}{3}\int_0^1 dz \frac{f(z)}{\{1-z\}_+} + \frac{4}{3}\int_0^1 dz (z^2-1) \left(\frac{f(z)}{1-z} \right)_+ \\
&= \frac{4}{3}\int_0^1 dz \left(\frac{z^2+1}{1-z} \right)_+ f(z) \ . \qquad (5.62)
\end{aligned}
$$

EXERCISE

5.3 The Bremsstrahlung Part of the GLAP Equation

Problem. The bremsstrahlung process of Fig. 5.1 does not change the number of quarks or antiquarks. Prove that this statement is also contained in the GLAP equation (5.58).

Solution. The number of quarks with a specific flavor is given by *Exercise 5.3*

$$N = \int_0^1 \mathrm{d}x \, f(x) \ . \tag{1}$$

One has to show that N does not depend on Q^2 and consequently not on t. Therefore

$$\frac{\mathrm{d}}{\mathrm{d}t} N = \int_0^1 \mathrm{d}x \frac{\mathrm{d}f}{\mathrm{d}t} = \frac{\alpha_s}{2\pi} \int_0^1 \mathrm{d}x \int_x^1 \frac{\mathrm{d}y}{y} f(y) P_{qq}\left(\frac{x}{y}\right) \tag{2}$$

must vanish. We exchange the x and y integrations,

$$\frac{\mathrm{d}}{\mathrm{d}t} N = \frac{\alpha_s}{2\pi} \int_0^1 \mathrm{d}y \int_0^y \frac{\mathrm{d}x}{y} f(y) P_{qq}\left(\frac{x}{y}\right) \ , \tag{3}$$

and introduce the new variable $z = x/y$,

$$\frac{\mathrm{d}}{\mathrm{d}t} N = \frac{\alpha_s}{2\pi} \int_0^1 \mathrm{d}y \ f(y) \int_0^1 \mathrm{d}z \, P_{qq}(z) = \frac{\alpha_s}{2\pi} N \int_0^1 \mathrm{d}z \, P_{qq}(z) \ . \tag{4}$$

By means of (5.49) we have

$$\int_0^1 \mathrm{d}z \frac{4}{3} \left(\frac{1+z^2}{1-z}\right)_+ = \int_0^1 \mathrm{d}z \left(\frac{4}{3} - \frac{4}{3}\right) \frac{1+z^2}{1-z} = 0 \tag{5}$$

and hence

$$\frac{\mathrm{d}}{\mathrm{d}t} N = 0 \ ,$$

which was to be shown.

It is no coincidence that the right-hand side of (5.58) integrated over x separates into two factors, namely the integrals over y and z. Since QCD is a dimensionless theory, P_{qq} can only depend on ratios of the momenta that occur, i.e., on x/y. Consequently all contributions to the GLAP equation exhibit the same feature. In the next section we shall discuss how this mathematical structure can be derived from very general assumptions.

The content of (5.58) can also be interpreted in the following way: owing to the QCD interaction a quark can split up into a couple of particles with the same total charge, for example, into a quark and a gluon or into two quarks, one antiquark, and a gluon.

Now the question arises whether the quark "fine structure" can be resolved by the exchanged photon. Clearly this depends on the photon momentum. For larger values of Q^2 smaller structures which might exist only for shorter times become visible (see Fig. 5.9). In this sense P_{qq} is also denoted a "*splitting function*" and interpreted as the probability that a quark shows the inner structure quark + gluon at a resolution of $t = \ln(Q^2/Q_0^2)$. Now it is easy to guess in what way one has to generalize (5.56) in order to describe the following processes also:

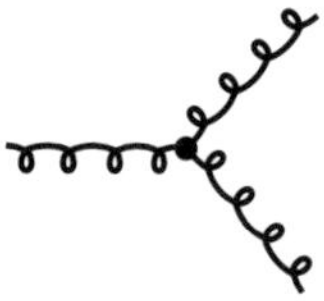

and

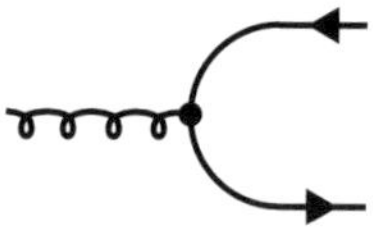

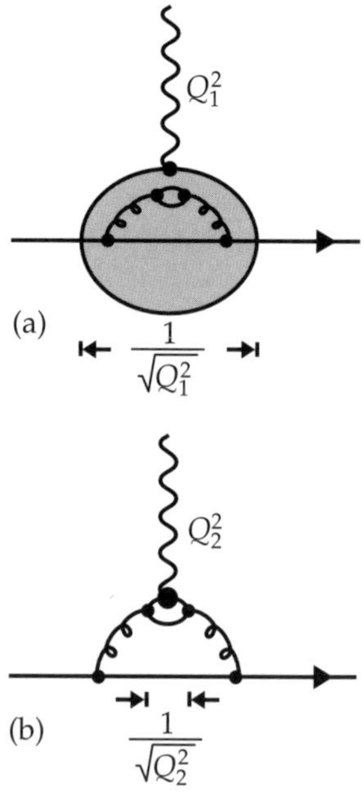

Fig. 5.9a,b. An illustration of the GLAP equation. Larger values of Q^2 $(Q_2^2 \gg Q_1^2)$ resolve smaller structures. (**a**) The non-resolved vertex at low $Q^2\,(= Q_1^2)$. (**b**) A particular vertex structure at high $Q^2\,(= Q_2^2)$

First we introduce new symbols. The single quark distribution functions $q_i(x)$ and $\overline{q}_i(x)$ are replaced by a total quark distribution,

$$\Sigma(x) = \sum_i \left(q_i(x) + \overline{q}_i(x)\right) \;, \quad i = \mathrm{u,\ d,\ s,\ c,} \ldots \;, \tag{5.63}$$

and by differences between the quark distributions,

$$\Delta_{ij}(x) = q_i(x) - q_j(x) \;, \tag{5.64}$$

$$\overline{\Delta}_{ij} = \overline{q}_j(x) - \overline{q}_j(x) \;, \tag{5.65}$$

$$V(x) = \sum_i (q_i(x) - \overline{q}_i(x)) \;. \tag{5.66}$$

Usually $\Sigma(x)$ is referred to as a *singlet distribution function*, since it is symmetric in all flavors (it transforms as a singlet in the flavor symmetry group). $V(x)$ measures the valence quark distribution. If we assume $q_s(x) = \overline{q}_s(x)$, $q_c(x) = \overline{q}_c(x)$, etc., we have for the proton $V(x) = q_u(x) - \overline{q}_u(x) + q_d(x) - \overline{q}_d(x) = u_{\mathrm{val}}(x) + d_{\mathrm{val}}(x)$, and due to the isospin of the proton the sum rules, $\int_0^1 u_{\mathrm{val}}(x)\,\mathrm{d}x = 2$

and $\int_0^1 d_{\mathrm{val}}(x)\,\mathrm{d}x = 1$ hold. A GLAP equation can be derived for each of these distribution functions. They read

$$\frac{\mathrm{d}\Delta_{ij}(x,t)}{\mathrm{d}t} = \frac{\alpha_s(t)}{2\pi}\int\limits_x^1 \frac{\mathrm{d}y}{y}\Delta_{ij}(y,t)P_{qq}\left(\frac{x}{y}\right) , \tag{5.67}$$

$$\frac{\mathrm{d}\Sigma(x,t)}{\mathrm{d}t} = \frac{\alpha_s(t)}{2\pi}\int\limits_x^1 \frac{\mathrm{d}y}{y}\left[\Sigma(y,t)P_{qq}\left(\frac{x}{y}\right) + 2N_{\mathrm{f}}G(y,t)P_{qG}\left(\frac{x}{y}\right)\right] , \tag{5.68}$$

$$\frac{\mathrm{d}G(x,t)}{\mathrm{d}t} = \frac{\alpha_s(t)}{2\pi}\int\limits_x^1 \frac{\mathrm{d}y}{y}\left[\Sigma(y,t)P_{Gq}\left(\frac{x}{y}\right) + G(y,t)P_{GG}\left(\frac{x}{y}\right)\right] . \tag{5.69}$$

Remember, $t = \ln(Q^2/Q_0^2)$ and $\Delta_{ij}(x, t=0)$ has been simply denoted as $\Delta_{ij}(x)$ in (5.64). The same holds for the other quantities. Here in (5.67)–(5.68), $G(x,t)$ and N_{f} denote the gluon distribution function and the number of flavors; P_{qG}, P_{Gq}, and P_{GG} correspond to the above graphs, respectively. A calculation analogous to the one discussed above yields

$$P_{qG}(z) = \frac{1}{2}\left[z^2 + (1-z)^2\right] , \tag{5.70}$$

$$P_{Gq}(z) = \frac{4}{3}\left[1 + (1-z)^2\right]\frac{1}{z} , \tag{5.71}$$

$$P_{GG}(z) = 6\left[z\left(\frac{1}{1-z}\right)_+ \frac{1-z}{z} + z(1-z) + \left(\frac{11}{12} - \frac{N_{\mathrm{f}}}{18}\right)\delta(1-z)\right] . \tag{5.72}$$

The derivation of the explicit expressions (5.70)–(5.72) is quite cumbersome, some of them will be treated in Examples 5.5–5.7[5] The general structure of (5.67)–(5.69) on the other hand is very simple. The bigger Q^2, the more partons with decreasing x are resolved. Therefore only such processes occur on the right-hand side, that increase the particle number.

It should be clear how to interpret the splitting functions P_{qq}, P_{qG}, P_{Gq}, and P_{GG}. It seems to be natural to interpret, for example, $P_{qq}(x/y)$ as the probability density of finding a quark around a quark with fraction x/y of the parent quark momentum y. This clearly leads to a change of the parent quark density $\sum(y,t)$ to $\sum(x,t)$. In a quite analogous way we can interpret $P_{qG}(x/y)$ as the probability of finding a quark around a gluon.

The splitting functions P_{Gq} and P_{GG} therefore describe the probability of finding a gluon around a quark or gluon, respectively, and, accordingly, they contribute to the gluon density $G(x,t)$.

[5] See also, G. Altarelli: Partons in Quantum Chromodynamics, Phys. Rep. C **81**, 1.

However, it should be clear by now that the GLAP equations are an approximation, valid only for large Q^2 and sufficiently large x. The first condition is obvious, since the GLAP equations are a perturbative expansion that becomes rather meaningless for too large α_s. Note, for example, the following process:

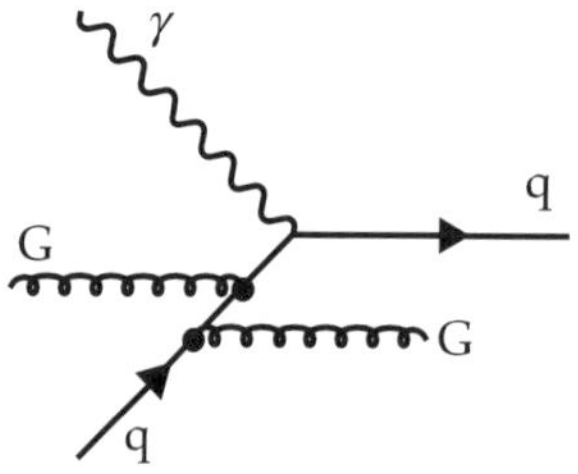

It represents a higher-order correction to the processes in Fig. 5.1; not one bremsstrahlung gluon is involved (as in Fig. 5.1) but two. Therefore it contributes to P_{qq} and P_{Gq} in proportion to

$$\alpha_s^2 \int G(x_1) q(x_2) \tilde{P}_{qq}(x_1, x_2, x) \frac{\mathrm{d}x_1}{x_1} \frac{\mathrm{d}x_2}{x_2} \;, \tag{5.73}$$

with a corresponding function $\tilde{P}_{qq}$. This equation describes recombination effects. At very small x, not only bremsstrahlung processes occur, which are described by the GLAP equations and one typical type of splitting function.

In the GLAP equation, one parent parton radiates a daughter parton and the corresponding splitting function depends only on the ratio of the corresponding momenta. If we want to treat recombination effects we have to introduce more complicated "splitting functions", which depend on three momenta: one parent parton, one radiated "daughter" parton, and an additional absorbed gluon from the surrounding gluon bath. Therefore one gets a type of equation which represents a convolution of the gluon bath structure function $G(x_1)$, the parent quark $q(x_2)$, and a complicated recombination splitting function $\tilde{P}_{qq}(x_1, x_2, x)$. The structure function $G(x_1)$ and the splitting function $\tilde{P}_{qq}(x_1, x_2, x)$ are difficult to determine. We shall not follow up this any further here.

The essential point now is the following: In the case of very small x_1 the gluon distribution function $G(x_1)$ can increase so much that even $\alpha_s\, G(x_1)$ remains larger than one (we shall find later that $G(x_1)$ is proportional to $1/x_1$ for small x_1). Hence there are kinematic regions where contributions like those of Fig. 5.9 are no longer negligible. With the new HERA accelerator at DESY in Hamburg, Germany, these regions have for the first time become accessible to experimental investigations. At HERA, an electron beam and a proton beam are collided with an invariant mass of

$$s = (p_\mathrm{e} + p_\mathrm{p})^2 \approx 4 E_\mathrm{e} E_\mathrm{p} \approx 10^5\ \mathrm{GeV}^2 \;. \tag{5.74}$$

Since the maximum momentum transfer $\nu_{\max}$ is just $s/2$, events with $Q^2 \geq 5\ \mathrm{GeV}^2$ and $x \geq 10^{-4}$ can be investigated (see Exercise 5.4). Currently much effort is being invested in the necessary generalizations of the GLAP equations for small values of x. To this end their structure must be modified, as can already be seen from (5.73). Terms have to be introduced, for example, that depend on products of two or more distribution functions. Up to now there has been no generally accepted solution to this problem (Sect. 6.2). Presumably such a solution will only be found in connection with experimental investigations. The HERA experiments have been producing data since 1992.

In this section we have learned how QCD corrections can be evaluated from the corresponding Feynman graphs. This procedure is of great clarity, but it requires extensive calculations. Therefore one frequently chooses another way to derive the GLAP equations that is much more elegant but unfortunately less obvious. The latter disadvantage occurs because a part of the result, namely the general structure of the GLAP equations, has to be more or less assumed. We shall discuss this alternative derivation in the following section, because it will shed additional light on the meaning of the GLAP equations.

5.2 An Alternative Approach to the GLAP Equations

In this section we shall be concerned with the derivation of the GLAP equations using only our knowledge about the properties of QCD.[6] We know already that scattering experiments using leptons as projectiles and hadrons as targets reveal information about the hadronic structure. The structure functions or momentum distributions of the partons inside the hadron depend on the four-momentum q_μ carried by the exchanged photon (in general the vector meson). If the momentum transfer Q^2 is increased, more and more details of the hadronic structure become visible (see Fig. 5.10).

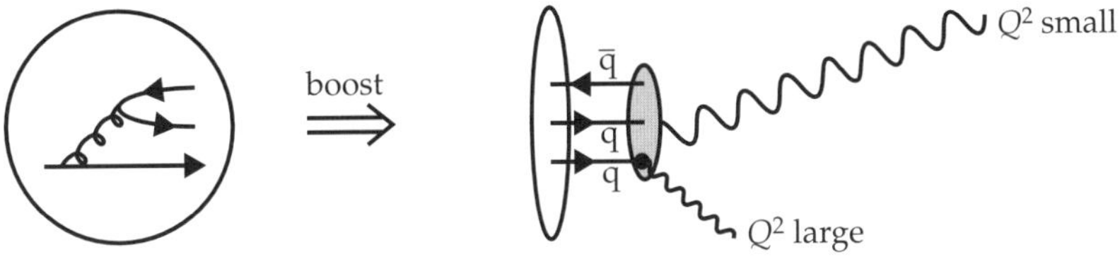

Fig. 5.10. The role of Q^2 as transverse resolving power

Q^2 is related to the maximum transverse momentum of a parton in the final state. This fact is easily understood by looking at Fig. 5.10. In the Breit frame, where the distribution functions are defined, the nucleon is strongly contracted in the direction of movement. Therefore the resolution with which quarks are seen depends only on the transverse momentum. We denote by $1/\sqrt{Q^2}$ the *resolving power*, since hadronic structures down to $1/\sqrt{Q^2}$ can be resolved owing to the Heisenberg inequality $\Delta x \Delta Q \sim 1$. Let us assume that we have chosen Q^2 high enough to enter the perturbative regime. We can resolve partons of size $1/\sqrt{Q^2}$ inside the hadron. If the momentum transfer is increased to $Q'^2 > Q^2$, one can

[6] O. Nachtmann: *Elementary Particle Physics* (Springer, Berlin, Heidelberg 1990).

resolve smaller constituents of size $1/\sqrt{Q'^2}$ inside the hadron. For a given resolving power $1/\sqrt{Q^2}$ the momentum distribution of the partons of kind a is denoted by $N_a(x_a, Q^2)$. The smaller constituents b at $1/\sqrt{Q'^2}$ carry some of the parton momenta x_a:

$$\frac{x_b}{x_a} < 1 \ . \tag{5.75}$$

This results in a decrease in the momentum distribution for large x and an increase for small momentum fractions. In other words $N_a(x_a, Q^2) < N_b(x_b, Q^2)$.

The variation of the distribution functions $\Delta N(x, Q^2)$ when Q^2 is increased by ΔQ^2 can now be treated by using perturbation theory. To lowest order, $\Delta N(x, Q^2)$ is proportional to the coupling parameter

$$\alpha_s(Q^2) = \frac{g^2(Q^2)}{4\pi} \ . \tag{5.76}$$

If the quark masses are neglected, QCD lacks any energy scale. Therefore only momentum ratios can appear that govern the development of $N_a(x_a, Q^2)$ to higher Q^2. This distribution function for partons of kind a, i.e. $N_a(x_a, Q^2)$, depends for $Q'^2 > Q^2$ on the distribution of partons of all kinds b. We assume this dependence to be linear. The probability of finding a parton a with momentum fraction x_a inside a parton b of momentum fraction x_b is therefore given by

$$\Delta N_b(x_b, Q^2) = \frac{\alpha_s(Q^2)}{2} P_{ba}\left(\frac{x_b}{x_a}\right) N_a(x_a, Q^2) \frac{\Delta Q^2}{Q^2} \ . \tag{5.77}$$

The quotient $\Delta Q^2/Q^2 = \Delta \ln Q^2$ is introduced to keep the change in momentum transfer dimensionless. In the same way it follows that P_{ba} can only depend on the ratio of the momentum fractions. The factor $\frac{1}{2}$ is convient.

To obtain the total change of N_b we have to sum over all partons of kind a that may "contain" the parton b, and we have to integrate over all parton momenta $x_a > x_b$. This leads us, finally, to the GLAP equations:

$$\Delta N_b(x_b, Q^2) = \frac{\alpha_s(Q^2)}{2\pi} \sum_b \int_{x_b}^{1} P_{ba}\left(\frac{x_b}{x_a}\right) N_a(x_a, Q^2) \frac{\mathrm{d}x_a}{x_a} \Delta \ln Q^2 \ . \tag{5.78}$$

Here we have introduced the ratio $\mathrm{d}x_a/x_a$ instead of $\mathrm{d}x_a$ alone; otherwise a scale would enter the GLAP equations. One might wonder why a dimensionless integration measure $\mathrm{d}x_a/x_b$ does not appear in (5.75). However, remember, the also dimensionless integration measure $\mathrm{d}x_a/\mathrm{d}x_b$ would lead to a splitting function with explicit dependence on single momenta and not only ratios. Equation (5.75) describes a set of equations for all types of partons, which in differential form becomes

$$\frac{\partial N_b(x_b, Q^2)}{\partial \ln Q^2} = \frac{\alpha_s(Q^2)}{2\pi} \sum_b \int_{x_b}^{1} P_{ba}\left(\frac{x_b}{x_a}\right) N_a(x_a, Q^2) \frac{\mathrm{d}x_a}{x_a} \ . \tag{5.79}$$

The functions P_{ba} are called *splitting functions* since they describe the breaking up of a parton of type a into partons of type b, when the momentum transfer Q^2 is increased.

In the following we will derive the splitting functions P_{ba} for several cases. Life is made easy by assuming that there are only two types of partons: quarks and gluons. Since we assume that all antiquarks are due to gluon splitting, $G \to \bar{q}q$, the increase in the number of quarks always equals the increase in the number of antiquarks. $P_{qG} = P_{uG} = P_{\bar{u}G} = P_{dG} = \ldots$. This is justified because of charge-conjugation symmetry (change in quark distribution equals change in antiquark distribution) and because of our treating quarks as massless objects (changes in u, d, s quarks proceeds in the same way). This reduces the number of splitting functions to four:

$$P_{qq} = -\frac{4}{3}(1+x) + 2\delta(1-x) + \frac{8}{3}\left(\frac{1}{1-x} - \delta(1-x)\int_0^1 \mathrm{d}x' \frac{1}{1-x'}\right) , \tag{5.80}$$

$$P_{qG} = \frac{1}{2}\left[(1-x)^2 + x^2\right] , \tag{5.81}$$

$$P_{Gq} = \frac{4}{3}\frac{1+(1-x)^2}{x} , \tag{5.82}$$

$$P_{GG} = 6\left[\frac{1}{x} - 2 + x(1-x)\right] + \left(\frac{11}{2} - \frac{N_\mathrm{f}}{3}\right)\delta(1-x) + 6\left(\frac{1}{1-x} - \delta(1-x)\int_0^1 \mathrm{d}x' \frac{1}{1-x'}\right) . \tag{5.83}$$

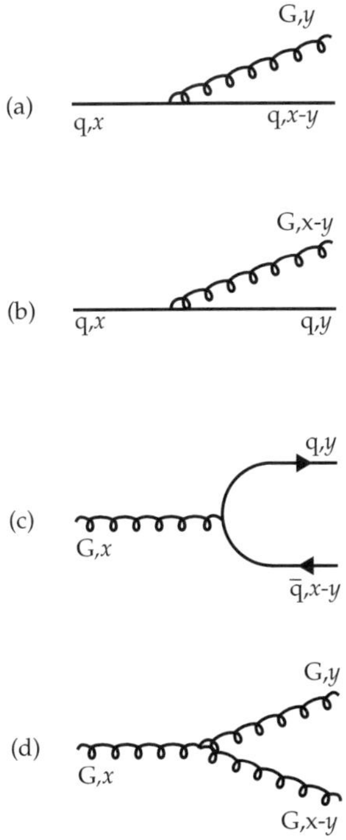

Fig. 5.11a–d. The processes described by the splitting functions **(a)** P_{Gq}, **(b)** P_{qq}, **(c)** P_{qG}, and **(d)** P_{GG}

We shall evaluate the functions P_{Gq}, P_{qq}, and P_{qG} in Examples 5.5–5.7. The effect of these functions is depicted graphically in Fig. 5.11. The evaluation of P_{GG} can be found in other papers.[7]

EXERCISE

5.4 The Maximum Transverse Momentum

Problem. (a) Determine the maximum transverse momentum of a parton occurring in the final state if the photon momentum before the collision is given in the Breit system (see Exercise 3.6) by

$$q_\mu = (0; 0, 0, -Q) \tag{1}$$

and the initial parton momentum in the Breit system is

$$p_\mu = (p; 0, 0, p) . \tag{2}$$

[7] See also, G. Altarelli: Partons in Quantum Chromodynamics, Phys. Rep. C **81**, 1.

Exercise 5.4

(b) Verify the relation $\nu_{\max} = s/2$ and investigate the kinematical region accessible at the HERA collider.

Solution. (a) We first assume that two partons escape after the collision. These partons carry momenta

$$p_1' = \left(\frac{p}{2}; \boldsymbol{p}_\perp, \frac{p-Q}{2}\right) ,$$
$$p_2' = \left(\frac{p}{2}; -\boldsymbol{p}_\perp, \frac{p-Q}{2}\right) . \tag{3}$$

Obviously the four-momentum is conserved:

$$q + p = p_1' + p_2' . \tag{4}$$

Furthermore squaring the parton momentum p_1' yields

$$\begin{aligned}(p_1')^2 = 0 &= \left(\frac{p}{2}\right)^2 - (\boldsymbol{p}_\perp)^2 - \left(\frac{p-Q}{2}\right)^2 \\ &= \left(\frac{p}{2}\right)^2 - (\boldsymbol{p}_\perp)^2 - \left(\frac{p}{2}\right)^2 + \frac{2p\cdot Q}{4} - \left(\frac{Q}{2}\right)^2 \\ &= -(\boldsymbol{p}_\perp)^2 + \frac{p\cdot Q}{2} - \left(\frac{Q}{2}\right)^2 ,\end{aligned} \tag{5}$$

because it is assumed to be of rest mass zero, i.e. ${p_1'}^2 = 0$. Introducing the momentum fraction $x = Q^2/(2p\cdot Q)$, one gets

$$-(\boldsymbol{p}_\perp)^2 - \left(\frac{Q}{2}\right)^2 + \frac{Q^2}{4x} = 0 \tag{6}$$

or

$$|\boldsymbol{p}_\perp| = \frac{Q}{2}\sqrt{\frac{1-x}{x}} . \tag{7}$$

For more than two partons in the final state this is the maximum transverse momentum that can be achieved.

(b) According to the definition (5.11), $\nu = q\cdot p_p$, we have in the rest system of the nucleon

$$\begin{aligned}\nu &\overset{\text{R.S.}}{=} (E_e - E_e')M_p ,\\ \nu_{\max} &= E_e M_p .\end{aligned} \tag{8}$$

For the center-of-mass energy

$$\begin{aligned}s &= (p_p + p_e)^2 = M_p^2 + 2p_p\cdot p_e + m_e^2 \sim 2p_p\cdot p_e \\ &\overset{\text{R.S.}}{=} 2E_e M_p = 2\nu_{\max} .\end{aligned} \tag{9}$$

Exercise 5.4

From this we get for x_{min}

$$x_{\text{min}} = \frac{Q^2}{2\nu_{\text{max}}} \simeq \frac{Q^2}{s} \ . \tag{10}$$

For HERA kinematics $E_e = 30\,\text{GeV}$, $E_p = 820\ \text{GeV}$, so that

$$s \simeq eE_eE_p \simeq 10^5\ \text{GeV}^2 \ . \tag{11}$$

Then we get for $Q^2 \sim 5\,\text{GeV}$ a minimal value $x_{\text{min}} \sim 10^{-4}$.

EXAMPLE

5.5 Derivation of the Splitting Function P_{Gq}

For given momentum transfer Q^2 – in the perturbative region – we define the Hamiltonian

$$\hat{H}_{Q^2} = \hat{H}^{\text{kin}}_{Q^2} + \hat{V} \ . \tag{1}$$

The potential $\hat{V}$ describes the binding of the partons. It is the nonperturbative part, including interactions with small Q^2. The kinetic part $\hat{H}^{\text{kin}}_{Q^2}$ yields – when applied to the parton wave functions – the kinetic parton energies

$$\begin{aligned} \hat{H}^{\text{kin}}_{Q^2}|q_{Q^2}(p)\rangle &= E|q_{Q^2}(p)\rangle = |\boldsymbol{p}||q_{Q^2}(p)\rangle \ , \\ \hat{H}^{\text{kin}}_{Q^2}|G_{Q^2}(k)\rangle &= \omega|G_{Q^2}(k)\rangle = |\boldsymbol{k}||G_{Q^2}(k)\rangle \ . \end{aligned} \tag{2}$$

Here $|q_{Q^2}(p)\rangle$ and $|G_{Q^2}(k)\rangle$ denote the quark and gluon wave functions, respectively, at momentum transfer Q^2. Again Q is related to the maximum transverse momentum by $|\boldsymbol{k}_\perp|_{\text{max}} \equiv |\boldsymbol{p}_\perp|$ of Exercise 5.4, (7), i.e.

$$|\boldsymbol{k}_\perp|_{\text{max}} = \frac{Q}{2}\sqrt{\frac{1-x}{x}} \equiv \varepsilon Q \ , \quad \varepsilon = \frac{1}{2}\sqrt{\frac{1-x}{x}} \ . \tag{3}$$

At $Q'^2 > Q^2$ the kinetic part of the Hamiltonian can be written as

$$\hat{H}^{\text{kin}}_{Q'^2} = \hat{H}^{\text{kin}}_{Q^2} + \hat{H}_{\Delta Q^2} \ . \tag{4}$$

Application of $\hat{H}^{\text{kin}}_{Q'^2}$ to the quark and gluon wave functions at Q'^2 again yields

$$\begin{aligned} \hat{H}^{\text{kin}}_{Q'^2}|q_{Q'^2}(p)\rangle &= |\boldsymbol{p}||q_{Q'^2}(p)\rangle \ , \\ \hat{H}^{\text{kin}}_{Q'^2}|G_{Q'^2}(k)\rangle &= |\boldsymbol{k}||G_{Q'^2}(k)\rangle \ . \end{aligned} \tag{5}$$

The additional operator $\hat{H}_{\Delta Q^2}$ just describes the interactions leading to transverse momenta $|\boldsymbol{k}_\perp|$ between εQ and $\varepsilon Q'$ (see Exercise 5.4).

Example 5.5

The interaction part of the QCD Hamiltonian can be derived in the temporal Weyl gauge, $A_0^\alpha = 0$ for all gluons. The Hamiltonian $\hat{H}_{\Delta Q^2}$ up to first order in the coupling parameter g is

$$\hat{H}_{\Delta Q^2} = -g(Q^2) \int d^3r \sum_{f,a} \hat{\bar{q}}_f(\boldsymbol{r}) \left[\boldsymbol{\gamma} \cdot \hat{\boldsymbol{A}}^a(\boldsymbol{r}) \frac{\lambda^a}{2} \right] \hat{q}_f(\boldsymbol{r})$$
$$+ g(Q^2) \int d^3r \, f_{abc} \hat{A}_j^a(\boldsymbol{r}) \hat{A}_k^b(\boldsymbol{r}) \frac{\partial}{\partial r_j} \hat{A}_k^c(\boldsymbol{r}) \ . \tag{6}$$

This interaction Hamiltonian may be directly read off from Example 4.2, where we derived the Feynman rules (see (1) and (6) of Example 4.2). The additional minus sign stems from the Legendre transformation that relates Lagrange and Hamilton density. The first part describes the interaction between quarks and gluons, the second part handles the 3-gluon interaction. In the first part the sum is over all quark flavors u, d, s, ...

We turn now to the quark and gluon wave functions. Since for higher resolution ("virtuality") Q'^2 smaller constituents of the hadron can be seen, the wave function describing big partons at Q^2 can be written as a superposition of wave functions for the constituents at Q'^2. For the quark and gluon wave functions at Q^2 we get the expansion

$$|q_{Q^2}(p)\rangle = C_q(p)|q_{Q'^2}(p)\rangle + \sum_{\boldsymbol{k}',\boldsymbol{p}'} C_{Gq}(k', p')|G_{Q'^2}(k')q_{Q'^2}(p')\rangle \ ,$$
$$|G_{Q^2}(k)\rangle = C_G(k)|G_{Q'^2}(k)\rangle + \sum_{\boldsymbol{k}',\boldsymbol{p}'} C_{qG}(k', p')|G_{Q'^2}(k')q_{Q'^2}(p')\rangle$$
$$+ \sum_{\boldsymbol{k}',\boldsymbol{k}''} C_{GG}(k', k'')|G_{Q'^2}(k')G_{Q'^2}(k'')\rangle \ . \tag{7}$$

The coefficients are given by

$$C_q(p) = \langle q_{Q'^2}(p)|q_{Q^2}(p)\rangle \ ,$$
$$C_G(k) = \langle G_{Q'^2}(k)|G_{Q^2}(k)\rangle \ ,$$
$$C_{Gq}(k', p') = \langle G_{Q'^2}(k')q_{Q'^2}(p')|q_{Q^2}(p)\rangle \ ,$$
$$C_{qG}(k', p') = \langle G_{Q'^2}(k')q_{Q'^2}(p')|G_{Q^2}(p)\rangle \ ,$$
$$C_{GG}(k', k'') = \langle G_{Q'^2}(k')G_{Q'^2}(k'')|G_{Q^2}(k)\rangle \ . \tag{8}$$

The squared coefficients can be related to the momentum distribution functions. Note again that (7) describe the evolution of the various distribution function from virtuality Q^2 to virtuality Q'^2. The whole evolution is calculated in first-order perturbation theory. Now we have all the ingredients to determine the splitting functions. We start with the calculation of the probability of resolving a gluon inside a quark (Fig. 5.11a). This will lead us to an expression for the splitting function P_{Gq}.

Example 5.5

We apply the kinetic part $\hat{H}^{\text{kin}}_{Q'^2}$ of the Hamiltonian to the quark wave function at resolution Q^2. This gives on the one hand

$$\begin{aligned}\hat{H}^{\text{kin}}_{Q'^2}|q_{Q^2}(p)\rangle &= \hat{H}^{\text{kin}}_{Q^2}|q_{Q^2}(p)\rangle + \hat{H}_{\Delta Q^2}|q_{Q^2}(p)\rangle \\ &= |\boldsymbol{p}|\,|q_{Q^2}(p)\rangle + \hat{H}_{\Delta Q^2}|q_{Q^2}(p)\rangle \ . \end{aligned} \tag{9}$$

The expansion (7) of the quark wave function yields on the other hand

$$\begin{aligned}&\hat{H}^{\text{kin}}_{Q'^2}|q_{Q^2}(p)\rangle \\ &\quad = C_q(p)\hat{H}^{\text{kin}}_{Q'^2}|q_{Q'^2}(p)\rangle + \sum_{k',p'} C_{Gq}(k',p')\hat{H}^{\text{kin}}_{Q'^2}|G_{Q'^2}(k')q_{Q'^2}(p')\rangle \\ &\quad = C_q(p)|\boldsymbol{p}|\,|q_{Q'^2}(p)\rangle + \sum_{k',p'}(|\boldsymbol{k}'|+|\boldsymbol{p}'|)C_{Gq}(k',p')|G_{Q'^2}(k')q_{Q'^2}(p')\rangle \ . \end{aligned} \tag{10}$$

We project (9) onto the final-state wave function $|G_{Q'^2}(k')q_{Q'^2}(p')\rangle$:

$$\begin{aligned}&|\boldsymbol{p}|\langle G_{Q'^2}(k')q_{Q'^2}(p')|q_{Q^2}(p)\rangle + \langle G_{Q'^2}(k')q_{Q'^2}(p')|\hat{H}_{\Delta Q^2}|q_{Q^2}(p)\rangle \\ &\quad = \langle G_{Q'^2}(k')q_{Q'^2}(p')|\hat{H}^{\text{kin}}_{Q'^2}|q_{Q^2}(p)\rangle \\ &\quad = (|\boldsymbol{k}'|+|\boldsymbol{p}'|)\langle G_{Q'^2}(k')q_{Q'^2}(p')|q_{Q^2}(p)\rangle \\ &\quad = (|\boldsymbol{k}'|+|\boldsymbol{p}'|)C_{Gq}(k',p') \ , \end{aligned} \tag{11}$$

employing (5) and (8) as well as the orthogonality between the wave functions. Thus we obtain an expression for C_{Gq}:

$$C_{Gq}(k',p') = \frac{\langle G_{Q'^2}(k')q_{Q'^2}(p')|\hat{H}_{\Delta Q^2}|q_{Q^2}(p)\rangle}{|\boldsymbol{k}'|+|\boldsymbol{p}'|-|\boldsymbol{p}|} \ . \tag{12}$$

This is just first-order perturbation theory. The absolute square of this coefficient is proportional to the probability of finding a gluon of longitudinal momentum k' inside a quark of momentum p when the resolution Q^2 is increased. The momentum fraction of the gluon relative to the quark is therefore

$$x = \frac{k'_\parallel}{|\boldsymbol{p}|} = \frac{\boldsymbol{k}'\cdot\boldsymbol{p}}{|\boldsymbol{p}|^2} \ . \tag{13}$$

Starting with a single quark $|q_Q(p)\rangle$ and no gluons at the resolution Q^2 we will have initially a gluon distribution $N_G(Q^2)=0$. Increasing the resolution from Q^2 to Q'^2 we get due to Bremsstrahlung $q\to qG$, at Q'^2 the following distribution function (see Fig. 5.12). Averaging over the incoming quark spin s yields $\frac{1}{2}\sum_s$, and therefore

$$\begin{aligned}N_G(x,Q'^2) = \sum_a \frac{1}{2}\sum_{ss'\varepsilon^a} \int \frac{\mathrm{d}^3p'}{(2\pi)^3 2E'} \int \frac{\mathrm{d}^3k'}{(2\pi)^3 2\omega'} \\ \times \delta\left(x - \frac{\boldsymbol{k}'\cdot\boldsymbol{p}}{|\boldsymbol{p}|^2}\right)|C_{Gq}(k',p')|^2\frac{1}{2E} \ , \end{aligned} \tag{14}$$

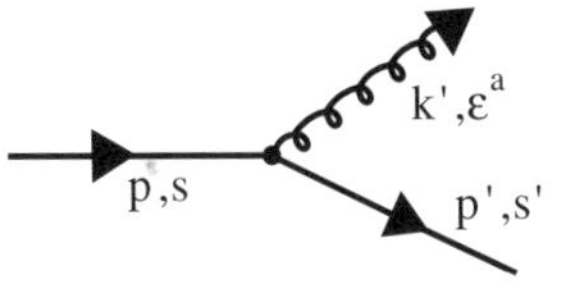

Fig. 5.12. Bremsstrahlung of a gluon by a quark. The quark wave function is normalized within the volume V according to $1/\sqrt{2EV}\times u(p,s)\mathrm{e}^{\mathrm{i}px}$. s, s' denote the spin of the quark before and after the scattering, respectively. ε^a is the polarization vector of the gluon

Example 5.5

where the $\boldsymbol{k}$ integration is performed for $|\boldsymbol{k}_\perp|$ in the range from εQ^2 to $\varepsilon Q'^2$. Here $\int \mathrm{d}^3p'/(2\pi)^3 2E'$ and $\int \mathrm{d}^3k'/(2\pi)^3 2\omega'$ are the invariant phase-space factors (see (2.107)). The factor $1/2E$ stems from the normalisation of the incoming quark wave function. The eight color degrees of freedom of the gluons are enumerated by a. We now evaluate the squared coefficient $|C_{Gq}(k', p')|^2$. To do this, only the first part of Hamiltonian (6) is required. Inserting (6) into (12) yields

$$\begin{aligned}
C_{Gq}(k', p') &= -g(Q^2)\int \mathrm{d}^3r \sum_f \langle G_{Q'^2}(k')q_{Q'^2}(p')|\hat{\bar{q}}_f(\boldsymbol{r})\\
&\quad\times\left(\boldsymbol{\gamma}\cdot\hat{\boldsymbol{A}}^a(\boldsymbol{r})\frac{\lambda^a}{2}\right)\hat{q}_f(\boldsymbol{r})|q_{Q^2}(p)\rangle\frac{1}{|\boldsymbol{k}'|+|\boldsymbol{p}'|-|\boldsymbol{p}|}\\
&= -g(Q^2)\delta^3(\boldsymbol{k}'+\boldsymbol{p}'-\boldsymbol{p})(2\pi)^3\overline{u(\boldsymbol{p}', s')}\boldsymbol{\varepsilon}^{a*}\cdot\boldsymbol{\gamma}\frac{\lambda^a}{2}u(\boldsymbol{p}, s)\\
&\quad\times\frac{1}{|\boldsymbol{k}'|+|\boldsymbol{p}'|-|\boldsymbol{p}|}\ .
\end{aligned}\tag{15}$$

Note that the quark spinors $u(p, s)$ are here defined without the normalization factor $1/\sqrt{2EV}$. The latter is explicitely denoted in (14) within the invariant phase space factors. Again it is obvious that $C_{Gq}(k', p')$ is calculated in first-order perturbation theory (see also Fig. 5.12). From this it follows that

$$\begin{aligned}
N_G(x, Q'^2) &= \int\frac{\mathrm{d}^3p'}{(2\pi)^3}\int\frac{\mathrm{d}^3k'}{(2\pi)^3}\frac{1}{8EE'\omega'}\delta\left(x-\frac{\boldsymbol{k}'\cdot\boldsymbol{p}}{|\boldsymbol{p}|^2}\right)\\
&\quad\times\delta^3(\boldsymbol{k}'+\boldsymbol{p}'-\boldsymbol{p})(2\pi)^3 g(Q^2)^2\frac{1}{(|\boldsymbol{k}'|+|\boldsymbol{p}'|-|\boldsymbol{p}|)^2}\\
&\quad\times\frac{1}{2}\sum_a\sum_{s,s',\varepsilon^a}\left|\overline{u(\boldsymbol{p}', s')}(\boldsymbol{\varepsilon}^{a*}\cdot\boldsymbol{\gamma})\frac{\lambda^a}{2}u(\boldsymbol{p}, s)\right|^2\ .
\end{aligned}\tag{16}$$

Here $[\delta^3(\boldsymbol{k}'+\boldsymbol{p}'-\boldsymbol{p})]^2 = (2\pi)^3 V\delta^3(\boldsymbol{k}'+\boldsymbol{p}'-\boldsymbol{p})$ has been used and the normalization volume V is set to $V = 1$. We consider the expression for the squared matrix element. Performing the summation over color and spin, where we have used the relation $\not{\varepsilon}^a = -\boldsymbol{\varepsilon}^a\cdot\boldsymbol{\gamma}$ for $\varepsilon^a = (0, \boldsymbol{\varepsilon}^a)$, we get

$$\begin{aligned}
&\frac{1}{2}\sum_s\frac{1}{3}\sum_c\sum_a\sum_{\varepsilon^a}\sum_{s',c'}\left|\overline{u_{c'}}(\boldsymbol{p}', s')\not{\varepsilon}^a\frac{\lambda^a_{cc'}}{2}u_c(\boldsymbol{p}, s)\right|^2\\
&= \frac{1}{6}\sum_{s,s'}\sum_{c,c'}\sum_{\varepsilon^a}\sum_{a,a'}\mathrm{tr}\left[\not{\varepsilon}^a\frac{\lambda^a_{cc'}}{2}u_{c'}(\boldsymbol{p}, s)\overline{u_{c'}}(\boldsymbol{p}, s)\frac{\lambda^{a'}_{c'c}}{2}\not{\varepsilon}^{a'}u_c(\boldsymbol{p}, s)\overline{u_c}(\boldsymbol{p}, s)\right],
\end{aligned}$$

where c, c' denote the colors of the quarks. Performing the sums over spins and over colors

$$\sum_s u_{c'}(\boldsymbol{p}, s)\overline{u_{c'}}(\boldsymbol{p}, s) = \not{p}+m \approx \not{p}\ ,$$

$$\sum_{c,c'}\frac{\lambda^a_{cc'}}{2}\frac{\lambda^{a'}_{c'c}}{2} = \mathrm{tr}\frac{\lambda^a}{2}\frac{\lambda^{a'}}{2} = \frac{1}{2}\delta^{aa'}\ ,$$

we obtain

Example 5.5

$$
\begin{aligned}
2\cdot\frac{1}{6}&\sum_{\varepsilon^a}\sum_{a}\mathrm{tr}\left\{\not p'\not\varepsilon^{a*}\not p\not\varepsilon^{a}\right\}\\
&=\frac{1}{6}\sum_{a}\sum_{\varepsilon^a}\frac{1}{2}\cdot 4\left[(p'\cdot\varepsilon^{a*})(p\cdot\varepsilon^{a})+(p'\cdot\varepsilon^{a})(p\cdot\varepsilon^{a*})-(p\cdot p')(\varepsilon^{a}\cdot\varepsilon^{a*})\right]\\
&=\frac{1}{3}\sum_{a}\sum_{\varepsilon^a}\left[(\boldsymbol{p}'\cdot\boldsymbol{\varepsilon}^{a*})(\boldsymbol{p}\cdot\boldsymbol{\varepsilon}^{a})\right.\\
&\quad\left.+(\boldsymbol{p}'\cdot\boldsymbol{\varepsilon}^{a})(\boldsymbol{p}\cdot\boldsymbol{\varepsilon}^{a*})-(p_0'p_0-(\boldsymbol{p}'\cdot\boldsymbol{p}))(\boldsymbol{\varepsilon}^{a}\cdot\boldsymbol{\varepsilon}^{a*})\right]\\
&=\frac{1}{3}\sum_{a}\sum_{\varepsilon^a}(\varepsilon^a)_i(\varepsilon^{a*})_j\left\{p_ip_j'+p_i'p_j-\left[(\boldsymbol{p}'\cdot\boldsymbol{p})-|\boldsymbol{p}||\boldsymbol{p}'|\right]\delta_{ij}\right\}\\
&=\frac{1}{3}8\cdot\left(\delta_{ij}-\frac{k_i'k_j'}{|\boldsymbol{k}'|^2}\right)\left\{p_ip_j'+p_i'p_j-\left[(\boldsymbol{p}'\cdot\boldsymbol{p})-|\boldsymbol{p}||\boldsymbol{p}'|\right]\delta_{ij}\right\}\\
&=\frac{16}{3}\frac{1}{|\boldsymbol{k}'|^2}\left[|\boldsymbol{p}||\boldsymbol{p}'||\boldsymbol{k}'|^2-(\boldsymbol{k}'\cdot\boldsymbol{p})(\boldsymbol{k}'\cdot\boldsymbol{p}')\right]\ ,
\end{aligned}
\tag{17}
$$

where we summed over transversal polarisations of the gluon with momentum k'

$$
\sum_{a}\sum_{\varepsilon^a}(\varepsilon^a)_i(\varepsilon^{a*})_j=8\sum_{\varepsilon}\varepsilon_i\varepsilon_j^*=8\left(\delta_{ij}-\frac{k_i'k_j'}{|k'|^2}\right)\ .
$$

Inserting this into (16) yields

$$
\begin{aligned}
N_G\left(x,Q'^2\right)=&\frac{2}{3}\frac{1}{(2\pi)^3}g^2(Q^2)\int \mathrm{d}^3p'\int\mathrm{d}^3k'\frac{1}{EE'\omega'}\delta\left(x-\frac{\boldsymbol{k}'\cdot\boldsymbol{p}}{|\boldsymbol{p}|^2}\right)\\
&\times\delta^3(\boldsymbol{k}'+\boldsymbol{p}'-\boldsymbol{p})\cdot\frac{1}{(|\boldsymbol{k}'|+|\boldsymbol{p}'|-|\boldsymbol{p}|)^2}\\
&\times\frac{1}{|\boldsymbol{k}'|^2}\left[|\boldsymbol{p}||\boldsymbol{p}'||\boldsymbol{k}'|^2-(\boldsymbol{k}'\cdot\boldsymbol{p})(\boldsymbol{k}'\cdot\boldsymbol{p}')\right]\ .
\end{aligned}
\tag{18}
$$

The momenta can be written as (see Fig. 5.11)

$$
\begin{aligned}
\boldsymbol{p}=&(0,0,p)\ ,\\
\boldsymbol{k}'=&(\boldsymbol{k}_\perp',xp)\ ,\\
\boldsymbol{p}'=&(-\boldsymbol{k}_\perp',(1-x)p)\ .
\end{aligned}
\tag{19}
$$

Obviously the three-momentum is conserved: $\boldsymbol{p}=\boldsymbol{k}'+\boldsymbol{p}'$. The transverse momentum is expected to be small compared to the longitudinal momenta:

$$
|\boldsymbol{k}_\perp'|\ll p\ ,\quad |\boldsymbol{k}_\perp'|\ll xp\ ,\quad |\boldsymbol{k}_\perp'|\ll(1-x)p\ .
\tag{20}
$$

Example 5.5

The term $\frac{1}{|\boldsymbol{k}'|+|\boldsymbol{p}'|-|\boldsymbol{p}|}$ in the expression $N_G(x,\,Q^2)$ of (18) can be approximated by

$$\begin{aligned}|\boldsymbol{k}'|+|\boldsymbol{p}'|-|\boldsymbol{p}| &\approx xp+\frac{|\boldsymbol{k}'_\perp|^2}{2x|\boldsymbol{p}|}+(1-x)p+\frac{|\boldsymbol{k}'_\perp|^2}{2(1-x)|\boldsymbol{p}|}-|\boldsymbol{p}|\\ &=\frac{|\boldsymbol{k}'_\perp|^2}{2p}\frac{1}{x(1-x)}\;, \end{aligned} \tag{21}$$

and therefore

$$\frac{1}{|\boldsymbol{k}'|+|\boldsymbol{p}'|-|\boldsymbol{p}|}\approx\frac{2p}{|\boldsymbol{k}'_\perp|^2}x(1-x)\;.$$

Furthermore

$$\begin{aligned}&\frac{1}{|\boldsymbol{k}'|^2}\left[|\boldsymbol{p}||\boldsymbol{p}'||\boldsymbol{k}'|^2-(\boldsymbol{k}'\cdot\boldsymbol{p})(\boldsymbol{k}'\cdot\boldsymbol{p}')\right]\\ &\approx\frac{1}{x^2p^2}\Bigg[p(1-x)p(|\boldsymbol{k}'_\perp|^2+x^2p^2)\\ &\quad+\frac{1}{2(1-x)p^2}|\boldsymbol{k}'_\perp|p(|\boldsymbol{k}'_\perp|^2+x^2p^2)+xp^2|\boldsymbol{k}'_\perp|^2-x^2p^2(1-x)p^2\Bigg]\\ &=\frac{|\boldsymbol{k}'_\perp|^2}{x^2}\left[1+\frac{x^2}{2(1-x)}\right] \end{aligned} \tag{22}$$

and

$$\frac{1}{EE'\omega'}=\frac{1}{|\boldsymbol{p}||\boldsymbol{p}'||\boldsymbol{k}'|}\approx\frac{1}{p^3x(1-x)}\;. \tag{23}$$

Inserting (21)–(23) into the gluon distribution (18) results in

$$\begin{aligned}N_G(x,\,Q'^2)=&\frac{2}{3}\frac{1}{(2\pi)^3}g^2(Q^2)\int\mathrm{d}^3k'\int\mathrm{d}^3p'\\ &\times\delta\left(x-\frac{\boldsymbol{k}'\cdot\boldsymbol{p}}{|\boldsymbol{p}|^2}\right)\delta^3(\boldsymbol{k}'+\boldsymbol{p}'-\boldsymbol{p})\frac{1}{p^3x(1-x)}\\ &\times\frac{4p^2}{|\boldsymbol{k}'_\perp|^4}x^2(1-x)^2\cdot\frac{|\boldsymbol{k}'_\perp|^2}{x^2}\left(1+\frac{x^2}{2(1-x)}\right)\\ =&\frac{4}{3}\frac{1}{(2\pi)^3}g^2(Q^2)\frac{1+(1-x)^2}{x}\int\frac{\mathrm{d}^2k'_\perp}{|\boldsymbol{k}'_\perp|^2}\;. \end{aligned} \tag{24}$$

In the last step we wrote $\mathrm{d}^3k'=\mathrm{d}^2k'_\perp\,\mathrm{d}k_\parallel$ and performed the integration over $\boldsymbol{p}'$ and $k_\parallel$, thus cancelling the δ functions.

Example 5.5

The last integration has to be performed for transverse momenta $|\boldsymbol{k}'_\perp|$ between εQ and $\varepsilon Q'$ (see (3)–(5) and Exercise 5.4):

$$\begin{aligned}\int_{\varepsilon Q}^{\varepsilon Q'} \frac{d^2k'_\perp}{|\boldsymbol{k}'_\perp|^2} &= 2\pi \int_{\varepsilon Q}^{\varepsilon Q'} \frac{dk'_\perp}{|\boldsymbol{k}'_\perp|} = 2\pi \ln |k'_\perp| \Bigg|_{\varepsilon Q}^{\varepsilon Q'} = 2\pi(\ln \varepsilon Q' - \ln \varepsilon Q) \\ &= \pi(\ln \varepsilon^2 Q'^2 - \ln \varepsilon^2 Q^2) \\ &= \pi\Delta \ln \varepsilon^2 Q^2 = \pi\Delta \ln Q^2 \ . \end{aligned} \tag{25}$$

Q^2 is preferred in the argument of the logarithm, because it is this quantity which enters the GLAP equations (5.75) and (5.76). Since ε was introduced as an arbitrary constant $\Delta \ln \varepsilon = 0$. We are left with

$$N_G(x, Q^2) = \frac{g^2(Q^2)}{8\pi^2} \frac{4}{3} \frac{1+(1-x)^2}{x} \Delta \ln Q^2 \ . \tag{26}$$

Here we introduced again Q^2 instead of Q'^2 in the argument of $N_G(x, Q^2)$, because Q'^2 is hidden in $\Delta \ln Q^2$ – see (25). Both notations will be used in the following. This result now has to be compared to the GLAP equation (5.76), where we have to impose the initial condition

$$\begin{aligned} &N_q(x, Q^2) = \delta(1-x) \ , \\ &N_G(x, Q^2) = 0 \ , \end{aligned} \tag{27}$$

which is in accordance with the initial quark state $|q_{Q^2}(p)\rangle$ in (12) where gluons are absent. From this condition we evolve the parton distribution function to $N_G(x, Q^2)$. The term "initial condition" has to be understood in that way that one has distribution functions at a virtuality Q^2, from which one determines the distribution functions at the higher virtuality $Q'^2 > Q^2$.

From the GLAP equations (5.76) it follows that

$$\begin{aligned} N_G(x, Q'^2) &= N_G(x, Q^2) + \Delta N_G(x, Q^2) \\ &= \Delta N_G(x, Q^2) \\ &= \frac{g^2}{8\pi^2} P_{Gq}(x) \Delta \ln Q^2 \ . \end{aligned} \tag{28}$$

Therefore we can conclude that

$$\Delta N_G(x, Q^2) = \frac{g^2}{8\pi^2} P_{Gq}(x) \Delta \ln Q^2 \tag{29}$$

with the P_{Gq} splitting function given by

$$P_{Gq}(x) = \frac{4}{3} \frac{1+(1-x)^2}{x} \ . \tag{30}$$

EXAMPLE

5.6 Derivation of the Splitting Function P_{qq}

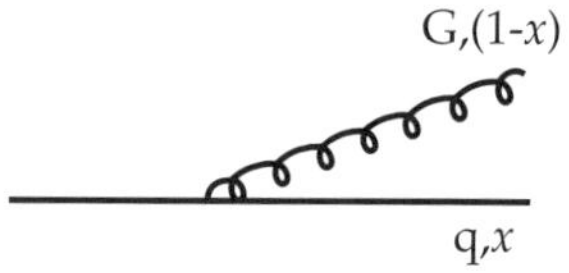

Fig. 5.13.

Up to now we have determined only the gluon distribution function $N_G(x, Q'^2)$ (see (28) of Example 5.5). To determine the quark momentum distribution $N_q(x, Q'^2)$ we have to keep in mind that the production of a quark of momentum fraction x corresponds to the production of a gluon of momentum fraction $1-x$ (see Fig. 5.13). For $x = 1$ the original quark is obtained with a probability $|C_q(p)|^2$. We can therefore write

$$\begin{aligned} N_q(x, Q'^2) &= N_q(x, Q^2) + \Delta N_q(x, Q^2) \\ &= \Delta N_G(1-x, Q^2) + \delta(1-x)|C_q(p)|^2 \ . \end{aligned} \tag{1}$$

The first term on the right-hand side denotes the probability for having a gluon with virtuality Q^2 and momentum fraction $1-x$; the second term describes the probability for having no gluon at all. The change in the quark distribution function is then given by (see (27) of Example 5.5)

$$\begin{aligned} \Delta N_q(x, Q^2) &= N_q(x, Q'^2) - N_q(x, Q^2) = N_q(x, Q'^2) - \delta(1-x) \\ &= \Delta N_G(1-x, Q^2) - \delta(1-x)\left[1 - |C_q(p)|^2\right] \ . \end{aligned} \tag{2}$$

$\Delta N_G(1-x, Q^2)$ is already known from Example 5.5. Furthermore the longitudinal momentum is conserved. In other words, the longitudinal momentum of the newly created quarks, $x\Delta N_q(x, Q^2)$, together with the total longitudinal momentum of the created gluons, $x\Delta N_G(x, Q^2)$, should vanish, i.e. the expectation value of the longitudinal momentum should equal zero:

$$\int_0^1 \mathrm{d}x\, x(\Delta N_q(x, Q^2) + \Delta N_G(x, Q^2)) = 0 \ . \tag{3}$$

If we use the result of the last section, that the change in the gluon distribution is given by the splitting function P_{Gq},

$$\begin{aligned} \Delta N_G(1-x, Q^2) &= \frac{g^2(Q^2)}{8\pi^2}\frac{4}{3}\frac{1+x^2}{1-x}\Delta \ln Q^2 \ , \\ \Delta N_G(x, Q^2) &= \frac{g^2(Q^2)}{8\pi^2}\frac{4}{3}\frac{1+(1-x)^2}{x}\Delta \ln Q^2 \ , \end{aligned} \tag{4}$$

momentum conservation allows us to determine the coefficient $|C_q(p)|^2$:

$$0 = \int_0^1 \mathrm{d}x\, x\left[\Delta N_G(1-x, Q^2) - \delta(1-x)(1-|C_q(p)|^2) + \Delta N_G(x, Q^2)\right] \tag{5}$$

or *Example 5.6*

$$|C_q(p)|^2 = 1 - \int_0^1 \mathrm{d}x\, x \left[\Delta N_G(1-x, Q^2) + \Delta N_G(x, Q^2)\right]$$

$$= 1 - \frac{g^2(Q^2)}{8\pi^2} \Delta \ln Q^2 \frac{4}{3} \int_0^1 \mathrm{d}x\, x \left(\frac{1+(1-x)^2}{x} + \frac{1+x^2}{1-x} \right)$$

$$= 1 - \frac{g^2(Q^2)}{8\pi^2} \Delta \ln Q^2 \left(\frac{8}{3} \int_0^1 \mathrm{d}x \frac{1}{1-x} - 2 \right) . \tag{6}$$

Finally we get for the change in the quark distribution (2)

$$\Delta N_q(x, Q^2) = \frac{g^2(Q^2)}{8\pi^2} \Delta \ln Q^2 \left[\frac{4}{3} \frac{1+x^2}{1-x} + 2\delta(1-x) \right.$$
$$\left. - \frac{8}{3}\, \delta(1-x) \int_0^1 \mathrm{d}x' \frac{1}{1-x'} \right] . \tag{7}$$

Comparing this to the GLAP equation (see (28) in the previous example, which can be directly translated into a similar equation for ΔN_q; see also (5.74)),

$$\Delta N_q(x, Q^2) = \frac{g^2(Q^2)}{8\pi^2} P_{qq}(x) \Delta \ln Q^2 , \tag{8}$$

allows us to extract the splitting function

$$P_{qq} = \frac{4}{3} \frac{1+x^2}{1-x} + 2\delta(1-x) - \frac{8}{3} \delta(1-x) \int_0^1 \mathrm{d}x' \frac{1}{1-x'}$$
$$= -\frac{4}{3}(x+1) + 2\delta(1-x) + \frac{8}{3} \left(\frac{1}{1-x} - \delta(1-x) \int_0^1 \mathrm{d}x' \frac{1}{1-x'} \right) . \tag{9}$$

EXAMPLE

5.7 Derivation of the Splitting Function P_{qG}

Next we calculate the splitting function that describes the quark fraction inside a gluon when the resolving power is increased from Q^2 to Q'^2 (Fig. 5.11c). The final-state wave function is given by

$$|q'_{Q'^2}(p) \overline{q}_{Q'^2}(p')\rangle . \tag{1}$$

Example 5.7

The initial gluon wave function at Q^2 can be expanded into these quarks states:

$$|G_{Q^2}(k)\rangle = \sum_{p,p'} C_{q\overline{q}}(p,p')|q_{Q'^2}(p)\overline{q}_{Q'^2}(p')\rangle \ , \tag{2}$$

where the coefficients are given by

$$C_{q\overline{q}}(p,p') = \langle q_{Q'^2}(p)\overline{q}_{Q'^2}(p')|G_{Q^2}(k)\rangle \ . \tag{3}$$

Applying the kinetic part of the Hamiltonian (see (4) of Example 5.5) yields

$$\hat{H}_{Q^2}|G_{Q^2}(k)\rangle + \Delta\hat{H}_{Q^2}|G_{Q^2}(k)\rangle = \sum_{p,p'} C_{q\overline{q}}(p,p')\hat{H}_{Q'^2}|q_{Q'^2}(p)\overline{q}_{Q'^2}(p')\rangle \tag{4}$$

or

$$\begin{aligned} &k|G_{Q^2}(k)\rangle + \Delta\hat{H}_{Q^2}|G_{Q^2}(k)\rangle \\ &\quad = \sum_{p,p'}(|\boldsymbol{p}|+|\boldsymbol{p}'|)\times C_{q\overline{q}}(p,p')|q_{Q'^2}(p)\overline{q}_{Q'^2}(p')\rangle \ . \end{aligned} \tag{5}$$

The coefficients can be obtained by projecting with $\langle q_{Q'^2}(p)\overline{q}_{Q'^2}(p')|$:

$$C_{q\overline{q}}(p,p') = \frac{\langle q_{Q'^2}(p)\overline{q}_{Q'^2}(p')|\Delta\hat{H}_{Q^2}|G_{Q^2}(k)\rangle}{|\boldsymbol{p}|+|\boldsymbol{p}'|-|\boldsymbol{k}|} \ . \tag{6}$$

To determine $C_{q\overline{q}}(pp')$ we need from the Hamiltonian $\Delta\hat{H}_{Q^2}$ (see (6) of Example 5.5) the part that describes the quark–gluon interaction:

$$\Delta\hat{H}_{Q^2} = -g^2(Q^2)\sum_a \int \mathrm{d}^3x \sum_f \overline{q}_f(\boldsymbol{r})\boldsymbol{A}^a(\boldsymbol{r})\boldsymbol{\gamma}\frac{\lambda^a}{2}q_f(\boldsymbol{r}) . \tag{7}$$

As in (15) in Example 5.5 it follows for the matrix element that

$$\begin{aligned} &\langle q_{Q'^2}(p)\overline{q}_{Q'^2}(p')|\Delta\hat{H}_{Q^2}|G_{Q^2}(k)\rangle \\ &\quad = -\sum_a g(Q^2)(2\pi)^3\delta^3(\boldsymbol{p}+\boldsymbol{p}'-\boldsymbol{k})\overline{u}(\boldsymbol{p},s)\boldsymbol{\varepsilon}^a\cdot\boldsymbol{\gamma}\frac{\lambda^a}{2}v(\boldsymbol{p}',s') \ . \end{aligned} \tag{8}$$

Again we can relate the squared coefficient $C_{q\overline{q}}(p,p')$ to the quark distribution function by calculating the "decay probability" of a gluon if the resolving power Q^2 is increased. Following our previous calculation (see (14) of Example 5.5) we start with a single gluon $|G_{Q^2}(k)\rangle$ and no quarks at the resolution Q^2. Increasing the resolution to Q'^2 will generate quarks through the process of pair production and we obtain the following quark distribution function:

$$\begin{aligned} \Delta N_q(c,Q'^2) = &\frac{1}{8}\sum_a \frac{1}{2}\sum_{\varepsilon^a}\sum_{cs}\sum_{c's'}\frac{1}{2\omega}\int\frac{\mathrm{d}^3p}{(2\pi)^3 2E}\int\frac{\mathrm{d}^3p'}{(2\pi)^3 2E'} \\ &\times\delta\left(x-\frac{\boldsymbol{p}\cdot\boldsymbol{k}}{|\boldsymbol{k}|^2}\right)|C_{q\overline{q}}(p,p')|^2 \ . \end{aligned} \tag{9}$$

Example 5.7

This is summed over spins (s, s') and colors (c, c') (factor 1/8) of the outgoing quarks and it is averaged over the spin ε^a (factor 1/2) and color of the initial gluon. Remember that the gluon, being a vector particle, has spin 1. However, being massless, the gluon has only two (transverse) spin directions, as is the case for the photon. The δ function guarantees that the quark longitudinal momentum is x times the gluon momentum:

$$x = \frac{p''}{|\boldsymbol{k}|} = \frac{\boldsymbol{p}\cdot\boldsymbol{k}}{|\boldsymbol{k}|^2} \ . \tag{10}$$

Inserting $C_{q\bar{q}}(p, p')$ from (6) into (9), we find that (9) becomes

$$\begin{aligned}\Delta N_q(c, Q^2) &= g(Q^2)^2 \frac{1}{16}(2\pi)^3 \int \frac{\mathrm{d}^3 p}{(2\pi)^3} \int \frac{\mathrm{d}^3 p'}{(2\pi)^3} \\ &\quad \times \delta\left(x - \frac{\boldsymbol{p}\cdot\boldsymbol{k}}{|\boldsymbol{k}|^2}\right) \delta^3(\boldsymbol{p}+\boldsymbol{p}'-\boldsymbol{k}) \frac{1}{8\omega E E'} \left(\frac{1}{|\boldsymbol{p}|+|\boldsymbol{p}'|-|\boldsymbol{k}|}\right)^2 \\ &\quad \times \sum_a \sum_{\varepsilon^a} \sum_{cs} \sum_{c's'} |\bar{u}_c(\boldsymbol{p}, s)\boldsymbol{\varepsilon}^a \cdot \boldsymbol{\gamma} \frac{\lambda^a}{2} v_{c'} A(\boldsymbol{p}', s')|^2 \ . \end{aligned} \tag{11}$$

We first perform the spin and color summations, again using $\not{\varepsilon}^a = -\boldsymbol{\gamma}\cdot\boldsymbol{\varepsilon}$ for $\varepsilon^a = (0, \boldsymbol{\varepsilon}^a)$:

$$\begin{aligned} &\sum_a \sum_{\varepsilon^a} \sum_{cs} \sum_{c's'} |\bar{u}_c(\boldsymbol{p}, s)\boldsymbol{\varepsilon}^a \cdot \boldsymbol{\gamma} \frac{\lambda^a}{2} v_{c'}(\boldsymbol{p}', s')|^2 \\ &= \sum_{a,b} \mathrm{tr}\left(\frac{\lambda^a \lambda^b}{4}\right) \sum_{\varepsilon^a} \mathrm{tr}[\not{p}(\boldsymbol{\varepsilon}^a\cdot\boldsymbol{\gamma})\not{p}'(\boldsymbol{\varepsilon}^{b*}\cdot\boldsymbol{\gamma})] \\ &= \sum_a \frac{1}{2} \sum_{\varepsilon^a} \mathrm{tr}\left(\not{p}\not{\varepsilon}^a \not{p}' \not{\varepsilon}^{a*}\right) \\ &= \sum_a \frac{1}{2} \sum_{\varepsilon^a} 4\cdot[(p\cdot\varepsilon^a)(p'\cdot\varepsilon^{a*}) + (p'\cdot\varepsilon^a)(p\cdot\varepsilon^{a*}) - (p\cdot p')(\varepsilon^a\cdot\varepsilon^{a*})] \\ &= 2\sum_a \sum_{\varepsilon^a} [(\boldsymbol{p}\cdot\boldsymbol{\varepsilon}^a)(\boldsymbol{p}'\cdot\boldsymbol{\varepsilon}^{a*}) \\ &\quad + (\boldsymbol{p}'\cdot\boldsymbol{\varepsilon}^a)(\boldsymbol{p}\cdot\boldsymbol{\varepsilon}^{a*}) - (p_0 p_0' - \boldsymbol{p}\cdot\boldsymbol{p}')(\boldsymbol{\varepsilon}^a\cdot\boldsymbol{\varepsilon}^{a*})] \\ &= 16\left(\delta_{ij} - \frac{k_i k_j}{|\boldsymbol{k}|^2}\right)(p_i p_j' + p_i' p_j + (|\boldsymbol{p}||\boldsymbol{p}'| - \boldsymbol{p}\cdot\boldsymbol{p}')\delta_{ij}) \\ &= \frac{32}{|\boldsymbol{k}|^2}\left[|\boldsymbol{k}|^2|\boldsymbol{p}||\boldsymbol{p}'| - (\boldsymbol{k}\cdot\boldsymbol{p})(\boldsymbol{k}\cdot\boldsymbol{p}')\right] \ . \end{aligned} \tag{12}$$

Example 5.7

The summation over the colors a is performed exactly the same way as was done in Example 5.5, (17). The quark distribution function can now be written as

$$\Delta N_q(c, Q^2) = g(Q^2)^2 \frac{1}{4}(2\pi)^3 \int \mathrm{d}^3 p \delta \left(x - \frac{\boldsymbol{p} \cdot \boldsymbol{k}}{|\boldsymbol{k}|^2} \right) \frac{1}{8\omega E E'} \times \left(\frac{1}{|\boldsymbol{p}| + |\boldsymbol{p}'| - |\boldsymbol{k}|} \right)^2 \frac{1}{|\boldsymbol{k}|^2} (|\boldsymbol{k}|^2 |\boldsymbol{p}||\boldsymbol{p}'| - (\boldsymbol{k} \cdot \boldsymbol{p})(\boldsymbol{k} \cdot \boldsymbol{p}')) \ . \tag{13}$$

Here $\boldsymbol{p}' = \boldsymbol{k} - \boldsymbol{p}$ is due to the $\delta^3(\boldsymbol{p} + \boldsymbol{p}' - \boldsymbol{k})$ function. This conservation of the three-momentum will now be handled explicitly. We introduce explicit forms for the gluon and quark momenta,

$$\begin{aligned} \boldsymbol{k} &= (0, 0, k) \ , \\ \boldsymbol{p} &= (\boldsymbol{k}_\perp, xk) \ , \\ \boldsymbol{p}' &= (-\boldsymbol{k}_\perp, (1-x)k) \ , \end{aligned} \tag{14}$$

so that the three-momentum is conserved: $\boldsymbol{k} = \boldsymbol{p} + \boldsymbol{p}'$. As in the last sections we take $|\boldsymbol{k}_\perp|$ to lie between εQ and $\varepsilon Q'$ and assume that all momenta are much greater than $|\boldsymbol{k}_\perp|^2$:

$$|\boldsymbol{k}| \gg |\boldsymbol{k}_\perp| \ , \quad |\boldsymbol{p}| \gg |\boldsymbol{k}_\perp| \ , \quad |\boldsymbol{p}'| \gg |\boldsymbol{k}_\perp| \ . \tag{15}$$

The terms that enter (13) can now be approximated as in previous sections (see (25) in Example 5.5):

$$\begin{aligned} \frac{1}{|\boldsymbol{p}| + |\boldsymbol{p}'| - |\boldsymbol{k}|} &= \frac{1}{\sqrt{\boldsymbol{k}_\perp^2 + x^2 \boldsymbol{k}^2} + \sqrt{\boldsymbol{k}_\perp^2 + (1-x)^2 \boldsymbol{k}^2} - |\boldsymbol{k}|} \\ &= \frac{2k}{|\boldsymbol{k}_\perp|} x(1-x) \ , \end{aligned} \tag{16}$$

$$\frac{1}{\omega E E'} = \frac{1}{x(1-x)k^3} \ , \qquad \omega = k \ , \tag{17}$$

$$\frac{1}{|\boldsymbol{k}|^2} (|\boldsymbol{k}|^2 |\boldsymbol{p}||\boldsymbol{p}'| - (\boldsymbol{k} \cdot \boldsymbol{p})(\boldsymbol{k} \cdot \boldsymbol{p}')) = \frac{1}{2} |\boldsymbol{k}_\perp|^2 \frac{(1-x)^2 + x^2}{x(1-x)} \ . \tag{18}$$

Inserting these expressions into (13) and rewriting

$$\mathrm{d}^3 p = \mathrm{d}^2 k_\perp \, \mathrm{d} p_\parallel \tag{19}$$

we can perform the integration over the longitudinal momentum, obtaining

$$\begin{aligned} \Delta N_q(x, Q'^2) &= \frac{g(Q^2)^2}{8\pi^3} \int \frac{\mathrm{d}^2 k_\perp}{|\boldsymbol{k}_\perp|^2} \frac{1}{2} \left((1-x)^2 + x^2 \right) \\ &= \frac{g(Q^2)^2}{4\pi^2} \int \frac{\mathrm{d} k_\perp}{|\boldsymbol{k}_\perp|} \frac{1}{2} \left((1-x)^2 + x^2 \right) \ . \end{aligned} \tag{20}$$

Example 5.7

Keeping in mind that $|\boldsymbol{k}_\perp|$ is in the range εQ to $\varepsilon Q'$, we can perform the last integration. The result is

$$\begin{aligned} \Delta N_q(x, Q'^2) &= \frac{g(Q^2)^2}{4\pi^2}\,\frac{1}{2}\left((1-x)^2+x^2\right)\ln|\boldsymbol{k}_\perp|\Big|_{\varepsilon Q}^{\varepsilon Q'} \\ &= \frac{g(Q^2)^2}{8\pi^2}\,\frac{1}{2}\left((1-x)^2+x^2\right)\Delta\ln Q^2\ , \end{aligned} \tag{21}$$

since ε is an arbitrary constant $\Delta\ln\epsilon = 0$. Comparing this result with the GLAP equation (see (5.74)) yields, with the initial condition

$$N_G(x, Q^2) = \delta(1-x)\ , \quad N_q(x, Q^2) = 0\ , \tag{22}$$

the expression for the splitting function:

$$P_{qG}(x) = \frac{1}{2}\left((1-x)^2+x^2\right)\ . \tag{23}$$

5.3 Common Parametrizations of the Distribution Functions and Anomalous Dimensions

Instead of dealing with the GLAP equations (5.76) we can transform the set of integral equations into a set of linear equations by introducing the moments of the structure functions. The nth moment is defined as[8]

$$M_a(n, Q^2) = \int_0^1 \mathrm{d}x\, x^{n-1} N_a(x, Q^2)\ , \tag{5.84}$$

where N_a is the structure function of the parton of kind a. Differentiating this with respect to $\ln(Q^2)$ and inserting the GLAP equations (5.76) results in

$$\begin{aligned} \frac{\mathrm{d}M_a(n, Q^2)}{\mathrm{d}\ln Q^2} &= \int_0^1 \mathrm{d}x\, x^{n-1}\frac{\mathrm{d}N_a(x, Q^2)}{\mathrm{d}\ln Q^2} \\ &= \int_0^1 \mathrm{d}x\, x^{n-1}\frac{g^2(Q^2)}{8\pi^2}\sum_b\int_x^1\frac{\mathrm{d}x'}{x'}P_{ab}\left(\frac{x}{x'}\right)N_b(x', Q^2) \\ &= \frac{g_s^2(Q^2)}{8\pi^2}\sum_b\int_0^1 \mathrm{d}x\, x^{n-1}\int_x^1\frac{\mathrm{d}x'}{x'}P_{ab}\left(\frac{x}{x'}\right)N_b(x', Q^2)\ . \end{aligned} \tag{5.85}$$

[8] Usually one defines the nth moment as the expectation value of x^n rather than x^{n-1}, which is commonly chosen in QCD.

The second integration over x' between x and 1 can be performed from $x' = 0$ to $x' = 1$ if one introduces the Heavyside function in the integrand. Thus the integrand vanishes for $x' < x$:

$$\frac{\mathrm{d}M_a(n,Q^2)}{\mathrm{d}\ln Q^2} = \frac{g^2(Q^2)}{8\pi^2} \sum_b \left(\int_0^1 \mathrm{d}x\, x^{n-1} \right.$$
$$\left. \times \int_0^1 \frac{\mathrm{d}x'}{x'} P_{ab}\left(\frac{x}{x'}\right) \Theta(x'-x) N_b(x',Q^2) \right) . \tag{5.86}$$

Now the integrations can be exchanged. Substituting $z = x/x'$ we get

$$\begin{aligned}
\frac{\mathrm{d}M_a(n,Q^2)}{\mathrm{d}\ln Q^2} &= \frac{g^2(Q^2)}{8\pi^2} \sum_b \left(\int_0^1 \frac{\mathrm{d}x'}{x'} N_b(x',Q^2) \right. \\
&\quad \left. \times \int_0^1 \mathrm{d}x\, x^{n-1} P_{ab}\left(\frac{x}{x'}\right) \Theta(x'-x) \right) \\
&= \frac{g^2(Q^2)}{8\pi^2} \sum_b \left(\int_0^1 \frac{\mathrm{d}x'}{x'} N_b(x',Q^2) \right. \\
&\quad \left. \times \int_0^{1/x'} \mathrm{d}z\, x' (zx')^{n-1} P_{ab}(z) \Theta(x'-x'z) \right) \\
&= \frac{g^2(Q^2)}{8\pi^2} \sum_b \left(\int_0^1 \mathrm{d}x' x'^{n-1}\, N_b(x',Q^2) \right. \\
&\quad \left. \times \int_0^{1/x'} \mathrm{d}z\, z^{n-1} P_{ab}(z) \Theta(1-z) \right) \\
&= \frac{g^2(Q^2)}{8\pi^2} \sum_b \int_0^1 \mathrm{d}x' x'^{n-1}\, N_b(x',Q^2) \int_0^1 \mathrm{d}z\, z^{n-1}\, P_{ab}(z) \; .
\end{aligned} \tag{5.87}$$

The last step can be performed since $1/x' \geq 1$ (because $x' \leq 1$) and the Θ function contributes only for $z = 1/x' \leq 1$. Therefore the upper bound of the integration can be taken as 1.

In the high-Q^2 regime $\alpha_s(Q^2) = g_s(Q^2)^2/4\pi$ is given in leading-log order (see (4.147) which has to be taken at $\mu^2 = Q^2$) by

$$\alpha_s(Q^2) = \frac{12}{33-2N_\mathrm{f}} \frac{\pi}{\ln(Q^2/\Lambda^2)} , \tag{5.88}$$

where N_f denotes the number of quark flavors and Λ is the QCD cut-off parameter. We obtain (see Example 4.4)

$$\begin{aligned} \frac{\mathrm{d}M_a(n,Q^2)}{\mathrm{d}\ln Q^2} &= \frac{6}{33-2N_\mathrm{f}} \frac{1}{\ln(Q^2/\Lambda^2)} \sum_b M_b(n,Q^2) \int\limits_0^1 \mathrm{d}z\, z^{n-1}\, P_{ab}(z) \\ &= -\frac{1}{\ln(Q^2/\Lambda^2)} \sum_b M_b(n,Q^2) d_{ab}(n) , \end{aligned} \tag{5.89}$$

where we have abbreviated the moment of the splitting function $P_{ab}(z)$ by

$$d_{ab}(n) = -\frac{6}{33-2N_\mathrm{f}} \int\limits_0^1 \mathrm{d}z\, z^{n-1}\, P_{ab}(z) . \tag{5.90}$$

Thus we have derived a system of linear equations between the moments of the structure functions and their derivatives. The moments $d_{ab}(n)$ of the splitting functions are the anomalous dimensions. We shall come back to that in the discussion of the renormalization group (Sect. 5.4). Once the splitting functions are given, the moments $d_{ab}(n)$ can easily be obtained (see Exercise 5.8):

$$\begin{aligned} d_{Gq}(n) &= -\frac{8}{33-2N_\mathrm{f}} \frac{n^2+n+2}{n(n^2-1)} , \\ d_{qq}(n) &= \frac{4}{33-2N_\mathrm{f}} \left(1 - \frac{2}{n(n+1)} + 4\sum_{j=2}^{n} \frac{1}{j}\right) , \\ d_{qG}(n) &= -\frac{3}{33-2N_\mathrm{f}} \frac{n^2+n+2}{n(n+1)(n+2)} , \\ d_{GG}(n) &= \frac{9}{33-2N_\mathrm{f}} \left(\frac{1}{3} + \frac{2N_\mathrm{f}}{9} - \frac{4}{n(n-1)} - \frac{4}{(n+1)(n+2)} + 4\sum_{j=2}^{n} \frac{1}{j}\right) . \end{aligned} \tag{5.91}$$

The moments $M(n,Q^2)$ of the parton distributions $N(x,Q^2)$ are given by experiment. However, it is more difficult to calculate the parton distributions $N(x)$ for given moments $M(n)$. Here we deleted the Q^2 dependence in the argument, because the relation between the distribution function $N(x,Q^2)$ and the momenta $M(n,Q^2)$ is shown at the renormalization point $Q^2=\mu^2$. At this special point the Q^2 dependence drops out. After $N(x,Q^2=\mu^2)$ is known one may obtain $N(x,Q^2)$ at any other Q^2 by applying the GLAP evolution; see, e.g., (5.86).

Since the polynomials x^n do not form an orthonormal system of functions in the range $0 \le x \le 1$, one has to transform the moments $M(n)$ into coefficients

$$c_i = f\big(M(n)\big) \ , \tag{5.92}$$

so that

$$N(x) = \sum_i c_i x^i \ . \tag{5.93}$$

The coefficients c_i are functions of the moments $N(n)$ and can be obtained from the relation

$$M(n) = \sum_i \frac{1}{i+n} c_i \ , \tag{5.94}$$

which is proven as follows.

Let $f(x)$ be the function that is to be approximated by $g(x) = \sum_i c_i x^i$. Again the moments of $f(x)$ are denoted by

$$f_n = \int\limits_0^1 \mathrm{d}x\, x^{n-1} f(x) \ . \tag{5.95}$$

We require that

$$\mathcal{M} = \int\limits_0^1 \mathrm{d}x \big(f(x) - g(x)\big)^2 \tag{5.96}$$

becomes minimal, in which case the variation of $\mathcal{M}$ with respect to the coefficients c_k vanishes:

$$0 = \delta\mathcal{M} = \frac{\partial \mathcal{M}}{\partial c_k} \delta c_k \ , \tag{5.97}$$

from which it follows that

$$\frac{\partial \mathcal{M}}{\partial c_k} = 0 \quad \text{for all } k \ . \tag{5.98}$$

Inserting the series expansion for $g(x)$ into (5.93), we can express $\mathcal{M}$ by the coefficients c_i and the moments f:

$$\begin{aligned}
\mathcal{M} &= \int\limits_0^1 \mathrm{d}x \Big(f(x)^2 + g(x)^2 - 2 f(x) g(x) \Big) \\
&= \int\limits_0^1 \mathrm{d}x\ f(x)^2 + \int\limits_0^1 \mathrm{d}x\ g(x)^2 - 2 \int\limits_0^1 \mathrm{d}x\ f(x) g(x)
\end{aligned}$$

$$= C + \sum_{i,j} c_i c_j \int_0^1 \mathrm{d}x\, x^{i+j} - 2\sum_i c_i \int_0^1 \mathrm{d}x\, x^i f(x)$$
$$= C + \sum_{i,j} \frac{c_i c_j}{i+j+1} - 2\sum_i c_i f_{i+1} \;. \tag{5.99}$$

$C = \int_0^1 \mathrm{d}x\, f(x)^2$ is a constant.

Performing the derivation with respect to c_k we get the desired relation between the coefficients c_i and the moments f_i of the function f:

$$\begin{aligned} 0 &= \frac{\partial \mathcal{M}}{\partial c_k} \\ &= \frac{\partial}{\partial c_k}\left(C + \sum_{i,j} \frac{c_i c_j}{i+j+1} - 2\sum_i c_i f_{i+1} \right) \\ &= \sum_{i,j} \frac{\partial}{\partial c_k} \frac{c_i c_j}{i+j+1} - 2\sum_i \frac{\partial c_i}{\partial c_k} f_{i+1} \\ &= \sum_{i,j} \frac{\delta_{ik} c_j + c_i \delta_{jk}}{i+j+1} - 2\sum_i \delta_{ik} f_{i+1} \\ &= 2\sum_i \frac{c_i}{i+k+1} - 2 f_{k+1} \;, \end{aligned} \tag{5.100}$$

or

$$f_k = \sum_i \frac{c_i}{i+k} \equiv \sum_i A_{ki} c_i \;. \tag{5.101}$$

In order to compute the coefficients c_i from the known moments f_k, one has to determine the inverse of the matrix A_{ki}:

$$c_i = \left(A^{-1}\right)_{ik} f_k \;. \tag{5.102}$$

Obviously the coefficients c_i change their values if one more moment f_k is added in the calculation.

It is possible to give an upper and lower bound to $f(x)$ by employing the Padé approximation method for the moments f_k.[9] Recent parametrizations for the parton distributions inside the nucleon are given by CTEQ. (The CTEQ group provides continuously updated sets of distribution functions, obtainable via WWW from http://www.phys.psu.edu/~cteq/, which is regularly updated.) Here we give another parametrization.[10] Writing for s in a leading-log approximation

$$s = \ln \frac{\ln\left[Q^2/(0.232\,\mathrm{GeV})^2\right]}{\ln\left[\mu^2/(0.232\,\mathrm{GeV})^2\right]} \;, \tag{5.103}$$

[9] W.W. Press, B.P. Flannery, S.A. Teukolsky, W.T. Vetterling: *Numerical Recipes* (Cambridge University Press 1992)
[10] M. Glück, E. Reya, A. Vogt: Z. Physik C **53**, 127 (1992)

where $\mu^2 = 0.25\ \text{GeV}^2$, the parton distributions are parametrized as follows. The distribution function for the u and d valence quarks in a proton are denoted by $v_p(x, Q^2)$. We use the parametrization

$$xv_p(x, Q^2) = Nx^a\big(1 + A\sqrt{x} + Bx\big)(1-x)^D \ , \tag{5.104}$$

with

$$\begin{aligned} N &= 0.663 + 0.191s - 0.041s^2 + 0.031s^3 \ , \\ a &= 0.326 \ , \\ A &= -1.97 + 6.74s - 1.96s^2 \ , \\ B &= 24.4 - 20.7s + 4.08s^2 \ , \\ D &= 2.86 + 0.70s - 0.02s^2 \ , \end{aligned} \tag{5.105}$$

for $v = u_v + d_v$ and

$$\begin{aligned} N &= 0.579 + 0.283s + 0.047s^2 \ , \\ a &= 0.523 - 0.015s \ , \\ A &= 2.22 - 0.59s - 0.27s^2 \ , \\ B &= 5.95 - 6.19s + 1.55s^2 \ , \\ D &= 3.57 + 0.94s - 0.16s^2 \ , \end{aligned} \tag{5.106}$$

for $v = d_v$. The distributions for a neutron are obtained by interchanging u and d quarks.

The gluon and *light-sea-quark distributions* are given by

$$\begin{aligned} xw_\mathrm{p}(x, Q^2) = \Bigg[& x^\alpha \left(A + Bx + Cx^2\right) \ln\left(\frac{1}{x}\right)^b \\ & + s^\alpha \exp\left(-E + \sqrt{E' s^\beta \ln\left(\frac{1}{x}\right)}\right)\Bigg] (1-x)^D \ , \end{aligned} \tag{5.107}$$

with

$$\begin{aligned} \alpha &= 0.558 \ , \\ \beta &= 1.218 \ , \\ a &= 1.00 - 0.17s \ , \\ b &= 0 \ , \\ A &= 4.879s - 1.383s^2 \ , \\ B &= 25.92 - 28.97s + 5.596s^2 \ , \\ C &= -25.69 + 23.68s - 1.975s^2 \ , \\ D &= 2.537 + 1.718s + 0.353s^2 \ , \\ E &= 0.595 + 2.138s \ , \\ E' &= 4.066 \end{aligned} \tag{5.108}$$

for $w = G$ and

$$
\begin{aligned}
\alpha &= 1.396 \ , \\
\beta &= 1.331 \ , \\
a &= 0.412 - 0.171s \ , \\
b &= 0.566 - 0.496s \ , \\
A &= 0.363 \ , \\
B &= -1.196 \ , \\
C &= 1.029 + 1.785s - 0.459s^2 \ , \\
D &= 4.696 + 2.109s \ , \\
E &= 3.838 + 1.944s \ , \\
E' &= 2.845
\end{aligned}
\tag{5.109}
$$

for $w = \bar{u}$ and $w = \bar{d}$, characterizing the light sea quarks. Finally for the s, c, b quark distributions, we employ the relation

$$
\begin{aligned}
xw'_{\mathrm{p}}(x, Q^2) = {} & \frac{(s - s_{w'})^\alpha}{[\ln(1/x)]^a} \left(1 + A\sqrt{x} + Bx\right)(1-x)^D \\
& \times \exp\left(-E + \sqrt{E' s^\beta \ln\left(\frac{1}{x}\right)}\right) \ ,
\end{aligned}
\tag{5.110}
$$

where the *s-quark distribution* is given by

$$
\begin{aligned}
\alpha &= 0.803 \ , \\
\beta &= 0.563 \ , \\
a &= 2.082 - 0.577s \ , \\
A &= -3.055 + 1.024s^{0.67} \ , \\
B &= 27.4 - 20.0s^{0.154} \ , \\
D &= 6.22 \ , \\
E &= 4.33 + 1.408s \ , \\
E' &= 8.27 - 0.437s \ , \\
s_{w'} &= s_{\mathrm{s}} = 0 \ .
\end{aligned}
\tag{5.111}
$$

The *c, $\bar{c}$ distribution* is parametrized by

$$
\begin{aligned}
\alpha &= 1.01 \ , \\
\beta &= 0.37 \ , \\
a &= 0 \ , \\
A &= 0 \ , \\
B &= 4.24 - 0.804s \ , \\
D &= 3.46 + 1.076s \ ,
\end{aligned}
$$

$$\begin{aligned} E &= 4.61 + 1.49s \ , \\ E' &= 2.555 + 1.961s \ , \\ s_{w'} &= s_c = 0.888 \ , \end{aligned} \tag{5.112}$$

and the $b,\bar{b}$ *distribution* by

$$\begin{aligned} \alpha &= 1.00 \ , \\ \beta &= 0.51 \ , \\ a &= 0 \ , \\ A &= 0 \ , \\ B &= 1.848 \ , \\ D &= 2.929 + 1.396s \ , \\ E &= 4.71 + 1.514s \ , \\ E' &= 4.02 + 1.239s \ , \\ s_{w'} &= s_b = 1.351 \ . \end{aligned} \tag{5.113}$$

The parametrizations are valid in the x range

$$10^{-5} \leq x < 1 \tag{5.114}$$

and the scale

$$\mu^2 \leq Q^2 \leq 10^8 \ \text{GeV}^2 \tag{5.115}$$

with an averaged inaccuracy of $2\% - 3\%$.

EXERCISE

5.8 Calculation of Moments of Splitting Functions (Anomalous Dimensions)

Problem. Calculate the moments of the splitting functions P_{Gq}, P_{qq}, P_{qG}, and P_{GG}. The splitting functions are explicitly (see (5.70)–(5.72))

(a) $$P_{Gq} = \frac{4}{3}\frac{1+(1-x)^2}{x} \ ,$$

(b) $$\begin{aligned} P_{qq} &= -\frac{4}{3}(1+x) + 2\,\delta(1-x) \\ &\quad + \lim_{\varepsilon\to 0}\frac{8}{3}\left[\frac{1}{1-x+\varepsilon} - \delta(1-x)\int\limits_0^1 \mathrm{d}y\,\frac{1}{1-y+\varepsilon}\right] \ , \end{aligned}$$

Exercise 5.8

(c) $P_{qG} = \frac{1}{2}\left[x^2 + (1-x)^2\right]$,

(d) $P_{GG} = 6\left[\frac{1}{x} - 2 + x(1-x)\right] + \left(\frac{11}{2} - \frac{N_f}{3}\right)\delta(1-x)$

$$+ \lim_{\varepsilon\to 0} 6\left[\frac{1}{1-x+\varepsilon} - \delta(1-x)\int_0^1 \mathrm{d}y\,\frac{1}{1-y+\varepsilon}\right] .$$

Solution. (a) Inserting the explicit expression for the splitting function P_{Gq} we get for the nth moment

$$\begin{aligned} d_{Gq}(n) &= -\frac{6}{33-2N_f}\int_0^1 \mathrm{d}x\, x^{n-1} P_{Gq}(x) \\ &= -\frac{6}{33-2N_f}\int_0^1 \mathrm{d}x\, x^{n-1}\,\frac{4}{3}\,\frac{1+(1-x)^2}{x} \\ &= -\frac{8}{33-2N_f}\int_0^1 \mathrm{d}x\,\left(2x^{n-2} - 2x^{n-1} + x^n\right) \\ &= -\frac{8}{33-2N_f}\left(\frac{2}{n-1}x^{n-1} - \frac{2}{n}x^n + \frac{1}{n+1}x^{n+1}\right)\Bigg|_0^1 . \end{aligned} \quad (1)$$

Since $n \geq 1$ we get

$$\begin{aligned} d_{Gq}(n) &= -\frac{8}{33-2N_f}\left(\frac{2}{n-1} - \frac{2}{n} + \frac{1}{n+1}\right) \\ &= -\frac{8}{33-2N_f}\,\frac{n^2+n+2}{n(n^2-1)} . \end{aligned} \quad (2)$$

(b) The nth moment of the splitting function P_{qq} is given by

$$\begin{aligned} d_{qq}(n) &= -\frac{6}{33-2N_f}\int_0^1 \mathrm{d}x\, x^{n-1} P_{qq}(x) \\ &= -\frac{6}{33-2N_f}\int_0^1 \mathrm{d}x\, x^{n-1}\left[-\frac{4}{3}(1+x) + 2\delta(1-x)\right. \\ &\quad \left. + \lim_{\varepsilon\to 0}\frac{8}{3}\left(\frac{1}{1-x+\varepsilon} - \delta(1-x)\int_0^1 \mathrm{d}y\,\frac{1}{1-y+\varepsilon}\right)\right] \end{aligned}$$

Exercise 5.8

$$= -\frac{6}{33-2N_\mathrm{f}}\left[-\frac{4}{3}\int_0^1 \mathrm{d}x\,\left(x^{n-1}+x^n\right)+2\right.$$
$$\left.+\lim_{\varepsilon\to 0}\frac{8}{3}\left(\int_0^1 \mathrm{d}x\,\frac{x^{n-1}}{1-x+\varepsilon}-\int_0^1 \mathrm{d}y\,\frac{1}{1-y+\varepsilon}\right)\right]\,. \tag{3}$$

With the expansion

$$\frac{x^{n-1}}{1-x+\varepsilon} = -\sum_{i=0}^{n-2} x^i + \frac{1}{1-x+\varepsilon} \tag{4}$$

this equation can be put into a simple form:

$$d_{qq}(n) = -\frac{6}{33-2N_\mathrm{f}}\left[-\frac{8}{3}\sum_{i=0}^{n}\int_0^1 \mathrm{d}x\,x^i+\frac{4}{3}\int_0^1 \mathrm{d}x\,\left(x^{n-1}+x^n\right)+2\right]$$
$$= -\frac{6}{33-2N_\mathrm{f}}\left[-\frac{8}{3}\sum_{i=0}^{n}\frac{1}{i+1}+\frac{4}{3}\left(\frac{1}{n}+\frac{1}{n+1}\right)+2\right]$$
$$= -\frac{6}{33-2N_\mathrm{f}}\left[-\frac{8}{3}\sum_{i=2}^{n}\frac{1}{i}-\frac{8}{3}\frac{1}{n+1}-\frac{8}{3}+\frac{4}{3}\frac{1}{n}+\frac{4}{3}\frac{1}{n+1}+2\right]$$
$$= \frac{4}{33-2N_\mathrm{f}}\left[4\sum_{i=2}^{n}\frac{1}{i}+1-\frac{2}{(n+1)n}\right]\,. \tag{5}$$

(c) For the moment $d_{qG}(n)$ we get

$$d_{qG}(n) = -\frac{6}{33-2N_\mathrm{f}}\int_0^1 \mathrm{d}x\,x^{n-1}P_{\mathrm{qG}}(x)$$
$$= -\frac{6}{33-2N_\mathrm{f}}\int_0^1 \mathrm{d}x\,x^{n-1}\,\frac{1}{2}\left[x^2+(1-x)^2\right]$$
$$= -\frac{3}{33-2N_\mathrm{f}}\int_0^1 \mathrm{d}x\,\left(2x^{n+1}-2x^n+x^{n-1}\right)$$
$$= -\frac{3}{33-2N_\mathrm{f}}\left(2n+2x^{n+2}-\frac{2}{n+1}x^{n+1}+\frac{1}{n}x^n\right)\Bigg|_0^1$$
$$= -\frac{3}{33-2N_\mathrm{f}}\,\frac{n^2+n+2}{(n+2)(n+1)n}\,. \tag{6}$$

Exercise 5.8

(d) The remaining moments $d_{GG}(n)$ of the splitting function P_{GG} are obtained in the same way as in (b):

$$
\begin{aligned}
d_{GG}(n) &= -\frac{6}{33-2N_\mathrm{f}} \int_0^1 \mathrm{d}x\, x^{n-1} P_{GG}(x) \\
&= -\frac{6}{33-2N_\mathrm{f}} \left[6 \int_0^1 \mathrm{d}x\, x^{n-1} \left(\frac{1}{x} - 2 + x(1-x) \right) + \left(\frac{11}{2} - \frac{f}{3} \right) \right. \\
&\quad \left. + 6 \lim_{\varepsilon \to 0} \left(\int_0^1 \mathrm{d}x\, \frac{x^{n-1}}{1-x+\varepsilon} - \int_0^1 \mathrm{d}y\, \frac{1}{1-y+\varepsilon} \right) \right] . \qquad (7)
\end{aligned}
$$

Again we employ the relation

$$
\frac{x^{n-1}}{1-x+\varepsilon} = -\sum_{i=0}^{n-2} x^i + \frac{1}{1-x+\varepsilon} \qquad (8)
$$

and get, after integrating and rearranging,

$$
\begin{aligned}
d_{GG}(n) &= -\frac{6}{33-2N_\mathrm{f}} \left[6 \left(\frac{1}{n-1} - \frac{2}{n} + \frac{1}{n+1} - \frac{1}{n+2} \right) \right. \\
&\quad \left. + \left(\frac{11}{2} - \frac{f}{3} \right) - 6 \sum_{i=0}^{n-2} \int_0^1 \mathrm{d}x\, x^i \right] \\
&= -\frac{6}{33-2N_\mathrm{f}} \left[6 \left(\frac{1}{n-1} - \frac{2}{n} + \frac{1}{n+1} - \frac{1}{n+2} - \sum_{i=0}^{n-2} \frac{1}{i+1} \right) \right. \\
&\quad \left. + \frac{11}{2} - \frac{N_\mathrm{f}}{3} \right] \\
&= -\frac{6}{33-2N_\mathrm{f}} \left[-6 \sum_{i=2}^{n} \frac{1}{i} - 6 \right. \\
&\quad \left. + 6 \left(\frac{1}{n-1} - \frac{1}{n} + \frac{1}{n+1} - \frac{1}{n+2} \right) + \frac{11}{2} - \frac{N_\mathrm{f}}{3} \right] \\
&= \frac{9}{33-2N_\mathrm{f}} \left[4 \sum_{i=2}^{n} \frac{1}{i} + \frac{1}{3} + \frac{2N_\mathrm{f}}{9} - 4 \left(\frac{1}{n-1} - \frac{1}{n} + \frac{1}{n+1} - \frac{1}{n+2} \right) \right] \\
&= \frac{9}{33-2N_\mathrm{f}} \left[4 \sum_{i=2}^{n} \frac{1}{i} + \frac{1}{3} + \frac{2N_\mathrm{f}}{9} - \frac{4}{n(n-1)} - \frac{4}{(n+2)(n+1)} \right] . \qquad (9)
\end{aligned}
$$

5.4 Renormalization and the Expansion Into Local Operators

The Q^2 dependence of the distribution functions in the previous sections was described with the help of the GLAP equations. Starting with an – experimentally measured – input distribution function at low Q^2 the evolution to a higher scale Q'^2 was determined by perturbative QCD which entered into the calculation of the four splitting functions. However, there exists a more general approach of wider applicability which allows us to rederive the GLAP equations in a more formal manner. This formalism will be discussed in this section. This particular approach, the so-called *operator product expansion* (OPE), is of general validity and not restricted to the context of deep inelastic lepton nucleon scattering. In fact, we shall encounter it again during the discussion of QCD sum rules in Chap. 7.

The following analysis again starts with (5.1):

$$W_{\mu\mu'} = \frac{1}{2\pi}\int \mathrm{d}^4x \exp\left(\mathrm{i}q\cdot x\right)\frac{1}{2}\sum_{\text{pol.}}\langle \mathrm{N}|\hat{J}_\mu(x)\hat{J}_{\mu'}(0)|\mathrm{N}\rangle \ . \tag{5.116}$$

In the rest system of the nucleon (see, e.g., Exercise 3.6, (2) and (8))

$$P_\mu = (M,0,0,0) \ , \quad q_\mu = (q_0,0,0,q_3) = \left(\frac{\nu}{M},0,0,\sqrt{\frac{\nu^2}{M^2}+Q^2}\right) \ , \tag{5.117}$$

where $\nu = q\cdot p = q_0 M$ and $Q^2 = -q^2 = q_3^2 - q_0^2 = q_3^2 - \nu^2/M^2$. Here we are interested in the asymptotic region (the Bjorken limit), ν, $Q^2 \gg M^2$, $Q^2/(2\nu M) = x$, and the exponential factor in (5.116) is

$$\begin{aligned}
\exp\left[\mathrm{i}q_0x^0 - q_3x^3\right] &= \exp\left[\mathrm{i}\frac{\nu}{M}x^0 - \sqrt{\frac{\nu^2}{M^2}+Q^2}x^3\right]\\
&= \exp\left[\mathrm{i}\frac{\nu}{M}\left(x^0 - \sqrt{1+\frac{Q^2M^2}{\nu^2}}x^3\right)\right]\\
&\approx \exp\left[\mathrm{i}\frac{\nu}{M}\left(x^0 - x^3 - \frac{1}{2}\frac{Q^2}{\nu^2}M^2x^3\right)\right]\\
&= \exp\left[\mathrm{i}\frac{\nu}{M}\left(x^0 - x^3 - x\frac{M^2}{\nu}x^3\right)\right]\\
&\approx \exp\left[\mathrm{i}\frac{\nu}{M}\left(x^0 - x^3\right)\right] \ .
\end{aligned} \tag{5.118}$$

Provided that $\left|x^0 - x^3\right|$ is much larger than M/ν, its contribution to (5.116) averages to zero:

$$\left|x^0 - x^3\right| \sim O\left(\frac{M}{\nu}\right) \ . \tag{5.119}$$

Therefore only x_μ values with

$$x^2 = x_0^2 - x_1^2 - x_2^2 - x_3^2 \le x_0^2 - x_3^2 \sim O\left(\frac{M}{\nu}\right) \tag{5.120}$$

contribute to the integral (5.116). On the other hand, one can prove that the matrix element $\langle N|J_\mu(x)J_{\mu'}(0)|N\rangle$ only contributes for $x^2 \ge 0$. To this end recall that this matrix element can be written as

$$\sum_X \int \mathrm{d}^4x \langle \mathrm{N}|\hat{J}_\mu(x)|X\rangle\langle X|\hat{J}_{\mu'}(0)|\mathrm{N}\rangle \exp(\mathrm{i}q\cdot x) \tag{5.121}$$

with *physical* final states X. The crucial point is that the expression with exchanged coordinates

$$\begin{aligned}&\sum_X \int \mathrm{d}^4x \langle \mathrm{N}|\hat{J}_\mu(0)|X\rangle\langle X|\hat{J}_{\mu'}(x)|\mathrm{N}\rangle \exp(\mathrm{i}q\cdot x)\\ &\quad\sim \sum_X \ldots \delta^4(P_X - P_N + q)\end{aligned} \tag{5.122}$$

is equal to zero, since there is no physical state $|X\rangle$ with baryon number one and energy $E_X = P_\mathrm{N}^0 - q^0 = M_N - \nu/M_\mathrm{N} < M_\mathrm{N}$. Hence the matrix element in (5.116) can be replaced by the matrix element of the commutator $\left[\hat{J}_\mu(x)\hat{J}_{\mu'}(0)\right]_-$:

$$\langle \mathrm{N}|\hat{J}_\mu(x)\hat{J}_{\mu'}(0)|\mathrm{N}\rangle - \langle \mathrm{N}|\hat{J}_\mu(0)\hat{J}_{\mu'}(x)|\mathrm{N}\rangle = \langle N|\left[\hat{J}_\mu(x)\hat{J}_{\mu'}(0)\right]_- |N\rangle\ . \tag{5.123}$$

The commutator vanishes for spacelike distances $x^2 < 0$ due to the requirements of causality: the commutator is proportional to the Green function, which has to vanish for space like distances. Therefore only time and lightlike x_μ contribute to (5.116):

$$W_{\mu\mu'} = \frac{1}{2\pi}\int\limits_{x^2\ge 0} \mathrm{d}^4x\ \exp(\mathrm{i}qx)\ \frac{1}{2}\sum_{\mathrm{pol.}}\langle \mathrm{N}|[\hat{J}_\mu(x)\hat{J}_{\mu'}(0)]_-|\mathrm{N}\rangle\ . \tag{5.124}$$

It follows with the help of (5.119) that $W_{\mu\mu'}$ is dominated by the matrix element of the current commutator for lightlike x values. Therefore the following expansion is also known as the *light-cone expansion*.

It is well known that a commutator of currents with a lightlike distance diverges. This divergence also occurs in the free case

$$\hat{J}_\mu(x) = \hat{\bar{\Psi}}(x)\gamma_\mu\hat{\Psi}(x)\ , \tag{5.125}$$

where it can be reduced to the divergence of the free propagator $S(x)$ for $x^2 \to 0$:

$$
\begin{aligned}
[\hat{J}_\mu(x)\hat{J}_\nu(0)]_- &= [\hat{\bar{\Psi}}(x)\gamma_\mu\hat{\Psi}(x),\, \hat{\bar{\Psi}}(0)\gamma_\nu\hat{\Psi}(0)]_- \\
&= \hat{\bar{\Psi}}(x)\gamma_\mu\hat{\Psi}(x)\hat{\bar{\Psi}}(0)\gamma_\nu\hat{\Psi}(0) \\
&\quad + [\hat{\bar{\Psi}}(x)\gamma_\mu]_\alpha\hat{\bar{\Psi}}_\beta(0)\hat{\Psi}_\alpha(x)[\gamma_\nu\hat{\Psi}(0)]_\beta \\
&\quad - \hat{\bar{\Psi}}(0)\gamma_\nu\hat{\Psi}(0)\hat{\bar{\Psi}}(x)\gamma_\mu\hat{\Psi}(x) \\
&\quad - \hat{\bar{\Psi}}_\beta(0)[\hat{\bar{\Psi}}(x)\gamma_\mu]_\alpha[\gamma_\nu\hat{\Psi}(0)]_\beta\hat{\Psi}_\alpha(x) \ .
\end{aligned} \tag{5.126}
$$

Here we have added the second term and substracted it again (fourth term). Indeed, owing to the anticommutation properties of $\hat{\Psi}$ and $\hat{\bar{\Psi}}$, the second and the fourth term cancel. The first two and the last two terms are combined and the anticommutators that occur are replaced by the propagator:

$$
\begin{aligned}
&[\hat{\bar{\Psi}}(x)\gamma_\mu]_\alpha S_{\alpha\beta}(x)[\gamma_\nu\hat{\Psi}(0)]_\beta - [\hat{\bar{\Psi}}(0)\gamma_\nu]_\beta S_{\beta\alpha}(-x)[\gamma_\mu\hat{\Psi}(x)]_\alpha \ , \\
&\qquad \{\hat{\Psi}_\alpha(x),\, \hat{\bar{\Psi}}_\beta(0)\} = S_{\alpha\beta}(x) = \left(\mathrm{i}\gamma_\mu\partial^\mu + m\right)_{\alpha\beta}\cdot \mathrm{i}\Delta(x) \ .
\end{aligned} \tag{5.127}
$$

In the case of massless fields and $x^2 \to 0$, $S(x)$ is given by

$$
S(x) \to \gamma_\mu \partial_x{}^\mu \frac{1}{2\pi}\varepsilon(x_0)\delta\left(x^2\right) \tag{5.128}
$$

with [11]

$$
\varepsilon(x_0) = \mathrm{sgn}(x_0) \ , \tag{5.129}
$$

$$
\begin{aligned}
[\hat{J}_\mu(x),\, \hat{J}_\nu(0)]_- &\to \left(\overline{\Psi}(x)\gamma_\mu\gamma_\lambda\gamma_\nu\Psi(0) - \overline{\Psi}(0)\gamma_\nu\gamma_\lambda\gamma_\mu\Psi(x)\right) \\
&\quad \times \frac{1}{2\pi}\partial^\lambda\varepsilon(x_0)\delta(x^2) \ .
\end{aligned} \tag{5.130}
$$

In the following, only the leading divergence of $S(x)$, i.e., the free propagator, is considered. The remaining terms are discussed later. For the electromagnetic interaction treated here $W_{\mu\nu}$ is symmetric in μ and ν (see (3.18)). Therefore we can write the current commutator as

$$
\begin{aligned}
&\frac{1}{2}\left(\left[\hat{J}_\mu\left(\frac{x}{2}\right), \hat{J}_\nu\left(-\frac{x}{2}\right)\right]_- + \left[\hat{J}_\nu\left(\frac{x}{2}\right), \hat{J}_\mu\left(-\frac{x}{2}\right)\right]_-\right) \\
&= \left[\hat{\bar{\Psi}}\left(\frac{x}{2}\right)\frac{1}{2}(\gamma_\mu\gamma_\lambda\gamma_\nu + \gamma_\nu\gamma_\lambda\gamma_\mu)\hat{\Psi}\left(-\frac{x}{2}\right)\right. \\
&\quad \left. - \hat{\bar{\Psi}}\left(-\frac{x}{2}\right)\frac{1}{2}(\gamma_\nu\gamma_\lambda\gamma_\mu + \gamma_\mu\gamma_\lambda\gamma_\nu)\hat{\Psi}\left(\frac{x}{2}\right)\right]\frac{1}{2\pi}\partial^\lambda\varepsilon(x_0)\delta(x^2) \\
&= s_{\mu\lambda\nu\beta}\left[\hat{\bar{\Psi}}\left(\frac{x}{2}\right)\gamma^\beta\hat{\Psi}\left(-\frac{x}{2}\right) - \hat{\bar{\Psi}}\left(-\frac{x}{2}\right)\gamma^\beta\hat{\Psi}\left(\frac{x}{2}\right)\right] \\
&\quad \times \frac{1}{2\pi}\partial^\lambda\varepsilon(x_0)\delta(x^2) \ ,
\end{aligned} \tag{5.131}
$$

[11] W. Greiner and J. Reinhardt: *Field Quantization* (Springer, Berlin, Heidelberg 1996).

with

$$s_{\mu\lambda\nu\beta} = g_{\mu\lambda}g_{\nu\beta} + g_{\mu\beta}g_{\nu\lambda} - g_{\mu\nu}g_{\lambda\beta} \ . \tag{5.132}$$

Here we have shifted the position arguments of the current operators from "x" and "0" to "$\pm x/2$". This operation does not change anything, because (5.124) is invariant under such translations, but it emphasizes the symmetry properties of the current operator. In the case of neutrino–nucleon scattering, i.e., for a parity-violating process, $W_{\mu\mu'}$ is no longer symmetric in μ and μ' (see Exercise 3.3). Let us see how the commutator $\left[\hat{J}_\mu(x), \hat{J}_\nu(0)\right]_-$ can be calculated in this case. Because

$$\gamma_\mu\gamma_\lambda\gamma_\nu = (s_{\mu\lambda\nu\beta} + \mathrm{i}\varepsilon_{\mu\lambda\nu\beta}\gamma_5)\gamma^\beta \tag{5.133}$$

(as we shall show in Exercise 5.9), we obtain the following additional terms:

$$\begin{aligned}
&\left[\hat{J}_\mu\left(\frac{x}{2}\right), \hat{J}_\nu\left(-\frac{x}{2}\right)\right]_- \\
&\quad = \left[\hat{\bar{\Psi}}\left(\frac{x}{2}\right)(s_{\mu\lambda\nu\beta} + \mathrm{i}\varepsilon_{\mu\lambda\nu\beta}\gamma_5)\gamma^\beta\hat{\Psi}\left(-\frac{x}{2}\right)\right. \\
&\qquad \left. - \hat{\bar{\Psi}}\left(-\frac{x}{2}\right)(s_{\nu\lambda\mu\beta} + \mathrm{i}\varepsilon_{\nu\lambda\mu\beta}\gamma_5)\gamma^\beta\hat{\Psi}\left(\frac{x}{2}\right)\right]\frac{1}{2\pi}\partial^\lambda\varepsilon(x_0)\delta(x^2) \\
&\quad = \left\{s_{\mu\lambda\nu\beta}\left[\hat{\bar{\Psi}}\left(\frac{x}{2}\right)\gamma^\beta\hat{\Psi}\left(-\frac{x}{2}\right) - \hat{\bar{\Psi}}\left(-\frac{x}{2}\right)\gamma^\beta\hat{\Psi}\left(\frac{x}{2}\right)\right]\right. \\
&\qquad \left. + \mathrm{i}\varepsilon_{\mu\lambda\nu\beta}\left[\hat{\bar{\Psi}}\left(\frac{x}{2}\right)\gamma_5\gamma^\beta\hat{\Psi}\left(-\frac{x}{2}\right) + \hat{\bar{\Psi}}\left(-\frac{x}{2}\right)\gamma_5\gamma^\beta\hat{\Psi}\left(\frac{x}{2}\right)\right]\right\} \\
&\qquad \times \frac{1}{2\pi}\partial^\lambda\varepsilon(x_0)\delta(x^2) \ .
\end{aligned} \tag{5.134}$$

Here the full antisymmetry of the ε tensor, i.e. $\varepsilon_{\mu\lambda\nu\beta} = -\varepsilon_{\nu\lambda\mu\beta}$, has been utilized. Inserting (5.134) into the matrix element (5.124) yields

$$\begin{aligned}
&\langle \mathrm{N}|\left[\hat{J}_\mu\left(\frac{x}{2}\right), \hat{J}_\nu\left(-\frac{x}{2}\right)\right]_- |\mathrm{N}\rangle \\
&\quad = \frac{1}{2\pi}\left[\partial^\lambda\varepsilon(x_0)\delta(x^2)\right] \\
&\qquad \times \left\{s_{\mu\lambda\nu\beta}\langle \mathrm{N}|\hat{\bar{\Psi}}\left(\frac{x}{2}\right)\gamma^\beta\hat{\Psi}\left(-\frac{x}{2}\right) - \hat{\bar{\Psi}}\left(-\frac{x}{2}\right)\gamma^\beta\hat{\Psi}\left(\frac{x}{2}\right)|\mathrm{N}\rangle\right. \\
&\qquad \left. + \mathrm{i}\varepsilon_{\mu\lambda\nu\beta}\langle \mathrm{N}|\hat{\bar{\Psi}}\left(\frac{x}{2}\right)\gamma_5\gamma^\beta\hat{\Psi}\left(-\frac{x}{2}\right) + \hat{\bar{\Psi}}\left(-\frac{x}{2}\right)\gamma_5\gamma^\beta\hat{\Psi}\left(\frac{x}{2}\right)|\mathrm{N}\rangle\right\} ,
\end{aligned} \tag{5.135}$$

which is a typical form for the *operator-product expansion* (OPE). The name OPE will become obvious in the following.

The first factor on the right-hand side of (5.132) is a divergent expression, which can be derived from perturbative QCD and which does not depend on the hadron considered. The terms in the curly brackets are finite matrix elements, which contain all the nonperturbative information. The divergent factors are called *Wilson coefficients*.

The *Wilson coefficients* can be calculated using perturbation theory and written as a systematic expansion in the coupling constant α_s. The determination of the matrix elements requires nonperturbative methods such as lattice calculations or QCD sum rules. This can be compared to the approach of the GLAP equations where the splitting functions are calculated within perturbative QCD but the input distribution functions – as pure nonperturbative quantities – have to be taken from models or from experiment.

EXERCISE

5.9 Decomposition Into Vector and Axial Vector Couplings

Problem. Prove relation (5.133),

$$\gamma_\mu \gamma_\alpha \gamma_\nu = (s_{\mu\alpha\nu\beta} + \mathrm{i}\varepsilon_{\mu\alpha\nu\beta}\gamma_5)\gamma^\beta \ . \tag{1}$$

Solution. $\gamma_\mu \gamma_\alpha \gamma_\nu$ anticommutes with γ_5. On the other hand, it must be a linear combination of the 16 matrices which span the Clifford space, namely 1, γ_5, γ_μ, $\gamma_\mu \gamma_5$, and $\sigma_{\mu\nu}$. Obviously this implies

$$\gamma_\mu \gamma_\alpha \gamma_\nu = a_{\mu\alpha\nu\beta}\gamma^\beta + b_{\mu\alpha\nu\beta}\gamma_5\gamma^\beta \ . \tag{2}$$

Multiplying by γ_δ and taking the trace we find with

$$\begin{aligned} \mathrm{tr}\{\gamma_\mu \gamma_\alpha \gamma_\nu \gamma_\delta\} &= 4(g_{\mu\alpha}g_{\nu\delta} + g_{\mu\delta}g_{\nu\alpha} - g_{\mu\nu}g_{\alpha\delta}) \\ &= 4a_{\mu\alpha\nu\delta} \end{aligned} \tag{3}$$

and

$$\mathrm{tr}\{\gamma^\beta \gamma_\delta\} = g^\beta_\delta \ , \quad \mathrm{tr}\{\gamma_5 \gamma^\beta \gamma^\delta\} = 0$$

that

$$g_{\mu\alpha}g_{\nu\delta} + g_{\mu\delta}g_{\nu\alpha} - g_{\mu\nu}g_{\alpha\delta} = a_{\mu\alpha\nu\delta} \ . \tag{4}$$

The second coefficient is determined by multiplying (2) by $\gamma_\delta \gamma_5$ and taking the trace

$$4b_{\mu\alpha\nu\delta} = 4\mathrm{i}\varepsilon_{\mu\alpha\nu\delta} \ . \tag{5}$$

Thus we have proven (1):

$$\gamma_\mu \gamma_\alpha \gamma_\nu = g_{\mu\alpha}\gamma_\nu + g_{\nu\alpha}\gamma_\mu - g_{\mu\nu}\gamma_\alpha + \mathrm{i}\varepsilon_{\mu\alpha\nu\delta}\gamma_5\gamma^\delta \ . \tag{6}$$

We have shown above that in the Bjorken limit short (or more precisely lightlike) distances give the dominant contribution to the scattering tensor. For later purposes it is convenient to expand the bilocal operator $\overline{\Psi}(x/2)\Psi(-x/2)$ into a sum of local operators. We easily find

$$\begin{aligned}
\hat{\overline{\Psi}}\left(\frac{x}{2}\right)\hat{\Psi}\left(-\frac{x}{2}\right) &= \hat{\overline{\Psi}}(0)\left[1+\overleftarrow{\partial}_{\mu_1}\frac{x^{\mu_1}}{2}+\frac{1}{2!}\overleftarrow{\partial}_{\mu_1}\overleftarrow{\partial}_{\mu_2}\frac{x^{\mu_1}}{2}\frac{x^{\mu_2}}{2}+\dots\right] \\
&\quad\times\left[1-\frac{x^{\nu_1}}{2}\overrightarrow{\partial}_{\nu_1}+\frac{x^{\nu_1}}{2}\frac{x^{\nu_2}}{2}\frac{1}{2!}\overrightarrow{\partial}_{\nu_1}\overrightarrow{\partial}_{\nu_2}-\dots\right]\hat{\Psi}(0) \\
&= \sum_n \frac{(-1)^n}{n!}\left(\frac{x}{2}\right)^{\mu_1}\left(\frac{x}{2}\right)^{\mu_2}\dots\left(\frac{x}{2}\right)^{\mu_n} \\
&\quad\times\hat{\overline{\Psi}}(0)\overleftrightarrow{\partial}_{\mu_1}\overleftrightarrow{\partial}_{\mu_2}\dots\overleftrightarrow{\partial}_{\mu_n}\hat{\Psi}(0)\ , \qquad (5.136)\\
\overleftrightarrow{\partial}_{\mu} &= \overrightarrow{\partial}_{\mu}-\overleftarrow{\partial}_{\mu}\ . \qquad (5.137)
\end{aligned}$$

With the free current operator, (5.121) therefore becomes

$$\begin{aligned}
W_{\mu\mu'} &= -\frac{1}{2\pi}\int_{x^2\geq 0} dx^4\,\exp(iq\cdot x)\sum_n\frac{(-1)^n}{n!}\left(\frac{x}{2}\right)^{\mu_1}\left(\frac{x}{2}\right)^{\mu_2}\dots\left(\frac{x}{2}\right)^{\mu_n} \\
&\quad\times[1-(-1)^n]\frac{1}{2}\sum_{\text{pol.}}\langle \mathrm{N}|\hat{O}^n_{\mu_1\dots\mu_n\beta}|\mathrm{N}\rangle s_{\mu\alpha\mu'}{}^{\beta}\frac{1}{2\pi}\partial^{\alpha}\left[\varepsilon(x^0)\delta(x^2)\right] \\
&= \frac{1}{2\pi}\int_{x^2\geq 0} dx^4\,\exp(iq\cdot x)\sum_{n\ \text{odd}}\frac{1}{n!}\left(\frac{x}{2}\right)^{\mu_1}\left(\frac{x}{2}\right)^{\mu_2}\dots\left(\frac{x}{2}\right)^{\mu_n}s_{\mu\alpha\mu'}{}^{\beta} \\
&\quad\times\frac{1}{2}\sum_{\text{pol.}}\langle \mathrm{N}|\hat{O}^n_{\mu_1\dots\mu_n\beta}|\mathrm{N}\rangle\frac{1}{\pi}\partial^{\alpha}\left[\varepsilon(x^0)\delta(x^2)\right]\ , \qquad (5.138)
\end{aligned}$$

where

$$\hat{O}^n_{\mu_1\dots\mu_n\beta}(0)=\hat{\overline{\Psi}}(0)\overleftrightarrow{\partial}_{\mu_1}\overleftrightarrow{\partial}_{\mu_2}\dots\overleftrightarrow{\partial}_{\mu_n}\gamma_{\beta}\,\hat{\Psi}(0)\ . \qquad (5.139)$$

Although the product of operator $\left[\hat{J}_{\mu}\left(\frac{x}{2}\right),\hat{J}_{\nu}\left(\frac{-x}{2}\right)\right]$ is singular for $x^2\to 0$, this is not the case for the operators $\hat{O}^n_{\mu_1\mu_2\dots\mu_n}(0)$, which are now well defined. All singularities of the current commutator are contained in the coefficient function $\partial^{\alpha}\left[\varepsilon(x^0)\delta(x^2)\right]$ (see (5.138)). To calculate the structure function we write

$$\begin{aligned}
\frac{1}{2}\sum_{\text{pol.}}\langle \mathrm{N}(P)|\hat{O}^n_{\mu_1\dots\mu_n\beta}(0)|\mathrm{N}(P)\rangle &= A^{(n)}\left[P_{\mu_1}P_{\mu_2}\dots P_{\mu_n}P_{\beta}\right] \\
&\quad+\text{trace terms}\ , \qquad (5.140)
\end{aligned}$$

where the $A^{(n)}$ are some constants. This equation follows simply from Lorentz covariance. Because the operator is a Lorentz tensor with $n+1$ indices we have to parameterize its matrix element accordingly. Furthermore, since we sum over the polarizations of the nucleon the only Lorentz vectors available for this purpose is P_μ, the momentum vector of the nucleon, and the metric tensor $g_{\mu_i\mu_j}$. Trace terms which contain one or more factors of $g_{\mu_i\mu_j}$ will produce powers of x^2 when contracted with $x^{\mu_i}x^{\mu_j}$ in (5.138) and will therefore be less singular near the light cone $x^2 \sim 0$. Consequently, such terms are less important in the Bjorken limit $Q^2 \to \infty$.

The x integration in (5.138) leads to a replacement of $x^{\mu_1}x^{\mu_2}\dots x^{\mu_n}$ by the corresponding q components. A Fourier transformation gives

$$x^{\mu_1}x^{\mu_2}\dots x^{\mu_n} \to (\dots)\cdot q^{\mu_1}q^{\mu_2}\dots q^{\mu_n} \ . \tag{5.141}$$

Hence (5.140) is contracted with $q^{\mu_1}q^{\mu_2}\dots q^{\mu_n}$. It then follows that

$$\frac{q^{\mu_i}q^{\mu_j}P_{\mu_i}P_{\mu_j}}{q^{\mu_i}q^{\mu_j}M^2 g_{\mu_i\mu_j}} = \frac{\nu^2}{-Q^2M^2} = -\frac{\nu}{2M^2}\frac{1}{x} \to \infty \tag{5.142}$$

for ν, $Q^2 \to \infty$, and x fixed. Thus all the remaining terms on the right-hand side of (5.140) are suppressed by powers of $1/Q^2$ (higher twist):

$$\frac{1}{2}\sum_{\text{pol.}}\langle \mathrm{N}(P)|\hat{O}^n_{\mu_1\dots\mu_n\beta}|\mathrm{N}(P)\rangle \to A^{(n)}P_{\mu_1}P_{\mu_2}\dots P_{\mu_n}P_\beta + \dots \ , \tag{5.143}$$

$$\begin{aligned} W_{\mu\mu'} &= \frac{1}{2\pi}\int\limits_{x^2\geq 0} \mathrm{d}^4x \exp(\mathrm{i}q\cdot x)\sum_{n\,\text{odd}}\frac{1}{n!}\left[\frac{x\cdot P}{2}\right]^n s_{\mu\alpha\mu'}{}^{\beta}P_\beta A^{(n)} \\ &\quad\times\frac{1}{\pi}\partial^\alpha\left[\varepsilon(x^0)\delta(x^2)\right] + \dots \\ &= \frac{1}{2\pi}\int\limits_{x^2\geq 0} \mathrm{d}^4x \exp(\mathrm{i}q\cdot x)\sum_{n\,\text{odd}}(\text{finite term})\times(\text{divergent term}) + \dots \ . \end{aligned} \tag{5.144}$$

We have succeeded in splitting up the contributions to $W_{\mu\mu'}$ into a finite and a divergent factor $\frac{1}{\pi}\left[\partial^\alpha\left(\varepsilon(x^0)\delta(x^2)\right)\right]$. The whole dependence on the specific process under consideration is related to the finite factor whereas the divergent factor always remains the same. This seperation of divergent current commutators into an operator with finite matrix elements specific for the hadron under consideration and a constant divergent function is the basic idea of the expansion into local operators.

One can ask what is the practical use of the formal expansion (5.141). As a first simple answer we wish to show that the main statements of the parton model can be immediately derived from this equation. The simple parton model assumes noninteracting constituents, which correspond exactly to the use of the free current commutator. Let us consider the relation

$$\sum_{n\,\text{odd}}\left[\frac{x\cdot P}{2}\right]^n\frac{A^{(n)}}{n!} = -\int \mathrm{d}\xi \exp(\mathrm{i}x\cdot\xi P)\, f(\xi) \ , \tag{5.145}$$

with

$$f(\eta) = -\frac{1}{2\pi} \int \mathrm{d}(x \cdot P) \sum_{n \text{ odd}} \left[\frac{x \cdot P}{2}\right]^n \frac{A^{(n)}}{n!} \exp(-\mathrm{i}x \cdot P\eta) \ . \tag{5.146}$$

Indeed, substituting (5.146) into the right-hand side of (5.145) yields exactly the verification of the identity (5.145). Inserting this into the hadron scattering tensor (5.144), one obtains

$$W_{\mu\mu'} = \frac{-1}{2\pi} \int \mathrm{d}^4x \int \mathrm{d}\xi \exp(\mathrm{i}x \cdot (q + \xi P))\, s_{\mu\alpha\mu'}{}^{\beta} \times P_\beta f(\xi) \frac{1}{\pi} \partial^\alpha \left[\varepsilon(x^0)\delta(x^2)\right] \ . \tag{5.147}$$

Using (see Exercise 5.10)

$$\int \mathrm{d}^4x \exp(\mathrm{i}x \cdot (q + \xi P))\, \delta(x^2)\varepsilon(x^0) = \mathrm{i}4\pi^2\delta\left[(q + \xi P)^2\right]\varepsilon(q_0 + \xi P_0) \tag{5.148}$$

and partial integration yields

$$W_{\mu\mu'} = -2 \int \mathrm{d}\xi\, s_{\mu\alpha\mu'}{}^{\beta} P_\beta f(\xi)(q + \xi P)^\alpha \delta\left[(q + \xi P)^2\right]\varepsilon(q_0 + \xi P_0) \ . \tag{5.149}$$

The argument of the δ function simplifies to

$$(q + \xi P)^2 = q^2 + \xi^2 M^2 + 2\xi q \cdot P \approx -Q^2 + \xi 2\nu = \frac{Q^2}{x}(\xi - x) \ , \tag{5.150}$$

$$\begin{aligned} W_{\mu\mu'} &= -2 s_{\mu\alpha\mu'}{}^{\beta} P_\beta f(x) \frac{x}{Q^2}(q + xP)^\alpha \\ &= -2(g_{\mu\alpha}g_{\mu'}{}^{\beta} + g_\mu{}^{\beta} g_{\mu'\alpha} - g_{\mu\mu'}g_\alpha{}^{\beta})P_\beta(q + xP)^\alpha \frac{x}{Q^2} f(x) \\ &= -2\frac{x}{Q^2} f(x)\left[(q + xP)_\mu P_{\mu'}\right. \\ &\qquad \left. + P_\mu(q + xP)_{\mu'} - g_{\mu\mu'}(xM^2 + P \cdot q)\right] \ . \end{aligned} \tag{5.151}$$

Note that $x = Q^2/2\nu$ denotes the Bjorken-variable. Since $W_{\mu\mu'}$ is multiplied by the lepton scattering tensor $L^{\mu\mu'}$, we can omit all terms proportional to q_μ and $q_{\mu'}$, because of

$$q^\mu L_{\mu\mu'} = q^{\mu'} L_{\mu\mu'} = 0 \ . \tag{5.152}$$

With $\nu = P \cdot q$ this yields

$$\begin{aligned} W_{\mu\mu'} &\to -2\frac{x}{Q^2} f(x)\left[-g_{\mu\mu'}(xM^2 + \nu) + 2xP_\mu P_{\mu'}\right] \\ &\approx -2\frac{x}{Q^2} f(x)\left[-g_{\mu\mu'}\nu + 2xP_\mu P_{\mu'}\right] \ . \end{aligned} \tag{5.153}$$

In an analogous way, i.e., by neglecting all terms proportional to q_μ or $q_{\mu'}$, we obtain from (3.18)

$$W_{\mu\mu'} \to -g_{\mu\mu'} W_1(Q^2, \nu) + P_\mu P_{\mu'} \frac{W_2(Q^2, \nu)}{M_\mathrm{N}^2} \ . \tag{5.154}$$

Comparison of (5.153) with (5.154) shows that $W_1(Q^2, x)$ and $W_2(Q^2, x)$ in fact depend only on x. Furthermore we get the Callan–Gross relation . For that purpose we also remember the definition (3.43) and identity

$$W_1(Q^2, \nu) = -2\frac{x}{Q^2} f(x)\nu =: F_1(x) \ , \tag{5.155}$$

$$\frac{W_2(Q^2, \nu)}{M_N^2} = -2\frac{x}{Q^2} f(x)\cdot 2x =: \frac{F_2(x)}{\nu}$$

$$\Rightarrow \quad F_2(x) = 2xF_1(x) \ . \tag{5.156}$$

This was just a simple example. In general the operator product expansion (OPE) provides two major results.

1. It gives a formal expression for all the terms that can occur. This allows us to analyze in a systematic way the terms contributing to a given power of $1/Q^2$. Furthermore it generates relations between various phenomenological expressions such as the structure functions.

2. For the individual terms appearing in the $1/Q^2$ expansion the OPE gives the corresponding correlators. This shows to what property of the (unknown) exact wave function a given observable is sensitive and allows us to calculate it in an appropriate model (i.e., a model which should describe the specific correlation in question realistically).

EXERCISE

5.10 The Proof of (5.148)

Problem. Prove (5.148) by replacing $(q+\xi P)_\mu$ by k_μ and evaluating the Fourier transformation of (5.148), i.e., by proving that

$$\frac{\mathrm{i}}{4\pi^2}\int \exp(-\mathrm{i}k\cdot x)\,\delta(k^2)\varepsilon(k_0)\,\mathrm{d}^4k = \delta(x^2)\varepsilon(x_0) \ . \tag{1}$$

Solution. Performing the k_0 integration turns the left-hand side into

$$\begin{aligned} I =& \frac{\mathrm{i}}{4\pi^2}\int \exp(-\mathrm{i}(k_0x_0 - \boldsymbol{k}\cdot\boldsymbol{x}))\,\delta(k_0^2 - \boldsymbol{k}^2)\varepsilon(k_0)\,\mathrm{d}k_0\,\mathrm{d}^3k \\ =& \frac{\mathrm{i}}{4\pi^2}\int \exp(-\mathrm{i}(k_0x_0 - \boldsymbol{k}\cdot\boldsymbol{x}))\,\delta\left((k_0-k)(k_0+k)\right)\varepsilon(k_0)\,\mathrm{d}k_0\,\mathrm{d}^3k \end{aligned}$$

Exercise 5.10

$$
\begin{aligned}
&= \frac{\mathrm{i}}{4\pi^2} \int \frac{1}{2k} \left[\exp\left(-\mathrm{i}k(x^0 - r\cos\theta)\right) \right. \\
&\quad \left. - \exp\left(+\mathrm{i}k(x^0 + r\cos\theta)\right) \right] \mathrm{d}^3k \qquad (2)
\end{aligned}
$$

with

$$
\begin{aligned}
k &:= \sqrt{\boldsymbol{k}^2} \ , \\
r &= \sqrt{\boldsymbol{x}^2} \ , \\
\boldsymbol{k}\cdot\boldsymbol{x} &= kr\cos\theta \ . \qquad (3)
\end{aligned}
$$

We evaluate the angular integration

$$
\begin{aligned}
I &= \frac{\mathrm{i}}{2\pi} \int_0^\infty \mathrm{d}k \frac{k^2}{2k} \left[\frac{1}{\mathrm{i}kr} \left(\exp\left(-\mathrm{i}k(x^0 - r)\right) - \exp\left(-\mathrm{i}k(x^0 + r)\right) \right) \right. \\
&\quad \left. - \frac{1}{\mathrm{i}kr} \left(\exp\left(\mathrm{i}k(x^0 + r)\right) - \exp\left(\mathrm{i}k(x^0 - r)\right) \right) \right] \\
&= \frac{1}{4\pi r} \int_{-\infty}^{\infty} \mathrm{d}k \left[\exp\left(-\mathrm{i}k(x^0 - r)\right) - \exp\left(\mathrm{i}k(x^0 + r)\right) \right] \qquad (4)
\end{aligned}
$$

and obtain a difference of δ functions, which yields the postulated result:

$$
\begin{aligned}
I &= \frac{2\pi}{4\pi r} \left[\delta(x^0 - r) - \delta(x^0 + r) \right] \\
&= \delta(x^2)\Big|_{x^0 = r} - \delta(x^2)\Big|_{x^0 = -r} \\
&= \delta(x^2)\varepsilon(x_0) \ . \qquad (5)
\end{aligned}
$$

Let us discuss next how the expansion of $W_{\mu\mu'}$ changes if in (5.144) we replace the free by the exact current commutator, i.e., if we take the quark interactions into account. This corresponds in lowest order to the transition from the simple parton model to the GLAP equations.

The bilocal operators $\hat{\bar{\Psi}}\left(\frac{x}{2}\right)\hat{\Psi}\left(-\frac{x}{2}\right)$ are not gauge invariant. In order to achieve that, the derivatives in $\hat{O}^n_{\mu_1\mu_2\ldots\mu_n\beta}$ have to be replaced by covariant derivatives $\hat{D}_\mu$. For the sake of simplicity all derivatives that operate to the left are partially integrated such that they act to the right:

$$
\left(\boldsymbol{\partial}_\mu - \overleftarrow{\partial_\mu}\right) \rightarrow 2\boldsymbol{\partial}_\mu \ , \qquad (5.157)
$$

$$
\hat{O}^n_{\mu_1\mu_2\ldots\mu_n\beta} \rightarrow 2^n \hat{D}_{\mu_1}\hat{D}_{\mu_2}\ldots\hat{D}_{\mu_n}\gamma_\beta \ . \qquad (5.158)
$$

In this way, i.e. by measuring gauge-invariant bilocal structures, the interactions with the gauge field A^a_μ are appearing. In the general interacting case, i.e. if instead of the simple handbag diagrams

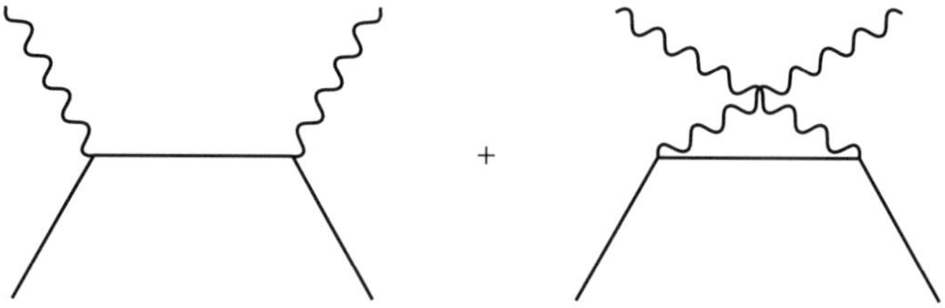

more complicated processes like

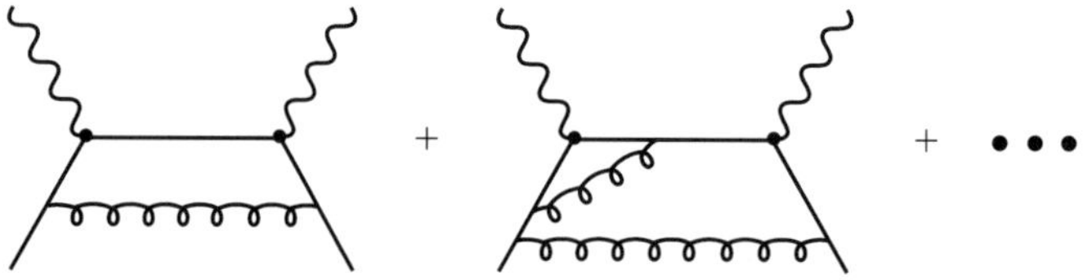

etc. are considered, some modified divergent function takes the place of $\partial^\alpha[\varepsilon(x_0)\delta(x^2)]$. The leading divergence remains the same, but additional terms occur. Also the appearance of $s_{\mu\alpha\mu'\beta}$ was due to the specific situation of the free case (see (5.130)). Now the general form of (5.138) is

$$W_{\mu\mu'} = \frac{1}{2\pi}\int \mathrm{d}^4x\ \exp(\mathrm{i}q\cdot x)\sum_j\sum_n\Big[-g_{\mu\mu'}\,x^{\mu_1}x^{\mu_2}\dots x^{\mu_n}\mathrm{i}^n C^{(n)}_{1,j}(x) + \left(g_{\mu\mu_1}g_{\mu'\mu_2}x^{\mu_3}x^{\mu_4}\dots x^{\mu_n}\right)\mathrm{i}^{n-2}C^{(n)}_{2,j}(x)\Big] \times\frac{1}{2}\sum_{\text{pol.}}\langle \mathrm{N}|\hat{O}^{(n),j}_{\mu_1\mu_2\dots\mu_n}|\mathrm{N}\rangle\ . \tag{5.159}$$

The general form again is determined by the requirements of Lorentz covariance and current conservation which lead in the unpolarized case to one structure function proportional to $g_{\mu\mu'}$ and another structure function proportional to $P_\mu P_{\mu'}$ (see (5.154)). Instead of the tensor $s_{\mu\alpha\mu'\beta}$ in (5.133) there are now two independent contributions which in coordinate space are proportional to $g_{\mu\mu'}$ and $x_\mu x_{\mu'}$, respectively. The $C^{(n)}_{1j}$ and $C^{(n)}_{2j}$ are divergent functions. They are independent Wilson coefficients corresponding to $F_1(x,Q^2)$ and $F_2(x,Q^2)$, respectively.

Again in leading order all terms containing g_{ij} can be neglected, since the matrix element of the operator $\hat{O}^{(n)j}$ is proportional to $P_{\mu_1}P_{\mu_2}\dots P_{\mu_n}$. Therefore all terms with $g_{\mu_i\mu_j}$ yield much smaller contributions than those with $x_{\mu_i}x_{\mu_j}$:

$$\frac{x^2 g_{\mu_i\mu_j}P^{\mu_i}P^{\mu_j}}{x_{\mu_i}x_{\mu_j}P^{\mu_i}P^{\mu_j}} \xrightarrow{\text{Fourier transformation}} \frac{q^2P^2}{(q\cdot P)^2} = -\frac{M^2Q^2}{\nu^2} = -\frac{2M^2x}{\nu} \to 0 \ . \qquad (5.160)$$

Since we do not know the expansion analogous to (5.138), we have to regard the divergent functions of the two terms $C_1^{(n)}$ and $C_2^{(n)}$ that occur as different. Furthermore we must take into account that the operator $\hat{O}^{(n)}$ can now change the flavor of a quark and can in addition act on the gluons. Consequently we have to distinguish between three operators:

$$j=q\ :\ \hat{O}^{(n),q}_{\mu_1\mu_2\dots\mu_n} = \frac{\mathrm{i}^{n-1}}{2n!}\{\gamma_{\mu_1}\hat{D}_{\mu_2}\hat{D}_{\mu_3}\dots\hat{D}_{\mu_n} + \text{permutations of vector indices}\} \qquad (5.161)$$

acts on quark states (SU(3) singlet);

$$j=\mathrm{NS}\ :\ \hat{O}^{(n),\mathrm{NS}}_{\mu_1\mu_2\dots\mu_n,a} = \frac{\mathrm{i}^{n-1}}{2n!}\left\{\frac{\hat{\lambda}^a}{2}\gamma_{\mu_1}\hat{D}_{\mu_2}\hat{D}_{\mu_3}\dots\hat{D}_{\mu_n} + \text{permutations}\right\} \qquad (5.162)$$

acts on quark states and can change the flavor (SU(3) octet); and

$$j=\mathrm{G}\quad :\quad \hat{O}^{(n),\mathrm{G}}_{\mu_1\mu_2\dots\mu_n} \qquad (5.163)$$

acts on the gluon states

$$\langle\bar{F}_{\mu\nu}|\hat{O}^{(n),\mathrm{G}}_{\mu_1\mu_2\dots\mu_n}|\bar{F}_{\mu\nu}\rangle = \frac{\mathrm{i}^{n-2}}{n!}\mathrm{tr}\{\bar{F}_{\mu\mu_1}\hat{D}_{\mu_2}\hat{D}_{\mu_3}\dots\hat{D}_{\mu_{n-1}}\bar{F}^{\mu}_{\mu_n} + \text{permutations}\}. \qquad (5.164)$$

Here we have chosen the standard normalizations for C_1 and C_2, which are called coefficient functions. Since P_μ is the only four-vector that occurs in

$$\frac{1}{2}\sum_{\text{pol.}}\langle N(P)|\hat{O}^{(n),j}_{\mu_1\mu_2\dots\mu_n}|N(P)\rangle\ , \qquad (5.165)$$

we can expand (5.165) into powers of P_μ. Again $g_{\mu_i\mu_j}P^2$ is negligible compared to $P_{\mu_i}P_{\mu_j}$ and therefore the leading term is

$$\frac{1}{2}\sum_{\text{pol.}}\langle N(P)|\hat{O}^{(n),j}_{\mu_1\mu_2\dots\mu_n}|N(P)\rangle = A^{(n),j}P_{\mu_1}P_{\mu_2}\dots P_{\mu_n}\ , \qquad (5.166)$$

and

$$W_{\mu\mu'} = \frac{1}{2\pi} \int d^4x \exp(iq\cdot x) \sum_j \sum_n \Big[-g_{\mu\mu'}(x\cdot P)^n i^n C_{1,j}^{(n)}(x) + P_\mu P_{\mu'}(x\cdot P)^{n-2} i^{n-2} C_{2,j}^{(n)}(x) \Big] A^{(n),j} \ . \tag{5.167}$$

Now we substitute for

$$x_\mu \exp(iq\cdot x) = -i\frac{\partial}{\partial q^\mu}\exp(iq\cdot x) = -2iq^\mu \frac{\partial}{\partial q^2}\exp(q\cdot x) \ , \tag{5.168}$$

and therefore

$$x_{\mu_1}\dots x_{\mu_n} \to (-i)^n \frac{\partial}{\partial q^{\mu_1}}\cdots\frac{\partial}{\partial q^{\mu_n}} = (-2i)^n q_{\mu_1}\dots q_{\mu_n}\left(\frac{\partial}{\partial q^2}\right)^n + \text{trace terms} \ , \tag{5.169}$$

giving

$$\begin{aligned} W_{\mu\mu'} &= \frac{1}{2\pi}\sum_{j,n}\Bigg[-g_{\mu\mu'}(2q\cdot P)^n \left(\frac{\partial}{\partial q^2}\right)^n \int d^4x \exp(iq\cdot x)\, C_{1,j}^{(n)}(x) \\ &\quad + P_\mu P_{\mu'}(2q\cdot P)^{n-2}\left(\frac{\partial}{\partial q^2}\right)^{n-2}\int d^4x \exp(iq\cdot x)\, C_{2,j}^{(n)}(x)\Bigg] A^{(n),j} \\ &= \frac{1}{2\pi}\sum_{j,n}\Bigg[-g_{\mu\mu'}\left(\frac{2q\cdot P}{Q^2}\right)^n (Q^2)^n \left(\frac{\partial}{\partial q^2}\right)^n \\ &\quad \times \int d^4 \exp(iq\cdot x)\, C_{i,j}^{(n)}(x) \quad + \frac{P_\mu P_{\mu'}}{2q\cdot P}\left(\frac{2q\cdot P}{Q^2}\right)^{n-1} \\ &\quad \times (Q^2)^{n-1}\left(\frac{\partial}{\partial q^2}\right)^{n-2}\int d^4x \exp(iq\cdot x)\, C_{2,j}^{(n)}(x)\Bigg] \ . \end{aligned} \tag{5.170}$$

Just as in the free case we can read off the structure functions $F_1(x)$ and $F_2(x)$, which in their most general form are now

$$F_1(x) = \frac{1}{2\pi}\sum_{j,n}\left[\left(\frac{2q\cdot p}{Q^2}\right)^n (Q^2)^n \left(\frac{\partial}{\partial q^2}\right)^n\right] \times \int d^4x \exp(iq\cdot x)\, C_{1,j}^{(n)}(x) A^{(n),j} \ , \tag{5.171}$$

$$F_2(x) = \frac{1}{4\pi}\sum_{j,n}\left[\left(\frac{2q\cdot p}{Q^2}\right)^{n-1} (Q^2)^{n-1} \left(\frac{\partial}{\partial q^2}\right)^{n-2}\right] \times \int d^4x \exp(iq\cdot x)\, C_{2,j}^{(n)}(x) A^{(n),j} \ . \tag{5.172}$$

To simplify these expressions we define the nth moment of the coefficient functions

$$\tilde{C}^{(n)}_{1,j}(Q^2) = \frac{1}{4\pi}(Q^2)^n \left(\frac{\partial}{\partial q^2}\right)^n \int \mathrm{d}^4x \exp(\mathrm{i}q\cdot x)\, C^{(n)}_{1,j}(x) \tag{5.173}$$

and

$$\tilde{C}^{(n)}_{2,j}(Q^2) = \frac{1}{16\pi}(Q^2)^{n-1} \left(\frac{\partial}{\partial q^2}\right)^{n-2} \int \mathrm{d}^4x \exp(\mathrm{i}q\cdot x)\, C^{(n)}_{2,j}(x) \ . \tag{5.174}$$

Now the structure functions are

$$F_1(x, Q^2) = 2\sum_{j,n} x^{-n}\tilde{C}^{(n)}_{1,j} A^{(n),j} \ , \tag{5.175}$$

$$F_2(x, Q^2) = 4\sum_{j,n} x^{-n+1}\tilde{C}^{(n)}_{2,j} A^{(n),j} \ . \tag{5.176}$$

Note that (5.175) and (5.176) now have the form of a Taylor expansion in $1/x$. However, physically $0 < x < 1$, so that a series in $1/x$ hardly makes sense. The key observation is, however, that this expansion which is mathematically correct in the unphysical region of the current commutator $1 < x < \infty$ can be analytically continued with the help of a dispersion relation. This will be done below in (5.199) to (5.205). What have we gained by this general formulation? We have expressed the structure functions by sums over products of divergent functions $\tilde{C}_{1,j}$, $\tilde{C}_{2,j}$ and unknown constants $A^{(n),j}$. The only thing we know is that the $C_{1,j}$, $C_{2,j}$ are independent of the hadron considered. Therefore the $\tilde{C}_{1,j}$, $\tilde{C}_{2,j}$ can only depend on Q^2 and the constants of the theory, which are the coupling constant g and the renormalization point μ:

$$\begin{aligned} \tilde{C}^{(n)}_{1,j} &= \tilde{C}^{(n)}_{1,j}(Q^2, g, \mu) \ , \\ \tilde{C}^{(n)}_{2,j} &= \tilde{C}^{(n)}_{2,j}(Q^2, g, \mu) \ . \end{aligned} \tag{5.177}$$

In addition, the constants $A^{(n),j}$ are matrix elements of certain operators sandwiched between nucleon states. Therefore all bound state complexities inherent in the $|N(p)\rangle$ state are buried in these matrix elements. Hence, we know that the constants $A^{(n),j}$ are characteristic for the hadronic state under consideration. Of course, all quantities can be approximately evaluated, the $\tilde{C}_{1,j}$, $\tilde{C}_{2,j}$ reliably by means of perturbation theory from the free current commutator $\left[\hat{J}_\mu(x), \hat{J}_\nu(0)\right]$ (see (5.126)) and the QCD Feynman rules. The $A^{(n),j}$, however, are truely nonperturbative and can only be obtained from a phenomenological model of the nucleon or from lattice QCD or from a sum-rule calculation, using (5.166). The finite matrix elements $A^{(n),j}$ thus contain information about the inner structure of the nucleons, for example. To obtain the perturbative expansion of the $\tilde{C}^{(n)}_{1,j}$, $\tilde{C}^{(n)}_{2,j}$ for the structure functions, we have to repeat the calculation in

Sect. 5.1, which again leads to the GLAP equations. An advantage of this general OPE formulation is that formal features of the $\tilde{C}_{1,j}, \tilde{C}_{2,j}$ can be deduced and relationships between different quantities can be established for the full nonperturbative expressions. The most important way to deduce information about the coefficient functions is to consider their renormalization behavior.

Here we only need to know that QCD is renormalizable and that for a given renormalization scheme carried out for a specific kinematic the bare coupling constants, Green functions, and so on are replaced by the renormalized ones. The renormalization schemes differ by the chosen kinematics, which can in general be characterized by a momentum parameter μ^2. Correspondingly the renormalized functions $\tilde{C}_1, \tilde{C}_2$ in (5.177) can also be written as

$$\tilde{C}_{i,j}^{(n)}(Q^2, g, \mu) = \sum_k Z_{jk}^{(n)}(g_0, \mu)\tilde{C}_{i,k}^{(n)}(Q^2, g_0)_{\text{unren.}} \; , \qquad (5.178)$$
$$g = g(g_0, \mu, Q^2) \; .$$

Equation (5.178) takes care of the fact that renormalization in general mixes different $\tilde{C}_j^{(n)}$, $j = q$, NS, G. This can be understood qualitatively by taking into account that not only do the graphs

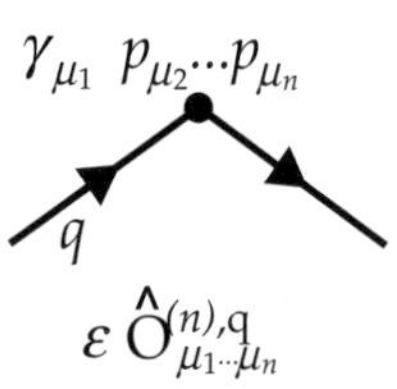

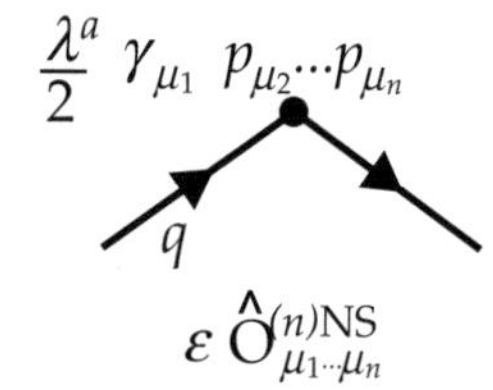

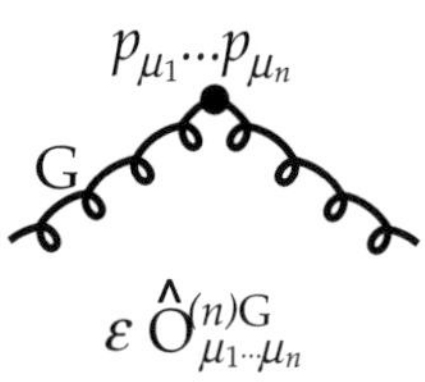

occur, but also, for example, the graph

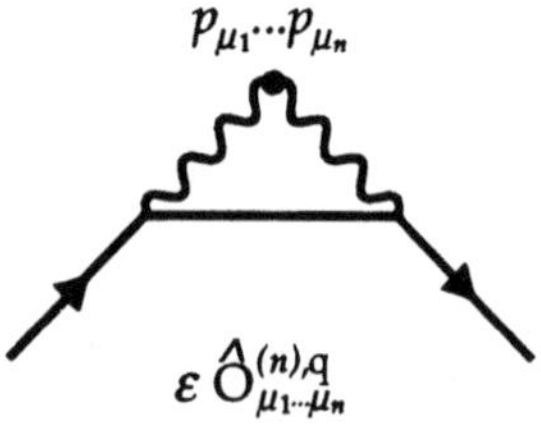

Hence the sum over $k = q$, NS, G occurs on the right-hand side of (5.178). Since the physical theory must not depend on the renormalization point μ, we can demand that

$$\frac{d\tilde{C}_{ij}^{(n)}(Q^2, g, \mu)}{d\mu} = 0 \; .$$

Now the total derivative is split up into partial derivatives.

$$\begin{aligned}
\frac{\mathrm{d}}{\mathrm{d}\mu}\tilde{C}^{(n)}_{i,j}(Q^2,g,\mu) &= \left(\frac{\partial g}{\partial\mu}\frac{\partial}{\partial g}+\frac{\partial}{\partial\mu}\right)\tilde{C}^{(n)}_{i,j}(Q^2,g,\mu) \\
&= \sum_k \frac{\partial Z^{(n)}_{jk}}{\partial\mu}\tilde{C}^{(n)}_{i,k}(Q^2,g_0)_{\text{unren.}} \\
&= \sum_{klm} \frac{\partial Z^{(n)}_{jk}}{\partial\mu}\left(Z^{(n)}\right)^{-1}_{kl}\left(Z^{(n)}\right)_{lm}\tilde{C}^{(n)}_{i,m}(Q^2,g_0)_{\text{unren.}} \\
&= \sum_l \left[\sum_k \frac{\partial Z^{(n)}_{jk}}{\partial\mu}\left(Z^{(n)}\right)^{-1}_{kl}\right]\tilde{C}^{(n)}_{i,l}(Q^2,g,\mu)\ .
\end{aligned}$$

Defining

$$\gamma^{(n)}_{jl} := -\mu\sum_k \frac{\partial Z^{(n)}_{jk}}{\partial\mu}\left(Z^{(n)}\right)^{-1}_{kl} = \gamma^{(n)}_{jl}(g,\mu) \tag{5.179}$$

yields

$$\sum_l \left[\left(\mu\frac{\partial g}{\partial\mu}\frac{\partial}{\partial g}+\mu\frac{\partial}{\partial\mu}\right)\delta_{jl}+\gamma^{(n)}_{jl}\right]\tilde{C}^{(n)}_{i,l}(Q^2,g,\mu)=0\ . \tag{5.180}$$

The term $\partial g/\partial\mu$, i.e., the dependence of the coupling constant on the renormalization point, is a fixed characteristic function for every field theory. It is referred to as the β *function*:

$$\beta(g)=\mu\frac{\partial g}{\partial\mu}\ . \tag{5.181}$$

In the case of QCD, perturbation theory yields (see Exercise 5.11)

$$\beta(g) = -\left(11-\frac{2}{3}N_{\mathrm{f}}\right)\frac{g^3}{(4\pi)^2} - \left(102-\frac{38}{3}N_{\mathrm{f}}\right)\frac{g^5}{(4\pi)^4}+O(g^7)\ , \tag{5.182}$$

where N_{f} denotes the number of quark flavors.

EXERCISE

5.11 The Lowest-Order Terms of the β Function

Problem. Show that (5) in Example 4.3 or (4.156) indeed yields the lowest-order contribution to the β function:

$$\alpha_s(-q^2) = \frac{\alpha_s(\mu^2)}{1+\frac{11-2N_\mathrm{f}(-q^2)/3}{4\pi}\alpha_s(\mu^2)\ln\left(-\frac{q^2}{\mu^2}\right)} . \tag{1}$$

Here μ^2 denotes the renormalization point where $\alpha_s(-q^2)$ assumes the renormalized value $\alpha_s(\mu^2)$. $N_\mathrm{f}(-q^2)$ is the number of light quarks with masses smaller than $-q^2$.

Solution. Equation (1) yields for $g(-q^2)$

$$g(-q^2) = \frac{g(\mu^2)}{\sqrt{1+\frac{11-2N_\mathrm{f}(-q^2)/3}{(4\pi)^2}g^2(\mu^2)\ln\left(-\frac{q^2}{\mu^2}\right)}} \tag{2}$$

with $\alpha_s = g^2/4\pi$. We multiply (2) by the demominator on the right-hand side and differentiate the resulting equation with respect to μ^2:

$$\begin{aligned} g'(\mu^2) &= g(-q^2)\frac{\partial}{\partial\mu^2}\sqrt{1+\frac{11-\frac{2}{3}N_\mathrm{f}(-q^2)}{(4\pi)^2}g^2(\mu^2)\ln\left(-\frac{q^2}{\mu^2}\right)} \\ &= g(-q^2)\frac{\frac{1}{2}\frac{11-2N_\mathrm{f}(-q^2)/3}{(4\pi)^2}}{\sqrt{1+\frac{11-2N_\mathrm{f}(-q^2)/3}{(4\pi)^2}g^2(\mu^2)\ln\left(-\frac{q^2}{\mu^2}\right)}} \\ &\quad\times\left[2g(\mu^2)g'(\mu^2)\ln\left(-\frac{q^2}{\mu^2}\right)-\frac{1}{\mu^2}g^2(\mu^2)\right] . \end{aligned} \tag{3}$$

Once more we insert (2), into this expression and collect the terms containing g':

$$\begin{aligned} & g'(\mu^2)\left[1+\frac{11-\frac{2}{3}N_\mathrm{f}(-q^2)}{(4\pi)^2}g^2(\mu^2)\ln\left(-\frac{q^2}{\mu^2}\right)\right] \\ &\quad = \frac{11-\frac{2}{3}N_\mathrm{f}(-q^2)}{(4\pi)^2}\left[g^2(\mu^2)g'(\mu^2)\ln\left(-\frac{g^2}{\mu^2}\right)-\frac{g^3(\mu^2)}{2\mu^2}\right] , \end{aligned} \tag{4}$$

$$g'(\mu^2) = -\frac{1}{2\mu^2}\frac{11-\frac{2}{3}N_\mathrm{f}(-q^2)}{(4\pi)^2}g^3(\mu^2) . \tag{5}$$

By means of (5) the β function now assumes the form

$$\beta = \mu\frac{\partial g(\mu)}{\partial\mu} = 2\mu^2\frac{\partial g(\mu^2)}{\partial\mu^2} = -\left[11-\frac{2}{3}N_\mathrm{f}(-q^2)\right]\frac{g^3(\mu^2)}{(4\pi)^2} . \tag{6}$$

Exercise 5.11

In fact this is identical with the first term in (5.182). We can see immediately where the higher terms come from. From Example 4.3, (1) is derived from vacuum polarization graphs. Consequently we obtain the higher-order contributions to the β function if higher graphs are taken into account in this calculation.

Finally, we summarize the basic ideas of this last calculation. The renormalization is carried out for a given value μ^2 of the momentum transfer, i.e., all divergencies are subtracted in that renormalization scheme by relating to the *measured* value of the coupling constant at the point, i.e. to $\alpha_s(\mu^2)$. This procedure yields a renormalized coupling constant $\alpha_s(\mu^2)$ which has to be identified with the physical coupling constant. However, we must decide on what we mean by "physical QCD coupling" and then we must express the effective (running) coupling $\alpha_s(Q^2)$ in terms of it. In QED the effective coupling is expressed in terms of the fine structure constant $\alpha \sim 1/137$ which is defined as the effective coupling in the limit $-q^2 \to 0$. In QCD $\alpha_s(-q^2)$ diverges as $-q^2 \to 0$. Hence we cannot define $\alpha_s(-q^2)$ in terms of its value at $-q^2 = 0$. Instead we choose some value of $-q^2$, say $-q^2 = \mu^2$, and define the experimental QCD coupling to be $\alpha_s \equiv \alpha_s(\mu^2)$. The *renormalization point* μ^2 is, of course, arbitrary. Had we instead chosen the point $\overline{\mu}^2$ then the two couplings would be related by

$$\alpha_s(\overline{\mu}^2) = \frac{\alpha_s(\mu^2)}{1 + \left(\frac{11 - 2N_f/3}{4\pi}\right) \ln\left(\frac{\overline{\mu}^2}{\mu^2}\right)} \, .$$

Obviously (5.180) yields a restriction for $\tilde{C}^{(n)}_{i,\ell}$, from which one can derive, using (5.175) and (5.176), a constraint for the observable structure functions. In this way our formal analysis leads to measurable predictions. Equations of the type (5.180) are called *renormalization group equations* or *Callan–Symanzik equations*. They play an important role in formal considerations of field theories, because they yield nonperturbative results.

We now wish to explain the meaning of the β function in some more detail. As shown in Exercise 5.11, the coupling constant g in fact depends on the dimensionless quantity $t = +\frac{1}{2} \ln Q^2/\mu^2$. This enables us to write the defining equation of the dimensionless β function as

$$\beta = \mu \frac{\partial g(\mu)}{\partial \mu} = \mu \frac{\partial t}{\partial \mu} \frac{\partial g(t)}{\partial t} = -\frac{\partial g(t)}{\partial t} \, . \tag{5.183}$$

The β function describes how the coupling depends on the momentum transfer, given a fixed renormalization point μ^2. In fact there are two definitions known in the literature. The first one is expression (5.183) and the second one is

$$\tilde{\beta}(g) = t \frac{\partial g(t)}{\partial t} = Q^2 \ln \frac{Q^2}{\mu^2} \frac{\partial g(Q^2)}{\partial Q^2} \, . \tag{5.184}$$

A great advantage of the latter definition is that its features are not dependent on the sign of t and they can be analyzed in a more general manner. Because of

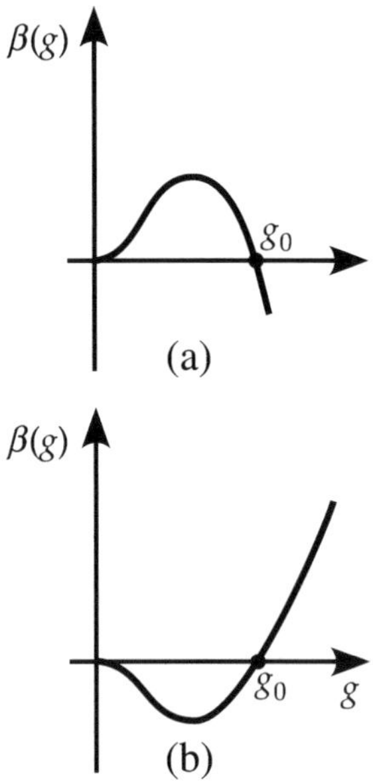

Fig. 5.14a,b. The behaviour of the β function as a function of the running coupling constant g.The arrows attached to the curve indicate the direction of movement along $g(Q^2)$ when $Q^2 \to \infty$

these two different definitions, the renormalization group equations found in the literature sometimes differ by a factor t in front of the β function.

The zeros of the β function are crucial for the general behavior of the coupling constant in the case of very large momentum transfers (the so-called *ultraviolet limit*) and very small momentum transfers (the so-called *infrared limit*) and thus for the most basic properties of the theory. This can be easily understood by analyzing the two examples in Fig. 5.14.

Case (a) If $\beta(g)$ is positive, then g becomes larger with increasing Q^2. In the region $g < g_0$, g therefore approaches the value g_0 with increasing Q^2. On the other hand $\beta(g)$ is negative for $g > g_0$ and here g becomes smaller with increasing Q^2. g_0 is therefore called a stable ultraviolet fixpoint. This point, however, is not stable in the infrared limit. As soon as g is different from g_0, and consequently $\tilde{\beta}(g)$ no longer vanishes, g moves away from the value g_0 for decreasing Q^2. Correspondingly the behavior at the point $g = 0$ is just the opposite. This is a stable infrared fixpoint and it is unstable in the ultraviolet limit. Hence Fig. 5.12a characterizes a theory whose coupling constant vanishes for small momentum transfers and assumes a constant value for large momentum transfers.

Case (b) In this case the points $g = 0$ and $g = g_0$ have exchanged their meanings compared to *Case (a)*. The coupling constant vanishes for large momentum transfers. This behavior is called *asymptotic freedom*. For small momentum transfers g assumes the constant value g_0. If g_0 is very large, which is known as *infrared slavery*, such a theory exhibits similar features to QCD. For our problem we want to evaluate (5.180) using perturbation theory. To this end we use

$$\beta = -\left(11 - \frac{2}{3}N_\mathrm{f}\right)\frac{g^3}{(4\pi)^2} = -bg^3 \tag{5.185}$$

as an approximation for $\beta(g)$. With $\bar{g}$ we now denote the running coupling constant as the solution of (5.185). Combining (5.183) with (5.185) we find

$$\bar{g}^2 = \frac{g^2}{1+2bg^2t} , \tag{5.186}$$

and (5.180) becomes

$$\sum_\ell \left[\left(-bg^3\frac{\partial}{\partial g} + \mu\frac{\partial}{\partial \mu}\right)\delta_{j\ell} + \gamma_{j\ell}^{(n)}\right]\tilde{C}_{i,\ell}^{(n)}(Q^2, g, \mu) = 0 . \tag{5.187}$$

Furthermore it holds that

$$\frac{\partial \bar{g}}{\partial t} = \frac{1}{2\bar{g}}\frac{\partial \bar{g}^2}{\partial t} = \frac{1}{2\bar{g}}\frac{-g^2 2bg^2}{(1+2bg^2t)^2} = -b\bar{g}^3 , \tag{5.188}$$

$$\frac{\partial \bar{g}}{\partial g} = \frac{g}{\bar{g}}\frac{\partial \bar{g}^2}{\partial g^2} = \frac{g}{\bar{g}}\frac{1}{(1+2bg^2t)^2} = \frac{\bar{g}^3}{g^3} , \tag{5.189}$$

and hence the derivatives in (5.187) acting on $\bar{g}$ yield zero:

$$\left(-bg^3\frac{\partial}{\partial g} - \frac{\partial}{\partial t}\right)\bar{g}(g,t) = -bg^3\frac{\bar{g}^3}{g^3} + b\bar{g}^3 = 0 . \tag{5.190}$$

The solution of (5.187) for $\gamma^{(n)}_{i,j\ell} = 0$ is therefore simply

$$\tilde{C}^{(n)}_{i,\ell}(Q^2, g, \mu) = f_{i,\ell}(\bar{g}(g,t)) \ , \tag{5.191}$$

with an unknown function $f_{i,\ell}(\bar{g}(g,t))$. Thus the renormalization group equation implies that, when $\gamma^{(n)}_{i,j\ell} = 0$, the entire Q^2 dependence arises from the running of the coupling constant $\overline{g}(g,t)$ which in turn is governed by (5.181). Therefore, if we replace g by the running $\overline{g}(t)$ we find with the help of (5.188) that (5.187) can be transformed into a total derivative

$$\left(-b\overline{g}^3 \frac{\partial}{\partial g} + \mu \frac{\partial}{\partial \mu}\right) = \frac{\partial \overline{g}}{\partial t}\frac{\partial}{\partial g} + \frac{\partial}{\partial t} = \frac{\mathrm{d}}{\mathrm{d}t} \ , \tag{5.192}$$

so that the differential equation

$$\sum_\ell \left(\frac{\mathrm{d}}{\mathrm{d}t}\delta_{j\ell} + \gamma^{(n)}_{j\ell}\right) \tilde{C}^{(n)}_{i,\ell}(\overline{g}(g,t)) = 0 \tag{5.193}$$

can be easily integrated to give

$$\tilde{C}^{(n)}_{i,\ell}(Q^2, g, \mu) = \sum_k f_{i,k}\left(\bar{g}(g,t)\right) \exp\left[-\int\limits_0^t \mathrm{d}t' \gamma^{(n)}_{k,\ell}\left(\bar{g}(g,t')\right)\right] \ . \tag{5.194}$$

Now we also expand the $\gamma^{(n)}_{k,\ell}$ into a power series

$$\gamma^{(n)}_{k,\ell} = b^{(n)}_{k,\ell} + c^{(n)}_{k,\ell}\bar{g} + d^{(n)}_{k,\ell}\bar{g}^2 + \cdots \ . \tag{5.195}$$

A perturbative calculation shows that both the constant and linear terms are equal to zero. The lowest-order contribution is proportional to $\bar{g}^2$. Hence the t' integral can be approximately evaluated:

$$-\int\limits_0^t \mathrm{d}t' \gamma^{(n)}_{k,\ell} \approx -d^{(n)}_{k,\ell} \int\limits_0^t \frac{g^2}{1+2bg^2t'}\mathrm{d}t' = -\frac{d^{(n)}_{k,\ell}}{2b}\ln(1+2bg^2t) \ , \tag{5.196}$$

$$\begin{aligned}\tilde{C}^{(n)}_{i,\ell}(t,g) &= \sum_k f_{i,k}\left(\bar{g}(g,t)\right)(1+2bg^2t)^{-d^{(n)}_{k,\ell}/2b} \\ &= \sum_k f_{i,k}\left(\bar{g}(g,t)\right)\left(\bar{g}^2(g,t)/g^2\right)^{d^{(n)}_{k,\ell}/2b} \ .\end{aligned} \tag{5.197}$$

In the last step we used the definition of the running coupling in (5.186). In the limiting case $t \gg 1$, $\bar{g} \to 0$ the function $f_{i,k}\left(\bar{g}(g,t)\right)$ assumes a constant value $a^{(n)}_{i,k}$:

$$t \gg 1\ , \quad x \text{ constant} \quad \Rightarrow \quad \tilde{C}^{(n)}_{i,\ell}(t,g) = \sum_k a^{(n)}_{i,k}\left(\bar{g}^2(g,t)/g^2\right)^{d^{(n)}_{k,\ell}/2b} \ . \tag{5.198}$$

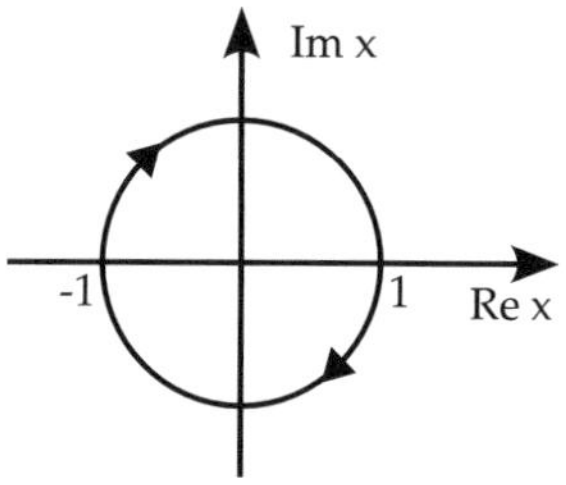

Fig. 5.15. Integration contour $\mathcal{C}$ in (5.199) and (5.200)

We insert this into (5.175) and (5.176), which we slightly modify for this purpose. Let $F_i(z, Q^2)$ be the complex continuation of $F_i(x, Q^2)$. Then we obviously have

$$\frac{1}{2\pi\mathrm{i}}\int\limits_{\mathcal{C}} F_1(z, Q^2)z^{n-1}\,\mathrm{d}z = 2\sum_j \tilde{C}^{(n)}_{1,j}A^{(n),j}\ , \tag{5.199}$$

$$\frac{1}{2\pi\mathrm{i}}\int\limits_{\mathcal{C}} F_2(z, Q^2)z^{n-2}\,\mathrm{d}z = 4\sum_j \tilde{C}^{(n)}_{2,j}A^{(n),j}\ , \tag{5.200}$$

with the integration contour C^n depicted in Fig. 5.15. Since $F_1(z)$ and $F_2(z)$ can become singular only in the case $z \to 0$, $\mathcal{C}$ can be deformed and replaced by an ε prescription:

$$\begin{aligned}\int\limits_{\mathcal{C}} F_1(z, Q^2)z^{n-1}\,\mathrm{d}z &= \int\limits_{-1}^{1} F_1(x+\mathrm{i}\varepsilon, Q^2)x^{n-1}\,\mathrm{d}x \\ &\quad - \int\limits_{-1}^{1} F_1(x-\mathrm{i}\varepsilon, Q^2)x^{n-1}\,\mathrm{d}x \\ &= \int\limits_{-1}^{1}\Big(F_1(x+\mathrm{i}\varepsilon, Q^2) - F_1^*(x+\mathrm{i}\varepsilon, Q^2)\Big)\,x^{n-1}\,\mathrm{d}x\ .\end{aligned} \tag{5.201}$$

Now we have to determine how the analytically continued structure function behaves under complex conjugation. Thus we have to take into account that $W_{\mu\nu}$ is the imaginary part of the scattering amplitude in the forward direction $T_{\mu\nu}$ (this is a special case of the so-called optical theorem):

$$W_{\mu\nu} = \frac{1}{\pi}\mathrm{Im}\left(T_{\mu\nu}\right)\ . \tag{5.202}$$

Correspondingly we have

$$F_1\left(x+\mathrm{i}\varepsilon, Q^2\right) - F_1^*\left(x+\mathrm{i}\varepsilon, Q^2\right) = 2\mathrm{Im}F_1\left(x, Q^2\right)\ , \tag{5.203}$$

$$\int\limits_{\mathcal{C}} F_1\left(z, Q^2\right)z^{n-1}\,\mathrm{d}z = 4\int\limits_{0}^{1} F_1\left(x, Q^2\right)x^{n-1}\,\mathrm{d}x =: 4M_1^{(n+1)}\left(Q^2\right)\ , \tag{5.204}$$

$$\int\limits_{\mathcal{C}} F_2\left(z, Q^2\right)z^{n-2}\,\mathrm{d}z = 4\int\limits_{0}^{1} F_2\left(x, Q^2\right)x^{n-2}\,\mathrm{d}x =: 4M_2^{(n)}\left(Q^2\right)\ . \tag{5.205}$$

This completes our derivation of (5.175) and (5.176) where the structure functions were given as expansions in $1/x$. Now we have suceeded in analytically continuing them to the physical region $0 < x < 1$. The quantities $M_1^{(n)}$ and $M_2^{(n)}$ are called *moments of the structure functions*. Owing to (5.198) they obey the evolution equations

$$M_1^{(n+1)}(t) = \frac{1}{2}\sum_{jk} a_{1,k}^{(n)} \left(\frac{\bar{g}^2(g,t)}{g^2}\right)^{d_{kj}^{(n)}/2b} A^{(n),j} \ , \tag{5.206}$$

$$M_2^{(n)}(t) = \sum_{jk} a_{2,k}^{(n)} \left(\frac{\bar{g}^2(g,t)}{g^2}\right)^{d_{kj}^{(n)}/2b} A^{(n),j} \ . \tag{5.207}$$

In fact these equations correspond to the GLAP equations. We shall prove this for the nonsinglet contributions to the moments of the structure functions. Since $d_{kj}^{(n)}$ is diagonal for $j = \mathrm{NS}$, the manipulations required are quite simple. The moments of the NS-quark distribution functions are defined as

$$M_{1,ij}^{(n+1)}(t) = \int_0^1 x^{n-1} \Delta_{ij}(x,t)\ \mathrm{d}x \ . \tag{5.208}$$

Equation (5.67) then yields (we substitute $x = yz$ in the second step)

$$\begin{aligned}
\frac{\mathrm{d}}{\mathrm{d}t} M_{1,ij}^{(n+1)}(t) &= \int_0^1 \mathrm{d}x\, x^{n-1} \frac{\alpha_s(t)}{2\pi} \int_x^1 \frac{\mathrm{d}y}{y} \Delta_{ij}(y,t) P_{qq}\left(\frac{x}{y}\right) \\
&= \frac{\alpha_s(t)}{2\pi} \int_0^1 \mathrm{d}y \int_0^y \frac{\mathrm{d}x}{y} x^{n-1} \Delta_{ij}(y,t) P_{qq}\left(\frac{x}{y}\right) \\
&= \frac{\alpha_s(t)}{2\pi} \int_0^1 \mathrm{d}y \Delta_{ij}(y,t) y^{n-1} \int_0^1 \mathrm{d}z P_{qq} z^{n-1} \\
&= \frac{\alpha_s(t)}{2\pi} D^{(n)} M_{1,ij}^{(n+1)} \ ,
\end{aligned} \tag{5.209}$$

with

$$D^{(n)} = \int_0^1 \mathrm{d}z P_{qq} z^{n-1} \ . \tag{5.210}$$

From (5.206) it follows that

$$M_{1,\mathrm{NS}}^{(n+1)}(t) = M_{1,\mathrm{NS}}^{(n+1)} \left(\frac{\bar{g}^2(g,t)}{g^2}\right)^{d_{\mathrm{NS}}^{(n)}/2b} \ , \tag{5.211}$$

where $M^{(n+1)}_{1,\mathrm{NS}} = M^{(n+1)}_{1,\mathrm{NS}}(t=0) = \frac{1}{2}\sum_j a^{(n)}_1 A^{(n),j}$ is the moment of the structure function at the renormalization point $\mu^2 = Q^2$, i.e. $t = \frac{1}{2}\ln(Q^2/\mu^2) = 0$ and correspondingly $g(t=0) = g$. We now differentiate this equation with respect to t, so that we can compare it with (5.209), especially the coefficients of both equations, and find a relation between the integral over the splitting function $D^{(n)}$ (5.210) and the anomalous dimensions $d^{(n)}_{\mathrm{NS}}$:

$$\begin{aligned}\frac{\partial M^{(n+1)}_{1,\mathrm{NS}}(t)}{\partial t} &= \frac{d^{(n)}_{\mathrm{NS}}}{2b}\frac{1}{\bar{g}^2}2\bar{g}\left(-b\bar{g}^3\right)M^{(n+1)}_{1,\mathrm{NS}}(t)\\ &= -\bar{g}^2 d^{(n)}_{\mathrm{NS}} M^{(n+1)}_{1,\mathrm{NS}}(t) = \frac{\alpha_s(t)}{2\pi}\left(-8\pi^2 d^{(n)}_{\mathrm{NS}}\right)M^{(n+1)}_{1,\mathrm{NS}}(t)\ . \end{aligned} \tag{5.212}$$

Clearly (5.212) and (5.209) are the same relations, provided $-8\pi^2 d^{(n)}_{\mathrm{NS}}$ is identified with $D^{(n)}$. If the $\gamma^{(n)}_{\mathrm{NS}}$ are evaluated by means of perturbation theory, we obtain the expression (5.210) for $d^{(n)}_{\mathrm{NS}}$ with $P_{qq}(z)$ given in Example 5.6.

Let us now reflect a little more on what we have learnt from the comparison of the OPE approach to evolution of structure functions and the GLAP approach. Equation (5.211) tells us how the QCD equations for the evolution of moments look like. Dropping unnecessary labels we have for the nonsinglet case

$$M^{(n)}_{\mathrm{NS}}(t) = \left[\frac{\bar{g}^2(t)}{g(t=0)}\right]^{d^{(n)}_{\mathrm{NS}}/2b} \cdot M^{(n)}_{\mathrm{NS}}(t=0)\ ,$$

while for the singlet case one has

$$\boldsymbol{M}^{(n)}_{\mathrm{S}}(t) = \left[\frac{\bar{g}^2(t)}{g(t=0)}\right]^{d^{(n)}_{\mathrm{S}}/2b} \cdot \boldsymbol{M}^{(n)}_{\mathrm{S}}(t=0)\ ,$$

where d_S now is to be considered as a matrix

$$d_S = \begin{pmatrix} d_{qq} & d_{qG} \\ d_{Gq} & d_{GG} \end{pmatrix}\ .$$

It only remains to calculate the anomalous dimensions $d^{(n)}_{\mathrm{NS}}$ and $d^{(n)}_{\mathrm{S}}$ to predict the moments of the structure functions at arbitrary $t = \frac{1}{2}\ln(Q^2/\mu^2)$ from those given at $t = 0$. In other words we obtain the moments at virtuality Q^2 if they are known at the renormalization point $Q^2 = \mu^2$. While this approach is fairly rigorous and not too difficult to handle, it may lack physical interpretation and, in particular, the connection to the simple parton model is not to fully apparent.

Taking the derivative on both sides of (5.211) with respect to t and comparing with (5.209) we have immediatly found that (5.211) can be cast into the form of the GLAP equations

$$\frac{\partial \Delta_{ij}}{\partial t} = \frac{\alpha_s}{2\pi}\int\limits_x^1 \frac{\mathrm{d}y}{y}\Delta_{ij}(y,t)P_{qq}\left(\frac{x}{y}\right)\ , \tag{5.213}$$

where we identified

$$\int_0^1 \mathrm{d}z\, P_{qq}(z) z^{n-1} = -8\pi^2 d_{\mathrm{NS}}^{(n)} \ .$$

Thus the integral over the splitting function can be identified with the anomalous dimension and (5.213) tells us how the structure function changes with t. Actually P_{qq} can be interpreted as governing the rate of change. However, this GLAP equation is a complicated integrodifferential equation. Its transformation to momentum space reduces it to a set of ordinary differential equations that can be easily solved.

EXAMPLE

5.12 The Moments of the Structure Functions

In this example we discuss the physical meaning of the moments of structure functions. Looking at (5.175), (5.176), (5.203), (5.204), (5.208), and (5.209) one could get the impression that knowing the structure function is equivalent to knowledge of all its moments. But this is only true in the limit $Q^2 \to \infty$. For every finite Q^2, deviations occur owing to the corrections that were neglected on the way from (5.142) to (5.143). One can prove that the corrections to $\hat{O}^n_{\mu_1\cdots\mu_n\mu_\beta}$, i.e., to $C^{(n)}$ and therefore to the nth moment of the structure function F_2 and to the $(n+1)$th moment of F_1, are suppressed by a factor

$$\frac{n\mu^2}{Q^2} \ , \tag{1}$$

where μ^2 is some mass scale. Equation (1) shows already that for every finite Q^2 the corrections become large at some value of n, i.e., each perturbative calculation of the moments is only valid up to a specific maximum number n. The factor (1) can be understood quite easily. Equation (5.166) is inserted into the expression for the scattering tensor and contracted with the leptonic scattering tensor. Then we count how often terms of the kind

$$q^{\mu_1} \cdots q^{\mu_n} g_{\mu_i\mu_j} p_{\mu_1} \cdots p_{\mu_{i-1}} p_{\mu_{i+1}} \cdots p_{\mu_{j-1}} p_{\mu_{j+1}} \cdots p_{\mu_n} \tag{2}$$

occur compared to the term

$$q^{\mu_1} \cdots q^{\mu_n} p_{\mu_1} \cdots p_{\mu_n} \ . \tag{3}$$

This procedure yields the factor n.[12] Thus a perturbative QCD calculation is well suited for the lower moments, for example, the integral over the structure functions, but fails in the case of the higher moments. From (5.204) and (5.205) these

[12] Since the explicit calculation is quite cumbersome, we refer to A. de Rujula, H. Georgi, and H.D. Politzer: Ann. Phys. **103**, 315 (1977)

Example 5.12

higher moments are determined by the structure functions near $x = 1$. The value $x = 1$, however, represents elastic scattering. Correspondingly here we find the elastic peak and for x slightly below 1 we find peaks due to the different resonances. Since these resonance structures depend on the individual features of the hadronic bound states, there is no way to describe this region by perturbative QCD. On the other hand, this analysis clearly shows that the nonperturbative effects must be included in higher-order terms of the operator product expansion.

The points just discussed (usually referred to as *precocious scaling*) are explained again in Fig. 5.16. This figure depicts the experimentally observed structure function $\nu W_2(x, Q^2)$ (full line), which converges to $F_2(x)$ for large Q^2. The Nachtmann variable

$$\xi = \frac{2x}{1 + \left(1 + 4x^2 \frac{M_N^2}{Q^2}\right)^{\frac{1}{2}}} \tag{4}$$

has been used instead of x. In the limit $Q^2 \to \infty$, ξ is equivalent to x. At finite Q^2 this variable takes the effects of the nucleon mass into account. The contributions

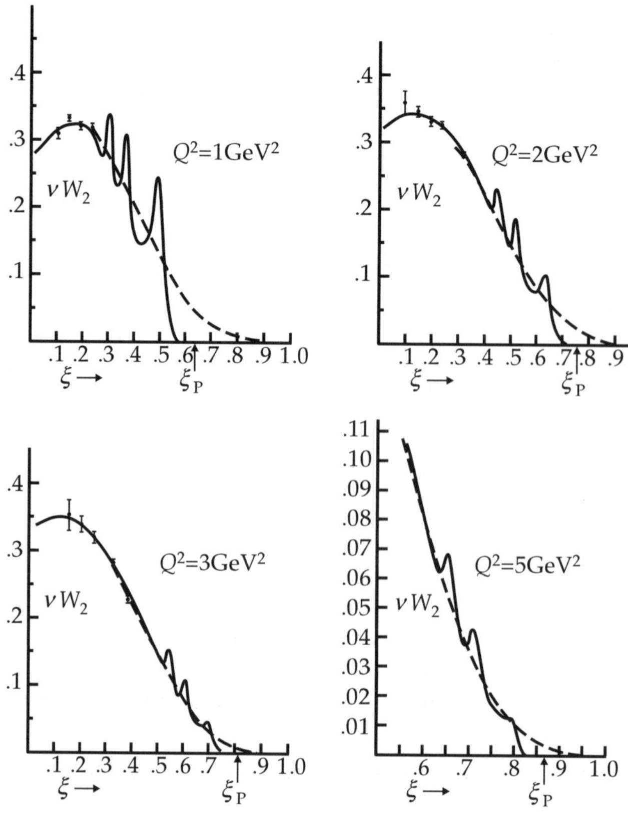

Fig. 5.16. Comparison of the experimental structure function νW_2 for electron–proton scattering with the result of a perturbative QCD calculation. ξ denotes the Nachtmann variableand ξ_P the position of the elastic peak. From A. de Rujula, H. Georgi, H.D. Politzer: Ann. Phys. **103**, 315 (1977)

Example 5.12

of the various resonances can be clearly identified. The elastic peak at ξ_p is not plotted.

The dashed curve results from a perturbative QCD calculation. Clearly the agreement becomes better for increasing Q^2 values, since here the effect of the individual hadron resonances can be neglected. But also for small values of Q^2 the lowest moments

$$M_2^{(1)}(Q^2) = \int_0^1 F_2(x, Q^2)\, x^{-1}\, \mathrm{d}x \tag{5}$$

and

$$M_2^{(2)}(Q^2) = \int_0^1 F_2(x, Q^2)\, \mathrm{d}x \tag{6}$$

are quite well described by the QCD calculation.

Before we continue our discussion of the OPE (operator product expansion) we briefly summarize the derivations carried out so far. Employing deep inelastic scattering as an example, we explained how scattering tensors can be expanded into products of divergent coefficient functions, which can be evaluated by means of perturbation theory and finite matrix elements not dependent on the momentum transfer, for example. In the case of $F_2(x, Q^2)$ we had

$$\int_0^1 \mathrm{d}x\, x^{n-2} F_2(x, Q^2) = 2\pi\mathrm{i} \sum_j \tilde{C}_{2,j}^{(n)}\left(Q^2, g, \mu\right) A^{(n),j} \; . \tag{5.214}$$

This so-called *factorization* into a Q^2-dependent perturbative part and fixed numbers $A^{(n),j}$ that contain information about the distribution functions at a given value of Q_0^2 is of fundamental importance to most QCD applications. We showed the possibility of factorization for deep inelastic scattering, but in fact one must prove that this method is not destroyed by higher-order terms. Since each *soft* (Q^2 is small of the order ~ 1 GeV) gluon line couples strongly ($\alpha_s(Q^2)$ is large), higher orders in QCD are in general as important as the lowest order. Therefore the validity of the expansion (5.214), or of analogous expansions for other processes, has to be investigated very carefully. Because of its fundamental importance we shall discuss this question in the following section. The nonperturbative matrix elements $A^{(n),j}$ can be treated as pure parameters or can be calculated with help of nonperturbative models, e.g., the MIT bag model. To clarify this statement we return to the starting point of our discussion, which was the QCD analysis of the commutator of hadronic currents,

$$\left[\hat{J}_\mu(x), \hat{J}_\nu(0)\right]_- \tag{5.215}$$

which, taken between nucleon (spin $\frac{1}{2}$) particles, yields the hadronic scattering tensor (see (5.1)):

$$\frac{1}{2\pi}\int \mathrm{d}^4x\, \mathrm{e}^{\mathrm{i}qx}\langle PS|\Big[\hat{J}_\mu(x),\, \hat{J}_\nu(0)\Big]_-|PS\rangle = W_{\mu\nu}\ . \tag{5.216}$$

Let us now formulate the basic ideas of the OPE by analysing the Compton scattering amplitude

$$T\left(\hat{J}_\mu(x)\hat{J}_\nu(0)\right)\ , \tag{5.217}$$

which is related by virtue of the optical theorem (5.202) to the cross section:

$$\begin{aligned}&\frac{1}{\pi}\mathrm{Im}\left\{\mathrm{i}\int \mathrm{d}^4x\, \mathrm{e}^{\mathrm{i}qx}\langle PS|T\left(\hat{J}_\mu(x)\hat{J}_\nu(0)\right)|PS\rangle\right\}\\&=\frac{1}{2\pi}\int \mathrm{d}^4x\, \mathrm{e}^{\mathrm{i}qx}\langle PS|\Big[\hat{J}_\mu(x),\, \hat{J}_\nu(0)\Big]_-|PS\rangle\end{aligned} \tag{5.218}$$

The basic idea of OPE now is to perform a separation of scales. Analysing deeply inelastic scattering we have to deal with two separate scales. One is the limit of high virtualities $Q^2\to\infty$, $pq\to\infty$, the Bjorken limit, which can be identified with physics on the light cone in coordinate space $x^2\to 0$. According to the behavior of the running coupling, which vanishes at $Q^2\to\infty$ we may treat this region of phase space using perturbative methods.

The other scale of typical hadronic size $M^2\sim 1\ \mathrm{GeV}^2$ comes about when we analyze the current commutator between nucleon states.

The OPE separates these two scales:

$$\begin{aligned}\mathrm{i}\int \mathrm{d}^4x\, \mathrm{e}^{\mathrm{i}qx}T\left(\hat{J}_\mu(x)\hat{J}_\nu(0)\right) &= \mathrm{i}\int \mathrm{d}^4x\, \mathrm{e}^{\mathrm{i}qx}\sum_i\Big[C_i(x,\mu^2)\hat{O}_i(\mu^2)\Big]_{\mu\nu}\\&=\sum_i\Big[\tilde{C}_i(Q^2,\mu^2)\hat{O}_i(\mu^2)\Big]_{\mu\nu}\ .\end{aligned} \tag{5.219}$$

The coefficient functions $\tilde{C}_i(Q^2,\mu^2)=\mathrm{i}\int \mathrm{d}^4x\, \mathrm{e}^{\mathrm{i}qx}C_i(x,\mu^2)$ (called *Wilson coefficients*) contain only virtualities that are larger than the separation scale μ^2, while the operators $\hat{O}_i(\mu^2)$ contain only contributions from momenta smaller than μ^2 (see Fig. 5.17).

Motivated by the behavior of the running coupling one now invokes the following assumption: The coefficient functions $\tilde{C}_i(Q^2,\mu^2)$ are treated completely perturbatively while the operators are assumed to be of nonperturbative nature. That means that the coefficient functions obey usual renormalization group equations and can be systematically expanded in powers of $\alpha_s(Q^2,\mu^2)$, while the matrix elements $\langle PS|\hat{O}_i(\mu^2)|PS\rangle$ in principle are unknown and parametrize our ignorance of bound-state complexities. However, in some special cases, matrix elements occuring in the analysis can be related to processes other than deep inelastic scattering.

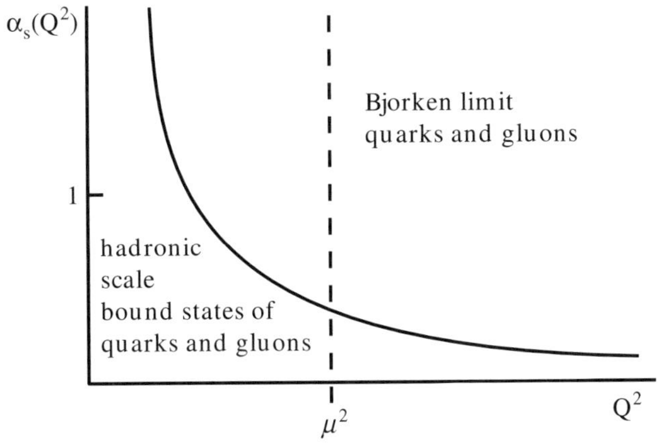

Fig. 5.17. Running QCD couplingconstant as a function of the virtuality Q^2. High virtuality, $Q^2 \to \infty$, corresponds to free quarks and gluons. $Q^2 < 1$ GeV is the region where quarks and gluons are bound to hadrons

For instance, the operator $\langle proton|\hat{\bar{\Psi}}\gamma_\mu\gamma_5\hat{T}_3\hat{\Psi}|proton\,\rangle$ occurs in spin-dependent deep inelastic scattering and can be related by isospin symmetry to an operator occuring in weak β decay:

$$\langle proton|\hat{\Psi}\gamma_\mu\gamma_5\hat{T}_+\hat{\Psi}|neutron\,\rangle\ . \tag{5.220}$$

Let us now derive the separation of scales, i.e. the separation into perturbative and nonperturbative contributions to the hadronic scattering tensor.

To be explicit we consider vector currents

$$\begin{aligned}\hat{J}_\mu(x) &= \sum_f e_f\hat{\bar{\Psi}}_f(x)\gamma_\mu\hat{\Psi}_f(x)\ ,\\ \hat{J}_\nu(0) &= \sum_{f'} e_{f'}\hat{\bar{\Psi}}_{f'}(0)\gamma_\mu\hat{\Psi}_{f'}(0)\ ,\end{aligned} \tag{5.221}$$

i.e. only purely electromagnetic interactions. The sum runs over the different quark flavors: u, d, c, ... Taking weak interactions into account imposes no further difficulties, we only have to take care of additional axial vector combinations.[13]

These currents have to be inserted into (5.215) to (5.219). Equation (5.217) then becomes

$$T\left(\hat{J}_\mu(x)\hat{J}_\nu(0)\right) = \sum_{ff'} T\left(e_f\hat{\bar{\Psi}}_f(x)\gamma_\mu\hat{\Psi}_f(x)e_{f'}\hat{\bar{\Psi}}_{f'}(0)\gamma_\mu\hat{\Psi}_{f'}\right)\ , \tag{5.222}$$

[13] See W. Greiner and B. Müller: *Gauge Theory of Weak Interactions*, 3rd ed. (Springer, Berlin, Heidelberg 2000).

which can further be simplified with Wick's theorem[14]

$$T\left(\hat{\Psi}_\alpha(x)\hat{\bar{\Psi}}_\beta(y)\right) = - : \hat{\bar{\Psi}}_\beta(y)\hat{\Psi}_\alpha(x) : + \langle 0|T\left(\hat{\Psi}_\alpha(x)\hat{\bar{\Psi}}_\beta(y)\right)|0\rangle \ , \tag{5.223}$$

where $\langle 0|T\left(\hat{\Psi}_\alpha(x)\hat{\bar{\Psi}}_\beta(y)\right)|0\rangle = \mathrm{i}S_{\alpha\beta}(x,y)$ is the usual Feynman propagator. The free propagator can be written in the usual way:

$$\mathrm{i}S(x,y) = \int \frac{\mathrm{d}^4 p}{(2\pi)^4} e^{\mathrm{i}q(x-y)} \frac{\mathrm{i}}{\not{p} - m + i\varepsilon} = \frac{\mathrm{i}}{2\pi^2} \frac{(\not{x} - \not{y})}{[(x-y)^2 - i\varepsilon]^2} + O(m) \ . \tag{5.224}$$

Applying (5.223) to (5.217) we thus obtain

$$\begin{aligned} T\left(\hat{J}_\mu(x)\hat{J}_\nu(0)\right) = &\sum_{ff'} : e_f \hat{\bar{\Psi}}_f(x)\gamma_\mu \hat{\Psi}_f(x) : \ : e_{f'} \hat{\bar{\Psi}}_{f'}(0)\gamma_\nu \hat{\Psi}_{f'}(0) : \\ &+ \sum_f e_f^2 \hat{\bar{\Psi}}_f(x)\gamma_\mu \mathrm{i}S(x,0)\gamma_\nu \hat{\Psi}_f(0) \\ &+ \sum_{f'} e_{f'}^2 \hat{\bar{\Psi}}_{f'}(0)\gamma_\nu \mathrm{i}S(0,x)\gamma_\mu \hat{\Psi}_{f'}(x) \\ &- \sum_f e_f^2 \mathrm{tr}\left\{\gamma_\mu \mathrm{i}S(x,0)\gamma_\nu \mathrm{i}S(0,x)\right\} \ . \end{aligned} \tag{5.225}$$

Inserting the free fermion propagator yields

$$\begin{aligned} T\left(\hat{J}_\mu(x)\hat{J}_\nu(0)\right) = &\sum_f e_f^2 \mathrm{tr}\left(\gamma_\mu \not{x} \gamma_\nu \not{x}\right) \cdot \frac{1}{4\pi^4} \frac{1}{(x^2 - \mathrm{i}\varepsilon)^4} \\ &- \frac{\mathrm{i}}{2\pi^2} \sum e_{f'}^2 \hat{\bar{\Psi}}_{f'}(x)\gamma_\mu \not{x} \gamma_\nu \hat{\Psi}_{f'}(0) \frac{1}{(x^2 - \mathrm{i}\varepsilon)^2} \\ &+ \frac{\mathrm{i}}{2\pi^2} \sum e_f^2 \hat{\bar{\Psi}}_f(0)\gamma_\nu \not{x} \gamma_\mu \hat{\Psi}_f(x) \frac{1}{(x^2 - \mathrm{i}\varepsilon)^2} \\ &+ \sum_{ff'} : e_f \hat{\bar{\Psi}}_f(x)\gamma_\mu \hat{\Psi}_f(x) : \ : e_{f'} \hat{\bar{\Psi}}_{f'}(0)\gamma_\nu \hat{\Psi}_{f'}(0) : \ . \end{aligned} \tag{5.226}$$

If we sandwich the time-ordered product (5.222) between proton states $\langle PS|T(\quad)|PS\rangle$ the process of deep inelastic scattering can be graphically depicted (see Fig. 5.17).

OPE separates the free quark and gluon propagation from bound-state complexities. While the upper part of the diagram may be treated with the tools of

[14] See W. Greiner and J. Reinhardt: *Field Quantization* (Springer, Berlin, Heidelberg 1996).

perturbation theory, the lower part is subject to nonperturbative methods. Moreover, these nonperturbative methods can now be applied to well defined hadronic matrix elements.

While the "hand bag" diagrams (b) and (c) contain one free quark propagator, i.e. they are singular as $x^2 \to 0$, the "cat ear" diagram (d) contains none and therefore we can safely neglect it. It will not give a dominant contribution on the light cone. Diagrams (b) and (c) can therefore be written as

$$T\left(\hat{J}_\mu(x)\hat{J}_\nu(0)\right) \overset{x^2\to 0}{=} \frac{1}{2\pi^2}\mathrm{i}\frac{x^\alpha}{(x^2-\mathrm{i}\varepsilon)^2}\sum_f \hat{\bar{\Psi}}_f(x)e_f^2\gamma_\mu\gamma_\alpha\gamma_\nu\hat{\Psi}(0)$$
$$-\hat{\bar{\Psi}}_f(0)e_f^2\gamma_\nu\gamma_\alpha\gamma_\mu\hat{\Psi}(x) \ . \tag{5.227}$$

Using the identy

$$\gamma_\mu\gamma_\alpha\gamma_\nu = (s_{\mu\alpha\nu\beta} + \mathrm{i}\varepsilon_{\mu\alpha\nu\beta}\gamma_5)\gamma^\beta \tag{5.228}$$

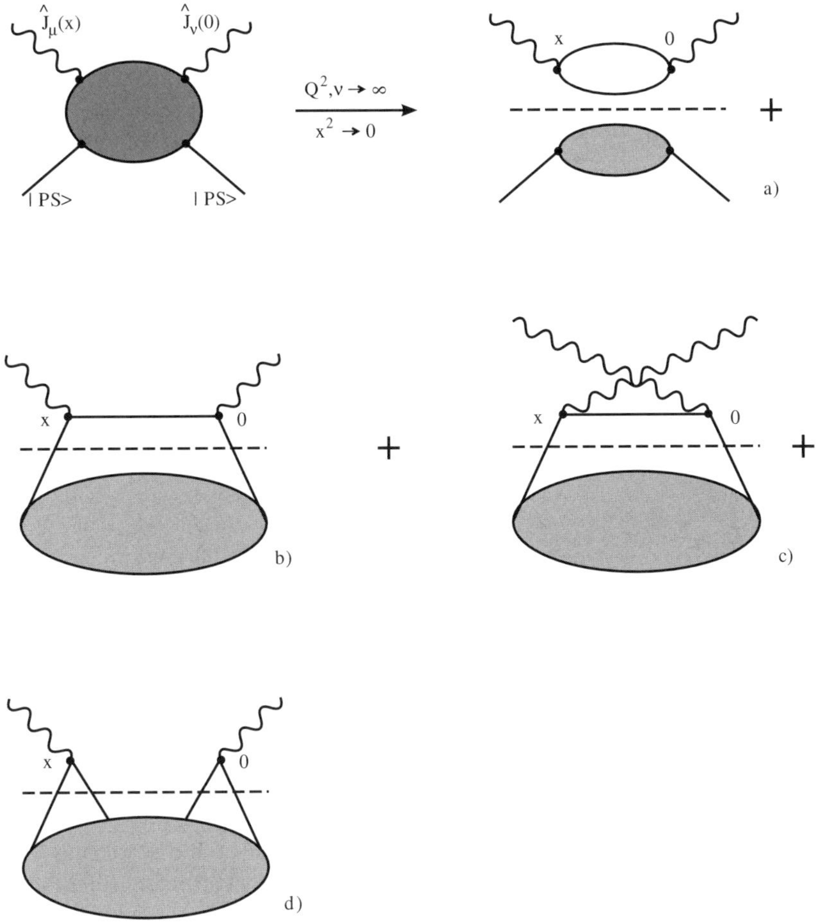

Fig. 5.18a–d. Diagrams occuring in the analysis of the time-ordered product on the light cone $x^2 \to 0$. Wavy lines denote the photon, straight lines the free massless quark propagator and the blob the proton state. Diagramm (**a**) will only contribute between vacuum states and will not contribute to lepton proton scattering since it is disconnected. The dashed line is a symbolic indication of the separation in perturbative and nonperturbative contributions

(see Excercise 5.9) we obtain

$$T\left(\hat{J}_\mu(x)\hat{J}_\nu(0)\right) \overset{x^2\to 0}{=} \frac{\mathrm{i}x^\alpha}{2\pi^2(x^2-\mathrm{i}\varepsilon)^2} s_{\mu\alpha\nu\beta} \times \left(\sum_f \hat{\bar{\Psi}}_f(x) e_f^2 \gamma^\beta \hat{\Psi}(0) - \hat{\bar{\Psi}}_f(0) e_f^2 \gamma^\beta \hat{\Psi}(x)\right) - \frac{x^\alpha}{2\pi^2(x^2-\mathrm{i}\varepsilon)^2} \varepsilon_{\mu\alpha\nu\beta} \times \left(\sum_f \hat{\bar{\Psi}}_f(x) e_f^2 \gamma^\beta \gamma_5 \hat{\Psi}(0) + \hat{\bar{\Psi}}_f(0) e_f^2 \gamma^\beta \gamma_5 \hat{\Psi}(x)\right) . \tag{5.229}$$

We can now expand the bilocal operator

$$\hat{\bar{\Psi}}(0) e_f^2 \gamma^\alpha \hat{\Psi}(x) = \sum_{n=0}^\infty \frac{x^{\mu_1}\dots x^{\mu_n}}{n!} \hat{\bar{\Psi}}(0) e_f^2 \gamma^\alpha \hat{D}_{\mu_1}\cdots\hat{D}_{\mu_n} \hat{\Psi}(0) ,$$
$$\hat{\bar{\Psi}}(x) e_f^2 \gamma^\alpha \hat{\Psi}(0) = \sum_{n=0}^\infty \frac{x^{\mu_1}\dots x^{\mu_n}}{n!} (-1)^n \hat{\bar{\Psi}}(0) e_f^2 \gamma^\alpha \hat{D}_{\mu_1}\cdots\hat{D}_{\mu_n} \hat{\Psi}(0) , \tag{5.230}$$

where we have brought all derivatives to act on the right-hand side and have neglected total derivative terms. Note that we have also included covariant derivative terms to take interactions into account. Rearranging the equations we get

$$T\left(\hat{J}_\mu(x)\hat{J}_\nu(0)\right) = \frac{\mathrm{i}x^\alpha}{\pi^2(x^2-\mathrm{i}\varepsilon)^2} s_{\mu\alpha\nu\beta} \times \left\{\sum_{n=1,3,5}^\infty \frac{x^{\mu_1}\dots x^{\mu_n}}{n!} \sum_f \hat{\bar{\Psi}}_f(0) e_f^2 \gamma^\beta \hat{D}_{\mu_1}\cdots\hat{D}_{\mu_n} \hat{\Psi}(0)\right\} - \frac{\mathrm{i}x^\alpha}{\pi^2(x^2-\mathrm{i}\varepsilon)^2} \varepsilon_{\mu\alpha\nu\beta} \times \left\{\sum_{n=0,2,}^\infty \frac{x^{\mu_1}\dots x^{\mu_n}}{n!} \sum_f \hat{\bar{\Psi}}_f(0) e_f^2 \gamma^\beta \gamma_5 \hat{D}_{\mu_1}\cdots\hat{D}_{\mu_n} \hat{\Psi}(0)\right\} . \tag{5.231}$$

This can be compared with the general structure of the scattering tensor (see (3.18)) (see also Sect. 5.6 where the spin-dependent structure functions are discussed extensively)

$$
\begin{aligned}
T_{\mu\nu} = & \left(-g_{\mu\nu} + \frac{q_\mu q_\nu}{q^2}\right) \tilde{F}_1(x, Q^2) \\
& + \frac{\tilde{F}_2(x, Q^2)}{p \cdot q} \left(p_\mu - \frac{p \cdot q}{q^2} q_\mu\right) \left(p_\nu - \frac{p \cdot q}{q^2} q_\nu\right) \\
& - \mathrm{i}\varepsilon_{\mu\nu\lambda\sigma} \frac{q^\lambda}{p \cdot q} \left(\tilde{g}_1(x, Q^2) S^\sigma + \tilde{g}_2(x, Q^2) \left(S^\sigma - p^\sigma \frac{S \cdot q}{p \cdot q}\right)\right) ,
\end{aligned}
\tag{5.232}
$$

where $\frac{1}{\pi}\mathrm{Im}\tilde{F}_1 = F_1$, etc., according to the optical theorem. This can also be used to identify the terms in the general representation of the time-ordered product. Indeed, we can conclude that the spin-independent structure functions F_1 and F_2 are somehow related to the vector operators

$$
\sim \sum_f e_f^2 \hat{\bar{\Psi}}_f(0) \gamma_\alpha \hat{D}_{\mu_1} \cdots \hat{D}_{\mu_n} \hat{\Psi}_f(0) ,
$$

while the spin-dependent structure functions $\tilde{g}_1$ and $\tilde{g}_2$ are related to the antisymmetric part of (5.227) and therefore to the axial vector operator

$$
\sim \sum_f e_f^2 \hat{\bar{\Psi}}_f(0) \gamma_\alpha \gamma_5 \hat{D}_{\mu_1} \cdots \hat{D}_{\mu_n} \hat{\Psi}_f(0) .
$$

To proceed we restrict ourselves to the antisymmetric part of (5.227), which contributes to the spin-dependent scattering.

$$
\begin{aligned}
& T\left(\hat{J}_\mu(x) \hat{J}_\nu(0)\right)_A = \\
& \frac{-\mathrm{i}}{\pi^2} \frac{x^\alpha}{(x^2 - \mathrm{i}\varepsilon)^2} \varepsilon_{\mu\alpha\nu\beta} \\
& \times \left\{ \sum_{n=0,2,} \frac{x^{\mu_1} \ldots x^{\mu_n}}{n!} \sum_f \hat{\bar{\Psi}}_f(0) e_f^2 \gamma^\beta \gamma_5 \hat{D}_{\mu_1} \cdots \hat{D}_{\mu_n} \hat{\Psi}_f(0) \right\} .
\end{aligned}
\tag{5.233}
$$

The time-ordered product is to be inserted between nucleon states and the Fourier transformation has to be carried out. In doing so we have to define matrix elements of local operators

$$
\begin{aligned}
& \mathrm{i}^{n-1} \langle PS | \hat{\bar{\Psi}}_f(0) \gamma^\beta \gamma_5 \hat{D}_{\mu_1} \cdots \hat{D}_{\mu_n} \hat{\Psi}_f(0) | PS \rangle \big|_{\text{twist-2}} \\
& = \mathcal{S}\left[S^\beta p_{\mu_1} \cdots p_{\mu_n}\right] \left(2 A_{n+1}^f\right) ,
\end{aligned}
\tag{5.234}
$$

where we proceeded completely analogously to (5.140) and (5.166). Here $\mathcal{S}$ is defined as $\mathcal{S}(\mu_1 \ldots \mu_n) = (1/n!) \sum a_{\text{permutations}(\mu_1 \ldots \mu_n)}$.

Again we have to take the completely symmetric part symbolized by $\mathcal{S}$. It corresponds to the leading twist contribution. We will explain the notion of twist in detail in Example 5.16. For the moment it is enough to know that leading twist operators are those that are completely symmetric in all indices.

Unlike the case of the symmetric (in μ, ν) contribution to the scattering tensor – due to the axial vector character of the operator – the spin vector S^β appears in the symmetrized product. This leads to additional complications.

Let us split up the symmetrized part (writing $S^\beta = S^{\mu_1}$ for convenience)

$$\mathcal{S}\left[S^{\mu_1} p^{\mu_2} \cdots p^{\mu_n}\right] = S^{\mu_1} p^{\mu_2} \cdots p^{\mu_n} + R^{\mu_1 \cdots \mu_n} \ , \tag{5.235}$$

where the residual tensor is

$$\begin{aligned} R^{\mu_1 \cdots \mu_n} =& \mathcal{S}\left[S^{\mu_1} p^{\mu_2} \cdots p^{\mu_n}\right] - S^{\mu_1} \cdots p^{\mu_n} \\ =& \frac{1}{n}\left[S^{\mu_1} p^{\mu_2} \cdots p^{\mu_n} + p^{\mu_1} S^{\mu_2} \cdots p^{\mu_n} + \cdots + p^{\mu_1} p^{\mu_2} \cdots S^{\mu_n}\right] \\ & - S^{\mu_1} \cdots p^{\mu_n} \\ =& -\frac{n-1}{n} S^{\mu_1} p^{\mu_2} \cdots p^{\mu_n} + \frac{1}{n} p^{\mu_1} S^{\mu_2} \cdots p^{\mu_n} + \cdots \\ & + \frac{1}{n} p^{\mu_1} p^{\mu_2} \cdots S^{\mu_n} \ . \end{aligned} \tag{5.236}$$

The tensor $R^{\mu_1 \cdots \mu_n}$ has no completely symmetric part, and so has one additional unit of twist. Thus the contribution of R to deep inelastic scattering is twist three, rather than twist two, even though it comes from the matrix element of a twist-two operator. This will contribute to the second spin-dependent structure function $\tilde{g}_2$. In this context we will omit R and refer the interested reader to the literature.[15]

Inserting (5.234) in (5.233) we arrive at

$$\begin{aligned} & \frac{-\mathrm{i}}{\pi^2} \frac{x^\alpha}{(x^2 - \mathrm{i}\varepsilon)^2} \varepsilon_{\mu\alpha\nu\beta} \\ & \times \left\{ \sum_{n=0,2,} \frac{x^{\mu_1} \cdots x^{\mu_n}}{n!} \sum_f \langle PS | \hat{\bar{\Psi}}_f(0) e_f^2 \gamma^\beta \gamma_5 \hat{D}_{\mu_1} \cdots \hat{D}_{\mu_n} \hat{\Psi}_f(0) | PS \rangle \right\} \\ & = \frac{(-\mathrm{i})^n}{\pi^2} \frac{x^\alpha}{(x^2 - \mathrm{i}\varepsilon)^2} \varepsilon_{\mu\alpha\nu\beta} \sum_{n=0,2,} \frac{x^{\mu_1} \cdots x^{\mu_n}}{n!} \mathcal{S}\left[S^\beta p_{\mu_1} \cdots p_{\mu_n}\right] \left(2A^f_{n+1}\right) \\ & = \mathrm{i} \int \mathrm{d}^4 x \, \mathrm{e}^{\mathrm{i}qx} \frac{x^\alpha}{\pi^2 (x^2 - \mathrm{i}\varepsilon)^2} \varepsilon_{\mu\alpha\nu\beta} S^\beta \sum_{n=0,2,} \frac{(-\mathrm{i}px)^n}{n!} \sum_f \left(2A^f_{n+1} e_f^2\right) \ . \end{aligned} \tag{5.237}$$

The final formula we require is

$$\frac{\partial}{\partial q^{\mu_1}} \cdots \frac{\partial}{\partial q^{\mu_n}} = 2^n q^{\mu_1} \cdots q^{\mu_n} \left(\frac{\partial}{\partial q^2}\right)^n + \text{trace terms} \ , \tag{5.238}$$

where we do not need the trace terms $g_{\mu_i \mu_j}$ explicitely since they will be supressed by powers of Q^2 (see (5.143)). By replacing $(\mathrm{i}x_\mu)$ by $\partial/\partial q^\mu$ we can write

$$\varepsilon_{\mu\alpha\nu\beta} q^\alpha S^\beta \sum_{n=0,2,} \frac{(-2p \cdot q)^n}{n!} \cdot 2 \sum_f e_f^2 2A^f_{n+1} \left(\frac{\partial}{\partial q^2}\right)^{n+1} \int \frac{\mathrm{d}^4 x}{\pi^2} \frac{\mathrm{e}^{\mathrm{i}qx}}{(x^2 - \mathrm{i}\varepsilon)^2} \ . \tag{5.239}$$

[15] See B. Ehrnsperger, et al.: *OPE Analysis for Polarized Deep Inelastic Scattering*, Phys. Lett. B **323**, 439 (1994) and references therein.

The Fourier transformation is performed by confining the integral to d dimensions and making use of the formula derived in Exercise 4.6. We obtain ($d = 4+2\varepsilon$)

$$\int \frac{\mathrm{d}^d x}{\pi^2} \frac{\mathrm{e}^{\mathrm{i}qx}}{(x^2-\mathrm{i}\varepsilon)^2} = -\mathrm{i}\Gamma(\varepsilon)\left(\frac{-4\pi\mu^2}{q^2}\right)^{\varepsilon} = -\mathrm{i}\ln q^2 + \text{const.} \tag{5.240}$$

Performing the derivatives with respect to q^2,

$$\left(\frac{\partial}{\partial q^2}\right)^{n+1} \ln(q^2) = n!(-1)^n \frac{1}{(q^2)^{n+1}} \,, \tag{5.241}$$

we get

$$-\mathrm{i}\varepsilon_{\mu\alpha\nu\beta} q^\alpha S^\beta \sum_{n=0,2,} \left(\frac{2p\cdot q}{q^2}\right)^n \frac{2}{q^2} \sum_f e_f^2 2A_{n+1}^f \,, \tag{5.242}$$

and by comparing with (5.232) we end up with

$$\begin{aligned} \tilde{g}_1 &= \sum_{n=0,2,} \left(\frac{2p\cdot q}{Q^2}\right)^{n+1} \sum_f e_f^2 2A_{n+1}^f \\ &= \sum_{n=1,3,\cdots} \omega^n \left(\sum_f e_f^2 2A_n^f\right) = \sum_{n=1,3,\cdots} \omega^n 2A_n \,, \\ \omega &\equiv \frac{1}{x} = \frac{2p\cdot q}{Q^2} \,, \quad A_n = \sum_f e_f^2 A_n^f \,. \end{aligned} \tag{5.243}$$

This gives us again, as in (5.175) and (5.176), an expansion in $\omega = 1/x$. A_n can be considered as a reduced matrix element. The nth term in this expansion is due to an operator of twist two and containing n symmterized indices. The radius of convergence of the series is $|\omega| = 1$ since this is the location of the first singularity in the complex ω plane. This is precisely where the physical region begins.

We can relate $\tilde{g}_1$ in the unphysical region to its value in the physical region using contour integration. The coefficient A_n can be extracted by the contour integral

$$2A_n = \frac{1}{2\pi\mathrm{i}} \oint_C \tilde{g}_1(\omega) \frac{\mathrm{d}\omega}{\omega^{n+1}} \,, \tag{5.244}$$

where the contour of integration is shown in Fig 5.19.

Now we have to assume that the contour at infinity does not contribute to the integral and can be neglected. This assumption should be testable by experiment. If the contour is important when $\omega \to \infty$ then this should be measurable as the small-x behavior of the structure functions. Neglecting the contour at infinity we

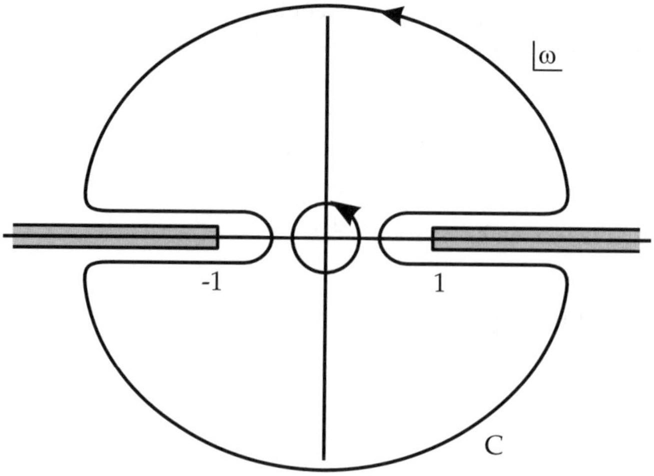

Fig. 5.19. Integration contour in (5.244). The circle around the origin is deformed to pick up the contributions along the real axis

can write

$$
\begin{aligned}
2A_n &= \frac{1}{2\pi\mathrm{i}}\left[\int_1^{\infty}\frac{\mathrm{d}\omega}{\omega^{n+1}}\left[\tilde{g}_1(\omega+\mathrm{i}\varepsilon)-\tilde{g}_1(\omega-\mathrm{i}\varepsilon)\right]\right.\\
&\quad\left.+\int_{-\infty}^{-1}\frac{\mathrm{d}\omega}{\omega^{n+1}}\left[\tilde{g}_1(\omega+\mathrm{i}\varepsilon)-\tilde{g}_1(\omega-\mathrm{i}\varepsilon)\right]\right]\\
&= \frac{1}{\pi}\int_1^{\infty}\mathrm{d}\omega\frac{1}{2\mathrm{i}}\left[\tilde{g}_1(\omega+\mathrm{i}\varepsilon)-\tilde{g}_1^*(\omega+\mathrm{i}\varepsilon)\right]\cdot\frac{1}{\omega^{n+1}}\\
&\quad+\frac{1}{\pi}\int_{-\infty}^{-1}\mathrm{d}\omega\frac{1}{2\mathrm{i}}\left[\tilde{g}_1(\omega+\mathrm{i}\varepsilon)-\tilde{g}_1^*(\omega+\mathrm{i}\varepsilon)\right]\cdot\frac{1}{\omega^{n+1}}\\
&= \int_1^{\infty}\frac{\mathrm{d}\omega}{\omega^{n+1}}\tilde{g}_1(\omega)+\int_{-\infty}^{1}\frac{\mathrm{d}\omega}{\omega^{n+1}}\tilde{g}_1(\omega)\\
&= \int_{-1}^{1}\mathrm{d}x\,g_1(x)x^{n-1}=2\int_0^1\mathrm{d}x\,g_1(x)x^{n-1}\,,\qquad n=\mathrm{odd}\,,
\end{aligned}
\tag{5.245}
$$

where $g_1(x)$ is now the physical measurable structure function. In deriving the above relation we made use of the optical theorem $W_{\mu\nu}=\frac{1}{\pi}T_{\mu\nu}$ and of $g_1(x)=g_1(-x)$, which follows from crossing symmetry $W_{\mu\nu}(q)=-W_{\nu\mu}(-q)$ and the explicit decomposition of the scattering tensor (5.232). To be explicit we write

$$
\begin{aligned}
W_{\mu\nu}(q) &= \frac{1}{2\pi}\int \mathrm{d}^4x\, \mathrm{e}^{\mathrm{i}qx}\langle PS\,|\,[\hat{J}_\mu(x), \hat{J}_\nu(0)]\,|\,PS\rangle \\
W_{\mu\nu}(-q) &= \frac{1}{2\pi}\int \mathrm{d}^4y\, \mathrm{e}^{-\mathrm{i}qy}\langle PS\,|\,[\hat{J}_\mu(y), \hat{J}_\nu(0)]\,|\,PS\rangle \\
&= \frac{1}{2\pi}\int \mathrm{d}^4y\, \mathrm{e}^{\mathrm{i}qy}\langle PS\,|\,[\hat{J}_\mu(-y), \hat{J}_\nu(0)]\,|\,PS\rangle \\
&= \frac{1}{2\pi}\int \mathrm{d}^4y\, \mathrm{e}^{\mathrm{i}qy}\langle PS\,|\,[\hat{J}_\mu(0), \hat{J}_\nu(y)]\,|\,PS\rangle \\
&= -\frac{1}{2\pi}\int \mathrm{d}^4y\, \mathrm{e}^{\mathrm{i}qy}\langle PS\,|\,[\hat{J}_\nu(y), \hat{J}_\mu(0)]\,|\,PS\rangle \\
&= -W_{\nu\mu}(q)\ .
\end{aligned}
$$

From the decomposition of the scattering tensor (5.232) it follows that $g_1(x) = g_1(-x)$ as well as $F_1(x) = -F_1(-x)$.

Thus we arrive at the final equation

$$
A_n = \int_0^1 \mathrm{d}x\, x^{n-1} g_1(x)\ , \tag{5.246}
$$

which relates the measurable moments of the structure function $g_1(x)$ to the matrix elements of operators defined in (5.234), which are calculable, at least in principle.

This is a prototype of a so called *moment sum rule*, which relates a quantity defined at high energy such as g_1 to a low-energy quantity, the zero momentum transfer matrix element of a local operator. One can derive sum rules in QCD only for even moments of F_1 and the odd moments of g_1; there are no sum rules for the other moments. This is due to the symmetry properties of the structure functions under crossing, $g_1(x) = g_1(-x)$ and $F_1(x) = -F_1(-x)$.

The only questionable assumption in the derivation of the sum rule is that the contour at infinity does not contribute to the integral. Indeed, the sum rules (5.246) become more convergent for the higher moments. Therefore any problem with convergence will hopefully occur only for the lowest few moments.

Finally, we remember that we performed the calculation above with the free quark propagator; $O(\alpha_s)$ corrections are neglected as well as higher twist operators. The full sum rule should read

$$
\int_0^1 \mathrm{d}x x^{n-1} g_1(x, Q^2) = C_n\left(\alpha_s(Q^2), \frac{Q^2}{\mu^2}\right) A_n(\mu^2) + O\left(\frac{1}{Q^2}\right)\ , \tag{5.247}
$$

where $C_n\left(\alpha_s(Q^2), Q^2/\mu^2\right)$ is the *coefficient function* which can be calculated order by order in perturbation theory. In the following, we will demonstrate how the coefficient functions are calculated.

5.5 Calculation of the Wilson Coefficients

In the previous section we recapitulated how the operator product expansion separates scales, thus allowing us to describe moments of structure functions through coefficient functions calculable in perturbation theory and to parametrize the nonperturbative contributions in terms of matrix elements of operators of well-defined twist.

These matrix elements then can be further investigated with nonperturbative techniques such as lattice QCD, QCD sum rules, or models like the MIT bag model. The Q^2 dependence of the Wilson coefficient in addition can be studied with the help of the renormalization group equation; cf (5.180). There we have shown how the Q^2 dependence of coefficients is described at leading order. To obtain the Q^2 dependence in next-to-leading order, we need the α_s corrections to the coefficient functions. Therefore, in this section some techniques for calculating Wilson coefficients in perturbation theory are explained.

The key observation is that the coefficient functions $\tilde{C}_i(Q^2, \mu^2)$ are *independent of the states* for which we are calculating the time-ordered product:

$$\begin{aligned} &\langle PS|\mathrm{i}\int \mathrm{d}^4x\,\mathrm{e}^{\mathrm{i}qx}T\left(\hat{J}_\mu(x)\hat{J}_\nu(0)\right)|PS\rangle \\ &= \sum_i \left[\tilde{C}_i(Q^2,\mu^2)\langle PS|\hat{O}_i(\mu^2)|PS\rangle\right]_{\mu\nu}\ ; \end{aligned} \tag{5.248}$$

cf (5.219). Hence we have the liberty to take which ever state is the simplest and, of course, we choose quark and gluon states. Calculation of the time-ordered product $\langle quark/gluon|\mathrm{i}\int \mathrm{d}^4x\,\mathrm{e}^{\mathrm{i}qx}T\left(\hat{J}_\mu(x)\hat{J}_\nu(0)\right)|quark/gluon\rangle$ and of the matrix elements $\langle quark/gluon|\hat{O}_i(\mu^2)|quark/gluon\rangle$ will give us the answer for $\tilde{C}_i(Q^2,\mu^2)$.

We shall examplify the idea by investigating unpolarized Compton forward scattering.

Let us write down the parametrization of the unpolarized part, which according to (3.18) and (3.43) is given by

$$\begin{aligned} T^{qq}_{\mu\nu}(p,q) &= \mathrm{i}\int \mathrm{d}^4x\,\mathrm{e}^{\mathrm{i}qx}\langle p, quark|T\left(\hat{J}_\mu(x)\hat{J}_\nu(0)\right)|p, quark\rangle \\ &= \left(-g_{\mu\nu}+\frac{q_\mu q_\nu}{q^2}\right)\tilde{F}_1 \\ &\quad+\frac{\tilde{F}_2}{p\cdot q}\left(p_\mu-\frac{p\cdot q}{q^2}q_\mu\right)\left(p_\nu-\frac{p\cdot q}{q^2}q_\nu\right)\ . \end{aligned} \tag{5.249}$$

Here the superscript qq refers to the fact that we are considering quarks as ingoing and outgoing hadronic states.

It is convenient to express the time-ordered product (5.249) in terms of the structure functions $\tilde{F}_L = \tilde{F}_2 - 2x\tilde{F}_1$ and $\tilde{F}_2$ instead of $\tilde{F}_1$ and $\tilde{F}_2$, which yields the decomposition

$$T^{qq}_{\mu\nu}(p,q) = e_{\mu\nu}\frac{1}{2x}\tilde{F}_L(x,Q^2) + d_{\mu\nu}\frac{1}{2x}\tilde{F}_2(x,Q^2) \tag{5.250}$$

with

$$e_{\mu\nu} = \left(g_{\mu\nu} - \frac{q_\mu q_\nu}{q^2}\right) \tag{5.251a}$$

and

$$d_{\mu\nu} = \left(-g_{\mu\nu} + p_\mu p_\nu \frac{4x^2}{q^2} - (p_\mu q_\nu + p_\nu q_\mu)\frac{2x}{q^2}\right) . \tag{5.251b}$$

Using the projectors $p_\mu p_\nu$ and $g_{\mu\nu}$, we find, for massless states $p^2 = 0$,

$$\begin{aligned}\frac{\tilde{F}_L}{2x} &= \frac{Q^2}{(p\cdot q)^2} p^\mu p^\nu T^{qq}_{\mu\nu} = \frac{4x^2}{Q^2} p^\mu p^\nu T^{qq}_{\mu\nu} ,\\ \frac{\tilde{F}_2}{2x} &= \left(\frac{Q^2}{(p\cdot q)^2} p^\mu p^\nu - \frac{g^{\mu\nu}}{2}\right) T^{qq}_{\mu\nu} = -\frac{\tilde{F}_L}{2x} - \frac{g^{\mu\nu}}{2} T^{qq}_{\mu\nu} .\end{aligned} \tag{5.252}$$

To set the stage, we calculate the 4-point Green function in zeroth-order perturbation theory, i.e. the forward scattering amplitude of a virtual photon on a quark target. The Feynman diagram is the same as we had in Fig. 5.18, only now the nucleon states are replaced by quark states (Fig. 5.20).

We can immediately denote the amplitude as

$$\begin{aligned}T^{qq(0)}_{\mu\nu} = \mathrm{i}\frac{1}{2}\sum_s \Bigg\{ &\bar{u}(p,s)\gamma_\mu \frac{\mathrm{i}}{\not p + \not q}\gamma_\nu u(p,s)\\ &+ \bar{u}(p,s)\gamma_\nu \frac{\mathrm{i}}{\not p - \not q}\gamma_\mu u(p,s)\Bigg\} .\end{aligned} \tag{5.253}$$

The superscript (0) to $T^{qq(0)}_{\mu\nu}$ indicates the zeroth-order approximation in α_s for $T^{qq}_{\mu\nu}$. Note that there is an overall factor of i because we are computing i times the time-ordered product in (5.219) or (5.249).

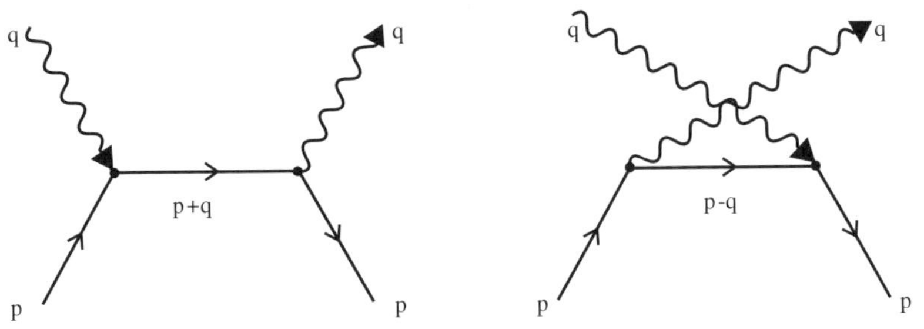

Fig. 5.20. Compton forward scattering amplitude for photon–quark scattering

The crossed diagram (second term) can be obtained by the replacement $\mu \leftrightarrow \nu$, $q \to -q$ from the direct (first) term. Hence it is sufficient to concentrate on the first term.

Applying the projector (5.251a) onto $\tilde{F}_L$ we find immediatly $\tilde{F}_L = 0$ in zeroth-order perturbation theory due to the equations of motions $\not{p}u(p,s) = 0$. This is a simple reflection of the celebrated Callan–Gross relation $\tilde{F}_2 = 2x\tilde{F}_1$. Turning to $\tilde{F}_2$ we project with $g_{\mu\nu}$ and find, for the direct term with $\gamma_\mu\gamma^\alpha\gamma^\mu = -2\gamma^\alpha$,

$$\begin{aligned} g^{\mu\nu}T^{qq}_{\mu\nu}(\text{direct}) &= -2\frac{\mathrm{i}^2}{2}\sum_s\left(\overline{u}(p,s)\frac{\not{p}+\not{q}}{(p+q)^2}u(p,s)\right) \\ &= -2\mathrm{i}^2\frac{1}{2}\sum_s \operatorname{tr}\left[\frac{\not{p}+\not{q}}{(p+q)^2}u(p,s)\overline{u}(p,s)\right] \\ &= -\mathrm{i}^2 4\frac{(p\cdot q)}{(p+q)^2} = 2\frac{(p\cdot q)}{q^2}\frac{2}{\left(1+\frac{2p\cdot q}{q^2}\right)} = -\omega\frac{2}{1-\omega} \ . \end{aligned} \tag{5.254}$$

Remember, $\omega \equiv 1/x = 2p\cdot q/Q^2 = -2p\cdot q/q^2$. Expanding the denominator gives

$$g^{\mu\nu}T^{qq}_{\mu\nu}(\text{direct}) = -2\omega\sum_{n=0}^{\infty}\omega^n \ . \tag{5.255}$$

Utilizing the second equation (5.252) with $\tilde{F}_L = 0$ and taking into account the exchange term we find, for $\tilde{F}_2$,

$$\frac{\tilde{F}_2}{2x} = \sum_{n=0}^{\infty}\omega^{n+1} + (-\omega)^{n+1} = 2\sum_{n=2,4}\omega^n \ , \quad \omega = \frac{1}{x} \ , \tag{5.256}$$

which is essentially the same expression as in (5.243), where we calculated $\tilde{g}_1$. The only difference is the lack of the reduced matrix elements A_n simply because those are normalized to one for free quark states:

$$\mathrm{i}^{n-1}\langle p, quark|\hat{\overline{\Psi}}\gamma_{\mu_1}\hat{D}_{\mu_2}\cdots\hat{D}_{\mu_n}\hat{\Psi}|p, quark\rangle = p_{\mu_1}\cdots p_{\mu_n} \ . \tag{5.257}$$

Had we done a similar calculation as previously we would have ended up with

$$\frac{\tilde{F}_2}{2x} = 2\sum_{n=2,4}\omega^n V_n \ , \tag{5.258}$$

where V_n are the reduced matrix elements of the vector operator (5.257) in an arbitrary state. Furthermore, by using dispersion relations as in the case of $\tilde{g}_1$ we finally would find the corresponding equation to (5.246) and (5.247):

$$\int_0^1 \mathrm{d}x x^{n-2}F_2(x,Q^2) = C_n\left(\alpha_s(Q^2), \frac{Q^2}{\mu^2}\right)V_n(\mu^2) \ , \tag{5.259}$$

which shows that the nth power of ω is directly related to the nth moment of F_2.

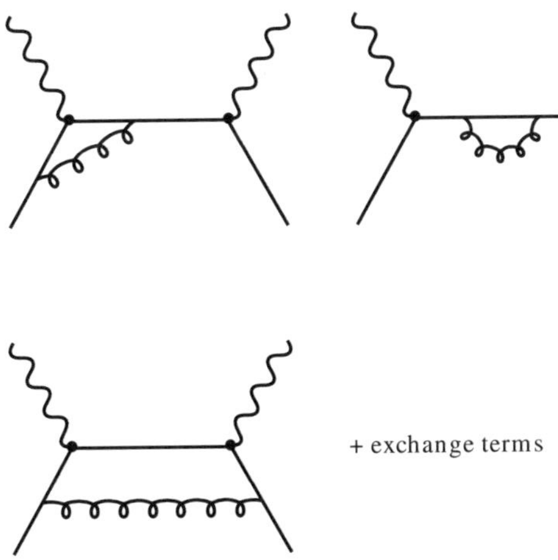

Fig. 5.21. Order α_s correction to the Compton forward scattering amplitude for photon–quark scattering

Keeping that in mind we proceed to our actual task, which is to calculate the Wilson coefficients at next-to-leading order, i.e. the order of α_s corrections to the forward scattering. The corresponding diagrams are depicted in Fig. 5.21.

We restrict ourselves to the calculation of $C_L(\alpha_s(Q^2))$, i.e. the Wilson coefficients for the longitudinal structure function, which measures the violation of the Callan–Gross relation. The calculation of $C_L(\alpha_s(Q^2))$ involves only the last diagram in Fig. 5.21, since all the others vanish by the equations of motion when we project with $p_\mu p_\nu$.

Using the Feynman rules for the first-order correction for the direct term in Fig. 5.22 we find

$$\begin{aligned} &T^{qq(1)}_{\mu\nu}(\text{direct}) \\ &= \mathrm{i}C_F\frac{1}{2}g^2\sum_s \overline{u}(p,s)\int\frac{\mathrm{d}^dk}{(2\pi)^d}\frac{\gamma_\alpha(\not{p}+\not{k})\gamma_\mu(\not{p}+\not{k}+\not{q})\gamma_\nu(\not{p}+\not{k})\gamma^\alpha}{(p+k)^4(p+k+q)^2k^2}u(p,s), \end{aligned} \tag{5.260}$$

where we have already performed the color sum

$$\frac{1}{N}\sum_{A,a}\sum_{b,i}\frac{\lambda^A_{ai}}{2}\frac{\lambda^A_{ib}}{2}\delta^{ab} = \frac{1}{N}\sum_A \mathrm{tr}\frac{\lambda^A}{2}\frac{\lambda^A}{2} = \frac{N^2-1}{2N} = C_F = \frac{4}{3} \tag{5.261}$$

stemming from the Gell-Mann matrices entering at the quark gluon vertices.

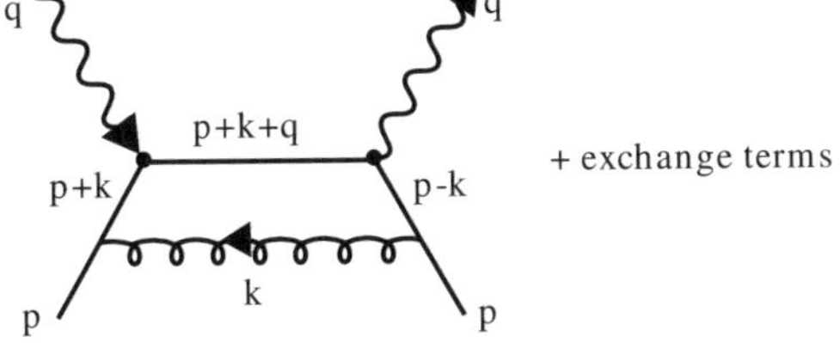

Fig. 5.22. Diagram that contributes to the $\mathcal{O}(\alpha_s)$ correction to the longitudinal structure function F_L(non-singlet)

In general we have to perform the calculation in $d=4+2\varepsilon$ dimensions to regularize divergencies. However, it turns out that for the projection onto F_L the integral is free of any divergence, which is not the case if we were to project with $g_{\mu\nu}$ onto F_2.

Performing the projection and using $p^2=0$ we obtain

$$\begin{aligned} p^\mu p^\nu T^{qq}_{\mu\nu}(\text{direct}) &= \mathrm{i}C_F g^2 \frac{1}{2}\sum_s \int \frac{\mathrm{d}^4k}{(2\pi)^4} \frac{\overline{u}(p,s)\gamma_\alpha \not{k}\not{p}(\not{k}+\not{q})\not{p}\not{k}\gamma^\alpha u(p,s)}{(p+k)^4(p+k+q)^2k^2} \\ &= -\mathrm{i}C_F g^2 \int \frac{\mathrm{d}^4k}{(2\pi)^4} \frac{\mathrm{tr}\,[\not{k}\not{p}(\not{k}+\not{q})\not{p}\not{k}\not{p}]}{(p+k)^4(p+k+q)^2k^2}\,, \end{aligned} \tag{5.262}$$

where we used $\sum_s \overline{u}(p,s)\hat{M}u(p,s)=\mathrm{tr}[\hat{M}\not{p}]$ and $\gamma_\alpha\not{p}\gamma^\alpha=-2\not{p}$. The trace can be further simplified by again using $p^2=0$:

$$\begin{aligned} \mathrm{tr}\,[\not{k}\not{p}(\not{k}+\not{q})\not{p}\not{k}\not{p}] &= 2kp\,\mathrm{tr}\,[\not{k}\not{p}(\not{k}+\not{q})\not{p}] \\ &= (2kp)^2\mathrm{tr}\,[(\not{k}+\not{q})\not{p}] \\ &= 16\left[(k\cdot p)^3+(k\cdot p)^2(p\cdot q)\right]\,. \end{aligned} \tag{5.263}$$

What remains is therefore to perform the following integral:

$$\begin{aligned} & p^\mu p^\nu T^{qq}_{\mu\nu}(\text{direct}) \\ &= -16\mathrm{i}C_F g^2 \int \frac{\mathrm{d}^4k}{(2\pi)^4} \frac{(k\cdot p)^3+(k\cdot p)^2(p\cdot q)}{(p+k)^4(p+k+q)^2k^2} \\ &= -16\mathrm{i}C_F g^2 \left[I_{\alpha\beta\varrho}(p,q)p^\alpha p^\beta p^\varrho + (p\cdot q)I_{\alpha\beta}(p,q)p^\alpha p^\beta\right]\,. \end{aligned} \tag{5.264}$$

This will be achieved by using the *Feynman parameter technique* (4.95). We begin with the integral

$$\begin{aligned} I_{\alpha\beta}(p,q) &= \int \frac{\mathrm{d}^4k}{(2\pi)^4} \frac{k_\alpha k_\beta}{(p+k)^4(p+k+q)^2k^2} \\ &= \frac{\Gamma(4)}{\Gamma(2)\Gamma(1)\Gamma(1)} \int_0^1 \mathrm{d}u \int_0^{\overline{u}} \mathrm{d}v\, u \\ &\quad\times \int \frac{\mathrm{d}^4k}{(2\pi)^4} \frac{k_\alpha k_\beta}{\left[(p+k)^2u+(p+k+q)^2v+k^2(1-u-v)\right]^4}\,, \end{aligned} \tag{5.265}$$

where we used the Feynman parametrization (see Exercise 4.7) and denote $\overline{u}=1-u$, $\overline{v}=1-v$. The denominator can be simplified:

$$\begin{aligned} & \left[u(p+k)^2+v(p+k+q)^2+k^2(1-u-v)\right]^4 \\ &\quad = k^2(1-u-v+u+v)+2k(up+v(p+q))+(p+q)^2v \\ &\quad = k^2+2k(up+v(p+q))+(p+q)^2v \end{aligned}$$

$$
\begin{aligned}
&= (k+up+v(p+q))^2 - (up+v(p+q))^2 + (p+q)^2 v \\
&= k'^2 - 2vu\,pq + (p+q)^2 v\bar{v} \\
&= k'^2 + v\bar{v}\left((p+q)^2 - \frac{u}{\bar{v}} 2pq\right) \\
&= k'^2 - v\bar{v}Q^2\left(1 - \frac{2pq(1-u')}{Q^2}\right) \\
&= k'^2 - v\bar{v}Q^2\left(1 - \omega\bar{u}'\right)
\end{aligned}
$$

with $u' = \frac{u}{\bar{v}}$ and $k' = k + up + v(p+q)$ and therefore

$$
k = k' - (u'\bar{v} + v)p - vq \ . \tag{5.266}
$$

With that we get

$$
I_{\alpha\beta}(p,q) = \Gamma(4)\int_0^1 \mathrm{d}u \int_0^1 \mathrm{d}v u\bar{v}^2 \int \frac{\mathrm{d}^4k}{(2\pi)^4} \frac{k'_\alpha k'_\beta}{\left[k'^2 - v\bar{v}Q^2(1-\omega\bar{u})\right]^4} \ , \tag{5.267}
$$

where we have rewritten the integration

$$
\int_0^1 \mathrm{d}u \int_0^{\bar{u}} \mathrm{d}vu\, f(u,v) = \int_0^1 \mathrm{d}v \int_0^{\bar{v}} \mathrm{d}uu\, f(u,v) = \int_0^1 \mathrm{d}v\bar{v}^2 \int_0^1 \mathrm{d}u'u'\, f(u'\bar{v},v) \ . \tag{5.268}
$$

The fact that the integral will be multiplied by $p^\alpha p^\beta$ again simplifies our calculation since due to the onshell condition $p^2 = 0$ we are allowed to retain only the terms $\sim q^\alpha q^\beta$ in the substitution $k'^\alpha k'^\beta$:

$$
\begin{aligned}
&I_{\alpha\beta}(p,q)p^\alpha p^\beta \\
&= \Gamma(4)(p\cdot q)^2 \int_0^1 \mathrm{d}v\bar{v}^2 v^2 \int_0^1 \mathrm{d}uu \int \frac{\mathrm{d}^4k}{(2\pi)^4} \frac{1}{\left[k^2 - v\bar{v}Q^2(1-\omega\bar{u})\right]^4} \ .
\end{aligned} \tag{5.269}
$$

Similarly we get for the other integral in (5.264)

$$
\begin{aligned}
&I_{\alpha\beta\varrho}(p,q)p^\alpha p^\beta p^\varrho \\
&= -\Gamma(4)(p\cdot q)^3 \int_0^1 \mathrm{d}v\bar{v}^2 v^3 \int_0^1 \mathrm{d}uu \int \frac{\mathrm{d}^4k}{(2\pi)^4} \frac{1}{\left[k^2 - v\bar{v}Q^2(1-\omega\bar{u})\right]^4} \ ,
\end{aligned} \tag{5.270}
$$

where the only difference is that instead of $k_\alpha k_\beta \to k'_\alpha k'_\beta \to v^2 q_\alpha q_\beta$ in (5.267) we have $k^\alpha k^\beta k^\varrho \to -q^\alpha q^\beta q^\varrho v^3$. The remaining k integration is readily performed with the help of the tables in (4.97) to (4.101):

$$\int \frac{\mathrm{d}^4 k}{(2\pi)^4} \frac{1}{\left[k^2 - v\bar{v}Q^2(1-\omega\bar{u})\right]^4} = \frac{\mathrm{i}}{(4\pi)^2} \frac{\Gamma(2)}{\left[v\bar{v}Q^2(1-\omega\bar{u})\right]^2} \frac{1}{\Gamma(4)} . \quad (5.271)$$

Since we simplified the denominator in such an intelligent way that the Feynman parameters *factorize*, we are now able to perform the $\int \mathrm{d}v$ integration:

$$\int_0^1 \mathrm{d}v \bar{v}^2 v^2 (v\bar{v})^{-2} = \frac{\Gamma(1)\Gamma(1)}{\Gamma(2)} = 1 ,$$

$$\int_0^1 \mathrm{d}v \bar{v}^2 v^3 (v\bar{v})^{-2} = \frac{\Gamma(1)\Gamma(2)}{\Gamma(3)} = \frac{1}{2} . \quad (5.272)$$

In passing, we note that the evaluation of the integrals within the Feynman parameter integral is one of the main obstacles in multiloop calculations. A factorization of Feynman parameters can always be enforced by expanding nonfactorized expressions like $(u+v)^a = \sum_{n=0}^a \binom{a}{n} u^n v^{a-n}$ for the price of introducing an additional sum. Summarizing we get

$$p^\mu p^\nu T^{qq(1)}_{\mu\nu}(\text{direct}) = \frac{16}{(4\pi)^2} C_F g^2 \frac{(p\cdot q)^3}{Q^4} \frac{1}{2} \int_0^1 \mathrm{d}u \frac{u}{(1-\omega\bar{u})^2} . \quad (5.273)$$

The superscript (1) to $T^{qq(1)}_{\mu\nu}$ indicates the first order in α_s approximation for $T^{qq}_{\mu\nu}$. To perform the remaining u integration we expand the denominator

$$\frac{1}{(1-\omega\bar{u})^2} = \sum_{n=0}^\infty (n+1)(\bar{u}\omega)^n \quad (5.274)$$

and with

$$\int_0^1 \mathrm{d}u \bar{u}^n u = \frac{\Gamma(n+1)\Gamma(2)}{\Gamma(n+3)} = \frac{1}{(n+2)(n+1)} \quad (5.275)$$

we end up with

$$p^\mu p^\nu T^{qq}_{\mu\nu(1)}(\text{direct}) = \frac{16}{(4\pi)^2} C_F g^2 \frac{(p\cdot q)^3}{Q^4} \frac{1}{2} \sum_{n=0}^\infty \frac{\omega^2}{(n+2)}$$

$$= C_F \frac{\alpha_s}{4\pi} \frac{Q^2}{x^2} \sum_{n=0}^\infty \frac{\omega^{n+1}}{(n+2)} . \quad (5.276)$$

The crossed diagram doubles even, and cancels odd, powers of ω so that we find

$$\frac{\tilde{F}_L^q}{2x} = 2C_F \frac{\alpha_s}{4\pi} \sum_{n=2,4,\ldots} \frac{4}{(n+1)} \omega^n \ . \tag{5.277}$$

The nth power of ω corresponds exactly to the nth moment of the structure function F_2 as we saw in the introductory discussion. Indeed, comparing with (5.256) to (5.259) leads us to the final result

$$\int_0^1 \mathrm{d}x x^{n-2} F_L^q(x, Q^2) = C_F \frac{\alpha_s}{4\pi} \frac{4}{(n+1)} \int_0^1 \mathrm{d}x x^{n-2} F_2^q(x, Q^2) \ . \tag{5.278}$$

Now we want to express the above relation not only on the level of moments of the structure function but also as a convolution in Bjorken-x space. With

$$\int_0^1 \mathrm{d}z z^{n-1} z = \frac{1}{n+1} \tag{5.279}$$

we can write

$$\begin{aligned}
\int_0^1 \mathrm{d}x x^{n-1} F_2^q(x, Q^2) &= 4C_F \frac{\alpha_s}{4\pi} \int_0^1 \mathrm{d}z z^{n-1} z \int_0^1 \mathrm{d}y y^{n-1} F_2(y, Q^2) \\
&= 4C_F \frac{\alpha_s}{4\pi} \int_0^1 \mathrm{d}y \int_0^1 \mathrm{d}z (zy)^{n-1} z F_2(y, Q^2) \\
&= 4C_F \frac{\alpha_s}{4\pi} \int_0^1 \frac{\mathrm{d}y}{y} \int_0^y \mathrm{d}x x^{n-1} \left(\frac{x}{y}\right) F_2(y, Q^2) \\
&= 4C_F \frac{\alpha_s}{4\pi} \int_0^1 \mathrm{d}x x^{n-1} \int_x^1 \frac{\mathrm{d}y}{y} \left(\frac{x}{y}\right) F_2(y, Q^2)
\end{aligned} \tag{5.280}$$

and therefore

$$F_L^q(x, Q^2) = 4C_F \frac{\alpha_s}{4\pi} \int_x^1 \frac{\mathrm{d}y}{y} \left(\frac{x}{y}\right) F_2^q(y, Q^2) \ . \tag{5.281}$$

So far we have only considered the $O(\alpha_s)$ correction coming from photon–quark scattering. However, to $O(\alpha_s)$ also diagrams from photon–gluon scattering containing a quark loop contribute (see Fig. 5.23).

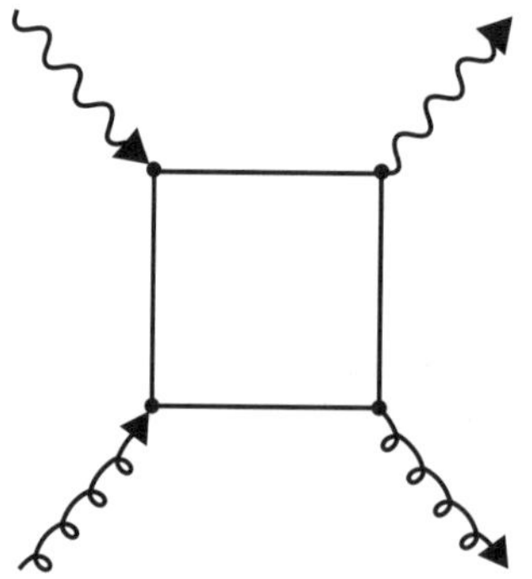

Fig. 5.23. $\mathcal{O}(\alpha_s)$ contribution to the Compton forward scattering amplitude stemming from photon–gluon scattering

We will calculate those diagrams that contribute only to singlet structure functions in Excercise 5.13. The final result is

$$F_L^g(x, Q^2) = T_\mathrm{f}\left(\frac{\alpha_s}{4\pi}\right) 16 \int\limits_x^1 \frac{\mathrm{d}y}{y}\left(\frac{x}{y}\right)\left(1-\frac{x}{y}\right) F_2^g(y) \; , \tag{5.282}$$

where $T_\mathrm{f} = \frac{1}{2}N_\mathrm{f}$ is the color factor times a factor N_f coming from the sum over the quark flavors running through the quark loop.

For F_L we therefore have calculated the complete $O(\alpha_s)$ corrections. As can be seen from (5.280) and (5.282) F_L is expressed in terms of the structure function $F_2(x, Q^2)$ times a factor that is calculable in perturbation theory. $F_2(x, Q^2)$ is simply the quark distribution function $F_2^q(x, Q^2) = xq(x, Q^2)$ and the gluon distribution function $F_2^g(x, Q^2) = xG(x, Q^2)$, respectively.

That F_L is completely expressed through F_2 is due to the fact that only *one* set of unpolarized operators $V_n(\mu^2)$ contributes to F_2 and F_L – see (5.175) and (5.176). (One has to keep in mind that the *vector* operators $V_n(\mu^2)$ contribute to the unpolarized structure functions while the *axialvector* operators $A_n(\mu^2)$ contribute to the polarized structure functions.)

We can therefore express F_L and F_2 as

$$\begin{aligned} \int\limits_0^1 \mathrm{d}x x^{n-2} F_L(x, Q^2) &= C_{n,L}\left(\alpha_s, \frac{Q^2}{\mu^2}\right) V_n(\mu^2) \; , \\ \int\limits_0^1 \mathrm{d}x x^{n-2} F_2(x, Q^2) &= C_{n,2}\left(\alpha_s, \frac{Q^2}{\mu^2}\right) V_n(\mu^2) \; , \end{aligned} \tag{5.283}$$

and solving the equations for F_L and F_2 by eliminating $V_n(\mu^2)$ we find

$$\int\limits_0^1 \mathrm{d}x x^{n-1} F_L(x, Q^2) = \frac{C_{n,L}\left(\alpha_s, \frac{Q^2}{\mu^2}\right)}{C_{n,2}\left(\alpha_s, \frac{Q^2}{\mu^2}\right)} \int\limits_0^1 \mathrm{d}x x^{n-1} F_2(x, Q^2)$$

$$
= \frac{\sum_{m=1} C_{n,L}^{(m)} \left(\frac{\alpha_s}{4\pi}\right)^m \int\limits_0^1 \mathrm{d}x x^{n-1} F_2(x, Q^2)}{\left(1 + \sum_{m=1} C_{n,2}^{(m)} \left(\frac{\alpha_s}{4\pi}\right)\right) \int\limits_0^1 \mathrm{d}x x^{n-1} F_2(x, Q^2)}
$$

$$
= \left\{ C_{n,L}^{1} \frac{\alpha_s}{4\pi} + \left[C_{n,L}^{(2)} - C_{n,L}^{(1)} C_{n,2}^{(1)} \right] \left(\frac{\alpha_s}{4\pi}\right)^2 + O\left(\left(\frac{\alpha_s}{4\pi}\right)^3\right) \right\} \times \int\limits_0^1 \mathrm{d}x x^{n-1} F_2(x, Q^2) \ . \quad (5.284)
$$

Here we have written the Wilson coefficients as an expansion in $\alpha_s/4\pi$ and made use of the fact that F_L is equal to zero at zeroth order and that usually the corresponding coefficient is normalized to 1 for F_2. This means $C_{n,L}^{(0)} = 0$ and $C_{n,2}^{(0)} = 1$. With this we have shown that to all orders in α_s, F_L is determined completely by F_2 up to higher-twist $\left(\frac{1}{Q^2}\right)$ corrections.

The expansion coefficients $C_{n,2}^{(m)}$ corresponding to F_2 can be calculated in a similar way as we did for F_L. The calculations, however, are more lengthy because now all diagrams of Fig. 5.21 contribute. An additional complication arises due to the fact that the calculations have to be carried through in d dimensions since for the projection onto F_2 the diagrams do not remain finite and have to be renormalized.

The details of such calculations can be found in the very comprehensive paper by Bardeen et al.[16] The coefficient functions are available in the literature for the first 10 moments up to the third order.[17] The coefficient functions for all moments i.e. the full x dependence are available up to the second order.[18] Such calculations cannot be done in the simple way presented here for the lowest order but depend heavily on the use of algebraic computer codes. Also it seems not to be feasible to push the calculations to even higher orders since at each order the number of contributing diagrams increases considerably and with each additional loop integration the necessary CPU time increases as well. Instead, the study of the perturbative series to all orders in certain approximations will be persued, e.g. for small n (n characterizes the nth moment of the structure function) or for large N_f.

---

[16] W.A. Bardeen, A.J. Buras, D.W. Duke and T. Muta: Phys. Rev. D **18**, 3998 (1978).

[17] S.A. Larin, P. Nogueira, T. van Ritbergen and J.A.M. Vermaseren: Nucl. Phys. B **492**,338 (1997).

[18] E.B. Zijlstra and W.L. van Neerven: Nucl. Phys. B **383**, 525 (1992).

EXERCISE

5.13 Calculation of the Gluonic Contribution to $F_L(x, Q^2)$

Problem. Calculate in the same way as done in the foregoing text the contribution to F_L due to photon–gluon scattering via a quark loop. The corresponding diagrams are shown in the figure. A usefull trick is to express the product of two propagators involving an onshell momentum $p^2 = 0$ as

$$\frac{1}{(k+p)^2}\frac{1}{k^2} = \int_0^1 du \frac{1}{(k+up)^4} \; .$$

Prove this formula and find its generalization.

Solution. The color factor for the diagrams coming from the Gell-Mann matrices at the quark–gluon vertices can be written as

$$\frac{1}{(N^2-1)}\sum_A\sum_B \mathrm{tr}\left[\frac{\lambda^A}{2}\frac{\lambda^B}{2}\right] = \frac{1}{(N^2-1)}\sum_A\sum_B \frac{1}{2}\delta^{AB} = \frac{1}{2} \; , \tag{1}$$

where we summed over outgoing gluon colors and averaged over the incoming ones. Together with a factor N_f, which is due to the sum over flavors in the quark loop, this yields an overall factor $T_\mathrm{f} = \frac{1}{2}N_\mathrm{f}$.

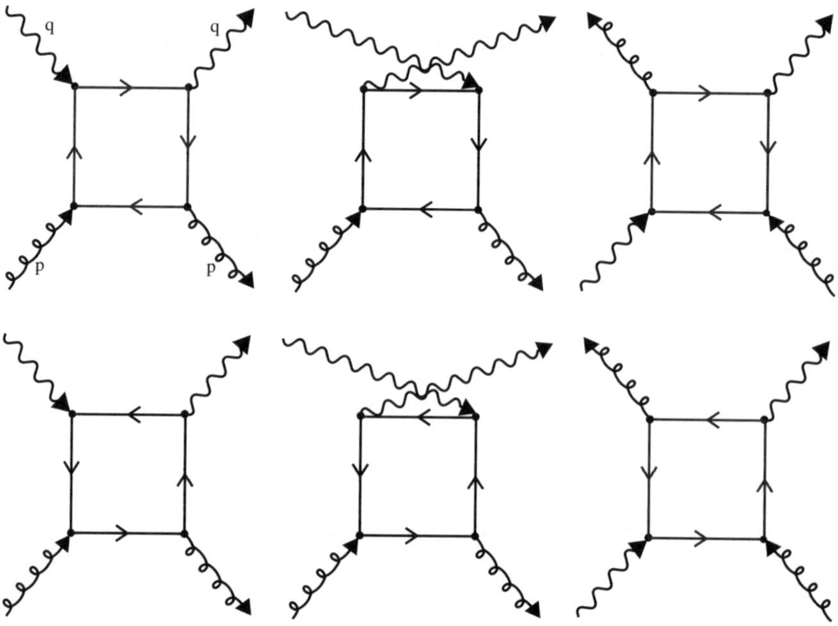

Fig. 5.24. Diagrams contributing in $\mathcal{O}(\alpha_s)$ to the virtual photon–gluon scattering.

Exercise 5.13

We start with the first diagram which we name $T^{(1)gg}_{\mu\nu}$(direct). In the usual way the exchange term can be found by substituting $\mu \leftrightarrow \nu$ and $q \leftrightarrow -q$. According to the Feynman rules we write

$$T^{(1)gg}_{\mu\nu}(\text{direct}) = \mathrm{i}\frac{g^2}{2}T_\mathrm{f}\int\frac{\mathrm{d}^4k}{(2\pi)^4}\mathrm{tr}\left[\gamma_\mu\frac{\mathrm{i}}{\not p+\not q+\not k}\gamma_\nu\frac{\mathrm{i}}{\not p+\not k}\gamma_\alpha\frac{\mathrm{i}}{\not k}\gamma_\beta\frac{i}{\not p+\not k}\right]\left(-g^{\alpha\beta}\right)\ , \tag{2}$$

where we have inserted an additional factor of $\frac{1}{2}$ due to the averaging over the gluon spin and used $\left(-g^{\alpha\beta}\right)$ for the polarization sum $\frac{1}{2}\sum\varepsilon^\alpha(k)\varepsilon^\beta(k) = -g^{\alpha\beta}/2$. Again we compute i times the time-ordered product. Projecting with $p^\mu p^\nu$ we find, with $\gamma_\alpha\not k\gamma^\alpha = -2\not k$,

$$\begin{aligned}p^\mu p^\nu T^{(1)gg}_{\mu\nu}(\text{direct}) &= \mathrm{i}g^2T_\mathrm{f}\int\frac{\mathrm{d}^4k}{(2\pi)^4}\frac{\mathrm{tr}\left[\not p(\not q+\not k)\not p\not k\not k\not k\right]}{(p+q+k)^2(p+k)^4k^2}\\ &= 8\mathrm{i}g^2T_\mathrm{f}\int\frac{\mathrm{d}^4k}{(2\pi)^4}\frac{p\cdot(q+k)(kp)}{(p+q+k)^2(p+k)^4}\ .\end{aligned} \tag{3}$$

The integral can be evaluated in the usual way

$$\begin{aligned}&\int\frac{\mathrm{d}^4k}{(2\pi)^4}\frac{(k+q)_\alpha k_\beta}{(p+k+q)^2(p+k)^4}\\ &= \frac{\Gamma(3)}{\Gamma(1)\Gamma(2)}\int_0^1\mathrm{d}u\bar{u}\int\frac{\mathrm{d}^4k}{(2\pi)^4}\frac{(k+q)_\alpha k_\beta}{\left[u(p+k+q)^2+\bar{u}(p+k)^2\right]^3}\\ &= \Gamma(3)\int_0^1\mathrm{d}u\bar{u}\int\frac{\mathrm{d}^4k}{(2\pi)^4}\frac{[k-(p+uq)]_\alpha[k-p+\bar{u}q)]_\beta}{\left[k^2-u\bar{u}Q^2\right]^3}\\ &= -q_\alpha q_\beta\Gamma(3)\int_0^1\mathrm{d}u\bar{u}\int\frac{\mathrm{d}^4k}{(2\pi)^4}\frac{1}{\left[k^2-u\bar{u}Q^2\right]^3}+\mathcal{O}\left(g_{\alpha\beta},\,p_\alpha p_\beta\right)\\ &= -q_\alpha q_\beta\int_0^1\mathrm{d}u\bar{u}u\frac{\mathrm{i}}{(4\pi)^{\frac{d}{2}}}\frac{1}{\left(u\bar{u}Q^2\right)}+\mathcal{O}\left(g_{\alpha\beta},\,p_\alpha p_\beta\right)\\ &= \frac{-q_\alpha q_\beta}{Q^2}\frac{\mathrm{i}}{(4\pi)^2}\cdot\frac{1}{2}+\mathcal{O}\left(g_{\alpha\beta},\,p_\alpha p_\beta\right)\ .\end{aligned} \tag{4}$$

We performed the usual Feynman parametrization (see Exercise 4.7) and wrote the denominator as $\left[u(p+k+q)^2+\bar{u}(p+k)^2\right] = \left[(k+p+uq)^2 + u\bar{u}q^2\right]$, where $\bar{u} = 1-u$. Inserting this into (3) yields

$$p^\mu p^\nu T^{(1)gg}_{\mu\nu}(\text{direct}) = \frac{4g^2T_\mathrm{f}}{(4\pi)^2}\frac{(p\cdot q)^2}{Q^2}\ . \tag{5}$$

Exercise 5.13

Taking the other four diagrams of the same type into account, i.e. the exchange term and the opposite direction of the fermion loop, we get a contribution to F_L (5.251) of

$$\frac{\tilde{F}_L^{(1)g}}{2x} = \frac{16g^2T_\mathrm{f}}{(4\pi)^2} \,, \tag{6}$$

which is, however, independent of ω, i.e. proportional to ω^0. However, the sum in (5.258) starts as ω^2, i.e. after transformation to Bjorken x these four diagrams do not contribute to the measurable structure function.

The remaining diagrams are sligthly more complicated to evaluate. They are shown in the last row of the figure at the beginnning of this exercise. We write according to the Feynman rules (see Fig. 5.25)

$$\begin{aligned} &T_{\mu\nu}^{(1)gg}(\text{direct}) \\ &= \frac{1}{2}T_\mathrm{f}g^2\mathrm{i}\int\frac{\mathrm{d}^4k}{(2\pi)^4}\mathrm{tr}\left[\gamma_\mu\frac{\mathrm{i}}{\not{q}+\not{k}}\gamma_\alpha\frac{\mathrm{i}}{\not{q}+\not{k}+\not{p}}\gamma_\nu\frac{\mathrm{i}}{\not{k}+\not{p}}\gamma_\beta\frac{\mathrm{i}}{\not{k}}\right]\left(-g^{\alpha\beta}\right) \,. \end{aligned} \tag{7}$$

Projecting onto F_L we write

$$\begin{aligned} p^\mu p^\nu T_{\mu\nu}^{(1)gg}(\text{direct}) &= -\frac{1}{2}T_\mathrm{f}g^2\mathrm{i} \\ &\quad\times\int\frac{\mathrm{d}^4k}{(2\pi)^4}\frac{\mathrm{tr}\,[\not{p}(\not{q}+\not{k})\gamma_\alpha(\not{q}+\not{k}+\not{p})\not{p}(\not{k}+\not{p})\gamma^\alpha\not{k}]}{(q+k)^2(q+k+p)^2(k+p)^2k^2} \\ &= +T_\mathrm{f}g^2\mathrm{i}\int\frac{\mathrm{d}^4k}{(2\pi)^4}\frac{\mathrm{tr}\,[\not{p}(\not{q}+\not{k})\not{k}\not{p}(\not{q}+\not{k})\not{k}]}{(q+k)^2(q+k+p)^2(k+p)^2k^2} \,, \end{aligned} \tag{8}$$

where we used $\not{p}^2 = p^2 = 0$ and $\gamma_\alpha\gamma^\mu\gamma^\nu\gamma^\varrho\gamma^\alpha = -2\gamma^\varrho\gamma^\nu\gamma^\mu$. The trace can be further simplified as

$$\begin{aligned} \mathrm{tr}\,[\not{p}(\not{q}+\not{k})\not{k}\not{p}(\not{q}+\not{k})\not{k}] &= 2k\cdot p\,\mathrm{tr}\,[\not{p}(\not{q}+\not{k})(\not{q}+\not{k})\not{k}] \\ &\quad-\mathrm{tr}\,[\not{p}(\not{q}+\not{k})\not{p}\not{k}(\not{q}+\not{k})\not{k}] \\ &= 8(k\cdot p)^2(q+k)^2 - 2p\cdot(q+k)\mathrm{tr}\,[\not{p}\not{k}(\not{q}+\not{k})\not{k}] \\ &= 8(k\cdot p)^2(q+k)^2 - 16p\cdot(q+k)(k\cdot q)(p\cdot k) \\ &\quad-16p(q+k)p\cdot k\,k^2 + 8[p\cdot(q+k)]^2k^2 \,, \end{aligned} \tag{9}$$

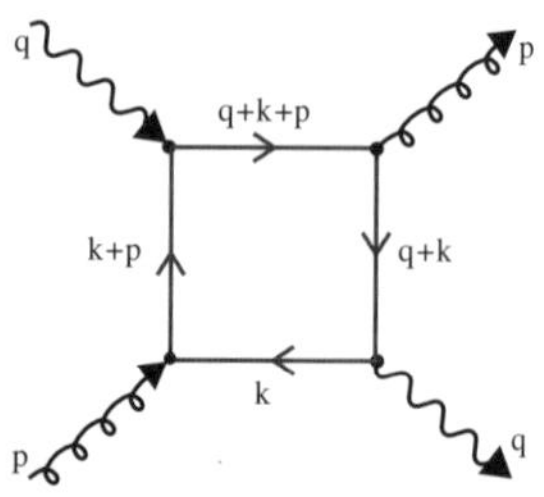

Fig. 5.25. Second type of diagrams that contribute to virtual photon–gluonscattering.

so that we get

$$\begin{aligned} p^\mu p^\nu T_{\mu\nu}^{(1)gg}(\text{direct}) &= 8g^2\mathrm{i}T_\mathrm{f}\,[I_1(q,\,p)+I_1(-q,\,p) \\ &\quad-2I_2(q,\,p)-2I_3(q,\,p)] \,, \end{aligned} \tag{10}$$

where *Exercise 5.13*

$$
\begin{aligned}
I_1(q,p) &= \int \frac{\mathrm{d}^4k}{(2\pi)^4} \frac{(k\cdot p)^2}{(q+k+p)^2(k+p)^2k^2} \ , \\
I_1(-q,p) &= \int \frac{\mathrm{d}^4k}{(2\pi)^4} \frac{(p\cdot(q+k))^2}{(q+k)^2(q+k+p)^2(k+p)^2} \\
&= \int \frac{\mathrm{d}^4k}{(2\pi)^4} \frac{(p\cdot k)^2}{(k+p)^2(k+p-q)^2k^2} \ , \\
I_2(q,p) &= \int \frac{\mathrm{d}^4k}{(2\pi)^4} \frac{p\cdot(q+k)p\cdot k}{(q+k)^2(q+k+p)^2(k+p)^2} \ , \\
I_3(q,p) &= \int \frac{\mathrm{d}^4k}{(2\pi)^4} \frac{p\cdot(q+k)k\cdot q p\cdot k}{(q+k)^2(q+k+p)^2(k+p)^2k^2} \ .
\end{aligned} \tag{11}
$$

The integrals $I_1(q,p)$ and $I_2(q,p)$ are of the same type as we had before (see (5.265)). Using suitable substitutions we can therefore deduce the corresponding results directly from the previous calculation of F_L^q. Comparing $I_1(q,p)$ with (5.265) we see that $(p+k)^2$ in the denominator comes now with a power of 1 instead of 2. Therefore we copy from (5.269) with evident changes and get

$$
I_1(q,p) = (p\cdot q)^2\Gamma(3)\int_0^1 \mathrm{d}u \int_0^1 \mathrm{d}v\bar{v}v^2 \int \frac{\mathrm{d}^4k}{(2\pi)^4} \frac{1}{(k^2-v\bar{v}Q^2(1-\omega\bar{u}))^3} \ . \tag{12}
$$

We changed $\Gamma(4)\to\Gamma(3)$ corresponding to the different power and we omitted the Feynman parameter $(u\bar{v})$, which came from the parametrization of $(p+k)^4$ in (5.265), which in our case now is $(p+k)^2$. Also we have taken care of the shift in the integration variable $k_\alpha \to k_\alpha - (u\bar{v}+v)p_\alpha - vq_\alpha$ where due to the contraction with $p^\alpha p^\beta$ only $v^2 q_\alpha q_\beta$ contributes.

For the $\int \mathrm{d}k$ integral we get

$$
\int \frac{\mathrm{d}^4k}{(2\pi)^4} \frac{1}{\left[k^2-v\bar{v}Q^2(1-\omega\bar{u})\right]^3} = \frac{-\mathrm{i}}{(4\pi)^2}\frac{1}{\Gamma(3)}\frac{1}{v\bar{v}Q^2(1-\omega\bar{u})} \ . \tag{13}
$$

The $\int \mathrm{d}v$ integration gives

$$
\int_0^1 \mathrm{d}v\bar{v}^2 v(v\bar{v})^{-1} = \frac{1}{2} \ , \tag{14}
$$

and therefore

$$
I_1(q,p) = \frac{-\mathrm{i}}{(4\pi)^2}\frac{(p\cdot q)^2}{Q^2}\cdot\frac{1}{2}\int_0^1 \mathrm{d}u\frac{1}{(1-\omega\bar{u})} \ . \tag{15}
$$

Exercise 5.13

Expanding

$$\frac{1}{1-\omega\overline{u}} = \sum_{n=0}^{\infty}(\omega\overline{u})^n \ , \tag{16}$$

we find, after trivial integration,

$$I_1(q,\,p) = \frac{-\mathrm{i}}{(4\pi)^2}\frac{(p\cdot q)^2}{Q^2}\frac{1}{2}\sum_{n=0}^{\infty}\frac{1}{(n+1)}\omega^n \ . \tag{17}$$

In the same way we can evaluate $I_2(q,\,p)$:

$$\begin{aligned} I_2(q,\,p) &= \int \frac{\mathrm{d}^4k}{(2\pi)^4}\frac{p\cdot(q+k)\ p\cdot k}{(q+k)^2(q+k+p)^2(k+p)^2} \\ &= \int \frac{\mathrm{d}^4k}{(2\pi)^4}\frac{(p\cdot k)\ p\cdot(k-q)}{k^2(k+p)^2(k+p-q)^2} \\ &= \Gamma(3)\int_0^1 \mathrm{d}u \int_0^1 \mathrm{d}v\overline{v}\int\frac{\mathrm{d}^4k}{(2\pi)^4}\frac{p\cdot(k-vq)\ p\cdot(k-(1-v)q)}{\left[k^2-v\overline{v}Q^2(1+\omega\overline{u})\right]^3} \ , \end{aligned} \tag{18}$$

where we again copied from (5.269) and performed in addition to the changes carried out in (12) the substitution $q \to -q$, i.e. $\omega \to -\omega$. Also the shift of the integration variable now reads $k \to k-(u\overline{v}+v)p+vq$.

I_2 therefore becomes

$$I_2(q,\,p) = -\Gamma(3)(p\cdot q)^2\int_0^1 \mathrm{d}u\int_0^1 \mathrm{d}v\overline{v}^2v\int\frac{\mathrm{d}^4k}{(2\pi)^4}\frac{1}{\left[k^2-v\overline{v}Q^2(1+\omega\overline{u})\right]^3} \tag{19}$$

$$= \frac{+\mathrm{i}}{(4\pi)^2}\frac{(p\cdot q)^2}{Q^2}\frac{1}{2}\sum_{n=0}^{\infty}\frac{1}{(n+1)}(-\omega)^n \ , \tag{20}$$

where we used (13) and obtained for the $\int \mathrm{d}v$ integration

$$\int_0^1 \mathrm{d}v\overline{v}^2v(\overline{v}v)^{-1} = \frac{1}{2} \ . \tag{21}$$

We also expanded

$$\frac{1}{(1+\omega\overline{u})} = \sum_{n=0}^{\infty}(-\omega\overline{u})^n \ . \tag{22}$$

Let us now proceed to calculate $I_3(q,\,p)$, which is a little bit more tricky since we deal with four propagators, two of which contain off-shell momenta

Exercise 5.13

$q^2 = -Q^2 \neq 0$. To circumvent this problem we note the following relation:

$$\frac{1}{(k+p)^{2A}} \cdot \frac{1}{(k^2)^B} = \frac{\Gamma(A+B)}{\Gamma(A)\Gamma(B)} \int_0^1 \mathrm{d}u \frac{u^{A-1}\bar{u}^{B-1}}{(k+up)^{2(A+B)}} , \tag{23}$$

which holds for arbitrary $k^2 \neq 0$ and $p^2 = 0$. It can be proven using Feynman parameters:

$$\frac{1}{(k+p)^{2A}} \cdot \frac{1}{(k^2)^B} = \frac{\Gamma(A+B)}{\Gamma(A)\Gamma(B)} \int_0^1 \mathrm{d}u \frac{u^{A-1}\bar{u}^{B-1}}{\left[u(k+p)^2+\bar{u}k^2\right]^{A+B}} , \tag{24}$$

where $\left[u(k+p)^2+\bar{u}k^2\right] = k^2+2kup = (k+up)^2$ completes the proof. Using this twice, the integral $I_3(q,p)$ can be written as

$$\begin{aligned}
I_3(q,p) &= \int \frac{\mathrm{d}^4k}{(2\pi)^4} \frac{p\cdot(q+k)\; k\cdot q\; p\cdot k}{(q+k)^2(q+k+p)^2(k+p)^2k^2} \\
&= \int_0^1 \mathrm{d}u \int_0^1 \mathrm{d}v \int \frac{\mathrm{d}^4k}{(2\pi)^4} \frac{p\cdot(q+k)\; k\cdot q\; p\cdot k}{(k+up)^4(q+k+vp)^4} \\
&= \Gamma(4) \int_0^1 \mathrm{d}u \int_0^1 \mathrm{d}v \int_0^1 \mathrm{d}y y\bar{y} \\
&\quad \times \int \frac{\mathrm{d}^4k}{(2\pi)^4} \frac{p\cdot(q+k)\; p\cdot k\; (k-up)\cdot q}{\left[\bar{y}k^2+y(k+q+(v-u)p)^2\right]^4} ,
\end{aligned} \tag{25}$$

where in the last step $k \to k-up$ is shifted and a third Feynman parameter is introduced. The denominator can now be simplified in the usual way

$$\begin{aligned}
\left[\bar{y}k^2+y(k+q+(v-u)p)^2\right] &= k^2+2k\cdot(yq+(v-u)yp) \\
&\quad + y(q+(v-u)p)^2 \\
&= \left[k+yq+(v-u)yp)^2+y\bar{y}(q+(v-u)p\right]^2 \\
&= k'^2+y\bar{y}\left[q^2+2q\cdot p(v-u)\right] \\
&= k'^2-y\bar{y}Q^2[1-\omega(v-u)]
\end{aligned} \tag{26}$$

The shift of the integration momenta $k = k'-yp-(v-u)yp$ has to be performed in a similar way in the numerator:

$$\begin{aligned}
&(q+k)_\alpha(k-up)_\beta k_\sigma \left[p^\alpha q^\beta p^\sigma\right] \\
&\quad = (k'+\bar{y}q)_\alpha(k'-yq-vyp-u\bar{y}p)_\beta(k'-yq)_\sigma \left[p^\alpha q^\beta p^\sigma\right] \\
&\quad = k'_\alpha k'_\beta(-yq)_\sigma + k'_\beta k'_\sigma(\bar{y}q)_\alpha + \bar{y}q_\alpha(yq+vyp+u\bar{y}p)_\beta yq_\sigma \left[p^\alpha q^\beta p^\sigma\right] \\
&\quad = \frac{k'^2}{4}(p\cdot q)^2[1-2y] + \bar{y}y(p\cdot q)^3(vy+u\bar{y}) + \bar{y}y^2(p\cdot q)^2q^2 .
\end{aligned} \tag{27}$$

Exercise 5.13

Here we used the fact that under integration, contributions like $\sim k_\alpha$ or $k_\alpha k_\beta k_\sigma$ vanish and that $k_\alpha k_\beta = \frac{1}{4}k^2 g_{\alpha\beta}$. (See our discussion on dimensional regularization before.)

With that we are left with two k' integrations:

$$\begin{aligned}\frac{1}{4}\int \frac{\mathrm{d}^4k'}{(2\pi)^4}\frac{k'^2}{\left[k'^2 - y\bar{y}Q^2(1-\omega(v-u))\right]^4} &= \frac{-\mathrm{i}}{4(4\pi)^2}\frac{\Gamma(3)}{\Gamma(2)\Gamma(4)}\frac{1}{\left(y\bar{y}Q^2(1-\omega(v-u))\right)} \\ &= \frac{-\mathrm{i}}{2(4\pi)^2}\frac{1}{\Gamma(4)}\frac{1}{\left(y\bar{y}Q^2(1-\omega(v-u))\right)}\end{aligned} \tag{28a}$$

and

$$\begin{aligned}\int \frac{\mathrm{d}^4k}{(2\pi)^4}\frac{1}{\left[k^2 - y\bar{y}Q^2(1-\omega(v-u))\right]^4} &= \frac{\mathrm{i}}{(4\pi)^2}\frac{\Gamma(2)}{\Gamma(4)\left[y\bar{y}Q^2(1-\omega)(v-u)\right]^2} \ .\end{aligned} \tag{28b}$$

Inserting everything in (25) and performing the trivial $\int \mathrm{d}y$ integration we get

$$\begin{aligned}I_3(q,p) &= \frac{\mathrm{i}}{2(4\pi)^2}\int_0^1 \mathrm{d}u\int_0^1 \mathrm{d}v\left\{\frac{(p\cdot q)^3}{Q^4}\frac{(v+u)}{[1-\omega(v-u)]^2}\right. \\ &\quad \left. + \frac{(p\cdot q)^2}{Q^4}q^2\frac{1}{[1-\omega(v-u)]^2}\right\} \\ &= \frac{\mathrm{i}}{4(4\pi)^2}\int_0^1 \mathrm{d}u\int_0^1 \mathrm{d}v\frac{(v+u)\omega-2}{[1-\omega(v-u)]^2}\frac{(p\cdot q)^2}{Q^2} \ .\end{aligned} \tag{29}$$

Expanding again in ω yields

$$\frac{1}{[1-\omega(v-u)]^2} = \sum_{n=0}^{\infty}\frac{\Gamma(n+2)}{\Gamma(2)\Gamma(n+1)}(v-u)^n\omega^n = \sum_{n=0}^{\infty}(1+n)(v-u)^n\omega^n \tag{30}$$

and the remaining integrations can be performed:

$$\begin{aligned}\int_0^1 \mathrm{d}u\int_0^1 \mathrm{d}v(v-u)^n &= \frac{1}{(1+n)}\int_0^1 \mathrm{d}u\left[(1-u)^{1+n} + (-)^n u^{1+n}\right] \\ &= \frac{1}{(1+n)(2+n)}\left(1+(-)^n\right) \ ,\end{aligned} \tag{31}$$

Exercise 5.13

$$\int_0^1 \mathrm{d}u \int_0^1 \mathrm{d}v (v+u)(v-u)^n$$

$$= \int_0^1 \mathrm{d}u \left[\frac{\overline{u}^{n+2}}{n+2} - (-)^n \frac{u^{n+2}}{n+2} + 2\frac{u\overline{u}^{n+1}}{(n+1)} - (-)^{n+1} \frac{2u^{n+2}}{n+1} \right]$$

$$= \frac{1}{(n+2)(n+3)} - \frac{(-)^n}{(n+2)(n+3)} + \frac{2}{(n+2)(n+3)(n+1)}$$

$$- (-)^{n+1} \frac{2}{(n+1)(n+3)} \ . \tag{32}$$

Inserting (30), (31), and (32) into (29) yields

$$I_3(q, p)$$

$$= \left\{ \frac{\mathrm{i}}{4(4\pi)^2} \sum_{n=0}^{\infty} \left\{ \frac{1+n}{(n+2)(n+3)} \left(1-(-)^n\right) + \frac{2}{(n+2)(n+3)} \right. \right.$$

$$\left. \left. + \frac{2(-)^n}{(n+3)} \right\} \omega^{n+1} - \frac{\mathrm{i}}{2(4\pi)^2} \sum_{n=0}^{\infty} \frac{1}{(2+n)} \left(1+(-)^n\right) \omega^n \right\} \frac{(p\cdot q)^2}{Q^2}$$

$$= \left\{ \frac{\mathrm{i}}{4(4\pi)^2} \sum_{n=0}^{\infty} \left\{ \left[\frac{2}{(3+n)} - \frac{1}{(2+n)} \right] \left(1-(-)^n\right) \right. \right.$$

$$\left. + 2\left[\frac{1}{(2+n)} - \frac{1}{(3+n)} \right] + \frac{2(-)^n}{(n+3)} \right\} \omega^{n+1}$$

$$\left. - \frac{\mathrm{i}}{2(4\pi)^2} \sum_{n=0}^{\infty} \frac{1}{(2+n)} \left(1+(-)^n\right) \omega^n \right\} \frac{(p\cdot q)^2}{Q^2}$$

$$= \left\{ \frac{\mathrm{i}}{4(4\pi)^2} \sum_{n=0}^{\infty} \frac{1}{(2+n)} \omega^{n+1} \left(1+(-)^n\right) \right.$$

$$\left. - \frac{\mathrm{i}}{2(4\pi)^2} \sum_{n=0}^{\infty} \frac{1}{(2+n)} \left(1+(-)^n\right) \omega^n \right\} \frac{(p\cdot q)^2}{Q^2}$$

$$= \left\{ \frac{\mathrm{i}}{2(4\pi)^2} \sum_{n=0,2,4,}^{\infty} \frac{1}{(2+n)} \omega^{n+1} - \frac{\mathrm{i}}{(4\pi)^2} \sum_{n=0,2,4,}^{\infty} \frac{1}{(2+n)} \omega^n \right\} \frac{(p\cdot q)^2}{Q^2} \ . \tag{33}$$

Taking into account the contribution coming from the exchange term $(q \to -q)$, which in this case cancels odd powers of ω and doubles even powers, we find

$$I_3(q, p) + I_3(-q, p) = \left(\frac{-2\mathrm{i}}{(4\pi)^2} \sum_{n=0,2,4,}^{\infty} \frac{1}{(2+n)} \omega^n \right) \frac{(p\cdot q)^2}{Q^2} \ . \tag{34}$$

Exercise 5.13

To get the final answer we have to insert (34), (17), and (20), into (10), which yields

$$\begin{aligned}
&p^\mu p^\nu T^{(2)gg}_{\mu\nu}(\text{direct}) + p^\mu p^\nu T^{(2)gg}_{\mu\nu}(\text{exchange}) \\
&= 8g^2 \mathrm{i} T_\mathrm{f}\, [2I_1(q,p) + 2I_1(-q,p) - 2(I_2(q,p) \\
&\quad + I_2(-q,p) + I_3(q,p) + I_3(-q,p))] \\
&= \frac{8}{(4\pi)^2} g^2 T_\mathrm{f} \sum_{n=0,2,4,} \left\{ \frac{2}{(n+1)} + \frac{2}{(n+1)} - \frac{4}{(n+2)} \right\} \omega^n \frac{(p\cdot q)^2}{Q^2} \\
&= \frac{32}{(4\pi)^2} g^2 T_\mathrm{f} \sum_{n=0,2,4,} \frac{1}{(1+n)(2+n)} \omega^n \left(\frac{(p\cdot q)^2}{Q^2} \right) . \qquad (35)
\end{aligned}$$

Finally we therefore get

$$\frac{\tilde{F}_2}{2x} = 32 \frac{\alpha_s}{4\pi} T_\mathrm{f} \cdot \sum_{n=0,2,4,}^{\infty} \frac{1}{(1+n)(2+n)} \omega^n . \qquad (36)$$

Repeating the steps we did after (5.277), and observing that

$$\begin{aligned}
\frac{1}{(1+n)(2+n)} = \frac{1}{(1+n)} - \frac{1}{(2+n)} &= \int_0^1 \mathrm{d}x x^{n-1} x - \int_0^1 \mathrm{d}x x^{n-1} x^2 \\
&= \int_0^1 \mathrm{d}x x^{n-1} x(1-x) , \qquad (37)
\end{aligned}$$

the desired result emerges as

$$F_L^g(x, Q^2) = T_\mathrm{f} \left(\frac{\alpha_s}{4\pi} \right) 16 \int_x^1 \frac{\mathrm{d}y}{y} \left(\frac{x}{y} \right) \left(1 - \frac{x}{y} \right) F_2^g(y) , \qquad (38)$$

which has been quoted in (5.282).

EXAMPLE

5.14 Calculation of Perturbative Corrections to Structure Functions with the Cross-Section Method

We have calculated in the previous sections the Wilson coefficients in $O(\alpha_s)$ using operator product expansion and a dispersion relation (the optical theorem), which relates the imaginary part of the forward scattering amplitude to the cross section. Instead of looking at the imaginary part it is sometimes convenient to calculate the cross section directly. This procedure is more general and has an

Example 5.14

wider applicablity. We would like to demonstrate this method, which we call the *cross-section method*, by essentially reproducing the results from the sections before, i.e. the perturbative corrections to F_L^q and F_L^g.

Following (5.281) and (5.282), we can denote the structure function $F_L(x, Q^2)$ and $F_2(x, Q^2)$ as

$$F_L^k(x, Q^2) = \int_x^1 \frac{dz}{z} \mathcal{F}_{L,k}(z, Q^2) F_2^k\left(\frac{x}{z}, Q^2\right) , \tag{1a}$$

$$F_2^k(x, Q^2) = \int_x^1 \frac{dz}{z} \mathcal{F}_{2,k}(z, Q^2) F_2^k\left(\frac{x}{z}, Q^2\right) , \tag{1b}$$

where the index k refers either to quarks q or gluons g. As usual, $Q^2 = -q^2$ describes the momentum of the virtual photon. $x = Q^2/2Pq$ refers to the Bjorken variable with respect to the nucleon state $|P\rangle$. The variable $z = Q^2/2pq$, on the other hand, corresponds to the Bjorken variable with respect to the parton momentum p.

At leading order, $F_2^{(0)k}$ is the structure function related to the pure parton density $F_2^{(0)q} = xq(x)$. $\mathcal{F}_{L,k}$ and $\mathcal{F}_{2,k}$ describe the radiative corrections to the parton subprocess

$$k + \gamma^* \rightarrow X ,$$

which are illustrated in the Fig. 5.26.

In the language of operator product expansion, the distribution functions $xq(x)$ and $xG(x)$ can be considered as the nonperturbative input parameters, while $\mathcal{F}_{L,k}$ and $\mathcal{F}_{2,k}$ are the coefficient functions, calculable order by order in perturbation theory. Then $\mathcal{F}_{L,k}$ is related to scattering of a longitudially polarized virtual photon off parton k and $\mathcal{F}_{2,k}$ to the scattering of a transversely polarized photon. Since the variable z corresponds to the Bjorken variable with respect to the parton momentum, the calligraphic structures $\mathcal{F}_{2,k}$ and $\mathcal{F}_{L,k}$ can be interpreted as *parton structure functions*.

In the following calculation, all problems related to renormalization will be neglected. Indeed, as we have seen before, in the special case we are interested in, the quantity $\mathcal{F}_{L,k}(x)$ is finite at $\mathcal{O}(\alpha_s)$. In general this function is divergent and we have to use a renormalization prescription to remove poles like $1/\varepsilon$ that arise when using dimensional regularization. However, for the projection onto F_L, poles do not arise. Therefore, we can carry out the calculation in $d = 4$ dimensions.

The parton structure functions can be obtained from the parton tensor, which we can define as

$$W_{\mu\nu} = \frac{1}{2}\frac{1}{2\pi} \sum_l \int dPS^{(l)} M_\mu(l) M_\nu^*(l) \tag{2}$$

Example 5.14

with

$$\int \mathrm{dPS}^{(l)} = \left(\prod_{j=1}^{l}\int \frac{\mathrm{d}^3 p_j}{(2\pi)^3 2p_0}\right)(2\pi)^4\delta\left(p+q-\sum_{j=1}^{l} p_j\right)$$
$$= \left(\prod_{j=1}^{l}\int \frac{\mathrm{d}^4 p_j}{(2\pi)^3}\delta(p_j^2)\Theta(p_0)\right)(2\pi)^4\delta\left(p+q-\sum_{j=1}^{l} p_j\right) , \quad (3)$$

and $M_\mu(l)$ denotes the amplitude for the photon–parton reaction

$$\gamma^* + p \longrightarrow p_1 + p_2 + \cdots + p_l \ . \quad (4)$$

Here p stands for the incoming parton and $p_1 \cdots p_l$ for the partons produced in the scattering reaction. The integral $\int \mathrm{dPS}^{(l)}$ is just the Lorentz invariant phase space for on-shell particles $p_j^2 = 0$. Remember that $\mathrm{d}^3 p/(2\pi)^3 2p_0$ is the Lorentz-invariant measure. Note that we have averaged over all spins in the initial state. The parton structure tensor can again be decomposed as in the general case in (5.249) in terms of $\mathcal{F}_L$ and $\mathcal{F}_2$

$$W_{\mu\nu} = \left(g_{\mu\nu} - \frac{q_\mu q_\nu}{q^2}\right)\mathcal{F}_L$$
$$+\left(p_\mu p_\nu - \frac{p\cdot q}{q^2}(p_\mu q_\nu + p_\nu q_\mu) + g_{\mu\nu}\frac{(p\cdot q)^2}{q^2}\right)\frac{\mathcal{F}_2}{p\cdot q} , \quad (5)$$

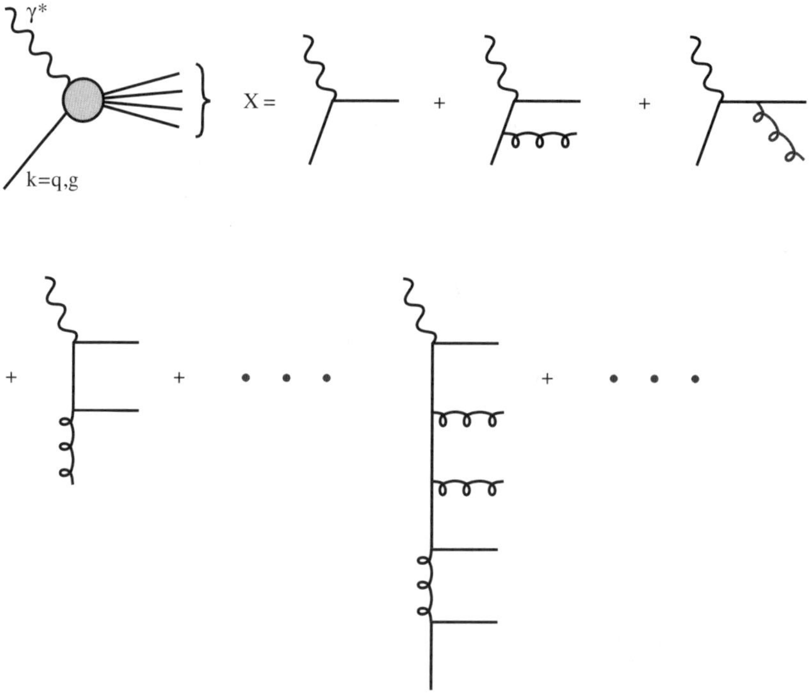

Fig. 5.26. Examples of deep inelastic lepton–parton ($k + \gamma^* \to X$) subprocesses

with the corresponding projection onto $\mathcal{F}_L$

$$\mathcal{F}_L = \frac{Q^2}{(p\cdot q)^2}\, p^\mu p^\nu W_{\mu\nu} \; . \tag{6}$$

To check wether our normalization is correct we first calculate the zeroth order, i.e. the tree-level contribution which is the first diagram on the right-hand side of the pictorial equation in Fig. 5.26, i.e. the contribution without additional gluons.

Inserting the tree-level amplitude

$$M_\mu(1) = \bar{u}(p,s)\gamma_\mu u(p_1,s_1) \tag{7}$$

into (2) yields

$$W_{\mu\nu} = \frac{1}{4\pi}\int \frac{\mathrm{d}^4 p_1}{(2\pi)^3}\delta(p_1^2)\Theta(p_0)(2\pi)^4\delta(p+q-p_1)\mathrm{tr}\left[\gamma_\mu \not{p}\gamma_\nu \not{p}_1\right] \; .$$

When projecting onto $\mathcal{F}_2$ with $-g_{\mu\nu}/2$ (see (5.252); the $p_\mu p_\nu$ term of the projector does not contribute!) we get

$$\begin{aligned}\frac{-g^{\mu\nu}}{2} W_{\mu\nu} &= \frac{1}{2}\delta\left((p+q)^2\right)\Theta(p_0)4p\cdot q \\ &= \delta\left(\frac{Q^2}{z}(1-z)\right)\frac{Q^2}{z} = \delta(1-z) = \mathcal{F}_2(z) \; ,\end{aligned} \tag{8}$$

where we have used $p^2 = 0$ and $2p\cdot q/Q^2 = 1/z$. This result has an obvious interpretation and – if inserted into (1) – shows that at zeroth order the structure function is equal to the bare parton density

$$F_2^q(x) = F_2^{(0)q}(x) = xq(x) \; , \tag{9}$$

as it must be. To obtain the corresponding first-order corrections to $\mathcal{F}_L$ we have to analyze the diagrams shown in Fig. 5.27.

This requires the evaluation of the real-gluon emission graphs (c) and (d) and also the interference of the lowest-order graph (a) with the virtual gluon exchange graph (b). Due to the equations of motion

$$\not{p}u(p) = \bar{u}(p)\not{p} = 0 \; , \tag{10}$$

only diagram (c) contributes to F_L. All other diagrams vanish under the projection (6). The real gluon emission graphs describe the reaction

$$\gamma^*(q) + q(p) \longrightarrow q(p') + G(k) \; ,$$

where the symbols in brackets are the momenta carried by the corresponding particles.

Note that the letter q in $\gamma^*(q)$ denotes the four-momentum of the photon while $q(p)$ denotes a quark with four-momentum p. To calculate the cross section we must evaluate the two-particle phase space

Example 5.14

$$\int \mathrm{dPS}^{(2)} = \int \frac{\mathrm{d}^4 p'}{(2\pi)^3} \int \frac{\mathrm{d}^4 k}{(2\pi)^3} (2\pi)^4 \delta(p+q-p'-k)\delta^+(p'^2)\delta^+(k^2) \;, \tag{11}$$

where we introduced the shorthand notation $\delta^+(p^2) = \delta(p^2)\Theta(p^0)$.

We will work in the centre-of-mass system (CMS) of the parton and virtual photon. The incoming momenta are directed along the z direction. Thus the kinematics is given by

$$\begin{aligned} p &= (|p|, 0, 0, |p|) \;, \\ q &= (q^0, 0, 0, -|p|) = \left(\sqrt{|p|^2 - Q^2}, 0, 0, -|p|\right) \;. \end{aligned} \tag{12}$$

The momenta of the produced particles are denoted as

$$\begin{aligned} k &= (|k|, 0, |k|\sin\Theta, |k|\cos\Theta) \;, \\ p' &= (|k|, 0, -|k|\sin\Theta, -|k|\cos\Theta) \;, \end{aligned} \tag{13}$$

so that $\boldsymbol{k}+\boldsymbol{p}' = 0$ and $k^2 = p'^2 = 0$. For later use we will also introduce the Mandelstam variables

$$s = (p+q)^2 \;, \quad t = (p-k)^2 \;, \quad u = (p-p')^2 \;. \tag{14}$$

The phase space then can be written as

$$\int \mathrm{dPS}^{(2)} = \int \frac{\mathrm{d}^4 k}{(2\pi)^2} \delta^+(k^2)\delta^+\left((p+q-k)^2\right)$$

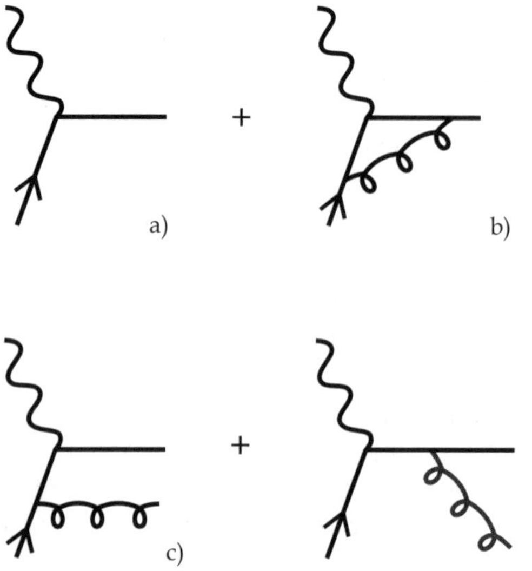

Fig. 5.27. Diagrams giving the correction of order $\mathcal{O}(\alpha_s)$ to the pointlike quark–photon cross section. The interference term of $|a+b|^2$ give rise to a contribution of order $\mathcal{O}(\alpha_s)$. For F_L with the projection $p^\mu p^\nu$ and $p^2 = 0$ only diagram (*c*) gives a nonzero contribution

Example 5.14

$$
\begin{aligned}
&= \int \frac{d^4k}{(2\pi)^2} \delta^+(k^2)\delta^+(s-2|k|\sqrt{s}) \\
&= \int \frac{d^3k}{(2\pi)^2(2|k|)} \delta^+(s-2|k|\sqrt{s}) \ ,
\end{aligned} \tag{15}
$$

where

$$
\int dk^0 \delta^+(k^2) = \int dk^0 \delta(k^{02} - |\boldsymbol{k}|^2)\Theta(k^0) = \frac{1}{2|\boldsymbol{k}|} = \frac{1}{2|k|} \ . \tag{16}
$$

has been used. Performing the trivial ϕ integration yields

$$
\int dPS^{(2)} = \frac{1}{4\pi} \int_0^\infty d|k|\,|k| \int_{-1}^{1} d\cos\Theta \delta(s - 2\sqrt{s}|k|) \ . \tag{17}
$$

To proceed, let us introduce the variable $y = \frac{1}{2}(1+\cos\Theta)$ and write the δ function as

$$
\delta(s - 2\sqrt{s}|k|) = \frac{1}{2\sqrt{s}} \delta\left(|k| - \frac{1}{2}\sqrt{s}\right) \tag{18}
$$

so that the simple expression

$$
\int dPS^{(2)} = \frac{1}{8\pi} \int_0^1 dy \tag{19}
$$

is obtained. Now we write the Mandelstam variables s, t, u (14) in terms of the variables Q^2, y, and $z = Q^2/2p\cdot q$ and find

$$
\begin{aligned}
s &= \frac{Q^2}{z}(1-z) \ , \\
t &= \frac{-Q^2}{z}(1-y) \ , \\
u &= \frac{-Q^2}{z} y \ .
\end{aligned} \tag{20}
$$

The last two equations are obtained by recognizing that in the CMS

$$
\begin{aligned}
t &= (p-k)^2 = -2p\cdot k = -2|p||k|(1-\cos\Theta) = -p\cdot q(1-\cos\Theta) \\
&= \frac{-Q^2}{z}(1-y) \ , \\
u &= (p-p')^2 = -2p\cdot p' = -2|p||k|(1+\cos\Theta) = -p\cdot q(1+\cos\Theta) \\
&= \frac{-Q^2}{z} y \ ,
\end{aligned} \tag{21}
$$

Example 5.14

with $z = Q^2/2p\cdot q$. Twice we have used here that in the CMS

$$p\cdot q = p^0 q^0 - \boldsymbol{pq} = |\boldsymbol{p}|q^0 + |\boldsymbol{p}||\boldsymbol{p}| = |\boldsymbol{p}|\left(q^0 + p^0\right) = |\boldsymbol{p}|(2|\boldsymbol{k}|) \ . \tag{22}$$

With these preliminaries we have everything at hand to calculate the real gluon emission process. For the amplitude in (c) of the last figure we write

$$M_\mu = g\bar{u}_a(p', s')\gamma_\mu \frac{\mathrm{i}}{\not{p} - \not{k}} \gamma_\alpha \frac{\lambda^A_{ab}}{2} u_b(p, s)\varepsilon^{*\alpha} \ , \tag{23}$$

so that we get for the spin-summed matrix element

$$\begin{aligned}
&\sum_{s',s,\varepsilon} M_\mu(2)M^*_\nu(2) \\
&= g^2 C_{\mathrm{F}} \sum_{s',s,\varepsilon} \bar{u}(p', s')\gamma_\mu \frac{1}{\not{p} - \not{k}} \gamma_\alpha u(p, s)\bar{u}(p, s)\gamma_\beta \frac{1}{\not{p} - \not{k}} \gamma_\nu u(p', s')\varepsilon^{*\alpha}\varepsilon^\beta \\
&= g^2 C_{\mathrm{F}}\, \mathrm{tr}\, \frac{\left[\not{p}'\gamma_\mu(\not{p} - \not{k})\gamma_\alpha \not{p}\gamma_\beta(\not{p} - \not{k})\gamma_\nu\right]\left(-g^{\alpha\beta}\right)}{(p-k)^4}
\end{aligned} \tag{24}$$

where the color factor is

$$\frac{1}{N}\sum_{a,b}\sum_{A,B} \frac{\lambda^A_{ab}}{2}\frac{\lambda^B_{ba}}{2}\delta^{AB} = \frac{N^2-1}{2N} = C_{\mathrm{F}} = \frac{4}{3} \tag{25}$$

and the sum over gluon polarisation in the Feynman gauge

$$\sum_\varepsilon \varepsilon^{*\alpha}\varepsilon^\beta = -g^{\alpha\beta} \ . \tag{26}$$

Projecting onto $\mathcal{F}_L$ with $p^\mu p^\nu$ we find

$$\begin{aligned}
p^\mu p^\nu \sum_{spins} M_\mu(2)M^*_\nu(2) &= \frac{g^2 C_{\mathrm{F}}}{(p-k)^4} \cdot 2\,\mathrm{tr}\left[\not{p}'\not{p}\not{k}\not{p}\not{k}\not{p}\right] \\
&= \frac{g^2 C_{\mathrm{F}}}{(p-k)^4} \cdot 8(p\cdot k)^2\,\mathrm{tr}\left[\not{p}'\not{p}\right] \\
&= \frac{-4g^2 C_{\mathrm{F}}}{(p-k)^4}(-2p\cdot k)^2(-2p'\cdot p) \\
&= \frac{-4g^2 C_{\mathrm{F}}}{z^2} z^2 u = -4g^2 C_{\mathrm{F}} u = +4g^2 C_{\mathrm{F}} \frac{Q^2}{z} y \ .
\end{aligned} \tag{27}$$

In the last step we inserted the Mandelstam variables according to (20). Inserting (27) and (19) into (2) yields

$$\begin{aligned}
p^\mu p^\nu W_{\mu\nu} &= \frac{1}{8\pi^2} g^2 C_{\mathrm{F}} \frac{Q^2}{z} \int_0^1 \mathrm{d}y\, y \\
&= \frac{1}{16\pi^2} g^2 C_{\mathrm{F}} \frac{Q^2}{z} \ .
\end{aligned} \tag{28}$$

Using (6)

Example 5.14

$$\mathcal{F}_L = \frac{Q^2}{(pq)^2} p^\mu p^\nu W_{\mu\nu} = \frac{4z^2}{Q^2} p^\mu p^\nu W_{\mu\nu} \tag{29}$$

gives

$$\mathcal{F}_L = \frac{g^2}{4\pi^2} C_{\mathrm{F}} \cdot z = 4C_{\mathrm{F}} \frac{\alpha_s}{4\pi} \cdot z \;, \tag{30}$$

and inserting that into (1) brings us finally to

$$\begin{aligned} F_L^q(x, Q^2) &= 4C_{\mathrm{F}} \frac{\alpha_s}{4\pi} \int\limits_x^1 \frac{\mathrm{d}z}{z} z F_2^{q0}\left(\frac{x}{z}, Q^2\right) \\ &= 4C_{\mathrm{F}} \frac{\alpha_s}{4\pi} \int\limits_x^1 \frac{\mathrm{d}y}{y} \left(\frac{x}{y}\right) F_2^{q0}\left(y, Q^2\right) \;. \end{aligned} \tag{31}$$

This confirms by use of a completely different method our former result of (5.281). The confirmation of the gluon contribution F_L^g will be left to Excercise 5.14.

Here we have presented a calculation of the α_s correction to F_L which apparently is much simpler than the more formal calculation we did before. Also it might appeal to the physical intuition to calculate directly the cross section and stay as close as possible to the parton model in each step of the calculation.

The calculations of the corrections to F_2 may be found in the work of G. Altarelli at al.[19] The complete 2-loop corrections were calculated by Zijlstra and van Neerven.[20] However, the cross-section method becomes increasingly difficult with 3-particle phase space integrals and, at least in the near future, calculations of 3-loop corrections do not seem to be feasible.

Finally, it is worthwhile to mention that 3-loop corrections are already available for the first 10 moments of F_L and F_2 with the operator product expansion method.[21]

[19] G. Altarelli, R.K. Ellis, G. Martinelli: Nucl. Phys. B **157**, 461 (1979).

[20] E.B. Zijlstra and W.L. van Neerven: Nucl. Phys. B **383**, 525 (1992) 525.

[21] S.A. Larin, P. Nogueira, T. van Ritbergen and J.A.M. Vermaseren: Nucl. Phys. B **492**, 338 (1997).

EXERCISE

5.15 Calculation of the Gluonic Contribution to F_L with the Cross-Section Method

Problem. Calculate the contribution of photon–gluon scattering to the longitudinal structure function. Use the cross-section method as in the previous example 5.14.

Solution. The necessary diagrams we encounter are shown in the Fig. 5.28.

Let us first collect the necessary definitions from Examples 5.13 and 5.14. The gluonic structure function is given by a convolution

$$F_L^g = \int\limits_x^1 \frac{dz}{z} \mathcal{F}_L(z, Q^2) F_2^{(0)g}\left(\frac{x}{z}, Q^2\right) \tag{1}$$

with a function $\mathcal{F}_L(z, Q^2)$ calculable in perturbation theory and the gluon density $F_2^{(0)g}(x, Q^2) = xG(x, Q^2)$. We calculate $\mathcal{F}_L$ by projecting onto the partonic scattering tensor $W_{\mu\nu}$ (see (6) and (29) of Example 5.14):

$$\mathcal{F}_L(z, Q^2) = \frac{4z^2}{Q^2} p^\mu p^\nu W_{\mu\nu} \ , \tag{2}$$

where the partonic scattering tensor to the order we are working is given by

$$W_{\mu\nu} = \frac{1}{(4\pi)} \frac{1}{(8\pi)} \int\limits_0^1 dy M_\mu(2) M_\nu^*(2) \ , \tag{3}$$

where we have used the representation (2) and (19) from Example 5.14 for the two-particle phase space ($l = 2$). The kinematics are the same as previously; i.e. for the reaction

$$\mathrm{g}(p) + \gamma^*(p) \to \bar{\mathrm{q}}(k) + \mathrm{q}(p') \tag{4}$$

we choose

$$\begin{aligned} p &= (|p|, 0, 0, |p|) \ , \\ k &= (|k|, 0, |k| \sin\theta, |k| \cos\Theta) \ , \\ p' &= (|k|, 0, -|k| \sin\theta, -|k| \cos\Theta) \ , \end{aligned} \tag{5}$$

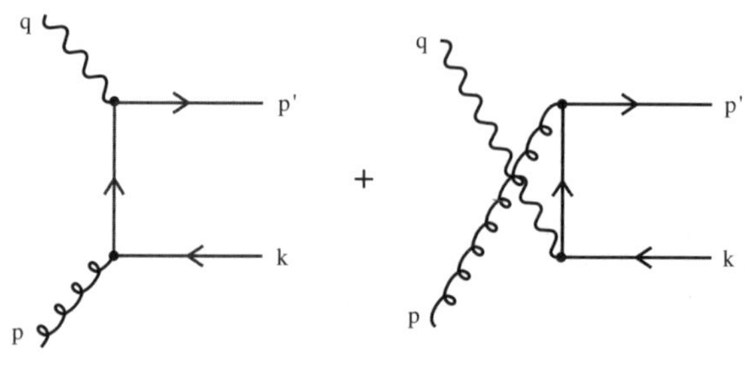

Fig. 5.28. Diagrams contributing to order α_s to photon–gluon scattering. The wavy line represents the virtual photon, the curled line the gluon

and define the Mandelstam variables as

$$
\begin{aligned}
s &= (p+q)^2 = \frac{Q^2}{z}(1-z) \ , \\
t &= (p-k)^2 = \frac{-Q^2}{z}(1-y) \ , \\
u &= (p-p')^2 = \frac{-Q^2}{z}y \ ,
\end{aligned} \tag{6}
$$

with $y = \frac{1}{2}(1+\cos\Theta)$. Remember also that $p'^2 = p^2 = 0 = k^2$, $q^2 = -Q^2$ and that the Bjorken variable with respect to the parton momentum p is defined as $z = Q^2/2p\cdot q$. With these definitions the calculation is straightforward. The amplitude of the process reads

$$
\begin{aligned}
M_\mu(2) = g \Bigg[& \bar{u}_a(p')\gamma_\mu \frac{\mathrm{i}}{\not{p}-\not{k}} \frac{\lambda^A_{ab}}{2} \gamma_\alpha u_b(k) \\
& + \bar{u}_\alpha(p')\gamma_\alpha \frac{\lambda^A_{ab}}{2} \frac{\mathrm{i}}{\not{q}-\not{k}} \gamma_\mu u_b(k) \Bigg] \cdot \varepsilon^\alpha \ .
\end{aligned} \tag{7}
$$

When squaring we find for the color sum

$$
\frac{1}{N^2-1} \sum_{A,B,a,b} \frac{\lambda^A_{ab}}{2} \frac{\lambda^A_{ba}}{2} = \frac{1}{N^2-1} \sum_{A,B} \mathrm{tr} \frac{\lambda^A}{2} \frac{\lambda^A}{2} = \frac{1}{N^2-1} \sum_{A,B} \frac{1}{2} \delta^{AB} = \frac{1}{2} \ . \tag{8}
$$

Together with the sum over quark flavors this gives a factor $T_\mathrm{f} = \frac{1}{2}N_\mathrm{f}$. With $p'-p = q-k$ the squared amplitude becomes

$$
\begin{aligned}
& p^\mu p^\nu \sum_{\mathrm{spins}} M_\mu(2) M^*_\nu(2) \\
&= T_\mathrm{f} g^2 \Bigg\{ \mathrm{tr}\left[\not{p}(\not{p}-\not{k})\gamma_\alpha \not{k} \gamma_\beta (\not{p}-\not{k})\not{p}\not{p}'\right] \cdot \frac{1}{(p-k)^4} \\
&\quad + \mathrm{tr}\left[\gamma_\alpha(\not{p}'-\not{p})\not{p}\not{k}\not{p}(\not{p}'-\not{p})\gamma_\beta\not{p}'\right] \cdot \frac{1}{(p-p')^4} \\
&\quad + \mathrm{tr}\left[\not{p}(\not{p}-\not{k})\gamma_\alpha\not{k}\not{p}(\not{p}'-\not{p})\gamma_\beta\not{p}'\right] \frac{1}{(p-p')^2(p-k)^2} \\
&\quad + \mathrm{tr}\left[\gamma_\alpha(\not{p}'-\not{p})\not{p}\not{k}\gamma_\beta(\not{p}-\not{k})\not{p}\not{p}'\right] \frac{1}{(p-p')^2(p-k)^2} \Bigg\} \left(-g^{\alpha\beta}\right)
\end{aligned} \tag{9}
$$

$$
\begin{aligned}
= T_\mathrm{f} g^2 2 \Bigg\{ & \mathrm{tr}\left[\not{p}\not{k}\not{k}\not{k}\not{p}\not{p}'\right] \cdot \frac{1}{t^2} \mathrm{tr}\left[\not{p}'\not{k}\not{p}\not{p}'\not{p}'\right] \cdot \frac{1}{u^2} \\
& - \mathrm{tr}\left[\not{p}\not{k}\not{p}'\not{p}\not{k}\not{p}'\right] \cdot \frac{1}{ut} \quad - \mathrm{tr}\left[\not{k}\not{p}\not{p}'\not{k}\not{p}\not{p}'\right] \cdot \frac{1}{ut} \Bigg\} \ .
\end{aligned} \tag{10}
$$

Exercise 5.15

The first two traces vanish due to $k^2 = p'^2 = 0$ and the second two are identical because $\mathrm{tr}\left(\gamma_\alpha \gamma_\beta \dots \gamma_\varrho \gamma_\sigma\right) = \mathrm{tr}\left(\gamma_\sigma \gamma_\varrho \dots \gamma_\beta \gamma_\alpha\right)$. The remaining two traces give

$$\begin{aligned} -\mathrm{tr}\left[\not{p}\not{k}\not{p}'\not{p}\not{k}\not{p}'\right] &= \mathrm{tr}\left[\not{p}\not{k}\not{p}'\not{k}\not{p}\not{p}'\right] \\ &= (2p\cdot p')(2p'\cdot k)\cdot \mathrm{tr}\left[\not{p}\not{k}\right] \\ &= 2u\cdot t(2p'\cdot k) = 2u\cdot t\cdot s\ . \end{aligned} \tag{11}$$

What remains is therefore

$$p^\mu p^\nu \sum_{spins} M_\mu(2) M_\nu^*(2) = T_{\mathrm{f}} g^2 \cdot 8s = 8\frac{Q^2}{z}(1-z) T_{\mathrm{f}} g^2\ . \tag{12}$$

Inserting this into (2) we finally find

$$\begin{aligned} \mathcal{F}_L(x, Q^2) &= \frac{g^2}{\pi} T_{\mathrm{f}} z(1-z) \\ &= 16\left(\frac{\alpha_{\mathrm{s}}}{4\pi}\right) T_{\mathrm{f}} z(1-z)\ , \end{aligned} \tag{13}$$

which inserted into (1) gives the final answer

$$\begin{aligned} F_L^g(x, Q^2) &= 16\left(\frac{\alpha_s}{4\pi}\right) T_{\mathrm{f}} \int\limits_x^1 \frac{\mathrm{d}z}{z}\,[z(1-z)]\, F_2^{(0)g}\left(\frac{x}{z}, Q^2\right) \\ &= 16\left(\frac{\alpha_{\mathrm{s}}}{4\pi}\right) T_{\mathrm{f}} \int\limits_x^1 \frac{\mathrm{d}y}{y}\left(\frac{x}{y}\left(1-\frac{x}{y}\right)\right) F_2^{(0)g}(y, Q^2)\ , \end{aligned} \tag{14}$$

which is completely in agreement with our previous result in (5.282).

5.6 The Spin-Dependent Structure Functions

In recent years it has become possible to measure, in addition to the structure functions $F_1(x, Q^2)$ and $F_2(x, Q^2)$, the so-called spin-dependent structure functions $g_1(x, Q^2)$ and $g_2(x, Q^2)$. The spin-dependent structure functions play a role only when the scattered leptons and hadrons are polarized. The origin of these additional structure functions is quite obvious. For a polarized hadron, e.g., a proton, there is an additional Lorentz vector available, namely the spin vector s^μ. Accordingly (3.6) has to be extended by some additional terms. Repeating the analysis of Chap. 3 with the additional spin vector, we are led to the following form of the scattering tensor:

$$\begin{aligned} W_{\mu\nu} &= \left(-g_{\mu\nu} + \frac{q_\mu q_\nu}{q^2}\right) W_1(x, Q^2) \\ &+ \left(P_\mu - q_\mu \frac{q \cdot P}{q^2}\right)\left(P_\nu - q_\nu \frac{q \cdot P}{q^2}\right) \frac{W_2(x, Q^2)}{M^2} \\ &+ \mathrm{i}\varepsilon_{\mu\nu\lambda\sigma} q^\lambda s^\sigma M G_1(x, Q^2) \\ &+ \mathrm{i}\varepsilon_{\mu\nu\lambda\sigma} q^\lambda (s^\sigma q \cdot P - P^\sigma q \cdot s) \frac{G_2(x, Q^2)}{M} \ . \end{aligned} \tag{5.285}$$

This equation is obviously an extension of (3.18). The factor i guarantees that the transition current is real (the relation $\Gamma^*_\mu(p \leftrightarrow p') = \Gamma_\mu$ must hold; see Exercise 3.3), while the ε tensor ensures that $q^\mu W_{\mu\nu} = q^\nu W_{\mu\nu} = 0$ holds. Since the additional contributions change their sign when the hadron spin direction is reversed, they cancel upon spin-averaging. Thus only spin-independent structure functions can be measured with an unpolarized target. Analogously to the unpolarized case, dimensionless functions are introduced according to

$$M G_1(x, Q^2) = \frac{g_1(x, Q^2)}{P \cdot q} \ , \qquad \frac{G_2(x, Q^2)}{M} = \frac{g_2(x, Q^2)}{(P \cdot q)^2} \ . \tag{5.286}$$

This choice is motivated by the fact that g_1 has a simple interpretation in terms of parton distributions and $g_2(x)$ thus occurs on the same footing as $g_1(x)$. The additional spin-dependent contributions to the hadron scattering tensor are antisymmetric; thus they cannot contribute when contracted with the symmetric, unpolarized lepton tensor of (3.20). In order to measure polarized structure functions in deep inelastic scattering, the lepton must also be polarized. The resulting leptonic scattering tensor is

$$\begin{aligned} L_{\mu\nu} &= \frac{1}{4} \mathrm{tr}[(\not{p} + m)(1 - \not{s}_\mathrm{e}\gamma_5)\gamma_\mu(\not{p}' + m)\gamma_\nu] \\ &= \frac{1}{4}[p_\mu p'_\nu + p_\nu p'_\mu - g_{\mu\nu}(p \cdot p' - m^2)] \\ &\quad - \frac{m}{4}\mathrm{tr}(\not{p}\not{s}_\mathrm{e}\gamma_5\gamma_\mu\gamma_\nu) - \frac{m}{4}\mathrm{tr}(\not{s}_\mathrm{e}\gamma_5\gamma_\mu\not{p}'\gamma_\nu) \\ &= \frac{1}{4}[p_\mu p'_\nu + p_\nu p'_\mu - g_{\mu\nu}(p \cdot p' - m^2)] + \mathrm{i} m \varepsilon_{\mu\nu\alpha\beta} q^\alpha s^\beta_\mathrm{e} \ , \end{aligned} \tag{5.287}$$

where $\frac{1}{2}(1 - \not{s}\gamma_5)$ is the standard spin projection operator, and s^β_e the spin vector of the electron. Thus the polarized lepton tensor also contains an antisymmetric part. Contracting both, we obtain the additional term

$$\begin{aligned} L_{\mu\nu} W^{\mu\nu} &= \cdots - 2mM(g_{\lambda\alpha}g_{\sigma\beta} - g_{\lambda\beta}g_{\sigma\alpha}) q^\lambda s^\sigma_\mathrm{p} q^\alpha s^\beta_\mathrm{e} G_1 \\ &\quad - \frac{m}{M}(g_{\lambda\alpha}g_{\sigma\beta} - g_{\lambda\beta}g_{\sigma\alpha}) q^\lambda (q \cdot P s^\sigma_\mathrm{p} - q \cdot s_\mathrm{p} P^\sigma) q^\alpha s^\beta_\mathrm{e} G_2 \ . \end{aligned} \tag{5.288}$$

This poses the question of what choice of the polarization of electron and hadron, i.e., nucleon, is most suitable. The relevant vectors in the centre-of-momentum frame are

$$
\begin{aligned}
p_\mu = (E_\mathrm{e}, 0, 0, p_\mathrm{e}) \quad &, \quad P_\mu = (E_\mathrm{p}, 0, 0, -P_\mathrm{p}) \\
s_\mathrm{e}^\mu(\text{longitudinal}) &= \pm(p_\mathrm{e}, 0, 0, E_\mathrm{e}) \cdot \frac{1}{m} \; , \\
s_\mathrm{p}^\mu(\text{longitudinal}) &= \pm(p_\mathrm{p}, 0, 0, -E_\mathrm{p}) \cdot \frac{1}{M} \; , \\
s_\mathrm{e}^\mu(\text{transverse}) = \pm(0,1,0,0) \quad &\text{and} \quad \pm(0,0,1,0) \; , \\
s_\mathrm{p}^\mu(\text{transverse}) = \pm(0,1,0,0) \quad &\text{and} \quad \pm(0,0,1,0) \; .
\end{aligned}
\tag{5.289}
$$

We recognize that the following products are large:

$$
s_\mathrm{e}(l) \cdot s_\mathrm{p}(l) \approx \pm 2 \frac{E_\mathrm{e} E_\mathrm{p}}{mM} \; , \quad s_\mathrm{e}(l) \cdot P \approx \pm 2 \frac{E_\mathrm{e} E_\mathrm{p}}{m} \; . \tag{5.290}
$$

All the other scalar products in (5.288) are considerably smaller. Thus to measure spin-dependent structure functions, longitudinally polarized leptons must be scattered off longitudinally or transversely polarized nucleons. This property can also be seen from projecting on physical degrees of freedom of the photon instead of summing over photon polarizations. In this case, the hadronic scattering tensor is contracted with $\varepsilon_\mu^* \varepsilon_\nu$. Here ε is the polarization vector of the virtual photon. In order for spin-dependent terms to be able to contribute, this expression must contain an antisymmetric part. As can easily be checked, this is the case only for longitudinal polarization vectors, e.g., it holds for

$$
\varepsilon_\mu = (0, 1, \mathrm{i}, 0)/\sqrt{2} \quad \text{that} \quad \varepsilon_1^* \varepsilon_2 = -\varepsilon_2^* \varepsilon_1 \; . \tag{5.291}
$$

A longitudinally polarized photon is most readily emitted by a longitudinally polarized lepton. For other lepton polarizations its coupling is suppressed by exactly the kinematic factors of (5.288).

In the experiments performed up to now, a longitudinally polarized proton target has been used, and the asymmetry

$$
A = \frac{\left[\frac{\mathrm{d}^2\sigma}{\mathrm{d}E'\,\mathrm{d}\Omega}(\uparrow\uparrow) - \frac{\mathrm{d}^2\sigma}{\mathrm{d}E'\,\mathrm{d}\Omega}(\uparrow\downarrow) \right]}{\left[\frac{\mathrm{d}^2\sigma}{\mathrm{d}E'\,\mathrm{d}\Omega}(\uparrow\uparrow) + \frac{\mathrm{d}^2\sigma}{\mathrm{d}E'\,\mathrm{d}\Omega}(\uparrow\downarrow) \right]} \tag{5.292}
$$

has been measured.

Inserting (5.289) into (5.288) one finds, for longitudinal polarization,

$$
A_1 = D \left\{ \frac{g_1(x) - \frac{2Mx}{Ey} g_2(x)}{F_1(x)} + \eta \sqrt{\frac{2Mx}{Ey}} \, \frac{g_1(x) + g_2(x)}{F_1(x)} \right\} \; , \tag{5.293}
$$

with

$$D = \frac{y(2-y)}{y^2+2(1-y)(1+R)} \,,$$
$$y = \frac{E-E'}{E} \,, \qquad R = \frac{\sigma_{\gamma^* L}}{\sigma_{\gamma^* T}} \,,$$
$$\eta = \frac{2\gamma(1-y)}{2-y} \,, \quad \text{and} \quad \gamma = \sqrt{\frac{Q^2 M^2}{\nu^2}} = \sqrt{\frac{2Mx}{Ey}} \,. \tag{5.294}$$

For transverse proton polarization the resulting asymmetry reads

$$\begin{aligned} A_2 &= \frac{\left[\frac{d^2\sigma}{dE'\,d\Omega}(\uparrow\rightarrow) - \frac{d^2\sigma}{dE'\,d\Omega}(\downarrow\rightarrow)\right]}{\left[\frac{d^2\sigma}{dE'\,d\Omega}(\uparrow\rightarrow) + \frac{d^2\sigma}{dE'\,d\Omega}(\downarrow\rightarrow)\right]} \\ &= D\sqrt{\frac{2\varepsilon}{1+\varepsilon}}\left(\sqrt{\frac{2Mx}{Ey}}\,\frac{g_1(x)+g_2(x)}{F_1(x)} - \eta\frac{1+\varepsilon}{2\varepsilon}\,\frac{g_1(x)-\frac{2Mx}{Ey}g_2(x)}{F_1(x)}\right) \\ &\longrightarrow \sqrt{\frac{2Mx}{E-E'}}\,\frac{g_1(x)+g_2(x)}{F_1(x)} \qquad \text{for } Q^2\to\infty,\ y\to 1 \end{aligned} \tag{5.295}$$

with $\varepsilon = (1-y)/(1-y+y^2/2)$. Since the cross sections for transverse and longitudinal polarization should be of approximately the same size, we deduce that $g_2(x)$ is certainly not larger than $g_1(x)$. Therefore the contribution of $g_2(x)$ at longitudinal polarization is suppressed by a factor Mx/E and constitutes only a small correction. The asymmetry A_2 therefore measures mainly the ratio of $g_1(x)$ to the unpolarized structure function $F_1(x)$. Next, there is the question of whether $g_1(x)$ can be given a practical meaning like $F_1(x)$? This is indeed the case. To see this, it is sufficient to compare the hadronic and the leptonic scattering tensor of (5.285) and (5.287), respectively.

Obviously each Dirac particle gives a contribution with the same form as that proportional to G_1 multiplied by the probability to find a quark with the fitting momentum fraction x and the right polarization. Accordingly, one obtains, in analogy with the parton interpretation of F_1,

$$F_1(x) = \frac{1}{2} \sum_{q=u,d,\bar{u},\bar{d}\ldots} q(x) Q_q^2 \,, \tag{5.296}$$

where $q(x)$ is the probability of finding a quark q with momentum fraction x, a similar expression for $g_1(x)$

$$g_1(x) = \frac{1}{2} \sum_{q=u,d,\bar{u},\bar{d}\ldots} Q_q^2 \left[q^\uparrow(x) - q^\downarrow(x)\right] \,. \tag{5.297}$$

Here $q^\uparrow(x)$ indicates the probability of finding a quark with a momentum fraction x polarized in the same direction as the whole proton. The asymmetry

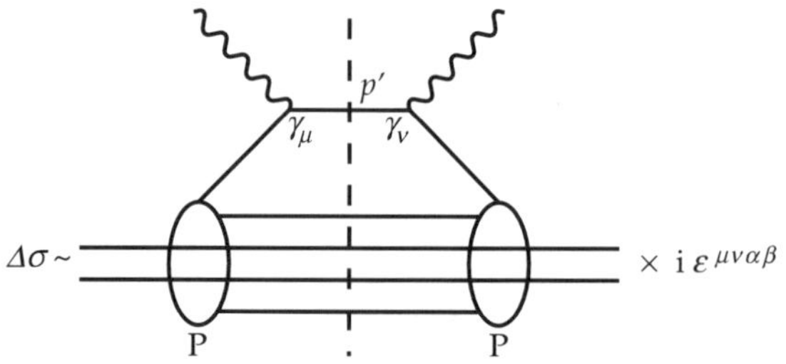

Fig. 5.29. The relationship between deep inelastic scattering and the forward matrix element

measured is a measure of the distribution of the proton spin among its constituents.

One of the most important properties of unpolarized structure functions is that the momentum fraction deduced from them, which is carried by the quarks, accounts for only half of the total momentum. This is striking evidence for the existence of gluons. Analogously, from polarized structure functions we can ask how much of the proton spin is carried by quarks, how much by gluons and how much is present in angular momentum. This question has, up to now, not been uniquely answered. The present experimental data have caused strong theoretical discussions and led to the design of much improved and completely new experiments. In the near future, experiments of the type

$$\begin{aligned}
&\mathrm{e}^{\uparrow}(\text{long.}) + \mathrm{p}^{\uparrow},\, {}^{3}\mathrm{He}^{\uparrow},\, \mathrm{d}^{\uparrow}(\text{long. or trans.}) \to \mathrm{e}' + X\ ,\\
&\mathrm{e}^{\uparrow}(\text{long.}) + \mathrm{p}^{\uparrow},\, {}^{3}\mathrm{He}^{\uparrow},\, \mathrm{d}^{\uparrow}(\text{long. or trans.}) \to \mathrm{e}' + \pi^{\pm} + X\ ,\\
&\mathrm{p} + \mathrm{p}^{\uparrow}(\text{trans.}) \to \gamma + X \ \text{ plus many more channels}
\end{aligned} \tag{5.298}$$

should be performed. In addition, polarized proton–proton collisions are a possibility for the future.

This field is extremely active right now and a more detailed discussion of the current situation is not suitable for a textbook. Instead, we consider two particular aspects where crucial concepts of QCD can be exemplified. As it turns out, nearly all techniques of QCD are, in a nearly singular manner, important in analyzing spin structure.

In particular, we consider the following aspects of the discussion.

1. The Bjorken and Ellis–Jaffe sum rule: From the knowledge of the axial vector coupling in weak interactions, predictions based on isospin and SU(3) flavor symmetry for $\int \mathrm{d}x(g_1^{\mathrm{p}}(x) - g_1^{\mathrm{n}}(x))$ and $\int \mathrm{d}x g_1^{\mathrm{p}}(x)$ can be obtained. This is a nice example that flavor symmetry continues to play an important role even after the introduction of color SU(3).

2. The axial anomaly and the gluonic contribution to $g_1(x)$: The axial anomaly plays a major role for hadronic physics in general. We met it in Sect. 4.2 when reviewing the foundations of QCD and will show in Exercise 7.2 its importance in understanding lattice QCD. As it turns out, the anomaly can contribute to the spin structure function in a subtle manner. The analysis of this effect gives exquisite insight into the physical meaning of the anomaly.

Let us start by discussing the Bjorken sum rule. The antisymmetric (spin-dependent) part of the hadronic scattering tensor can be written as an axial-vector forward matrix element of the proton (see Fig. 5.29). As the proton couples to quarks we find that the spin-asymmetric part of the cross section $\Delta\sigma$ is proportional to

$$\begin{aligned} \mathrm{i}\gamma_\mu \not{p}' \gamma_\nu \, \varepsilon^{\mu\nu\alpha\beta} &= p'^\sigma \left(g_{\mu\sigma}\,\gamma_\nu + g_{\sigma\nu}\,\gamma_\mu - g_{\mu\nu}\,\gamma_\sigma + \mathrm{i}\varepsilon_{\mu\sigma\nu\lambda}\,\gamma^\lambda\,\gamma_5 \right) \mathrm{i}\,\varepsilon^{\mu\nu\alpha\beta} \\ &= p'^\sigma \, \varepsilon_{\mu\nu\sigma\lambda}\,\varepsilon^{\mu\nu\alpha\beta}\,\gamma^\lambda\,\gamma_5 \\ &= -2p'^\sigma \left(g_\sigma{}^\alpha g_\lambda{}^\beta - g_\sigma{}^\beta g_\lambda{}^\alpha \right) \gamma^\lambda\,\gamma_5 \\ &= -2p'^\alpha\,\gamma^\beta\,\gamma_5 + 2p'^\beta\,\gamma^\alpha\,\gamma_5 \; , \end{aligned} \tag{5.299}$$

where we have used the well-known decomposition of the product of three gamma matrices. Thus we find that

$$\Delta\sigma \sim \sum_q \langle \mathrm{p}\, s \,|\, Q_q^2 \bar{q}\,\gamma^\alpha\,\gamma_5\, q \,|\, \mathrm{p}\, s \rangle \; . \tag{5.300}$$

It is important to note that, owing to the optical theorem, we have obtained a forward cross section, i.e., the momentum transfer is zero. In contrast, the momentum transfer in the lepton–hadron scattering reaction q_μ is very large, but squaring this diagram to obtain the cross section leads to a graph in which the second photon removes the momentum transferred by the first. (This is just a description of the content of the optical theorem.) Thus this forward matrix element can be related by isospin symmetry to a corresponding matrix element between neutron and proton. More precisely we write

$$Q^2 = \frac{15}{18} + \frac{1}{3}\tau_3 = \begin{cases} \frac{4}{9} & \text{for the up quark} \\ \frac{1}{9} & \text{for the down quark} \end{cases} \; , \tag{5.301}$$

implying that

$$\Delta\sigma^\mathrm{p} - \Delta\sigma^\mathrm{n} \sim \frac{1}{3} \langle \mathrm{N}\, s \,|\tau_3\, \bar{q}\,\gamma^\alpha\,\gamma_5\, q\, |\mathrm{N}\, s \rangle \; , \tag{5.302}$$

where $\mathrm{N} = \binom{\mathrm{p}}{\mathrm{n}}$ is the usual nucleon doublet. Now we can introduce $\tau_+\tau_- + \tau_-\tau_+ = \mathbb{1}$; we obtain

$$\begin{aligned} \Delta\sigma^\mathrm{p} - \Delta\sigma^\mathrm{n} &\sim \frac{1}{3} \Big\langle \mathrm{N}\, s \,\Big| \left(\tau_+\tau_- + \tau_-\tau_+ \right) \tau_3\, \bar{q}\,\gamma^\alpha\,\gamma_5\, q \,\Big| \mathrm{N}\, s \Big\rangle \\ &= \frac{1}{3} \Big\langle \mathrm{N}\, s \Big| \left(\tau_+\tau_- - \tau_-\tau_+ \right) \bar{q}\,\gamma^\alpha\,\gamma_5\, q \,\Big| \mathrm{N}\, s \Big\rangle \\ &= \frac{1}{3} \langle \mathrm{p}\, s \,|\tau_+\, \bar{q}\,\gamma^\alpha\,\gamma_5\, q\, |\mathrm{n}\, s \rangle + \text{h.c.} \, . \end{aligned} \tag{5.303}$$

This, however, is just the weak-interaction matrix element of neutron beta decay. The corresponding coupling is just g_A/g_V. Putting all the constants together, we end up with

$$\int_0^1 \mathrm{d}x \left[g_1^\mathrm{p}(x) - g_1^\mathrm{n}(x) \right] = \frac{1}{6} \frac{g_A}{g_V} \; . \tag{5.304}$$

This is the famous *Bjorken sum rule*, which allows us to connect the proton and the neutron results. It is strictly valid for $Q^2 \to \infty$. Various perturbative and higher-twist corrections have been calculated:

$$\begin{aligned}
&\int_0^1 \mathrm{d}x \left[g_1^{\mathrm{p}}(x, Q^2) - g_1^{\mathrm{n}}(x, Q^2)\right] \\
&\quad = \frac{1}{6}\frac{g_A}{g_V}\left[1 - \frac{\alpha_s}{\pi} - \frac{43}{12}\left(\frac{\alpha_s}{\pi}\right)^2\right] \\
&\quad + \frac{M^2}{Q^2}\int_0^1 \mathrm{d}x\, x^2 \left[\frac{2}{9}(g_1^{\mathrm{p}} - g_1^{\mathrm{n}})(x, Q^2) + \frac{4}{3}(g_2^{\mathrm{p}} - g_2^{\mathrm{n}})(x, Q^2)\right] \\
&\quad + \frac{4}{9}\frac{M^2}{Q^2}(f_{2\mathrm{u}} - f_{2\mathrm{d}}) + \frac{4}{9}\frac{M^2}{Q^2}(d_{2\mathrm{u}} - d_{2\mathrm{d}})
\end{aligned} \tag{5.305}$$

with

$$2(M^2)^2 f_{2q} s^\sigma = \left\langle \mathrm{P}\, S \left| g\, \bar{q}\, \tilde{G}^{\sigma\lambda}\, \gamma_\lambda\, q \right| \mathrm{P}\, S \right\rangle , \quad q = u,\, d\, . \tag{5.306}$$

The matrix element d_2 is defined in Example 5.16. The f_{2q} can be calculated, for example, by QCD sum rule techniques and turn out to be small. Also the d_{2q} have been estimated and they are expected to give measurable contributions.[22] The mass correction is small, basically because of the factor x^2 which suppresses the small x contributions.

Thus the Q^2 dependence of the Bjorken sum rule is well under control. It is especially noteworthy that no anomalous dimension, i.e., no factor of the form $[\alpha(Q^2)\,/\,\alpha(Q_0^2)]^{-d/2b}$, occurs on the right-hand side of (5.305). This fact has a simple physical reason. The main part of the Q^2 evolution of unpolarized structure functions is due to the many quark–antiquark pairs that contribute at high Q^2 (see Fig. 5.23). Because the vector coupling to gluons conserves helicity, these quark pairs are, however, predominantly unpolarized. In fact a double spin flip is needed to obtain a polarized $\mathrm{q}\bar{\mathrm{q}}$ pair, and except for very small values of x this is completely suppressed. Consequently the zeroth moment of g_1^{p}, g_1^{n}, g_2^{p}, g_2^{n} should show only a very mild Q^2 dependence. (Actually there is good reason to believe that $\int g_2(x)\, \mathrm{d}x \equiv 0$.) Now, assuming that the strange quarks are unpolarized, which is actually a very controversial assumption, we can obtain the Ellis–Jaffe sum rule from (5.305). From the definition of $g_1(x)$ we have, assuming strict isospin symmetry at the quark level, which is also controversial,

$$\int_0^1 \mathrm{d}x\, g_1^{\mathrm{p}}(x) = \frac{1}{2}\left(\frac{4}{9}\Delta u + \frac{1}{9}\Delta d\right) ,$$

[22] E. Stein, P. Gornicki, L. Mankiewicz, A. Schäfer and W. Greiner, Phys. Lett. B **343**, 369 (1995).

$$\int_0^1 dx\, g_1^{n}(x) = \frac{1}{2}\left(\frac{4}{9}\Delta d + \frac{1}{9}\Delta u\right) , \tag{5.307}$$

where Δu, Δd are the fractions of the proton spin carried by the u and d quarks, respectively.

The Bjorken sum rule (5.304) implies that

$$\int_0^1 dx\, g_1^{p}(x) - \int_0^1 dx\, g_1^{n}(x) = \frac{1}{6}\big(\Delta u - \Delta d\big) = \frac{1}{6}\frac{g_A}{g_V} ,$$
$$\Delta u = \Delta d + \frac{g_A}{g_V} , \tag{5.308}$$

which is already one constraint. Still another constraint comes from the coupling of s and u quarks in hyperon decays. To understand this we have to review a little group theory. The general axial vectorial flavor SU(3) matrix element $\langle B|S^5_{j\sigma}|B'\rangle$ can be analyzed with the equivalent of the Wigner–Eckhardt theorem. Here B, B' are baryon states from the flavor octet, σ is the Lorentz index, and $j = 1, \dots , 8$ is the index of the SU(3) generator. We find that

$$\begin{aligned}\langle \mathrm{p} \mid S^5_{j\sigma} \mid \mathrm{p}\rangle = c\cdot n_\sigma \Bigg[&\left(\begin{matrix} 8 & 8 \\ I=\frac{1}{2}\ I_3=-\frac{1}{2} & I=\frac{1}{2}\ I_3=\frac{1}{2} \\ Y=-1 & Y=1 \end{matrix}\ \middle|\ \begin{matrix} 8_1 \\ I_j\ \ -I_{3j} \\ -Y_j \end{matrix}\right) A_1 \\ &+ \left(\begin{matrix} 8 & 8 \\ I=\frac{1}{2}\ I_3=-\frac{1}{2} & I=\frac{1}{2}\ I_3=\frac{1}{2} \\ Y=-1 & Y=1 \end{matrix}\ \middle|\ \begin{matrix} 8_2 \\ I_j\ \ -I_{3j} \\ -Y_j \end{matrix}\right) A_2 \Bigg] \\ &\times \left(\begin{matrix} 8 & 8 \\ I_j\ \ -I_{3j} & I_j\ \ I_{3j} \\ -Y_{-j} & Y_j \end{matrix}\ \middle|\ \begin{matrix} 1 \\ 0 \end{matrix}\right) .\end{aligned} \tag{5.309}$$

The isoscalar factors are listed.[23] A peculiarity of the SU(3) group is the appearance of two unequivalent octet representations. This is related to the fact that three octets can be coupled either by the symmetric d_{abc} or the antisymmetric f_{abc} structure constants. It leads to the definition of two constants, D and F:

$$\big\langle N\big|S^5_{3\sigma}\big|N\big\rangle = c\cdot n_\sigma \left(\frac{\sqrt{30}}{40}A_1 + \frac{\sqrt{6}}{24}A_2\right) =: c\cdot n_\sigma\big(D+F\big) ,$$
$$D = \frac{\sqrt{30}}{40}A_1 , \quad F = \frac{\sqrt{6}}{24}A_2 . \tag{5.310}$$

Every axial-vector matrix element coupling two baryon states from the octet is thus proportional to a specific combination of D and F and can therefore be used

[23] See J.J. de Swart: Rev. Mod. Phys. **35** (1963) 916.

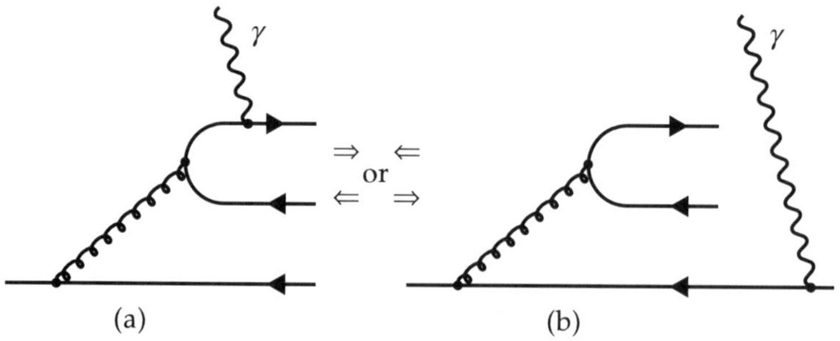

Fig. 5.30. Graphs contributing to the Q^2 dependence of the zeroth moment of the unpolarized structure functions. For the spin-dependent structure functions only (**b**) contributes, since the quark–antiquark pairs produced are unpolarized

to measure F/D. However, the combined analysis of all the hyperon decays described by different groups gave conflicting results. A conservative estimate is $F/D = 0.55$–0.60. From (5.310) we read off

$$\Delta u - \Delta d = F + D \ . \tag{5.311}$$

Similarly by inserting the SU(3) isoscalar factors we get

$$\langle N | S^5_{8\sigma} | N \rangle = c \cdot n_\sigma \left(-\frac{1}{4\sqrt{10}} A_1 + \frac{\sqrt{2}}{8} A_2 \right) = c \cdot n_\sigma \left(-\frac{1}{\sqrt{3}} D + \sqrt{3} F \right) \tag{5.312}$$

$$\Rightarrow \frac{1}{\sqrt{3}} (\Delta u + \Delta d - 2\Delta s) = \sqrt{3} F - \frac{1}{\sqrt{3}} D \ . \tag{5.313}$$

Assuming again that $\Delta s = 0$ we obtain another independent combination,

$$\frac{\Delta u + \Delta d}{\Delta u - \Delta d} = \frac{3F - D}{F + D} = \frac{3F/D - 1}{F/D + 1} \ , \tag{5.314}$$

which together with (5.309) allows us to determine $\int_0^1 g_1^{\mathrm{p}}(x)\, \mathrm{d}x$. A complete analysis along these lines gives

$$\begin{aligned} \int_0^1 \mathrm{d}x\, g_1^{\mathrm{p}}(x, Q^2) &= \frac{g_A}{g_V} \left[\frac{1}{18} \frac{3F/D + 1}{F/D + 1} \left(1 - \frac{\alpha_s}{\pi} \right) + \frac{1}{9} \frac{3F/D - 1}{F/D + 1} \left(1 - \frac{\alpha_s}{\pi} \right) \right] \\ &\quad + \frac{M^2}{Q^2} \int_0^1 \mathrm{d}x\, x^2 \left[\frac{2}{9} g_1^{\mathrm{p}}(x, Q^2) + \frac{4}{3} g_2^{\mathrm{p}}(x, Q^2) \right] \\ &\quad - \frac{4}{9} \frac{1}{Q^2} \left(\frac{4}{9} f_{2\mathrm{u}} + \frac{1}{9} f_{2\mathrm{d}} + \frac{1}{9} f_{2\mathrm{s}} \right) \\ &\approx 0.172 \pm 0.009 \quad \text{for } Q^2 = 5\,\mathrm{GeV}^2 \ . \end{aligned} \tag{5.315}$$

This is the Ellis–Jaffe sum rule. As mentioned several times already, its validity is a matter of dispute. We have presented it here to illustrate the potential usefulness of flavor SU(3) for the analysis of deep inelastic scattering cross sections.

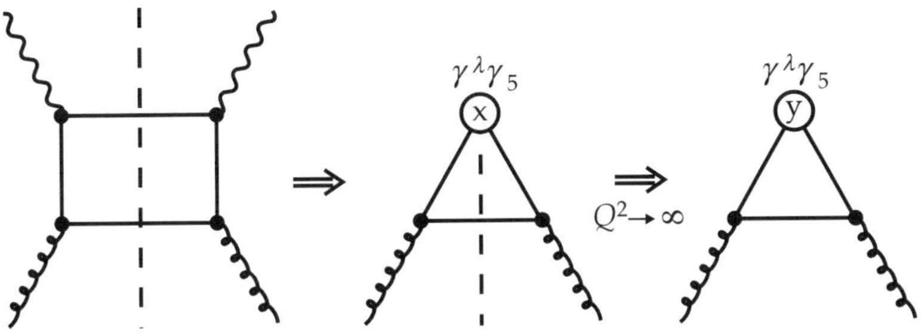

Fig. 5.31. The perturbative anomalous gluon contribution to g_1

Next let us discuss the role of the anomaly for the isosinglet axial-vector current. To illustrate the point consider first the perturbative graph in Fig. 5.31.

It is obvious that for $Q^2 \to \infty$ one indeed obtains the triangle anomaly, implying pointlike photon–gluon coupling. The basic problem of this interpretation is also obvious from Fig. 5.29, namely that there is no unambiguous way to separate the quark contribution and the anomalous gluonic contribution. It is unclear whether the quarks of the fermion line should be absorbed into Δu, Δd, etc. or into a contribution of the structure

$$\Delta g_1^{\mathrm{p}}(x) = \left\langle \frac{Q_q^2}{2} \right\rangle \int_0^1 \frac{\mathrm{d}z}{z} A\left(\frac{x}{z}\right) \Delta G(z) \ . \tag{5.316}$$

In other words, the Δqs appearing in the analysis we have presented so far should be split up according to

$$\Delta q = \Delta \tilde{q} - \frac{\alpha_s}{2\pi} \Delta G \cdot c \ . \tag{5.317}$$

It turns out that, for variety reasons, such a separation is very problematic.

1. Performing the perturbative calculation for the graph in Fig. 5.29 we get constants c that depend critically on the chosen infrared regulators indicating that we are not looking at an infrared safe quantity.
2. From the operator-product-expansion point of view, we find that there is no gauge-independent local definition of ΔG.
3. A so-called large gauge transformation, i.e., a gauge transformation with a nonzero topological quantum number, shifts contributions from $\Delta \tilde{q}$ to ΔG. A detailed discussion of this rather complicated issue is beyond the framework of this book.

On the other hand, there exists a practical argument for the decomposition (5.317). If we construct a phenomenological model for Δq, we will most probably miss the highly virtual quark components from Fig. 5.29. Therefore it might be easier to model $\Delta \tilde{q}$ and ΔG instead.

We will not discuss these still very much disputed questions further.

EXAMPLE

5.16 Higher Twist in Deep Inelastic Scattering

In the framework of operator product expansion we found that structure functions can be represented as

$$\int_0^1 \mathrm{d}x x^{n-2} F(x, Q^2) = C_n(\alpha_s, Q^2/\mu^2) V_n(\mu^2) + \mathcal{O}(1/Q^2) , \tag{1}$$

where $V_n(\mu^2)$ are the reduced matrix elements of the twist-2 operators (see (5.257)–(5.259)). The Wilson coefficients $C_n(\alpha_s, Q^2/\mu^2)$ can be evaluated in perturbation theory as an expansion in α_s, which is actually an expansion in $1/\ln(Q^2)$ because $\alpha_s \sim 1/\ln(Q^2)$.

We shall now discuss the $(1/Q^2)$ corrections that contribute to deep inelastic scattering (DIS). This will be done in a quite heuristic way and, in order to exemplify the main ideas, simple examples shall be used without going through the operator product expansion, which would be necessary to obtain the $(1/Q^2)$ corrections rigorously. The interested reader is referred to the literature and references given there.[24] In general, power-suppressed contributions are referred to as higher-twist contributions. This is, of course, a quite loose way of speaking. We distinguish power corrections that are of pure kinematic origin, the so-called target mass corrections (TMC), which we have already discussed in Example 5.12, and true higher-twist corrections. True higher-twist corrections are of dynamical origin related to quark–gluon interactions in the nucleon. TMC consist of power-suppressed twist-2 operators that occur in the expansion when the target mass is not set to zero, i.e. assumed to be negligible as compared to Q^2. Those corrections can be exactly taken into account by not expressing the structure functions in terms of the Bjorken variable x, but instead in terms of the Nachtmann variable ξ:[25]

$$\xi = \frac{2x}{1 + \left(1 + 4x^2 \frac{M_N^2}{Q^2}\right)^{1/2}} . \tag{2}$$

Setting M_N^2 to zero, one recovers the normal Bjorken variable. The structure functions expressed in terms of ξ depend analytically on M_N^2/Q^2 and can be expanded in that variable so that one recovers the $1/Q^2$ corrections in the form of (1). The classical papers of De Rujuly et al.[26] give some more insight into this procedure. Here we will not dwell on this issue, but instead restrict ourselves to

---

[24] B. Ehrnsperger et al.: Phys. Lett. B **150**, 439 (1994).

[25] O. Nachtmann: Nucl. Phys. B **63**, 237 (1973).

[26] A. De Rujula, W. Georgi, W.D. Politzer: Phys. Rev. D **15** (1997) 2495 and Ann. Phys. (1997) 315.

Example 5.16

the more interesting case of power suppressed twist-3 and twist-4 operators. The rigorous definition of twist is

$$\text{twist} = \text{dimension} - \text{spin} \ , \tag{3}$$

where dimension is the naive canonical dimension of the operator (see the following table), and spin relates to its transformation properties under the Lorentz group. For example, the vector operator $\overline{\psi}\gamma_\mu\psi$ has spin 1, the scalar operator $\overline{\psi}\psi$ has spin 0, etc. Using the fact that the action $\int \mathrm{d}^4x\mathcal{L} = \mathcal{S}$ is dimensionless, it is quite easy to derive the canonical dimensions of the field operators from the knowledge of the Lagrangian. In the table we give the mass dimensions of various operators.

operator	$\mathcal{S}$	P_μ, D_μ, A_μ	ψ	$G_{\mu\nu}, \tilde{G}_{\mu\nu}$	$\overline{\psi}\gamma_\mu\psi$	$\mathcal{L}$
dimension	0	1	3/2	2	3	4

How the decomposition of an operator into its irreducible representation under the Lorentz group (well-defined spin) is done will be demonstrated by the simple example of $\overline{\psi}iD_\mu\gamma_\nu\psi$. It has two vector indices, i.e. it is the direct product of two vectors, that means the maximal spin is 2. According to our knowledge from coupling of angular momentum with Clebsch–Gordan coefficients[27] we know that we can decompose it into a spin-0, a spin-1 and a spin-2 part. We can write

$$\begin{aligned}
\overline{\psi}i\hat{D}_\mu\gamma_\nu\psi = & \underbrace{\frac{1}{2}\left(\overline{\psi}i\hat{D}_\mu\gamma_\nu\psi + \overline{\psi}\gamma_\mu i\hat{D}_\nu\psi\right) - \frac{g_{\mu\nu}}{4}\overline{\psi}\not{D}\psi}_{\text{twist}-2} \\
& + \underbrace{\frac{1}{2}\left(\overline{\psi}i\hat{D}_\mu\gamma_\nu\psi - \overline{\psi}\gamma_\mu i\hat{D}_\nu\psi\right)}_{\text{twist}-3} \\
& + \underbrace{\frac{g_{\mu\nu}}{4}\overline{\psi}\hat{\not{D}}\psi}_{\text{twist}-4} \ ,
\end{aligned} \tag{4}$$

where the first line on the right-hand-side gives the twist-2, the second line the twist-3, and the third line the twist-4 contribution. Since the operator is of dimension-4 this corresponds to a spin-2, spin-1 and spin-0 contribution. The spin-2 part is completely symmetric and traceless, the spin-0 part is the trace we subtracted, and the remaining antisymmetric part has spin-1. In principle, such a decomposition is possible for even higher operators but becomes increasingly tedious and is not even unique, i.e. three spin-1 operators can be coupled to spin-2 in two different ways. In (4) we have found one twist-2 operator, one twist-3

[27] W. Greiner and B. Müller: *Quantum Mechanics – Symmetries*, 2nd ed., (Springer, Berlin, Heidelberg 1994).

Example 5.16

operator, and one twist-4 operator. However, sometimes some operators vanish by equations of motions, i.e. $i\hat{\not{D}}\psi = m\psi \sim 0$ for massless quarks. For instance the twist-4 operator is actually zero for a massless quark state. Sometimes, matrix elements of operators are zero due to trivial arguments. For example, if we were to sandwich the twist-3 operator between nucleon states

$$\left\langle PS\left|\overline{\psi}\left(i\hat{D}_\mu\gamma_\nu - i\hat{D}_\nu\gamma_\mu\right)\psi\right|PS\right\rangle = V_2\left(P_\mu P_\nu - P_\nu P_\mu\right) = 0 \tag{5}$$

it is evident that we would not be able to parametrize it in terms of a reduced matrix element. Due to the vector character of the object we have to parametrize it with $P_\mu P_\nu$ (due to Lorentz covariance this is the only vector at hand), which obviously cannot be antisymmetrized. On the other hand, the operator

$$\left\langle PS\left|\overline{\psi}\left(i\hat{D}_\mu\gamma_\nu\gamma_5 - i\hat{D}_\nu\gamma_\mu\gamma_5\right)\psi\right|PS\right\rangle = A_2\left(P_\mu S_\nu - P_\nu S_\mu\right) \tag{6}$$

exists. Due to the presence of γ_5 we can parametrize it with the spin vector S_μ which is an axial vector.

Now we take a look at a physical example. Performing the operator product expansion as we did before, but taking all trace terms carefully into account (and not only the free quark propagator) we could derive the first moment of the spin dependent structure function g_1 and obtain the following formula[28]

$$\int_0^1 \mathrm{d}x g_1(x, Q^2) = \frac{1}{2}a^{(0)} + \frac{M_\mathrm{N}^2}{9Q^2}\left(a^{(2)} + 4d^{(2)} + 4f^{(2)}\right) + \mathcal{O}\left(\frac{M_\mathrm{N}^4}{Q^4}\right) . \tag{7}$$

Here the reduced matrix elements are defined as:
twist-2

$$\left\langle PS\left|\overline{\psi}\gamma^\sigma\gamma^5\psi\right|PS\right\rangle = 2S^\sigma a^{(0)} , \tag{8}$$

twist-3

$$\begin{aligned}
&\frac{1}{6}\langle PS|\overline{\psi}\left[\gamma^\alpha g\tilde{G}^{\beta\sigma} + \gamma^\beta\tilde{G}^{\alpha\sigma}\right]\psi|PS\rangle - \text{traces} = \\
&2d^{(2)}\left[\frac{1}{6}\left(2P^\alpha P^\beta S^\sigma + 2P^\beta P^\alpha S^\sigma - P^\beta P^\sigma S^\alpha - P^\alpha P^\sigma S^\beta\right.\right. \\
&\qquad\left.\left. - P^\sigma P^\alpha S^\beta - P^\sigma P^\beta S^\alpha\right) - \text{traces}\right] ,
\end{aligned} \tag{9}$$

twist-4

$$\langle PS|\overline{\psi}g\tilde{G}_{\alpha\beta}\gamma^\beta\psi|PS\rangle = 2M_\mathrm{N}^2 f^{(2)} S_\alpha . \tag{10}$$

[28] B. Ehrnsperger et al.: Phys. Lett. B **150**, 439 (1994)

Example 5.16

Here we have a list of nontrivial twist-2, twist-3, and twist-4 operators. For the twist-3 operator one has to subtract traces so that the operator has well-defined spin. Explicitely they look like the following

$$
\begin{aligned}
\text{traces (l.h.s. of (9))} &= +\frac{1}{6}\langle PS|\overline{\psi}\left[\frac{4}{9}g^{\alpha\beta}\tilde{G}^{\sigma\mu}\gamma_\mu\right. \\
&\quad \left. +\frac{1}{9}g^{\alpha\sigma}\tilde{G}^{\beta\mu}\gamma_\mu+\frac{1}{9}g^{\beta\sigma}\tilde{G}^{\alpha\mu}\gamma_\mu\right]\psi|PS\rangle\ , \\
\text{traces (r.h.s. of (9))} &= +\frac{1}{18}M^2\left(g^{\alpha\beta}S^\sigma+g^{\beta\sigma}S^\alpha+g^{\alpha\sigma}S^\beta\right)\ .
\end{aligned}
$$

The twist-2 operator is a two-particle correlator, i.e. it can be interpreted as a probability distribution; see Fig. 5.32. Schematically it can be represented as the square of an amplitude.

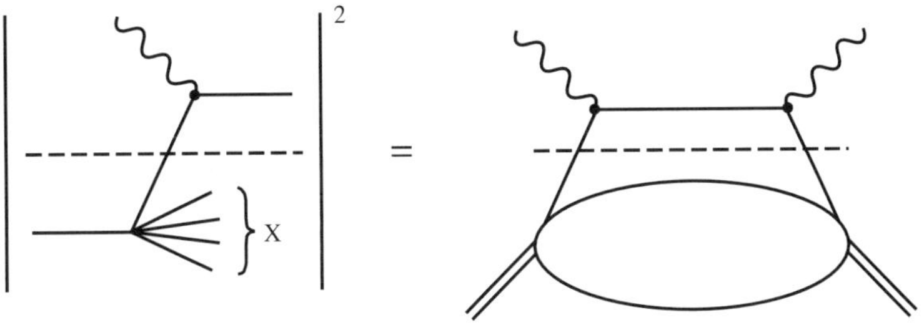

Fig. 5.32. Very schematic graphical representation of a twist-2 operator. See also Fig. 5.17, which we discussed previously. The blob represents the nucleon state $\langle PS|\ |PS\rangle$, the dashed lines the factorization scale

For twist-3 and twist-4 operator s this is not the case. Higher twists in general are correlations of more than two particles. In our case we have a quark–gluon–quark correlation. The gluon ($g\tilde{G}_{\mu\nu}$) is of different origin compared to the perturbative gluonic contribution that lead to the α_s expansion we discussed before. It is of nonperturbative nature and depends on the structure of the nucleon.

In the special case of the higher twist contributions to the first moment of g_1 we can make the physical meaning of higher twist even more explicit. In the rest frame of the nucleon $P_\mu = (M_\mathrm{N}, \mathbf{0})$, $S_\mu = (0, \mathbf{S})$ we can write the operator

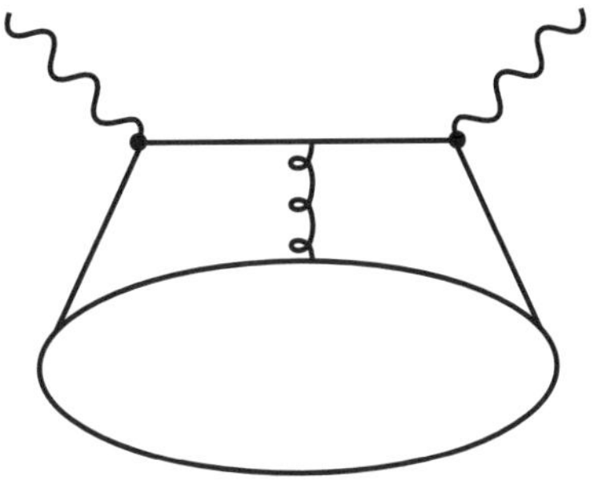

Fig. 5.33. Schematic graphical representation of a higher twist operator in DIS. The gluon, represented by the curled line, couples to the nucleon state and not to an other perturbative quark line

Example 5.16

determining $f^{(2)}$ and $d^{(2)}$ in components using the dual field strength tensor

$$\tilde{G}^{\mu\nu} = \frac{1}{2}\epsilon^{\mu\nu\varrho\sigma}G_{\varrho\sigma} = \tilde{G}^{a\mu\nu}\frac{\lambda_a}{2} = \begin{pmatrix} 0 & -B^1 & -B^2 & -B^3 \\ B^1 & 0 & E^3 & -E^2 \\ B^2 & -E^3 & 0 & E^1 \\ B^3 & E^2 & -E^1 & 0 \end{pmatrix} \tag{11}$$

as

$$\left\langle PS \middle| -B^\sigma_a j^0_a + (j_a \times \boldsymbol{E}_a)^\sigma \middle| PS \right\rangle = 2M_\mathrm{N}^2 f^{(2)} S^\sigma \tag{12}$$

and

$$\left\langle PS \middle| 2B^\sigma_a j^0_a + (j_a \times \boldsymbol{E}_a)^\sigma \middle| PS \right\rangle = 8M_\mathrm{N}^2 d^{(2)} S^\sigma \tag{13}$$

with the quark current $j^\mu_a = -g\overline{\psi}\gamma^\mu\lambda_a/2\psi$ and color-electric and color-magnetic field strength $\boldsymbol{E}^a$ and $\boldsymbol{B}^a$. The quark current in the nucleon induces an electric field. This induced field contributes to the spin of the nucleon. Clearly, correlations like that are well beyond the simple parton model. That is one of the reasons why theoreticians like to deal with higher twist, because of the insight gained into the structure of the nucleon. Moreover, present experiments are reaching such precision that even $1/Q^2$ corrections in DIS are accessible. Recent progress has been made to analyze higher twist beyond the level of the lowest moments. We will discuss this approach in the next example.

EXAMPLE

5.17 Perturbation Theory in Higher Orders and Renormalons

In the previous section we learned how to calculate perturbative corrections to structure functions measured in deep inelastic scattering (DIS). While this was quite an easy task for the first-order correction to F_L, higher-order corrections become increasingly difficult to handle. Up to now, third-order corrections for the lowest moments of some structure functions are available. As an example, we quote the corrections to the first moment of the nonsinglet polarized structure function g_1, the Bjorken sum rule, i.e. the difference between the proton structure function g_1^p and the neutron structure function g_1^n [29] (see also Sect. 5.6 on the spin-dependent structure functions):

$$\begin{aligned}\int\limits_0^1 \mathrm{d}x g_1^{p-n}(x, Q^2) =& \frac{1}{6}\frac{g_A}{g_V}\left(1 - \frac{\alpha_s}{\pi} - 3.5833\left(\frac{\alpha_s}{\pi}\right)^2\right.\\ &\left.-20.2153\left(\frac{\alpha_s}{\pi}\right)^3 + \dots\right) + \mathcal{O}\left(\frac{1}{Q^2}\right) .\end{aligned} \tag{1}$$

[29] S.A. Larin, J.A.M. Vermaseren: Phys. Lett. B **259** (1991) 345.

Example 5.17

First we notice that the numerical values of the expansion coefficient increase, a phenomenon not only observed in this special case.[30] In general, perturbative series in interacting quantum field theories are regarded as divergent rather than as convergent series. Observing the phenomenological success of perturbation theory in QED, for instance the calculation of the anomalous magnetic moment of electron and muon, one interpretes the divergent series as an asymptotic one. That means, even if the series does not converge it still makes sense up to a certain order. This order naturally depends on the numerical magnitude of the expansion parameter. While in QED $\alpha = 1/137$ is a rather small number even a factorial divergent coefficient, e.g. like $n!\alpha^n$, would lead to the explosion of the terms only for $n \gtrsim 136$ (check this!). The situation is different in QCD, where $\alpha_s(Q^2 = 1\,\text{GeV}^2) \approx 0.4$ is much larger. One therefore has to be careful to decide up to which order the expansion really makes sense and whether the divergence of the series may be seen already at low orders. An asymptotic expansion is only meaningful up to a certain order, which usually is the smallest term in the expansion. This minimal term is defined as the term of order n_0 in the expansion. Terms of higher order, i.e. terms with $n > n_0$ of the series, start to increase:

$$\left|r_{n_0+1}\alpha_s^{n_0+1}\right| > \left|r_{n_0}\alpha_s^{n_0}\right| \;,$$

The truncated series may be written as

$$\mathcal{R}(Q^2) = \sum_{n=0}^{n_0-1} r_n\alpha_s^n \pm r_{n_0}\alpha_s^{n_0} = \sum_{n=0}^{n_0-1} r_n\alpha_s^n \pm \Delta\mathcal{R}(Q^2) \;, \tag{2}$$

where $\alpha_s = \alpha_s(Q^2)$. The uncertainty in the approximation, characterized by the quantity $\mathcal{R}(Q^2)$ that we would like to calculate in perturbation theory, is given by the minimal term. As we will see in the following, it can be shown that the uncertainty does not depend logarithmically on Q^2 but has a power behavior

$$\Delta\mathcal{R}(Q^2) \sim \frac{1}{\left(Q^2\right)^s} \;, \tag{3}$$

where s is some integer number.

We normally attribute the same $1/Q^2$ dependence to higher-twist corrections. Let us consider the quantity $\mathcal{R}(Q^2)$ in some more detail. In general we can write it in the general form

$$\mathcal{R} - \mathcal{R}_{tree} = r_0\alpha_s + r_1\alpha_s^2 + r_2\alpha_s^3 + \cdots + r_k\alpha_s^{k+1} + \ldots \;, \tag{4}$$

[30] G. 't Hooft in *The Whys of Subnuclear Physics*, ed. by A. Zichichi (Plenum, New York 1977).
A.H. Müller: *The QCD Perturbation Series in QCD – Twenty Years Later*, ed. by P.M. Zerwas and H.A. Kastrup (World Scientific, Singapore, 1922), p. 162.

Example 5.17

where we have already subtracted the contribution of the tree-level term $\sim \alpha_s^0$ so that the expansion starts at order α_s. Using the integral representation of $k!$

$$\int_0^\infty \mathrm{d}t e^{-t} t^k = k! \tag{5}$$

we can write

$$\mathcal{R} - \mathcal{R}_{tree} = \sum_{m=0}^\infty \alpha_s^{m+1} \frac{r_m}{m!} \int_0^\infty \mathrm{d}t e^{-t} t^m \ . \tag{6}$$

Interchanging integration and summation gives

$$\begin{aligned} \mathcal{R} - \mathcal{R}_{tree} &= \int_0^\infty \mathrm{d}t e^{-t} \alpha_s \sum_{m=0}^\infty (\alpha_s t)^m \frac{r_m}{m!} \\ &= \int_0^\infty \mathrm{d}u e^{-u/\alpha_s} \sum_{m=0}^\infty u^m \frac{r_m}{m!} \\ &= \int_0^\infty \mathrm{d}u e^{-u/\alpha_s} \mathcal{B}\left[\mathcal{R}\right](u) \end{aligned} \tag{7}$$

with $u = \alpha_s t$. In the last step we *defined* the *Borel representation* of the series as

$$\mathcal{B}\left[\mathcal{R}\right](u) = \sum_{m=0}^\infty u^m \frac{r_m}{m!} \ . \tag{8}$$

The Borel representation can be used as a *generating function* for the fixed-order coefficients

$$r_n = \frac{\mathrm{d}^n}{\mathrm{d}u^n} \mathcal{B}\left[\mathcal{R}\right](u)|_{u=0} \ . \tag{9}$$

But, more importantly, (7) is now taken as the definition of the summed series. Even if the series in the expansion (4) is, in fact, divergent, with the help of the Borel transformation we may be able to give it a unique value. This is not always possible, of course. But if this is possible the series is said to be *Borel summable*. As a simple example we consider the series which diverges factorially but with alternating sign $r_n = (-)^n n!$. The Borel transformed series reads

$$1 - u + u^2 - u^3 \pm \cdots = \frac{1}{1+u} \tag{10}$$

and its integral representation is well defined,

$$\int_0^\infty \mathrm{d}u \frac{e^{-u/\alpha_s}}{1+u} \widehat{=} \sum_{m=0}^\infty (-)^n n! \alpha_s^n \ . \tag{11}$$

Example 5.17

If, on the other hand, we have a series with a *fixed* sign factorial growth $r_n \sim n!$ the transformed series reads

$$1+u+u^2+u^3+\cdots = \frac{1}{1-u} \tag{12}$$

and in the integral representation we have to integrate over the pole at $u = 1$. The series therefore acquires an imaginary part and the integral is defined only via its principal value. Such a series as in (12) is referred to as *not Borel summable*. In general a series exhibiting factorial growth of the coefficients leads to singularities in the complex Borel plane. Singularities on the negative real axis are integrable, singularities on the positive real axis are not. The QCD perturbation theory is not Borel summable: singularities on the positive real axis are present. One finds singularities at the following positions.

• Instanton–antiinstanton singularities (instantons–antiinstantons are classical solutions of the QCD equations of motion[31]) They can be found at $u = 4\pi, 8\pi, \ldots$ and are simply due to a classical effect. The number of diagrams grows factorially with the order of the perturbative expansion.

• Renormalon singularities. As opposed to the instanton–antiinstanton singularities that are due to the overall number of diagrams, there are single diagrams that contribute a factor $n!$ at order n to the amplitude. Diagrams of this kind are those where a simple gluon line is replaced by a fermion bubble chain (see Fig. 5.34). In QED these are exactly the diagrams that lead to the running of the coupling constant. In general, if we substitute for the fixed coupling at the vertex the running coupling

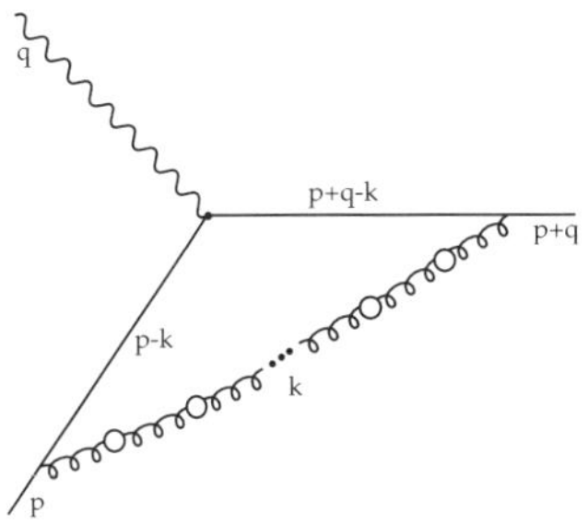

Fig. 5.34. Diagram that leads to corrections to the photon–quark vertex and diverges as $n!$

$$\alpha_s(k^2) = \frac{\alpha_s(Q^2)}{1+\beta_0\alpha_s(Q^2)\ln(k^2/Q^2)} , \tag{13}$$

we find that integration over the loop momenta k leads to a factorial growth of the expansion coefficient. In (13) β_0 is given by

$$\beta_0 = \frac{1}{4\pi}\left(\frac{11}{3}N - \frac{2}{3}N_\mathrm{f}\right) \tag{14}$$

where $N = 3$ for QCD (SU(3)) and zero for QED. N_f is the number of flavors.

To show that the running coupling constant leads to $n!$ growth we consider a dimensionless quantity that is calculated from a loop integration over $\int \mathrm{d}k_E^2/k_E^2$ with Euclidean momentum $k_E^2 = -k^2$ (see Sect. 4.3):

$$I(Q^2) = \int \frac{\mathrm{d}k_E^2}{k_E^2} f(k_E^2, Q^2)\alpha_s(k_E^2) , \tag{15}$$

[31] G. 't Hooft in *The Whys of Subnuclear Physics*, ed. by A. Zichichi (Plenum, New York 1977).
A.H. Müller: *The QCD Perturbation Series in QCD – Twenty Years Later*, ed. by P.M. Zerwas and H.A. Kastrup (World Scientific, Singapore, 1922), p. 162.

Example 5.17

where Q^2 is the external momentum and $f(k_E^2, Q^2)$ some function. Assuming $k_E^2 > Q^2$ we can expand $f(k_E^2, Q^2)$ in terms of Q^2/k_E^2 and find

$$\begin{aligned}
I_{\mathrm{UV}} &\sim \int_{Q^2}^{\infty} \frac{\mathrm{d}k_E^2}{k_E^2} \left(\frac{Q^2}{k_E^2}\right)^m \alpha_s(k_E^2) \\
&= \sum_n \int_{Q^2}^{\infty} \frac{\mathrm{d}k_E^2}{k_E^2} \left(\frac{Q^2}{k_E^2}\right)^m (-\beta_0)^n \ln^n \left(\frac{k_E^2}{Q^2}\right) \alpha_s(Q^2)^{n+1} \\
&= \sum_n \int_1^{\infty} \frac{\mathrm{d}t}{t} (t)^{-m} (-\beta_0)^n \ln^n(t) \alpha_s(Q^2)^{n+1} \\
&= \sum_n \int_0^{\infty} \mathrm{d}y e^{-my} (-\beta_0 y)^n \alpha_s(Q^2)^{n+1} \\
&= \sum_n \left(-\frac{\beta_0}{m}\right) \frac{1}{m} \underbrace{\int_0^{\infty} \mathrm{d}x e^{-x} x^n}_{n!} \alpha_s(Q^2)^{n+1} \\
&\sim \sum_n \left(\frac{-\beta_0}{m}\right)^n n! \alpha_s(Q^2)^{n+1} \; .
\end{aligned} \tag{16}$$

The alternating sign factorial growth ($\beta_0 > 0$ in QCD) leads to singularities in the complex Borel plane at

$$u = -\frac{m}{\beta_0}; \quad m = 1, 2, \dots \; . \tag{17}$$

Due to its origin from high virtualities ($k_E^2 \gg Q^2$) these singularities are called *ultraviolet renormalons*.

Considering the case of small loop momenta $k_E^2 < Q^2$ and expanding the integrand in k_E^2/Q^2 we get

$$\begin{aligned}
I_{\mathrm{IR}} &\sim \int_0^{Q^2} \frac{\mathrm{d}k_E^2}{k_E^2} \left(\frac{k_E^2}{Q^2}\right)^m \alpha_s(k_E^2) \\
&= \sum_n \int_0^{Q^2} \frac{\mathrm{d}k_E^2}{k_E^2} \left(\frac{k_E^2}{Q^2}\right)^m (-\beta_0)^n \ln^n \left(\frac{k_E^2}{Q^2}\right) \alpha_s(Q^2)^{n+1} \\
&\sim \sum_n \left(\frac{\beta_0}{m}\right)^n n! \alpha_s(Q^2)^{n+1} \; .
\end{aligned} \tag{18}$$

The corresponding singularities at the positive axis in the Borel plane are called *infrared renormalons*. They lie at $u = +m/\beta_0$, $m = 1, 2, \dots$ (see Fig. 5.35).

Example 5.17

Having explained what renormalons are, we turn to the calculation of the poles in the Borel plane, for instance for the perturbative expansion of the Bjorken sum rule in (1). For that purpose let us consider once more the expansion in (4). Each coefficient r_k can be written as an expansion in N_f, the number of active flavors:

$$r_k = r_k^{(0)} + r_k^{(1)} N_\mathrm{f} + \cdots + r_k^{(k)} N_\mathrm{f}^k \; . \qquad (19)$$

The different terms in the N_f expansion are due to Feynman diagrams that contain fermion bubbles . Each fermion bubble contributes a factor N_f.

The leading term in the expansion (19), i.e. the term with the highest power of N_f, comes from an *uncut* bubble chain. Two fermion bubble chains, containing k_1 and k_2 bubbles respectively, sublead in α_s, i.e. they contribute at order $r_{k-1}^{(k)} N_\mathrm{f}^{k-1}$ (see the Fig. 5.36). In our example both diagrams are of order N_f^2 but the one on the right-hand side with two bubble chains contains one additional factor α_s as compared to the left-hand diagram.

Diagrams that contain only *one* bubble chain are relatively straightforward to evaluate and are those that lead to the renormalon. Such a calculation – taking only fermion bubbles into account and neglecting all gluon–gluon interactions – is a simple QED calculation, and the exactly calculated

$$r_k^{(k)} N_\mathrm{f}^k \qquad (20)$$

coefficient can be used as the starting point to approximate a QCD calculation.

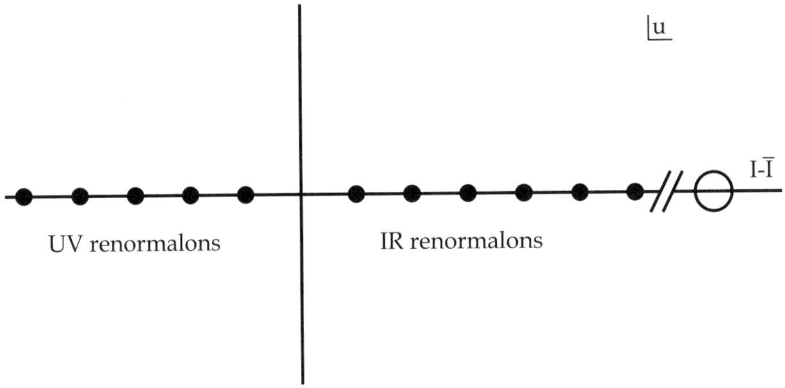

Fig. 5.35. Singularities in the Borel plane. For asymptotically free theories ($\beta_0 > 0$) the IR renormalons can be found on the positive axis, the UV renormalons on the negative axis. Instanton–antiinstanton singularities ($I\bar{I}$) are on the positive axis

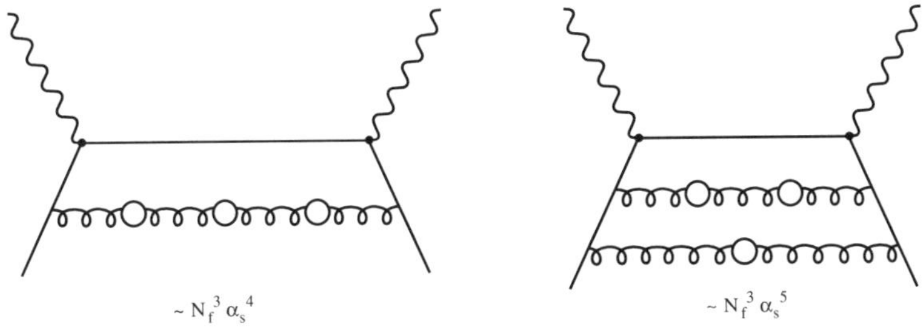

Fig. 5.36. Corrections to the Compton forward scattering amplitude: Double fermion lines containing in total the same number of bubbles are subleading in N_f as compared to uncut fermion lines

Example 5.17

In QED the summed fermion bubble chains are the only diagrams that contribute to the running of the coupling. Ward identities connect vertex correction $\Gamma_\mu(p,p)$ and self-energy $\sum(p)$ via the relation $-\partial/\partial p^\mu \sum(p) = \Gamma_\mu(p,p)$ so that the sum of all contributions their divergencies cancel and therefore do not have to be renormalized.

In QCD also gluon and ghost bubbles contribute so that the one-loop β function gets a positive sign as opposed to the QED β function. The idea is then simply to perform a QED calculation with the QED β function and, afterwards, to substitute the QCD β function for the corresponding QED piece. The latter is essentially given by (14) with $N=0$. In the QCD case $N=3$ and we substitute for N_f in (19)

$$N_\mathrm{f} = \frac{3}{2}\,(11 - 4\pi\beta_0) \tag{21}$$

and write the expansion (19) in terms of β_0. This gives

$$r_k = \tilde{r}_k^{(0)} + \tilde{r}_k^{(1)}\beta_0 + \dots + \tilde{r}_k^{(k)}\beta_0^k\ , \tag{22}$$

where the leading term of the β_0 expansion is given by

$$\tilde{r}_k^{(k)} = r_k^{(k)}\,(-6\pi)^k\,\beta_0^k\ . \tag{23}$$

Now we set all subleading (exact) coefficients of the β_0 expansion (22) to zero and get, in leading order β_0 (and therefore also in leading order N_f),

$$\begin{aligned} r_k &\approx \tilde{r}_k^{(k)}\beta_0^k = r_k^{(k)}(-6\pi\beta_0)^4 = r_k^{(k)}\left(N_\mathrm{f} - \frac{33}{2}\right)^k \\ &= r_k^{(k)}N_\mathrm{f}^k + r_k^{(k)}N_\mathrm{f}^{k-1}\left(-\frac{33}{2}\right)\cdot k + \dots + r_k^{(k)}\left(-\frac{33}{2}\right)^k\ , \end{aligned} \tag{24}$$

where we substituted $(-6\pi\beta_0) = N_\mathrm{f} - 33/2$ according to (21). In that way we obtained approximations for the unknown subleading coefficients that are much harder to calculate exactly as compared to $r_k^{(k)}$. This procedure is called *Naive Non-Abelianization* (NNA).[32] To test how well this procedure works we write the Bjorken sum rule with explicit N_f dependence.[33] For example, $\left[N_\mathrm{f}^{k-1}(-33/2)kr_k^{(k)}\right]$ approximates the term $r_k^{(k-1)}$ in (19). One should emphasize that there is no deep justification for this procedure. Instead it should be considered as a toy model to study the behavior of perturbation theory in QCD

[32] M. Beneke: Nucl. Phys. B **405**, 424 (1993), D.J. Broadhurst: Z. Phys. C **58**, 339 (1993).

[33] S.A. Larin, J.A.M. Vermaseren: Phys. Lett. B **259** (1991) 345.

at large orders. *Example 5.17*

$$\begin{aligned}\int_0^1 \mathrm{d}x g_1^{p-n}(x,Q^2) &= \frac{1}{6}\frac{g_A}{g_V}\left\{1-\frac{\alpha_s}{\pi}+\left(\frac{\alpha_s}{\pi}\right)^2\left(-4.58+\frac{1}{3}N_\mathrm{f}\right)\right.\\ &\quad\left.+\left(\frac{\alpha_s}{\pi}\right)^3\left(-41.439+7.61N_\mathrm{f}-0.177N_\mathrm{f}^2\right)+\dots\right\}\\ &= \frac{1}{6}\frac{g_A}{g_V}\left\{1-\frac{\alpha_s}{\pi}-3.5833\left(\frac{\alpha_s}{\pi}\right)^2\right.\\ &\quad\left.-20.215\left(\frac{\alpha_s}{\pi}\right)^3+\dots\right\}\,,\end{aligned}\tag{25}$$

where the last line was derived for $N_\mathrm{f}=3$. This agrees with (1). Substituting for the highest power of N_f, ($N_\mathrm{f}\alpha_s^2$ and $N_\mathrm{f}^2\alpha_s^3$) $N_\mathrm{f}\to N_\mathrm{f}-33/2$, we find

$$\begin{aligned}\int_0^1 \mathrm{d}x g_1^{p-n}(x,Q^2)|_\mathrm{NNA} &= \frac{1}{6}\frac{g_A}{g_V}\left\{1-\frac{\alpha_s}{\pi}+\left(\frac{\alpha_s}{\pi}\right)^2\left(-5.5+\frac{1}{3}N_\mathrm{f}\right)+\left(\frac{\alpha_s}{\pi}\right)^3\right.\\ &\quad\left.\times\left(-48.188+5.841N_\mathrm{f}-0.177N_\mathrm{f}^2\right)+\dots\right\}\\ &= \frac{1}{6}\frac{g_A}{g_V}\left\{1-\frac{\alpha_s}{\pi}-4.5\left(\frac{\alpha_s}{\pi}\right)^2\right.\\ &\quad\left.-32.26\left(\frac{\alpha_s}{\pi}\right)^3+\dots\right\}\,,\end{aligned}\tag{26}$$

which reproduces the sign and the order of magnitude of the exactly calculated coefficients amazingly well.

In the following we explain a simple way to calculate the contributions from bubble chains. The idea is extremly simple and rests on the observation that the Borel transform of the bubble chain (or running coupling) is an extremly simple function.

We therefore do not calculate each bubble contribution separately but instead calculate the Borel transform of the whole bubble chain. The effective gluon propagator with the insertion of fermion, gluon, and ghost bubbles reads

$$\alpha_s(k^2)D_{\mu\nu}^{AB}(k^2)=\mathrm{i}\delta^{AB}\left(\frac{k_\mu k_\nu-k^2g_{\mu\nu}}{k^4}\right)\frac{\alpha_s(\mu^2)}{1+\beta_0\alpha_s(\mu^2)\ln\left(-k^2/\mu^2e^{-C}\right)}\,,\tag{27}$$

where we multiplied the gluon propagator, given in Landau gauge , with the running QCD coupling at scale k^2, the momentum which runs through the gluon line. The reference scale is μ^2 and will only at the end be set to Q^2 as in (13). The factor e^{-C} accounts for scheme dependence and is $-5/3$ in the $\overline{\mathrm{MS}}$ scheme ($\overline{\mathrm{MS}}$ stands for minimal subtraction).[34] In that scheme not only the $1/\epsilon$ pole but also the contribution $1/\epsilon-\gamma_E+\ln(4\pi\mu^2)$ is subtracted.

[34] J. Brodsky, G.P. Lepage, P.B. Mackenzie: Phys. Rev. D **28**, 228 (1983).

Example 5.17

Expanding the series

$$\frac{\alpha_s(\mu^2)}{1+\beta_0\alpha_s(\mu^2)\ln\left(-k^2/\mu^2e^{-C}\right)} = \sum_{n=0}^{\infty}(\alpha_s)^{n+1}(-\beta_0)^n\ln^n\left(-\frac{k^2}{\mu^2e^{-C}}\right) \tag{28}$$

and performing the Borel transformation $\alpha_s^{n+1} \to u^n/n!$ we find

$$\begin{aligned}\sum_{n=0}^{\infty}(\alpha_s)^{n+1}(-\beta_0)^n\ln^n\left(-\frac{k^2}{\mu^2e^{-C}}\right) &\Rightarrow \sum_{n=0}^{\infty}\frac{u^n}{n!}(-\beta_0)^n\ln^n\left(-\frac{k^2}{\mu^2e^{-C}}\right)\\ = \exp\left(-\beta_0 u\ln\left(-\frac{k^2}{\mu^2e^{-C}}\right)\right) &= \left(\frac{\mu^2e^{-C}}{-k^2}\right)^{\beta_0 u}\end{aligned} \tag{29}$$

and get for the transformed propagator

$$\mathcal{B}T\left[\alpha_s(k^2)D^{AB}_{\mu\nu}(k)\right](u) = \frac{\mathrm{i}\delta^{AB}}{-k^2}\left(g_{\mu\nu}+\frac{k_\mu k_\nu}{-k^2}\right)\left(\frac{\mu^2e^{-C}}{-k^2}\right)^{\beta_0 u} . \tag{30}$$

The result means the following: To obtain an all-order expression for a diagram that only contains one bubble chain one only has to calculate the normal α_s correction but with the simple gluon propagator replaced by the Borel transformed propagator in (30). From the technical point of view one has to change the normal power $1/(-k^2)$ to $1/(-k^2)^{1+\beta_0 u}$. This simple fact was first observed in.[35] To obtain fixed-order contributions at the end one can apply (9). As a result for such a calculation we will give the perturbative expansion of the Bjorken sum rule to all orders in the Borel representation. Let us write

$$\frac{\int_0^1 \mathrm{d}x g_1^{p-n}(x,Q^2)}{(1/6)g_A/g_v} = 1+\sum_n a_n\alpha_s^{n+1}+\sum_n \delta_n\alpha_s^{n+1} . \tag{31}$$

Here the coefficients a_n stem from the insertion of n bubbles and the δ_n, which shall be neglectded in the following, can only be obtained from an exact all-order calculation. The generating function for the a_n is obtained by the calculation of all the diagrams in Fig. 5.21 and using the propagator in (30). The calculation is straightforward by following the general procedure explained in the section on Wilson coefficients. An important simplification arises, however, since we are only interested in the first momentum of the structure function g_1. This is the contribution $\sim \omega^0 = \left(2p\cdot q/Q^2\right)^0$ and therefore does not depend on the target momentum p. Therefore, in the following we can set $p=0$ without any loss of generality. This will simplify the calculations tremendously. Some intermediate steps of the calculation will be given in the following: For the self-energy one

[35] M. Beneke: Nucl. Phys. B **405**, 424 (1993).

gets

$$S(q,d,s) = (-\mathrm{i}g)^2(\mu^2)^{2-d/2} \times \int \frac{\mathrm{d}^dk}{(2\pi)^d}\gamma_\mu \frac{\mathrm{i}}{\not{q}-\not{k}}\gamma_\nu \frac{-\mathrm{i}}{k^2}\left(g^{\mu\nu}+\frac{k^\mu k^\nu}{(-k^2)}\right)\left(\frac{\mu^2 e^{-C}}{(-k^2)}\right)^s , \tag{32}$$

where the propagator (30) was used with the substitution $s = \beta_0 u$. In principle such calculations have to be performed in d dimensions to regulate divergencies. On the other hand, using the special propagator (30) the additional power of s serves as a regulating device. Divergencies that commonly appear at $d = 4$ will not appear at $s = 0$. The final result, i.e. the sum of all diagrams, which is finite, will not depend on the used procedure of regulation. At intermediate steps, however, results for single diagrams will differ for different regularization procedures.

Performing the integrals one is led to the following expression:

$$S(q,d,s=0) = 0 ,$$
$$S(q,d=4,s) = \frac{3\mathrm{i}}{(4\pi)^2}g^2\left(\frac{\mu^2 e^{-C}}{-q^2}\right)^s \not{q}\left(\frac{\Gamma(1+s)\Gamma(1-s)}{\Gamma(3-s)\Gamma(2+s)}\right) . \tag{33}$$

The first equation reflects the fact that the self-energy vanishes in the Landau gauge. The second equation shows that the limits $s \to 0$ and $d \to 4$ do not commute. For the vertex correction one gets

$$\Gamma_\mu(q,d,s) = -\mathrm{i}g^2(\mu^2)^{2-d/2}(\mu^2 e^{-C})^s \times \int \frac{\mathrm{d}^dk}{(2\pi)^d}\frac{\gamma_\alpha(\not{q}-\not{k})\gamma_\mu\not{k}\gamma_\beta}{(-(q-k)^2)(-k^2)^{3+s}}\left(g^{\alpha\beta}(-k^2)+k^\alpha k^\beta\right) , \tag{34}$$

which can be evaluated to give

$$\Gamma_\mu(q,d\to 4,s=0) = -\frac{g^2}{8\pi^2}\left(\frac{q_\mu\not{q}}{(-q^2)}+\gamma_\mu\right)$$
$$\Gamma_\mu(q,d=4,s) = -\frac{g^2}{(4\pi)^2}\left(\frac{\mu^2 \mathrm{e}^{-C}}{Q^2}\right)^s \times\left(\frac{\Gamma(1+s)\Gamma(1-s)}{\Gamma(3-s)\Gamma(2+s)}\right)\gamma_\mu + \mathcal{O}(q_\mu\not{q}) . \tag{35}$$

Terms of the order $q_\mu\not{q}$ do not contribute to the antisymmetric part of the Compton amplitude and therefore can be neglected. For the diagram with self-energy insertion one finds

$$\begin{aligned} t^{(1)}_{\mu\nu}\Big|^{S(q,4,s)}_{p=0} g^2 &= -\frac{3}{(4\pi)^2}g^2\left(\frac{\mu^2 e^{-C}}{Q^2}\right)^s\left(\frac{\Gamma(1+s)\Gamma(1-s)}{\Gamma(3-s)\Gamma(2+s)}\right) \\ &\quad\times\left(\gamma_\mu\frac{\mathrm{i}}{\not{q}}\gamma_\nu+(\mu\to\nu, q\to -q)\right) \\ &= -\frac{3}{4\pi}\alpha_s\left(\frac{\mu^2 e^{-C}}{Q^2}\right)^s\left(\frac{\Gamma(1+s)\Gamma(1-s)}{\Gamma(3-s)\Gamma(2+s)}\right)\left(t^{(0)}_{\mu\nu}\Big|_{p=0}\right) . \end{aligned} \tag{36}$$

Example 5.17

Here we have introduced the notion

$$t^{(0)}_{\mu\nu}\Big|_{p=0} = \gamma_\mu \frac{\mathrm{i}}{\not q}\gamma_\nu + (\mu \to \nu, q \to -q) \tag{37}$$

for the tree-level, i.e. the $\mathcal{O}(g^2=0)$ contribution. The result for diagrams with vertex corrections gives, according to (35),

$$t^{(1)}_{\mu\nu}\Big|^{\Gamma(q,4,s)}_{p=0} g^2 = -\frac{2}{4\pi}\alpha_s \left(\frac{\mu^2 e^{-C}}{Q^2}\right)^s \left(\frac{\Gamma(1+s)\Gamma(1-s)}{\Gamma(3-s)\Gamma(2+s)}\right) t^{(0)}_{\mu\nu}\Big|_{p=0}\Bigg) . \tag{38}$$

Calculation of the remaining box diagram

$$\begin{aligned} t^{(1)}_{\mu\nu}\Big|^{B(q,d,s)}_{p=0} g^2 &= (-\mathrm{i}g)^2(\mu^2)^{2-d/2}\left(\mu^2 e^{-C}\right)^s \\ &\quad \times \int \frac{\mathrm{d}^d k}{(2\pi)^d}\gamma_\varrho \frac{\mathrm{i}}{\not k}\gamma_\nu \frac{\mathrm{i}}{\not k+\not q}\gamma_\mu \frac{\mathrm{i}}{\not k}\gamma_\sigma \frac{\mathrm{i}(g^{\varrho\sigma}(-k^2)+k^\varrho k^\sigma)}{(-k^2)^{2+s}} , \end{aligned} \tag{39}$$

yields for $d=4$, $s \neq 0$ the following result:

$$t^{(1)}_{\mu\nu}\Big|^{B(q,4,s)}_{p=0} g^2 = -\frac{1}{4\pi}\alpha_s \left(\frac{\mu^2 e^{-C}}{Q^2}\right)\left(\frac{\Gamma(1+s)\Gamma(1-s)}{\Gamma(3+s)\Gamma(2-s)}\right)\left(t^{(0)}_{\mu\nu}\Big|_{p=0}\right) . \tag{40}$$

The sum of all contributions (36), (38), and (40) therefore gives

$$\begin{aligned} t^{(1)}_{\mu\nu}\Big|_{p=0} g^2 &= -\alpha_s \frac{C_\mathrm{F}}{\pi}\left(\frac{\mu^2 e^{-C}}{Q^2}\right)^s \left(\frac{\Gamma(1+s)\Gamma(1-s)}{\Gamma(3-s)\Gamma(3+s)}(3+s)\right)\left(t^{(0)}_{\mu\nu}\Big|_{p=0}\right) \\ &= -\alpha_s \frac{C_\mathrm{F}}{\pi}\left(\frac{\mu^2 e^{-C}}{Q^2}\right)^s \frac{3+s}{(1-s)(1+s)(2-s)(2+s)}\left(t^{(0)}_{\mu\nu}\Big|_{p=0}\right) \\ &\overset{s=0}{=} -\frac{\alpha_s}{\pi}\left(t^{(0)}_{\mu\nu}\Big|_{p=0}\right) , \end{aligned} \tag{41}$$

where we have inserted the color factor $C_\mathrm{F}=4/3$. Also we used the property of the Γ-function $x\Gamma(x)=\Gamma(x+1)$. For $s=0$ we reproduce the one-loop correction to the Bjorken sum rule.

$$C^A(1,\alpha_s(Q^2/\mu^2)) = 1 - \frac{\alpha_s(Q^2)}{\pi} + \cdots \tag{42}$$

For $s \neq 0$ one gets from (41) the generating functional for the a_n:

$$\begin{aligned} B[\mathcal{R}](s) &= -\frac{C_\mathrm{F}}{\pi}\left(\frac{\mu^2 e^{-C}}{Q^2}\right)^s \frac{3+s}{(1-s^2)(4-s^2)} \\ &= -\frac{C_\mathrm{F}}{3\pi}\left(\frac{\mu^2 e^{-C}}{Q^2}\right)^s \left(\frac{2}{1-s}+\frac{1}{1+s}+\frac{5}{4(2-s)}-\frac{1}{4(2+s)}\right) . \end{aligned} \tag{43}$$

Example 5.17

With $s = \beta_0 u$ the relevant equations now read

$$\mathcal{R} - \mathcal{R}_{\text{tree}} = \frac{1}{\beta_0} \int_0^\infty \mathrm{d}s \, e^{-s/\beta_0 \alpha_s} \mathcal{B}[\mathcal{R}](s) \tag{44}$$

$$\mathcal{B}[\mathcal{R}](s) = \sum_m \left(\frac{s}{\beta_0} \right)^m \frac{a_m}{m!} \tag{45}$$

$$a_m = \beta_0^m \frac{\mathrm{d}^m}{\mathrm{d}s^m} \mathcal{B}[\mathcal{R}](s)|_{s=0} \tag{46}$$

(compare with (7), (8), and (9)).

Applying (46) and setting $\mu^2 = Q^2$ one generates a series in $\beta_0 \alpha_s$:

$$\frac{\int_0^1 \mathrm{d}x g_1^{p-n}(x, Q^2)|_{\text{NNA}}}{(1/6) g_A/g_V} = 1 - \frac{\alpha_s}{\pi} \left(1 + 2\beta_0 \alpha_s + \frac{115}{18} \beta_0^2 \alpha_s^2 + \frac{605}{27} \beta_0^3 \alpha_s^3 + \ldots \right) , \tag{47}$$

which reproduces the expansion in (26) with the corresponding insertion of β_0. In Fig. 5.37 we have plotted the NNA approximation to the perturbative expansion of the Bjorken sum rule. One recognizes the typical behavior of an asymptotic series: up to a maximal order of expansion the terms converge to a minimal value characterized by n_0, but from there on $n > n_0$ the series starts to diverge. Analyzing (43) we see that the Borel representation exhibits four renormalons. With $u = s/\beta_0$ we find infrared renormalons at $1/\beta_0$ and $2/\beta_0$, while the ultraviolet renormalons are situated at $-1/\beta_0$, and $-2/\beta_0$.

The first IR renormalon leads to a fixed sign factorial divergence

$$\frac{\mathrm{d}^n}{\mathrm{d}s^n} \frac{1}{1-s} = n! \, . \tag{48}$$

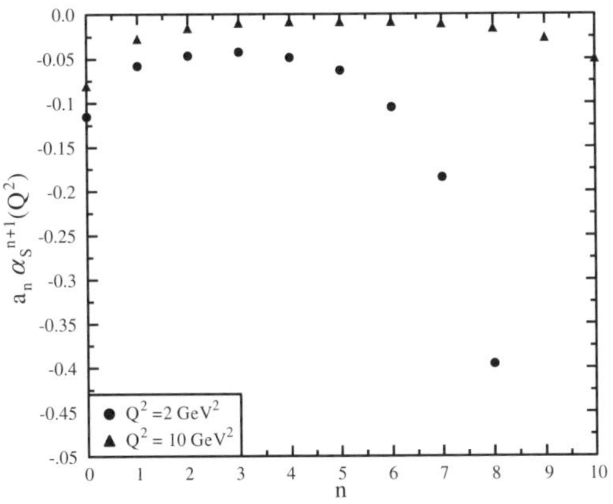

Fig. 5.37. NNA corrections to the Bjorken sum rule. We have chosen $\Lambda_{\overline{\text{MS}}} = 200\,\text{MeV}$ and $Q^2 = 2\,\text{GeV}$ (full data) and $Q^2 = 10\,\text{GeV}$ (triangles). On the n axis we have drawn the order of expansion

Example 5.17

The pole next to the origin dominates the asymptotic expansion. Applying the Leibnitz rule of differentation we are able to give closed formulas for the asymptotic behavior resulting from (43).

With a being some number (which can be read off from the first, second, ... fourth term in (43)) we find

$$\begin{aligned}
&\frac{\mathrm{d}^n}{\mathrm{d}s^n}\left(\frac{e^{-C}\mu^2}{Q^2}\right)^s \frac{1}{1+as} \\
&= \sum_{k=0}^{n}\binom{n}{k}\ln^k\left(\frac{e^{-C}\mu^2}{Q^2}\right)\left(\frac{e^{-C}\mu^2}{Q^2}\right)^s (n-k)!\frac{(-a)^{n-k}}{(1-s)^{n-k+1}}\Bigg|_{s=0} \\
&= n!(-a)^n\sum_{k=0}^{n}\frac{1}{k!}\left(\frac{1}{-a}\right)^k \ln^k\left(\frac{e^{-C}\mu^2}{Q^2}\right) \\
&= n!(-a)^n\sum_{k=0}^{n}\frac{1}{k!}\left[\ln\left(\frac{e^{-C}\mu^2}{Q^2}\right)^{\frac{1}{-a}}\right]^k \\
&= n!(-a)^n\left(\frac{e^{-C}\mu^2}{Q^2}\right)^{-1/a} ,
\end{aligned} \tag{49}$$

where $n \to \infty$ was used in the last line. Let us now combine (49) with (46) and determine the number a in (49) from the four terms in (43). Then, asymptotically the series derived from (43) receives four distinct contributions which behave as

$$\begin{aligned}
a_n^{IR1} &\sim -\frac{2C_\mathrm{F}}{3\pi}\left(\frac{e^{-C}\mu^2}{Q^2}\right) n!\beta_0^n , \\
a_n^{IR2} &\sim -\frac{5C_\mathrm{F}}{24\pi}\left(\frac{e^{-C}\mu^2}{Q^2}\right)\left(\frac{1}{2}\right)^n n!\beta_0^n , \\
a_n^{UV1} &\sim -\frac{C_\mathrm{F}}{3\pi}\left(\frac{e^{-C}\mu^2}{Q^2}\right)^{-1} n!(-)^n\beta_0^n , \\
a_n^{UV2} &\sim \frac{C_\mathrm{F}}{24\pi}\left(\frac{e^{-C}\mu^2}{Q^2}\right)^{-2} n!\left(-\frac{1}{2}\right)^n \beta_0^n ,
\end{aligned} \tag{50}$$

where the superscripts refer to the first (IR1) and second (IR2) infrared renormalons and the ultraviolet (UV1, UV2) renormalons.

The obvious question now is: Which term gives the dominant contribution for $n \to \infty$? Immediately one finds that the second renormalons (UV2 and IR2) are suppressed by a factor $(1/2)^n$. On the other hand, the contribution from the UV renormalon (UV1) is smaller than the first IR renormalon by a factor 2. Thus, the IR renormalon yields the most important contribution at $n \to \infty$. Asymptotically the coefficients therefore behave as

$$\frac{a_{n+1}}{a_n} \sim \beta_0(n+1) . \tag{51}$$

Example 5.17

The maximal order n_0 which corresponds to the minimal term in the series (see Fig. 5.38) follows from the condition

$$\left|\frac{a_{n+1}\alpha_s}{a_n}\right| \geq 1, \tag{52}$$

so that

$$n_0 = \frac{1}{\beta_0 \alpha_s} - 1 = \ln\left(Q^2/\Lambda_C^2\right) - 1 \ , \tag{53}$$

where we have used the one-loop definition of $\Lambda_{\rm QCD}$,

$$\alpha_s = \frac{1}{\beta_0 \ln\left(Q^2/\Lambda_{QCD}^2\right)} \ , \tag{54}$$

and introduced the index C ($\Lambda_{\rm QCD}^2 = \Lambda_C^2$) to remind the reader that $\Lambda_{\rm QCD}$ depends on the chosen renormalization scheme.

Setting $Q^2 = \mu^2$ and inserting (57) into (2) one finds the inherent uncertainty $\Delta\mathcal{R}$ of the asymptotic expansion:

$$\begin{aligned}
\Delta\mathcal{R} &= \alpha_s^{n+1} a_{n_0} \\
&= -\frac{2}{3}\frac{C_{\rm F}}{\pi}\beta_0^{\ln(Q^2/\Lambda_C^2)-1}\,\Gamma\left(\ln(Q^2/\Lambda_C^2)\right)\,{\rm e}^{-C}\alpha_s^{\ln(Q^2/\Lambda_C^2)} \\
&= -\frac{2}{3}\frac{C_{\rm F}}{\pi}\frac{1}{\beta_0}\Gamma\left(\ln(Q^2/\Lambda_C^2)\right)\,{\rm e}^{-C}\ln(Q^2/\Lambda_C^2)^{-\ln(Q^2/\Lambda_C^2)} \ .
\end{aligned} \tag{55}$$

Using Sterling's formula for the Γ function,

$$\Gamma(z) = {\rm e}^{-z} z^{-1/2} (2\pi)^{1/2}\left(1 + \frac{1}{12z} + \mathcal{O}\left(1/z^2\right)\right) \ , \tag{56}$$

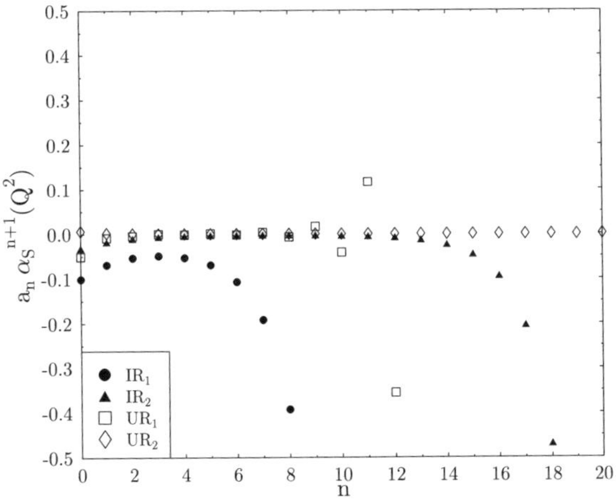

Fig. 5.38. Contributions of *UV* and *IR* renormalons to the Bjorken sum rule, cf. (50). The *n* axis shows the order of expansion. The perturbative expansion is completely dominated by the first *IR* renormalon (marked by the full bullets)

Example 5.17

and therefore

$$\Gamma(z)z^{-z} = \mathrm{e}^{-z}z^{-1/2}(2\pi)^{1/2} + \mathcal{O}\left(1/z^2\right) \ , \tag{57}$$

we obtain for $\Delta\mathcal{R}$

$$\Delta\mathcal{R} = -\frac{2}{3}\frac{C_\mathrm{F}}{\pi}\frac{1}{\beta_0}(2\pi)^{1/2}\left(\frac{\Lambda_C^2\,\mathrm{e}^{-C}}{Q^2}\right)\frac{1}{\sqrt{\ln Q^2/\Lambda_C^2}} + \mathcal{O}\left(\frac{1}{Q^2\ln Q^2/\Lambda_C^2}\right) . \tag{58}$$

As promised at the beginning, the uncertainty in the asymptotic expansion has no logarithmic dependences on the external momentum Q^2 but a power dependence. For large Q^2 it vanishes like $1/Q^2$.

Another more elegant way to find the uncertainty in the expansion is to take the imaginary part of the Borel transform (divided by π). From (44) we get

$$\begin{aligned}
&\frac{Im}{\pi}\int\limits_0^\infty \mathrm{d}s e^{-s/\beta_0\alpha_s}\mathcal{B}\left[\mathcal{R}\right](s) \\
&= \pm\frac{1}{\beta_0}\int\limits_0^\infty \mathrm{d}s e^{-s/\beta_0\alpha_s}(-)\frac{2C_\mathrm{F}}{3\pi}\left(\frac{\mu^2 e^{-C}}{Q^2}\right)^s \delta(1-s) \\
&= \pm\frac{2C_\mathrm{F}}{3\pi}\frac{1}{\beta_0}\left(\frac{\Lambda_C^2 e^{-C}}{Q^2}\right) ,
\end{aligned} \tag{59}$$

which reproduces (58) up to logarithmic corrections.

The ambiguity in the sign of the IR renormalon contribution is due to the two possible contour deformations above or below the pole at $s = 1$. The IR renormalon pole at $s = 2$ leads to $1/Q^4$ corrections:

$$\frac{5C_\mathrm{F}}{12\pi}\frac{1}{\beta_0}\left(\frac{\Lambda_C^2 e^{-C}}{Q^2}\right)^2 . \tag{60}$$

Let us summarize our findings. For a physical quantity like the Bjorken sum rule the perturbative QCD series is not summable, even in the Borel sense. This is due to the appearance of fixed sign factorial growth of its coefficients. It results in a power-suppressed ambiguity of the magnitude $\Lambda_\mathrm{QCD}^2/Q^2$. The ambiguity therefore has exactly the same power behavior as that known from higher-twist corrections.

Since the Bjorken sum rule represents a physical quantity it must have a unique value. The ambiguity apparent in our calculation shows the need to include additional nonperturbative corrections which cancel the ambiguity and give a meaning to the summed perturbative series. These additional nonperturbative corrections are higher-twist corrections.

It can be shown that the higher-twist corrections themselves are ill defined.[36,37] The ambiguity in the definition of the higher-twist corrections can-

[36] E. Stein, M. Meyer-Hermann, et al.: Phys. Lett. B **376**, 177 (1996); E. Stein, M. Maul et al.: Nucl. Phys. B **536**, 318 (1999).

cels exactly the IR renormalon ambiguity in the perturbative series of the twist-2 term. In turn, the investigation of the ambiguities in the definition of the perturbative series of the leading twist shows which higher-twist corrections are needed for the unambigious definition of a physical quantity.

Finally we would like to note that the calculation we have presented here for the Bjorken sum rule can be easily extended to all other structure functions. Such calculations can be easily done for all moments and hence give the IR renormalon ambiguity as a function of x. Since a rigorous operator product expansion for *all* moments of a structure function up to twist 4, or the calculation of the corresponding nucleon matrix elements, is beyond present-day capabilities, the easily calculable IR renormalon ambiguity has proven to be a phenomenological successful model for power corrections. We will not dwell on this issue here but refer the reader to the literature.[36,37]

[37] For a general overview on renormalons see: M. Beneke: Phys. Rept. **317**, 1 (1999).

6. Perturbative QCD II: The Drell–Yan Process and Small-x Physics

6.1 The Drell–Yan Process

Deep inelastic scattering is investigated by shooting electrons or muons at different fixed targets or by colliding them with a proton beam. From these experiments only information about nucleon structure functions can be extracted, while little is learned about the inner structure of pions, for example. There are, however, different options for high-energy collisions. Electrons can be made to collide with positrons (as is done at Stanford, DESY, and CERN), and protons can be made to collide with protons, pions, kaons, or antiprotons. The latter is done at sites such as Fermilab, where secondary beams of pions, antiprotons, etc. are created and collide with a fixed target.

To determine the internal structure of hadrons in these reactions, it is again preferable to consider reactions as hard as possible. For such reactions there is hope that perturbation theory will yield good results. To obtain information about the inner structure of hadrons, these must be in the initial channel. We therefore consider the collision of two hadrons. The process most similar to deep inelastic scattering in these reactions is the annihilation of a quark and an antiquark, each deriving from a different hadron, into a lepton pair (see Fig. 6.1). This is called the Drell–Yan process. Bear in mind that the creation of hadrons in e^+e^- scattering is described by the very same, time-reversed graph. While quark annihilation into leptons tells us about the initial momentum distribution of the quarks inside the hadrons involved, e^+e^- annihilation is used to investigate how full-blown hadrons are created from just two dissociating quarks, that is, how the momentum originally carried by the quark–antiquark pair that has been created is distributed between real and virtual constituents of all created hadrons. This is mainly a dynamical problem in which nonperturbative effects are crucial. Its description in the framework of QCD is of the utmost difficulty and has consequently not yet proceeded beyond phenomenological models. Keywords characterizing this area of research are "hadronic strings" and "jet physics".

Fig. 6.1. The Drell–Yan process

We shall first address the Drell–Yan process of Fig. 6.1. By carrying over our experience of deep inelastic scattering to the Drell–Yan process, we expect the cross section to behave like

$$\begin{aligned} d\sigma(\mathrm{DY}) =& \left[q_a(x_a)\bar{q}_b(x_b) + q_b(x_b)\bar{q}_a(x_a)\right] \\ & \times \hat{\sigma}(\mathrm{q}+\bar{\mathrm{q}} \to \gamma^* \to \mu^+\mu^-)\, dx_a\, dx_b \ . \end{aligned} \tag{6.1}$$

Here a designates the first and b the second hadron. Consequently $q_a(x_a)$ gives the probability of finding a quark of flavor q carrying the momentum fraction x_a, in the first hadron. This is based on the expectation that at sufficiently high energies all color interactions will be suppressed by the large denominator of a propagator. However, this argument has to be treated with great caution, since there are also soft processes that cannot be neglected (see below). Another problematic point is the kinematic difference between Drell–Yan and deep inelastic scattering. While in deep inelastic scattering the photon momentum is $q^2 = -Q^2 < 0$ with a large Q^2, now $q^2 = M^2 > 0$ holds with M the invariant mass of the lepton pair. An extended formal analysis has shown that such formal continuation is possible and (6.1) can be justified. Basically Q^2 is substituted by $-M^2$ in all formulas. This, however, quite substantially changes all logarithmic terms: $\ln(Q^2/m) \to \ln(-M^2/m^2) = \mathrm{i}\pi + \ln(M^2/m^2)$. The running coupling constant still has its usual form with $\alpha(Q^2) \to \alpha(M^2)$, but in practical calculations the perturbative expansion does not converge well owing to the term proportional to $\mathrm{i}\pi$. It must be conceded, therefore, that the analysis of Drell–Yan reactions is afflicted with far more uncertainties than deep inelastic scattering. All this will be discussed in more detail below.

First we shall evaluate (6.1) further while skipping the more involved questions. In doing this, it is customary to define some new quantities. Let $s = (P_a + P_b)^2$ be the invariant mass of the colliding hadrons. The momenta of the partons participating in the Drell–Yan process are $p_a = x_a P_a$, $p_b = x_b P_b$. Neglecting hadron and quark masses, the invariant mass of the lepton pair is then

$$\begin{aligned} M^2 &= (p_a + p_b)^2 = 2\, p_a \cdot p_b = 2x_a x_b\, P_a \cdot P_b = x_a x_b (P_a + P_b)^2 \\ &= x_a x_b s \ . \end{aligned} \tag{6.2}$$

It is furthermore standard to change the variables x_a and x_a to

$$\tau = x_a x_b = \frac{M^2}{s} \ , \qquad y = \frac{1}{2} \ln\left(\frac{x_a}{x_b}\right) \ . \tag{6.3}$$

Keep in mind that for the energy and longitudinal momentum of the lepton pair

$$E = x_a E_a + x_b E_b \ , \qquad P_L = x_a E_a - x_b E_b$$

are valid. Since the incoming partons are parallel to the indicent protons and therefore do not have transverse momenta, we obtain in the center-of-momentum system of the two hadrons, $E_a = E_b = \sqrt{s}/2$,

$$y = \frac{1}{2} \ln\left(\frac{E + P_L}{E - P_L}\right) \ .$$

Thus y is the rapidity of the lepton pair in the center-of-momentum system. To rewrite (6.1) in the new variables, we need the functional determinant

$$\frac{\partial(\tau, y)}{\partial(x_a, x_b)} = 1 \ . \tag{6.4}$$

Finally we must know the elementary cross section $\mathrm{q}+\bar{\mathrm{q}} \rightarrow \mu^{+}+\mu^{-}$. Calculating this is very easy and will be done in Exercise 6.1. We obtain

$$\sigma(\mathrm{q}+\bar{\mathrm{q}} \rightarrow \mu^{+}+\mu^{-})=\frac{1}{3} \frac{4 \pi \alpha^{2}}{3 M^{2}} . \tag{6.5}$$

The additional factor $1/3$ stems from averaging over quark colors, since the structure functions of the nucleon are already known.

The Drell–Yan cross section for quark–antiquark annihilation is thus to lowest order

$$\frac{\mathrm{d}^{2} \sigma_{\mathrm{DY}}}{\mathrm{d} \tau\, \mathrm{d} y}=\frac{4 \pi \alpha^{2}}{9 M^{2}} \sum_{q} e_{q}^{2}\left[q_{a}(x_{a}) \bar{q}_{b}(x_{b})+q_{b}(x_{b}) \bar{q}_{a}(x_{a})\right] \tag{6.6}$$

or, with

$$\frac{\mathrm{d} y}{\mathrm{d} x_{a}}=\frac{1}{2 x_{a}}, \quad M^{2}=\tau s ,$$

$$\sigma_{\mathrm{DY}}=\frac{4 \pi \alpha^{2}}{9} \int_{0}^{1} \mathrm{d} \tau \int_{\tau}^{1} \frac{\mathrm{d} x_{a}}{2 x_{a}} \frac{1}{\tau s} \sum_{q} e_{q}^{2}\left[q_{a}(x_{a}) \bar{q}_{b}\left(\frac{\tau}{x_{a}}\right)+q_{b}\left(\frac{\tau}{x_{a}}\right) \bar{q}_{a}(x_{a})\right] . \tag{6.7}$$

To calculate Drell–Yan cross sections one has in addition to calculate the Compton process $q+G \rightarrow \gamma^{*}+q \rightarrow \mu^{+} \mu^{-}+q$, which is usually even more important. We shall not do this, since all the problems encountered are exactly the same for both processes. We shall simply state the corresponding result for the Compton process at the very end.

Equation (6.7) and the corresponding Compton cross section yield unique prediction for lepton-pair production in high-energy proton–proton collisions. When this was measured, it turned out that the experimental cross section exceeded the prediction by a factor of about two. This missing factor, i.e., the quotient of the measured cross section over the calculated one, is termed the "K factor". Surprisingly it is independent of x_a and x_b, i.e., a true constant. This can, in the framework of perturbative QCD, only be due to higher-order corrections. Since such higher-order perturbative corrections are of the same order as the lowest-order contribution, convergence of the perturbation series is exceedingly questionable.

To gain a deeper understanding of the problems in Drell–Yan reactions, in the following we shall discuss the calculation of some corrections to the annihilation graph $\mathrm{q}+\bar{\mathrm{q}} \rightarrow \gamma^{*}$. The relevant processes are displayed in Fig. 6.2. However, we can avoid much work by using some results from Chap. 5. The graphs in Fig. 6.3 differ from those in Fig. 6.2 only by the exchange of the Mandelstam variables s and u. But the graphs in Fig. 6.3 are just those we calculated in Chap. 5, namely the term with $\mu=\mu'$. Accordingly one gets for the processes in Fig. 6.3 the result (see (5.12))

$$F^{2}=S_{\mu}^{\mu}=-8\left(\frac{s}{t}+\frac{t}{s}+\frac{2 M^{2} u}{s t}\right) . \tag{6.8}$$

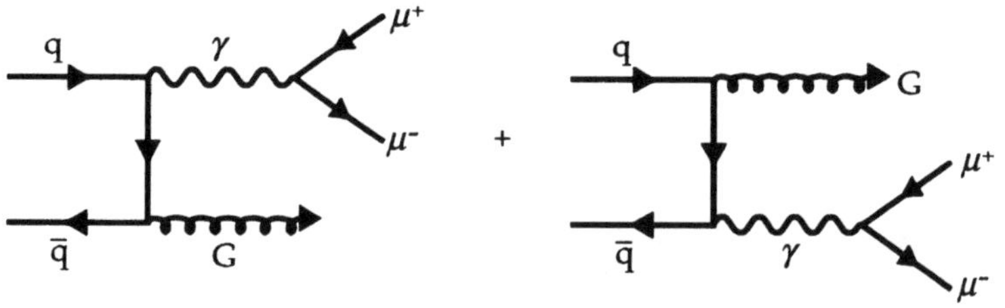

Fig. 6.2. First-order gluonic corrections to the Drell–Yan process

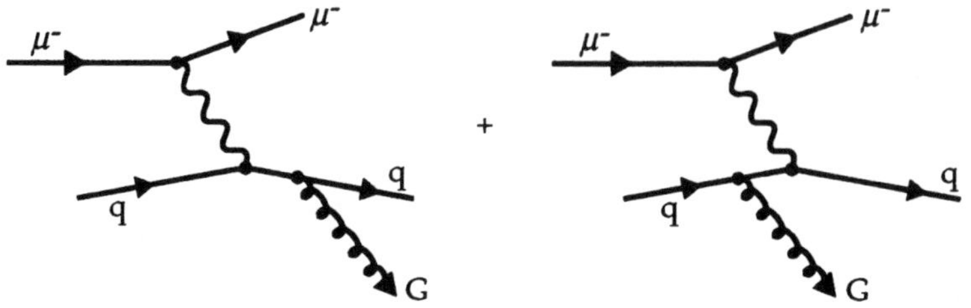

Fig. 6.3. First-order gluonic corrections to the muon–quark scattering

Here we have replaced $-Q^2$ by M^2. We must then, according to the "crossing symmetry", exchange s and u:

$$F^2 = 8\left(\frac{u}{t}+\frac{t}{u}+\frac{2M^2 s}{ut}\right) . \tag{6.9}$$

Here an additional minus sign had to be taken into account. This phase is related to the definition of particle states (we replaced a quark by an antiquark). Now the combinatorial factors remain to be determined. The different charges give $e^2 e_q^2 g^2$, spin averaging yields 1/4, and color averaging

$$\sum_a \frac{1}{9}\sum_{c,c'}\left(\frac{\lambda^a}{2}\right)_{cc'}\left(\frac{\lambda^a}{2}\right)_{c'c} = \frac{8}{9}\frac{1}{2} = \frac{4}{9} .$$

We therefore have

$$\mathrm{d}\sigma' = \frac{1}{2s}\frac{\mathrm{d}^3 q_1}{(2\pi)^2 s} e^2 e_q^2 g^2 \frac{8}{9}\delta\left(q_1^0+q_2^0-p_1^0-p_2^0\right)\left(\frac{u}{t}+\frac{t}{u}+\frac{2M^2 s}{ut}\right) . \tag{6.10}$$

With $\mathrm{d}t = \mathrm{d}\Omega|\boldsymbol{q}|^2/\pi$ (Exercise 2.6, (26)), we obtain

$$\begin{aligned}\mathrm{d}\sigma' &= \frac{1}{2s}\frac{\mathrm{d}t}{8\pi s}\frac{8}{9}\alpha\alpha_s e_q^2\, 16\pi^2\left(\frac{u}{t}+\frac{t}{u}+\frac{2M^2 s}{ut}\right)\\ &= \frac{\alpha\alpha_s e_q^2\pi}{s^2}\frac{8}{9}\left(\frac{u}{t}+\frac{t}{u}+\frac{2M^2 s}{ut}\right)\mathrm{d}t .\end{aligned} \tag{6.11}$$

It should be noted that the relation between the Mandelstam variables connects u, s, and t:

$$u = M^2 - s - t .$$

The additional contribution to the cross section is thus, at given s, a function of t only. We find that

$$M^2 - s \leq t \leq 0$$

holds and that the integral diverges at the upper bound of the integration, i.e., for $t \to 0$. The appearance of infrared divergences is typical for all calculations of perturbative QCD. Its reason is that in the framework of the parton model all small dimensional quantities like quark masses and gluon virtualities are neglected. These quantities normally inhibit the appearance of infrared divergences. The simplest solution to the problem would therefore be to introduce finite masses and virtualities for all particles. But then the results would exhibit a strong dependence on these parameters and thus be more or less meaningless. The only chance of getting reliable predictions lies in the hope that the infrared divergences of different contributions cancel. This is indeed what happens.

At this point it is useful to recall QED bremsstrahlung, where similar problems surface owing to the vanishing mass of the photon. It turned out there that the infrared divergence canceled with the diverging radiative corrections. This effect, known as the Bloch–Nordsieck theorem[1], is of basic relevance for the consistency of QED, and even more so for that of QCD. Its physical background is related to the fact that massless particles with arbitrarily small energy, e.g., photons of infinitely long wavelengths, are, strictly speaking, unphysical since they cannot be detected by any means. The transition from some state to the same state with an additional undetectable photon is not well defined. In perturbation theory, however, all states can be classified by their occupation numbers and such states can be strictly distinguished. Since this distinction is unphysical, it can very well lead to spurious divergences in different contributions that cancel each other.

Next we face the question of how a radiative correction whose amplitude is proportional to g^2 can cancel an amplitude that is of first order in the coupling. This is explained by noting that all radiative correction graphs interfere with the lowest-order graph since they have the same initial and final states. Fig. 6.4 depicts this.

The validity of the Bloch–Nordsieck theorem is of great relevance for the whole of perturbative QCD. Nonetheless a general complete proof seems still to be missing and is clearly confronted with problems for the Drell–Yan process in particular. We are unable to explore this theoretical question in more detail here, but the Bloch–Nordsieck theorem holds for the lowest-order term, investigated in the following, as will be shown.

The procedure to show that two divergences cancel is as follows. First a regulator is introduced to render the contributions finite. Then, after summing the contibutions, we send the regulator to zero or infinity depending on the nature of the divergences (infrared or ultraviolet).

In our case we choose a finite gluon mass m_{G} as a regulator. Then the calculation of the graphs in Fig. 6.2 must be repeated with a finite gluon mass. We

[1] F. Block and A. Nordsieck: Phys. Rev. **52**, 54 (1937)

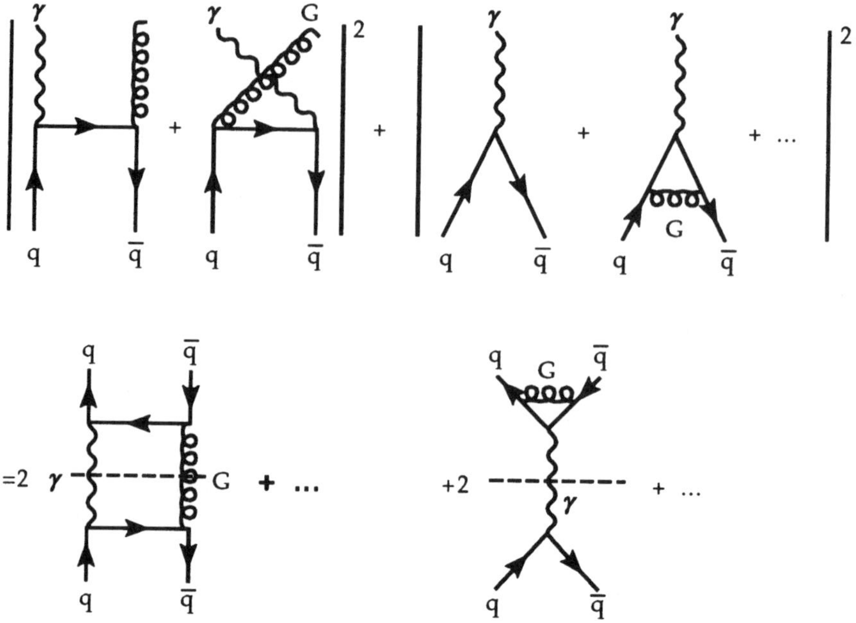

Fig. 6.4. All graphs leading to the same final state interfere with one another. Therefore both photon + gluon graphs and radiative corrections contribute to the order α_s. The Bloch–Nordsieck theorem states that infrared divergencies of these contributions cancel

shall skip this and give the final result only. The result is

$$\begin{aligned} & \mathrm{d}\sigma'(\mathrm{q}+\bar{\mathrm{q}} \to \gamma^* + G) \\ & = \frac{\alpha\alpha_s e_q^2 \pi}{s^2} \frac{8}{9} \left[\frac{u}{t} + \frac{t}{u} + \frac{2(M^2 + m_{\mathrm{G}}^2)s}{ut} - M^2 m_{\mathrm{G}}^2 \left(\frac{1}{t^2} + \frac{1}{u^2} \right) \right] \mathrm{d}t \end{aligned} \tag{6.12}$$

with

$$u = M^2 + m_{\mathrm{G}}^2 - s - t$$

and

$$\mathrm{d}^2\sigma''_{\mathrm{DY}}(\mathrm{q}+\bar{\mathrm{q}} \to \mu^+ + \mu^- + G) = \frac{\alpha}{3\pi M^2} \, \mathrm{d}\sigma'(\mathrm{q}+\bar{\mathrm{q}} \to \gamma^* + G) \, \mathrm{d}M^2 \; .$$

Obviously the changes in the cross section are minor. In particular, the troublesome terms with $1/t$ are unchanged, and an even more troublesome term in $1/t^2$ has appeared. The crucial question is thus: what are the bounds of the integral for finite m_{G}? We choose the center-of-momentum frame to calculate these bounds. Here $\boldsymbol{k}_\gamma = -\boldsymbol{k}_{\mathrm{G}} = \boldsymbol{k}$, and thus

$$s = (E_\gamma + E_{\mathrm{G}})^2 = k^2 + M^2 + 2\sqrt{k^2 + M^2}\sqrt{k^2 + m_{\mathrm{G}}^2} + k^2 + m_{\mathrm{G}}^2 \; , \tag{6.13}$$

$$(s - M^2 - m_{\mathrm{G}}^2 - 2k^2)^2 = 4(k^2 + M^2)(k^2 + m_{\mathrm{G}}^2) \; . \tag{6.14}$$

From this, k is calculated:

$$k = \sqrt{\frac{(s - M^2 - m_{\mathrm{G}}^2)^2 - 4M^2 m_{\mathrm{G}}^2}{4s}} \ . \tag{6.15}$$

Bounds for t are calculated from

$$t = (p_q - k_\gamma)^2 = M^2 - 2(E_\gamma E_q - E_q k \cos\theta) \ , \tag{6.16}$$

$$t_{\mathrm{min/max}} = M^2 - 2E_\gamma E_q \mp 2E_q k \ . \tag{6.17}$$

We put $2E_q = \sqrt{s}$ and we use (6.15):

$$\begin{aligned} t_{\mathrm{min/max}} &= M^2 - \sqrt{\frac{s}{4s}} \left(\sqrt{(s - M^2 - m_{\mathrm{G}}^2)^2 - 4M^2 m_{\mathrm{G}}^2 + 4sM^2} \right. \\ &\qquad \left. \pm \sqrt{(s - M^2 - m_{\mathrm{G}}^2)^2 - 4M^2 m_{\mathrm{G}}^2} \right) \\ &= M^2 - \frac{1}{2} \left(\sqrt{(s + M^2 - m_{\mathrm{G}}^2)^2} \pm \sqrt{(s - M^2 - m_{\mathrm{G}}^2)^2 - 4M^2 m_{\mathrm{G}}^2} \right) \\ &= -\frac{1}{2} \left((s - M^2 - m_{\mathrm{G}}^2) \pm \sqrt{(s - M^2 - m_{\mathrm{G}}^2)^2 - 4M^2 m_{\mathrm{G}}^2} \right) \ . \end{aligned} \tag{6.18}$$

Obviously t_{max} is now nonzero, so the integral is finite. As claimed above, t_{max} goes to zero as $m_{\mathrm{G}} \to 0$. This result can be rewritten by introducing the quantities $a = M^2/s$ and $b = m_{\mathrm{G}}^2/M^2$:

$$t_{\mathrm{min/max}} = -\frac{M^2}{2a} \left(1 - a - ab \pm \sqrt{(1-a)^2 + ba(ba - 2 - 2a)} \right) \ . \tag{6.19}$$

The integral over t is explicitly calculated in Exercise 6.2, leading to

$$\begin{aligned} \sigma' = \frac{8\pi\alpha\alpha_s e_q^2}{9s} &\left[2\frac{1 + B^2/s^2}{1 - B/s} \ln \left(\frac{s - B + \sqrt{(s-B)^2 - 4M^2 m_{\mathrm{G}}^2}}{s - B - \sqrt{(s-B)^2 - 4M^2 m_{\mathrm{G}}^2}} \right) \right. \\ &\left. - 4\sqrt{\left(1 - \frac{B}{s}\right)^2 - 4\frac{M^2 m_{\mathrm{G}}^2}{s^2}} \right] \end{aligned} \tag{6.20}$$

with

$$B = M^2 + m_{\mathrm{G}}^2 \ .$$

To get the total Drell–Yan contribution we have to substitute $\tau s = x_a x_b s$ for s, where s is now the total four-momentum squared of the colliding hadrons. We also have to insert the relation between $\sigma(\mathrm{q} + \bar{\mathrm{q}} \to \gamma^*)$ and $\mathrm{d}\sigma/\mathrm{d}M^2(\mathrm{q} + \bar{\mathrm{q}} \to \gamma^* \to \mu^+\mu^-)$, which is derived in Exercise 6.3. For negligible lepton masses this is

$$\frac{\mathrm{d}\sigma}{\mathrm{d}M^2}(\mathrm{q} + \bar{\mathrm{q}} \to \mu^+\mu^-) = \sigma(\mathrm{q} + \bar{\mathrm{q}} \to \gamma^*) \frac{\alpha}{3\pi M^2} \mathrm{d}M^2 \ . \tag{6.21}$$

In our problem, i.e., the first-order gluon correction, M^2 corresponds to the invariant mass of the γ^* and gluon. We therefore replace it by τs. Thus the correction to the annihilation part of the Drell–Yan cross section is

$$\begin{aligned}\frac{\mathrm{d}\Delta\sigma_{\mathrm{DY}}}{\mathrm{d}M^2} &= \frac{\alpha}{3\pi}\int \mathrm{d}x_b \int \mathrm{d}x_a \frac{\sigma'(s\to\tau s)}{\tau s}\sum_q [q_a(x_a)\bar{q}_b(x_b)+\bar{q}_a(x_a)q_b(x_b)] \\ &= \frac{\alpha}{3\pi}\int \mathrm{d}\tau \int \frac{\mathrm{d}x_a}{x_a}\frac{\sigma'(\tau s)}{\tau s}\sum_q [q_a(x_a)\bar{q}_b(x_b)+\bar{q}_a(x_a)q_b(x_b)] \ .\end{aligned} \tag{6.22}$$

We still have to specify the integration boundaries. Obviously it holds that

$$\tau s \geq (M+m_{\mathrm{G}})^2 \approx M^2\left(1+2\frac{m_{\mathrm{G}}}{M}\right) \ . \tag{6.23}$$

On the other hand τs can be very large. We write

$$\tau s \leq M^2 N \tag{6.24}$$

and later on set $1/N \approx 0$. Because $q_a(x_a)$, $\bar{q}_b(x_b)$, and so on are smooth functions of x_a and x_b, we can neglect the small shifts in x_a and x_b induced by the difference between τs and M^2, i.e., we can neglect the τ dependence of the $x_a x_b$:

$$\begin{aligned}\frac{\mathrm{d}\Delta\sigma_{\mathrm{DY}}}{\mathrm{d}M^2} &= \frac{\alpha}{3\pi}\int_0^1 \frac{\mathrm{d}x_a}{x_a}\sum_q \left[q_a(x_a)\bar{q}_b(x_b)+\bar{q}_a(x_a)q_b(x_b)\right] \\ &\times \int\limits_{M^2(1+2m_{\mathrm{G}}/M)/s}^{M^2N/s} \mathrm{d}\tau\, \frac{8\pi\alpha\alpha_s e_q^2}{9(\tau s)^2}\left[2\frac{1+\left(\frac{M^2+m_{\mathrm{G}}^2}{\tau s}\right)^2}{1-\frac{M^2+m_{\mathrm{G}}^2}{\tau s}}\right. \\ &\times \ln\left(\frac{1-\frac{M^2+m_{\mathrm{G}}^2}{s\tau}+\sqrt{\left(1-\frac{M^2+m_{\mathrm{G}}^2}{s\tau}\right)^2-4\frac{M^2 m_{\mathrm{G}}^2}{s^2\tau^2}}}{1-\frac{M^2+m_{\mathrm{G}}^2}{s\tau}-\sqrt{\left(1-\frac{M^2+m_{\mathrm{G}}^2}{s\tau}\right)^2-4\frac{M^2 m_{\mathrm{G}}^2}{s^2\tau^2}}}\right) \\ &\left. -4\sqrt{\left(1-\frac{M^2+m_{\mathrm{G}}^2}{s\tau}\right)^2-4\frac{M^2 m_{\mathrm{G}}^2}{s^2\tau^2}}\,\right] \ .\end{aligned} \tag{6.25}$$

The τ integral is expanded in $\beta = m_0^2/M^2$. To this end we substitute $\tau = M^2/sr$:

$$I = \frac{8\pi\alpha\alpha_s e_q^2}{9M^2} \int\limits_{\frac{1}{N}\sim 0}^{1-2\sqrt{\beta}} \mathrm{d}r \left[2\,\frac{1+r^2(1+\beta)^2}{1-r(1+\beta)} \right.$$
$$\times \ln\left(\frac{1-r(1+\beta)+\sqrt{[1-r(1+\beta)]^2-4\beta r^2}}{1-r(1+\beta)-\sqrt{[1-r(1+\beta)]^2-4\beta r^2}} \right)$$
$$\left. -4\sqrt{[1-r(1+\beta)]^2-4\beta r^2} \right] . \tag{6.26}$$

We study the limit $\beta \to \infty$. We shall now take this limit, and in doing so we shall make sure that we keep all the constant terms as well as those proportional to logarithms of β. Powers of β are, however, neglected.

We make the substitution $r \to r/(1+\beta)$. The factors $1/(1+\beta)$ lead only to corrections proportional to β, which we neglect. The leading term, which we have to take care to treat correctly, is that proportional to $2\sqrt{\beta}$.

$$I = \frac{8\pi\alpha\alpha_s e_q^2}{9M^2} \int\limits_{0}^{1-2\sqrt{\beta}} \mathrm{d}r \left[2\,\frac{1+r^2}{1-r} \ln\left(\frac{2(1-r)}{2\beta r^2/(1-r)} \right) \right.$$
$$-2\,\frac{1+r^2}{1-r} \ln\left(\frac{2(1-r)}{1-r+\sqrt{(1-r)^2-4\beta r^2}} \right)$$
$$+2\,\frac{1+r^2}{1-r} \ln\left(\frac{2\beta r^2}{(1-r)^2\left[1-\sqrt{1-4\beta r^2/(1-r)^2}\right]} \right)$$
$$\left. -4\sqrt{(1-r)^2-4\beta r^2} \right] . \tag{6.27}$$

In the terms on the right-hand side it is tempting to neglect $4\beta r^2$. This would be incorrect, however, since it is of the same order as $(1-r)^2$ at the upper integration boundary. To treat this carefully we have split up the logarithmic term into three parts. The first contains the approximation valid for $(1-r)^2 > 4\beta r^2$. The other two terms contain the ratio between the approximate expressions and the exact ones. We investigate these first:

$$I_1 = 2 \int\limits_{0}^{1-2\sqrt{\beta}} \mathrm{d}r\, \frac{1+r^2}{1-r} \left[\ln\left(\frac{2\beta r^2}{(1-r)^2\left[1-\sqrt{1-4\beta r^2/(1-r)^2}\right]} \right) \right.$$
$$\left. -\ln\left(\frac{2(1-r)}{(1-r)+\sqrt{(1-r)^2-4\beta r^2}} \right) \right] . \tag{6.28}$$

Obviously these terms vanish unless $r \approx 1$. Thus we set $r^2 = 1$ except in $(1-r)$. Next we substitute $1-r=t$:

$$I_1 = 4 \int\limits_{2\sqrt{\beta}}^{1} \mathrm{d}t \, \frac{1}{t} \left[\ln\left(\frac{2\beta}{t^2\left(1-\sqrt{1-4\beta/t^2}\right)} \right) - \ln\left(\frac{2}{1+\sqrt{1-4\beta/t^2}} \right) \right] . \tag{6.29}$$

Now we substitute $t = 2\sqrt{\beta}/u$, $\mathrm{d}t = -2\sqrt{\beta}\,\mathrm{d}u\,/u^2 = -t\,\mathrm{d}u\,/u$:

$$I_1 = -4 \int\limits_{1}^{2\sqrt{\beta}} \frac{\mathrm{d}u}{u} \left[\ln\left(\frac{u^2}{2\left(1-\sqrt{1-u^2}\right)} \right) - \ln\left(\frac{2}{1+\sqrt{1-u^2}} \right) \right] . \tag{6.30}$$

In this expression it is now safe to let $\sqrt{\beta}$ go to zero. We are left with finite integrals only. To do these we further substitute $u = \sqrt{1-z^2}$, $\mathrm{d}u = -z\,\mathrm{d}z\,/u$:

$$\begin{aligned} I_1 &= 4 \int\limits_0^1 \mathrm{d}z \left[\ln\left(\frac{1-z^2}{2(1-z)} \right) - \ln\left(\frac{2}{1+z} \right) \right] \frac{z}{1+z^2} \\ &= 2 \int\limits_0^1 \mathrm{d}z \; 2 \ln\left[\frac{1}{2}(1+z) \right] \left(\frac{1}{1-z} - \frac{1}{1+z} \right) \\ &= -2 \ln^2\left(\frac{1+z}{2} \right) \Bigg|_0^1 + 4 \int\limits_0^1 \mathrm{d}z \, \frac{1}{1-z} \ln\left(\frac{1+z}{2} \right) . \end{aligned} \tag{6.31}$$

We finally substitute $z = 2u-1$ and use the fact that

$$-\int\limits_{1-x}^{1} \mathrm{d}t \, \frac{\ln(t)}{1-t} = \mathrm{Li}_2(x) \;, \quad \mathrm{Li}_2\left(\frac{1}{2} \right) = \frac{\pi^2}{12} - \frac{1}{2} \ln^2(2) . \tag{6.32}$$

Here $\mathrm{Li}_2(x) = \sum_{n=1}^{\infty} x^2/n^2$, $|x| < 1$ is the dilogarithm, also called the Struve function,

$$\begin{aligned} I_1 &= 2 \ln^2(2) + 8 \int\limits_{1/2}^{1} \mathrm{d}u \, \frac{1}{2(1-u)} \ln(u) \\ &= 2 \ln^2(2) - 4 \left(\frac{\pi^2}{12} - \frac{1}{2} \ln^2(2) \right) = -\frac{\pi^2}{3} + 4 \ln^2(2) . \end{aligned} \tag{6.33}$$

The last integral in (6.27) is trivial, since one can directly set $\beta = 0$:

$$I_2 = -4 \int\limits_0^{1-2\sqrt{\beta}} \mathrm{d}r \sqrt{(1-r)^2 - 4\beta r^2} \to -4 \int\limits_0^1 \mathrm{d}r \, (1-r) = -2 . \tag{6.34}$$

The remaining integral leads to an expression found in good integration tables:

$$
\begin{aligned}
I_3 &= -2 \int_0^{1-2\sqrt{\beta}} \mathrm{d}r \, \frac{1+r^2}{1-r} \left[\ln(\beta) + 2\ln(r) - 2\ln(1-r) \right] \\
&= -2\ln(\beta) \int_0^{1-2\sqrt{\beta}} \mathrm{d}r \, \frac{(r-1)^2 + 2(r-1) + 2}{1-r} \\
&\quad - 4 \int_0^{1-2\sqrt{\beta}} \mathrm{d}r \, \frac{1+r^2}{1-r} \ln(r) + 4 \int_0^{1-2\sqrt{\beta}} \mathrm{d}r \, \frac{1+r^2}{1-r} \ln(1-r) \\
&= -2\ln(\beta) \left[\frac{1}{2} - 2 - 2\ln\left(2\sqrt{\beta}\right) + O\left(\sqrt{\beta}\right) \right] \\
&\quad - 4\left(-\frac{\pi^2}{3} + \frac{5}{4} \right) + 4\left[-\ln^2\left(2\sqrt{\beta}\right) + \frac{7}{4} \right] \\
&= 2 - 4\ln^2\left(2\sqrt{\beta}\right) + \frac{4}{3}\pi^2 + 3\ln(\beta) + 4\ln(\beta)\,\ln\left(2\sqrt{\beta}\right) \\
&= \frac{4}{3}\pi^2 + 2 + 3\ln(\beta) + 4\ln\left(2\sqrt{\beta}\right)\left[\ln(\beta) - \ln\left(2\sqrt{\beta}\right) \right] \\
&= \frac{4}{3}\pi^2 + 2 + 3\ln(\beta) + 4\left[\ln\left(\sqrt{\beta}\right) + \ln(2) \right]\left[\ln\left(\sqrt{\beta}\right) - \ln(2) \right] \\
&= \frac{4}{3}\pi^2 + 2 + 3\ln(\beta) + 4\left[\ln^2\left(\sqrt{\beta}\right) - \ln^2(2) \right] \\
&= \ln^2(\beta) + 3\ln(\beta) + \frac{4}{3}\pi^2 + 2 - 4\ln^2(2) \, .
\end{aligned} \tag{6.35}
$$

Finally we put the results (6.33), (6.34), and (6.35) together to get

$$
I = \left(\ln^2(\beta) + 3\ln(\beta) + \pi^2 \right) \frac{8\pi\alpha\alpha_s e_q^2}{9M} \, , \tag{6.36}
$$

or

$$
\begin{aligned}
\frac{\mathrm{d}\Delta\sigma_{\mathrm{DY}}}{\mathrm{d}M^2} &= \frac{\alpha}{3\pi} \int_0^1 \frac{\mathrm{d}x_a}{x_a} \sum_q \left[q_a(x_a)\bar{q}_b(x_b) + \bar{q}_a(x_a) q_b(x_b) \right] \\
&\quad \times \frac{8\pi\alpha\alpha_s e_q^2}{9M^2} \left[\ln^2(\beta) + 3\ln(\beta) + \pi^2 \right] \, .
\end{aligned} \tag{6.37}
$$

We hope that this explicit example shows how the logarithms associated with the infrared divergences can be isolated and where one has to be careful to avoid errors. We shall next discuss how the Block–Nordsieck theorem works for this specific example, i.e., how the logarithmic terms in (6.37) are cancelled by those of the radiative corrections. However, we shall give only an overview of the explicit calculation of the radiative corrections, since it contains nothing new.

Fig. 6.5a,b. First-order radiative vertex corrections

According to Fig. 6.4 the graphs we have to calculate are those contributing to the vertex correction to first order (see Fig. 6.5). If we introduce a momentum cutoff Λ, after the introduction of Feynman parameters the two relevant integrals are, for the quark self-energy,

$$\Sigma(p) = -g_s^2 \int\limits_{m_g^2}^{\Lambda^2} \mathrm{d}s \int\limits_0^1 \mathrm{d}x \int \frac{\mathrm{d}^4 k}{(2\pi)^4} \frac{2(1-x)\left[-2(1-x)\right]}{\left[k^2 + p^2 x(1-x) - s(1-x)\right]^3}$$

$$= \ldots = -\frac{\alpha_s}{4\pi} \ln\left(\frac{\Lambda^2}{m_g^2}\right) \tag{6.38}$$

and, for the vertex correction,

$$\alpha \to \alpha \left(1 + \frac{2\alpha_s}{3\pi} \int\limits_{m_g^2}^{\Lambda^2} \mathrm{d}s \int\limits_0^1 \mathrm{d}y \int\limits_0^1 \mathrm{d}x \left\{ \frac{x(1-x)q^2\left[2 + 2x^2 y(1-y) - 2x\right]}{\left[-y(1-y)x^2 q^2 + s(1-x)\right]^2} \right.\right.$$

$$\left.\left. + \frac{2x(1-x)}{\left[-y(1-y)x^2 q^2 + s(1-x)\right]} \right\}\right)$$

$$= \alpha \left\{ 1 + \frac{2\alpha_s}{3\pi} \left[-\ln^2\left(\frac{m_g^2}{-q^2}\right) - 3\ln\left(\frac{m_g^2}{-q^2}\right) - \frac{7}{2} - \frac{2\pi^2}{3} \right] \right\} . \tag{6.39}$$

In contrast to deep inelastic scattering the four-momentum squared $\left(q^2 = M^2\right)$ is now positive:

$$\ln\left(\frac{m_g^2}{-q^2}\right) = \mathrm{Re}\left[\ln\left(-\frac{m_g^2}{M^2}\right)\right] = \mathrm{Re}\left[\ln(\beta) - \mathrm{i}\pi\right] = \ln(\beta) \tag{6.40}$$

$$\ln^2\left(\frac{m_g^2}{-q^2}\right) = \mathrm{Re}\left[\ln^2\left(-\frac{m_g^2}{M^2}\right)\right] = \mathrm{Re}\left[\ln^2(\beta) - 2\mathrm{i}\pi\ln(\beta) - \pi^2\right]$$

$$= \ln^2(\beta) - \pi^2 , \tag{6.41}$$

so the correction to α can be written as

$$\alpha \to \alpha\left\{1+\frac{2\alpha_s}{3\pi}\left[-\ln^2(\beta)-3\ln(\beta)-\frac{7}{2}+\frac{\pi^2}{3}\right]\right\} . \tag{6.42}$$

The corresponding correction to the Drell–Yan cross section is

$$\frac{\mathrm{d}\Delta\sigma_{\mathrm{DY}}(\mathrm{rad.-corr.})}{\mathrm{d}M^2} = \frac{\alpha}{3\pi}\int_0^1 \frac{\mathrm{d}x_a}{x_a}\sum_q \left[q_a(x_a)\bar{q}_b(x_b)+\bar{q}_a(x_a)q_b(x_b)\right] \times\frac{8\pi\alpha\alpha_s e_q^2}{9M^2}\left(-\ln^2(\beta)-3\ln(\beta)-\frac{7}{2}+\frac{\pi^2}{3}\right) . \tag{6.43}$$

Adding this result to (6.37) gives the final result:

$$\frac{\mathrm{d}\Delta\sigma_{\mathrm{DY}}(\mathrm{total})}{\mathrm{d}M^2} = \frac{2\alpha_s}{3\pi}\int_0^1 \frac{\mathrm{d}x_a}{x_a}\sum_q \left[q_a(x_a)\bar{q}_b(x_b)+\bar{q}_a(x_a)q_b(x_b)\right] \times\frac{4\pi\alpha^2 e_q^2}{9M^2}\left(\frac{4}{3}\pi^2-\frac{7}{2}\right) , \tag{6.44}$$

which implies that the Drell–Yan cross section is just multiplied by a constant factor, which can be identified with the K factor

$$K(\text{1st order}) = 1+\frac{2\alpha_s}{3\pi}\left(\frac{4\pi^2}{3}-\frac{7}{2}\right) = 1+2.05\,\alpha_s . \tag{6.45}$$

If we insert $\alpha_s \approx 0.3$, we get $K \approx 1.6$, which is already a good step towards the experimental value $K \approx 2$. Thus it can be hoped that the perturbative expression will indeed converge to the correct K factor. One can continue by pursuing this tedious calculation order by order. Instead we wish to address a much more fundamental approach, namely that it is often possible to sum up certain radiative corrections exactly, i.e., to all orders in the coupling constant. Such techniques are called "resummation techniques" and are a central issue of current research in QCD. The idea is very simple. Sometimes corrections can be iterated and it can be proven that the iteration has a very simple form, i.e. assumes the form of an exponential or a geometric series. A geometric series we know already from QED, where we resummed fermion bubbles. We have depicted the resummation of fermion bubbles in Fig. 6.6. In the following, we will resum an exponential series. The result will be that K is proportional to $\exp[2\pi\alpha_s(M^2)/3]$, implying that

$$\begin{aligned} K &= \exp\left[\frac{2}{3}\pi\alpha_s\left(M^2\right)\right]\left[1-\frac{2}{3}\pi\alpha_s+\frac{2\alpha_s}{3\pi}\left(\frac{4\pi^2}{3}-\frac{7}{2}\right)\right] \\ &= \exp\left[\frac{2}{3}\pi\alpha_s\left(M^2\right)\right]\left[1-\alpha_s\left(\frac{2\pi}{3}-\frac{8\pi}{9}+\frac{7}{3\pi}\right)\right] \\ &= \exp\left[\frac{2}{3}\pi\alpha_s\left(M^2\right)\right]\left(1-0.0446\,\alpha_s\right) . \end{aligned} \tag{6.46}$$

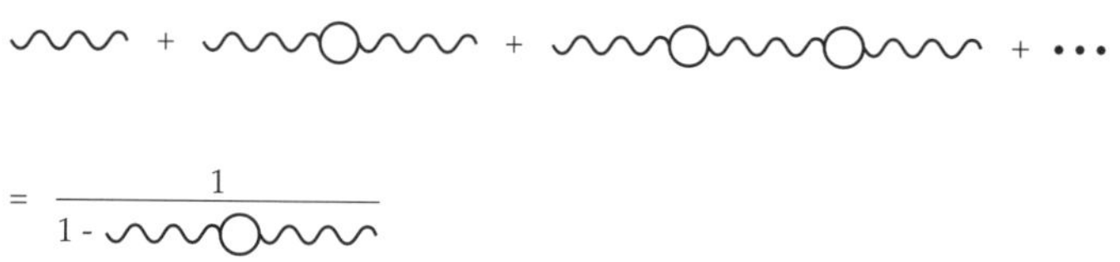

Fig. 6.6. Resummation of fermion bubblesin QED

This suggests an excellent convergence of the perturbative expansion *after* the resummed exponential factor is isolated. Furthermore the numerical result for $\alpha_s \approx 0.25-0.35$,

$$\begin{aligned} K(\text{1st order + resummation}) &= \exp\left[\frac{2}{3}\pi\alpha_s\left(M^2\right)\right](1-0.0446\,\alpha_s) \\ &= 1.7-2.0\ , \end{aligned} \tag{6.47}$$

agrees well with the empirical values. In fact (6.47) is further improved by also taking into account the second order and serves then to determine $\alpha_s(M^2)$ or Λ_{QCD}. The starting point of the argument is as follows. The terms absorbed into the exponent are the π^2 terms in (6.43) and (6.44). According to (6.40) and (6.41) these arise from the analytic continuation of the logarithm to negative arguments. Thus it is sufficient to keep only the logarithmic terms to high orders, to sum them up, and to continue them then to negative arguments. This procedure is called "leading-logarithm approximation" (LLA) and is the standard procedure for resummation. Next we shall discuss the correction due to gluon emission in the production of a quark–antiquark pair by a massive photon. The resummed corrections for the Drell–Yan process can then be obtained from the result by analytic continuation. The calculation can be found in Exercise 6.4. Here we simply give the result for the decay rate Γ of a massive photon to a quark–antiquark pair + photon:

$$\frac{\mathrm{d}^2\Gamma}{\mathrm{d}x_1\,\mathrm{d}x_2} = 3\alpha e_q^2 M\frac{32\alpha_s}{3\pi}\frac{x_1^2+x_2^2}{(1-2x_1)(1-2x_2)}\ , \tag{6.48}$$

where x_1 and x_2 are the energy fractions carried by the quark and antiquark. The gluon energy fraction is $x_3 = 1-x_1-x_2$. We treat the decay of the massive photon in its rest frame, i.e., $k_\gamma = (M,0,0,0)$. Obviously this expression diverges if x_1 or x_2 approaches 1/2. This is obvious from Fig. 6.7 for the case $x_1 \to 1/2$. The intermediate propagator becomes on-mass-shell, if

$$\begin{aligned} (k-p_1)^2 &= (M(1-x_1),0,0,-x_1M)^2 = M^2((1-x_1)^2-x_1^2) \\ &= M^2(1-2x_1) = 0\ . \end{aligned} \tag{6.49}$$

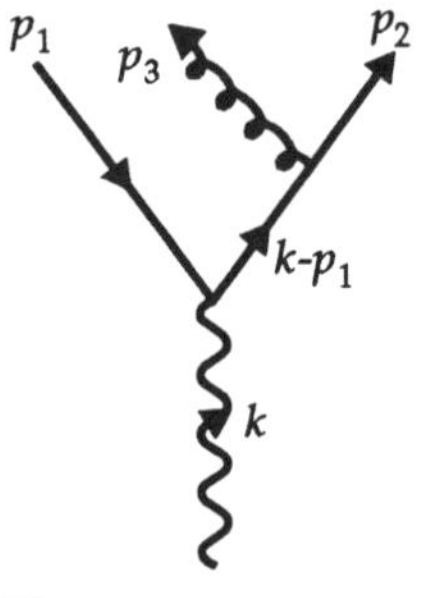

Fig. 6.7. Illustration of the origin of divergences in (6.48)

For a finite gluon energy only x_1 or x_2 can be equal to 1/2. Thus (6.48) has two leading, logarithmically diverging contributions: one for $x_1 \to 1/2$ and one for $x_2 \to 1/2$. We change to the coordinates

$$\begin{aligned} t &= (p_2+p_3)^2 = (k-p_1)^2 = M^2(1-2x_1)\ , \\ z &= \left.\frac{E_2}{E_2+E_3}\right|_{x_1\to\frac{1}{2}} = \frac{E_2}{\frac{1}{2}M} = 2x_2\ , \end{aligned} \tag{6.50}$$

$$4M^2 \frac{d^2}{dx_1\, dx_2} \Rightarrow \frac{d^2}{dt\, dz} \ , \tag{6.51}$$

$$\begin{aligned} \frac{d^2\Gamma}{dt\, dz} &= 3\alpha e_q^2 \frac{2\alpha_s}{3\pi} \frac{1+z^2}{t(1-z)} \\ &= 3\alpha e_q^2 \frac{\alpha_s}{2\pi t}\left(\frac{4}{3}\frac{1+z^2}{1-z}\right) \ . \end{aligned} \tag{6.52}$$

The new variable z has obviously the meaning of the fraction of the energy of the original quark state carried by the final quark state after gluon emission. Therefore it is not suprising that we recover in (6.51) again the quark-splitting function $P_{qq}(z)$ known from our derivation of the GLAP equations. As usual the virtual corrrection adds effectively the '+' prescription to the bracket in (6.52). Let us now turn to multigluon emission. An important point to know is that for the specific gauge we used in Exercise 6.4 only a single amplitude contributes, namely S_{22}. The sum of the three terms $S_{11}+S_{22}+S_{12}$ is of course independent of the choice of gauge. However, choosing an appropriate gauge we can move the contributions of each single diagram around, so that, as done in our case, only S_{22} contains a logarithmic divergence, i.e. we have chosen a gauge where interference terms are absent (see figure).

The t integration leads to a logarithmic divergence. With (6.50) we get

$$\begin{aligned} S_{11} &= 8c\frac{1-2x_1}{1-x_2} = 8c\frac{t}{M^2(1-z)} && \text{(no logarithmic divergence)}, \\ S_{22} &= 8c\frac{(1+z^2)M^2}{t(1-z)} && \text{(logarithmic divergence)}, \\ S_{12} &= -16c\frac{1}{1-2x_2} = -\frac{16c}{1-z} && \text{(no logarithmic divergence)}, \\ c &= 4g^2e^2e_{\mathrm{q}}^2 \ . \end{aligned} \tag{6.53}$$

Thus for the leading term the two-gluon emission probability is just the product of two one-gluon emission probabilities. There are no interference effects. This fact is intimately connected to the form of the GLAP equations. If interference effects were important, one would not get such a simple equation on the level of distribution functions, i.e. probabilities. Therefore in leading logarithmic approximation one simply gets

$$\frac{d^n\Gamma}{dz_1 \dots dz_n\, dt_1 \dots dt_n} = 3\alpha e_q^2 \left(\frac{\alpha_s(t_1)}{2\pi t_1} P_{qq}(z_1)\right) \dots \left(\frac{\alpha_s(t_n)}{2\pi t_n} P_{qq}(z_n)\right) \ . \tag{6.54}$$

We define the t_i such that $t_1 \geq t_2 \geq \dots \geq t_n$. Note that for our specific gauge only those graphs in which the quark lines couple to gluons contribute in the LLA. Graphs in which the antiquark couples to the gluons give no leading-log contribution. This fact has no deep physical meaning but is exclusively due to the specific gauge used. Equation (12) in Exercise 6.4 contains p_2/x_2 and not p_1/x_1.

This asymmetry generates the asymmetry in the contributing processes. Now let us define $S(\Theta)$ to be the probability that after n-gluon emissions the outgoing quark still has a scattering angle $p_\perp/p_\parallel < \Theta$, with $p_\parallel$ and $p_\perp$ being defined with respect to the jet axis, and the original quark momentum, respectively. For the following we need the relation between t and the scattering angle θ. With

$$p_2 = \left(z(1-x_1)M,\, p_\perp,\, \sqrt{z^2(1-x_1)^2M^2 - p_\perp^2} \right) , \tag{6.55}$$

$$p_3 = \left((1-z)(1-x_1)M,\, -p_\perp,\, \sqrt{(1-z)^2(1-x_1)^2M^2 - p_\perp^2} \right) \tag{6.56}$$

we find that [using $p_\perp^2 \ll (1-z)^2(1-x_1)^2M^2 \sim (1-z_1)^2M^2/4$]

$$\begin{aligned} t = 2\, p_2 \cdot p_3 = 2 \Bigg\{ & z(1-z)(1-x_1)^2M^2 + p_\perp^2 \\ & - \left[z(1-x_1)M - \frac{p_\perp^2}{2z(1-x_1)M} \right] \\ & \times \left[(1-z)(1-x_1)M - \frac{p_\perp^2}{2(1-z)(1-x_1)M} \right] \Bigg\} \\ \approx 2 & \left(p_\perp^2 + \frac{p_\perp^2(1-z)}{2z} + \frac{p_\perp^2 z}{2(1-z)} \right) \approx \frac{p_\perp^2}{(1-z)} . \end{aligned} \tag{6.57}$$

We are interested in typical bremsstrahlung processes, i.e., $z \approx 1$. For $z \to 0$, multigluon emission plays no role. The angle is related to $p_\perp$ by

$$\vartheta \approx \frac{p_\perp}{p_\parallel} \simeq \frac{p_\perp}{\left(z(1-x_1)M - \frac{p_\perp^2}{2z(1-x_1)M} \right)} \simeq \frac{p_\perp}{z(1-x_1)M} \simeq \frac{2p_\perp}{M} < \Theta \tag{6.58}$$

$$\Rightarrow p_\perp^2 < \frac{\Theta^2}{4} M^2 . \tag{6.59}$$

Remember that the leading-logarithm approximation (LLA) we are investigating is only valid for $x_1 \approx 1/2$. As a consequence of (6.58) we find that

$$(1-z) < \frac{\Theta^2 M^2}{4t} . \tag{6.60}$$

While $S(\Theta)$ relates to the probability that the quark is diverted by an angle less than Θ we define the conjugate probability $T(\Theta)$ that the quark is diverted by an angle greater than Θ, i.e. $S(\Theta) + T(\Theta) = 1$. The probability for the emission of a gluon is given by the function $\mathrm{d}^2\Gamma/\mathrm{d}t\,\mathrm{d}z$ in (6.52). To find the total "decay rate" we integrate over z and t with the integration bounds we can deduce from (6.60). It holds that

$$\left. \frac{\Theta^2 M^2}{4(1-z)} \right|_{z=0} < t < M^2 , \tag{6.61}$$

where the maximal momentum transfer t is bounded by the total mass of the gluon. From (6.60) it follows that

$$1-\frac{\Theta^2 M^2}{4z} < z < 1 \quad : \quad S(\Theta) \tag{6.62}$$

which relates to the probability $S(\Theta)$ that the emission takes place with an angle smaller than Θ. The conjugate probability $T(\Theta)$ that the emission occurs with angle greater than Θ is given by the other values of z,

$$0 < z < 1-\frac{\Theta^2 M^2}{4z} \quad : \quad T(\Theta) \ , \tag{6.63}$$

so that we find

$$\begin{aligned}
S_1(\Theta) = 1 - T_1(\Theta) &= 1 - \int\limits_{\Theta^2 M^2/4}^{M^2} \mathrm{d}t \int\limits_0^{1-\Theta^2 M^2/4t} \mathrm{d}z\, \frac{2\alpha_s}{3\pi} \frac{1+z^2}{t(1-z)} \\
&\overset{z\to 1}{\approx} 1 - \frac{4\alpha_s}{3\pi} \int\limits_{\Theta^2 M^2/4}^{M^2} \mathrm{d}t \int\limits_0^{1-\Theta^2 M^2/4t} \mathrm{d}z\, \frac{1}{t(1-z)} \\
&= 1 - \frac{4\alpha_s}{3\pi} \int\limits_{\Theta^2 M^2/4}^{M^2} \mathrm{d}t\, \frac{1}{t}\left[-\ln\left(\frac{\Theta^2 M^2}{4t}\right)\right] \\
&= 1 - \frac{4\alpha_s}{3\pi}\left[-\ln\left(\frac{\Theta^2}{4}\right)\ln(t) + \frac{1}{2}\ln^2\left(\frac{t}{M^2}\right)\right]_{\Theta^2 M^2/4}^{M^2} \\
&= 1 - \frac{4\alpha_s}{3\pi}\left[\ln^2\left(\frac{\Theta^2}{4}\right) - \frac{1}{2}\ln^2\left(\frac{\Theta^2}{4}\right)\right] \\
&= 1 - \frac{2\alpha_s}{3\pi}\ln^2\left(\frac{\Theta^2}{4}\right) \ .
\end{aligned} \tag{6.64}$$

For $T(\Theta)$ one does not have to treat the subtleties of the radiative corrections canceling the infrared divergencies for $t \to 0$, $z \to 1$. For finite Θ the integrals for $T_1(\Theta)$ are finite. The probability is strongly peaked toward $\theta \ll \Theta$. Therefore we can approximate

$$\frac{S_{n+1}(\Theta)}{S_n(\Theta)} \sim 1 - T_1(\Theta) \ , \tag{6.65}$$

which results in

$$S_n(\Theta) \sim \left[1 - T_1(\Theta)\right]^n \ . \tag{6.66}$$

We still have to correct for the fact that the t_j were ordered according to

$$t_1 \geq t_2 \geq t_3 \ldots \geq t_n \ , \tag{6.67}$$

so that only one of the $n!$ permutations contained in (6.65) is chosen:

$$S_n(\Theta) \sim \frac{1}{n!}\left[1 - T_1(\Theta)\right]^n . \tag{6.68}$$

With (6.67) we can now easily sum the gluon emission to all orders:

$$S(\Theta) \sim \sum_n S_n(\Theta) = \mathrm{e}^{1-T_1(\Theta)} = \mathrm{e} \times \exp\left[-\frac{2}{3}\frac{\alpha_s}{\pi}\ln^2\left(\frac{\Theta^2}{4}\right)\right] . \tag{6.69}$$

Obviously this still lacks the proper normalization factor. For $T_1(\Theta) \equiv 0$, $S(\Theta)$ must clearly be equal to one. Thus the missing normalisation factor is e^{-1} and the final expression reads

$$S(\Theta) = \exp\left[-\frac{2}{3}\frac{\alpha_s}{\pi}\ln^2\left(\frac{\Theta^2}{4}\right)\right] = \exp\left[-\frac{2}{3}\frac{\alpha_s}{\pi}\ln^2\left(\frac{p_{\perp\min}}{M}\right)^2\right] . \tag{6.70}$$

This expression has the structure of a form factor and is usually interpreted as such. It is then called "Sudakov form factor". Clearly (6.69) suppresses the emission of collinear soft gluons, rendering the cross sections finite. The Sudakov form factor is probably the best-known example of resummation in QCD and is of great phenomenological importance in describing jet formation correctly. As we shall discuss at the end of this section the renormalization group equation imposes the condition that the exponential must always be of the form (6.69). The only problem remaining is to find the correct analytic continuation for the given dynamics. In our case this is relatively easy. Figure 6.8 compares the graph for which we have calculated (6.68) with the gluon corrections to the Drell–Yan cross section. Obviously the outgoing gluons just have to be substituted by incoming ones. This corresponds to the analytic continuation to $z \geq 1$ rather than $z \leq 1$. According to (6.59) this is equivalent to the continuation $\Theta^2 \to -\Theta^2$, which in turn implies that

$$\begin{aligned}\exp\left\{-\frac{2}{3}\frac{\alpha_s}{\pi}\mathrm{Re}\left[\ln^2\left(\frac{\Theta^2}{4}\right)\right]\right\} &\Rightarrow \exp\left\{-\frac{2}{3}\frac{\alpha_s}{\pi}\left[-\pi^2 + \ln^2\left(\frac{\Theta^2}{4}\right)\right]\right\} \\ &\Rightarrow \exp\left(\frac{2}{3}\alpha_s\pi\right)\exp\left[-\frac{2}{3}\frac{\alpha_s}{\pi}\ln^2\left(\frac{\Theta^2}{4}\right)\right] .\end{aligned} \tag{6.71}$$

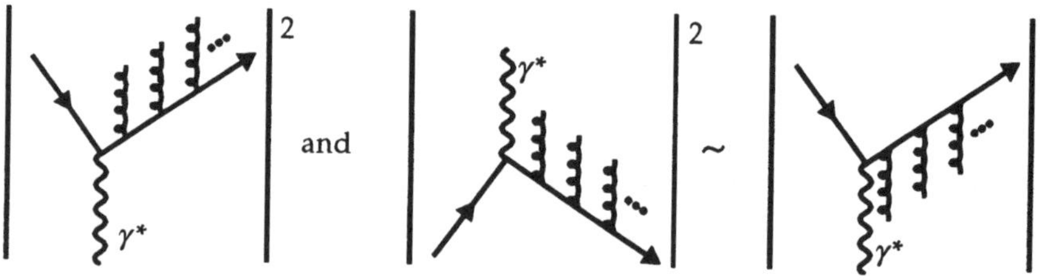

Fig. 6.8. Comparison of the gluonic corrections for two Drell–Yan processes

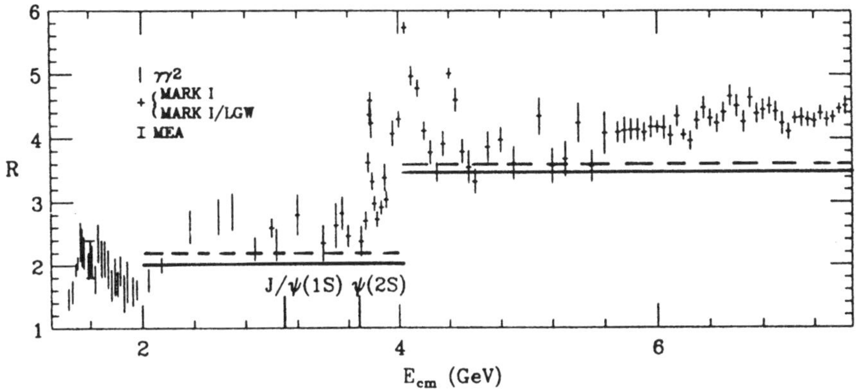

Fig. 6.9. The prediction of the parton model for the Drell–Yan process including QCD corrections. The data are taken from Review of Particle Properties, Phys. Rev. D **45** (1992)

As we have integrated over t respectively Θ the second factor is contained in the α_s corrections in the last bracket in (6.46). The remaining exponential factor is indeed $\exp(2\pi\alpha_s/3)$ as stated in (6.46).

The appearence of this factor is a typical property of higher-order corrections and the standard result of resummation. The same result can be obtained with the more formal apparatus of the renormalization group equation.

To complete this chapter let us add a few brief remarks on the phenomenology of the Drell–Yan process. As discussed in Sect. 4.2, the parton model predicts quite simple behavior for the total cross section:

$$R = \frac{\sigma\left(e^+e^- \to q\bar{q}\right)}{\sigma\left(e^+e^- \to \mu^+\mu^-\right)} = \begin{cases} 2 & E_{cm} > 2\,\text{GeV} \\ 10/3 & E_{cm} > 4\,\text{GeV} \\ 11/3 & E_{cm} > 10\,\text{GeV} \end{cases} . \tag{6.72}$$

We have just discussed the K factor that describes the resummed QCD corrections to this process. We sketch in Fig. 6.9 how these corrections improve the agreement with the data. The QCD corrections increase R, i.e., the K factor is always larger than 1. The change in the theoretical prediction for R has been included in Fig. 6.9 (at $\Lambda = 2$ GeV). To obtain a truly precise prediction, QED/radiative corrections, higher QCD corrections, mass corrections, and so on must also be included. These increase R somewhat more, in particular for small E_{cm}.

Figure 6.9 shows that the parton-model predictions are in general slightly too low while the QCD-corrected results clearly give a better description of the data. In this comparison one must take into account that in some energy regions the presence of resonances can induce large variations of R.

Choosing a special subgroup of the Drell–Yan processes, one can perform a far more specific test of QCD. This has been done most successfully for 3-jet events, which are characterized by the property that most of the energy is carried by hadrons, leading to three narrow and clearly distinguished angular regions. The simplest process that can lead to a 3-jet event is the gluon bremsstrahlung whose graph is shown in Fig. 6.7. However, not all bremsstrahlung events are identified as 3-jet events, but only those in which the angle between gluon jet and quark jets is sufficiently large. Conversely, the graph

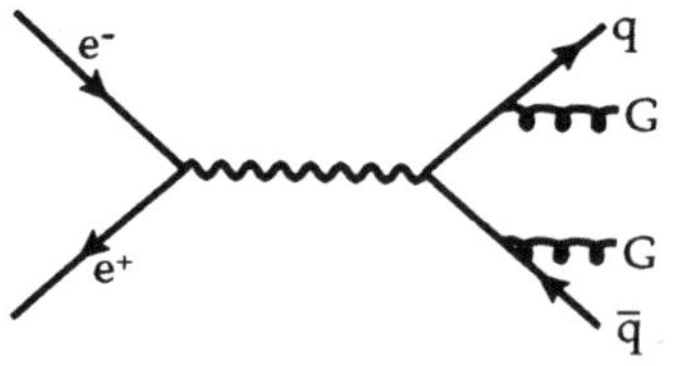

also contributes to 3-jet events, as long as one gluon jet is not resolved from the others. Finally the simplest Drell–Yan process,

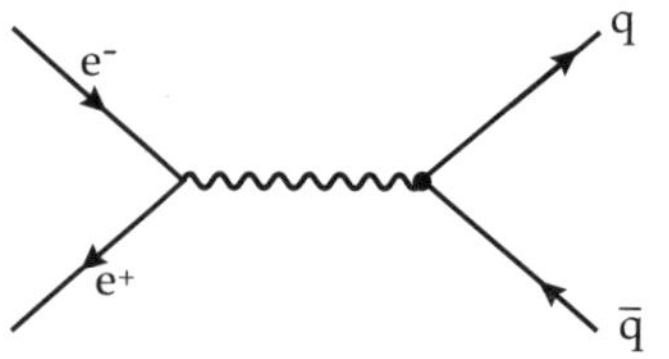

contains, with a certain statistical weight, events with unbalanced momentum distributions generated by the nonperturbative process of hadronization. A calculation of the 3-jet probability and a determination of the strong coupling constant α_s is therefore an extensive and model-dependent venture. See Fig. 6.10 for a typical 3-jet event.

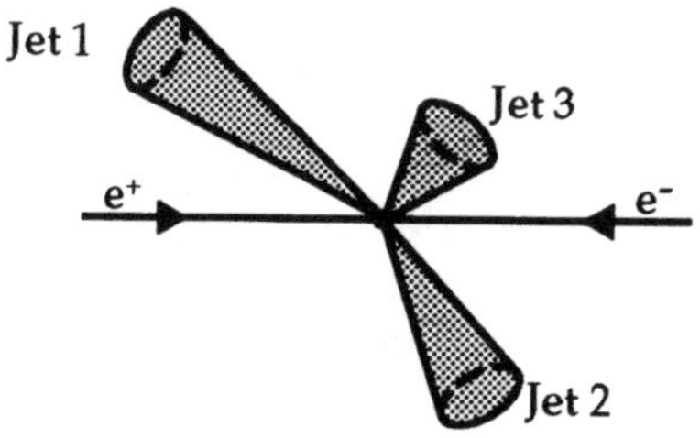

Fig. 6.10. A typical 3-jet event in the laboratory system

A typical calculation goes like this.

1. One calculates elementary QCD processes in perturbation theory. To order α_s^2, these are the graphs of Fig. 6.11, for example.
2. One calculates hadronization using one or more of the special computer codes developed for this purpose. These programs are based on some basic assumptions about the creation and decay of color strings between separating quarks (see Fig. 6.12). A typical code of this type depends on a substantial number of parameters and gives good fits for a large number of processes.
3. Finally one determines by the experimentally used definition of 3-jet events the corresponding contribution to this class of events. Here also experimental sensitivities enter.

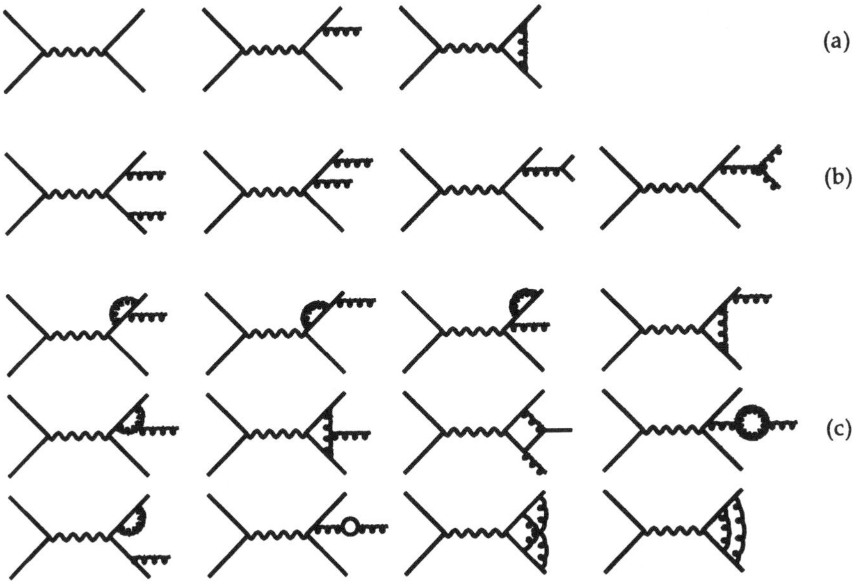

Fig. 6.11a–c. Elementary graphs of the Drell–Yan process up to order α_s^2

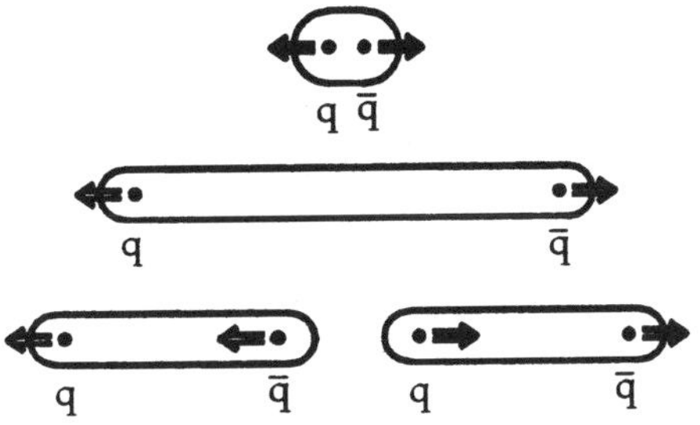

Fig. 6.12. Creation and break-up of a color string

The result of Fig. 6.13 was obtained in this way.[2] Here the energy asymmetry (whose exact definition is not important to us) is shown against the total center-of-mass energy. The two curves result from different hadronization programs. The coupling constant used was

$$\alpha_s = \frac{12\pi}{(33-2N_\mathrm{f})\ln\left(\frac{Q^2}{\Lambda^2}\right)}\left\{1-\frac{6(153-19N_\mathrm{f})}{(33-2N_\mathrm{f})^2}\frac{\ln\left[\ln\left(\frac{Q^2}{\Lambda^2}\right)\right]}{\ln\left(\frac{Q^2}{\Lambda^2}\right)}\right\} \tag{6.73}$$

with $N_\mathrm{f}=5$ and $\Lambda = 100\,\mathrm{MeV}$.

Equation (6.73) is the correct form of the running coupling constant when two-loop processes are taken into account. As the Drell–Yan process is calculated to order α_s^2, the coupling constant must also be determined to this order, to be consistent.

[2] See M. Chen: Int. J. Mod. Phys. A **1**, 669 (1986).

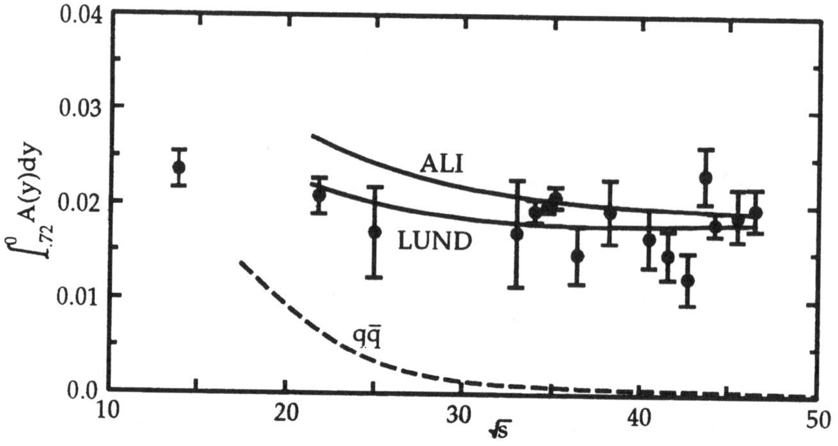

Fig. 6.13. The fit of experimental 3-jet events with $\Lambda = 100\,\text{MeV}$. LUND and ALI denote the two hadronization routines used

Using this fit to determine Λ gives

$$\Lambda = 100 \pm 30 \; {}^{+60}_{-45} \; \text{MeV} \quad \text{or} \quad \alpha_s(E_{\text{cm}} = 44\,\text{GeV}) = 0.12 \pm 0.02 \; . \quad (6.74)$$

Other analyses give somewhat different values for Λ. We do not want to enter this controversy here; we simply wanted to illustrate the procedure used.

In a similar way one attempts to describe semi-exclusive hadron–hadron scattering processes with hadrons in the final state. But here the uncertainties and model-dependence is even larger, so that frequently nothing more than rather general statements result. Such statements may concern, for example, the power of Q^2 with which a specific cross section decays. These processes are therefore less important as tests of QCD, with the exception of specific reactions like charmonium production (see Fig. 6.14).

The Drell–Yan process offers a number of possible ways to tests QCD. Although the results are quite convincing as a whole, their quality bears no comparison with tests of QED or of the Glashow–Salam–Weinberg model. Another

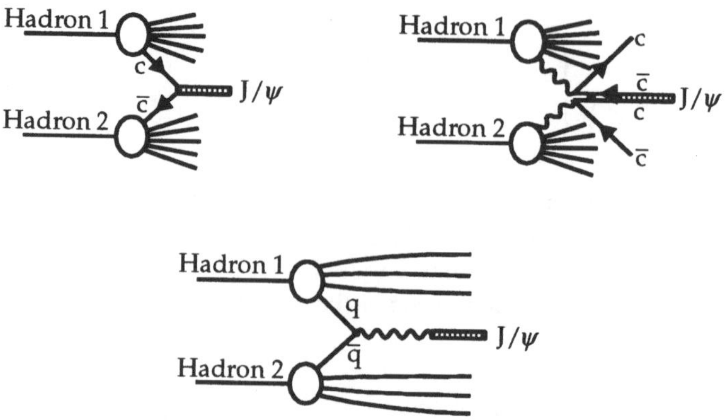

Fig. 6.14. Charmonium creation in hadron–hadron reactions

problem is that there is no competing model to QCD at this time. One does not therefore know how specific the predictions of QCD really are. There may be other models that describe scaling violations, 3-jet events, and so on equally well. Conversely many theoreticians consider the fact that there is only one theoretical candidate for describing quark–quark interactions to be the strongest argument for the correctness of QCD.

As a result of the analysis described here we conclude that Λ lies in a region between 100 and 200 MeV. (Λ depends on the renormalization procedure. The values given are for the $\overline{MS}$ scheme; see Sect. 5.1.)

EXERCISE

6.1 The Drell–Yan Cross Section

Problem. Derive the elementary Drell–Yan cross section in (6.5).

Solution. According to the techniques given in Chap. 2, we get

$$\mathrm{d}\sigma = \frac{e^4}{2M^2} \frac{\mathrm{d}^3 q_1}{(2\pi)^3 2E} \frac{\mathrm{d}^3 q_2}{(2\pi)^3} \frac{1}{4} \frac{F^2}{M^4} (2\pi)^4 \delta^4(p_1 + p_2 - q_1 - q_2) \ . \tag{1}$$

Here we have simply neglected the quark and gluon masses, thus getting the flux factor to $1/M^2$. (M is the invariant mass of the two quarks.) Spin-averaging yields a factor of 1/4, and F^2 is calculated from

$$\begin{aligned} F^2 &= \mathrm{tr}(\not{p}_1 \gamma_\mu \not{p}_2 \gamma_\nu)\, \mathrm{tr}(\not{q}_1 \gamma^\mu \not{q}_2 \gamma^\nu) \\ &= 16(p_1^\mu p_2^\nu + p_2^\mu p_1^\nu - g^{\mu\nu}\, p_1 \cdot p_2)(q_{1\mu} q_{2\nu} + q_{2\mu} q_{1\nu} - g_{\mu\nu}\, q_1 \cdot q_2) \\ &= 32(q_1 \cdot p_1\, q_2 \cdot p_2 + q_1 \cdot p_2\, q_2 \cdot p_1) \ . \end{aligned} \tag{2}$$

This is simplest in the center-of-momentum frame, where

$$\mathrm{d}\sigma = \frac{e^4}{8M^2} \frac{\mathrm{d}^3 q_1}{(2\pi)^2 M^6} F^2\, \delta(2q_1^0 - 2p_1^0) \ . \tag{3}$$

With $p_1 = (E, 0, 0, E)$, $p_2 = (E, 0, 0, -E)$, $q_1 = (E, E\sin\theta\cos\phi, E\sin\theta \times \sin\phi, E\cos\theta)$ and $q_2 = (E, -E\sin\theta\cos\phi, -E\sin\theta\sin\phi, -E\cos\theta)$, it follows that

$$\begin{aligned} F^2 &= 32\left[E^4(1-\cos\theta)^2 + E^4(1+\cos\theta)^2\right] \\ &= 64E^4(1+\cos^2\theta) \ , \end{aligned} \tag{4}$$

and therefore

$$\mathrm{d}\sigma = \frac{e^4}{8M^2} \frac{\mathrm{d}(\cos\theta)\, E^2}{4\pi M^6} 64E^4(1+\cos^2\theta) \tag{5}$$

Exercise 6.1

$$\sigma = \frac{e^4}{8M^2}\frac{1}{4\pi}\frac{8}{3} = \frac{4\pi\alpha^2}{3M^2} \ . \tag{6}$$

This is the relation we used in (6.5).

EXERCISE

6.2 The One-Gluon Contribution to the Drell–Yan Cross Section

Problem. Calculate the total Drell–Yan cross section for one-gluon emission, i.e., integrate (6.12) with bounds (6.17).

Solution. As s, u, and t are related by (introducing B for simplicity)

$$s+t+u = M^2+m_{\mathrm{G}}^2 = B \ , \tag{1}$$

we have to choose which variables we want to treat as independent. We use t and s, thus the integral becomes

$$\begin{aligned}
&\frac{8\pi\alpha_s\alpha e_q^2}{9s^2}\int\limits_{t_{\min}}^{t_{\max}} dt\left[\frac{B-s}{t}-1+\frac{t+s-B-s-B}{B-s-t}\right.\\
&\qquad\left.+\frac{2Bs}{t(B-s-t)}-M^2m_{\mathrm{G}}^2\left(\frac{1}{t^2}+\frac{1}{(B-s-t)^2}\right)\right]\\
&=\frac{8\pi\alpha_s\alpha e_q^2}{9s^2}\Bigg[(B-s)\ln t-2t+(s-B)\ln(B-s-t)\\
&\qquad+\frac{2Bs}{B-s}\left[\ln(t)-\ln(B-s-t)\right]-M^2m_{G}^2\left(-\frac{1}{t}+\frac{1}{B-s-t}\right)\Bigg]_{t_{\min}}^{t_{\max}} .
\end{aligned} \tag{2}$$

We now write

$$t_{\max/\min} = -\frac{1}{2}\left[s-B\mp\sqrt{(s-B)^2-4M^2m_{\mathrm{G}}^2}\right] \ , \tag{3}$$

$$\begin{aligned}
\sigma' &= \frac{8\pi\alpha_s\alpha e_q^2}{9s^2}\left[\frac{B^2+s^2}{B-s}\ln\left|\frac{t}{B-s-t}\right|-2t+M^2m_{\mathrm{G}}^2\left(\frac{1}{B-s-t}-\frac{1}{t}\right)\right]_{t_{\min}}^{t_{\max}}\\
&= \frac{8\pi\alpha_s\alpha e_q^2}{9s^2}\left[2\frac{B^2+s^2}{B-s}\ln\left(\frac{s-B-\sqrt{(s-B)^2-4M^2m_{\mathrm{G}}^2}}{s-B+\sqrt{(s-B)^2-4M^2m_{\mathrm{G}}^2}}\right)\right.\\
&\qquad -2\sqrt{(s-B)^2-4M^2m_{\mathrm{G}}^2}\\
&\qquad \left.+M^2m_{\mathrm{G}}^2(-2)\frac{2\sqrt{(s-B)^2-4M^2m_{\mathrm{G}}^2}}{(s-B)^2-(s-B)^2+4M^2m_{\mathrm{G}}^2}\,2\right]
\end{aligned} \tag{4}$$

and obtain the final expression

Exercise 6.2

$$\sigma' = \frac{8\pi\alpha\alpha_s e_q^2}{9s}\left[2\frac{1+B^2/s^2}{1-B/s}\ln\left(\frac{s-B+\sqrt{(s-B)^2-4M^2m_G^2}}{s-B-\sqrt{(s-B)^2-4M^2m_G^2}}\right)\right.$$
$$\left. -4\sqrt{\frac{(s-B)^2-4M^2m_G^2}{s^2}}\right] , \tag{5}$$
$$B = M^2+m_G^2 .$$

Obviously for $m_G \to 0$ the logarithmic divergence is recovered.

EXERCISE

6.3 The Drell–Yan Process as Decay of a Heavy Photon

Problem. Derive the relation between the cross sections for $q+\bar{q}\to\gamma^*\to \mu^+ + \mu^-$ and $q+\bar{q}\to\gamma^*$:

$$\frac{d\sigma}{dM^2}(q+\bar{q}\to\mu^+ + \mu^-) = \sigma(q+\bar{q}\to\gamma^*)\,\frac{\alpha}{3\pi M^2}\left(1-\frac{4m^2}{M^2}\right)^{3/2} dM^2 .$$

Solution. The decay cross section for $q+\bar{q}\to\gamma^*\to\mu^+ + \mu^-$ is

$$d\sigma = \frac{Q_q^2 e^4}{M^4}\,\frac{1}{2p_0 2p_0'}\,W^{\mu\nu}L_{\mu\nu}(2\pi)^{4-6}\delta(p+p'-k-k')\,\frac{d^3k\,d^3k'}{4k_0k_0'} , \tag{1}$$

where p, p', k, and k' are the quark, antiquark, μ^-, and μ^+ momenta. The virtual photon has the four-momentum q with $q^2 = M^2$. We now multiply the right-hand side of (1) by

$$\delta^4(q-k-k')\,\frac{d^3q}{2q_0}\,dM^2 = \delta^4(q-k-k')\,d^4q = 1 , \tag{2}$$

giving

$$d\sigma = \frac{Q_q^2 e^4}{M^4}\,\frac{1}{16k_0k_0'p_0p_0'\,2q_0}\,(2\pi)^{-2}\,W^{\mu\nu}L_{\mu\nu}$$
$$\times\,\delta^4(p+p'-q)\,\delta^4(q-k-k')\,d^3k\,d^3k'\,d^3q\,dM^2 . \tag{3}$$

We perform the k and k' integrations

$$I^{\mu\nu} = \int d^3k\,d^3k'\,(2\pi)^{-2}L^{\mu\nu}\,\delta^4(q-k-k') \tag{4}$$
$$= \frac{4}{(2\pi)^2}\int d^3k\,d^3k'\left[k^\mu k'^\nu + k^\nu k'^\mu - g^{\mu\nu}(k\cdot k'+m^2)\right]\delta^4(q-k-k') .$$

Exercise 6.3

We then evaluate this expression in the rest frame of the virtual photon, $q = (M, 0, 0, 0)$, $k^0 = M/2$:

$$\begin{aligned} I^{\mu\nu} &= \frac{1}{\pi^2} \int d^3k\, d^3k' \big[-2k^\mu k^\nu + k^\mu q^\nu \\ &\quad + q^\mu k^\nu - g^{\mu\nu}(-k^2 + k\cdot q + m^2)\big]\delta(q_0 - 2k_0)\, \delta^3(\boldsymbol{k}-\boldsymbol{k}') \\ &= \frac{1}{\pi^2} \int d^3k \big[2k^\mu k^0 \delta^{\nu 0} + 2k^\nu k^0 \delta^{\mu 0} - 2k^\mu k^\nu \\ &\quad - g^{\mu\nu}(k^0 q^0)\big]\delta(q^0 - 2k^0)\ . \end{aligned} \tag{5}$$

Because the integral over just one component of $\boldsymbol{k}$ vanishes, this expression is diagonal. For $\mu = \nu = 0$ we get

$$I^{00} = \frac{1}{\pi^2} 4\pi \int dk\, k^2 [2(k^0)^2 - 2(k^0)^2]\delta\left(q^0 - 2\sqrt{\boldsymbol{k}^2 + m^2}\right) = 0\ . \tag{6}$$

This is also required by current conservation:

$$q_\nu I^{\mu\nu} = q_\mu I^{\mu\nu} = 0\ . \tag{7}$$

For $\mu = \nu = j$ we find that

$$I^{jj} = \frac{1}{\pi^2} \int d^3k \left[-2(k^j)^2 + \frac{M^2}{2} \right] \delta\left(q^0 - 2\sqrt{\boldsymbol{k}^2 + m^2}\right)\ . \tag{8}$$

We substitute $k^j k^j \to 1/3\, \boldsymbol{k}^2$ and get

$$\begin{aligned} I^{jj} &= \frac{4}{\pi} \int dk\, k^2 \left(-\frac{2}{3}\boldsymbol{k}^2 + 2\boldsymbol{k}^2 \right) \delta\left(M - 2\sqrt{\boldsymbol{k}^2 + m^2}\right) \\ &= \frac{4}{\pi}\frac{4}{3}\left(\frac{M^2}{4} - m^2\right) \frac{\sqrt{\frac{M^2}{4} - m^2}\ \frac{M}{2}}{2} \\ &= \frac{2}{3\pi}\frac{M^4}{4}\left(1 - \frac{4m^2}{M^2}\right)^{3/2}\ . \end{aligned} \tag{9}$$

As $I^{\mu\nu}$ must be proportional to $g^{\mu\nu} - q^\mu q^\nu / q^2$, owing to current conservation, it is easy to obtain (note that $g_{\mu\nu} I^{\mu\nu} = I^{00} - 3I^{jj}$)

$$I^{\mu\nu} = \left(-g^{\mu\nu} + \frac{q^\mu q^\nu}{q^2}\right) \frac{M^4}{6\pi} \left(1 - \frac{4m^2}{M^2}\right)^{3/2}\ . \tag{10}$$

Substituting this result back into (3) gives

$$\begin{aligned} d\sigma &= \frac{Q_q^2 e^4}{M^4} \frac{1}{4M^2 p_0 p_0' 2q_0} W_{\mu\nu}\, \delta^4(p + p' - q) \\ &\quad \times \left(-g^{\mu\nu} + \frac{q^\mu q^\nu}{q^2}\right) \frac{M^4}{6\pi}\left(1 - \frac{4m^2}{M^2}\right)^{3/2} d^3q\, dM^2\ . \end{aligned} \tag{11}$$

To bring this into the form of $d\sigma(q+\bar{q} \to \gamma^*)$, we substitute

Exercise 6.3

$$\left(-g^{\mu\nu} + \frac{q^\mu q^\nu}{q^2}\right) = \sum_\varepsilon \varepsilon^\mu \varepsilon^{*\nu} \tag{12}$$

and get

$$\begin{aligned} d\sigma &= \frac{e^2}{6\pi M^2}\left[Q_q^2 e^2 W_{\mu\nu} \sum_\varepsilon \varepsilon^\mu \varepsilon^{*\nu} (2\pi)^{4-3} \delta^4(p+p'-q) \frac{d^3q}{4p_0 p_0' \cdot 2q_0}\right] \\ &\quad \times dM^2 \frac{1}{2\pi}\left(1-\frac{4m^2}{M^2}\right)^{3/2} \\ &= \frac{\alpha}{3\pi M^2}\left(1-\frac{4m^2}{M^2}\right)^{3/2} \sigma(q+\bar{q} \to \gamma^*)\, dM^2 \ . \end{aligned} \tag{13}$$

This completes our proof.

EXERCISE

6.4 Heavy Photon Decay Into Quark, Antiquark, and Gluon

Problem. Derive (6.48).

Solution. The differential decay rate for a one-body decay in its rest frame is simply

$$\begin{aligned} d\Gamma &= \frac{1}{2M}|M_1 + M_2|^2 \frac{d^3p_1}{(2\pi)^3 2E_1} \frac{d^3p_2}{(2\pi)^3 2E_2} \frac{d^3p_3}{(2\pi)^3 2E_3} \\ &\quad \times (2\pi)^4\, \delta^4(k-p_1-p_2-p_3) \end{aligned} \tag{1}$$

with

$$\begin{aligned} M_2 &= \bar{u}(p_2,s_2)\left(-\mathrm{i}g\gamma_\nu \frac{\lambda^a}{2}\right) \frac{\mathrm{i}(\not{p}_2+\not{p}_3)}{(p_2+p_3)^2} \\ &\quad \times (-\mathrm{i}\, e\, e_q\, \gamma_\mu)\, v(p_1,s_1)\, \tilde{\varepsilon}_\mu(k)\, \varepsilon_\nu(p_3) \end{aligned} \tag{2}$$

and

$$\begin{aligned} M_1 &= -\bar{u}(p_2,s_2)\,(-\mathrm{i}\, e\, e_q\, \gamma_\mu) \frac{(-\mathrm{i})(\not{p}_1+\not{p}_3)}{(p_1+p_3)^2} \\ &\quad \times \left(-\mathrm{i}g\gamma_\nu \frac{\lambda^a}{2}\right) v(p_1,s_1)\, \tilde{\varepsilon}_\mu(k)\, \varepsilon_\nu(p_3) \ . \end{aligned} \tag{3}$$

Exercise 6.4

The factor $1/2M$ in (1) is just the photon field normalization. In the photon rest system all four-products are rather simple for zero masses:

$$\begin{aligned}
2\,p_1\cdot p_3 &= (p_1+p_3)^2 = (k_\gamma - p_2)^2 = k_\gamma^2 - 2\,k_\gamma\cdot p_2 \\
&= M^2 - 2M^2x_2 = M^2(1-2x_2)\ , \\
2\,p_2\cdot p_3 &= M^2(1-2x_1)\ , \\
2\,p_1\cdot p_2 &= M^2(1-2x_3)\ ,
\end{aligned} \tag{4}$$

where x_i is the energy fraction carried by the parton i. Due to (4), $|M_1+M_2|^2$ depends only on x_1, x_2, and $x_3 = 1-x_1-x_2$, such that the phase-space integral is very easy.

$$\begin{aligned}
&\int \frac{\mathrm{d}^3p_1}{(2\pi)^3 2E_1}\,\frac{\mathrm{d}^3p_2}{(2\pi)^3 2E_2}\,\frac{\mathrm{d}^3p_3}{(2\pi)^3 2E_3}(2\pi)^4\,\delta^4(k-p_1-p_2-p_3)\cdots \\
&= \int \frac{\mathrm{d}^3p_1\,\mathrm{d}^3p_2}{(2\pi)^5 2^3 E_1E_2E_3}\delta\,(M-E_1-E_2-E_3)\cdots\ .
\end{aligned}$$

Using $E_1 = x_1M$, $E_2 = x_2M$ and $E_3 = (1-x_1-x_2)M$ we can simplify the formula above. In the δ function we express

$$E_3^2 = \boldsymbol{p}_3^2 = (\boldsymbol{k}-\boldsymbol{p}_1-\boldsymbol{p}_2)^2 = (\boldsymbol{p}_1+\boldsymbol{p}_2)^2$$

and get

$$\begin{aligned}
&\int \frac{\mathrm{d}^3p_1}{(2\pi)^3 2E_1}\,\frac{\mathrm{d}^3p_2}{(2\pi)^3 2E_2}\,\frac{\mathrm{d}^3p_3}{(2\pi)^3 2E_3}(2\pi)^4\,\delta^4(k-p_1-p_2-p_3)\cdots \\
&= \int \frac{\mathrm{d}^3p_1\,\mathrm{d}^3p_2}{2^5\pi^5 2^3 x_1x_2(1-x_1-x_2)M^3} \\
&\quad \times\delta\left(M-x_1M-x_2M-\sqrt{(\boldsymbol{p}_1-\boldsymbol{p}_2)^2}\right)\cdots \\
&= \int \frac{4\pi E_1^2\,\mathrm{d}E_1\,2\pi E_2^2\,\mathrm{d}E_2\,\mathrm{d}\cos(\theta_{12})}{2^8\pi^5 x_1x_2(1-x_1-x_2)M^3} \\
&\quad \times\delta\left(M(1-x_1-x_2)-M\sqrt{2x_1x_2\cos(\theta_{12})}\right)\cdots \\
&= \int \frac{x_1^2M^3x_2^2M^3\,\mathrm{d}x_1\,\mathrm{d}x_2}{2^5\pi^3 x_1x_2(1-x_1-x_2)M^3}\,\frac{\sqrt{2x_1x_2\cos(\theta_{12})}}{Mx_1x_2}\Bigg|_{\sqrt{2x_1x_2\cos(\theta_{12})}=1-x_1-x_2}\cdots \\
&= \int \frac{M^2\,\mathrm{d}x_1\,\mathrm{d}x_2}{2^5\pi^3}\cdots
\end{aligned} \tag{5}$$

such that the decay rate reads

$$\mathrm{d}\Gamma = \frac{M}{64\pi^3}\,\mathrm{d}x_1\,\mathrm{d}x_2\,|M_1+M_2|^2\ . \tag{6}$$

We now make a special choice for the polarization vectors of the gluon. The reason for this will become clear later:

$$\sum_{\varepsilon} \varepsilon_\mu \varepsilon^*_{\mu'} = -g_{\mu\mu'} + \frac{n_\mu p_{3\mu'} + n_{\mu'} p_{3\mu}}{n \cdot p_3} = \Sigma_{\mu\mu'} \ , \tag{7}$$

$$\sum_{\tilde{\varepsilon}} \tilde{\varepsilon}_\nu \tilde{\varepsilon}^*_{\nu'} = -g_{\nu\nu'} \ . \tag{8}$$

The choice (7) is equivalent to choosing the axial gauge $n \cdot A = 0$. n_μ is an arbitrary four-vector with $n^2 = 0$ and $n \cdot \varepsilon = 0$, while for the photon we choose the covariant Feynman gauge.

$$\begin{aligned} S_{11} &= \sum_{\text{spins}} M_1 M_1^* \\ &= -g^2 e^2 e_q^2 \,\mathrm{tr}\left\{\frac{\lambda^a \lambda^a}{4}\right\} \Sigma^{\mu\mu'} \\ &\quad \times \mathrm{tr}\left\{ \not{p}_2 \gamma_\nu \frac{\not{p}_1 + \not{p}_3}{(p_1+p_3)^2} \gamma_\mu \not{p}_1 \gamma_{\mu'} \frac{\not{p}_1 + \not{p}_3}{(p_1+p_3)^2} \gamma^\nu \right\} \ , \end{aligned} \tag{9}$$

$$\begin{aligned} S_{12} &= + \sum_{\text{spins}} (M_1 M_2^* + M_2 M_1^*) \\ &= 4\, g^2 e^2 e_q^2 \, \Sigma^{\mu\mu'} \left[\mathrm{tr}\left\{ \not{p}_2 \gamma_\nu \frac{\not{p}_1 + \not{p}_3}{(p_1+p_3)^2} \gamma_\mu \not{p}_1 \gamma^\nu \frac{\not{p}_2 + \not{p}_3}{(p_2+p_3)^2} \gamma_{\mu'} \right\} \right. \\ &\quad \left. + \mathrm{tr}\left\{ \not{p}_2 \gamma_\mu \frac{\not{p}_2 + \not{p}_3}{(p_2+p_3)^2} \gamma_\nu \not{p}_1 \gamma_{\mu'} \frac{\not{p}_1 + \not{p}_3}{(p_1+p_3)^2} \gamma_\nu \right\} \right] \ , \end{aligned} \tag{10}$$

$$\begin{aligned} S_{22} &= \sum_{\text{spins}} M_2 M_2^* \\ &= -4\, g^2 e^2 e_q^2 \, \Sigma^{\mu\mu'} \mathrm{tr}\left\{ \not{p}_2 \gamma_\mu \frac{\not{p}_2 + \not{p}_3}{(p_2+p_3)^2} \gamma_\nu \not{p}_1 \gamma^\nu \frac{\not{p}_2 + \not{p}_3}{(p_2+p_3)^2} \gamma_{\mu'} \right\} \ . \end{aligned} \tag{11}$$

We shall use the gauge

$$n_\mu = \left(k_\gamma - \frac{p_2}{2x_2} \right)_\mu \ , \tag{12}$$

$$n^2 = M^2 - 2\,\frac{x_2 M^2}{2x_2} + 0 = 0 \ . \tag{13}$$

This choice will result in a particularly simple result and will thus be justified only a posteriori. From

$$n_\mu \Sigma^{\mu\mu'} = n_{\mu'} \Sigma^{\mu\mu'} = 0 \quad \text{and} \quad p_{3\mu} \Sigma^{\mu\mu'} = 0 \tag{14}$$

we find that

$$\left(p_1 + p_2 + p_3 - \frac{p_2}{2x_2} \right)_\mu \Sigma^{\mu\mu'} = \left(p_1 - p_2 \frac{1 - 2x_2}{2x_2} \right)_\mu \Sigma^{\mu\mu'} = 0 \ . \tag{15}$$

Exercise 6.4

This allows us to substitute all momentum vectors occurring in the trace according to

$$\begin{aligned} (p_3)_{\mu'}\,,\ (p_3)_{\mu'} &\to 0\ , \\ (p_1)_{\mu} &\to (p_2)_{\mu}\frac{1-2x_2}{2x_2}\ , \\ (p_1)_{\mu'} &\to (p_2)_{\mu'}\frac{1-2x_2}{2x_2}\ , \end{aligned} \tag{16}$$

which simplifies the calculation substantially. We start with S_{12} and introduce the notation $c = 4g^2e^2e_q^2$:

$$\begin{aligned} S_{11} &= \frac{2c\,\Sigma^{\mu\mu'}}{M^4(1-2x_2)^2}\mathrm{tr}\{\not p_2\,(\not p_1+\not p_3)\,\gamma_\mu\,\not p_1\,\gamma_{\mu'}\,(\not p_1+\not p_3)\} \\ &= \frac{2c\,\Sigma^{\mu\mu'}}{M^4(1-2x_2)^2}\Big[M^2(1-2x_3+1-2x_1)\mathrm{tr}\{\gamma_\mu\,\not p_1\,\gamma_{\mu'}\,(\not p_1+\not p_3)\} \\ &\quad -2\,p_1\cdot p_3\,\mathrm{tr}\{\not p_2\,\gamma_\mu\,\not p_1\gamma_{\mu'}\}\Big]\ , \\ &= \frac{2c\,\Sigma^{\mu\mu'}}{M^2(1-2x_2)^2}\Big[2(1-x_3-x_1)\,4\big(p_{1\mu}\,(p_1+p_3)_{\mu'}+p_{1\mu'}\,(p_1+p_3)_\mu \\ &\quad -g_{\mu\mu'}\,p_1\cdot p_3\big)-\frac{2\,p_1\cdot p_3}{M^2}\,4\big(p_{2\mu}\,p_{1\mu'}+p_{2\mu'}\,p_{1\mu}-g_{\mu\mu'}\,p_2\cdot p_1\big)\Big]\ . \end{aligned} \tag{17}$$

With (14) this reads

$$\begin{aligned} S_{11} &= \frac{2c\,\Sigma^{\mu\mu'}}{M^2(1-2x_2)^2}\Big[16x_2\,p_{1\mu}\,p_{1\mu'}-8(1-2x_2)\,p_{2\mu'}\,p_{1\mu} \\ &\quad -4x_2(1-2x_2)M^2g_{\mu\mu'}+2(1-2x_2)M^2(1-2x_3)g_{\mu\mu'}\Big]\ . \end{aligned} \tag{18}$$

With (16) and

$$\Sigma^{\mu\mu'}g_{\mu\mu'} = -4+2 = -2 \tag{19}$$

this gives

$$\begin{aligned} S_{11} &= \frac{2c}{1-2x_2}\,(-2)\,(-4x_2+2-4x_3) \\ &= -\frac{4c}{1-2x_2}\,(-2+4x_1) \\ &= 8c\,\frac{1-2x_1}{1-2x_2}\ . \end{aligned} \tag{20}$$

Analogously S_{22} gives

$$S_{22} = \frac{2c\,\Sigma^{\mu\mu'}}{M^4(1-2x_1)^2}\mathrm{tr}\Big\{\not p_2\,\gamma_\mu(\not p_2+\not p_3)\not p_1(\not p_2+\not p_3)\gamma_{\mu'}\Big\}\ , \tag{21}$$

Exercise 6.4

which is the same as S_{11} with μ and μ' interchanged and p_1 interchanged with p_2. Thus the analogy to (18) now reads

$$S_{22} = \frac{2c}{M^2(1-2x_1)^2}\left\{\left[16\,x_1\, p_{2\mu}\, p_{2\mu'} - 8(1-2x_1)p_{1\mu'}\, p_{2\mu}\right]\Sigma^{\mu\mu'} + 8x_1(1-2x_1)M^2 - 4(1-2x_1)M^2(1-2x_3)\right\} . \tag{22}$$

Again using (16) this simplifies substantially to

$$S_{22} = \frac{2c}{M^2(1-2x_1)^2}\left\{8\, p_{2\mu}\, p_{2\mu'}\left[2x_1 - \frac{(1-2x_2)(1-2x_1)}{2x_2}\right] + 4M^2(1-2x_1)(1-2x_2)\right\} . \tag{23}$$

With

$$\begin{aligned}\Sigma^{\mu\mu'} p_{2\mu}p_{2\mu'} &= 2\,\frac{n\cdot p_2\, 2\, p_3\cdot p_2}{2\, n\cdot p_3}\\ &= 2\,\frac{M^2x_2\, M^2(1-2x_1)}{M^2(1-2x_2)\frac{1-2x_3}{2x_2}}\\ &= M^2\,\frac{4x_2^2(1-2x_1)}{(1-2x_2)(1-2x_3)} ,\end{aligned} \tag{24}$$

$$\begin{aligned}2\, n\cdot p_3 &= 2\, p_1\cdot p_3 - (1-2x_2)\frac{2\, p_2\cdot p_3}{2x_2}\\ &= M^2(1-2x_2)\left(1 - \frac{1-2x_1}{2x_2}\right)\\ &= M^2\,\frac{1-2x_2}{2x_2}(2x_2+2x_1-1)\\ &= M^2\,\frac{(1-2x_2)(1-2x_3)}{2x_2} ,\end{aligned} \tag{25}$$

we obtain

$$\begin{aligned}S_{22} &= \frac{2c}{(1-2x_1)^2}\left\{8\frac{4x_2^2(1-2x_1)}{(1-2x_2)(1-2x_3)}\left[2x_1 - \frac{(1-2x_2)(1-2x_1)}{2x_2}\right] + 4(1-2x_1)(1-2x_2)\right\}\\ &= \frac{8c}{(1-2x_1)(1-2x_2)(1-2x_3)}\left[16x_1\, x_2^2 - 4(1-2x_2)(1-2x_1)x_2 + (1-2x_2)^2(1-2x_3)\right]\\ &= \frac{8c}{(1-2x_1)(1-2x_2)(1-2x_3)}\left[16x_2^2 - 16x_2^3 - 16x_3x_2^2 + 4x_2(1-2x_2)(1-2x_2-2x_3) + (1-2x_2)^2(1-2x_3)\right]\end{aligned}$$

Exercise 6.4

$$
\begin{aligned}
&= \frac{8c}{(1-2x_1)(1-2x_2)(1-2x_3)}\left[8x_2^2 - 16x_3\,x_2^2 + 4x_2(1-2x_2)(1-2x_3)\right.\\
&\quad \left. + (1-2x_2)^2(1-2x_3)\right]\\
&= \frac{8c}{(1-2x_1)(1-2x_2)}\left[8x_2^2 + 4x_2 - 8x_2^2 + (1-2x_2)^2\right]\\
&= \frac{8c}{(1-2x_1)(1-2x_2)}\left(1+4x_2^2\right)\\
&= 8c\,\frac{1+4x_2^2}{(1-2x_1)(1-2x_2)}\ .
\end{aligned}
\qquad (26)
$$

Finally we calculate S_{12}.

$$
\begin{aligned}
S_{12} &= -\frac{2c\,\Sigma^{\mu\mu'}}{M^4(1-2x_1)(1-2x_2)}\,2\,\mathrm{tr}\left\{\not p_1\not p_2\gamma_{\mu'}(\not p_2+\not p_3)(\not p_1+\not p_3)\gamma_\mu\right\}\\
&= -\frac{4c\,\Sigma^{\mu\mu'}}{M^4(1-2x_1)(1-2x_2)}\,\mathrm{tr}\left\{\not p_1\not p_2\gamma_{\mu'}(\not p_2\not p_1+\not p_3\not p_1+\not p_2\not p_3)\gamma_\mu\right\}\\
&= -\frac{4c\,\Sigma^{\mu\mu'}}{M^4(1-2x_1)(1-2x_2)}\left[2(p_2)_{\mu'}\mathrm{tr}\{\not p_1\not p_2\not p_1\gamma_\mu\}\right.\\
&\quad \left. +2(p_1)_\mu\mathrm{tr}\{\not p_1\not p_2\gamma_{\mu'}\not p_3\} + 2(p_2)_{\mu'}\mathrm{tr}\{\not p_1\not p_2\not p_3\gamma_\mu\}\right]\\
&= -\frac{4c\,\Sigma^{\mu\mu'}}{M^4(1-2x_1)(1-2x_2)}\,4\left[4\,p_{2\mu'}\,p_1\cdot p_2\,p_{1\mu} + 2\,p_{1\mu}\,p_1\cdot p_3\,p_{2\mu'}\right.\\
&\quad \left. +2\,p_{1\mu}\,p_{1\mu'}\,p_2\cdot p_3 + 2\,p_{2\mu'}\,p_{1\mu}\,p_2\cdot p_3 - 2\,p_{2\mu'}\,p_1\cdot p_3\,p_{2\mu}\right]\\
&= -\frac{16c\,\Sigma^{\mu\mu'}}{M^2(1-2x_1)(1-2x_2)}\,p_{2\mu}\,p_{2\mu'}\left[2\,\frac{(1-2x_2)(1-2x_3)}{2x_2} + \frac{(1-2x_2)^2}{2x_2}\right.\\
&\quad \left. -(1-2x_1)\,\frac{(1-2x_2)^2}{(2x_2)^2} + (1-2x_1)\,\frac{1-2x_2}{2x_2} - (1-2x_2)\right]\\
&= -\frac{16c}{(1-2x_1)(1-2x_2)}\,\frac{4x_2^2(1-2x_1)}{(1-2x_2)(1-2x_3)}\,\frac{1-2x_2}{4x_2^2}\\
&\quad \times\left[4x_2(1-2x_3) + 2x_2(1-2x_2) - (1-2x_1)(1-2x_2)\right.\\
&\quad \left. + 2x_2(1-2x_1) - 4x_2^2\right]\\
&= -\frac{16c}{(1-2x_2)(1-2x_3)}\\
&\quad \times\left[4x_2(1-2x_3) + (1-2x_2)(1-2x_3) + 2x_2(1-2x_1-2x_2)\right]\\
&= -\frac{16c}{(1-2x_2)(1-2x_3)}\,(1-2x_3)\left(4x_2+1-2x_2-2x_2\right)\\
&= -\frac{16c}{1-2x_2}\ .
\end{aligned}
\qquad (27)
$$

Exercise 6.4

This completes our calculation:

$$\begin{aligned} \mathrm{d}\Gamma &= \frac{M}{64\pi^3}\,\mathrm{d}x_1\,\mathrm{d}x_2\left(S_{11}+S_{12}+S_{22}\right) \\ &= \frac{M}{64\pi^3}\,\mathrm{d}x_1\,\mathrm{d}x_2\,16\pi\alpha_s\,4\pi\alpha\,e_q^2 \\ &\quad\times\left[8\,\frac{1-2x_1}{1-2x_2}+8\,\frac{1+4x_2^2}{(1-2x_2)(1-2x_1)}-16\,\frac{1}{1-2x_2}\right] \\ &= \frac{8M}{\pi}\alpha_s\alpha e_q^2\,\mathrm{d}x_1\,\mathrm{d}x_2\frac{-(1+2x_1)(1-2x_1)+1+4x_2^2}{(1-2x_2)(1-2x_1)} \\ &= 3\alpha e_q^2 M\,\frac{32}{3\pi}\,\alpha_s\,\frac{x_1^2+x_2^2}{(1-2x_2)(1-2x_1)}\,\mathrm{d}x_1\,\mathrm{d}x_2\;. \end{aligned} \tag{28}$$

EXAMPLE

6.5 Factorization in Drell–Yan

In this example, we derive the cross section for the Drell–Yan process

$$N+N\to\mu^+\mu^-+X \tag{1}$$

in the most general form (Fig. 6.15). N represents the nucleon, μ^+ and μ^- the charged leptons and X the unobserved hadrons. In (6.6) we assumed that the cross section can be written as a convolution of the elementary parton–parton cross section and the structure functions known from deep inelastic lepton–nucleon scattering. In the following it will be shown that such a factorization can actually be derived from QCD.

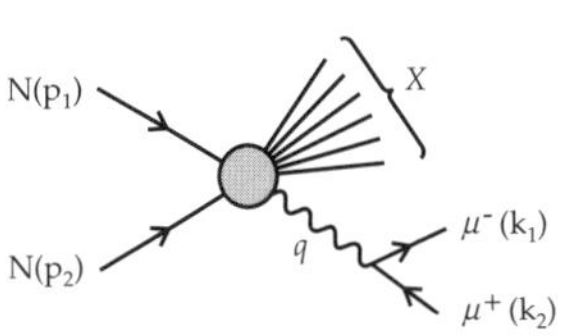

Fig. 6.15. Drell–Yan $\mu^+\mu^-$ pair production in nucleon–nucleon collisions

The cross section in general can be written as

$$\begin{aligned} \mathrm{d}\sigma &= \frac{1}{2(p_1\cdot p_2)}\frac{1}{4}\sum_{\mathrm{pol}}\frac{\mathrm{d}^3k_1}{(2\pi)^3 2k_1^0}\frac{\mathrm{d}^3k_2}{(2\pi)^3 2k_2^0} \\ &\quad\times\sum_X(2\pi)^4\delta^4(p_1+p_2-k_1-k_2-p_X)\left|\left\langle\mu^+\mu^-X\right|T\left|NN\right\rangle\right|^2\;, \end{aligned} \tag{2}$$

where p_1 and p_2 are the momenta of the nucleons, k_1, k_2 the momenta of the produced leptons, and the center-of-mass energy $s=(p_1+p_2)^2=2p_1\cdot p_2$. We neglect the masses of the leptons and nucleons. The sum runs over all polarizations of initial nucleons and final leptons. The factor 1/4 stands for the average over the initial polarizations. The matrix element in the one-photon approximation can be written as

$$\left\langle\mu^+\mu^-X\right|T\left|NN\right\rangle=\bar{u}(k_1)e\gamma_\mu v(k_2)\frac{g^{\mu\nu}}{(k_1+k_2)^2}\left\langle X\left|e j_\nu(0)\right|NN\right\rangle \tag{3}$$

Example 6.5

with $j_\nu(0)$ the electromagnetic current (we neglect Z and W exchange). Squaring the matrix element and performing the sum over lepton polarizations we arrive at

$$\frac{1}{4}\sum_{\text{pol}}\left|\langle\mu^+\mu^-|\,T\,|NN\rangle\right|^2 = \frac{e^4}{Q^4}\frac{1}{4}\mathrm{tr}\left[\gamma^\mu \not{k}_2\gamma^\nu \not{k}_1\right] \times \sum_X \langle NN\,|j_\nu(0)|\,X\rangle\left\langle X\left|j_\mu(0)\right|NN\right\rangle \tag{4}$$

with the timelike momentum $q=(k_1+k_2)$, $q^2=Q^2$. The sum over final hadrons can be performed by making use of the δ function in (2):

$$\begin{aligned}
&(2\pi)^4\delta^4(p_1+p_2-P_X-q)\left\langle X\left|j_\mu(0)\right|NN\right\rangle \\
&= \int \mathrm{d}^4x\,\mathrm{e}^{-\mathrm{i}(p_1+p_2-P_X-q)\cdot x}\left\langle X\left|j_\mu(0)\right|NN\right\rangle \\
&= \int \mathrm{d}^4x\,\mathrm{e}^{\mathrm{i}qx}\,\mathrm{e}^{\mathrm{i}P_X\cdot x}\left\langle X\left|j_\mu(0)\right|NN\right\rangle \mathrm{e}^{-\mathrm{i}(p_1+p_2)\cdot x} \\
&= \int \mathrm{d}^4x\,\mathrm{e}^{\mathrm{i}q\cdot x}\left\langle X\left|\mathrm{e}^{\mathrm{i}\hat{P}\cdot x}j_\mu(0)\,\mathrm{e}^{-\mathrm{i}\hat{P}\cdot x}\right|NN\right\rangle \\
&= \int \mathrm{d}^4x\,\mathrm{e}^{\mathrm{i}q\cdot x}\left\langle X\left|j_\mu(x)\right|NN\right\rangle\;.
\end{aligned} \tag{5}$$

Insertion into (2) finally leads to

$$\begin{aligned}
\mathrm{d}\sigma =& \frac{1}{2s}\frac{e^4}{Q^4}\frac{1}{4}\mathrm{tr}\left[\gamma^\mu \not{k}_2\gamma^\nu \not{k}_1\right] \\
&\times \int \mathrm{d}^4x\,\mathrm{e}^{\mathrm{i}qx}\left\langle NN\left|j_\nu(0)\,j_\mu(x)\right|NN\right\rangle \frac{\mathrm{d}^3k_1}{(2\pi)^3 2k_1^0}\frac{\mathrm{d}^3k_2}{(2\pi)^3 2k_2^0}\;,
\end{aligned}$$

where a complete set of states has been summed:

$$\sum_X \langle NN\,|j_\nu(0)|\,X\rangle\left\langle X\left|j_\mu(x)\right|NN\right\rangle = \left\langle NN\left|j_\nu(0)\,j_\mu(x)\right|NN\right\rangle\;. \tag{6}$$

Defining the leptonic tensor as

$$\frac{1}{4}\mathrm{tr}\left[\gamma_\mu \not{k}_2\gamma_\nu \not{k}_1\right] = \frac{1}{2}L_{\mu\nu} \tag{7}$$

and the hadronic tensor as

$$W_{\mu\nu} = \int \mathrm{d}^4x\,\mathrm{e}^{\mathrm{i}q\cdot x}\left\langle NN\left|j_\nu(0)\,j_\mu(x)\right|NN\right\rangle\;, \tag{8}$$

we get as a result

$$\begin{aligned}
\mathrm{d}\sigma &= \frac{1}{2s}\frac{e^4}{Q^4}\frac{L_{\mu\nu}}{2}W^{\mu\nu}\frac{\mathrm{d}^3k_1}{(2\pi)^3 2k_1^0}\frac{\mathrm{d}^3k_2}{(2\pi)^3 2k_2^0} \\
&= \frac{\alpha^2}{2s}\frac{1}{Q^4}2L_{\mu\nu}W^{\mu\nu}\frac{1}{(2\pi)^4}\frac{\mathrm{d}^3k_1}{2k_1^0}\frac{\mathrm{d}^3k_2}{2k_2^0}
\end{aligned} \tag{9}$$

with $\alpha = e^2/4\pi$ the electromagnetic fine-structure constant.

Example 6.5

The expression can be further simplified by introducing coordinates $q = k_1 + k_2$ and $k = k_1 - k_2$, which yields

$$\frac{\mathrm{d}^3 k_1}{2k_1^0}\frac{\mathrm{d}^3 k_2}{2k_2^0} = \frac{1}{2}\frac{1}{(q_0^2 - k_0^2)}\mathrm{d}^3 q\,\mathrm{d}^3 k\ . \tag{10}$$

In the center-of-mass system of the lepton pair $\boldsymbol{k}_1 + \boldsymbol{k}_2 = 0 = \boldsymbol{q}$ we have $Q^2 = q_0^2$, $k_0^2 = 0$ so that we can write

$$\begin{aligned}\frac{1}{2}\frac{\mathrm{d}^3 q}{Q^2}\mathrm{d}^3 k &= \frac{1}{2}\frac{\mathrm{d}^3 q}{Q^2}\mathrm{d}\Omega_k\,|\boldsymbol{k}|^2\,\mathrm{d}\,|\boldsymbol{k}|\\ &= \frac{1}{2}\mathrm{d}^3 q\,\mathrm{d}\,|\boldsymbol{k}|\,\mathrm{d}\Omega_k\\ &= \frac{1}{2}\mathrm{d}^4 q\,\mathrm{d}\Omega_k\ .\end{aligned} \tag{11}$$

The angle $\mathrm{d}\Omega_k = \mathrm{d}\phi_k\,\mathrm{d}\cos\theta_k$ corresponds to the angle relative to the appropriately chosen axis in the lepton-pair center-of-mass system. Putting everything together we find

$$\mathrm{d}\sigma = \frac{\alpha^2}{2s}\frac{1}{Q^4}L_{\mu\nu}W^{\mu\nu}\frac{1}{(2\pi)^4}\mathrm{d}^4 q\,\mathrm{d}\Omega_k\ . \tag{12}$$

As seen from (8) the Drell–Yan cross section is described by the hadronic tensor in a quite analoguos way to the hadronic tensor in the case of deep inelastic lepton–nucleon scattering. In the case of Drell–Yan the current product $j_\nu(0)\,j_\mu(x)$ is sandwiched not in a one-nucleon state but in the two-nucleon state $|NN\rangle \equiv |\,p_1 p_2\rangle$. Hence a similar procedure as in the case of operator product expansion is not applicable here. However, we will use a procedure a little bit more general to derive a factorization of the short-distance part from the long-distance part. In leading order in the strong coupling constant we insert into the hadronic tensor the vector currents

$$j_\mu(x)\,j_\nu(0) = e_q^2\bar{q}_i(x)\gamma_\mu q_j(x)\bar{q}_l(0)\gamma_\nu q_k(0)\delta_{ij}\delta_{lk}\ , \tag{13}$$

where we have made the dependence on the color denoted by the latin indices i, j, k, l explicit. Performing a Fierz transformation[3] on the quark fields we can write

$$\begin{aligned}\left(q_j(x)\right)_\alpha\left(q_l(0)\right)_\beta = &-\frac{1}{4}\left[\bar{q}_l(0)q_j(x)\right]\delta_{\alpha\beta} - \frac{1}{4}\left[\bar{q}_l(0)\gamma_\varrho q_j(x)\right]\gamma^\varrho_{\alpha\beta}\\ &-\frac{1}{8}\left[\bar{q}_l(0)\sigma_{\varrho\sigma}q_j(x)\right]\sigma^{\varrho\sigma}_{\alpha\beta} - \frac{1}{4}\left[\bar{q}_l(0)\gamma_5\gamma_\varrho q_j(x)\right]\gamma^\varrho\gamma_5\\ &-\frac{1}{4}\left[\bar{q}_l(0)\gamma_5 q_j(x)\right]\gamma_{5\alpha\beta}\end{aligned} \tag{14}$$

[3] See W. Greiner and B. Müller: *Gauge Theory of Weak Interactions*, 3rd ed. (Springer, Berlin, Heidelberg, New York 2000).

Example 6.5

and analogously for $(-)q_k(0)\overline{q}_i(x)$, where an additional minus sign is encountered due to the anticommuting fermion field operators. Note that the indices α, β refer to the Dirac indices of the γ matrices. Thus (13) is transformed to

$$\begin{aligned}
j_\mu(x)j_\nu(0) = e_q^2\delta_{ij}\delta_{lk}\left(-\frac{1}{16}\right)\{&\mathrm{tr}\left[\gamma_\mu\gamma_\nu\right]\left[\overline{q}_l(0)q_j(x)\right]\left[\overline{q}_i(x)q_k(0)\right]\\
&+\mathrm{tr}\left[\gamma_\mu\gamma_\varrho\gamma_\nu\gamma_\sigma\right]\left[\overline{q}_l(0)\gamma^\varrho q_j(x)\right]\left[\overline{q}_i(x)\gamma^\sigma q_k(0)\right]\\
&+\mathrm{tr}\left[\gamma_\mu\gamma_5\gamma_\varrho\gamma_\nu\gamma_5\gamma_\sigma\right]\left[\overline{q}_l(0)\gamma^\varrho\gamma_5 q_j(x)\right]\left[\overline{q}_i(x)\gamma^\sigma\gamma_5 q_k(0)\right]\\
&+\frac{1}{2}\mathrm{tr}\left[\gamma_\mu\sigma_{\varrho\sigma}\gamma_\nu\sigma_{\lambda\eta}\right]\left[\overline{q}_l(0)\sigma^{\varrho\sigma}q_j(x)\right]\left[\overline{q}_i(x)\sigma^{\lambda\eta}q_k(0)\right]\\
&+\mathrm{tr}\left[\gamma_\mu\gamma_5\gamma_\nu\gamma_5\right]\left[\overline{q}_l(0)\gamma_5 q_j(x)\right]\left[\overline{q}_i(x)\gamma_5 q_k(0)\right]\}\ ,
\end{aligned} \tag{15}$$

where all other combinations lead to an odd number of γ matrices under the trace and thus to zero.

Now the quark operators have to be factorized into the nucleon states. Only color-less combinations will contribute. A Fierz transformation in color space will achieve that. Using the relation

$$\delta_{ij}\delta_{lk} = \frac{1}{N}\delta_{jl}\delta_{ik} - 2t^A_{jl}\,t^A_{ik}\ , \tag{16}$$

with the SU(3) Gell-Mann matrices $\lambda^A/2 = t^A$ and $N = 3$ the number of colors, we find

$$\begin{aligned}
W_{\mu\nu} = &\left(-\frac{1}{16}\right)e_q^2\frac{1}{N}\mathrm{tr}\left[\gamma_\mu\gamma_\varrho\gamma_\nu\gamma_\sigma\right]\\
&\times\int \mathrm{d}^4x e^{iq\cdot x}\left\langle p_1\left|\overline{q}(0)\gamma^\varrho q(x)\right|p_1\right\rangle\left\langle p_2\left|\overline{q}(x)\gamma^\sigma q(0)\right|p_2\right\rangle\\
&+\text{other terms}\ .
\end{aligned} \tag{17}$$

In this equation we have used that only the first part in (16) will give color-less matrix elements and have not written down explicitely the other structures stemming from the expansion in (15). Only the vector matrix elements $\langle p_1\left|\overline{q}(0)\gamma^\varrho q(x)\right|p_1\rangle$ will contribute at leading twist to *unpolarized* scattering.

To proceed further we specify our coordinate system. We choose two lightlike vectors n_μ and $\overline{n}_\mu$ with $n^2 = \overline{n}^2 = 0$:

$$\begin{aligned}
n^\mu &= \frac{1}{\sqrt{2}}(1,0,0,-1)\ ,\\
\overline{n}^\mu &= \frac{1}{\sqrt{2}}(1,0,0,1)\,,\quad n\cdot\overline{n} = 1\ ,
\end{aligned} \tag{18}$$

so that in light-cone coordinates for any vector k_μ

$$\begin{aligned}
n\cdot k &= \frac{1}{\sqrt{2}}\left(k^0+k^3\right) = k^+\ ,\\
\overline{n}\cdot k &= \frac{1}{\sqrt{2}}\left(k^0-k^3\right) = k^-\ .
\end{aligned} \tag{19}$$

Example 6.5

The volume element in such coordinates reads $\mathrm{d}^4k = \mathrm{d}k^- \, \mathrm{d}k^+ \, \mathrm{d}^2k_\perp$, where $k_\perp = (k_x, k_y)$ is the transverse component of the vector k_μ. We use the two lightlike vectors n_μ and $\overline{n}_\mu$ to fix the incoming momenta:

$$\begin{aligned} p_1^\mu &= p_1^+ \overline{n}^\mu \ , \\ p_2^\mu &= p_2^- n^\mu \ , \end{aligned} \tag{20}$$

and the virtual photon's momentum q as

$$q^\mu = x_1 p_1^\mu + x_2 p_2^\mu + q_\perp^\mu \ . \tag{21}$$

In the same way as the Lorentz vector q^μ may be expanded in terms of two light-cone vectors and a transverse component any γ matrix can be decomposed into "plus" and "minus" and transverse directions:

$$\begin{aligned} \gamma^\mu &= \not{n} \cdot \overline{n}^\mu + \not{\overline{n}} \cdot n^\mu + \gamma_\perp^\mu \\ &= \not{n} \frac{p_1^\mu}{p_1^+} + \not{\overline{n}} \frac{p_2^\mu}{p_2^-} + \gamma_\perp^\mu \ . \end{aligned} \tag{22}$$

Such a decomposition normally is referred to as *Sudakov decomposition*. Assuming that in (17) proton 1 travels in the "plus" direction and proton 2 in the "minus" direction, we can approximate the matrix element as

$$\left\langle p_1 \left| \overline{q}(0) \gamma_\varrho q(x) \right| p_1 \right\rangle \sim \frac{p_{1\varrho}}{p_1^+} \left\langle p_1 \left| \overline{q}(0) \not{n} q(x) \right| p_1 \right\rangle \ , \tag{23a}$$

$$\left\langle p_2 \left| \overline{q}(x) \gamma_\sigma q(0) \right| p_2 \right\rangle \sim \frac{p_{2\sigma}}{p_2^-} \left\langle p_2 \left| \overline{q}(x) \not{\overline{n}} q(0) \right| p_2 \right\rangle \ , \tag{23b}$$

where all omitted contributions are of higher twist.

Further on we Fourier-transform the matrix elements

$$\left\langle p_1 \left| \overline{q}(0) \not{n} q(x) \right| p_1 \right\rangle = \int \frac{\mathrm{d}^4 r_1}{(2\pi)^4} \, \mathrm{e}^{-\mathrm{i} x \cdot r_1} S(r_1, p_1) \ , \tag{24a}$$

$$\left\langle p_2 \left| \overline{q}(x) \not{n} q(0) \right| p_2 \right\rangle = \int \frac{\mathrm{d}^4 r_2}{(2\pi)^4} \, \mathrm{e}^{-\mathrm{i} x \cdot r_2} \overline{S}(r_2, p_2) \ , \tag{24b}$$

which after insertion into (17) leads to

$$\begin{aligned} W_{\mu\nu} = \int \frac{\mathrm{d}^4 r_1}{(2\pi)^4} \int \frac{\mathrm{d}^4 r_2}{(2\pi)^4} \frac{e_q^2}{N} \frac{1}{4} \mathrm{tr} \left[\gamma_\mu \not{p}_1 \gamma_\nu \not{p}_2 \right] (2\pi)^4 \delta^4 (q - r_1 - r_2) \\ \times (-) \frac{1}{p_2^-} \overline{S}(r_2, p_2) \frac{1}{p_1^+} S(r_1, p_1) \ . \end{aligned} \tag{25}$$

Here we have succeded in writing the lepton pair production in NN scattering in terms of a product of a part calculable in perturbation theory (the trace) together with the δ function and the nonperturbative matrix elements $\overline{S}(r_2, p_2)$, $S(r_1, p_1)$, which symbolically can be written as

$$W_{\mu\nu} = \int \frac{\mathrm{d}^4 r_1}{(2\pi)^4} \int \frac{\mathrm{d}^4 r_2}{(2\pi)^4} H_{\mu\nu}(r_1, r_2, q) \tilde{S}(r_1, r_2) \ . \tag{26}$$

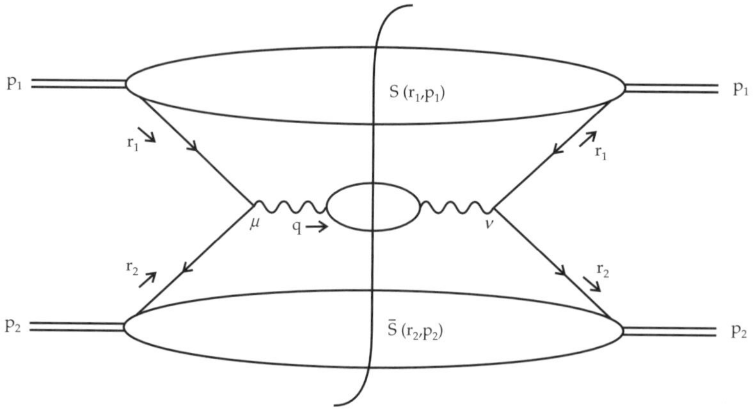

Fig. 6.16. Diagram relevant for the Drell–Yan process at leading order QCD

Here $H_{\mu\nu}(r_1, r_2, q)$ is the part of the diagramm consisting of lines that carry the "hard" momentum q^μ while $\tilde{S}(r_1, r_2)$ is the part that does not. See Fig. 6.16.

The leading contribution of the diagram arises when the momenta of the quark lines entering into $H_{\mu\nu}(r_1, r_2, q)$ are on shell and collinear to p_1 and p_2. This can systematically be accounted for by expanding the hard part around its collinear components. Approximating

$$\begin{aligned} r_1 &\approx \zeta_1 p_1; \quad \zeta_1 = r_1^+/p_1^+ \ , \\ r_2 &\approx \zeta_2 p_2; \quad \zeta_2 = r_2^-/p_2^- \ , \end{aligned} \tag{27}$$

the hard part can be written as

$$\begin{aligned} H(r_1, r_2) = {} & H(\zeta_1 p_1, \zeta_2 p_2) + (r_1 - \zeta_1 p_1)_\varrho \frac{\partial}{\partial r_1^\varrho} H(\zeta_1 p_1, \zeta_2 p_2) \\ & + (r_2 - \zeta_2 p_2)_\varrho \frac{\partial}{\partial r_2^\varrho} H(\zeta_1 p_1, \zeta_2 p_2) \\ & + \frac{1}{2!}(r_1 - \zeta_1 p_1)_\varrho (r_2 - \zeta_2 p_2)_\sigma \frac{\partial^2}{\partial r_1^\varrho \partial r_2^\sigma} H(\zeta_1 p_1, \zeta_2 p_2) \\ & + \dots \ . \end{aligned} \tag{28}$$

This procedure is known as collinear expansion[4] and in general is used to calculate the $1/Q^2$ correction terms to the parton model results. Here it will be sufficient to obtain from the leading term in the expansion in (28) the parton model already implemented in (6.6).

---

[4] R.K. Ellis, W. Furmanski, and R. Petronzio: Nucl. Phys. B **207**, 1 (1982); Nucl. Phys. B **212**, 19 (1983)
J. Qiu, G. Sterman: Nucl. Phys. B **353**, 105 (1991).

Example 6.5

Restricting ourselves to the leading term in (28) means approximating the δ function in (25) by

$$\begin{aligned}\delta^4(q-r_1-r_2) &\approx \delta^4(q-\zeta_1 p_1-\zeta_2 p_2)\\ &= \delta^2(q_\perp)\delta(x_1-\zeta_1)\frac{1}{p_1^+}\delta(x_2-\zeta_2)\frac{1}{p_2^-}\ .\end{aligned} \tag{29}$$

In the second step we have inserted the parametrization of the momentum q (see (21)). Noting that

$$\begin{aligned}&\int\frac{d^4r_1}{(2\pi)^4}\int\frac{d^4r_2}{(2\pi)^4}(2\pi)^4\delta^4(q-\zeta_1p_1-\zeta_2p_2)\\ &= \int d\zeta_1 p_1^+\, d\zeta_2 p_2^-\,\delta^2(q_\perp)\frac{1}{p_1^+}\delta(x_1-\zeta_1)\frac{1}{p_2^-}\delta(x_2-\zeta_2)\\ &\quad\times\int\frac{d^2r_{1\perp}}{(2\pi)^2}\int\frac{d^2r_{2\perp}}{(2\pi)^2}\int\frac{dr_1^-}{(2\pi)^2}\int\frac{dr_2^+}{(2\pi)^2}\ ,\end{aligned} \tag{30}$$

i.e. all but two integrations now are independent from integrations over the hard part, we can combine the six other integrations over the soft part

$$S(r_1)=\int d^4x\, e^{ir_1\cdot x}\left\langle p_1\left|\overline{q}(0)\not{n}q(x)\right|p_1\right\rangle\ , \tag{31a}$$

$$\overline{S}(r_2)=\int d^4x\, e^{ir_2\cdot x}\left\langle p_2\left|\overline{q}(x)\not{\overline{n}}q(0)\right|p_2\right\rangle\ , \tag{31b}$$

to get

$$\begin{aligned}W_{\mu\nu} &= \int d\zeta_1\int d\zeta_2\frac{e_q^2}{N}\frac{1}{4}\mathrm{tr}\left[\gamma_\mu\not{p}_1\gamma_\nu\not{p}_2\right]\frac{1}{p_1^+p_2^-}(2\pi)^4\delta(q_\perp^2)\delta(\zeta_1-x_1)\\ &\quad\times\delta(\zeta_2-x_2)\frac{1}{2}\int\frac{dx^-}{(2\pi)}e^{+ip_1^+\zeta_1x^-}\left\langle p_1\left|\overline{q}(0)\not{n}q(x^-)\right|p_1\right\rangle\\ &\quad\times\left(-\frac{1}{2}\right)\int\frac{dy^+}{(2\pi)}e^{-ip_2^-\zeta_2y^+}\left\langle p_2\left|\overline{q}(0)\not{\overline{n}}q(y^+)\right|p_2\right\rangle\ ,\end{aligned} \tag{32}$$

where in the last step it was noted that

$$\begin{aligned}&\int\frac{dy^+}{(2\pi)}e^{+ip_2^-x_2y^+}\left\langle p_2\left|\overline{q}(y^+)\not{\overline{n}}q(0)\right|p_2\right\rangle\\ &= \int\frac{dy^+}{(2\pi)}e^{-ip_2^-x_2y^+}\left\langle p_2\left|\overline{q}(0)\not{\overline{n}}q(y^+)\right|p_2\right\rangle\end{aligned} \tag{33}$$

due to translation invariance of the matrix elements

$$\left\langle p_2\left|\overline{q}(y^+)\not{\overline{n}}q(0)\right|p_2\right\rangle=\left\langle p_2\left|\overline{q}(0)\not{\overline{n}}q(-y^+)\right|p_2\right\rangle\ . \tag{34}$$

Example 6.5

To make the relationship with the previous parton model predictions more apparent we define

$$
\begin{aligned}
q(\zeta_1) &= \frac{1}{2}\int \frac{\mathrm{d}x^-}{(2\pi)} \mathrm{e}^{\mathrm{i}p_1 x^- \zeta_1} \left\langle p_1 \left| \overline{q}(0) \not{n} q(x^-) \right| p_1 \right\rangle , \\
\overline{q}(\zeta_2) &= \left(-\frac{1}{2}\right)\int \frac{\mathrm{d}y^+}{(2\pi)} \mathrm{e}^{-\mathrm{i}p_2 y^+ \zeta_2} \left\langle p_2 \left| \overline{q}(0) \not{\overline{n}} q(y^+) \right| p_2 \right\rangle .
\end{aligned} \tag{35}
$$

Apparently the bilocal operators along the light cone can be identified with the quark distribution function we discussed earlier. Further $p_1^+ p_2^- = p_1 p_2 = s/2$, where s is the overall center-of-mass energy. Hence (33) can be given as

$$
\begin{aligned}
W_{\mu\nu} = \int \mathrm{d}\zeta_1 \int \mathrm{d}\zeta_2 [&\hat{\tau}_{\mu\nu}(\zeta_1, \zeta_2) q(\zeta_1) \overline{q}(\zeta_2) \\
&+ \hat{\tau}_{\mu\nu}(\zeta_2, \zeta_1) \overline{q}(\zeta_1) q(\zeta_2)] ,
\end{aligned} \tag{36}
$$

i.e. a convolution of some partonic cross section with quark distributions. Note that we accounted for the fact that the antiquark can be picked up from either proton 1 or proton 2. In our derivation we analyzed only the case in which the antiparticle is picked up from proton 2. Comparing coefficients we see that $\hat{\tau}_{\mu\nu}$ is given by

$$
\begin{aligned}
\hat{\tau}_{\mu\nu} &= \frac{2e_q^2}{NS} \frac{1}{4} \mathrm{tr}\left[\gamma_\mu \not{p}_1 \gamma_\nu \not{p}_2\right] (2\pi)^4 \delta(q_\perp^2) \delta(\zeta_1 - x_1) \delta(\zeta_2 - x_2) \\
&= \frac{2e_q^2}{N(\zeta_1\zeta_2 S)} \hat{w}_{\mu\nu} (2\pi)^4 \delta(q_\perp^2) \delta(\zeta_1 - x_1) \delta(\zeta_2 - x_2) ,
\end{aligned} \tag{37}
$$

with the tree-level contribution from quark–antiquark scattering

$$
\begin{aligned}
\hat{w}_{\mu\nu} &= \frac{1}{4} \sum_{\mathrm{pol}} \left| \overline{v}(\zeta_2 p_2) \gamma_\mu u(\zeta_1 p_1) \right|^2 \\
&= \frac{1}{4} \mathrm{tr}\left[\gamma_\mu (\zeta_1 \not{p}_1) \gamma_\nu (\zeta_2 \not{p}_2)\right] .
\end{aligned} \tag{38}
$$

This completes our proof that the cross section for Drell–Yan pair production can be factorized into a partonic part, which is calculable order by order in perturbation theory, and a soft hadronic part, which can be parametrized in terms of quark distribution functions.

EXERCISE

6.6 Collinear Expansion and Structure Functions in Deep Inelastic Lepton–Nucleon Scattering

Problem. Apply the technique of the collinear expansion to the physical process of deep inelastic lepton–nucleon scattering. Analyze the hadronic tensor

$$\begin{aligned} W_{\mu\nu} &= \frac{1}{(2\pi)}\int \mathrm{d}^4x\,\mathrm{e}^{-\mathrm{i}q\cdot x}\langle pS\,|j_\nu(0)j_\mu(x)|\,pS\rangle \\ &= -F_1(x,Q^2)g_{\mu\nu} + F_2(x,Q^2)\frac{p_\mu p_\nu}{p\cdot q} \\ &\quad -\mathrm{i}\epsilon_{\mu\nu\lambda\sigma}\frac{q^\lambda}{p\cdot q}S^\sigma g_1(x,Q^2) + \mathcal{O}(q_\mu q_\nu) \end{aligned} \tag{1}$$

and express the structure functions F_1 and g_1 through bilocal operators defined along the light cone.

Solution. As in the previous example we first define the kinematics by choosing two lightlike vectors n_μ and $\overline{n}_\mu$ so that $n^2 = \overline{n}^2 = 0$ and $n\cdot\overline{n} = 1$.

The momenta of the photon, q_μ, and the target hadron, p_μ, can then be parametrized as

$$q^\mu = -xp^\mu + \frac{Q^2}{2xp^+}n^\mu\ , \tag{2a}$$

$$p^\mu = p^+\overline{n}^\mu\ . \tag{2b}$$

It is easy to convince oneself that indeed $p^2 = 0$, $q^2 = -Q^2$, and $x = Q^2/2pq$ refers to the usual definition of the Bjorken scaling variable. The proton is chosen to travel along the light cone in the + direction. We insert vector currents

$$j_\mu(x) = \sum_f e_f\overline{q}_f(x)\gamma_\mu q_f(x)\ , \tag{3a}$$

$$j_\nu(0) = \sum_{f'} e_{f'}\overline{q}_{f'}(0)\gamma_\nu q_{f'}(0) \tag{3b}$$

into the hadronic tensor $W_{\mu\nu}$ and formally apply the Wick theorem. With this we find

$$\begin{aligned} W_{\mu\nu} &= \frac{e^2}{2\pi}\int \mathrm{d}^4x\,\mathrm{e}^{-\mathrm{i}q\cdot x}\langle pS\,|\overline{q}(0)\gamma_\nu\mathrm{i}S(-x)\gamma_\mu q(x) \\ &\quad +\overline{q}(x)\gamma_\mu\mathrm{i}S(x)\gamma_\nu q(0)|\,pS\rangle \\ &= \frac{e^2}{2\pi}\int \mathrm{d}^4x\int\frac{\mathrm{d}^4k}{(2\pi)^4}\left[\mathrm{e}^{-\mathrm{i}(q-k)\cdot x}\left\langle pS\left|\overline{q}(0)\gamma_\nu\frac{\mathrm{i}}{\not k}\gamma_\mu q(x)\right|pS\right\rangle\right. \\ &\quad \left. +\,\mathrm{e}^{-\mathrm{i}(q+k)\cdot x}\left\langle pS\left|\overline{q}(x)\gamma_\mu\frac{\mathrm{i}}{\not k}\gamma_\nu q(0)\right|pS\right\rangle\right] \end{aligned} \tag{4}$$

Exercise 6.6

where we have introduced the free-quark propagator

$$\mathrm{i}S(-x) = \int \frac{\mathrm{d}^4k}{(2\pi)^4}\mathrm{e}^{\mathrm{i}k\cdot x}\frac{\mathrm{i}}{\not{k}} = \int \frac{\mathrm{d}^4k}{(2\pi)^4}\mathrm{e}^{\mathrm{i}k\cdot x}\frac{\mathrm{i}\not{k}}{k^2+\mathrm{i}\epsilon} \tag{5}$$

and suppressed flavor indices. Shifting the variable of integration $k \to k+q$ and performing a Fierz transformation as explained in the previous example we arrive at

$$\begin{aligned} W_{\mu\nu} = &\frac{e^2}{2\pi}\int\frac{\mathrm{d}^4k}{(2\pi)^4}\frac{1}{4}\mathrm{tr}\left[\gamma_\nu\frac{\mathrm{i}}{\not{k}+\not{q}}\gamma_\mu\gamma_\alpha\right]\int \mathrm{d}^4x\,\mathrm{e}^{\mathrm{i}k\cdot x}\langle\, p\,|\overline{q}(0)\gamma_\alpha q(x)|\, p\rangle \\ &+\frac{e^2}{2\pi}\int\frac{\mathrm{d}^4k}{(2\pi)^4}\frac{1}{4}\mathrm{tr}\left[\gamma_\nu\frac{\mathrm{i}}{\not{k}+\not{q}}\gamma_\mu\gamma_\alpha\gamma_5\right]\int \mathrm{d}^4x\,\mathrm{e}^{\mathrm{i}k\cdot x}\langle\, p\,|\overline{q}(0)\gamma_5\gamma_\alpha q(x)|\, p\rangle \\ &+(q\leftrightarrow -q;\,\mu\leftrightarrow\nu)\ . \end{aligned} \tag{6}$$

Note that these terms are the only ones that contribute. The other terms from the Fierz transformation ~ 1, γ_5, $\sigma_{\alpha\beta}$ vanish due to an odd number of γ matrices under the trace. Performing a *Sudakov decomposition* (see (22) of Example 6.5) of the γ matrix,

$$\begin{aligned} \gamma_\alpha &= \not{n}\overline{n}_\alpha + \not{\overline{n}}n_\alpha + \gamma_{\alpha\perp} \\ &= \not{n}\frac{p_\alpha}{p^+} + \not{\overline{n}}n_\alpha + \gamma_{\alpha\perp} \\ &\approx \not{n}\frac{p_\alpha}{p^+}\ , \end{aligned} \tag{7}$$

the Dirac indices can be factorized in (6). In addition to that, we assume that the parton momentum k_μ is collinear to the parent hadron's momentum:

$$k_\mu = x_1 p_\mu = x_1 p^+\overline{n}_\mu \tag{8}$$

so that $x_1 = k^+/p^+$.

With that all but one integration over k can be performed to give

$$\begin{aligned} W_{\mu\nu} = &\left[\frac{e^2}{2\pi}\int \mathrm{d}x_1\,p^+\frac{1}{4}\mathrm{tr}\left[\gamma_\nu\frac{\mathrm{i}}{x_1\not{p}+\not{q}}\gamma_\mu\frac{\not{p}}{p^+}\right]\right. \\ &\times\int\frac{\mathrm{d}x^-}{2\pi}\mathrm{e}^{\mathrm{i}k^+x^-}\langle p\,|\overline{q}(0)\not{n}q(x^-)|\,p\rangle \\ &+\frac{e^2}{2\pi}\int \mathrm{d}x_1\,p^+\frac{1}{4}\mathrm{tr}\left[\gamma_\nu\frac{\mathrm{i}}{x_1\not{p}+\not{q}}\gamma_\mu\frac{\not{p}}{p^+}\gamma_5\right] \\ &\left.\times\int\frac{\mathrm{d}x^-}{2\pi}\mathrm{e}^{\mathrm{i}k^+x^-}\langle pS|\overline{q}(0)\gamma_5\not{n}q(x^-)|pS\rangle\right] \\ &+(q\leftrightarrow -q;\,\mu\leftrightarrow\nu)\ . \end{aligned} \tag{9}$$

As explained in more detail in the chapter on the GLAP equation we are interested in that part where intermediate particles are on shell. Thus instead of the

Exercise 6.6

full propagator we retain only its imaginary part:

$$\frac{\mathrm{i}}{x_1 \not{p} + \not{q}} = \frac{\mathrm{i}(x_1 \not{p} + \not{q})}{(x_1 p + q)^2 + \mathrm{i}\epsilon} \to (x_1 \not{p} + \not{q})(2\pi)\delta^+((x_1 p + q)^2) \ . \tag{10}$$

The general rules for obtaining from a Feynman amplitude the discontinuity across a cut were first obtained by Cutkosky.[5] As soon as more than one propagator has to be cut the rules are not as straightforward as in our case. The one-dimensional δ function becomes

$$\begin{aligned}(2\pi)\delta((x_1 p + q)^2) &= \left(2\pi)\delta(-Q^2 + 2x_1 p^+ \frac{Q^2}{2xp^+}\right) \\ &= (2\pi)\delta\left(\frac{Q^2}{x}(x_1 - x)\right) = (2\pi)\frac{x}{Q^2}\delta(x_1 - x) \ ,\end{aligned} \tag{11}$$

so that (9) simplifies to

$$\begin{aligned}W_{\mu\nu} = {}& e^2 \frac{x}{Q^2}\frac{1}{2}\mathrm{tr}\left[\gamma_\nu(x\not{p} + \not{q})\gamma_\mu \not{p}\right]\frac{1}{2}\int \frac{\mathrm{d}x^-}{2\pi}\,\mathrm{e}^{\mathrm{i}x(p^+x^-)}\left\langle p\left|\bar{q}(0)\not{n}q(x^-)\right| p\right\rangle \\ &+ e^2\frac{1}{2}\frac{x}{Q^2}\mathrm{tr}\left[\gamma_\nu(x\not{p} + \not{q})\gamma_\mu \not{p}\gamma_5\right]\frac{1}{2} \\ &\times \int \frac{\mathrm{d}x^-}{2\pi}\,\mathrm{e}^{\mathrm{i}x(p^+x^-)}\left\langle pS\left|\bar{q}(0)\gamma_5\not{n}q(x^-)\right| pS\right\rangle \\ &+ (q \leftrightarrow -q;\, \mu \leftrightarrow \nu) \ .\end{aligned} \tag{12}$$

To complete our task we calculate the traces. For the trace without γ_5 we obtain

$$\begin{aligned}\frac{x}{Q^2}\frac{1}{2}\mathrm{tr}\left[\gamma_\nu(x\not{p} + \not{q})\gamma_\mu \not{p}\right] &= \frac{x}{Q^2}\cdot 2\left[2xp_\nu p_\mu - g_{\mu\nu}p\cdot q + \mathcal{O}(q_\mu, q_\nu)\right] \\ &= 2x\frac{p_\mu p_\nu}{p\cdot q} - g_{\mu\nu} + \mathcal{O}(q_\mu, q_\nu) \ .\end{aligned} \tag{13}$$

Note that we have neglected all terms that lead to zero after contracting with the leptonic tensor. Comparing with the parametrization of the hadronic tensor we can identify

$$\begin{aligned}F_1(x) &= \frac{1}{2}\int \frac{\mathrm{d}x^-}{2\pi}\,\mathrm{e}^{+\mathrm{i}x(p^+x^-)}\sum_f e_f\left\langle p\left|\bar{q}_f(0)\not{n}q_f(x^-)\right| p\right\rangle \ , \\ F_2(x) &= 2xF_1(x) \ .\end{aligned} \tag{14}$$

Note that we have taken into account the exchange term ($q \leftrightarrow -q;\, \mu \leftrightarrow \nu$) and made explicit the dependence on the quark flavor. In a similar way we find for

---

[5] R.E. Cutkosky, J. Math. Phys. **1**, 1344 (1958). See also C. Itzykson and J.-B. Zuber: *Quantum Field Theory* (McGraw-Hill, New York, 1978).

Exercise 6.6

the spin dependent function

$$\frac{x}{Q^2}\frac{1}{2}\mathrm{tr}\left[\gamma_\nu(x\not{p}+\not{q})\gamma_\mu\not{p}\gamma_5\right] = \frac{2x}{Q^2}\mathrm{i}\epsilon_{\nu\alpha\mu\beta}q^\alpha p^\beta$$
$$= -\mathrm{i}\epsilon_{\mu\nu\lambda\sigma}\frac{q^\lambda p^\sigma}{p\cdot q} \ . \qquad (15)$$

For longitudinal polarization $p^\sigma \sim S^\sigma$ we can identify in a similar way

$$g_1(x) = \frac{1}{2}\int \frac{\mathrm{d}x^-}{2\pi}\,\mathrm{e}^{+\mathrm{i}x(p^+x^-)}\sum_f e_f \left\langle pS\left|\bar{q}_f(0)\gamma_5\not{n}q_f(x^-)\right|pS\right\rangle \ . \qquad (16)$$

Hence we have proven that the hadronic tensor can be parametrized in terms of bilocal twist-2 operators along the light cone. In our analysis we confirmed once more the validity of the Callan–Gross relation

$$F_2(x) = +2xF_1(x). \qquad (17)$$

6.2 Small-x Physics

In the previous sections we acquainted ourselves with the leading-log approximation, which enables us to resum all terms which contain factors $\log(t)$ and $t = Q^2/\mu^2$ and are of lowest order in α_s. We now turn, however, to a kinematical range in which the leading-log approximation (LLA) is not sufficient, namely to deep inelastic scattering at very small x, more precisely, for x so small that e.g. $(\alpha_s/\pi)\log(1/x) > 1$. With the completion of the HERA accelerator at DESY this region (down to $x \sim 10^{-5}$) became accessible for the first time, generating an intense theoretical effort to isolate, understand, and resum the relevant graphs. In these reactions Q^2 is still rather large, so that perturbation theory should be applicable.

However, the small-x physics is a very complicated subject with a variety of theoretical approaches. In the literature there exist comprehensive theoretical reviews addressing the advanced reader.[6]

We will restrict ourselves to an overview of existing theoretical approaches and refer the reader to the more detailed original literature. In our overview we follow the presentation by Badelek et al.[7] We shall start by discussing the

[6] L.V. Gribov, E.M. Levin, M.G. Ryskin: Phys. Rept. **100**, 1 (1983), E.M. Levin, M.G. Ryskin: Phys. Rep. **189**, 267 (1990).

[7] B. Badelek, M. Krawczyk, K. Charchula, J. Kwiecinski, DESY 91-124: Rev. Mod. Phys. **64**, 927 (1992).

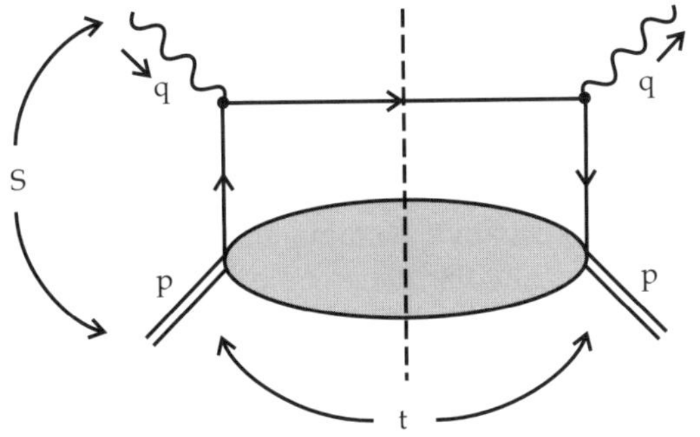

Fig. 6.17. Virtual Compton scattering of a photon with momentum q on a nucleon target

elements of Regge theory and its consequences for deep inelastic scattering on a nucleon. Let us consider once more the DIS process

$$\mathrm{l\,N} \to \mathrm{l'\,X} \ , \tag{6.75}$$

where lepton l scatters off nucleon N and produces the final hadronic state X. l′ denotes the scattered lepton. The interaction is mediated by the exchange of a vector boson, γ or Z^0 for the neutral current, $W^{\pm}$ for charged current interactions. At fixed energy the kinematics of inelastic lepton–nucleon scattering is determined by two independent variables. As usual we can choose the Bjorken parameter $x = Q^2/2p\cdot q$ and $Q^2 = -q^2$, the "mass" of the exchanged boson. q is the four-momentum transfer between the leptons, p the momentum of the proton. As will become apparent, small-x behavior of structure functions for fixed Q^2 reflects the high-energy behavior of the virtual Compton scattering amplitude.

Recall that the Bjorken limit is defined as the limit where Q^2 and $\nu = p\cdot q$ both go to infinty while their ratio $Q^2/2p\cdot q = x$ stays fixed.

The total center-of-mass energy in the process depicted in Fig. 6.17 is

$$s = (p+q)^2 \approx 2pq - Q^2 = \frac{Q^2}{x}(1-x) \ , \tag{6.76}$$

expressed in terms of x and Q^2. The limit $s \to \infty$ corresponds therefore to the limit where $x \to 0$,

$$s \gg Q^2 \leftrightarrow x \ll 1 \ . \tag{6.77}$$

The *small-x* limit of deep inelastic scattering is therefore the limit where the scattering energy is kept much larger than all external masses and momentum tranfers. This is by definition the *Regge limit*. In deep inelastic scattering Q^2 by definition is kept large, the limit where in addition s is large also, is the Regge limit of deep inelastic scattering. The old concepts of Regge theory and Regge phenomenology acquire a new content within the modern concept of QCD. It may be helpful to recapitulate some elements of Regge theory.[8]

[8] See, for example, P.D.B. Collins: *An Introduction to Regge Theory and High-Energy Physics* (Cambridge University Press, Cambridge 1977).

It is known that two-body scattering of hadrons is strongly dominated by small squared momentum transfer t or equivalently by small scattering angles. This behavior is modeled by assuming the exchange of particles with appropriate quantum numbers. Regge exchange is a generalization of this concept. In this description Regge poles rather than particles are exchanged. The Regge poles are characterized by quantum numbers like charge, isospin, C parity, parity, etc.

Especially the Regge pole that carries the quantum numbers of the vacuum ($J^{PC} = 0^{++}$) is called the *pomeron*. Pomeron exchange dominates, for instance, the total cross section of proton–proton and proton–antiproton scattering.

Other Regge poles are called *reggeons*. For example, ϱ-meson exchange can be considered as reggeon exchange. Instead of talking about ϱ exchange in terms of meson exchange it is useful to consider this at the level of quarks and gluons. Thus the exchange of $\mathrm{q\bar{q}}$ pairs that carry quantum numbers different from those of the vacuum can be classified by reggeons; exchange of gluons with quantum numbers of the vacuum can be considered as pomeron exchange (see Fig. 6.18).

Regge pole exchange describes the exchange of states with appropriate quantum numbers and different virtuality t and spin α. The relation between t and α is called the *Regge trajectory*, $\alpha(t)$. Whenever this function passes through an integer (for bosonic Regge poles) or half-integer (for fermionic Regge poles), i.e. $\alpha(t) = n$, $n = 1, 2, \cdots$ or $n = \frac{1}{2}, \frac{3}{2}, \cdots$, there should exist a particle of spin n and mass $M_n = \sqrt{t}$. The trajectory $\alpha(t)$ thus interpolates between particles of different spins.

It can be shown that the exchange of states as described above leads to a pole in the scattering amplitude, or more precisely a pole in the partial wave amplitude. The trajectory $\alpha(t)$ describes the t dependence of the pole of the partial wave amplitude in the complex angular momentum plane.

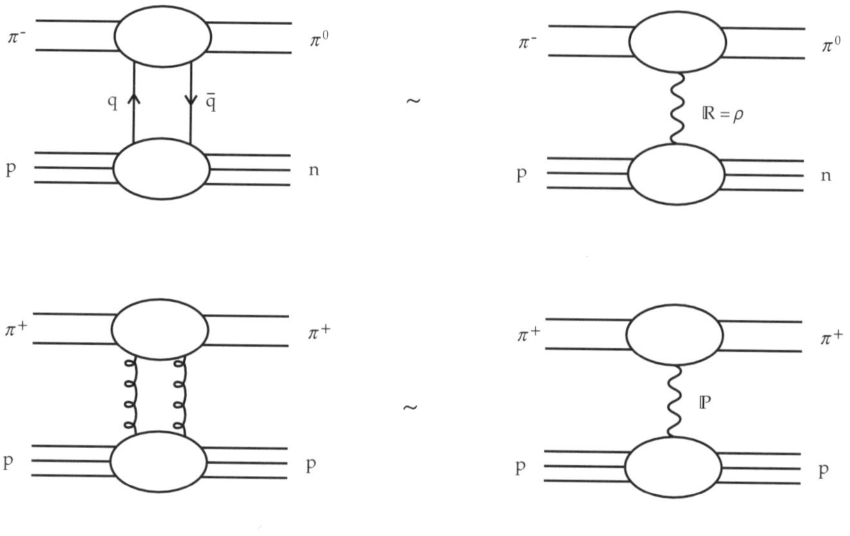

Fig. 6.18. Exchange of a $\mathrm{q\bar{q}}$ pair viewed as reggeon exchange: $\pi^-\mathrm{p} \to \pi^0\mathrm{n}$ scattering at high energies can be parametrized as ϱ exchange. Exchange of a gluon pair corresponds to pomeron exchange in high energy $\pi^+\mathrm{p} \to \pi^+\mathrm{p}$

At high energies the asymptotic behavior of a two-body amplitude can be parametrized as

$$A(s,t) \sim s^{\alpha(t)} \ . \tag{6.78}$$

The optical theorem relates the imaginary part of the forward scattering amplitude to the total cross section:

$$\mathrm{Im}A(s,t=0) = s\sigma^{\mathrm{tot}} \ . \tag{6.79}$$

Regge theory therefore predicts that the total cross section behaves as

$$\sigma^{\mathrm{tot}} = s^{\alpha(0)-1} \ , \tag{6.80}$$

where $\alpha(t=0) = \alpha(0)$ is called the *intercept*.

Non-vacuum trajectories of Regge poles associated with the known meson have intercepts smaller than one ($\alpha_{\mathrm{IR}} \approx \frac{1}{2}$ or less). Thus the total cross section decreases with increasing energy. However, in $\mathrm{p}\bar{\mathrm{p}}$ and pp scattering an increase in the total cross section with energy has been observed and no known reggeon can be attributed to this behavior. High-energy $\mathrm{p}\bar{\mathrm{p}}$ and pp scattering is therefore associated with the *pomeron*, which is assumed to have an intercept close to one, $\alpha_{\mathrm{IP}}(0) \approx 1$. It should be emphasized that the particular nature of the pomeron is not yet completely understood. The name pomeron is in general a name for the mechanism that leads to an increase in the cross section with increasing energy. Note, however, that the increase in the cross section is assymptotically constrained by the Froissart bound;[9] i.e. based on unitarity and analyticity of the amplitude it can be shown that the total cross section cannot increase faster than $\log^2 s$.

It should be stressed that this bound is an asymptotic one: for finite energies the total cross section can still behave like

$$\alpha_{\mathrm{IP}} > 1, \ \sigma^{\mathrm{tot}} \sim s^{\alpha_{\mathrm{IP}}-1} \ .$$

How can Regge theory be applied to deep inelastic lepton nucleon scattering? The natural quantities to consider are the structure functions F_1 and F_2, which are proportional to the total virtual photon nucleon cross section $\gamma^*\mathrm{N} \to \mathrm{X}$. This cross section is expected to show Regge behavior for $s \to \infty$.

We have shown above (see (6.76)) that the high-energy limit $s \to \infty$ corresponds to $x \to 0$, $s \sim \frac{1}{x}Q^2$. In the parton model the structure functions are related to quark antiquark distributions in the nucleon. The Regge behavior of the cross section is reflected in the small-x behavior of the structure function. Due to the quantum numbers of the operators that determine the distribution function, the sea quark distribution and the gluon distribution is expected to reflect the Regge behavior of the pomeron:

$$\begin{aligned} xq^{\mathrm{sea}}(x, Q_0^2) &\sim x^{1-\alpha_{\mathrm{IP}}} \ , \\ xG(x, Q_0^2) &\sim x^{1-\alpha_{\mathrm{IP}}} \ . \end{aligned} \tag{6.81}$$

[9] M. Froissart: Phys. Rev. **123**, 1053 (1961).

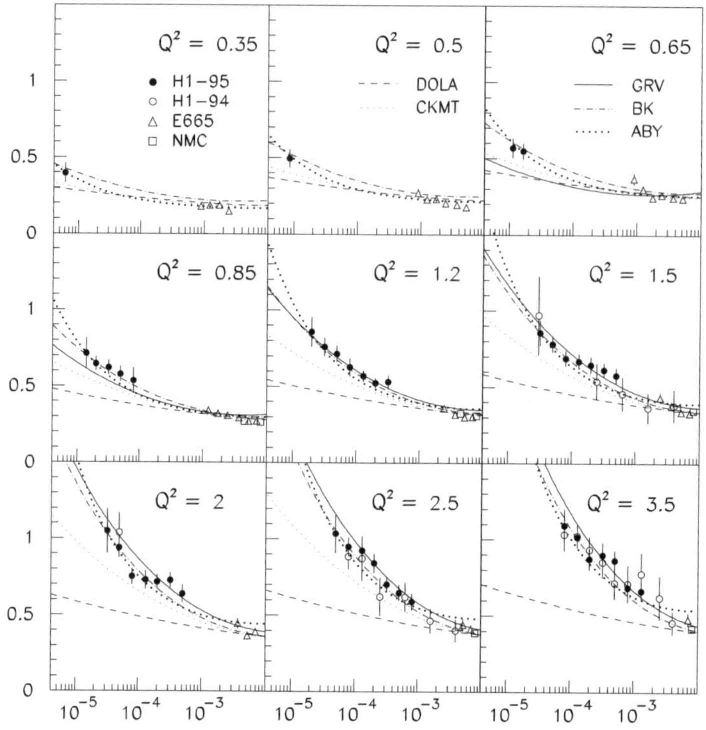

Fig. 6.19. Measurement of the proton structure function $F_2(x, Q^2)$ in the low-Q^2 region by H1 (full points), together with previously published results from H1 (open circles), E665 (open triangles), NMC (open squares). The Q^2 values are given in GeV^2. Various predictions for F_2 are compared with the data: the model of Donnachie and Landshoff (dashed line), the model of Capella et al. (dotted line/small), the model of Badelek, and Kwiecinski (dashed–dotted line), the model of Glück, Reya, and Vogt (full line) and the model of Adel et al. (dotted line/large). Global normalization uncertainties are not included[10]

On the other hand, the behavior of the valence quarks follows the Regge exchange of mesonic poles:

$$xq^{\mathrm{val}}(x, Q_0^2) \sim x^{1-\alpha_{\mathrm{IR}}} \; . \tag{6.82}$$

Assuming $\alpha_{\mathrm{IP}} \sim 1$, $\alpha_{\mathrm{IR}} \sim \frac{1}{2}$ we obtain the Regge prediction:

$$\begin{aligned} xq^{\mathrm{sea}}(x, Q_0^2) &\sim x^0 \; , \\ xG(x, Q_0^2) &\sim x^0 \; , \\ xq^{\mathrm{val}}(x, Q_0^2) &\sim \sqrt{x} \; . \end{aligned} \tag{6.83}$$

A detailed analysis of small-x structure function measurements shows that they are indeed consistent with predictions of Regge theory. For example, the behavior $xG(x, Q_0^2) \sim x^0$, $xq^{\mathrm{sea}}(x, Q_0^2) \sim x^0$ at small x can be observed in Fig. 6.19.[10]

But a modification of this behavior is still seen from Fig. 6.19 if Q^2 increases. The steepening of the behavior at small x as Q^2 increases may be attributed to the perturbative evolution of the structure functions discussed in previous chapters. Indeed HERA data are well described by NLO GLAP-based fits. The analysis

[10] H1 collaboration (C. Adloff et al.): Nucl. Phys. B **497**, 3 (1997).

shows that the $\sim x^0$ behavior of the sea quark and gluon distributions is unstable for $Q^2 > Q_0^2$. The evolution equations generate steeper behavior.

In Chap. 5 we have discussed these evolution equations. In the leading-log approximation, which keeps only the leading power of $\ln(Q^2)$, i.e. $\alpha_s^n \ln^n(Q^2)$, we get the well-known GLAP equations. When a physical gauge is chosen, this approximation corresponds to resumming ladder diagrams with gluon and quark exchange (see Fig. 6.20). When terms with higher powers of the coupling $\alpha_s(Q^2)$ are included, one obtains the next-to-leading-logarithmic approximation. Instead of the next-to-leading-log approximation we are concerned with the small-x limit of the GLAP evolution. To this end we note that the gluon splitting function $P_{\mathrm{GG}}(z)$ behaves as $6/z$ at small z (see, e.g., (5.72)), which is relevant at small x. (Remember that $P_{\mathrm{GG}}(z)$ enters as $P_{\mathrm{GG}}(x/y)$ in the evolution equations.)

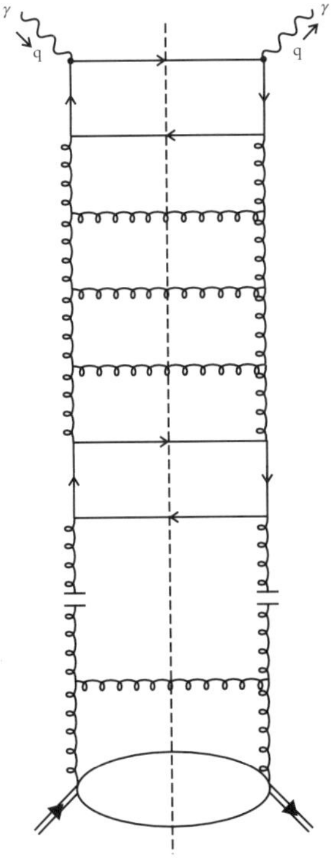

Fig. 6.20. The ladder diagram that contributes to DIS in leading-log approximation

Retaining only these terms in the GLAP equations, one gets maximal powers of both large logarithms $\ln(Q^2)$ and $\ln(1/x)$. This approximation is called the *double logarithmic approximation*. The powers of $\ln(1/x)$ come from the fact that the integrations over the longitudinal momentum fraction become also logarithmic, and, therfore, in the nth order given by the nth iteration of the evolution equations one finds[11]

$$G(x, Q^2) \sim \eta^n \left(Q^2, Q_0^2\right) \frac{\ln^{n-1}\left(\frac{1}{x}\right)}{x(n!)} \left(1 + \mathcal{O}\left(\frac{1}{\ln\left(\frac{1}{x}\right)}\right)\right) , \tag{6.84}$$

where

$$\eta\left(Q^2, Q_0^2\right) = \int\limits_{Q_0^2}^{Q^2} \frac{\mathrm{d}k^2}{k^2} \frac{3\alpha_s(k^2)}{\pi} . \tag{6.85}$$

The iterations can be summed to give

$$xG(x, Q^2) \sim \exp\left(2\sqrt{\eta\left(Q^2, Q_0^2\right) \ln\left(\frac{1}{x}\right)}\right) \tag{6.86}$$

at small x and large Q^2.

The gluon and sea quark distributions are therefore found to grow faster than any power of $\ln(1/x)$ in the small-x limit. Note that the dominant contribution to the sea quarks comes from $\mathrm{q\bar{q}}$ pairs emitted from gluons.

Sometimes it is convenient to discuss the behavior of the structure functions in terms of moments instead of the functions itself. Introducing

$$\tilde{G}_n(Q^2) = \int\limits_0^1 \mathrm{d}x x^{n-1} G(x, Q^2) , \tag{6.87}$$

[11] R.K. Ellis, W.J. Stirling, B.R. Webber: *QCD and Collider Physics* (Cambridge University Press, 1996).

the evolution equations can be solved in closed form (see similar discussion for (5.194)) to give

$$\tilde{G}_n(Q^2) = \tilde{G}_n(Q_0^2) \exp\left(d_{\mathrm{GG}}(n)\eta\left(Q^2, Q_0^2\right)\right) \tag{6.88}$$

where

$$d_{\mathrm{GG}}(n) = \frac{1}{6}\int_0^1 \mathrm{d}z z^{n-1} P_{\mathrm{GG}}(z) \tag{6.89}$$

is the moment of the splitting function and $\tilde{G}_n(Q_0^2)$ corresponds to the starting contribution of Q_0^2. If we keep only the most singular term in P_{GG} then $d_{\mathrm{GG}}(n)$ becomes

$$d_{\mathrm{GG}}(n) \approx \frac{1}{(n-1)} \,. \tag{6.90}$$

From (6.90) it follows that the moment $\tilde{G}_n(Q^2)$ has an essential singularity at $n = 1$. This is the leading singularity and it generates the small-x behavior given by (6.86).

The solution of the evolution equations for the moments $q^{\mathrm{val}}(Q^2)$ of the valence quark distribution is accordingly

$$\tilde{q}_n^{\mathrm{val}}(Q^2) = \tilde{q}_n^{\mathrm{val}}(Q_0^2) \exp\left(\gamma_{qq}(n)\eta\left(Q^2, Q_0^2\right)\right) \tag{6.91}$$

with

$$d_{qq}(n) = \frac{1}{6}\int_0^1 \mathrm{d}z z^{n-1} P_{qq} \tag{6.92}$$

and the leading pole of the anomalous dimensions in $d_{qq}(n)$ is at $n = 0$ (see Example 5.8). The moment of the starting distribution will have a pole at $n = 1/2$ due to the general Regge arguments given above. Thus this pole will remain the leading singularity of the moment $q_n^{\mathrm{val}}(Q^2)$ and the small-x behavior of the valence quark distribution will consequently remain unchanged by QCD evolution.

Attempts have been made to go beyond the double logarithmic approximation. To explain this it should be noted that the leading logs can be traced to those contributions in the diagram (Fig. 6.20) where the momenta are strongly ordered. In those diagrams the longitudinal momenta $\sim x_i$ are ordered along the chain ($x_i \geq x_{i+1}$) and the transverse momenta are strongly ordered as well, (i.e. $k_{\perp i}^2 \ll k_{\perp i+1}^2$). It is the strong ordering of transverse momenta towards Q^2 which gives the maximal power of $\ln(Q^2)$, since the integration over transverse momentum in each cell is logarithmic.

Going beyond the double logarithmic approximation means summing those terms which contain the leading $\ln(1/x)$ and retain the full Q^2 dependence.

This approximation leads to the Balitsky–Fadin–Kuraev–Lipatov (BFKL) equation.[12] In general the summation of $\ln(1/x)$ terms is refered to as the *leading* $\ln(1/x)$ *approximation*. This approximation gives the bare pomeron in QCD. The diagrams are the same as mentioned above, yet the transverse momenta of the gluons are no longer ordered. The equation that sums these diagrams is

$$f(x,k^2) = f0(x,k^2) + \frac{3\alpha_s(k^2)}{\pi}k^2 \times \int_x^1 \frac{\mathrm{d}x'}{x'} \int_{k_0^2}^{\infty} \frac{\mathrm{d}k'^2}{k'^2} \left\{ \frac{f(x',k'^2) - f(x',k^2)}{|k'^2 - k^2|} + \frac{f(x',k^2)}{\sqrt{4k'^4 + k^4}} \right\} , \tag{6.93}$$

where the function $f(x,k^2)$ is the nonintegrated gluon distribution, i.e.

$$f(x,k^2) = \frac{\partial xG(x,k^2)}{\partial \ln(k^2)} , \tag{6.94}$$

where k^2, k'^2 are the transverse momenta squared of the gluon in the initial and final states respectively; k_0^2 is a cutoff.

When effects of the running coupling are neglected, i.e. if a fixed coupling $\alpha_s(k^2) = \alpha_0$ is used, the BFKL equation can be solved analytically. And at small x one obtains

$$xG(x,Q^2) \sim \frac{x^{1-\alpha_{\mathrm{IP}}^{\mathrm{B}}}}{[\ln(x)]^{\frac{1}{2}}} \left(1 + \mathcal{O}\left(\frac{1}{\ln(x)}\right)\right) \tag{6.95}$$

with

$$\alpha_{\mathrm{IP}}^{\mathrm{B}} = 1 + \frac{12\alpha_0}{\pi}\ln(2) = 1 + \omega_{\mathrm{IP}}^{\mathrm{B}} , \tag{6.96}$$

which corresponds to the intercept of the bare QCD pomeron. It should be noticed that this leads to a large number $\alpha_{\mathrm{IP}}^{\mathrm{B}} > \frac{3}{2}$ for typical values of α_0. Taking the running of the coupling into account this picture does not change very much. Still the pomeron intercept is quite large, $\alpha_{\mathrm{IP}}^{\mathrm{B}} \sim \frac{3}{2}$. This is in contrast to the nonperturbative "soft" pomeron, which is used to fit data in high-energy pp collisions. This intercept is 1.08.[13]

Note that (6.95) and (6.96) suggest that the gluon distribution (multiplied by x), i.e. the function $xG(x,Q^2)$, can grow arbitrarily in the small-x limit. Quite obviously such behavior is forbidden by the finite size of the nucleon.

At a certain stage the gluons can no longer be treated as free particles. They begin to interact with each other. This interaction leads to screening and shadowing effects so that an infinite increase in the number density is tamed. These

[12] See e.g. L. Lipatov, in *Perturbative QCD*, ed. by A.H. Mueller (World Scientific, Singapore 1989), p. 411 and references therein.

[13] A. Donnachie and P.V. Landshoff: Nucl. Phys. B **231**, 189 (1984).

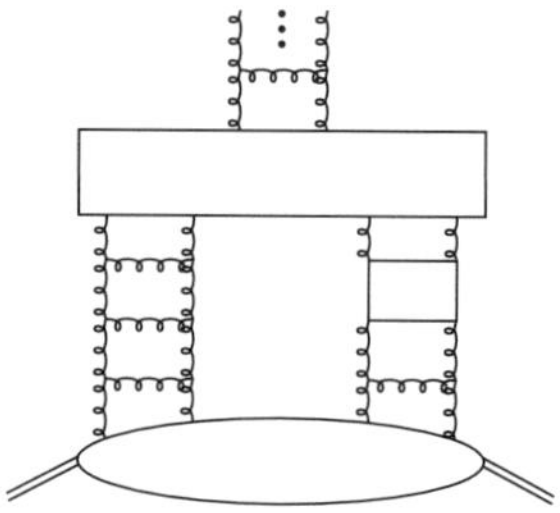

Fig. 6.21. Graphical representation of the quadratic shadowing term in the evolution equation (6.97). The box represents all possible pertubative QCD diagrams which couple four gluons to two gluons. The lower blob represents the nucleon

shadowing effects modify the evolution equations as well as the BFKL equation by nonlinear terms. Including shadowing corrections, the BFKL equation assumes the following form

$$-x\frac{\partial f(x,k^2)}{\partial x} = \frac{3\alpha_s(k^2)}{\pi}k^2 \int_{k_0^2}^{\infty} \frac{\mathrm{d}k'^2}{k'^2}\left\{\frac{f(x,k'^2)-f(x,k^2)}{|k'^2-k^2|} + \frac{f(x,k^2)}{\left[4k'^4+k^4\right]^{\frac{1}{2}}}\right\}$$
$$-\frac{81}{16}\frac{\alpha_s^2(k^2)}{R^2k^2}\left[xG(x,k^2)\right]^2 . \qquad (6.97)$$

This equation is called the Gribov–Levin–Ryskin (GLR) equation.[14] The second term on the right-hand side describes the shadowing effects. Note that this term is quadratic in the gluon distributions. The parameter R describes the size of the region within which the gluons are concentrated (see Fig. 6.21). In Figure 6.22 we have summarized the various regions in the $\frac{1}{x}$, Q^2 plane where the different equations and phenomenological descriptions might be applicable and in which direction the evolution manifests itself.

The complete discussion of BFKL and GLR equations would be far too lengthy, and, furthermore, the validity of these equations is presently very much under discussion. In particular, doubt was cast on the validity of the BFKL equation since NLO corrections were calculated which turned out to be larger

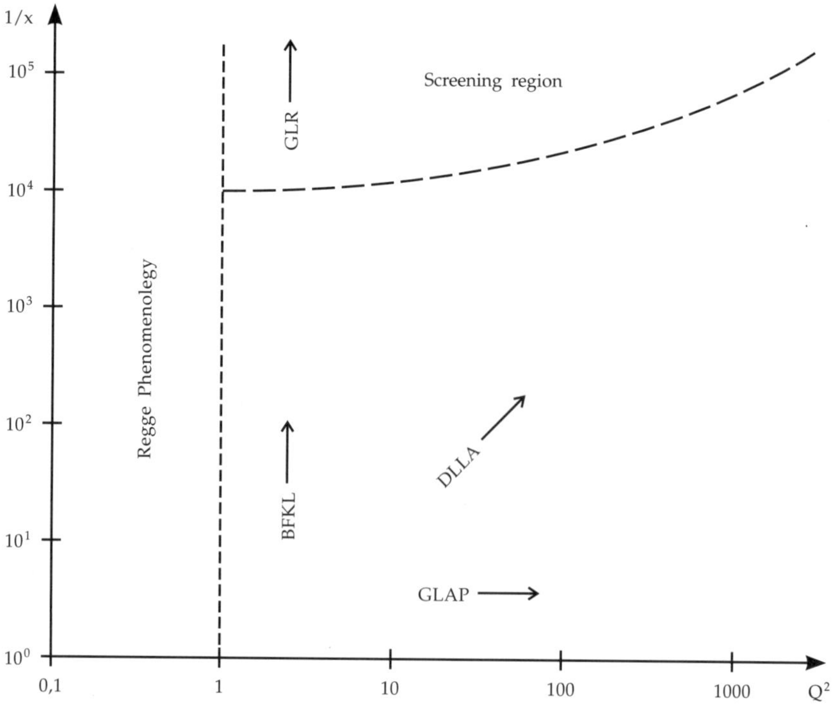

Fig. 6.22. The various evolution equations depicted in the $(1/x, Q^2)$ plane as discussed in the text

[14] L.V. Gribov, E.M. Levin, M.G. Ryskin: Phys. Rep. **100**, 1 (1983).

than the leading-order result.[15] This situation is not yet understood. Arguments exist that a complete resummation of various terms would lead to a more acceptable picture or that additional nonperturbative input is needed.

Refering to the leading-order result for the pomeron intercept as it follows from the BFKL equation

$$\omega_{\mathrm{IP}}^{\mathrm{B}} = \frac{12\alpha_0}{\pi} \ln(2) \ , \tag{6.98}$$

the corrected (NLO) intercept ω_{IP} can be written as

$$\omega_{\mathrm{IP}} = \omega_{\mathrm{IP}}^{\mathrm{B}} \left(1 - \frac{r\left(\frac{1}{2}\right)}{4\ln(2)} \omega_{\mathrm{IP}}^{\mathrm{B}} \right) = \omega_{\mathrm{IP}}^{\mathrm{B}} \left(1 - 2.4\omega_{\mathrm{IP}}^{\mathrm{B}} \right) \ , \tag{6.99}$$

where $r\left(\frac{1}{2}\right)$ is the eigenfunction of the BFKL kernel with the largest eigenvalue (for details see the original literature[15]). Discarding details, even if the coefficient 2.4 looks not very large, since $\omega_{\mathrm{IP}}^{\mathrm{B}}$ itself is not small, the NLO result completely cancels the leading-order one. For example, if $\alpha_0 = 0.15$, where the Born intercept is $\omega_{\mathrm{IP}}^{\mathrm{B}} = 12\alpha_0 \ln(2) = 0.397$, the relative correction for $n_f = 0$ is big:

$$\frac{\omega_{\mathrm{IP}}}{\omega_{\mathrm{B}}} = 1 - r\frac{1}{2}\frac{\alpha_0}{\pi} \cdot 3 = 0.0747 \ . \tag{6.100}$$

Whether the NNLO calculation or partial resummation of big corrections will be able to cure this problem remains to be seen.

[15] V. Fadin and L. Lipatov: Phys. Lett. B **429**, 127 (1998),
G. Camici and M. Ciafaloni: Phys. Lett. B **430**, 349 (1998).

7. Nonperturbative QCD

In the previous chapters we discussed mainly QCD effects as calculated in perturbation theory. We found reasonable agreement between the general theory of strong interaction and experimental results in high-energy scattering. However, there are other regimes in QCD that are much harder to treat theoretically. This problem had already been illustrated in Example 4.4. In fact, as can be seen in Fig. 4.8, the running coupling constant $\alpha(q^2)$ increases at small momentum transfers, reaching a value comparable to 1 at momenta q around $\sqrt{|q^2|} \sim 500$ MeV. Obviously a power-series expansion in a quantity of the order of one does not converge: $\alpha \approx \alpha^n$. Feynman graphs with many vertices (Fig. 7.1a) have an expansion coefficient of the same order as the most simple ones that carry only one vertex.

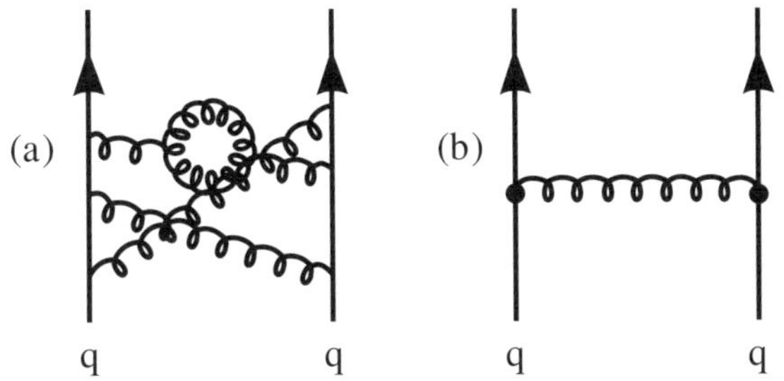

Fig. 7.1a,b. Feynman diagram for quark–quark interaction: (**a**) complicated gluonic exchange; (**b**) simple one-gluon exchange. In the low-momentum regime process (**a**) is not suppressed compared to (**b**) due to the break downof the perturbation expansion, yielding an infinite number of relevant diagrams that (in principle) need to be computed

This leads us to the general problem that QCD at small momenta or energies $E \leq 1$ GeV has to be treated nonperturbatively. The properties of a proton, for instance, with a mass of $M_{\mathrm{P}} = 938.3\ \mathrm{MeV}/c^2$ are in this regime. Therefore, if we want to understand protons (and, in fact, all hadrons), perturbation theory won't be of help and another approach has to be found. Nonperturbative systems are inherently difficult to treat, and there are generally no simple systematic methods available for solving them. In the course of this chapter we will look at two very different ways to tackle this problem: lattice gauge theory and sum rule techniques.

7.1 Lattice QCD Calculations

The idea of calculating QCD numerically on a lattice is directly tied to the path-integral or functional-integral representation of quantum mechanics and quantum field theory. Before continuing further with the derivation of the QCD Lagrangian on the lattice we will first discuss some basic notions of the method of path integration in order to provide the reader with some understanding of the background of the lattice formulation. A much more detailed discussion of this topic can be found in the book on field quantization.[1]

7.1.1 The Path Integral Method

Let us consider a particle propagating from a starting point x_i at time t_i to a point $x_\mathrm{f}(t_\mathrm{f})$. For a given (not necessarily classical) trajectory $x(t)$ and $\dot{x}(t)$ of the particle connecting the initial and final points we may determine the corresponding action S given its Lagrangian L:

$$S = \int_{t_\mathrm{i}}^{t_\mathrm{f}} \mathrm{d}t\, L\,[x(t), \dot{x}(t)] \ . \tag{7.1}$$

The action S depends on the position $x(t)$ and the velocity $\dot{x}(t)$ along the path. As we are discussing quantum mechanics the connecting path does not have to be the solution of the classical equations of motion. The path integral method, largely developed and propagated by Feynman following earlier ideas of Dirac,[2] states that the total probability amplitude of the transition of the particle from $x_\mathrm{i}(t_\mathrm{i})$ to $x_\mathrm{f}(t_\mathrm{f})$ can be described by a weighted sum over all possible trajectories (paths P) x_P between those two points:

$$\langle x_\mathrm{f}(t_\mathrm{f})|x_\mathrm{i}(t_\mathrm{i})\rangle \sim \sum_{\mathrm{paths}(x_\mathrm{P},\dot{x}_\mathrm{P})} \exp(\mathrm{i}S(\{x(t), \dot{x}(t)\})) \ . \tag{7.2}$$

We approximate the time evolution of the system, discretizing the time coordinate into $N-1$ intermediate time slices t_n ($n = 1, \ldots, N-1$):

$$\begin{aligned} \Delta t &= \frac{t_\mathrm{f} - t_\mathrm{i}}{N} \ , \\ t_n &= t_\mathrm{i} + n\Delta t \ . \end{aligned} \tag{7.3}$$

---

[1] W. Greiner and J. Reinhardt: *Field Quantization* (Springer, Berlin, Heidelberg 1996).
[2] P.A.M. Dirac: *Physikalische Zeitschrift der Sowjetunion* **3**, 64 (1933). Also see the monograph by R. P. Feynman and A. R. Hibbs: *Quantum Mechanics and Path Integrals* (McGraw-Hill, 1965).

One can now 'sum' over the trajectories at the different time slices. The sum over the paths P at a specific time t_n is given by

$$\sum_{P(t=t_n)} \exp(\mathrm{i}S(\{x(t_\mathrm{i}), \dot{x}(t_\mathrm{i})\})) \sim \int\limits_{-\infty}^{\infty} \mathrm{d}x(t_n) \exp(\mathrm{i}S(\{x(t_\mathrm{i}), \dot{x}(t_\mathrm{i})\})) \; ,$$

varying over all possible positions $x(t_n)$ of the particle at this time step. The corresponding velocity can be determined from the trajectory of the particle. This procedure can be used for every time step (see Fig. 7.2), yielding the transition amplitude

$$\langle x_\mathrm{f}(t_\mathrm{f})|x_\mathrm{i}(t_\mathrm{i})\rangle \sim \int \prod_{n=1}^{N-1} \mathrm{d}x(t_n) \exp(\mathrm{i}S(\{x(t_n), \dot{x}(t_n)\})) \; . \tag{7.4}$$

Strictly speaking, one has to perform a careful limit for an infinite number ($N \to \infty$) of time slices and corresponding integrations in this formula for the exact expression:

$$\langle x_\mathrm{f}(t_\mathrm{f})|x_\mathrm{i}(t_\mathrm{i})\rangle = \mathcal{N} \lim_{N\to\infty} \int \prod_{n=1}^{N-1} \mathrm{d}x(t_n) \exp(\mathrm{i}S(\{x(t_n), \dot{x}(t_n)\})) \tag{7.5}$$

with a normalization factor $\mathcal{N}$. As can be seen from (7.5), the discretization of the time variable (a *time lattice*) is quite useful for formulating the idea of summing over all paths in a practical manner. After studying the dynamics of a single particle, i.e. one-body quantum mechanics, we can apply the same general approach to fields. For simplicity we consider a real scalar field $\Phi(x, t)$ in one spatial dimension. The extension to higher dimensions is straightforward. Again we study the transition amplitude for the field having a value $\Phi_\mathrm{i}(x, t_\mathrm{i})$ at the beginning and $\Phi_\mathrm{f}(x, t_\mathrm{f})$ at the end. The path integral method states that the

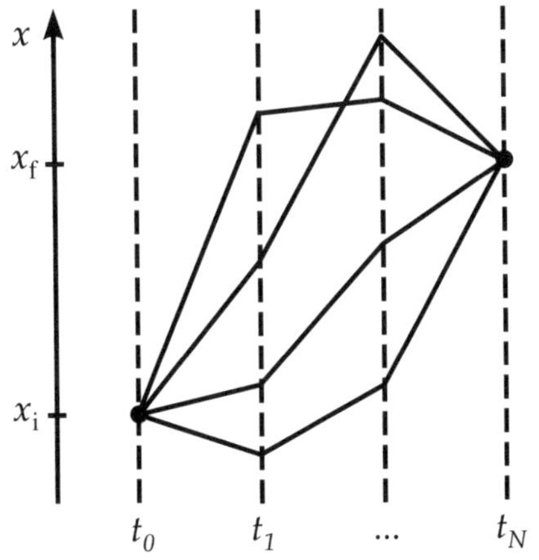

Fig. 7.2. Paths in space–time between $x_\mathrm{i}(t_0)$ and $x_\mathrm{f}(t_N)$ for discrete time steps. Every possible path is taken into account

amplitude

$$\langle \Phi_{\rm f}(x,t_{\rm f})|\Phi_{\rm i}(x,t_{\rm i})\rangle \sim \sum_{\Phi_{\rm P}} \exp\left(\mathrm{i}S\left(\{\Phi_{\rm P}\},\{\partial_\mu \Phi_{\rm P}\}\right)\right) \tag{7.6}$$

is a weighted sum over all connecting paths P. We subdivide the time evolution into $N-1$ steps as before using (7.3). In addition, as we consider space-dependent fields, we will also discretize the space coordinates x into N_x equally spaced points $x_1,\ldots,x_{N_x}$. Given this system of a finite number of points in time and space directions as an approximation of the real continuous system, the transition amplitude reads

$$\begin{aligned}\langle \Phi_{\rm f}(x,t_{\rm f})|\Phi_{\rm i}(x,t_{\rm i})\rangle = \mathcal{N} \lim_{N_x\to\infty, N\to\infty} \int \prod_{m=1}^{N_x}\prod_{n=1}^{N-1} \mathrm{d}\Phi(x_m,t_n)\\ \times \exp\left(\mathrm{i}S\left[\left\{\Phi(x_m,t_n),\partial_\mu\Phi(x_m,t_n)\right\}\right]\right)\ .\end{aligned} \tag{7.7}$$

This is the formulation of the *transition* matrix element within the path integral formalism in quantum field theory. Another typical feature of lattice gauge calculations is that one prefers to calculate quantities in the imaginary time formalism (the reason for using imaginary time will become clearer in Sect. 7.1.8 on Monte Carlo methods). Formally this is done by replacing the real time t by imaginary time τ using the equation $t=-\mathrm{i}\tau$. This corresponds to a Wick rotation by $\pi/2$ in the complex t plane (see Sect. 4.3 for a Wick rotation in Fourier space). In principle one can rotate results back to real time after finishing the calculation. However, in practice this turns out to be rather difficult. For the scalar field just discussed one introduces slices in the imaginary time coordinate τ. Take the action of a scalar field:

$$S=\int \mathrm{d}t\,\mathrm{d}^3x\left[\frac{1}{2}\partial_\mu\Phi\partial^\mu\Phi - V(\Phi)\right] \tag{7.8}$$

with some potential or mass term summarized in $V(\Phi)$. We get

$$S=-\mathrm{i}\int \mathrm{d}\tau\,\mathrm{d}^3x\left[\frac{1}{2}(-\partial_\tau\Phi\partial_\tau\Phi - \nabla\Phi\nabla\Phi) - V(\Phi)\right]\ . \tag{7.9}$$

With $\tau \equiv x_4$ the action can be written as

$$S=\mathrm{i}\int \mathrm{d}^4x\left[\frac{1}{2}\partial_\mu\Phi\partial_\mu\Phi + V(\Phi)\right]\ , \tag{7.10}$$

where the summation over μ runs over $\mu=1,\ldots,4$. Note that in this formulation the metric tensor now is simply $g_{\mu\nu}\equiv\delta_{\mu\nu}=\mathrm{diag}(1,1,1,1)$ so there is no difference between subscripts and superscripts, i.e. space–time is Euclidean.[3]

[3] It is customary to remove the factor i in front of (7.10) from the definition of the action. The Euclidean action $S^{\rm E}$ is defined as $S^{\rm E}=-\mathrm{i}S(t=-\mathrm{i}\tau)$.

The phase factor in the path integral formula (7.7) now reads

$$\exp(\mathrm{i}S) \to \exp(-S^{\mathrm{E}}) \ . \tag{7.11}$$

Thus for real and positive actions (or at least for actions bounded from below) the phase factor has become a weighting factor so that one can adopt importance sampling techniques for the evaluation of the expression. This will be very important for actual numerical calculations (see Sect. 7.1.8)

The Euclidean formulation allows a very direct connection of (7.7) with the partition function Z of the theory of statistical mechanics. The partition function is given by

$$Z = \sum_{\Phi} \langle \Phi(x) | \exp(-\beta H) | \Phi(x) \rangle \ , \tag{7.12}$$

summing over all possible states $|\Phi(x)\rangle$. $\beta = 1/(kT)$ is the inverse temperature of the system. We rewrite this expression as

$$\begin{aligned} Z &= \sum_{\Phi} \langle \exp(-H\beta)\Phi(x) | \Phi(x) \rangle \\ &= \sum_{\Phi} \langle \Phi(x, t = -\mathrm{i}\beta) | \Phi(x, t = 0) \rangle \ , \end{aligned} \tag{7.13}$$

which corresponds to the transition amplitude (7.7). The difference is that now the initial and final states Φ_{f}, Φ_{i} are the same in (7.12) and the variables over which the sum is performed, which can be incorporated by imposing periodic boundary conditions on Φ in time direction with a period $\tau = \beta$. Thus we have

$$\begin{aligned} Z &= \mathcal{N} \int \lim_{N_x \to \infty, N \to \infty} \prod_{n=1}^{N} \prod_{m=1}^{N_x} \mathrm{d}\Phi(x_n, \tau_m) \exp(-S^{\mathrm{E}}(\{x_n(t_m)\})) \\ &= \mathcal{N} \int [\mathrm{d}\Phi] \exp(-S^{\mathrm{E}}) \ . \end{aligned} \tag{7.14}$$

The bracketed term $[\mathrm{d}\Phi]$ is a short-hand notation for the product of the whole integration measure. We will see later that the normalization factor $\mathcal{N}$ drops out when observables are calculated.

EXERCISE

7.1 Derivation of the Transition Amplitude (7.5)

Problem. Derive the transition amplitude (7.5) from the concept of path integrals.

Solution. We extract the time dependence of the state vectors

$$T \equiv \langle x_{\mathrm{f}}(t_{\mathrm{f}}) | x(t_{\mathrm{i}}) \rangle = \langle x_{\mathrm{f}} | \exp\left(-\mathrm{i}\hat{H}(t_{\mathrm{i}} - t_{\mathrm{f}})\right) | x \rangle \ . \tag{1}$$

Exercise 7.1

In the next step we divide the time interval $(t_\mathrm{f}-t_\mathrm{i})$ into $N-1$ intermediate time steps with $\Delta t=(t_\mathrm{f}-t_\mathrm{i})/N$ as done before in (7.3):

$$T=\langle x_\mathrm{f}\equiv x_N|\underbrace{\exp\left(-\mathrm{i}\hat{H}\Delta t\right)\exp\left(-\mathrm{i}\hat{H}\Delta t\right)\cdots}_{N\text{ times}}|x_\mathrm{i}\equiv x_0\rangle \tag{2}$$

and insert complete sets of eigenstates of the positions between each step

$$\begin{aligned}T=&\langle x_N|\exp\left(-\mathrm{i}\hat{H}\Delta t\right)\int \mathrm{d}x_{N-1}|x_{N-1}\rangle\langle x_{N-1}|\exp\left(-\mathrm{i}\hat{H}\Delta t\right)\\&\times\int \mathrm{d}x_{N-2}|x_{N-2}\rangle\langle x_{N-2}|\cdots\exp\left(-\mathrm{i}\hat{H}\Delta t\right)|x_0\rangle\ .\end{aligned} \tag{3}$$

In addition we also insert complete sets of momentum eigenstates into (3)

$$\begin{aligned}T=&\int \mathrm{d}x_1\,\mathrm{d}x_2\,\mathrm{d}x_3\cdots\mathrm{d}x_{N-1}\frac{\mathrm{d}p_0}{2\pi}\cdots\frac{\mathrm{d}p_{N-1}}{2\pi}\\&\times\langle x_N|p_{N-1}\rangle\langle p_{N-1}|\exp(-\mathrm{i}\hat{H}\Delta t)|x_{N-1}\rangle\\&\times\langle x_{N-1}|p_{N-2}\rangle\langle p_{N-2}|\exp(-\mathrm{i}\hat{H}\Delta t)|x_{N-2}\rangle\\&\times\cdots\times\langle p_0|x_0\rangle\ .\end{aligned} \tag{4}$$

The main purpose of the whole procedure is that for sufficiently large N and small Δt one can expand the exponential

$$\exp(-\mathrm{i}\hat{H}\Delta t)\sim 1-\mathrm{i}\Delta t\hat{H}\ . \tag{5}$$

In (4) the operator $\hat{H}$ is sandwiched between eigenstates of the position and momentum operator. Thus, using $\hat{x}|x\rangle=x|x\rangle$ and $\hat{p}|p\rangle=p|p\rangle$ we can replace the operators in $\hat{H}$ by their eigenvalues:[4] $\langle p|\hat{H}(\hat{p},\hat{x})|x\rangle=\langle p|H(p,x)|x\rangle=\langle p|x\rangle H(p,x)$. This is the main trick used in the path integral formalism: operators are replaced by c numbers at the cost of (infinitely) many sets of basis states representing the possible paths of the particle. Using this results and $\langle x|p\rangle=\exp(\mathrm{i}px)$ we obtain the expression

$$\begin{aligned}T=&\int\prod_{n=1}^{N-1}\mathrm{d}x_n\prod_{m=0}^{N-1}\frac{\mathrm{d}p_m}{2\pi}\exp\left(\mathrm{i}\sum_{n=0}^{N-1}p_n(x_{n+1}-x_n)\right)\\&\times\prod_{n=0}^{N-1}(1-\mathrm{i}\Delta tH(p_n,x_n))\\\underset{N\to\infty}{=}&\ \mathcal{N}\int[\mathrm{d}x][\mathrm{d}p]\exp(\mathrm{i}\int_t^{t'}\mathrm{d}\tilde{t}(p\dot{x}-H(p,x)))\ .\end{aligned} \tag{6}$$

[4] Some care has to be taken with the ordering of the operators when the Hamiltonian contains mixed products of position and momentum operators.

Exercise 7.1

Equation (6) is the path integral in phase space. Depending on the specific structure of H it is rather straightforward to get rid of the momentum integration if H is quadratic in the momentum, e.g.

$$H = \frac{p^2}{2m} + U(x) \ . \tag{7}$$

Using this expression we see that we can complete the square with respect to the momentum variable in (6):

$$T = \mathcal{N} \int [\mathrm{d}x][\mathrm{d}p] \exp\left(-\mathrm{i} \int_t^{t'} \mathrm{d}\tilde{t} \left(\frac{1}{2m}(p - m\dot{x})^2 - \frac{m\dot{x}^2}{2} + U(x) \right) \right) \ . \tag{8}$$

Now one can transform the variable $p' = p - m\dot{x}$, which amounts to a Legendre transformation, and integrate the Gaussian integral of p', which generates a constant that can be absorbed in the normalization factor:

$$\begin{aligned} &= \mathcal{N}' \int [\mathrm{d}x] \exp\left(\mathrm{i} \int_t^{t'} \left[\mathrm{d}\tilde{t} \frac{m\dot{x}^2}{2} - U(x) \right] \right) \\ &= \mathcal{N}' \int [\mathrm{d}x] \exp(\mathrm{i}S) \ . \end{aligned} \tag{9}$$

Equation (9) is the path integral formula in the Lagrange formalism (7.5), here in Minkowski space–time.

7.1.2 Expectation Values

One can calculate expectation values of operators in the path integral method in a very straightforward way. One important observable in quantum mechanics and field theory is the two-point correlation function or propagator, defined as

$$G(t_1, t_2) = \langle 0 | T(\hat{q}(t_1)\hat{q}(t_2)) | 0 \rangle \tag{7.15}$$

in the case of a one-body quantum-mechanical system. Considering the quantum-mechanical case it describes the correlation of the position of a particle at different times t_1, t_2, where $|0\rangle$ is the ground state of the system. The time evolution of a general state is

$$\begin{aligned}|q(t)\rangle &= \exp(-\mathrm{i}Ht)|q(0)\rangle \\ &= \exp(-\mathrm{i}Ht)\sum_n |n\rangle\langle n|q(0)\rangle \\ &= \sum_n \exp(-\mathrm{i}E_n t)|n\rangle\langle n|q(0)\rangle \ ,\end{aligned} \tag{7.16}$$

where a complete set of energy eigenstates with $H|n\rangle = E_n|n\rangle$ has been inserted. Let us now consider the path integral with $t < t_1, t_2, t' > t_1, t_2$, and $t_1 > t_2$. Using (7.16) we get

$$\begin{aligned}\langle q(t')|\hat{q}(t_1)\hat{q}(t_2)|q(t)\rangle = &\sum_{n,m} \exp(+\mathrm{i}(E_n t - E_{n'}t'))\varrho_n^*(q)\varrho_{n'}(q') \\ &\times \langle n'|\hat{q}(t_1)\hat{q}(t_2)|n\rangle\end{aligned} \tag{7.17}$$

with $\varrho_n(q) \equiv \langle q|n\rangle$. We switch to imaginary time $\tau = \mathrm{i}t$ as discussed before:

$$\begin{aligned}\langle q(\tau')|\hat{q}(\tau_1)\hat{q}(\tau_2)|q(\tau)\rangle = &\sum_{n,m} \exp(+E_n(\tau - \tau_2) - E_{n'}(\tau' - \tau_1)) \\ &\times \varrho_n^*(q)\varrho_{n'}(q')\langle n'|\hat{q}(\tau_1)\hat{q}(\tau_2)|n\rangle \ .\end{aligned} \tag{7.18}$$

With $\tau \to -\infty$ and $\tau' \to +\infty$ the sum is dominated by the groundstate contribution $n = m = 0$

$$\begin{aligned}\langle q(+\infty)|\hat{q}(\tau_1)\hat{q}(\tau_2)|q(-\infty)\rangle = &\exp(+E_0(\tau - \tau'))\varrho_0(q)\varrho_0(q') \\ &\times \langle 0|\hat{q}(\tau_1)\hat{q}(\tau_2)|0\rangle \ .\end{aligned} \tag{7.19}$$

By considering the ratio

$$\frac{\langle q'(\tau')|\hat{q}(\tau_1)\hat{q}(\tau_2)|q(\tau)\rangle}{\langle q'(\tau')|q(\tau)\rangle} \tag{7.20}$$

and using (7.19) for the nominator and denominator we get

$$\frac{\langle q'(\tau')|\hat{q}(\tau_1)\hat{q}(\tau_2)|q(\tau)\rangle}{\langle q'(\tau')|q(\tau)\rangle} = \langle 0|\hat{q}(\tau_1)\hat{q}(\tau_2)|0\rangle \qquad (\tau \to -\infty, \tau' \to \infty) \ . \tag{7.21}$$

The path integral (7.4) is just the transition amplitude between different states, i.e. the denominator of expression (7.21). Following the earlier argument one can again split the time interval into many intermediate steps and replace the operators $\hat{q}(t_1)$, $\hat{q}(t_2)$ by their eigenvalues.

Note that by inserting a sequence of intermediate states one automatically generates the time-ordered product $T(\hat{q}(t_1)\hat{q}(t_2))$. Altogether we have (in real time)

$$\langle 0|T\left(\hat{q}(t_1)\hat{q}(t_2)\right)|0\rangle = \frac{\int [\mathrm{d}q]\, q(t_1)q(t_2)\exp(\mathrm{i}S(q,\dot{q}))}{\int [\mathrm{d}q]\exp(\mathrm{i}S(q,\dot{q}))} \ . \tag{7.22}$$

In the same way as for the extension of the path integral from quantum mechanics to quantum field theory, (7.22) can be used in a field-theoretical context, too. For general operators $\hat{O}$ containing s field operators $\hat{\Phi}$ one obtains the vacuum expectation value

$$\langle 0|\hat{O}\left(\hat{\Phi}\right)|0\rangle = \frac{\int [\mathrm{d}\Phi]\, O(\Phi)\exp(\mathrm{i}S(\{\Phi\},\{\partial_\mu\Phi\}))}{\int [\mathrm{d}\Phi]\exp(\mathrm{i}S(\{\Phi\},\{\partial_\mu\Phi\}))} \ , \tag{7.23}$$

where $O(\Phi)$ is the c-number expression with respect to the fields Φ corresponding to the operator $\hat{O}$. Equation (7.23) is the basic formula for calculating observables on the lattice. The straightforward possibility of discretization of the expression and the fact that there are no operators in the path integral formalism renders this approach very suitable for numerical calculations. However, first the theory of QCD has to be formulated on a lattice, which will be discussed in the following section.

7.1.3 QCD on the Lattice

The basic idea of a lattice calculation is to reduce the infinite number (more precisely, the continuum) of field variables to a finite tractable number by discretizing space and time. In the following we (and most of the current state-of-the-art calculations) will adopt the most simple discretization. We introduce a hypercubic equally spaced lattice in space and time with coordinates (time being the 4th coordinate x_4)

$$x_\mu \to x(i,j,k,t) = (i\boldsymbol{e}_1 + j\boldsymbol{e}_2 + k\boldsymbol{e}_3 + t\boldsymbol{e}_4)a \ , \tag{7.24}$$

where a is the lattice spacing, the distance between neighboring lattice points. In the next step we have to formulate the QCD gauge theory on this lattice structure, i.e. find a discretized version of (4.57). The guideline for describing a gauge theory on the lattice is the implementation of the *gauge principle* on the lattice structure. The Lagrangian should be gauge invariant, that is, invariant with respect to arbitrary local phase rotations of the fields in color space. This was also one of the essential requirements (the conceptual basis of a gauge theory) in the case of continuous space–time. In Example 4.1 we discussed the geometric properties of a gauge theory and introduced the notion of *parallel transport*. This is exactly what is needed in this context. One has to construct a method to transfer the value of the gauge field at one point of the lattice to another. To do this we have to generalize the parallel transport by an infinitesimal amount in space–time $\mathrm{d}x_\mu$ to a finite step from one lattice point to a neighboring one. The infinitesimally shifted quark field $\Psi(x)$ reads ((8) in Example 4.1)

$$\Psi(x+\mathrm{d}x) = (\mathbb{1} + \mathrm{d}x^\mu D_\mu)\Psi(x) \ , \tag{7.25}$$

where D_μ is the usual covariant derivative $D_\mu = \mathbb{1}\,\partial_\mu - \mathrm{i}\hat{A}_\mu(x)$. Here and in the rest of the chapter we will use the matrix notation of the gluon fields following (4.49), i.e. $\hat{A}_\mu(x) \equiv \sum \frac{\lambda^a}{2} g A^a_\mu(x)$. In contrast to the case of (4.54) we also absorb the coupling constant into the field in order to follow standard notation of lattice gauge theory. Except for the matrix structure the formulae look like expressions in QED (by keeping in mind that the covariant derivative for electrons in QED reads $D_\mu = \partial_\mu + \mathrm{i}eA_\mu(x)$ as the charge of the electron is defined to be negative). In fact, although the phenomenology is very different, most of the relations that will be derived in this chapter can be directly applied to the U(1) gauge group of QED. The first term in D_μ is the standard translational operator which is diagonal in color space. The second term, containing the gluon field, describes the actual color transport between the infinitesimally close points x and $x+\mathrm{d}x$. We concentrate on the parallel transport of the color orientation. We have a factor

$$(1+\mathrm{i}\hat{A}_\mu(x)\,\mathrm{d}x^\mu) \tag{7.26}$$

in (7.25). When we look at color transport between two neighboring points x and $x+ae_\mu$ in some direction μ on the lattice, we apply the infinitesimal color translation infinitely many times along a straight path connecting the points. We get (no summation over μ) an infinite product of small displacement factors (7.3) along the line between x and $x+ae_\mu$:

$$\begin{aligned} &\lim_{N\to\infty} \prod_{n=0}^{N-1} \left[1+\mathrm{i}\hat{A}_\mu(x+n\delta x)\Delta x^\mu\right] \hat{A}_\mu \\ &= P\exp\left(\mathrm{i}\int\limits_x^{x+ae_\mu} \mathrm{d}s^\mu\,\hat{A}_\mu(x)\right) \equiv U_\mu(x)\ , \end{aligned} \tag{7.27}$$

with small steps $\Delta x^\mu = ae^\mu/N$, letting the number of steps N go to infinity. A line integral connecting the initial and final points enters the expression (7.27). The symbol P denotes path ordering – the multiplication of the gauge field matrices is ordered along the path as is also obvious from the ordering of the terms in the product of the first line of the equation. The quantity $U_\mu(x)$, which connects two neighboring points on the lattice, is called the *link variable*. From (7.27) it becomes clear that the link variable that transports color from $x+ae_\mu$ to x, $U_{-\mu}(x+ae_\mu)$ is directly related to $U_\mu(x)$ via the relation

$$U_{-\mu}(x+ae_\mu) = U_\mu^\dagger(x)\ . \tag{7.28}$$

In the standard formulation of lattice QCD the link variables are chosen as the basic variables for the gluons instead of the usual gauge potentials $\hat{A}_\mu(x)$. For every point on the lattice there are four link variables. Therefore the number of degrees of freedom of the fields remains unchanged when switching to the new variables. We will see in the course of this chapter why this change in variables is useful for lattice formulations.

7.1.4 Gluons on the Lattice

In the following, we will derive the discretized counterpart of the gluonic action in the continuum[5]

$$S^{\mathrm{E}} = \frac{1}{2g^2}\mathrm{tr}\left[\int \mathrm{d}^4x F_{\mu\nu}(x) F_{\mu\nu}(x)\right] , \tag{7.29}$$

where, as in the previous section, we adopt the Euclidean formulation of the theory. As we discussed in Chap. 4 the action (and the Lagrange density) is gauge invariant. Adopting gauge invariance as the construction principle for the action, we will formulate the most simple nontrivial gauge invariant expression on the lattice by using the link variables (7.27), which we introduced previously. In order to perform this task we first have to determine the gauge properties of the link variable itself. If the lattice spacing a is very small one can approximate a single link $U_\mu(x)$, pointing in the direction μ, by

$$U_\mu(x) \approx 1 + \mathrm{i}aA_\mu(x) . \tag{7.30}$$

However, even if a is not small enough for (7.30) to be a good approximation, one can divide the integral over the path between x and $x+ae_\mu$, where e_μ is the unit vector in the μ direction, into many (N) small steps $\epsilon = a/N$ and repeatedly apply (7.30):

$$U_\mu(x) \approx \exp(\mathrm{i}\epsilon A_\mu(x))\exp(\mathrm{i}\epsilon A_\mu(x+\epsilon e_\mu))\cdots\exp(\mathrm{i}\epsilon A_\mu(x+(a-\epsilon)e_\mu)) . \tag{7.31}$$

Note that we have exploited the fact that the definition of U_μ contains path ordering, which is the reason why we could write (7.31) as successive products of exponentials along the path. For a single factor in (7.31) we can now use the form (7.30):

$$\delta U_\mu(x) \equiv \exp\left(\mathrm{i}\frac{a}{N}A_\mu(x)\right) \approx 1 + \mathrm{i}\epsilon A_\mu(x) . \tag{7.32}$$

For the gauge-transformed $\delta U'_\mu(x)$ we have

$$\delta U'_\mu(x) \approx 1 + \mathrm{i}\epsilon A'_\mu(x) , \tag{7.33}$$

which, according to (4.55), reads

$$\delta U'_\mu(x) \approx 1 + \mathrm{i}\epsilon g(x)A_\mu(x)g^+(x) + \epsilon(\partial_\mu g(x))g^+(x) , \tag{7.34}$$

where $g(x)$ is a SU(3) gauge transformation at point x. Equation (7.34) can be rewritten as

$$\begin{aligned}&\left[g(x)+\epsilon\partial_\mu g(x)\right]\left[1+\mathrm{i}\epsilon A_\mu(x)\right]g^+(x) + O(\epsilon^2)\\ &\quad = g(x+\epsilon e_\mu)\exp(\mathrm{i}\epsilon A_\mu(x))g^+(x) .\end{aligned} \tag{7.35}$$

[5] Note that all the colored quantities, the gauge field A_μ, the field strength $F_{\mu\nu}$, the link variable U_μ, and the gauge rotation g are 3×3 matrix-valued fields. To prevent cluttering of the text with too many symbols, we will suppress the ^ symbol in the formulae.

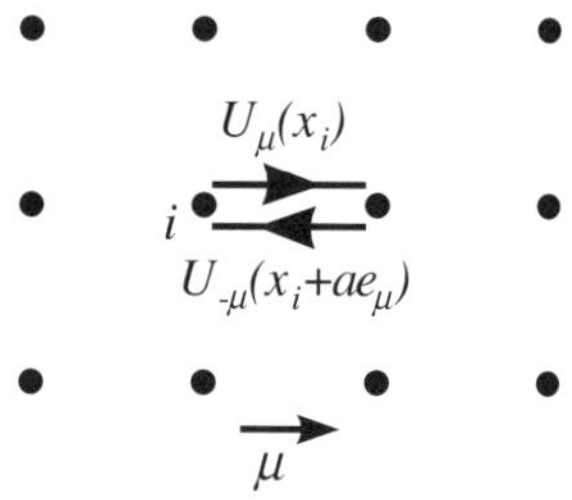

Fig. 7.3. The most simple (and trivial) two-link gauge-invariant object

Inserting this expression into (7.33), we finally get

$$U'_\mu(x) = g(x+e_\mu)U_\mu(x)g^+(x) \ . \tag{7.36}$$

This transformation property of the link variable is quite plausible (much more so than (4.55)). The link variable transports color from one point to the next and it transforms via a color rotation at the initial point and another (inverse) rotation at the final point. This form of gauge transformation also suggests the construction principle of a gauge invariant term. One has to build a product of link variables that are connected in space–time and do not carry an open color index (color singlet). The most simple object is shown in Fig. 7.3. It can be written as

$$\mathrm{tr}\left[U_\mu(x_\mathrm{i})U_{-\mu}(x_\mathrm{i}+ae_\mu)\right] \ . \tag{7.37}$$

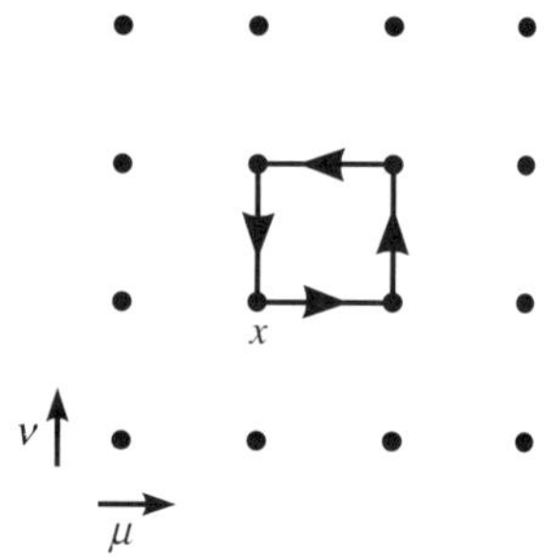

Fig. 7.4. Plaquette $P_{\mu\nu}(x)$ in the plane $(\mu\nu)$ first nontrivial gauge-invariant term

Unfortunately, using (7.28) we get

$$\mathrm{tr}\left[U_\mu(x_\mathrm{i})U_\mu(x_\mathrm{i}+ae_\mu)\right] = \mathrm{tr}\left[U_\mu(x_\mathrm{i})U^+_\mu(x_\mathrm{i})\right] = \mathrm{tr}\,[\mathbb{1}] = 3 \ , \tag{7.38}$$

which is a (trivial) constant. The first nontrivial expression is shown in Fig. 7.4. It is a product of four links encircling an elementary square of four neighboring points in a plane on the lattice. It is also called *plaquette* $P_{\mu\nu}$:

$$\begin{aligned} P_{\mu\nu}(x) &= \left[U_\mu(x)U_{-\nu}(x+ae_\nu)U_{-\mu}(x+ae_\mu+ae_\nu)U_\nu(x+ae_\mu)\right] \\ &= \left[U_\mu(x)U^+_\nu(x)U^+_\mu(x+ae_\nu)U_\nu(x+ae_\mu)\right] \ . \end{aligned} \tag{7.39}$$

One can directly check that the expression $\mathrm{tr}(P_{\mu\nu})$ is gauge invariant by inserting the gauge-transformed links into (7.39):

$$\begin{aligned} \mathrm{tr}\left[P'_{\mu\nu}(x)\right] &= \mathrm{tr}\left[U'_\mu(x)\left(U'_\nu(x)\right)^+\left(U'_\mu(x+e_\nu)\right)^+U'_\nu(x+e_\mu)\right] \\ &= \mathrm{tr}\left[g(x+e_\mu)U_\mu(x)g^+(x)g(x)U^+_\nu(x)g^+(x+e_\nu)\right. \\ &\quad \times g(x+e_\nu)U^+_\mu(x+e_\nu)g^+(x+e_\nu+e_\mu) \\ &\quad \left.\times g(x+e_\mu+e_\nu)\ U_\nu(x+e_\mu)g^+(x+e_\mu)\right] \end{aligned} \tag{7.40}$$

and by permuting the terms in the trace. This immediately yields

$$\mathrm{tr}\left[P'_{\mu\nu}(x)\right] = \mathrm{tr}\left[P_{\mu\nu}(x)\right] \ . \tag{7.41}$$

In the next step we have to find the continuum analogue of the plaquette, i.e. we calculate

$$P_{\mu\nu}(x) \ , \quad a \to 0 \ . \tag{7.42}$$

Let us first consider the links in the plaquette separately. Assuming that a is sufficiently small we approximate the line integral in the links as in (7.32). We take the value of $A_\mu(x)$ in the middle of the line

$$\begin{aligned} U_\mu(x) &\approx \exp\left(\mathrm{i}aA_\mu(x+ae_\mu/2)\right. \\ &\approx \exp\left(\mathrm{i}a\left[A_\mu(x)+a/2\partial_\mu A_\mu(x)\right]\right) \\ &\quad \text{(no summation over } \mu!) \ , \end{aligned} \tag{7.43}$$

$$
\begin{aligned}
U_\nu^+(x) &\approx \exp\left(-\mathrm{i}a\left[A_\nu(x)+a/2\partial_\nu A_\nu(x)\right]\right) , \\
U_\mu^+(x+e_\nu) &\approx \exp\left(-\mathrm{i}aA_\mu(x+a(e_\nu+e_\mu/2))\right) , \\
&\approx \exp\left(-\mathrm{i}a\left[A_\mu(x)+a\partial_\nu A_\mu(x)+a/2\partial_\mu A_\mu(x)\right]\right) , \\
U_\nu(x+e_\mu) &\approx \exp\left(\mathrm{i}a\left[A_\nu(x)+a\partial_\mu A_\nu(x)+a/2\partial_\nu A_\nu(x)\right]\right) .
\end{aligned}
\tag{7.44}
$$

The products of the links yield

$$
\begin{aligned}
&U_\mu(x)U_\nu^+(x)U_\mu^+(x+e_\nu)U_\nu(x+e_\mu) \\
&\approx \exp\left(\mathrm{i}a^2\left[\partial_\mu A_\nu(x)-\partial_\nu A_\mu(x)\right]+a^2\left[A_\mu(x),\, A_\nu(x)\right]\right) .
\end{aligned}
\tag{7.45}
$$

Comparing the exponent of this expression with the definition of the non-Abelian field strength tensor $F_{\mu\nu}$, in (4.58), we have

$$
P_{\mu\nu}(x) \overset{a\to 0}{=} \left[\exp\left(\mathrm{i}a^2 F_{\mu\nu}(x)\right)\right] . \tag{7.46}
$$

The continuum action for the gluons is (7.29)

$$
S_\mathrm{C}^E = \frac{1}{2g^2}\mathrm{tr}\left[\int \mathrm{d}^4x F_{\mu\nu}(x)F_{\mu\nu}(x)\right] . \tag{7.47}
$$

Comparing this expression with (7.46), we can construct the corresponding lattice version S_W^E (W for Wilson action) that coincides with the continuum action in the limit of vanishing lattice spacing a:

$$
S_\mathrm{W}^E = \frac{6}{g^2}\sum_\Box\left(1-\frac{1}{3}\mathrm{Re}\,\mathrm{tr}(\,P_\Box)\right) . \tag{7.48}
$$

The shortcut symbol $\Box$ denotes a single elementary plaquette on the lattice. The sum extends over all possible plaquettes of the four-dimensional lattice. There are 6 planes in a 4-dimensional lattice,[6] thus there are $6(N_x)^3N_t$ elementary squares. Note that you can replace the real part of the trace by $\mathrm{Re}\,\mathrm{tr}\,(P_\Box) = 1/2\mathrm{tr}\,(P_\Box + P_\Box^+)$. The Hermitean conjugate of the plaquette can be seen as the plaquette $P_\Box$ with reversed orientation of the links (see Fig. 7.4). Expression (7.48) is called the *Wilson action*.[7] One can easily generalize (7.48) to an arbitrary group SU(N)

$$
S_\mathrm{W}^{E,\,\mathrm{SU}(N)} = \beta\sum_\Box\left(1-\frac{1}{N}\mathrm{Re}\,\mathrm{tr}\,(P_\Box)\right) , \tag{7.49}
$$

where $P_\Box$ now contains the appropriate unitary $N\times N$ matrix-valued link variables. Many QCD lattice calculations use the SU(2) instead of the SU(3) group.

[6] A plane is defined by two axes. There are four directions on the lattice, therefore there exist six independent planes.

[7] The coefficient in front of the sum is usually denoted as $\beta = 6/g^2$ for SU(3), or in general $\beta = 2N/g^2$ for a SU(N) gauge group. It is the only free parameter of the theory.

The advantage is that there are far fewer SU(2) gluons (3 compared to 8 in SU(3)), which reduces the computational efforts considerably. Many features, like asymptotic freedom and color confinement, exist already for SU(2). In the following, we will also consider SU(2) several times for illustration.

7.1.5 Integration in SU(2)

In the expression for the path integral and for the observables calculated in the path-integral formalism, (7.23), one has to perform the integration over the group elements for every link on the lattice. We will show how this can be done for the group SU(2), which is much more intuitive (and simpler) than in the case of SU(3). The SU(2) link variables are unitary 2×2 matrices. One can parameterize them in the form

$$U = \begin{pmatrix} a & b \\ -b^* & a^* \end{pmatrix} . \tag{7.50}$$

The matrix fulfills the unitarity condition

$$U^+ U = 1 \tag{7.51}$$

if we demand that

$$|a|^2 + |b|^2 = 1 \tag{7.52}$$

holds. Instead of using complex parameters a,b we can also take real ones with $a = x_4 + \mathrm{i}x_3$ and $b = x_2 + \mathrm{i}x_1$. Then (7.52) reads

$$x_1^2 + x_2^2 + x_3^2 + x_4^2 = 1 . \tag{7.53}$$

Obviously, with this choice one can rewrite U also in a slightly different form. Inserting the x components into (7.50) explicitly yields

$$U = \begin{pmatrix} x_4 + \mathrm{i}x_3 & x_2 + \mathrm{i}x_1 \\ -x_2 + \mathrm{i}x_1 & x_4 - \mathrm{i}x_3 \end{pmatrix} , \tag{7.54}$$

which one can rewrite with the help of the Pauli matrices ($i = 1, 2, 3$) as

$$U = x_4 \mathbb{1} + \mathrm{i}x_i \sigma_i . \tag{7.55}$$

Now, using (7.53) we write the group integration (integration over the group elements) in the form

$$\int \mathrm{d}U = N \int \mathrm{d}^4 x \delta \left(x_1^2 + x_2^2 + x_3^2 + x_4^2 - 1 \right) . \tag{7.56}$$

Thus the SU(2) integration can be rewritten as an integration over the surface of the unit sphere S_3 in four dimensions. The normalization N is chosen such that the basic integration over the group space returns 1:

$$\int \mathrm{d}U = 1 \quad \Rightarrow \quad N = \frac{1}{\int \mathrm{d}^4 x \delta\, (x_i x_i - 1)} = \frac{1}{\Omega_3} , \tag{7.57}$$

where the surface Ω_3 of the hypersphere is $\Omega_3 = 2\pi^2$. Going a step further, consider the integration over one link. We take the single matrix element U_{11} as an example:

$$\int \mathrm{d}U\, U_{11} = \frac{1}{2\pi^2} \int \mathrm{d}^4x \delta\left(|x|^2 - 1\right)(x_4 + \mathrm{i}x_3) = 0\ . \tag{7.58}$$

The integration over odd functions of x_i vanishes:

$$\int \mathrm{d}U\, U_{ij} = 0\ . \tag{7.59}$$

An important integral is

$$\int \mathrm{d}U\, U_{ij}\left(U^+\right)_{kl}\ . \tag{7.60}$$

From the previous integral it is clear that integrals over functions of the type $U_{11}(U^+)_{12} = (x_4 + \mathrm{i}x_3)(x_2 + \mathrm{i}x_1)$ vanish. The following integral is zero, too:

$$\int \mathrm{d}U\, U_{11}\left(U^+_{22}\right) = \int \mathrm{d}U\, a^2 = \int \mathrm{d}U\, \left(x_4^2 - x_3^2 + 2\mathrm{i}x_3x_4\right) = 0\ . \tag{7.61}$$

The integrals over x_3^2 and x_4^2 cancel. The third term vanishes following (7.59). The only nonvanishing terms are proportional to $|a|^2$, $|b|^2$:

$$\begin{aligned}\int \mathrm{d}U\, U_{11}\left(U^+\right)_{11} &= \int \mathrm{d}U\, \left(x_3^2 + x_4^2\right)\\ &= \frac{1}{2\pi^2}\frac{1}{2}\int \mathrm{d}^4x\, \delta\left(x_\mathrm{i}^2 - 1\right)\left(x_1^2 + x_2^2 + x_3^2 + x_4^2\right)\\ &= \frac{1}{2}\end{aligned} \tag{7.62}$$

and also $\int \mathrm{d}U\, U_{12}\left(U^+\right)_{21} = \frac{1}{2}$. Thus we have

$$\int \mathrm{d}U\, U_{ij}\left(U^+\right)_{kl} = \frac{1}{2}\delta_{il}\delta_{jk}\ . \tag{7.63}$$

The factor $\frac{1}{2}$ normalizes the integral over the trace:

$$\int \mathrm{d}U\, \mathrm{tr}\left(UU^+\right) = 1\ , \tag{7.64}$$

which holds for all SU(N). In (7.63) one has to replace $\frac{1}{2}$ by $\frac{1}{N}$ for the general SU(N) result, accordingly. This is evident from (7.64). Rewriting the equation in the form

$$\int \mathrm{d}U\, \mathrm{tr}\left(UU^+\right) = \sum_i \int \mathrm{d}U\, U_{ii}U^\dagger_{ii} = 1\ ,$$

it necessarily follows that

$$\int \mathrm{d}U\, U_{ii} U^{\dagger}_{ii} = \frac{1}{N} \,,$$

since no direction in the group space is preferred.

7.1.6 Discretization: Scalar and Fermionic Fields

After introducing the idea of discretizing space and time on a rectangular equally spaced lattice and introducing the link variables for the gauge fields, we will now study the implementation of other field variables in such a system. We will first look at the equations of motion for scalar particles and then discuss the important case of spin-$\frac{1}{2}$ particles that we require in order to treat quarks in the lattice calculations. There are important differences between those cases which also have to be dealt with in a lattice description of QCD. First, let us introduce the Fourier transform of functions on our lattice. In one dimension the Fourier transform of a function $f(x_\mathrm{i})$ defined on the lattice is

$$\tilde{f}(p) = \sum_{n=-\infty}^{+\infty} a f(x_n) \exp(-\mathrm{i} p x_n) \quad \text{with the discrete points} \quad x_n = na \,. \tag{7.65}$$

(The spacing between neighboring points is again given by a.) As you can see from this formula $\tilde{f}(p)$ is periodic:

$$\tilde{f}(p) = \tilde{f}\left(p + 2\frac{\pi}{a}\right) \,. \tag{7.66}$$

We therefore restrict momentum integration to the range $(-\pi/a, \pi/a]$, which is well known as the first Brillouin zone in solid-state physics. The inverse transform is then

$$f(x_n) = \int\limits_{-\pi/a}^{\pi/a} \frac{\mathrm{d}p}{2\pi} \tilde{f}(p) \exp(\mathrm{i} p x_n) \,. \tag{7.67}$$

The Kronecker symbol δ_{nm} follows directly as

$$\delta_{nm} = \frac{a}{2\pi} \int\limits_{-\pi/a}^{\pi/a} \mathrm{d}p \exp(\mathrm{i} p (x_n - x_m)) \,. \tag{7.68}$$

After having defined these preliminaries we can study the discretization of the field equations of motion.

The (Euclidean) Klein–Gordon equation for a scalar particle reads

$$\left(\Box - m^2\right) \Phi(x) = 0 \,. \tag{7.69}$$

We have to discretize the Laplace operator $\Box$, i.e. the second derivatives acting on the field. Again, as mentioned before, we study the system in the Euclidean-time formalism, setting the $g^{\mu\nu}$ tensor to be $\delta^{\mu\nu}$. Using a 3-point formula for the second derivative we get the equation

$$\frac{1}{a^2}\sum_{\mu=1}^{4}\left[\Phi(\boldsymbol{x}_n-\boldsymbol{e}_\mu)+\Phi(\boldsymbol{x}_n+\boldsymbol{e}_\mu)\right]-\left(\frac{8}{a^2}+m^2\right)\Phi(\boldsymbol{x}_n)=0 \tag{7.70}$$

for the field at point x_n. We can rewrite the equation in matrix form:

$$\left(G^{-1}\right)_{nm}\Phi_m=0\ , \tag{7.71}$$

where the matrix (G^{-1}) is given by

$$\left(G^{-1}\right)_{nm}=\frac{1}{a^2}\left\{\sum_\mu\left[\delta_{n-\mu,m}+\delta_{n+\mu,m}\right]-\left(8+(ma)^2\right)\delta_{n,m}\right\}\ . \tag{7.72}$$

This tridiagonal matrix is the inverse of the propagator of the particle. Using the Fourier representation of the Kronecker δ_{nm}, (7.68), we can calculate the propagator of the Φ field in momentum space. Inserting (7.68) into (7.72) we read off the result

$$\begin{aligned}\left(G^{-1}\right)(p)&=a^4\sum_{n,m}\left(G^{-1}\right)_{nm}\exp(-\mathrm{i}p_\mu(x_n-x_m)_\mu)\\&=a^2\left[4\sum_\mu\sin^2\left(\frac{p_\mu a}{2}\right)+(ma)^2\right]\ .\end{aligned} \tag{7.73}$$

Remember, that the momenta can lie in the range $-\pi/a\leq p_\mu<\pi/a$. One important point in the lattice description is the continuum limit of the theory. The propagator has poles for momenta corresponding to the propagation of real particles. Except for the overall scale factor a^4 we can perform the continuum limit of (7.73). We get

$$a^{-4}\left(G^{-1}\right)(p_\mu)\stackrel{a\to0}{=}\left[p_\mu^2+m^2\right]\ . \tag{7.74}$$

The poles are at $p_\mu^2=-m^2$ (in Euclidean space–time!) as it should be. We can write the propagator in space–time accordingly:

$$G^{-1}(x_n-x_m)=a^2\int\frac{\mathrm{d}^4k}{(2\pi)^4}\frac{\exp(\mathrm{i}k(x_n-x_m))}{4\sum_\mu\sin^2\left(\frac{ka}{2}\right)+(ma)^2}\ , \tag{7.75}$$

with the obvious continuum limit $a\to0$,

$$G^{-1}(x)=\int\frac{\mathrm{d}^4k}{(2\pi)^4}\frac{\exp(\mathrm{i}kx)}{k^2+m^2}\ . \tag{7.76}$$

Let us now move over to the discretization of the equations of motion for a spin-$\frac{1}{2}$ field (a quark, for instance). The main difference from the previous case is that the Dirac equation is a first-order differential equation. We have to discretize the equation

$$(\not{p} - \mathrm{i}m)\,\psi = 0 \ . \tag{7.77}$$

The γ matrices fulfill the standard anticommutation relations

$$\{\gamma_\mu, \gamma_\nu\} = 2\delta_{\mu\nu} \ , \tag{7.78}$$

where we used the metric tensor in Euclidean space $\delta_{\mu\nu}$ instead of the usual $g_{\mu\nu}$. We adopt the standard (but not unique) choice of γ matrices in Euclidean space:

$$\gamma_4 \to \gamma^0 \ , \quad \boldsymbol{\gamma} \to \mathrm{i}\boldsymbol{\gamma}_i \ . \tag{7.79}$$

As γ_0 is hermitian and the original $\boldsymbol{\gamma}$ are antihermitean, the newly defined γ matrices are all hermitean. Inserting those matrices into the Dirac equation we obtain

$$\begin{aligned}
&(\gamma_\mu \partial_\mu + m)\,\psi = 0 \\
\Longleftrightarrow &\left[\sum_m \gamma_\mu \frac{1}{2a}\left[\psi(x_n + e_\mu) - \psi(x_n - e_\mu)\right] + m\delta_{mn}\right]\psi_m = 0 \ .
\end{aligned} \tag{7.80}$$

In matrix notation we get the inverse propagator

$$\left(G^{-1}\right)^{\alpha\beta}_{n,m} = \frac{1}{2a}\left[\sum_\mu \left(\delta_{n,m+\mu} - \delta_{n,m-\mu}\right)\gamma_\mu\right] + m\delta_{n,m}\delta_{\alpha,\beta} \ . \tag{7.81}$$

Using expression (7.68) for δ_{nm} we can derive the Fourier transform of the inverse fermion propagator

$$\left(G^{-1}\right)^{\alpha\beta}(p) = \frac{1}{a}\left[\mathrm{i}\sum_\mu \sin(p_\mu a)\gamma_\mu + ma\right]\delta^{\alpha\beta} \ . \tag{7.82}$$

Now, the continuum limit is more tricky in the case of spin-$\frac{1}{2}$ particles compared to the spin-0 case, or in general, for fermions as opposed to bosons. In the limit $a \to 0$, we get the standard result from (7.82):

$$\left(G^{-1}\right) \overset{a\to 0}{\sim} (\mathrm{i}p_\mu\gamma_\mu + m)\delta^{\alpha\beta} \ , \tag{7.83}$$

which is just the usual (Euclidean) Dirac equation. However, this is not the end of the story. The momenta p_μ vary between $-\pi/a$ and π/a. Usually, assuming a massless field for simplicity, the root of (7.83) is the single on-shell condition $p_\mu^2 = 0$. However, $\sin(p_\mu a)$ also vanishes at the edge of the Brillouin zone with $p_\mu = \pm\pi/a$. We end up with an additional pole, or particle, at π/a (if the Brillouin zone is defined to be a closed interval at that point). This effect is the so-called *fermion-doubling* problem. In one dimension we have two poles in the propagator. In the case of two dimensions we therefore have $2^4 = 16$ "particles" instead of 1. Where did this problem originate from? The trouble started with the symmetric first-order discretized derivative, (7.80), which leads to the fact that the argument of the sine is $(p_\mu a)$ instead of $(p_\mu a/2)$ for spin-0 particles. It is the linear nature of the Dirac equation that generates doublers in its discretized version.

There are a number of methods to deal with fermion doubling. We will discuss the two most common methods to remove the superfluous solutions. The first method was originally proposed by Wilson, like most of the first concepts in lattice gauge theory. One adds an additional term to the equations of motion $-\frac{r}{2}\Box\psi(x)$ to the Dirac equation:

$$\left[\gamma_\mu \partial_\mu + m - \frac{r}{2}\partial_\mu \partial_\mu\right]\psi(x) = 0 \tag{7.84}$$

with some parameter r, also often called the Wilson parameter. This equation is discretized on the lattice as usual, where we take the second derivative in (7.81) from (7.70). Thus we get the matrix equation

$$\left(G^{-1}\right)_{mn} \psi_n = 0 \tag{7.85}$$

with

$$\begin{aligned}\left(G^{-1}\right)_{mn} =& \frac{1}{2a}\sum_\mu \left[(\gamma_\mu - r)\delta_{m,n+\mu} - (\gamma_\mu + r)\delta_{m,n-\mu}\right] \\ &+ \left(m + 4\frac{r}{a}\right)\delta_{m,n}\mathbb{1} \ .\end{aligned} \tag{7.86}$$

Fourier transforming (7.86) the inverse propagator reads

$$\left(G^{-1}\right)(p) = \frac{\mathrm{i}}{a}\gamma_\mu \sin(p_\mu a) + m + 2\frac{r}{a}\sum_\mu \sin^2\left(\frac{p_\mu a}{2}\right) \ . \tag{7.87}$$

As you can see from (7.87) in the continuum limit, $a \to 0$, the additional term vanishes like a^1. However, at the corner of the Brillouin zone $p = \pm\pi/a$ the factor does not vanish because of the $\frac{1}{2}$ in the argument of the sine function. The term therefore diverges like $1/a$ effectively generating an infinite mass for the fermion doublers. Including the gluons, (7.86) then reads

$$\begin{aligned}M^F_{mn}(U) \equiv& \left(G^{-1}(U)\right)_{mn} \\ =& \frac{1}{2a}\sum_\mu \left[(\gamma_\mu - r)\, U_m u(x_n)\delta_{m,n+\mu} - (\gamma_\mu + r)U^+_\mu(x_m)\delta_{m,n-\mu}\right] \\ &+ \left(m + 4\frac{r}{a}\right)\delta_{m,n}\mathbb{1} \ .\end{aligned} \tag{7.88}$$

Fermions which obey (7.86) are called Wilson fermions. Wilson fermions are still widely used in lattice gauge calculations of QCD. One problem connected with this approach is that even for massless particles, i.e. a chirally symmetric theory, the extra term always introduces an additional mass term. This generates troubles in the case that one wants to study chiral symmetry restoration on the lattice.

Another approach to deal with fermion doubles is due to J. Kogut and L. Susskind.[8] The fermions treated this way are accordingly called Kogut–Susskind, or also staggered, fermions. The idea is to effectively define the fermion fields on a lattice with twice the lattice spacing. The corners for the Brillouin zone of this lattice are then not at $\pm\pi/a$ anymore. Staggered fermions are introduced by transforming the fermion fields $\psi(x_n)$ introducing new fields $\tilde{\psi}(x_n)$: The transformation $T_{(x_n)}$ is chosen to diagonalize the γ matrices in such a way that

$$T^+(x_n)\gamma_\mu T(x_n+\boldsymbol{e}_\mu) = \eta_\mu(x_n)\mathbb{1} \ . \tag{7.89}$$

Since in the Dirac equation (7.80) fields at two neighboring points are connected via the γ matrix, the transformation T at two neighboring points is involved in (7.89). This procedure generates an x-dependent field $\eta_\mu(x_n)$ without spin indices.

The standard choice for $T(x_n)$ to fulfill (7.89) is

$$T(x_{1,n_1}, x_{2,n_2}, x_{3,n_3}, x_{4,n_4}) = \gamma_1^{n_1}\gamma_2^{n_2}\gamma_3^{n_3}\gamma_4^{n_4} \ , \tag{7.90}$$

where we have used the four components of the lattice point x_n explicitly.

Inserting (7.90) into (7.89), we get the following expression for the field $\eta_\mu(x_n)$:

$$\begin{aligned} \eta_1(x_n) &= 1 \ , \\ \eta_2(x_n) &= (-1)^{n_1} \ , \\ \eta_3(x_n) &= (-1)^{n_1+n_2} \ , \\ \eta_4(x_n) &= (-1)^{n_1+n_2+n_3} \end{aligned} \tag{7.91}$$

or, more compactly,

$$\eta_\mu(x_n) = (-1)^{\sum_{i=1}^{\mu-1} n_i} \ . \tag{7.92}$$

Inserting $\psi = T\tilde{\psi}$ into the Dirac equation, we obtain

$$\eta_\mu(x_m)\partial_\mu\tilde{\psi}_\alpha(x_m) + m\tilde{\psi}_\alpha(x_m) = 0 \ . \tag{7.93}$$

Due to the diagonalization procedure (7.89), we now have a "Dirac equation" with decoupled spinor indices α. We therefore restrict ourselves to *one* species of staggered fermions, setting α to 1 or, simpler, dropping the index altogether.

$$\eta_\mu(x_m)\partial_\mu\tilde{\psi}(x_m) + m\tilde{\psi}_\alpha(x_m) = 0 \ . \tag{7.94}$$

[8] The relevant original references are T. Banks, S. Raby, L. Susskind, J. Kogut, D.R.T. Jones, P.N. Scharbach, and D.K. Sinclair: Phys. Rev. D **15**, 1111 (1977); L. Susskind, Phys. Rev. D **16**, 3031 (1977).

In order to understand what it means to throw out 3 of the 4 components we have to look back at the original fermion fields. We mentioned before that the basic idea of staggered fermions is to distribute fermionic degrees of freedom on a coarser lattice. In this spirit we introduce new space–time vectors $\tilde{n}_\mu$ with integer-valued components $n_\mu = 2\tilde{n}_\mu + \epsilon_\mu$. The components of ϵ_μ are either 0 or 1, depending on whether n_μ is an even or odd number, respectively. Thus we have reinterpreted the original space–time coordinates as the coordinates on a lattice with spacing $2a$ plus a displacement vector ϵ_μ. We therefore get the following Dirac equation

$$\eta_\mu^{(2\tilde{n}_\nu+\epsilon_\nu)} \partial_\mu \tilde{\psi}(2\tilde{n}_\nu + \epsilon_\nu) + m\tilde{\psi}(2\tilde{n}_\nu + \epsilon_\nu) = 0 \ . \tag{7.95}$$

One can write the field $\tilde{\psi}$ as

$$\tilde{\psi}(2\tilde{n}_\nu + \epsilon_\nu) = \tilde{\tilde{\psi}}_\epsilon(\tilde{n}) \ . \tag{7.96}$$

Note that this new field $\tilde{\tilde{\psi}}_\epsilon$ has 16 components ($\epsilon_\mu = 0$ or 1 for each μ). Another simplification comes from the fact that following the definition of η, (7.92), we get $\eta_\mu(2\tilde{n}_\nu + \epsilon_\nu) = \eta_\nu(\epsilon_\nu)$ since $(\gamma_\mu)^2 = 1$. The discretized Dirac equation contains the slight complication that in the kinetic terms there is a space–time point shifted by e_μ. Translating this into the new coordinates, one has

$$n_\mu = 2\tilde{n}_\mu + \epsilon_\mu$$

and

$$n_\mu + e_\nu = \begin{cases} 2\tilde{n}_\mu + (\epsilon + e)_\mu & \text{for } \epsilon_\mu e_\mu = 0 \ , \\ 2(\tilde{n} + e)_\mu + \epsilon_\mu & \text{for } \epsilon_\mu e_\mu = 1 \ . \end{cases} \tag{7.97}$$

Keeping in mind this subtlety, the Dirac equation for staggered fermions reads

$$\begin{aligned} &\frac{1}{2a} \sum_{\mu,\epsilon'} \eta_\mu(\epsilon) \Big[\delta_{\epsilon+\mu,\epsilon'} \left(\tilde{\tilde{\psi}}_{\epsilon'}(\tilde{n}) - \tilde{\tilde{\psi}}_{\epsilon'}(\tilde{n} - e_\mu) \right) \\ &\quad + \delta_{\epsilon-\mu,\epsilon'} \left(\tilde{\tilde{\psi}}_{\epsilon'}(\tilde{n} + e_\mu) - \tilde{\tilde{\psi}}_{\epsilon'}(\tilde{n}) \right) \Big] + m\tilde{\tilde{\psi}}_\epsilon(\tilde{n}) = 0 \ . \end{aligned} \tag{7.98}$$

Thus we end up with a single fermionic component which, according to the doubling argument, generates 4 particle modes in the continuum. Therefore incorporating staggered fermions, one ends up with a 4-flavor QCD.

7.1.7 Fermionic Path Integral

One additional complication arises when treating fermions on the lattice. Fermionic fields cannot be treated as c numbers in the path integral. They follow a so-called Grassmann algebra.[9] Fermionic field operators anticommute, which

[9] For a more extended discussion of this point, see W. Greiner and J. Reinhardt: *Field Quantization* (Springer, Berlin, Heidelberg 1996).

ensures that not more than one fermion can reside in any given state (the Pauli principle)

$$\hat{\psi}(x)\hat{\psi}(x) = 0 \ , \tag{7.99}$$

which is clear from the general fermionic anticommutator $\{\hat{\psi}(x), \hat{\psi}(y)\} = 0$. The fields in the path integral $\psi(x_n)$, which are not operators, obey the analogous algebra

$$\left\{\psi(x_i), \psi(x_j)\right\} = \psi(x_i)\psi(x_j) + \psi(x_j)\psi(x_i) = 0 \ . \tag{7.100}$$

Commutativity no longer holds for these fields. This behavior can also be obtained by assigning a 2×2 matrix to every fermionic variable:

$$\hat{\psi} = \psi\sigma^+ = \begin{pmatrix} 0 & \psi \\ 0 & 0 \end{pmatrix} \ . \tag{7.101}$$

Obviously, $\left(\hat{\psi}\right)^2 \sim (\sigma^+)^2 = 0$, fulfilling (7.100). Considering two variables ψ_1 and ψ_2, one can achieve the same behavior by choosing

$$\hat{\psi}_1 = \psi_1 \left(\sigma^+\right) \otimes (\mathbb{1}) \ , \qquad \hat{\psi}_2 = \psi_2 \left(\sigma^z\right) \otimes \left(\sigma^+\right) \ , \tag{7.102}$$

where we have introduced the direct product of two matrices. Again one has

$$\left(\hat{\psi}_1\right)^2 = \left(\hat{\psi}_2\right)^2 = 0 \ ,$$

and the combination of σ^+ and σ^z generates anticommutation:

$$\begin{aligned} \hat{\psi}_1\hat{\psi}_2 + \hat{\psi}_2\hat{\psi}_1 &= \psi_1\psi_2 \left[\left(\sigma^+\sigma^z\right) \otimes \sigma^+ + \left(\sigma^z\sigma^+\right) \otimes \sigma^+\right] \\ &= \left(-\sigma^+ + \sigma^+\right) \otimes \sigma^+ = 0 \ . \end{aligned} \tag{7.103}$$

One can implement the same procedure for many variables, too. However, this is not really a practical approach. On a standard-sized lattice one might have about one million quark degrees of freedom, thus one would need a direct product of a million 2×2 matrices to represent the fermionic variables. In the following, we will integrate out the fermionic variables occurring in the path integral by using standard integration rules for Grassmann variables. There are only two basic rules to observe:

$$\begin{aligned} &\int \mathrm{d}\psi = 0 \ , \\ &\int \mathrm{d}\psi\psi = 1 \ . \end{aligned} \tag{7.104}$$

Higher powers in ψ cannot occur following (7.99). The fact that one cannot pile up more than one fermion in a state makes integration of Grassmann variables much simpler than standard integration. Coming back to our original problem, we have an integral of the form

$$Z_\psi = N \int [\mathrm{d}\psi]\left[\mathrm{d}\bar{\psi}\right] \exp\left(-\int \mathrm{d}^4x\,\mathrm{d}^4y\,\bar{\psi}(x)G^{-1}(x,y)\psi(y)\right) . \quad (7.105)$$

$\psi(x)$ and $\bar{\psi}(x)$ (the real and imaginary parts of the field, respectively) are two independent Grassmann variables. Discretizing (7.105), one gets

$$Z_\psi = N \int \prod_{\mathrm{i}} \mathrm{d}\psi(x_\mathrm{i}) \prod_j \mathrm{d}\bar{\psi}(x_j) \exp\left(-\sum_{ij} \bar{\psi}(x_\mathrm{i})G^{-1}(x_\mathrm{i}, x_j)\psi(x_j)\right) , \quad (7.106)$$

where we have suppressed the lattice spacing and the spinor, color, and flavor indices of the quark fields in order not to confuse the reader with too many symbols. If $\left(G^{-1}\right)$ is a hermitean matrix one can diagonalize the matrix with some unitary transformation, which does not change the integration measure. We then have

$$Z_\psi = \int \prod_{\mathrm{i}} \mathrm{d}\tilde{\bar{\psi}}(x_\mathrm{i})\,\mathrm{d}\tilde{\psi}(x_\mathrm{i}) \exp\left(-\sum_{\mathrm{i}} \tilde{\bar{\psi}}(x_\mathrm{i}) \left(G_{\mathrm{D}}^{-1}\right)(x_\mathrm{i}, x_\mathrm{i})\tilde{\psi}(x_\mathrm{i})\right) . \quad (7.107)$$

From (7.104) we know that only the linear term in each variable contributes to the integral. Thus we get

$$Z_\psi = N \int \prod_{\mathrm{i}} \mathrm{d}\tilde{\bar{\psi}}(x_\mathrm{i})\,\mathrm{d}\tilde{\psi}(x_\mathrm{i}) \prod_{\mathrm{i}} \tilde{\bar{\psi}}(x_\mathrm{i}) \left(G_{\mathrm{D}}^{-1}\right)(x_\mathrm{i}, x_\mathrm{i})\tilde{\psi}(x_\mathrm{i}) . \quad (7.108)$$

Using (7.104), we then end up with

$$Z_\psi = N \prod \left(G_{\mathrm{D}}^{-1}\right)(x_\mathrm{i} x_\mathrm{i}) = N \det G_{\mathrm{D}}^{-1} = N \det G^{-1} , \quad (7.109)$$

where the last step followed from the fact that unitary transformations do not change the value of the determinant: $\det\left(G_{\mathrm{D}}^{-1}\right) = \det U \left(G_{\mathrm{D}}^{-1}\right) U^{+} = \det G^{-1}$. In total we could remove the unpleasant fermionic degrees of freedom from the path integral expression, but at the price of having to compute a determinant of a huge matrix.

7.1.8 Monte Carlo Methods

Using discretization of space and time we can reduce the number of variables from a continuous to a discrete, albeit infinite, set of degrees of freedom. In addition we put the system into a finite box, ending up with a finite number of variables. However, a direct calculation of observables using (7.23) is still out of reach. Take a standard lattice with a size of 16^4 lattice sites. At each site, counting gluons only, we have 32 variables (4 directions times 8 colors). Thus one has to calculate a 2 097 152-fold integral in (7.23). Luckily, for most quantities it turns out that one needs only a very small selected number (depending on the observable a few hundred to a few million) of points in this high-dimensional space of degrees of freedom. The sampling of this space is done via Monte Carlo methods.

As an example, think of the one-dimensional integration of a function $\varphi(x)$

$$I = \int_{x_\mathrm{i}}^{x_\mathrm{f}} \mathrm{d}x \varphi(x) \ . \tag{7.110}$$

One can introduce a weight function $w(x)$

$$I = \int_{x_\mathrm{i}}^{x_\mathrm{f}} \mathrm{d}x \underbrace{\left(\frac{\varphi(x)}{w(x)}\right)}_{\varphi'(x)} w(x) \ , \tag{7.111}$$

which we assume to be normalized $\int_{x_\mathrm{i}}^{x_\mathrm{f}} \mathrm{d}x w(x) = 1$.

One can view $w(x)$ also as a distribution that defines the probability that x attains a specific value. With a change of variables

$$y = \int_{x_\mathrm{i}}^{x} \mathrm{d}x' w(x') \ , \qquad \mathrm{d}y = w(x)\,\mathrm{d}x \ , \tag{7.112}$$

one gets

$$I = \int_0^1 \mathrm{d}y \varphi'(x(y)) \ . \tag{7.113}$$

The difference between (7.110) and (7.113) is that a good choice of the weight function $w(x)$ can reduce φ' to an easily integrable function. One can now sample the range of y by calculating a set of N uniformly distributed random numbers between 0 and 1 (i.e. all numbers between 0 and 1 have the same probability):

$$I_{(\{y_\mathrm{i}\})} \approx \frac{1}{N} \sum_{i=1}^{N} \varphi'(x(y_\mathrm{i})) \ . \tag{7.114}$$

Of course, one needs the function $x(y)$ in either analytical or numerical form. As can be guessed from our discussion, so far, this method can directly be extended to a multi-dimensional problem.

Hence the general principle of the algorithm is to generate a number of random numbers corresponding to some weight function and summing up the values of the the integrand for this set of numbers.

Often it is not trivial to determine $x(y)$ or its equivalent in many dimensions. Therefore one defines a stochastic process generating a sequence of variables following the weight function.

For that we construct a transition amplitude $P(x,x')$ for the variable x' to attain the value x in the next step of the stochastic process (we assume ergodicity, i.e. every value of x can be reached from any starting value x_{i} in the process). The succession of values is called a *Markov chain*. Within the theory of Markov chains it can be shown that $P(x,x')$ will converge to a unique *equilibrium distribution* of values of x, $w_{\mathrm{eq}}(x)$, i.e.

$$w_{\mathrm{eq}}(x) = \int \mathrm{d}x'\, P(x,x') w_{\mathrm{eq}}(x') \ . \tag{7.115}$$

This holds if we demand *detailed balance* (a so-called "reversible Markov chain"), which requires that

$$w_{\mathrm{eq}}(x') P(x,x') = w_{\mathrm{eq}}(x) P(x',x) \ . \tag{7.116}$$

Inserting (7.115) into (7.116), it follows that

$$w_{\mathrm{eq}}(x) = \int \mathrm{d}x' \left[\frac{w_{\mathrm{eq}}(x) P(x',x)}{P(x,x')} \right] P(x,x') \ , \tag{7.117}$$

and because of the normalization $\int \mathrm{d}x'\, P(x',x) = 1$ the equation is fulfilled. The value $I(\{y_{\mathrm{i}}\})$ converges to I with a statistical error $\sim 1/\sqrt{N}$ (N number of samples) following the central limit theorem. Note that this feature is independent of the dimensionality of the underlying integral, which we assumed to be one-dimensional for the sake of simplicity. In the case of lattice gauge theory we obviously have to deal with many dimensions.

7.1.9 Metropolis Algorithm

A widely used definition for the transition amplitude $P(x,x')$ was given by Metropolis as early as 1953.[10] It is defined in two steps. First one introduces a *general transition matrix* $\tilde{P}(x,x')$, which should obey ergodicity but otherwise is arbitrary. The new value x' is accepted with some probability $P(x,x')$ or it is discarded. Thus we have

$$P(x',x) = P(x',x)\tilde{P}(x',x) + \delta(x-x') \int \mathrm{d}x''\, \tilde{P}(x'',x)(1 - P(x'',x)) \ . \tag{7.118}$$

[10] N. Metropolis, A.W. Rosenbluth, M.N. Rosenbluth, A.H. Teller, and E. Teller: *J. Chem. Phys.* **21**, 1087 (1953).

Using detailed balance the equations

$$w_{\text{eq}}(x') = \left[\tilde{P}(x,x')P(x,x') + \delta(x-x')\int \mathrm{d}x\,\tilde{P}(x',x'')(1-P(x',x''))\right] ,$$
$$w_{\text{eq}}(x) = \left[\tilde{P}(x',x)P(x',x) + \delta(x-x')\int \mathrm{d}x''\,\tilde{P}(x,x'')(1-P(x,x''))\right] \tag{7.119}$$

or

$$\frac{P(x,x')}{P(x',x)} = \frac{w_{\text{eq}}(x)}{w_{\text{eq}}(x')}\frac{\tilde{P}(x',x)}{\tilde{P}(x,x')} \tag{7.120}$$

have to hold. This relation can be fulfilled by the choice (which is not unique) of Metropolis et al.

$$P(x,x') = \min\left[\frac{w_{\text{eq}}(x)}{w_{\text{eq}}(x')}\frac{\tilde{P}(x',x)}{\tilde{P}(x,x')}, 1\right] , \tag{7.121}$$

which clearly fulfils (7.120). If one takes the simple choice of $\tilde{P}(x,x')$ being a constant we get

$$P(x,x') = \min\left[\frac{w_{\text{eq}}(x)}{w_{\text{eq}}(x')}, 1\right] , \tag{7.122}$$

which can be computed numerically quite easily in many cases.

In practice, a QCD lattice calculation utilizing the Metropolis algorithm is done in the following general manner. Step through the lattice from link variable $U_\mu(x)$ to the next one; generate a new link variable $U'_\mu(x)$ by multiplying the old value of the link with a random SU(3) group element. Calculate the weight functions $w_{\text{eq}}(x) \equiv \mathrm{e}^{-S(\{U\})}$ for the former value $U_\mu(x)$ and the new value $U'_\mu(x)$; compare the weights according to (7.122) and either accept or reject the new choice. When one has performed this procedure for every link on the lattice (a so-called "lattice sweep") one has generated a configuration $\{U\}$. This procedure has to be repeated many times and observables can be calculated by using (7.23).

7.1.10 Langevin Algorithms

Another method to generate a set of configurations is based on the so-called Langevin algorithm. We again exploit the analogy of (7.14) to the canonical ensemble of a statistical system, by identifying $\mathrm{e}^{-S} \approx \mathrm{e}^{-\beta\hat{H}}$. We simulate this system by looking at the "time evolution" of the system, i.e. replacing configuration averaging by time averaging. For our field-theoretical system this means we introduce a fictitious time variable t and solve the dynamical equations in a heat bath in this $(4+1)$-dimensional pseudo system. Looking at a specific link variable $U \equiv U_\mu(x_n)$ we have the equation

$$\dot{U} = -\frac{\delta S}{\delta U} + \eta(t) . \tag{7.123}$$

The first term on the right-hand side drives the system to a minimum of the action, whereas the random noise $\eta(t)$ generates the fluctuations of the heat bath.

The noise term η is chosen to be a Gaussian with the requirement

$$\begin{aligned}\langle \eta(t_i)\eta(t_j)\rangle &= \int \Pi_k \, \mathrm{d}\eta(t_k) \; P\left(\{\eta(t_k)\}\right) \; \eta(t_i)\eta(t_j) \\ &= \frac{2}{\Delta t}\delta_{ij} \ ,\end{aligned} \tag{7.124}$$

assuming discrete "time" steps Δt. Here, P is the probability distribution for the η variables. $\eta(t_i)$ obeys the Gaussian distribution

$$P(\eta(t_i)) = \frac{1}{2}\sqrt{\frac{\Delta t}{2\pi}} \exp\left(-\eta^2(t_i)\frac{\Delta t}{4}\right) \ . \tag{7.125}$$

Discretizing (7.123) accordingly, we get

$$U(t_i+1) = U(t_i) - \Delta t \left[\frac{\delta S\left(\{U(t_i)\}\right)}{\delta U} - \eta(t_i)\right] \ . \tag{7.126}$$

Now let us check whether this procedure generates configurations according to a weight factor $\exp(-S)$. Solving (7.126) for η, the expression

$$\eta(t_i) = \frac{U(t_{i+1}) - U(t_i)}{\Delta t} + \frac{\delta S}{\delta U} \tag{7.127}$$

follows and therefore

$$P[U(t_i) \to U(t_{i+1})] = \frac{1}{2}\sqrt{\frac{\Delta t}{\pi}} \; \exp\left(-\left\{\frac{U(t_{i+1}) - U(t_i)}{\Delta t} + \frac{\delta S}{\delta U}\right\}^2 \Delta t/4\right) \ . \tag{7.128}$$

In the limit of an infinitesimal time step Δt we look at detailed balance, (7.116):

$$\frac{\mathrm{e}^{-S(U)}}{\mathrm{e}^{-S(U')}} = \frac{P(U' \to U)}{P(U \to U')} \ . \tag{7.129}$$

In our case we get

$$\begin{aligned}\frac{P(U' \to U)}{P(U \to U')} &= \frac{\exp\left(-\left\{\frac{U-U'}{\Delta t} + \frac{\delta S(U')}{\delta U}\right\}^2 \frac{\Delta t}{4}\right)}{\exp\left(-\left\{\frac{U'-U}{\Delta t} + \frac{\delta S(U)}{\delta U}\right\}^2 \frac{\Delta t}{4}\right)} \\ &\underset{\Delta t \to 0}{\longrightarrow} \exp\left(-\frac{\delta S}{\delta U}(U - U')\right) = \frac{\mathrm{e}^{-S(U)}}{\mathrm{e}^{-S(U')}} \ ,\end{aligned} \tag{7.130}$$

i.e. detailed balance holds.

As one can infer from this derivation, an additional scale Δt has entered the discussion. Δt has to be reasonably small for this algorithm to work. This is a disadvantage compared to Metropolis. An advantage of this method is that one always generates new updated fields per evolution step and one does not have to bother about the acceptance/rejection ratio of the Metropolis procedure that might keep the system stuck in configuration space for a while.

7.1.11 The Microcanonical Algorithm

Finally we want to discuss a microcanonical way of generating configurations. This procedure is again based on the thermodynamical analogue of our problem. The Langevin method describes an evolution of the system in a fictitious "time", generating a canonical ensemble of configurations. As we know from statistical mechanics in the thermodynamic limits of an infinite number of degrees of freedom it does not matter for averaged values of observables which ensemble one considers. Therefore one can generate configurations using the microcanonical ensemble at constant "energy" (which is *not* the real energy of the system). Let us look at a scalar field theory again. The partition function reads

$$Z = \int [\mathrm{d}\phi] \mathrm{e}^{-S[\{\phi\}]} \ . \tag{7.131}$$

One can introduce momentum variables π_i in the form

$$Z = \mathrm{const} \int [\mathrm{d}\phi][\mathrm{d}\pi] \ \exp\left(-\sum_i \frac{\pi_i^2}{2} - S[\{\phi\}]\right) \ . \tag{7.132}$$

The Gaussian integration over π_i generates only an (unimportant) constant. The exponent in (7.132) can be interpreted as a (fictitious) Hamiltonian

$$H[\pi, \phi] = \sum_i \frac{\pi_i^2}{2} + S[\{\phi\}] \ . \tag{7.133}$$

Keep in mind that this is not the real Hamiltonian of the scalar field theory.

In the thermodynamic limit one can replace (7.132) by the microcanonical partition sum

$$Z_{\mathrm{micro}}^{(E)} = \mathrm{const} \int [\mathrm{d}\phi][\mathrm{d}\pi] \, \delta(H[\pi, \phi] - E) \ . \tag{7.134}$$

The time evolution of the system occurs within the energy hypersurface of a given energy E of the system.

Using the equipartition theorem, stating that the mean kinetic energy per degree of freedom is just $T/2$,

$$\frac{1}{N}\left\langle \sum_{\mathrm{i}} \frac{\pi_{\mathrm{i}}^2}{2} \right\rangle = \frac{T}{2} = 1/2 \ , \tag{7.135}$$

where $\beta = 1/T = 1$ in our case, as can be seen from (7.131). That is, the microcanonical ensemble has to start with the initial conditions fulfilling (7.135). From

(7.134) and the assumption of ergodicity it follows that one only has to calculate classical equations of motion:

$$\dot{\phi}_i = \frac{\partial H}{\partial \pi_i} = \pi_i \ ,$$
$$\dot{\pi}_i = -\frac{\partial H}{\partial \phi_i} \ . \tag{7.136}$$

Equations (7.136) are solved by discretizing the "time variable" in the same way as in the Langevin method. It is amusing to note that one can actually solve quantum-field-theoretical problems by integrating classical equations of motion (in higher dimensions).

One problem with this procedure is that one has to ensure that (7.135) holds. In practice this is done by thermalizing the system first by using one of the methods discussed earlier. Another caveat arises from making use of the thermodynamic limit, which is naturally not exactly fulfilled on a finite lattice. Therefore finite-volume corrections enter which have to be studied separately.

There are a number of other updating schemes. Many so-called hybrid algorithms use some mixture of Langevin and microcanonical approaches.

As we derived in (7.109) one can integrate out the fermionic degrees of freedom leaving us with a determinant det$M$ in the path integral. When considering the full QCD problem we still have the gluon integration left. In addition the matrix M, which is the inverse Green's function of the quarks, contains a dependence on the gluons. The full partition function reads

$$Z = \int [\mathrm{d}U] \det M^{\mathrm{F}}(\{U\}) \, \mathrm{e}^{-S^{\mathrm{F}}(\{U\})} \tag{7.137}$$

where M^{F} is the fermion Green's function, for instance for Wilson fermions (7.88). The solution of (7.137) constitutes the main problem in solving full QCD on the lattice. There are various algorithms in use to solve (7.137). It would be beyond the scope of this book to go into a detailed discussion of those methods. Since the determinant in the equation depends nonlocally on the link variables, every updating step requires the calculation of the determinant of a huge matrix. This part of the numerical calculation is therefore extremely time-critical and has to be designed very carefully.

The basis of most algorithms is the transformation of the determinant to an additional path integral. The determinant in (7.137) is the result of a Gaussian integration over Grassmann variables as is discussed in Sect. 7.1.7. A similar relation holds for bosonic fields. Given a complex scalar field ϕ, the discretized path integral is of the form

$$\int [\mathrm{d}\phi^*][\mathrm{d}\phi] \, \mathrm{e}^{-\phi^*(x)M(x,x')\phi(x')} \tag{7.138}$$

with a Hermitean matrix M. Equation (7.138) can be solved directly after diagonalizing M with a unitary transformation $U(x,x')$:

$$\phi M \phi = \tilde{\phi} M^D \tilde{\phi} \quad \text{with} \quad \tilde{\phi} = U\phi \ . \tag{7.139}$$

The integral measure is invariant under the unitary transformation:

$$I = \int [\mathrm{d}\tilde{\phi}][\mathrm{d}\tilde{\phi}^*]\,\mathrm{e}^{-\tilde{\phi}^* M^D \tilde{\phi}} = \int [\mathrm{d}\tilde{\phi}_{\mathrm{re}}][\mathrm{d}\tilde{\phi}_{\mathrm{im}}]\,\mathrm{e}^{-(\tilde{\phi}_{\mathrm{re}} M^D \tilde{\phi}_{\mathrm{re}} + \tilde{\phi}_{\mathrm{im}} M^D \tilde{\phi}_{\mathrm{im}})}$$
$$= \left[\int [\mathrm{d}\tilde{\phi}_{\mathrm{re}}]\,\mathrm{e}^{-\tilde{\phi}_{\mathrm{re}} M^D \tilde{\phi}_{\mathrm{re}}}\right]^2 . \tag{7.140}$$

In the last line we switched from the variables ϕ, ϕ^* to the real and the imaginary ϕ_{re}, ϕ_{im}. Assuming, for simplicity, that the system is discretized, one can integrate over each variable $\phi_i = \phi(x_i)$ separately:

$$I = N\left(\frac{1}{\prod_i \sqrt{M^D_{ii}}}\right)^2 = \frac{N}{\det M} , \tag{7.141}$$

where we absorbed all constants coming from the definition of the measure and the integration in a factor N.

Considering this result and (7.137), we can write the partition function as

$$Z = N \int [\mathrm{d}\phi][\mathrm{d}\phi^*][\mathrm{d}U]\,\mathrm{e}^{-S^{\mathrm{E}}(\{U\}) - \phi^*(M^{\mathrm{F}}(\{U\}))^{-1}\phi} . \tag{7.142}$$

One has to perform a path integral over the link variables and the auxiliary complex field which replaces the fermion part (ϕ is therefore often called the "pseudo fermion"). There still remains the tough numerical challenge to evaluate the elements of the inverse $(M^{\mathrm{F}})^{-1}\phi$ during the updating process of the field ϕ.

EXERCISE

7.2 The Average Link Value

Problem. Calculate the average link variable

$$\overline{U} = \frac{\int \mathrm{d}U U\,\mathrm{e}^{\frac{\beta}{N}\mathrm{Re\,tr}\,UW}}{\int \mathrm{d}U\,\mathrm{e}^{\frac{\beta}{N}\mathrm{Re\,tr}\,UW}} \tag{1}$$

for the gauge group SU(2).

Solution. Since the trace of an SU(2) matrix is real, which can be understood directly from (7.54), parameterize the SU(2) group elements with the help of the Pauli matrices:

$$U = x_1 \mathbb{1} + \mathrm{i} \sum_i x_i \sigma_i . \tag{2}$$

Exercise 7.2

Exploiting this property we can write the exponent in the exponentials of (1) simply as $\beta/2\mathrm{tr}\, UW$. A sum of SU(2) matrices is again proportional to a SU(2) matrix:

$$\sum_i U_i = c\tilde{U} \ . \tag{3}$$

This behavior becomes clear when remembering our discussion of Sect. 7.1.5. There we showed that a SU(2) matrix corresponds to a unit vector on the hypersphere S_3. Adding several SU(2) matrices together is like adding several vectors, the resulting vector pointing to the hypersurface, albeit with a different length. Note that this behavior is very specific to the SU(2) group and cannot be applied to SU(3). Using this feature we get

$$W = \lambda V \ , \quad V \in \mathrm{SU(2)} \ , \quad \lambda = \sqrt{\det W} \ . \tag{4}$$

The invariance of the measure with respect to the variable change $U \to UV^+$ yields

$$\begin{aligned}
\overline{U} &= \frac{\int \mathrm{d}UU\, \mathrm{e}^{\frac{\beta}{2}\lambda\, \mathrm{tr}\, UV}}{\int \mathrm{d}U\, \mathrm{e}^{\frac{\beta}{2}\lambda\, \mathrm{tr}\, UV}} \\
&= \frac{\int \mathrm{d}UUV^\dagger\, \mathrm{e}^{\frac{\beta}{2}\lambda\, \mathrm{tr}\, U}}{\int \mathrm{d}U\, \mathrm{e}^{\frac{\beta}{2}\lambda\, \mathrm{tr}\, U}} \\
&= V^\dagger \frac{\int \mathrm{d}UU\, \mathrm{e}^{\frac{\beta}{2}\lambda\, \mathrm{tr}\, U}}{\int \mathrm{d}U\, \mathrm{e}^{\frac{\beta}{2}\lambda\, \mathrm{tr}\, U}} \ .
\end{aligned} \tag{5}$$

Let us first evaluate the denominator of (1):

$$\int \mathrm{d}U\, \mathrm{e}^{\frac{\beta}{2}\lambda\, \mathrm{tr}\, U} = \frac{1}{4\pi^2} \int \mathrm{d}\varphi\, \mathrm{d}\Omega(\boldsymbol{n}) \sin^2\frac{\varphi}{2} \exp\left(\frac{\alpha \sin\varphi}{2\sin\frac{\varphi}{2}}\right) \equiv I \ , \quad \alpha \equiv \beta\lambda \ . \tag{6}$$

The integration over the solid angle $\mathrm{d}\Omega$ can be directly performed, yielding a value of 4π. Using the relation

$$\sin\varphi = 2\sin\frac{\varphi}{2}\cos\frac{\varphi}{2} \tag{7}$$

we get

$$\begin{aligned}
I &= \frac{1}{\pi} \int \mathrm{d}\varphi \sin^2\frac{\varphi}{2}\, \mathrm{e}^{\alpha\cos\frac{\varphi}{2}} \\
&= \frac{1}{\pi}\left(1 - \frac{\partial^2}{\partial\alpha^2}\right) \underbrace{\int_0^{2\pi} \mathrm{d}\varphi\, \mathrm{e}^{\alpha\cos\frac{\varphi}{2}}}_{\equiv A} \ .
\end{aligned} \tag{8}$$

Exercise 7.2

We can evaluate the integral A explicitly:

$$\begin{aligned} A &= \sum_{k=0}^{\infty} \frac{\alpha^k}{k!} \int_0^{2\pi} \mathrm{d}\varphi \cos^k \frac{\varphi}{2} \\ &= \sum_{k=0}^{\infty} \frac{\alpha^k}{k!} 2 \int_0^{\pi} \mathrm{d}\phi \cos^k \phi \\ &= 2 \sum_{k=0}^{\infty} \frac{\alpha^k}{k!} \frac{1}{2^k} \sum_{l=0}^{\infty} \binom{k}{l} \int_0^{\pi} \mathrm{d}\phi \cos((k-2l)\phi) \\ &= 2 \sum_{k=0}^{\infty} \frac{\alpha^k}{k!} \frac{1}{2^k} \sum_{l=0}^{\infty} \binom{k}{l} \begin{cases} \pi & \text{if } k=2l \\ 0 & \text{otherwise} \end{cases} . \end{aligned} \tag{9}$$

Thus only terms with k even contribute to the sum:

$$\begin{aligned} A &= 2 \sum_{\mu=0}^{\infty} \frac{\alpha^{2\mu}}{(2\mu)!} \frac{1}{2^{2\mu}} \binom{2\mu}{\mu} \pi \\ &= 2\pi \sum_{\mu=0}^{\infty} \frac{\alpha^{2\mu}}{(2\mu)!} \frac{1}{2^{2\mu}} \frac{(2\mu)!}{(\mu!)^2} \\ &= 2\pi \sum_{\mu=0}^{\infty} \frac{\alpha^{2\mu}}{(\mu!)^2 2^{2\mu}} = 2\pi I_0(\alpha) . \end{aligned} \tag{10}$$

The sum corresponds to the series representation of the Bessel function $I_0(x)$ of the 2nd kind and 0th order. In general, the following relation for Bessel functions holds:[11]

$$I_m(x) = \sum_{k=0}^{\infty} \frac{1}{k!(k+m)!} \left(\frac{x}{2}\right)^{2k+m} . \tag{11}$$

For the next steps we can make use of the following recursion relations for Bessel functions:

$$\begin{aligned} I_{m-1}(x) - I_{m+1}(x) &= \frac{2m}{x} I_m(x) , \\ I_{m-1}(x) + I_{m+1}(x) &= 2I'_m(x) , \\ I'_0(x) &= I_1(x) . \end{aligned} \tag{12}$$

It follows that

$$I = 2\left(1 - \frac{\partial^2}{\partial \alpha^2}\right) I_0(\alpha)$$

[11] e.g. see M. Abramowitz, I.A. Stegun: *Handbook of Mathematical Functions* (Dover 1968) Formula (9.6.10).

Exercise 7.2

$$
\begin{aligned}
&= 2\left(I_0(\alpha) - \frac{\partial}{\partial\alpha} I_1(\alpha)\right) \\
&= 2\left(I_0(\alpha) - \frac{I_0(\alpha)+I_2(\alpha)}{2}\right) \\
&= I_0(\alpha) - I_2(\alpha) \\
&= \frac{2}{\alpha} I_1(\alpha) \ . \qquad (13)
\end{aligned}
$$

Now we have to compute the numerator of (1)

$$
\begin{aligned}
&\int dU U \, e^{\frac{\beta\lambda}{2}\operatorname{tr} U} \qquad (14) \\
&= \frac{1}{4\pi^2} \int d\varphi \, d\Omega(\boldsymbol{n}) \sin^2\frac{\varphi}{2} \left(\cos\frac{\varphi}{2} + \frac{i}{2}\sin\frac{\varphi}{2}\boldsymbol{n}\cdot\boldsymbol{\tau}\right) \exp\left(\frac{\alpha}{2}\frac{\sin\varphi}{\sin\frac{\varphi}{2}}\right) \ .
\end{aligned}
$$

Because

$$
\int d\Omega(\boldsymbol{n})\boldsymbol{n} = 0 \ , \qquad (15)
$$

only the first term in the sum contributes:

$$
\begin{aligned}
&\frac{1}{\pi}\int_0^{2\pi} d\varphi \sin^2\frac{\varphi}{2}\cos\frac{\varphi}{2} \, e^{\alpha\cos\frac{\varphi}{2}} \\
&= \frac{1}{\pi}\left(\frac{\partial}{\partial\alpha} - \frac{\partial}{\partial\alpha^3}\right)\int_0^{2\pi} d\varphi \, e^{\alpha\cos\frac{\varphi}{2}} \\
&= \frac{1}{\pi}\left(\frac{\partial}{\partial\alpha} - \frac{\partial}{\partial\alpha^3}\right) 2\pi I_0(\alpha) \\
&= 2\left(I_1(\alpha) - \frac{\partial}{\partial\alpha}\left\{\frac{I_0(\alpha)+I_2(\alpha)}{2}\right\}\right) \\
&= 2I_1(\alpha) - I_1(\alpha) - \frac{I_1(\alpha)+I_3(\alpha)}{2} = \frac{2}{\alpha} I_2(\alpha) \ . \qquad (16)
\end{aligned}
$$

Putting the results together we have determined the value of the average link:

$$
\begin{aligned}
\overline{U} &= V^\dagger \frac{\frac{2}{\alpha} I_2(\alpha)}{\frac{2}{\alpha} I_1(\alpha)} = V^\dagger \frac{I_2(\alpha)}{I_1(\alpha)} \\
&= (\det W)^{-1/2} W^\dagger \frac{I_2(\beta\sqrt{\det W})}{I_1(\beta\sqrt{\det W})} \ . \qquad (17)
\end{aligned}
$$

7.1.12 Strong and Weak Coupling Expansions

In the real world of low-energy QCD, lattice or continuum, analytical or semi-analytical studies are usually not manageable. One can still get more understanding of the behavior of the theory by tuning parameters to unphysical regimes. In the lattice QCD action (7.48), neglecting fermionic effects for now, we have one parameter β proportional to the inverse coupling squared. What happens to observables when β is tuned to very small (strong coupling) or large (weak coupling) values?

For that we consider one basic and useful quantity: the averaged plaquette $\langle P\rangle$, (7.39), which is directly connected to the QCD action as can be seen from (7.48). In the limit of infinite coupling, $\beta = 0$, the weighting factor e^{-S} vanishes, which means that all link values occur with equal probabilities and the value of the plaquette averages out to zero. In the opposite limit $\beta \to \infty$ the exponential weighting factor suppresses all contributions but the one with minimum action $S = 0$, which means all U_μ are equal to 1 up to a gauge transformaton. Thus $\langle P\rangle = 1$ attains the maximum value.

Now let us go a step further and first calculate deviations from $\langle P\rangle = 0$ for small values of β analytically:

$$\bar{P} \equiv \frac{1}{3}\mathrm{Re}[\langle \mathrm{tr} P\rangle] = Z^{-1}\frac{1}{n_\mathrm{P}}\sum_{\Box}\int [\mathrm{d}U]\,\frac{1}{3}\mathrm{Re}\,\mathrm{tr}_{\Box}(U_1U_2U_3U_4)\,\mathrm{e}^{-S[U]} \tag{7.143}$$

n_P is the number of plaquettes on the lattice, which is the number of independent planes in the hypervolume (i.e. $[(x,y),(x,z),(x,t),(y,z),(y,t),(z,t)]$) times the number of lattice points $n_\mathrm{P} = 6N_x^3N_t$. The shortcut $\sum_\Box$ denotes summation over all plaquettes on the lattice and the product of links encircles an elementary square.

We perform a Taylor expansion of (7.143) to first order in β:

$$\begin{aligned}\bar{P} = {} & \frac{1}{3n_\mathrm{P}}\int [\mathrm{d}U]\sum_{\Box}\mathrm{Re}\,\mathrm{tr}\underbrace{(U_1U_2U_3U_4)}_{P_\Box}\left(1-\beta\sum_{\Box'}\left[\frac{1}{3}\mathrm{Re}\,\mathrm{tr}P_{\Box'}\right]\right)\\ & \times\left[\int [\mathrm{d}U]\left(1-\beta\sum_{\Box'}\left[1-\frac{1}{3}\mathrm{Re}\,\mathrm{tr}P_{\Box'}\right]\right)\right]^{-1}\,,\end{aligned} \tag{7.144}$$

where we inserted the normalization factor in lowest-order β expansion.

In order to evaluate (7.144), one needs some basic integrals over the group elements. The following relations hold (see Sect. 7.1.5):

$$\int [\mathrm{d}U] = 1\,,$$

$$\int [\mathrm{d}U]U_{ij} = 0\,,$$

$$\int [\mathrm{d}U]U_{ij}U_{kl} = 0\,,$$

$$\int [\mathrm{d}U] U_{ij} U^*_{kl} = \frac{1}{3}\delta_{ik}\delta_{jl} \; . \tag{7.145}$$

The first relation in (7.145) is the choice of normalization for the group integration. The second equation expresses the fact that a group element varied over the whole group space vanishes. The first nontrivial relation is the last equation in (7.145). You see that the nonvanishing contribution does not have an open color index. This relates to our earlier discussion about gauge invariance, integrating over quantities with open color averages out to zero. The factor $1/3$ reads in general $1/N_\mathrm{c}$, the inverse number of colors.

With this set of integrals the integration of constants in the expression (7.144) is trivial. The next terms of type

$$\int [\mathrm{d}U] \mathrm{tr} P_\Box = \int [\mathrm{d}U_1\, \mathrm{d}U_2\, \mathrm{d}U_3\, \mathrm{d}U_4] \mathrm{tr}(U_1 U_2 U_3 U_4) = 0 \tag{7.146}$$

vanish because according to (7.145) the integrals over terms linear in a given link variable are zero.

Nontrivial terms arise from the product of two plaquettes:

$$\int [\mathrm{d}U] \mathrm{Re\; tr} P_\Box \mathrm{Re\; tr} P_{\Box'} \; . \tag{7.147}$$

If the two plaquettes $\Box$ and $\Box'$ were different, there would be link variables in linear order contained in (7.147) and the integral vanishes. Therefore one obtains a δ function $\delta_{\Box,\Box'}$ (see Fig. 7.5):

$$\begin{aligned} \int [\mathrm{d}U] \mathrm{Re\; tr} P_\Box \mathrm{Re\; tr} P_{\Box'} = & \frac{\delta_{\Box,\Box'}}{4} \int \left[\mathrm{d}U_{(1)}\, \mathrm{d}U_{(2)}\, \mathrm{d}U_{(3)}\, \mathrm{d}U_{(4)}\right] \\ & \times \left[U_{(1)ij} U_{(2)jk} U_{(3)kl} U_{(4)li} + \mathrm{h.c.}\right] \\ & \times \left[U_{(1)i'j'} U_{(2)j'k'} U_{(3)k'l'} U_{(4)l'i'} + \mathrm{h.c.}\right] \; . \end{aligned} \tag{7.148}$$

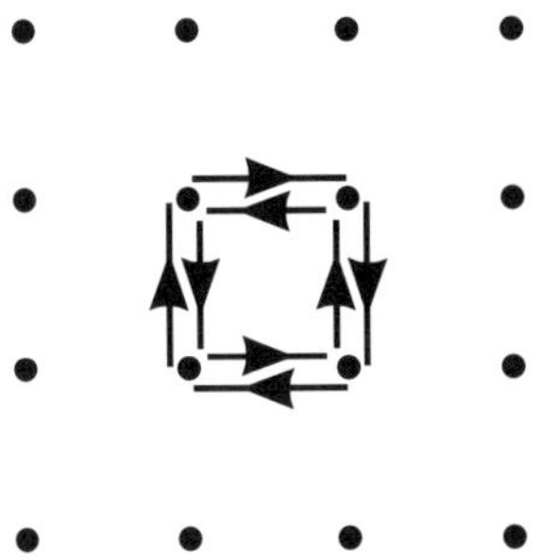

Fig. 7.5. The only nonvanishing contribution for the value of the plaquette operator in lowest order of the coupling β: the product of two elementary plaquettes with opposite orientation

Using the relations (7.145) we get

$$\begin{aligned} \int [\mathrm{d}U] \mathrm{Re\; tr} P_\Box \mathrm{Re\; tr} P_{\Box'} &= \frac{\delta_{\Box,\Box'}}{4} 2 \sum_{\substack{ijkl \\ i'j'k'l'}} \left(\frac{1}{3}\right)^4 \delta_{ii'}\delta_{jj'}\delta_{kk'}\delta_{jj'} \\ &= \frac{1}{2}\delta_{\Box,\Box'} \; . \end{aligned} \tag{7.149}$$

Putting this all together we get the result for the denominator Z of (7.143),

$$Z = 1 - \beta \sum_\Box = 1 - \beta n_\mathrm{P} \; , \tag{7.150}$$

and the expectation value of the plaquette in strong coupling approximation reads

$$\begin{aligned}\bar{P} &= \frac{Z^{-1}}{3n_{\mathrm{P}}}\left\{\frac{\beta}{3}-\frac{1}{2}\sum_{\Box,\Box'}\delta_{\Box,\Box'}\right\}\\ &\equiv \frac{Z^{-1}}{18}\beta \sim \frac{\beta}{18}+\mathcal{O}(\beta^2)\ .\end{aligned} \tag{7.151}$$

7.1.13 Weak-Coupling Approximation

Let us repeat the calculation of the plaquette value in the opposite regime of weak coupling or large β. As we mentioned before, at large β the weight factor in the path integral $\exp(-S(\{U\}))$ picks out the contributions with small actions, and in the extreme limit $\beta \to \infty$ the configuration with $S=0$ survives since the Wilson action is positive, $S \geq 0$. With the definition of S, (7.48), this means all link variables $U_\mu = 1$ (plus a possible gauge transformation; gauge transformations do not change the value of the action) throughout the lattice and therefore we also get $\langle \mathrm{tr} P\rangle = 1$. Calculating corrections in lowest order, $1/\beta$, we proceed by expanding the link variables about $U = \mathbb{1}$:

$$U = \mathrm{e}^{\mathrm{i}\lambda^a\alpha^a} \approx \mathbb{1} + \mathrm{i}\lambda^a\alpha^a \tag{7.152}$$

and the product of links in the plaquette yields

$$\begin{aligned}P_\Box &= U_1U_2U_3U_4 = \mathrm{e}^{\mathrm{i}\lambda^a\alpha_1^a}\cdot\mathrm{e}^{\mathrm{i}\lambda^a\alpha_2^a}\cdot\mathrm{e}^{\mathrm{i}\lambda^a\alpha_3^a}\cdot\mathrm{e}^{\mathrm{i}\lambda^a\alpha_4^a}\\ &= \mathrm{e}^{\mathrm{i}\lambda^a\epsilon^a} + \text{higher orders}\\ &= 1+\mathrm{i}\lambda^a\epsilon^a - \frac{1}{2}\left(\left(\lambda^a\epsilon^a\right)^2+\mathcal{O}(\epsilon^3)\right)\end{aligned} \tag{7.153}$$

with $\epsilon^a = \sum_{i=1}^4 \alpha_i^a$. The normalizing factor (or partition function) Z in (7.143) then gives

$$\begin{aligned}Z &= \int[\mathrm{d}U]\exp\left(-\beta\sum_\Box\left(1-\frac{1}{3}\mathrm{Re}\,\mathrm{tr}\left(1-\mathrm{i}\lambda^a\epsilon^a-\frac{1}{2}(\lambda^a\epsilon^a)^2\right)\right)\right)\\ &= \int[\mathrm{d}U]\exp\left(-\frac{\beta}{6}\sum_\Box \mathrm{tr}(\lambda^a\epsilon^a)^2\right)\ .\end{aligned} \tag{7.154}$$

Using the relation $\mathrm{tr}\,\lambda^a\lambda^b = \frac{1}{2}\,\mathrm{tr}\,\{\lambda^a,\lambda^b\} = 2\delta_{ab}$, we get

$$Z = \int[\mathrm{d}U]\exp\left(-\frac{\beta}{3}\sum_\Box \epsilon^a\epsilon^a\right)\ . \tag{7.155}$$

We do not really need the exact values of this expression but merely its β dependence. Let us rescale all the parameters in the link variables $\alpha^a \to \tilde{\alpha}^a/\sqrt{\beta}$. [dU] contains 8 colors times 4 directions times $(N_x)^3 N_t$ space–time points. Following the discussion in the continuum case (see Sect. 2.2.5) one can reduce the number of gluonic degrees of freedom from 4 to 3 directions by specifying a gauge. Gauge degrees of freedom leave the action invariant and therefore would not contribute to the Gaussian integration in (7.155), i.e. they do not generate a β-dependent term. Altogether we then get

$$\begin{aligned} Z &= \beta^{-12N_x^3 N_t} \int [\mathrm{d}\tilde{U}]\mathrm{e}^{-\frac{1}{3}\tilde{\varepsilon}^a\tilde{\varepsilon}^a} \\ &= \text{const. } \beta^{-12N_x^3 N_t} \ . \end{aligned} \tag{7.156}$$

The plaquette can be calculated by taking a derivative of Z with respect to β:

$$N_c^{-1}\mathrm{tr}P = 1 - \frac{1}{6N_x^3 N_t}\frac{\frac{\partial}{\partial\beta}Z}{Z} = 1 - \frac{2}{\beta} + \mathcal{O}\left(\frac{1}{\beta^2}\right) \ . \tag{7.157}$$

7.1.14 Larger Loops in the Limit of Weak and Strong Coupling

In addition to calculating the plaquette analytically in the case of extreme coupling, we can do a similar calculation in the case of larger Wilson loops. This is especially interesting because of the connection between the Wilson loops and the quark potential, which will be discussed in Sect. 7.1.15. The calculation is performed in the same way as in the previous section. The value of the rectangular loop with sides of length L, L' is given by

$$w(l = (L, L')) = \frac{1}{Z}\int [\mathrm{d}U]\mathrm{e}^{-S}\frac{1}{3}\mathrm{tr}\prod_l U_l \ . \tag{7.158}$$

where $\prod_l U_l$ symbolizes the product of link variables around a closed loop l. Obviously the plaquette and w are connected via $\mathrm{tr}P = w(1, 1)$. Again we will consider only lowest-order contributions in β. From the set of integrals (7.145), we see that the only nonvanishing terms after integration contain for every link U in the loop the oppositely oriented link $U^\dagger$, which comes from the plaquette contained in the action S.

Now we can construct the relevant term diagramatically in similar ways as shown in Fig. 7.4. Let us have a look at Fig. 7.7. The big loop is the Wilson loop whose expectation value we want to calculate. By expanding the action in powers of β we generate factors proportional to the plaquette operator (or 1×1 Wilson loop). As is clear from the integrals (7.145) and (7.148) the plaquettes have to compensate for the links in the loop by the conjugate links. Therefore you get the inner circle of plaquettes (one wants to use the least number of elementary loops because each one contains a factor β).

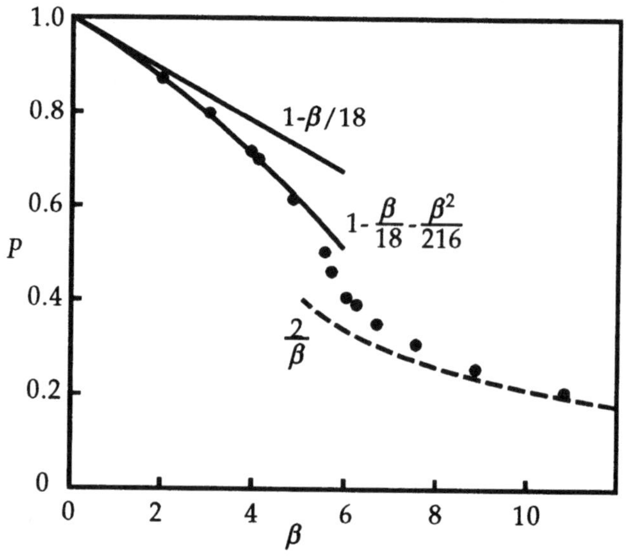

Fig. 7.6. Monte Carlo results for the expectation value of the plaquette. One can identify the results of strong (7.151) and weak coupling (7.157). The numerical data are taken from M. Creutz: *Quarks, Gluons, and Lattices*, Cambridge University Press, Cambridge, 1983

Since there are in general some additional unsaturated links from the plaquettes, one has to fill up the remaining holes with additional plaquettes. This then finally yields the nonvanishing contribution to the Wilson loop in lowest order in β.

As we mentioned before, every plaquette contains a factor β. One can infer from Fig. 7.7 that the first contribution to $w(L, L')$ is proportional to $\beta^{L\cdot L'}$. Thus, we are left calculating the coefficient in front of $\beta^{L\cdot L'}$.

From a perturbation theory point of view the introduction of a cutoff scale a is just another method of regularization of QCD. Gauge invariance arguments yielded a general formula for the running coupling constant in lowest-order loop calculation (see Sect. 4.4):

$$g^2(Q^2) = \frac{16\pi^2}{(11 - \frac{2}{3}N_\mathrm{f}) \ln \frac{Q^2}{\Lambda^2}} \ . \tag{7.159}$$

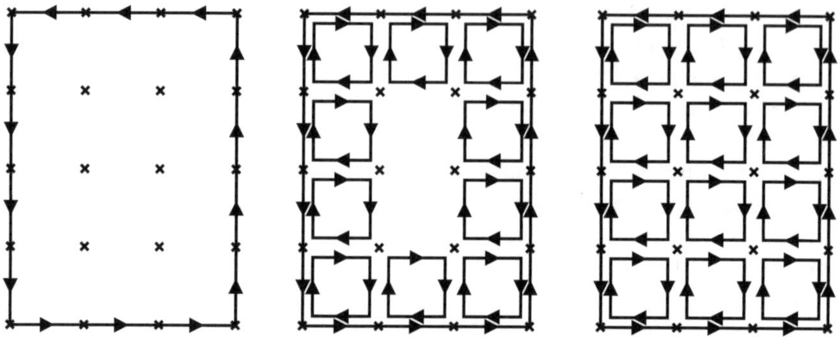

Fig. 7.7. The 3×4 Wilson loop $w(3, 4)$. The interior is filled up with plaquettes successively, matching each occurring link variable with an analogous link pointing in the opposite direction

The asymptotic scale parameter Λ depends on the regularization scheme. In lattice perturbation theory the momentum scale is therefore $Q = 1/a$. Thus for sufficiently dense lattices the coupling strength depends on the lattice spacing as

$$g^2(a) = -\frac{16\pi^2}{(11 - \frac{2}{3}N_\mathrm{f})\ln(a^2\Lambda_\mathrm{lat}^2)} , \qquad (7.160)$$

where Λ_lat is the scale parameter corresponding to a renormalisation scheme based on an Euclidean lattice. This parameter can be obtained only by actually solving the corresponding Feynman diagrams in the lattice formulation. This is a very complex calculation.[12] Compared to continuum renormalisation schemes, e.g. the momentum-subtraction scheme (MOM), we have $\Lambda_\mathrm{lat} \sim 0.012\,\Lambda_\mathrm{MOM}$, a very small number.

This has the consequence that one needs lattices with very small a (i.e. with many lattice points) to be in a regime where quantities measured on a lattice change only perturbatively when one changes the lattice spacing (the so-called scaling limit). So far lattice calculations have not yet reached this regime.

7.1.15 The String Tension

A very important quantity is the string tension σ. Its value of about 1 GeV/fm is often used as a scale or measure of calculated quantities in QCD. As will be argued in this section the potential between a quark and an antiquark $V_{\mathrm{Q\overline{Q}}}(R)$ is expected to grow linearly with distance. This can be readily understood by assuming a flux tube model connecting the quark and antiquark, where the color-electric field lines are confined in a tubelike connection between the two quarks. No field lines leak out of the tube, therefore the field strengths remain constant over the extension of the tube and the potential rises linearly with the distance of the quarks. Thus we have

$$V_{\mathrm{Q\overline{Q}}}(R) \overset{R\to\infty}{\longrightarrow} \sigma R . \qquad (7.161)$$

Here one additionally assumes heavy quarks in order to apply the nonrelativistic concept of a quark potential. This expected behavior is a basic but crucial testing ground for lattice QCD.

As mentioned, we assume static, i.e. very heavy, quarks. Then their kinetic energy vanishes. They sit at some given point and propagate in time only. Keeping in mind that for the Euclidean action the potential enters with a positive sign, the action is then merely

$$S^\mathrm{E}_{\mathrm{Q\overline{Q}}} = \int\limits_{\tau_\mathrm{i}}^{\tau_\mathrm{f}} \mathrm{d}\tau V_{\mathrm{Q\overline{Q}}}(R_{\mathrm{Q\overline{Q}}}) = TV(R) , \qquad (7.162)$$

[12] see e.g. P. Weisz: Phys. Lett. B **100**, 330 (1981).

where R is the distance between quark and antiquark. For simplicity, we do not consider the color matrix structure of the fields in the following argument. This does not change any results discussed here. We determine the quark–gluon interaction term

$$j_\mu A_\mu = \varrho A_4 \tag{7.163}$$

since the quarks are static. ϱ is just the color charge density, which reads

$$\varrho(x) = \delta^{(3)}(x) - \delta^{(3)}(x - R) \; , \tag{7.164}$$

where the positive charge sits at site $x = 0$ and the antiquark is put on a lattice site with distance $R = |\boldsymbol{R}|$. The action per time generated by this interaction is

$$\begin{aligned} S^{\mathrm{E}} &= -\mathrm{i} \int \mathrm{d}^4 x \varrho(x) A_4(x) = -\mathrm{i} \int \mathrm{d}^3 x \varrho(x) A_4(x) \\ &= -\mathrm{i}\,(A_4(0) - A_4(\boldsymbol{R})) \; . \end{aligned} \tag{7.165}$$

As this result is time-independent, we can write

$$S^{\mathrm{E}} = -\mathrm{i} \left[\int\limits_0^T \mathrm{d}\tau \big(A_4(0, \tau) - A_4(\boldsymbol{R}, \tau) \big) \right] \; . \tag{7.166}$$

If we assume that the time interval T is much larger than the distance $|\boldsymbol{R}|$ between the quarks we can approximate S^{E} by a closed-loop integration on the lattice:

$$\begin{aligned} S^{\mathrm{E}} \overset{T \gg |\boldsymbol{R}|}{\approx} & -\mathrm{i} \left[\int\limits_0^T \mathrm{d}\tau A_4(0, \tau) + \int\limits_{x=0}^{x=\boldsymbol{R}} \mathrm{d}\boldsymbol{s} \boldsymbol{A}(\boldsymbol{s}, T) \right. \\ & \left. + \int\limits_T^0 \mathrm{d}\tau A_4(\boldsymbol{R}, \tau) + \int\limits_{x=\boldsymbol{R}}^{x=0} \mathrm{d}\boldsymbol{s} \boldsymbol{A}(\boldsymbol{s}, 0) \right] \\ = & -\mathrm{i} \oint\limits_l \mathrm{d}x_\mu A_\mu(x) \\ = & \, T \cdot V(R) \; . \end{aligned} \tag{7.167}$$

Now, looking back to our discussion of Wilson loops, a rectangular loop with sides R and T (see Fig. 7.8)

$$w(R, T) = \mathrm{tr}\,[U_1 U_2 \ldots U_N] \tag{7.168}$$

is just the product of the link variables around the contour l. That is, in the continuum limit we have

$$\langle w(R, T) \rangle = \left\langle \mathrm{e}^{-\mathrm{i}g \oint_l \mathrm{d}x_\mu A_\mu(x)} \right\rangle = \mathrm{e}^{-T \cdot V(R)} \; , \tag{7.169}$$

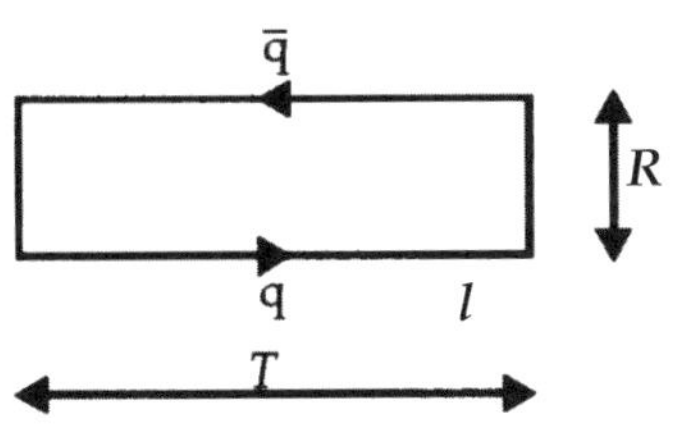

Fig. 7.8. The Wilson loop to calculate the quark potential. A quark–antiquark pair or their color fields, propagate along the time links

where we have averaged over all possible gluonic configurations. As we discussed in the beginning, the assumption is that for large distances the potential between quark and antiquark behaves like $V_{\mathrm{Q\overline{Q}}} = \sigma r$. From that we get

$$\sigma \overset{R\,\text{large}, T \gg R \gg a}{=} -\frac{1}{RT} \ln\langle w(R,T)\rangle \ . \tag{7.170}$$

In general the potential is similarly given by

$$V(R) = -\frac{1}{T} \ln\langle w(R,T)\rangle \ . \tag{7.171}$$

Figure 7.9 shows the results of a lattice calculation of the potential. One clearly recognizes the linear slope of the potential at large R. In fact, the total

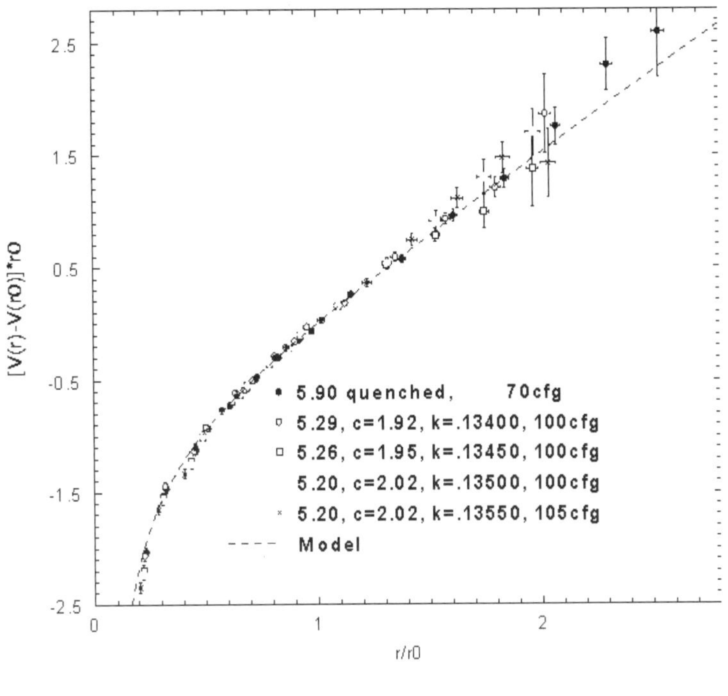

Fig. 7.9. Lattice calculation of the quark–antiquark potential. One can clearly see the linearly rising potential indicating confinement (from: J. Garden et al. (UKQCD collaboration): Nucl. Phys. Proc. Suppl. **83**, 165 (2000))

potential can be fitted by the expression

$$V(R) = -\frac{\alpha}{r} + \sigma r \ . \tag{7.172}$$

Another related method to determine the string tension from the value of the Wilson loops is to determine the so-called Creutz ratio Cr of loops:[13]

$$Cr \sim \sigma = -\ln\left[\frac{w(R,T)w(R-1,T-1)}{w(R,T-1)w(R-1,T)}\right] \ . \tag{7.173}$$

One can easily check that (7.173) will reduce to the value of the string tension in the limit of large R and T:

$$\begin{aligned} Cr &\sim -\ln\left[\frac{\mathrm{e}^{-\sigma RT-\sigma(R-1)(T-1)}}{\mathrm{e}^{-\sigma R(T-1)-\sigma(R-1)T}}\right] \\ &= -\ln\left[\frac{\mathrm{e}^{-\sigma(2RT-R-T+1)}}{\mathrm{e}^{-\sigma(2RT-R-T)}}\right] \\ &= -\ln\left[\mathrm{e}^{-\sigma}\right] = \sigma \ . \end{aligned} \tag{7.174}$$

The advantage of this expression is that finite-size effects due to the usually rather small values of R, T in a realistic calculation can be suppressed by taking ratios of observables.

7.1.16 The Lattice at Finite Temperature

One of the major research topics for lattice calculations, especially of QCD, is to study strong interactions at finite temperatures ("finite" means larger than zero here).

As we have discussed in the introduction to the path-integral formalism (Sect. 7.1.1), a transition from real to imaginary time, i.e. an analytical continuation into the complex t-plane, led us directly to the partition function of the theory

$$Z(T,V) = \mathcal{N}\int [\mathrm{d}\phi_i]\langle\phi_i|\,\mathrm{e}^{-\beta H}|\phi_i\rangle \ , \tag{7.175}$$

where the initial and final states are identical.

In order to study vacuum properties one adopts the limit $\beta \to \infty$ (or at least very large in actual numerical calculations). It is now straightforward to study finite temperatures within the same formalism. We just take a finite value of $\beta = 1/T$.

For imaginary times $\tau = \mathrm{i}t$ the time evolution operator and the Boltzmann factor look alike

$$\mathrm{e}^{-\mathrm{i}H(-\mathrm{i}\tau)} = \mathrm{e}^{-H\tau} \longleftrightarrow \mathrm{e}^{-\beta H} \ . \tag{7.176}$$

[13] M. Creutz: Phys. Rev. D **21**, 2308 (1980).

This means that (7.175) can be seen as a transition amplitude for some finite time interval $T = 1/\beta$. A finite extension in time together with periodic boundary conditions has the consequence that the possible energy values ω_n are discrete:

$$\mathrm{e}^{\mathrm{i}\omega_n^{\mathrm{B}} T} = \mathrm{e}^{\mathrm{i}\omega_n^{\mathrm{B}} 0} \longrightarrow \omega_n^{\mathrm{B}} = \frac{2\pi}{T} n \ , \quad n = 0, 1, \dots \ . \tag{7.177}$$

The ω_n are the *Matsubara frequencies*, which are commonly used in finite-temperature quantum field theory. Note that so far our discussion has not relied on any lattice discretization and is also valid for the continuum theory.

For fermions there is the difference that one must assume antiperiodic boundary conditions due to their anticommutation properties:

$$\mathrm{e}^{\mathrm{i}\omega_n^{\mathrm{F}} T} = -\mathrm{e}^{\mathrm{i}\omega_n^{\mathrm{F}} 0} \longrightarrow \omega_n^{\mathrm{F}} = \frac{2\pi}{T} n + \frac{\pi}{T} \quad n = 0, 1, \dots \tag{7.178}$$

N.B. As we have just discussed, the temperature of the system is equal to the inverse length of the system in time direction. When one performs a lattice calculation the lattice necessarily has a finite size in the time direction due to the lack of infinite computer power. This means there is *always* a temperature larger than zero in every lattice calculation. So one has to ensure the temperature is kept low enough by taking a big lattice so that the energy is much lower than the energy of excited modes in the system.

In studies of finite-temperature properties of QCD matter, the values of the energy density ϵ and the pressure density P are of central importance. Let us therefore outline how to determine ϵ and P on the lattice.

Using the partition function Z we have[14]

$$\begin{aligned} \epsilon &= \frac{T^2}{V} \frac{\partial}{\partial T} (\ln Z)_{V=\mathrm{const}} \ , \\ P &= T \frac{\partial}{\partial V} (\ln Z)_{T=\mathrm{const}} \ . \end{aligned} \tag{7.179}$$

We see from these equations that we have to treat space and time separately in order to keep V constant and vary T and vice versa.

There are actually two possibilities to vary T, for instance, as we discussed, the temperature is given by the temporal extent L_t of the lattice:

$$T = \frac{1}{L_t} = \frac{1}{N_t a} \ , \tag{7.180}$$

where N_t is the number of points of the lattice in time direction. One can either vary T by changing the lattice spacing a, or one could keep a but change the number of points in the time direction. The latter method has the advantage that it would change T without altering the volume of the system $V = (N_x a)^3$, which is just what one needs in (7.179). There is a major disadvantage, however. N_t is

[14] W. Greiner, H. Stöcker, L. Neise: *Thermodynamics and Statistical Mechanics* (Springer Berlin, Heidelberg 1995).

an integer (in addition many lattice gauge codes require even values of N_t), so that the variation possibilities of temperature

$$\frac{1}{N_t a} - \frac{1}{(N_t+1)a} \approx \frac{1}{(N_t)^2 a} = \frac{T}{N_t} \tag{7.181}$$

is rather coarse. Therefore it is customary to stick with the former approach, varying a. This one has its own problem since the volume would change too with changing scale parameter. So we formally have to introduce two scales a_x, a_t for the space and time directions (this is only needed to perform the derivatives; afterwards one can set $a_x = a_t = a$ again). In this case one has to rewrite the Wilson action (7.48) (including only gluons here):

$$S^{\mathrm{W}} = \frac{\beta_x}{r} \sum_{P_x} \left[1 - \frac{1}{3} \mathrm{Re}\, \mathrm{tr} P_x \right] + r\beta_t \sum_{P_t} \left[1 - \frac{1}{3} \mathrm{Re}\, \mathrm{tr} P_t \right] , \tag{7.182}$$

where P_x are plaquettes which lie in a plane with constant time, i.e. the (xy), (xz), or (yz) planes, whereas P_t contains links in the time direction ((xt), (yt), (zt) planes).

β_x and β_t are the inverse coupling strength defined via

$$\beta_x = \frac{6}{g_s^2(g_0, r)} , \qquad \beta_t = \frac{6}{g_t^2(g_0, r)} , \tag{7.183}$$

where the anisotropy factor is given by $r = a_x/a_t$.

The spatial and temporal coupling constants g_s, g_t generally depend on the coupling g_0, assuming an isotropic lattice $r = 1$, and the anisotropy in some complicated fashion. In the isotropic limit $r = 1$ we have $g_s^2(g_0, 1) = g_t^2(g_0, 1) = g_0$. The factors r and $1/r$ can be intuitively explained by considering the following argument. The volume element of the action a^4 becomes $a_x^3 a_t = r^{-1} a^4$. The spacelike part of the action is not further modified. The space–time plaquettes contain two links in the time direction $U_4 \sim \mathrm{e}^{\mathrm{i} A_4 a_\tau}$. For small lattice spacing $1 - U$ is proportional to a_τ compared to a_x for spacelike links, altogether there is an additional factor of r^2 for the contributions from P_τ. Neglecting corrections arising from $g_s^2 \neq g_x^2$ for $r \neq 1$, we now have the following expression for the action:

$$S \sim r S_\tau + \frac{1}{r} S_x + \text{corrections} , \tag{7.184}$$

where S_τ and S_x contain space–time plaquettes P_τ or space–space plaquettes P_x, respectively.

Using this result, we can return to calculate the energy and pressure of the system. For the energy we get, using (7.179) and $T = 1/(N_t a_t) = r/(N_t a_x)$,

$$\epsilon = \frac{1}{N_x^3 N_t a_x^4} \frac{\partial}{\partial r} \ln Z(r) \bigg|_{r=1} . \tag{7.185}$$

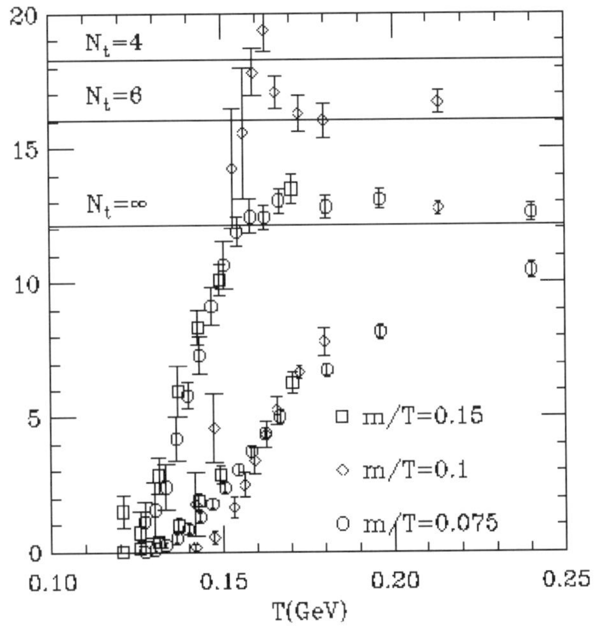

Fig. 7.10. Energy density and pressure in a lattice calculation including the fermion determinant (for 2 flavors). The graph is taken from C. Bernard et al. (MILC collaboration): Nucl. Phys. Proc. Suppl. **53**, 442 (1997)

Using the definition of Z and the expression for the action (7.184) we get the result

$$\begin{aligned}\epsilon &= \frac{1}{N_x^3 N_t a_x^4}\left\langle -\frac{\partial S}{\partial r}\right\rangle\Big|_{r=1} \\ &= \frac{1}{N_x^3 N_t a_x^4}\langle S_\tau - S_x\rangle + \epsilon_\mathrm{i}\end{aligned} \tag{7.186}$$

with

$$\epsilon_\mathrm{i} = \frac{1}{N_x^3 N_t a_x^4}\left[\frac{\partial \beta_x}{\partial r}P_x + \frac{\partial \beta_\tau}{\partial r}P_\tau\right]\Big|_{r=1}\,, \tag{7.187}$$

where ϵ_i is the contribution to the energy arising from the dependence of the coupling on a scale change, i.e. quantum corrections.

In Fig. 7.10 the results of a calculation of the energy and pressure on the lattice as a function of temperature are shown. One can see a clear dramatic rise in the energy density at a temperature of about 150 MeV, signaling a transition from hadronic QCD states to a gaslike state, the experimentally much sought-after quark–gluon plasma (QGP).

7.1.17 The Quark Condensate

Another order parameter which signals the phase transition is the quark condensate $\langle\overline{q}q\rangle$. Now $\overline{q}q = \overline{q}_L q_R + \overline{q}_R q_L$, where we have introduced the left-handed

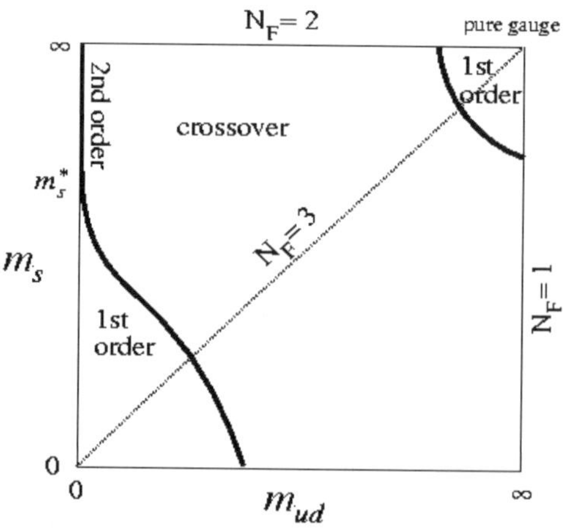

Fig. 7.11. A still much debated problem is the order of the phase transition to the plasma phase. The figure shows a rough sketch of the situation. For very heavy or very light up, down (m_{ud}, horizontal axis) and strange (m_s, vertical axis) quarks the transition is presumably first-order. The first-order regions and the smooth cross-over region are separated by a 2nd-order phase transition line. In the range of the physical masses the result could be first or second-order or even a smooth transition between hadronic and quark phase (from: K. Kanaya: Nucl. Phys. Proc. Suppl. **47**, 144 (1996))

(q_L) and right-handed (q_R) quark states, which are obtained by acting with the projectors on states with definite handedness $q_{\frac{L}{R}} = \frac{1}{2}(1 \mp \gamma_5)\psi$. The condensate measures the amount of coupling between left- and right-handed particles in the vacuum. Therefore it is also called the "chiral" condensate. Beyond a critical temperature its value vanishes, signaling the transition to a state with (nearly) massless quarks and gluons. On the lattice, $\langle \overline{q}q \rangle$ can be calculated as a path integral of the trace of the propagator

$$\langle \overline{q}q \rangle = \langle \mathrm{tr} G(x,x) \rangle \ , \tag{7.188}$$

using the master equation (7.23). Figure 7.12 shows the values of the Polyakov loop and $\langle \overline{q}q \rangle$ in a small temperature range around T_c. One can observe the

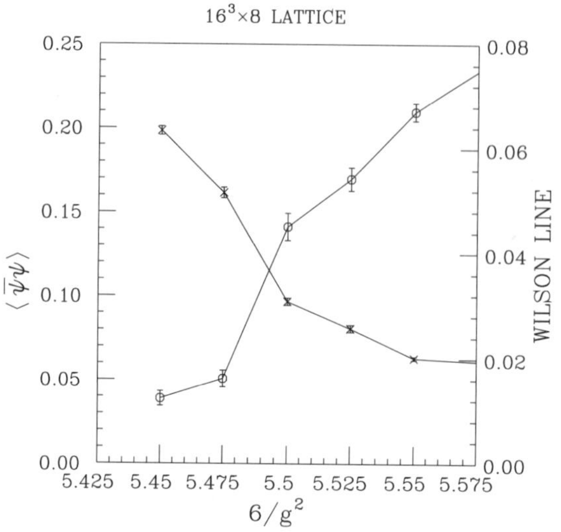

Fig. 7.12. Lattice calculation of the Polyakov loop (or Wilson line, which will be explained in the following section) and the chiral condensate as function of temperature around the critical temperature T_c. The results are taken from S. Gottlieb et al. (MILC collaboration): Phys. Rev. D **55**, 6852 (1997)

interesting behavior (also seen in many other calculations) that the critical temperatures for the deconfinement phase transition and the chiral transition seem to be identical. This is a nontrivial result as both phenomena relate to different physical effects. The theoretical explanation of this behavior is still under discussion.

7.1.18 The Polyakov Loop

In Sect. 7.1.15 on the string tension we have shown how to extract the static potential between quarks from the expectation of Wilson loops. There is an alternative method to do this. As mentioned, it is crucial to construct gauge-invariant quantities (if we do not want to fix the gauge on the lattice). Any quantity with some unsaturated color indices like the untraced plaquette P_{ab} will exactly vanish if we average it over all gauge rotations $\langle P_{ab}\rangle = 0$. There is a particularly useful type of gauge-invariant object we can specify on the lattice, the so-called "Polyakov loop" (sometimes also called the "Wilson line"), given by the following expression:

$$P(\boldsymbol{n}) = \operatorname{tr} \prod_{n_\tau=1}^{N_\tau} U_4(\boldsymbol{n}, n_\tau) \ . \tag{7.189}$$

$P(\boldsymbol{n})$ is defined at every space point and is the product of timelike links in the time direction (see Fig. 7.13). Since we have periodic boundary conditions in the time direction it forms a "loop" on the lattice. The trace guarantees gauge invariance:

$$\begin{aligned} \tilde{P}(\boldsymbol{n}) &= \operatorname{tr} G(\boldsymbol{n}, 1) U_4(\boldsymbol{n}, 1) G^\dagger(\boldsymbol{n}, 2) G(\boldsymbol{n}, 2) U_4(\boldsymbol{n}, 2) G^\dagger(\boldsymbol{n}, 3) \dots \\ &\qquad G(\boldsymbol{n}, N_\tau) U_4(\boldsymbol{n}, N_\tau) G^\dagger(\boldsymbol{n}, N\tau + 1) \\ &= \operatorname{tr} G(\boldsymbol{n}, 1) \left(\prod_{n_\tau 01}^{N_\tau} U_4(\boldsymbol{n}, n\tau) \right) G^\dagger(\boldsymbol{n}, N_\tau + 1) \\ &= P(\boldsymbol{n}) \end{aligned} \tag{7.190}$$

because of the periodicity $G(\boldsymbol{n}, 1) = G(\boldsymbol{n}, N_\tau + 1)$.

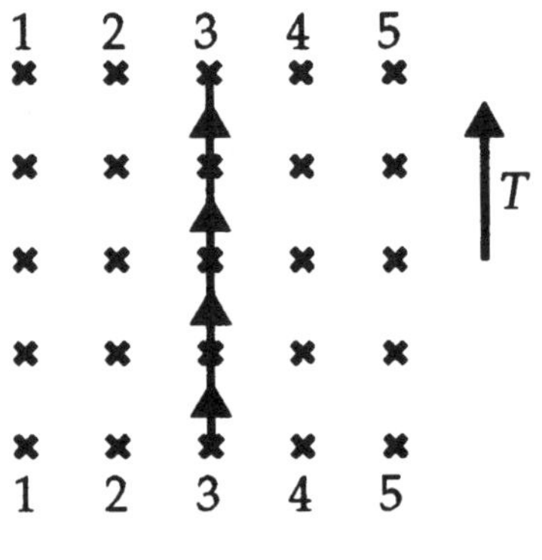

Fig. 7.13. Graphical representation of the Polyakov loop. The links along the time direction are multiplied, which yields a loop structure because of the periodic boundary conditions in time

The Polyakov loop has a rather simple physical meaning. If one considers an infinitely heavy quark sitting at a given point $\boldsymbol{x}$ in space and only propagating in the time direction, the Euclidean Schrödinger equation, $\varphi_Q(\boldsymbol{x}, \tau)$, is given by

$$\partial_\tau \varphi_Q(\boldsymbol{x}, \tau) = \mathrm{i} A_4(\boldsymbol{x}, \tau) \varphi_Q(\boldsymbol{x}, \tau) \ . \tag{7.191}$$

Thus we get a time-evolution operator $S(\tau)$ for the quark:

$$S(\tau) = P \exp \left[\mathrm{i} \int^{\tau} \mathrm{d}\tau' A_4(\boldsymbol{x}, \tau') \right] \ . \tag{7.192}$$

Expression (7.192) is just the continuum version of the Polyakov loop. It describes the propagation of a static color charge on the lattice. Thus calculating the expectation value of the Polyakov loop at finite temperature T, one calculates the partition function of the system Z_Q including an open charge:

$$\langle P(\boldsymbol{n}) \rangle = Z_Q = \exp\left\{ -F_Q/T \right\} \tag{7.193}$$

with the free energy of the system F_Q. $\langle P \rangle$ serves as a signal for the deconfinement phase transition. In the confinement phase, color charges cannot propagate. Therefore, adding a color charge to the system by hand generates an infinite (or very large, in the case of a finite-size lattice calculation) energy. Thus, according to (7.193) the expectation value vanishes: $\langle P \rangle = 0$. However, beyond a critical temperature one can observe a clear phase transition to finite values of $\langle P \rangle$, signalling the possibility that open color charges can propagate. Figure 7.12 shows the result of a lattice calculation. One can see that the Polyakov loop becomes finite for temperatures $T_\mathrm{c} \sim 250$ MeV in a sharp transition that has been clearly identified as a first-order transition. This is the result if one neglects the fermionic determinant as discussed in Sect. 7.1.7. If the effects of quarks are included, the temperature drops to $T_\mathrm{c} \sim 140$ MeV. Here the order of the transition is still subject to debate.

7.1.19 The Center Symmetry

There exists an interesting symmetry of lattice QCD in connection with the Polyakov loop. Take the link variables in a hyperplane for some fixed time, e.g. $n_\tau = 1$. One might transform all timelike links in this hyperplane by a unitary 3×3 matrix $\hat{Z}$

$$\tilde{U}_4(\boldsymbol{n}, n_\tau = 1) = C U_4(\boldsymbol{n}, n_\tau = 1) \ . \tag{7.194}$$

Z belongs to the center of the gauge group, here SU(3). The center of a group is defined as the class of elements C which leave all group elements invariant under transformation, i.e. which commute with all group elements: $[C, U] = 0$.

It is clear that C has to be proportional to the unit matrix in order to fulfil this requirement

$$C = \mathrm{e}^{\mathrm{i}\alpha \mathbb{1}} \quad \text{with} \quad \alpha \in [0, 2\pi] \; . \tag{7.195}$$

In addition, SU(N) requires the determinant of the group element to be 1. Therefore we get

$$\det C = 1 \Rightarrow \mathrm{tr}\, \alpha \mathbb{1} = 0 \bmod 2\pi \; , \tag{7.196}$$

which yields

$$\alpha_n = \frac{2\pi n}{3} \; , \quad n = 0, 1, 2 \; . \tag{7.197}$$

The three elements C_1 to C_3 form an own group, the Z(3). In general, the Z(N) group (replace 3 by N in (7.197)) is the group of discrete rotations around the unit circle in the complex plane.

In the limit $N \to \infty$ we get continuous rotations, which is just the usual U(1) of electrodynamics, for instance.

From the definition of the action for the gluons, (7.48), we see that under a transformation (7.195) the action does not change. Factors C only enter in the S_τ the part of the interaction containing space–time plaquettes. A plaquette in the 41 plane, for example, is

$$P_{41} = U_4(\boldsymbol{n}, 1) U_1(\boldsymbol{n}, 2) U_4^\dagger(\boldsymbol{n} + a\boldsymbol{e}_1, 1) U_1^\dagger(\boldsymbol{n}, 1) \; . \tag{7.198}$$

Transforming the links by using (7.194), we get

$$\tilde{P}_{41} = C U_4 U_1 U_4^\dagger C^\dagger U_1^\dagger \; . \tag{7.199}$$

Since C commutes with all group elements one can shift C to the right and use $CC^\dagger = 1$. Thus $P_{41} = \tilde{P}_{41}$.

Although the action is invariant under the center transformation, the Polyakov loop is not, since it contains only *one* link belonging to the hyperplane with $n_\tau = 1$. There we have the transformation property

$$\tilde{P}(\boldsymbol{n}) = C P(\boldsymbol{n}) \; . \tag{7.200}$$

As we have seen in the confined phase, $\langle P \rangle = 0$ because an isolated color field is infinitely heavy. In the deconfined phase one can have $\langle P \rangle \neq 0$. This is also a signal of spontaneous breaking of the center symmetry: the system clusters around one of the three values of C_n, i.e. $\langle P \rangle$ attains the corresponding phase factor $\mathrm{e}^{\mathrm{i}2\pi n/3}$. Otherwise, an averaging over the values of C_n $\langle C \rangle = 0$ would again generate a vanishing Polyakov loop. This intimate relation between the Z(3) structure of the Polyakov loops and the confinement/deconfinement phase transition led to the idea that the critical behavior of the whole SU(3) theory can be reduced to the one of a Z(3) spin model in three dimensions (as the Polyakov loops $P(\boldsymbol{n})$ are defined in three dimensions).[15] The fact that both systems exhibit a first-order phase transition supports this idea.

[15] See B. Svetitsky, L.G. Yaffe: Nucl. Phys. B **210**, 423 (1982).

7.2 QCD Sum Rules

In this section we discuss the techniques of QCD sum rules, which allow an approximate phenomenological treatment of nonperturbative effects in QCD. It turns out that this method is extremely useful in calculating the lowest-mass hadronic bound states or determining effective coupling constants. In addition, QCD sum rules offer some surprising insights into the internal wave functions of nucleons and pions.

The basic ideas of QCD sum rules are met frequently in QCD. As the full QCD interaction is strongly nonperturbative and can be solved exactly only in special cases (e.g., using lattice gauge theories; see Sect. 7.1), one attempts to separate perturbative and nonperturbative contributions, describing the latter by a set of phenomenologically effective Feyman rules.

The starting point for QCD sum rules, first formulated by M.A. Shifmann, A.I. Vainshtein, and V.I. Zakharow in 1979,[16] is the operator product expansion we met in Sect. 5.4. The OPE gives a general form for the quantities of interest, and QCD sum rules are a phenomenological procedure for evaluating the matrix elements of the operators that occur. This is a general procedure with many applications. We shall introduce it by discussing a specific application: the determination of hadron masses. For this purpose the operator to be expanded is the general time-ordered product of two currents:

$$\mathrm{i}\int \mathrm{d}x\, \mathrm{e}^{\mathrm{i}q\cdot x} T[j_\Gamma(x) j_\Gamma(0)] = C_{\mathrm{i}}^\Gamma \mathcal{I} + \sum_n C_n^\Gamma(q)\mathcal{O}_n \ . \tag{7.201}$$

The operators $\mathcal{O}_n$ can be ordered by their naive dimension ($\dim[q] = 3/2$, $\dim[G_{\mu\nu}] = 2$, $\dim[m] = 1$). As we shall be interested only in vacuum expectation values, we can restrict ourselves to spin-0 operators, which are

$$\begin{aligned}
\mathcal{I}\ , &\quad d=0\ ,\\
\mathcal{O}_m = m\bar{q}q\ , &\quad d=4\ ,\\
\mathcal{O}_G = G^a_{\mu\nu}\, G_{a\mu\nu}\ , &\quad d=4\ ,\\
\mathcal{O}_\Gamma = \bar{q}\Gamma q\bar{q}\Gamma q\ , &\quad d=6\ ,\\
\mathcal{O}_\sigma = m'\bar{q}\sigma_{\mu\nu}\frac{\lambda^a}{2} q G^a_{\mu\nu}\ , &\quad d=6\ ,\\
\mathcal{O}_{\mathrm{f}} = f_{abc} G^a_{\mu\nu} G^b_{\nu\gamma} G^c_{\gamma\mu}\ , &\quad d=6\ , \quad \text{etc.}
\end{aligned} \tag{7.202}$$

To evaluate (7.149) one must know the vacuum expectation values of all operators and the corresponding Wilson coefficients, which can be calculated perturbatively and will be discussed in detail below. But first we shall present the general technique of QCD sum rules.

QCD sum rules start from the fact that the vacuum polarization tensor can be described at the hadronic level, i.e., in terms of hadrons and hadronic resonances. The optical theorem states that the total cross section of a certain reaction

[16] See M.A. Shifmann, A.I. Vainshtein, and V.I. Zakharow: Nucl. Phys. B **147**, 385 and 448 (1979).

equals the imaginary part of its forward-scattering amplitude. This can be understood from the fact that the cross section is proportional to the square of some amplitude, graphically depicted by doubling the corresponding graph (see Fig. 7.14).

The forward-scattering amplitude corresponds to the vacuum polarization tensor. Hence for vector coupling we can write

$$\mathrm{Im}\,\Pi^V\left(s=q^2\right)=\frac{9}{64\pi^2\alpha^2}s\sigma\left(\mathrm{e}^+ + \mathrm{e}^- \to \gamma^* \to \text{hadrons}\right)\ . \tag{7.203}$$

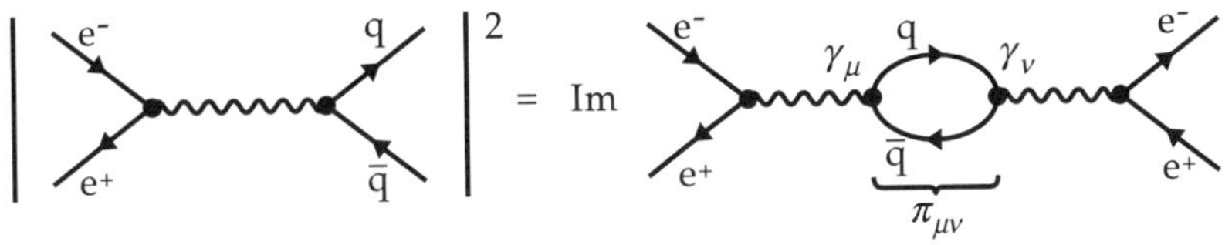

Fig. 7.14. A graphical demonstration of the optical theorem

The cross section of virtual photons decaying into hadrons has been extensively studied in electron positron annihilation, e.g., at DESY in Hamburg and at SLAC in Stanford. For low virtuality of the photon there is a contribution only when q^2 is just the mass squared of a hadronic resonance with photon quantum numbers, i.e., a 1^{--} resonance. (Here 1 is the spin, the first "−" represents the internal parity, and the second "−" C parity. Only neutral particles have a C parity.) Therefore the vacuum polarization tensor can be thought of as a sum of δ functions of the type $\delta(q^2 - m_{\mathrm{R}}^2)$. In reality the hadronic states are rather broad (see Fig. 7.15), but here they can be approximated by delta functions. Bear in mind that, in principle, we can imagine inserting a Breit–Wigner form instead; however, the precision achieved is such that it does not really make a difference. As Π^V is dimensionless, this must be multiplied by the only dimensional quantity $s = m_{\mathrm{R}}^2$. All other constants appearing here are absorbed in a parameter g_{R}, and we have

$$\mathrm{Im}\left[\Pi^V(s=q^2)\right]=\sum_q \frac{\pi}{e_q^2}\sum_R \frac{m_R^2}{g_{qR}^2}\delta\left(s-m_R^2\right)\ . \tag{7.204}$$

Here the sum over resonances has been split into sums of resonances that require the creation of a $\mathrm{q\bar{q}}$ pair of flavor q. For $q=c$ these are, e.g., $J/\psi,\ \psi',\ \psi'',\cdots$. For sufficiently high virtualities of the photon, more than one particle can be created, and discrete resonances are replaced by continua of states. For virtualities just above the continuum threshold, the cross section cannot be expressed in a general simple form. Above the continuum threshold this is possible, since just the elementary $\mathrm{e}^+ + \mathrm{e}^- \to \mathrm{q} + \bar{\mathrm{q}}$ cross section can be inserted.

$$\sigma(\mathrm{e}^+ + \mathrm{e}^- \to \text{hadrons}) = \frac{4\pi\alpha^2}{3s}\sum_q e_q^2\ . \tag{7.205}$$

The ratio of the cross sections for multipion production and (7.153) are plotted in Fig. 7.15.

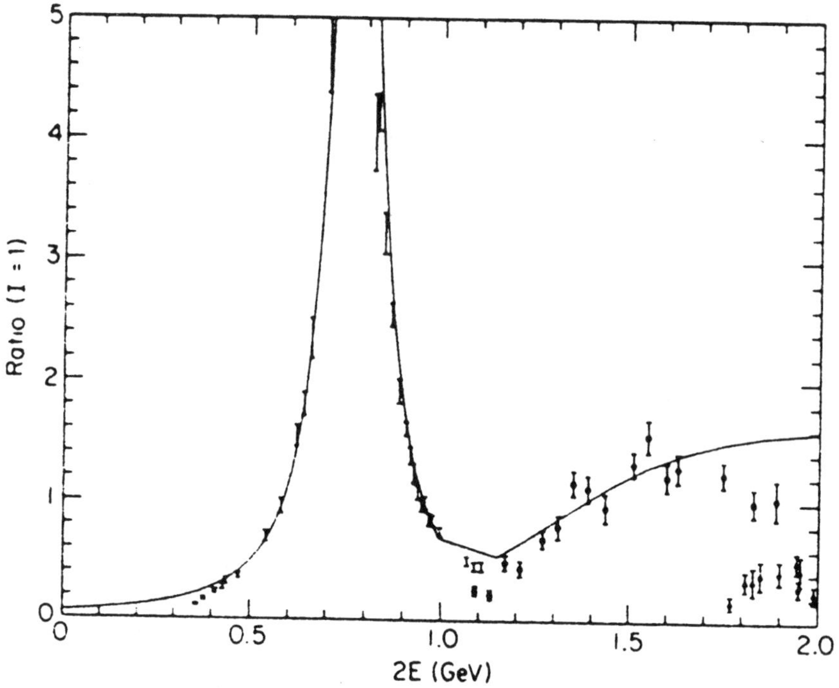

Fig. 7.15. The ratio $\sigma(e^+e^- \to n\pi)/\sigma(e^+e^- \to \mu^+\mu^-)$

The rho resonance at 770 MeV is very prominent, while the higher resonances merge with the continuum. The continuum threshold s_0 is at about 1.3 GeV.

$$\mathrm{Im}\left[\Pi^V(s=q^2)\right] = \frac{3\pi}{e_\mathrm{f}^2}\sum_R \frac{m_R^2}{g_R^2}\,\delta\left(s-m_R^2\right) + \sum_f \frac{9}{16\pi}e_\mathrm{f}^2\,\Theta\left(s-s_{0f}\right) . \tag{7.206}$$

Here we have counted each flavor f three times, since there are three colors. The continuum thresholds $s_{0\mathrm{f}}$ for the different flavors f can be taken to be free-fitting parameters. In our case we take only up and down quarks into account and choose $s_0 = s_{0u} = s_{0d} \approx 1.3$ GeV and neglect all except the ϱ resonance in the first term on the right-hand side of (7.154). As we have already mentioned, its approximation by the free-quark cross section is poor near the continuum threshold, and one attempts to minimize the errors resulting from this by suitably adjusting s_{0f}. In the following, we shall expand the correlator in (7.149) in the framework of OPE and in this way obtain a representation of Π^V as a function of expectation values of certain operators. On the other hand, the imaginary part of Π^V can, as discussed above, be parametrized by discrete resonances and a continuum contribution. This will enable us to express the mass of the lightest resonance in terms of the expectation values mentioned above. This approach can be repeated for each J^{PC} channel so that many masses (and coupling constants) are fitted essentially by three expectation values. Our trust in this procedure is based primarily on its success. QCD sum rules consistently describe a plethora of hadronic data, all of them with a typical accuracy of about 20%. On the other hand, the fitted expectation values

$$\begin{aligned}
\left\langle 0 \left| \frac{\alpha_s}{\pi} G^a_{\mu\nu} G^{a\mu\nu} \right| 0 \right\rangle &= (360 \pm 20\,\mathrm{MeV})^4 \ , \\
\langle 0 | \bar{u}u | 0 \rangle = \left\langle 0 \left| \bar{d}d \right| 0 \right\rangle &= -(225 \pm 25\,\mathrm{MeV})^3 \ , \\
\langle 0 | m_\mathrm{s} \bar{s}s | 0 \rangle &= -(210 \pm 5\,\mathrm{MeV})^4 \ , \\
\langle 0 | m\bar{q}q | 0 \rangle &= -(100\,\mathrm{MeV})^4 \ , \quad m = \frac{1}{2}(m_\mathrm{u} + m_\mathrm{d}) \ , \\
\langle 0 | \bar{q} g \sigma_{\mu\nu} G^{a\mu\nu} q | 0 \rangle &= 0.436\,\mathrm{GeV}^5
\end{aligned} \tag{7.207}$$

have far-reaching physical implications. They describe physical properties of the nonperturbative QCD ground state and give, for example, the vacuum energy density. To illustrate this, let us discuss the trace of the energy momentum tensor of QCD,

$$\Theta^\mu_\mu = \frac{\beta(g)}{2g} G^a_{\mu\nu} G^{a\mu\nu} + \sum m\bar{q}q \ , \tag{7.208}$$

with the QCD beta function

$$\beta(g) = -\frac{g^3}{(4\pi)^2} \left(11 - \frac{2}{3} N_\mathrm{f} \right) + \dots \ . \tag{7.209}$$

This gives, on inserting (7.209),

$$\Theta^\mu_\mu = -\frac{1}{8} \left(11 - \frac{2}{3} N_\mathrm{f} \right) \frac{\alpha_s}{\pi} G^a_{\mu\nu} G^{a\mu\nu} + \sum_q m_q \bar{q}q \ , \tag{7.210}$$

$$\left\langle \Theta^\mu_\mu \right\rangle \approx -(370\,\mathrm{MeV})^4 \ . \tag{7.211}$$

In the bag model Θ^μ_μ is simply (ϱ is the energy density, p is the pressure)

$$\Theta^\mu_\mu \big|_\mathrm{bag} = \varrho - 3p = -4B \ , \tag{7.212}$$

which suggests that the bag constant is ($N_\mathrm{F} = 3$)

$$B = \frac{9}{32} \left\langle \frac{\alpha_s}{\pi} G^a_{\mu\nu} G^{a\mu\nu} \right\rangle \approx (260\,\mathrm{MeV})^4 \ , \tag{7.213}$$

which is substantially larger than the value fitted in the simple bag model (see (3.151)) $B = (146\,\mathrm{MeV})^4$. There do exist a number of modified bag models with a smaller bag radius and correspondingly larger B values. Some of them use B as large as $(250\,\mathrm{MeV})^4$ and their authors claim that the sum-rule value (7.213) would support their model. Such claims are, however, not very sensible. Additional interactions are introduced in each of these models. The energy and pressure associated with these additional interactions and that due to the QCD vacuum state are interwoven, such that is not so clear which quantity should be compared to (7.211). Most probably B should be regarded only as an effective parameter and the bag model as a simple model with limited connections to QCD.

On the other hand, (7.211) is a clear statement of the fact that the vacuum of particle physics is far from being a simple empty state, but is instead a complicated structure with a definite energy. To illustrate this fact let us calculate the total energy content of a cubic meter of QCD vacuum:

$$\begin{aligned} 1\,\mathrm{m}^3 \times (260\,\mathrm{MeV})^4 &= \mathrm{MeV} \times 260 \times \left(10^{15}\right)^3 \left(\frac{260\,\mathrm{MeV\,fm}}{197\,\mathrm{MeV\,fm}}\right)^3 \\ &= 6.0 \times 10^{47}\,\mathrm{MeV} = 10^{18}\,\mathrm{kg} = 10^{15}\,\mathrm{t}\ . \end{aligned} \tag{7.214}$$

This corresponds to a mass energy of a million billions tonnes of matter. These numbers become much more exotic if one studies astronomical distances.

Chiral symmetry, i.e. SU(2), plays an important and still not fully understood role in full QCD. The normal isospin symmetry transformation is

$$\psi(x) \to \exp\left(-\frac{\mathrm{i}}{2}\alpha_a(x)\tau^a\right)\psi(x)\ . \tag{7.215}$$

In the absence of quark mass terms a second symmetry follows from this, namely

$$\psi(x) \to \exp\left(-\frac{\mathrm{i}}{2}\alpha_a(x)\tau^a\gamma_5\right)\psi(x)\ . \tag{7.216}$$

The reason that $\bar{\psi}(x)\gamma_\mu\psi(x)$ is invariant is that γ_5 anticommutes with every γ matrix, and thus any bilinear form with an even number of γ matrices between $\psi^\dagger(x)$ and $\psi(x)$ is chirally symmetric, and all bilinear forms with an odd number are not:

$$\begin{aligned} \psi^\dagger(x)\gamma_0\gamma_\mu\psi(x) \to\ & \psi^\dagger(x)\exp\left(\frac{\mathrm{i}}{2}\alpha_a(x)\tau^a\gamma_5\right)\gamma_0\gamma_\mu\exp\left(-\frac{\mathrm{i}}{2}\alpha_a\tau^a\gamma_5\right)\psi(x) \\ &= \psi^\dagger(x)\gamma_0\exp\left(-\frac{\mathrm{i}}{2}\alpha_a(x)\tau^a\gamma_5\right)\gamma_\mu\exp\left(-\frac{\mathrm{i}}{2}\alpha_a\tau^a\gamma_5\right)\psi(x) \\ &= \psi^\dagger(x)\gamma_0\gamma_\mu\exp\left(\frac{\mathrm{i}}{2}\alpha_a(x)\tau^a\gamma_5\right)\exp\left(-\frac{\mathrm{i}}{2}\alpha_a\tau^a\gamma_5\right)\psi(x) \\ &= \psi^\dagger(x)\gamma_0\gamma_\mu\psi(x)\ . \end{aligned} \tag{7.217}$$

Consequently the quark-mass term $m_q\bar{\psi}\psi$ violates the symmetry (7.216), but since the QCD up and down quark masses are small, their symmetry-breaking effects should be weak, such that a symmetric theory should give a good description. Surprisingly this is not the case. While the up and down quark masses are only about $m_\mathrm{u} = 5.6 \pm 1.1$ MeV and $m_\mathrm{d} \approx 9.9 \pm 1.1$ MeV, the chiral partners ϱ and a_1 have vastly different masses

$$m_{\mathrm{a}_1} - m_\varrho = 1260\,\mathrm{MeV} - 770\,\mathrm{MeV} = 490\,\mathrm{MeV}\ , \tag{7.218}$$

which implies that in addition to the small explicit breaking of chiral symmetry there must also be a spontaneous symmetry violation. “Spontaneous symmetry breaking" (SSB) designates all cases in which a symmetry present in the Lagrangian is not present in the actual physical ground state. Let us add that ϱ has the J^{PC} quantum numbers 1^{--}, while a_1 has 1^{++}. Both are the lightest mesons with these quantum numbers. As (7.216) mixes vector states with axial-vector states, ϱ and a_1 would have the same mass if (7.216) were a symmetry operation. Similarily $\omega(m_\omega = 782\,\text{MeV}, J^{PC} = 1^{--})$ and $f_1(m_{f_1} = 1282\,\text{MeV}, J^{PC} = 1^{++})$ should be degenerate, etc. The pion, however, should play a special role; see below. Let us also note that in lattice calculations (see Sect. 7.1) the restoration of chiral symmetry at the phase transition manifests itself by m_ϱ and m_{a_1} converging to the same value. If (7.216) is a symmetry operation, it follows that

$$Q_5^a = \int \mathrm{d}^3x\, \bar{\psi}(x)\gamma_0\gamma_5 \frac{1}{2}\tau^a \psi(x) \tag{7.219}$$

is an additional conserved quantity, similar to the total isospin

$$Q^a = \int \mathrm{d}^3x\, \bar{\psi}(x)\gamma_0 \frac{1}{2}\tau^a \psi(x) \ . \tag{7.220}$$

As (7.216) is a symmetry of the Lagrangian (neglecting quark masses for the time being), $|0\rangle$, $\hat{Q}_5|0\rangle$, $\hat{Q}_5^2|0\rangle, \dots$ all have the same energy. On the other hand, we have just argued that this symmetry is spontaneously broken for the ground state; thus

$$\hat{Q}_5|0\rangle = |v_1\rangle \neq |0\rangle \ , \quad \hat{Q}_5^2|0\rangle = |v_2\rangle \neq |0\rangle \ . \tag{7.221}$$

Owing to all the other symmetries of the theory, $|0\rangle, |v_1\rangle, |v_2\rangle, \dots$ can differ only by the presence of $\bar{q}q$ pairs. To lowest order

$$\left\langle 0 \left| \bar{\psi}\gamma_\mu\gamma_5 \frac{\tau^a}{2}\psi \right| v_1 \right\rangle \neq 0 \ . \tag{7.222}$$

We substitute $\phi_\mu^a(x) = \bar{\psi}(x)\gamma_\mu\gamma_5\frac{\tau^a}{2}\psi(x)$. As $\phi_\mu^a(x)$ transforms as an isospin vector we find that

$$\begin{aligned}
\phi_\mu^a(x) &\to \phi_\mu^a(x) - \mathrm{i}\delta\alpha^A(x)(T^A)_{ab}\phi_\mu^b(x) \ , \\
\phi_\mu^a(x) &\to \exp\left(\mathrm{i}\delta\alpha^b(x)Q_5^b\right)\phi_\mu^a(x)\exp\left(-\mathrm{i}\delta\alpha^c(x)Q_5^c\right) \\
&\Rightarrow \phi_\mu^a(x) + \mathrm{i}\delta_\alpha^A(x)\left[Q_5^A, \phi_\mu^a(x)\right] \\
&\Rightarrow \left(T^A\right)_{ab}\phi_\mu^b(x) = -\left[Q_5^A, \phi_\mu^a\right] \ .
\end{aligned} \tag{7.223}$$

Taking the vacuum expectation value of (7.223) yields

$$\begin{aligned}
\left\langle 0 \left| \left(T^A\right)_{ab}\phi_\mu^b(x) \right| 0 \right\rangle &= -\langle 0|Q_5^A \underbrace{\phi_\mu^a|0\rangle}_{=0} + \langle 0|\phi_\mu^a Q_5^A|0\rangle \\
&= \langle 0|\phi_\mu^a|v_1\rangle \neq 0 \ .
\end{aligned} \tag{7.224}$$

Obviously there must be at least one linear combination of the $\phi^b(x)$ with a nonvanishing vacuum expectation value. For SU(2) the number of generators is three, just like the number of vector components. In fact the pions are a regular representation of the SU(2). Therefore we shall not distinguish bestween A and a indices. Let us define

$$\delta^{ab}\sigma = \mathrm{i}\left[Q_5^a, \phi^b\right] . \tag{7.225}$$

Equation (7.224) then implies that

$$\mathrm{i}\left\langle 0\left|\left[Q_5^a, \phi^b\right]\right| 0\right\rangle = \langle 0|\sigma|0\rangle \neq 0 . \tag{7.226}$$

To describe SSB, one usually introduces an effective field σ with a potential that is minimal for a finite value σ. Here we have shown that even without specifying an explicit model one must always have an effective σ term. Next we introduce the pion in our scheme. Fourier transforming $\phi_\mu^a(x)$, we get

$$\left\langle 0\left|\bar{\psi}(p')\gamma_\mu\gamma_5\frac{\tau^a}{2}\psi(p)\right| v_1\right\rangle \neq 0 . \tag{7.227}$$

We assume that $\bar{\psi}$, ψ are incoming or outgoing on-shell particles (either quarks or nucleons). We then multiply this expression by $q_\mu = p'_\mu - p_\mu$. On the other hand, we know that $|v_1\rangle$ should be identified with some field. In principle

$$Q_5|0\rangle = |v_1\rangle$$

could be related to any pseudoscalar, but for $[H_0, Q_5] = 0$ it follows that

$$H_0|v_1\rangle = +Q_5 H_0|0\rangle = 0 , \tag{7.228}$$

such that in the absence of chiral symmetry breaking the particle must be massless (it is the corresponding, so-called Goldstone boson). The pion, with its extremely small mass (compared with other hadrons), is the only natural candidate. Equation (7.227) thus implies that

$$\begin{aligned}
&\left\langle 0\left|\bar{\psi}(p')(\not{p}' - \not{p})\gamma_5\frac{\tau^a}{2}\psi(p)\right|\pi^b\right\rangle \\
&= \left\langle 0\left|\bar{\psi}(p')(m\gamma_5 + \gamma_5\not{p})\frac{\tau^a}{2}\psi(p)\right|\pi^b\right\rangle \\
&= -\mathrm{i}m\left\langle 0\left|\bar{\psi}\mathrm{i}\gamma_5\tau^a\psi\right|\pi^b\right\rangle \sim \delta_{ab} \neq 0 .
\end{aligned} \tag{7.229}$$

Transforming this back into coordinate space yields

$$\begin{aligned}
&\left\langle 0\left|\partial^\mu\left(\bar{\psi}(x)\gamma_\mu\gamma_5\frac{\tau^a}{2}\psi\right)\right|\pi^b\right\rangle = \left\langle 0\left|\partial^\mu A_\mu^a\right|\pi^b\right\rangle \\
&= \delta^{ab}\cdot\text{const.} = \delta^{ab} f_\pi m_\pi^2 ,
\end{aligned} \tag{7.230}$$

with the usual definition of the pion decay constant f_π. This relationship is postulated to hold at the operator level to give what is called a PCAC (partially conserved axial vector current)

$$\partial_\mu A^{a\mu} = f_\pi m_\pi^2 \pi^a \ . \tag{7.231}$$

This equation implies that the violation of axial-vector current conservation can be described as being due to the coupling to pions. Many interesting relations can be deduced from this starting point. We are especially interested in the following:

$$\begin{aligned} I &= \int \mathrm{d}^4x\ \mathrm{e}^{-\mathrm{i}k\cdot x} \left\langle 0 \left| T\left(\partial^\mu A^a_\mu(0)\partial^\nu A^b_\nu(x)\right)\right| 0\right\rangle \\ &= f_\pi^2 m_\pi^4 \int \mathrm{d}^4x\ \mathrm{e}^{-\mathrm{i}k\cdot x} \left\langle 0 \left| T\left(\pi^a(0)\pi^b(x)\right)\right| 0\right\rangle \\ &= f_\pi^2 m_\pi^4 \delta^{ab}\cdot \mathrm{i}D_\pi(k) \ . \end{aligned} \tag{7.232}$$

We look at this in the limit $k \to 0$ $\left(D_\pi(k) = 1/\left(k^2 - m_\pi^2\right) \to -1/m_\pi^2\right)$ and obtain

$$I = -\mathrm{i} f_\pi^2 m_\pi^2 \delta^{ab} \ . \tag{7.233}$$

On the other hand, by partial integration we have

$$\begin{aligned} I(k\to 0) = &\int \mathrm{d}^4x\ \left(-\partial^\nu \mathrm{e}^{-\mathrm{i}k\cdot x}\right) \left\langle 0 \left| T\left(\partial^\mu A^a_\mu(0) A^b_\nu(x)\right)\right| 0\right\rangle \\ &- \int \mathrm{d}^4x\ \mathrm{e}^{-\mathrm{i}k\cdot x} \Big\langle 0 \Big| \Big[\partial^\nu \Theta(x_0) A^b_\nu(x) \partial^\mu A^a_\mu(0) \\ &+ \partial^\nu \Theta(-x_0)\partial^\mu A^a_\mu(0) A^b_\nu(x)\Big]\Big| 0\Big\rangle \ , \end{aligned} \tag{7.234}$$

$$\begin{aligned} I(k\to 0) &= 0 - \int \mathrm{d}^4x\ \delta(x_0) \left\langle 0 \left| \left[A^b_0(x),\ \partial^\mu A^a_\mu(0)\right]\right| 0\right\rangle \\ &= -\int \mathrm{d}^4x\ \delta(x_0) \left\langle 0 \left| \left[A^b_0(0),\ \partial^\mu A^b_\mu(-x)\right]\right| 0\right\rangle \\ &= \int \mathrm{d}^3x\ \left\langle 0 \left| \left[A^b_0(0,\boldsymbol{x}),\ \partial^0 A^a_0(0,0)\right]\right| 0\right\rangle \ , \end{aligned} \tag{7.235}$$

because only terms even in $\boldsymbol{x}$ contribute. As A_0^b is not explicitly time dependent, this becomes

$$I(k\to 0) = \mathrm{i}\int \mathrm{d}^3x\ \left\langle 0 \left| \left[A^b_0(0,\boldsymbol{x}),\ \left[A^a_0(0,0),\ H(0)\right]\right]\right| 0\right\rangle \ . \tag{7.236}$$

This we shall evaluate for

$$H = H_0 + H_{\text{int}} = H_0 + \int \mathrm{d}^3y\ \left[m_u \bar{u}(0,\boldsymbol{y})u(0,\boldsymbol{y}) + m_d \bar{d}(0,\boldsymbol{y})d(0,\boldsymbol{y})\right] \tag{7.237}$$

with

$$\left[q_\alpha^+(\boldsymbol{x},t),\ q_\beta(\boldsymbol{y},t)\right] = \delta_{\alpha\beta}\delta^3(\boldsymbol{x}-\boldsymbol{y}) \ . \tag{7.238}$$

By straightforward calculation (see Exercise 7.3) we get

$$\left\langle 0 \left| \left[A_0^b(0,\boldsymbol{x}), \left[A_0^b(0,0), H_{\text{int}}(0) \right] \right] \right| 0 \right\rangle = \left\langle 0 \left| \frac{m_{\text{u}} + m_{\text{d}}}{2} (\bar{u}u + \bar{d}d) \right| 0 \right\rangle \tag{7.239}$$

and thus

$$(m_{\text{u}} + m_{\text{d}}) \langle 0 | \bar{u}u + \bar{d}d | 0 \rangle = -f_\pi^2 m_\pi^2 \approx -2 \times (100\ \text{MeV})^4 \ . \tag{7.240}$$

We have seen how the nonperturbative effects leading to the breaking of chiral symmetry and thus to hadronic mass correction lead to the introduction of a quark condensate (7.240). QCD sum rules invoke similar condensates to describe the nonperturbative effects of QCD. In general, here one has to allow for the appearance of all possible condensates, of all the operators O_n appearing in (7.201). Let us now return to this equation.

EXERCISE

7.3 PCAC and the Quark Condensate

Problem. Calculate

$$\int\int \left\langle 0 \left| \left[A_0^b(0,\boldsymbol{x}), \left[A_0^a(0,\boldsymbol{0}), m_{\text{u}}\bar{u}(y)u(y) + m_{\text{d}}\bar{d}(y)d(y) \right] \right] \right| 0 \right\rangle \text{d}^3x\ \text{d}^3y \tag{1}$$

with

$$A_0^b(x) = \bar{q}(x)\gamma_0\gamma_5 \frac{\tau^b}{2} q(x) \ , \quad q(x) = \begin{pmatrix} u(x) \\ d(x) \end{pmatrix} \ . \tag{2}$$

Solution. First we write

$$m_{\text{u}}\bar{u}(y)u(y) + m_{\text{d}}\bar{d}(y)d(y) = \frac{m_{\text{u}} + m_{\text{d}}}{2} \bar{q}(y)q(y) + \frac{m_{\text{u}} - m_{\text{d}}}{2} \bar{q}(y)\tau^3 q(y) \tag{3}$$

and then we write, with simultaneous Lorentz and isospin indices $\alpha, \beta, \gamma, \delta$,

$$\begin{aligned} &\left[\bar{q}(0) \frac{\tau^a}{2} \gamma_0\gamma_5 q(0), \bar{q}(y)q(y) \right] \\ &= \left[q_\alpha^+(0) q_\beta(0), q_\gamma^+(y) q_\delta(y) \right] \left(\frac{\tau^a}{2}\gamma_5 \right)_{\alpha\beta} (\gamma_0)_{\gamma\delta} \\ &= \left(-q_\alpha^+ q_\gamma^+ \left\{ q_\beta, q_\delta \right\} + q_\alpha^+ \left\{ q_\beta q_\gamma^+ \right\} q_\delta - q_\gamma^+ \left\{ q_\alpha^+, q_\delta \right\} q_\beta \right. \\ &\quad \left. + \left\{ q_\alpha^+, q_\gamma^+ \right\} q_\delta q_\beta \right) \left(\frac{\tau^a}{2}\gamma_5 \right)_{\alpha\beta} (\gamma_0)_{\gamma\delta} \end{aligned} \tag{4}$$

Exercise 7.3

as

$$
\begin{aligned}
(\dots) &= -q^+_\alpha q^+_\gamma q_\beta q_\delta - q^+_\alpha q^+_\gamma q_\delta q_\beta + q^+_\alpha q_\beta q^+_\gamma q_\delta + q^+_\alpha q^+_\gamma q_\beta q_\delta \\
&\quad - q^+_\gamma q^+_\alpha q_\delta q_\beta - q^+_\gamma q_\delta q^+_\alpha q_\beta + q^+_\alpha q^+_\gamma q_\delta q_\beta + q^+_\gamma q^+_\alpha q_\delta q_\beta \\
&= q^+_\alpha q^+_\gamma q_\delta q_\beta (+1-1-1+1+1-1) + \left[q^+_\alpha q_\beta , q^+_\gamma q_\delta \right] \ .
\end{aligned} \tag{5}
$$

Now we use

$$
\begin{aligned}
&\left\{ q_\beta(0,\mathbf{0}), q_\delta(0,\mathbf{y}) \right\} = 0 = \left\{ q^+_\alpha(0,\mathbf{0}), q^+_\gamma(0,\mathbf{y}) \right\} \ , \\
&\left\{ q_\beta(0,\mathbf{0}), q^+_\gamma(0,\mathbf{y}) \right\} = \delta_{\beta\gamma}\delta^3(\mathbf{y}) \ , \\
&\left\{ q^+_\alpha(0,\mathbf{0}), q_\delta(0,\mathbf{y}) \right\} = \delta_{\alpha\delta}\delta^3(\mathbf{y})
\end{aligned} \tag{6}
$$

to get

$$
\begin{aligned}
[\dots, \dots] &= \delta^3(\mathbf{y}) \left[q^+_\alpha(0) q_\delta(0) \left(\frac{\tau^a}{2}\gamma_5\gamma_0 \right)_{\alpha\delta} - q^+_\gamma(0) q_\beta(0) \left(\gamma_0 \frac{\tau^a}{2}\gamma_5 \right)_{\gamma\beta} \right] \\
&= \delta^3(\mathbf{y}) q^+_\alpha(0) q_\delta(0) \left(\tau^a \gamma_5 \gamma_0 \right)_{\alpha\delta} \ .
\end{aligned} \tag{7}
$$

Similarly we obtain

$$
\begin{aligned}
&\left[\bar{q}(0) \frac{\tau^a}{2} \gamma_0\gamma_5 q(0), \bar{q}(0,\mathbf{y}) \tau^3 q(0,\mathbf{y}) \right] \\
&= \delta^3(\mathbf{y}) \left[q^+_\alpha(0) q_\delta(0) \left(\frac{\tau^a}{2}\gamma_5\gamma_0\tau^3 \right)_{\alpha\delta} - q^+_\gamma(0) q_\beta(0) \left(\gamma_0 \tau^3 \frac{\tau^a}{2}\gamma_5 \right)_{\gamma\beta} \right] \\
&= \delta^3(\mathbf{y}) q^+_\alpha(0) q_\delta(0) \Bigg(\gamma_5\gamma_0 \underbrace{\frac{1}{2}\left\{ \tau^a, \tau^3 \right\}}_{\delta_{a3}} \Bigg)_{\alpha\delta}
\end{aligned} \tag{8}
$$

and thus

$$
\begin{aligned}
&\int d^3y \left[A^a_0(0,\mathbf{0}), \frac{m_\mathrm{u}+m_\mathrm{d}}{2} \bar{q}(0,\mathbf{y}) \ \mathbb{1} \ q(0,\mathbf{y}) + \frac{m_\mathrm{u}-m_\mathrm{d}}{2} \bar{q}(0,\mathbf{y}) \tau^3 q(0,\mathbf{y}) \right] \\
&= \frac{m_\mathrm{u}+m_\mathrm{d}}{2} q^+_\alpha(0) q_\delta(0) \left(\gamma_5\gamma_0\tau^a \right)_{\alpha\delta} \\
&\quad + \frac{m_\mathrm{u}-m_\mathrm{d}}{2} q^+_\alpha(0) q_\delta(0) \left(\gamma_5\gamma_0 \right)_{\alpha\delta} \delta_{a3} = J(0) \ .
\end{aligned} \tag{9}
$$

The next step just repeats the preceding one:

$$
\begin{aligned}
&\int d^3x \left[A^b_0(0,\mathbf{x}), J(0) \right] \\
&= \int d^3r \, \delta^3(\mathbf{x}) \left[\frac{m_\mathrm{u}+m_\mathrm{d}}{2} q^\dagger_\alpha(0) q_\delta(0) \left(\frac{\tau^b}{2}\gamma_5\gamma_5\gamma_0\tau^a - \gamma_5\gamma_0\tau^a \frac{\tau^b}{2}\gamma_5 \right)_{\alpha\delta} \right. \\
&\quad \left. + \frac{m_\mathrm{u}-m_\mathrm{d}}{2} q^\dagger_\alpha(0) q_\delta(0) \left(\frac{\tau^b}{2}\gamma_5\gamma_5\gamma_0\delta_{a3} - \gamma_5\gamma_0\delta_{a3} \frac{\tau^b}{2}\gamma_5 \right)_{\alpha\delta} \right]
\end{aligned}
$$

Exercise 7.3

$$= q_\alpha^\dagger(0) q_\delta(0)\left[\frac{m_\mathrm{u}+m_\mathrm{d}}{2}\left(\gamma_0 \frac{1}{2}\{\tau^b,\tau^a\}\right)_{\alpha\delta} + \frac{m_\mathrm{u}-m_\mathrm{d}}{2}\left(\gamma_0 \tau^b \delta_{a3}\right)_{\alpha\delta}\right]$$
$$= \frac{m_\mathrm{u}+m_\mathrm{d}}{2}\bar{q}(0)\, q(0)\, \delta_{ab} + \frac{m_\mathrm{u}-m_\mathrm{d}}{2}\bar{q}(0)\, \tau^b\, q(0)\, \delta_{a3} \;. \tag{10}$$

Finally we take the vacuum expectation value of this expression. As the vacuum cannot carry any quantum number, such as isospin, the second term does not contribute:

$$\int \mathrm{d}^3x\ \mathrm{d}^3y$$
$$\times \left\langle 0 \left| \left[A_0^b(0,\boldsymbol{x}), \left[A_0^a(0,\boldsymbol{O}), m_\mathrm{u}\bar{u}(0,\boldsymbol{y})u(0,\boldsymbol{y}) + m_\mathrm{d}\bar{d}(0,\boldsymbol{y})d(0,\boldsymbol{y})\right]\right] \right| 0 \right\rangle$$
$$= \frac{m_\mathrm{u}+m_\mathrm{d}}{2}\langle 0|\bar{u}(0)u(0)+\bar{d}(0)d(0)|0\rangle\, \delta_{ab} \;. \tag{11}$$

Let us now return to the QCD sum-rule calculation. We illustrate this method by presenting all the detailed steps for one specific example: the calculation of the ϱ mass. We wish to evaluate the left-hand side of (7.201) at the quark level, using the condensates (7.202). The result will be compared with (7.206) and from this comparison we shall extract an estimate for the ϱ mass.

The simplest contribution to (7.201) is just the perturbative graph. The flavor structure of ϱ is simply

$$\varrho^+ = \frac{1}{\sqrt{2}}(u_1\bar{d}_2 - \bar{d}_1 u_2) \;,$$
$$\varrho^- = -\frac{1}{\sqrt{2}}(d_1\bar{u}_2 - \bar{u}_1 d_2) \;,$$
$$\varrho^0 = \frac{1}{2}\left[(d_1\bar{d}_2 - u_1\bar{u}_2) - (\bar{d}_1 d_2 - \bar{u}_1 u_2)\right] \;, \tag{7.241}$$

implying that

$$\langle | j_\mu(x) j_\nu(0) | \rangle_{\varrho^+} = -\frac{1}{2}\langle u_1(x)\gamma_\mu \bar{d}_2(x)\big(d_1\gamma_\nu\bar{u}_2 - \bar{u}_1\gamma_\nu d_2\big)(0)\rangle \cdot 2$$
$$= \langle u_1(x)\gamma_\mu\bar{d}_2(x)\bar{u}_1(0)\gamma_\nu d_2(0)\rangle$$
$$= \langle u(x)\gamma_\mu\bar{d}(x)\bar{u}(0)\gamma_\nu d(0)\rangle \;,$$
$$\langle | j_\mu(x) j_\nu(0) | \rangle_{\varrho^0} = \langle \bar{q}(x)\gamma_\mu q(x)\bar{q}(0)\gamma_\nu q(0)\rangle \;. \tag{7.242}$$

Thus we do not have to distinguish between up and down quarks, at least not in the usual limit $m_\mathrm{u}, m_\mathrm{d} \to 0$. In the following we discuss the ϱ^0 case. For the perturbative part (without any additive interactions or vacuum insertions) we get

$$\Pi_{\mu\nu}(q) = \mathrm{i}\int \mathrm{d}^4x\ \mathrm{e}^{\mathrm{i}qx}\langle 0|T\{\bar{q}(x)\gamma_\mu q(x)\bar{q}(0)\gamma_\nu q(0)\}|0\rangle \;. \tag{7.243}$$

Wick's theorem simply gives

$$
\begin{aligned}
&\langle 0|T\{\bar q(x)\gamma_\mu q(x)\bar q(0)\gamma_\nu q(0)\}|0\rangle \\
&\quad = (\gamma_\mu)_{i_1 i_2}(\gamma_\nu)_{i_3 i_4}\langle 0|T\{\bar q_{i_1}(x)q_{i_2}(x)\bar q_{i_3}(0)q_{i_4}(0)\}|0\rangle \\
&\quad = (\gamma_\mu)_{i_1 i_2}(\gamma_\nu)_{i_3 i_4}\langle 0|T\{\bar q_{i_1}(x)q_{i_4}(0)\}|0\rangle\langle 0|T\{q_{i_2}(x)\bar q_{i_3}(0)\}|0\rangle \\
&\quad = (\gamma_\mu)_{i_1 i_2}(\gamma_\nu)_{i_3 i_4}\,\mathrm{i}S_{\mathrm{F},i_4,i_1}(-x)\,\mathrm{i}S_{\mathrm{F},i_2,i_3}(x) \\
&\quad = \mathrm{tr}\left(\gamma_\mu\,\frac{\not x}{2\pi^2(x^2)^2}\,\gamma_\nu\,\frac{-\not x}{2\pi^2(x^2)^2}\right) \\
&\quad = \frac{1}{\pi^4}\frac{x^2 g_{\mu\nu}-2x_\mu x_\nu}{(x^2)^4}\ .
\end{aligned} \tag{7.244}
$$

Using (1) from Exercise 4.6,

$$
\int \mathrm{d}^d x \frac{\mathrm{e}^{\mathrm{i}p\cdot x}}{(-x^2)^\nu} = -\mathrm{i}\pi^2\frac{\Gamma(2-\nu+\varepsilon)}{\Gamma(\nu)}\left(\frac{-4\pi^2\mu^2}{p^2}\right)^\varepsilon\left(\frac{-p^2}{4}\right)^{\nu-2}\ , \tag{7.245}
$$

it is easy to calculate the Fourier transformation of (7.244):

$$
\begin{aligned}
&\mathrm{i}\int \mathrm{d}^4 x\ \mathrm{e}^{\mathrm{i}p\cdot x}\langle 0|T\{\bar q(x)\gamma_\mu q(x)\bar q(0)\gamma_\nu q(0)\}|0\rangle \\
&\quad = \mathrm{i}\int \mathrm{d}^4 x_\mathrm{E}\ \mathrm{e}^{-\mathrm{i}p_\mathrm{E}\cdot x_\mathrm{E}}\frac{-1}{\pi^4}\frac{x_\mathrm{E}^2 g_{\mu\nu}+2x_{\mathrm{E}\mu}x_{\mathrm{E}\nu}}{\left(x_\mathrm{E}^2\right)^4} \\
&\quad = \frac{-\mathrm{i}}{\pi^4}g_{\mu\nu}(+\mathrm{i}\pi^2)\frac{\Gamma(\varepsilon-1)}{\Gamma(3)}\left(\frac{-4\pi^2\mu^2}{p_\mathrm{E}^2}\right)^\varepsilon\frac{p_\mathrm{E}^2}{4} \\
&\qquad -\frac{\mathrm{i}}{\pi^4}2\left(-\frac{\partial}{\partial p_{\mathrm{E}\mu}}\frac{\partial}{\partial p_{\mathrm{E}\nu}}\right)\int \mathrm{d}^d x_\mathrm{E}\ \mathrm{e}^{-\mathrm{i}p_\mathrm{E}\cdot x_\mathrm{E}}\ \frac{1}{\left(x_\mathrm{E}^2\right)^4} \\
&\quad = \frac{1}{8\pi^2}\Gamma(\varepsilon-1)\ p_\mathrm{E}^2\left(\frac{-4\pi^2\mu^2}{p_\mathrm{E}^2}\right)^\varepsilon g_{\mu\nu} \\
&\qquad +\frac{2\mathrm{i}}{\pi^4}(-\mathrm{i}\pi^2)\frac{\Gamma(\varepsilon-2)}{\Gamma(4)}\left(-4\pi^2\mu^2\right)^\varepsilon\ \frac{1}{16}\frac{\partial}{\partial p_{\mathrm{E}\mu}}\frac{\partial}{\partial p_{\mathrm{E}\nu}}\left(p_\mathrm{E}^2\right)^{2-\varepsilon} \\
&\quad = \frac{1}{8\pi^2}\Gamma(\varepsilon-1)\ p_\mathrm{E}^2\ g_{\mu\nu}\left(\frac{-4\pi^2\mu^2}{p_\mathrm{E}^2}\right)^\varepsilon \\
&\qquad +\frac{1}{8\pi^2}\frac{\Gamma(\varepsilon-2)}{6}\left(\frac{-4\pi^2\mu^2}{p_\mathrm{E}^2}\right)^\varepsilon \\
&\qquad \times\left[(2-\varepsilon)(2-\varepsilon)4\ p_{\mathrm{E}\mu}\ p_{\mathrm{E}\nu}+(2-\varepsilon)\,2\,g_{\mu\nu}\ p_\mathrm{E}^2\right] \\
&\quad = \frac{1}{8\pi^2}\ \Gamma(\varepsilon-1)\left(\frac{-4\pi^2\mu^2}{p_\mathrm{E}^2}\right)^\varepsilon\left[\frac{2}{3}\ p_\mathrm{E}^2\ g_{\mu\nu}+(1-\varepsilon)\ \frac{2}{3}\ p_{\mathrm{E}\mu}\ p_{\mathrm{E}\nu}\right] \\
&\quad = \frac{1}{8\pi^2}\left(\frac{1}{\varepsilon}+\Gamma'(1)\right)\left(\frac{4\pi^2\mu^2}{p^2}\right)^\varepsilon\frac{2}{3}\left(p^2 g_{\mu\nu}-p_\mu p_\nu\right)+\ldots
\end{aligned}
$$

$$= \frac{1}{8\pi^2} \ln\left(\frac{4\pi^2\mu^2}{p^2}\right) \frac{2}{3} \left(p^2 g_{\mu\nu} - p_\mu p_\nu\right)$$
$$+ \text{terms without an imaginary part} \ . \tag{7.246}$$

Every quark color gives the same contribution such that we get finally

$$\Pi_{\mu\nu} = \frac{1}{4\pi^2} \left(p^2 g_{\mu\nu} - p_\mu p_\nu\right) \ln\left(\frac{4\pi^2\mu^2}{p^2}\right) + \dots \ . \tag{7.247}$$

Thus we have calculated the perturbative contribution to lowest order of α_s. Clearly these perturbative contributions alone cannot be sufficient to completely obtain nonperturbative quantities such as hadron masses. The fundamental assumption of the QCD sum-rule approach is that these nonperturbative effects can, to a good approximation, be described by vacuum properties. As the vacuum is supposed to be homogeneous, those fields coupling to it must have zero momentum.

In the presence of background particles, the propagator of, for example, a Dirac field is

$$S_{\mathrm{F}}(k) = (\not{k} + m) \left[\frac{1}{k^2 - m^2 + \mathrm{i}\varepsilon} + 2\pi\mathrm{i}\, \delta(k^2 - m^2)\, \Theta(k_0)\, n(\boldsymbol{k})\right] \ . \tag{7.248}$$

Assuming that $n(\boldsymbol{k}) = c\, m\, \delta^3(\boldsymbol{k})$, in coordinate space for $m \to 0$ this becomes

$$S_{\mathrm{F}}(x) = \frac{1}{(2\pi)^4} \int \mathrm{d}^4k\ \mathrm{e}^{\mathrm{i}k\cdot x}\, S(k)$$
$$= \frac{-\mathrm{i}}{2\pi^2} \frac{\not{x}}{(x^2)^2} + \frac{1}{(2\pi)^3} \int \mathrm{d}^4k\, \frac{c\,m}{2m}\, \delta^4(k)\, \mathrm{e}^{\mathrm{i}k\cdot x}$$
$$= \frac{-\mathrm{i}}{2\pi^2} \frac{\not{x}}{(x^2)^2} + \frac{c}{2(2\pi)^3} \ . \tag{7.249}$$

Thus it would seem natural to describe the complex vacuum structure by adding constant real numbers to the field propagators:

$$\left[S_{\mathrm{F}}(x)\right]_{ij}^{ab} = \frac{-\mathrm{i}}{2\pi^2} \frac{(\not{x})_{ij}}{(x^2)^2}\, \delta_{ab} - \frac{1}{12} \langle \bar{q}q \rangle\, \delta_{ab}\, \delta_{ij} \ . \tag{7.250}$$

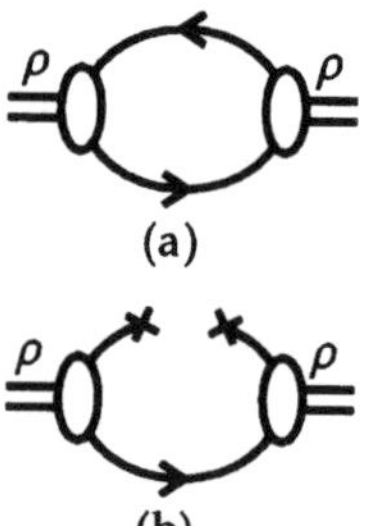

Fig. 7.16a,b. A graphical representation of (a) the lowest-order, purely perturbative contribution, and (b) the lowest quark condensate contribution

This form is also suggested by Wick's theorem if the expectation values of normal ordered products are interpreted as condensates.

$$T\left[\psi_i^a(x)\bar{\psi}_j^b(x')\right] = \,:\psi_i^a(x)\bar{\psi}_j^b(x'):\, + \langle 0|T\left[\psi_i^a(x)\bar{\psi}_j^b(x')\right]|0\rangle \ , \tag{7.251}$$

$$\langle \mathrm{vac}|T\{\psi_i^a(x)\bar{\psi}_j^b(x')\}|\mathrm{vac}\rangle$$
$$= \langle \mathrm{vac}|\,:\psi_i^a(x)\bar{\psi}_j^b(x'):\,|\mathrm{vac}\rangle + \langle 0|T\{\psi_i^a(x)\bar{\psi}_j^b(x')\}|0\rangle$$
$$= -\frac{1}{12}\, \delta_{ab}\, \delta_{ij} \langle \bar{q}q \rangle + \mathrm{i}S_{\mathrm{F}ij}(x - x')\, \delta_{ab} \ . \tag{7.252}$$

Using (7.250) we shall next calculate the quark-condensate contribution to the ϱ sum rule. It is represented by Fig. 7.16; Fig. 7.17 shows the lowest-order perturbative contribution.

With (7.252) we simply get

$$\begin{aligned}\Pi_{\mu\nu} &= \sum_{a,b} \mathrm{i} \int \mathrm{d}^4x\, \mathrm{e}^{\mathrm{i}qx}\, (\gamma_\mu)_{i_1 i_2} (\gamma_\nu)_{i_3 i_4}\, \mathrm{i} S_{\mathrm{F} i_4 i_1}(-x) \left(-\frac{1}{12} \langle \bar{q} q \rangle \right) \delta_{ab}\, \delta_{i_2 i_3} \\ &= \frac{-3\mathrm{i}}{12} \langle \bar{q} q \rangle \int \mathrm{d}^4x\, \mathrm{e}^{\mathrm{i}qx}\, \mathrm{tr}\big(\gamma_\mu \gamma_\nu S_\mathrm{F}(-x)\big) \ . \end{aligned} \tag{7.253}$$

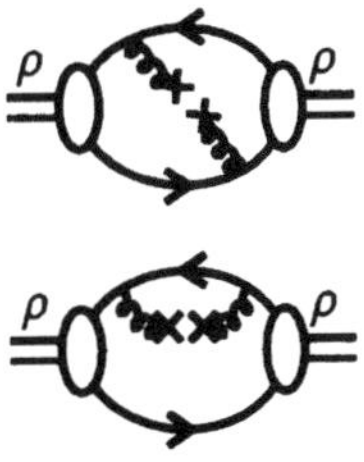

Fig. 7.17. Lowest-order couplings to the gluon condensate

For massless quarks the trace vanishes and we get strictly zero. Therefore we expand S_F in m and also keep the term linear in m:

$$\begin{aligned}-\mathrm{i} S_\mathrm{F}(-x) &= \frac{1}{(2\pi)^4} \int \mathrm{d}^4k\, \mathrm{e}^{\mathrm{i}kx} \frac{\not{k} + m}{k^2 - m^2 + \mathrm{i}\varepsilon} \\ &= \frac{-\mathrm{i}}{2\pi^2} \frac{\not{x}}{(x^2)^2} + \frac{m}{4\pi^2} \frac{1}{x^2} + \dots \ . \end{aligned} \tag{7.254}$$

We thus get

$$\begin{aligned}\Pi_{\mu\nu} &= -\frac{\mathrm{i}}{4} \langle \bar{q} q \rangle\, m \int \mathrm{d}^4x\, \mathrm{e}^{\mathrm{i}qx} \frac{1}{4\pi^2} 4 g_{\mu\nu} \frac{1}{x^2} \\ &= \frac{\pi^2}{4\pi^2} \frac{\Gamma(1)}{\Gamma(1)} \left(\frac{q^2}{4} \right)^{-1} \left(\frac{-4\pi^2\mu^2}{q^2} \right)^{\varepsilon} m\, \langle \bar{q} q \rangle\, g_{\mu\nu} \\ &= \frac{1}{q^2}\, m\, \langle \bar{q} q \rangle\, g_{\mu\nu} \\ &= \frac{1}{q^2 + \mathrm{i}\eta}\, m\, \langle \bar{q} q \rangle\, g_{\mu\nu} \ . \end{aligned} \tag{7.255}$$

In the last step we have reintroduced the usual $\mathrm{i}\eta$ prescription, which we did not write out explicitly during the calculation. As two graphs contribute, we have to multiply this by a factor of 2. Also we know that $\Pi_{\mu\nu}$ is transverse, and we can make this explicit by adding the $q_\mu q_\nu$ term:

$$\Pi_{\mu\nu} = \left(q^2 g_{\mu\nu} - q_\mu q_\nu \right) 2 \frac{m \langle \bar{q} q \rangle}{q^2 + \mathrm{i}\eta} \ . \tag{7.256}$$

Next we calculate the lowest-order gluon condensate contributions, corresponding to the graphs in Fig. 7.17.

The fields coupling to condensates all have zero momentum, since the vacuum condensates are time and space independent. Thus the gluon field can be easily expressed in terms of the constant gluon field strengths $G^a_{\mu\nu}$,

$$A^a_\mu(z) = \frac{1}{2} z_\varrho G^a_{\varrho\mu}(0) \tag{7.257}$$

$$\begin{aligned}\Rightarrow \quad & \partial_\nu A^a_\mu(z) - \partial_\mu A^a_\nu(z) + g f_{abc} A^b_\mu(z) A^c_\nu(z) \\ &= \frac{1}{2} G^a_{\nu\mu}(0) - \frac{1}{2} G^a_{\mu\nu}(0) + g \frac{1}{4} z_\alpha z_\beta\, f_{abc}\, G^b_{\alpha\mu}(0)\, G^c_{\beta\nu}(0) \\ &= G^a_{\nu\mu}(0) \ , \end{aligned} \tag{7.258}$$

where we have used the fact that $G^a_{\nu\mu}(0)$ is the same for all colors, since the vacuum does not distinguish any color state $G^b_{\alpha\mu}(0) = G^c_{\beta\nu}(0)$.

The quark propagator in coordinate space in the gluon background fields of the vacuum condensates is then

$$
\begin{aligned}
-\mathrm{i}S(x, y; A) = {} & \frac{1}{2\pi^2}\frac{(x-y)\cdot\gamma}{(x-y)^4} - \frac{gG^{\varrho\mu}(0)}{4\pi^4} \\
& \times \frac{1}{2}\int \mathrm{d}^4 z\, \frac{\not{x}-\not{z}}{(x-z)^4} z_\varrho \gamma_\mu \frac{\not{z}-\not{y}}{(z-y)^4} \\
& + \frac{g^2 G^{\varrho\mu}(0) G^{\varrho'\mu'}(0)}{8\pi^6} \\
& \times \frac{1}{4}\int \mathrm{d}^4 z_1\, \mathrm{d}^4 z_2\, \frac{\not{x}-\not{z}_1}{(x-z_1)^4}(z_1)_\varrho \gamma_\mu \frac{\not{z}_1-\not{z}_2}{(z_1-z_2)^4} \\
& \times (z_2)_{\varrho'}\gamma_{\mu'}\frac{\not{z}_2-\not{y}}{(z_2-y)^4} \\
& + \dots\,.
\end{aligned}
\tag{7.259}
$$

We have to perform these integrals using dimensional regularization. In Exercise 7.4 we do this for the first integral on the right-hand side. The double integral and all higher ones vanish. The result is then

$$
\begin{aligned}
S(x, y; A) = \mathrm{i}\Bigg[& \frac{1}{2\pi^2}\frac{\not{r}}{(r^2)^2} + \frac{\mathrm{i}g}{32\pi^2 r^2}(\not{r}\gamma_\mu\gamma_\varrho - \gamma_\varrho\gamma_\mu\not{r})G_{\varrho\mu}(0) \\
& + \frac{\mathrm{i}g}{4\pi^2 r^4}\not{r}\gamma_\varrho x_\mu G_{\varrho\mu}(0)\Bigg]
\end{aligned}
\tag{7.260}
$$

with

$$
r_\mu = x_\mu - y_\mu\,. \tag{7.261}
$$

EXERCISE

7.4 Calculation of QCD Sum-Rule Graphs with Dimensional Regularization

Problem. Calculate the first integral in (7.259) using the techniques of dimensional regularization introduced in Sect. 4.3.

Solution. We first introduce the usual Feynman parameters:

$$
\begin{aligned}
I &= \int \mathrm{d}^4 z\, \frac{(\not{x}-\not{z})\gamma_\mu(\not{z}-\not{y})z_\varrho}{(x-z)^4(z-y)^4} \\
&= \Gamma(4)\int \mathrm{d}u\, u(1-u)\int \mathrm{d}^4 z\, \frac{(\not{x}-\not{z})\gamma_\mu(\not{z}-\not{y})z_\varrho}{\left[u(x-z)^2+(1-u)(z-y)^2\right]^4}\,.
\end{aligned}
\tag{1}
$$

Then, as usual, we bring the denominator into the quadratic normal form: *Exercise 7.4*

$$\begin{aligned} & u(x-z)^2+(1-u)(z-y)^2 \\ & \quad = z^2-2z\big[ux+(1-u)y\big]+ux^2+(1-u)y^2 \\ & \quad = \Big\{z-\big[ux+(1-u)y\big]\Big\}^2+u(1-u)x^2+u(1-u)y^2-2u(1-u)x\cdot y \\ & \quad = \Big\{z-\big[ux+(1-u)y\big]\Big\}^2+u(1-u)(x-y)^2 \ . \end{aligned} \tag{2}$$

Substituting $z \rightarrow z+ux+(1-u)y$ gives

$$\begin{aligned} I = {} & \Gamma(4)\int \mathrm{d}u\ u(1-u)\int \mathrm{d}^4z \frac{\big[(1-u)(\not{x}-\not{y})-\not{z}\big]\gamma_\mu}{\big[z^2+u(1-u)(x-y)^2\big]^4} \\ & \times\big[\not{z}+u(\not{x}-\not{y})\big]\big[z_\varrho+ux_\varrho+(1-u)y_\varrho\big] \ . \end{aligned} \tag{3}$$

Only the terms even in z contribute to the integral, and $z_\varrho z_\alpha$ contribute $z^2 g_{\varrho\alpha}/4$:

$$\begin{aligned} I = {} & 6\int_0^1 \mathrm{d}u\ u(1-u)\int \mathrm{d}^4z\ \frac{(1-u)u(\not{x}-\not{y})\gamma_\mu(\not{x}-\not{y})}{\big[z^2+u(1-u)(x-y)^2\big]^4} \\ & \times\big[ux_\varrho+(1-u)y_\varrho\big]+6\int_0^1 \mathrm{d}u\ u(1-u) \\ & \times\int \frac{\mathrm{d}^4z}{\big[z^2+u(1-u)(x-y)^2\big]^4}\left\{\frac{1}{2}z^2\gamma_\mu\big[ux_\varrho+(1-u)y_\varrho\big]\right. \\ & \left.+\frac{1}{4}z^2(1-u)(\not{x}-\not{y})\gamma_\mu\gamma_\varrho-\frac{1}{4}z^2\gamma_\varrho\gamma_\mu u(\not{x}-\not{y})\right\} \ . \end{aligned} \tag{4}$$

The nominator of the second integral is abbreviated by writing $z^2\cdot f(u)$. We now integrate using (4.97) and (4.99):

$$\begin{aligned} I = {} & 6\int_0^1 \mathrm{d}u\ u(1-u)\Big\{\big[u(1-u)(x-y)^2\big]^{-2} \\ & \times u(1-u)(\not{x}-\not{y})\gamma_\mu(\not{x}-\not{y})\big[ux_\varrho+(1-u)y_\varrho\big]\frac{\mathrm{i}\pi^2\Gamma(2)}{\Gamma(4)} \\ & +f(u)\frac{\mathrm{i}\pi^2\Gamma(1)}{\Gamma(4)}\,2\big[u(1-u)(x-y)^2\big]^{-1}\Big\} \\ = {} & \mathrm{i}\pi^2\frac{1}{(x-y)^4}\left[\int_0^1 \mathrm{d}u\ (\not{x}-\not{y})\gamma_\mu(\not{x}-\not{y})\big[ux_\varrho+(1-u)y_\varrho\big]\right. \end{aligned}$$

Exercise 7.4

$$+2\int_0^1 \mathrm{d}u\,\left\{\frac{1}{2}\gamma_\mu\left[ux_\varrho+(1-u)y_\varrho\right]+\frac{1}{4}(1-u)(\not{x}-\not{y})\gamma_\mu\gamma_\varrho\right.$$
$$\left.\left.-\frac{1}{4}\gamma_\varrho\gamma_\mu u(\not{x}-\not{y})\right\}(x-y)^2\right]\,. \tag{5}$$

Both the integral $\int u\ \mathrm{d}u$ and $\int(1-u)\ \mathrm{d}u$ give $1/2$

$$\begin{aligned}
I=&\frac{\mathrm{i}\pi^2}{2}\frac{1}{(x-y)^4}\left\{(\not{x}-\not{y})\gamma_\mu(\not{x}-\not{y})(x+y)_\varrho+\gamma_\mu(x_\varrho+y_\varrho)(x-y)^2\right.\\
&\left.+\frac{1}{2}\left[(\not{x}-\not{y})\gamma_\mu\gamma_\varrho-\gamma_\varrho\gamma_\mu(\not{x}-\not{y})\right](x-y)^2\right\}\\
=&\frac{\mathrm{i}\pi^2}{4}\frac{1}{(x-y)^2}\left[(\not{x}-\not{y})\gamma_\mu\gamma_\varrho-\gamma_\varrho\gamma_\mu(\not{x}-\not{y})\right]\\
&+\frac{\mathrm{i}\pi^2}{2}\frac{1}{(x-y)^4}\left[2(x-y)_\mu(\not{x}-\not{y})(x+y)_\varrho\right]\\
=&\frac{\mathrm{i}\pi^2}{4}\frac{1}{(x-y)^2}\left[(\not{x}-\not{y})\gamma_\mu\gamma_\varrho-\gamma_\varrho\gamma_\mu(\not{x}-\not{y})\right]\\
&+\mathrm{i}\pi^2\frac{1}{(x-y)^4}\left[x_\mu x_\varrho+x_\mu y_\varrho-y_\mu x_\varrho-y_\mu y_\varrho\right](\not{x}-\not{y})\,.
\end{aligned}\tag{6}$$

This still has to be multiplied by $G_{\varrho\mu}(0)$, which is antisymmetric in ϱ and μ, leading to

$$\begin{aligned}
\Delta S(x,y;A)=&-\frac{\mathrm{i}g}{32\pi^2}\frac{1}{(x-y)^2}\left[(\not{x}-\not{y})\gamma_\mu\gamma_\varrho-\gamma_\varrho\gamma_\mu(\not{x}-\not{y})\right]G^{\varrho\mu}(0)\\
&-\frac{\mathrm{i}g}{8\pi^2}\frac{1}{(x-y)^2}2x_\mu y_\varrho(\not{x}-\not{y})G^{\varrho\mu}(0)\,.
\end{aligned}\tag{7}$$

Only the first graphs in Fig. 7.17 have to be calculated:

$$\begin{aligned}
&\langle 0|T\{\bar{q}(x)\gamma_\mu q(x)\bar{q}(0)\gamma_\nu q(0)\}|0\rangle\\
&\Rightarrow(\gamma_\mu)_{i_1i_2}(\gamma_\nu)_{i_3i_4}\mathrm{i}\left(-\frac{\mathrm{i}}{32\pi^2}\right)[\not{x}\gamma_{\mu''}\gamma_{\varrho''}-\gamma_{\varrho''}\gamma_{\mu''}\not{x}]_{i_4i_1}\frac{G^{\varrho''\mu''}(0)}{(x-y)^2}\\
&\times\mathrm{i}\left(+\frac{\mathrm{i}}{32\pi^2}\right)[\not{x}\gamma_{\mu'}\gamma_{\varrho'}-\gamma_{\varrho'}\gamma_{\mu'}\not{x}]_{i_2i_3}\frac{G^{\varrho'\mu'}(0)}{(x-y)^2}\\
&=-G^{\varrho''\mu''}(0)G^{\varrho'\mu'}(0)\frac{g^2}{32\cdot 32\pi^4}\frac{1}{(x-y)^4}\\
&\times\mathrm{tr}\{[\not{x}\gamma_{\mu''}\gamma_{\varrho''}-\gamma_{\varrho''}\gamma_{\mu''}\not{x}]\gamma_\mu[\not{x}\gamma_{\mu'}\gamma_{\varrho'}-\gamma_{\varrho'}\gamma_{\mu'}\not{x}]\gamma_\nu\}=:T_{\mu\nu}\,.
\end{aligned}\tag{7.262}$$

Now we also have to insert the SU(3) matrices to express $G_{\varrho''\mu''}G_{\varrho'\mu'}$ by the vacuum condensate:

$$\begin{aligned}\frac{\alpha_s}{\pi}G_{\varrho''\mu''}G_{\varrho'\mu'} &\to \mathrm{tr}\left\{\frac{\lambda^a}{2}\frac{\lambda^b}{2}\right\}\frac{\alpha_s}{\pi}G^a_{\varrho''\mu''}G^b_{\varrho'\mu'}\\ &= \frac{1}{2}\frac{\alpha_s}{\pi}G^a_{\varrho''\mu''}G^b_{\varrho'\mu'}\\ &= \mathrm{const}\cdot\left\langle 0\left|\frac{\alpha_s}{\pi}G^a_{\alpha\beta}G^{a\alpha\beta}\right|0\right\rangle\left(g_{\varrho''\varrho'}g_{\mu''\mu'}-g_{\varrho''\mu'}g_{\mu''\varrho'}\right) \ .\end{aligned} \tag{7.263}$$

To determine the constant we contract with $g^{\varrho''\varrho'}g^{\mu''\mu'}$:

$$\begin{aligned}&\frac{1}{2}\frac{\alpha_s}{\pi}G^a_{\varrho'\mu'}G^{a\varrho'\mu'} = \mathrm{const}\cdot\left\langle 0\left|\frac{\alpha_s}{\pi}G^a_{\alpha\beta}G^{a\alpha\beta}\right|0\right\rangle(16-4)\\ &\Rightarrow \mathrm{const} = \frac{1}{24} \ .\end{aligned} \tag{7.264}$$

Inserting this into (7.262) yields

$$\begin{aligned}&-\left\langle 0\left|\frac{\alpha}{\pi}G^a_{\alpha\beta}G^{a\alpha\beta}\right|0\right\rangle\frac{1}{32\cdot 8\pi^2}\cdot\frac{1}{24}\cdot\frac{1}{\left(x^2\right)^7}\\ &\quad\times\Big[\mathrm{tr}\left\{\left[\not{x}\gamma_{\mu'}\gamma_{\varrho'}-\gamma_{\varrho'}\gamma_{\mu'}\not{x}\right]\gamma_\mu\left[\not{x}\gamma^{\mu'}\gamma^{\varrho'}-\gamma^{\varrho'}\gamma^{\mu'}\not{x}\right]\gamma_\nu\right\}\\ &\quad-\mathrm{tr}\left\{\left[\not{x}\gamma_{\varrho'}\gamma_{\mu'}-\gamma_{\mu'}\gamma_{\varrho'}\not{x}\right]\gamma_\mu\left[\not{x}\gamma^{\mu'}\gamma^{\varrho'}-\gamma^{\varrho'}\gamma^{\mu'}\not{x}\right]\gamma_\nu\right\}\Big] \ .\end{aligned} \tag{7.265}$$

The traces give

$$\begin{aligned}&\mathrm{tr}\{-2\not{x}\gamma^{\mu'}\gamma_{\mu'}\not{x}\gamma_\mu\gamma_\nu+2\not{x}\gamma^{\mu'}\gamma_\mu\gamma_{\mu'}\not{x}\gamma_\nu+2\gamma^{\mu'}\not{x}\gamma_\mu\not{x}\gamma_{\mu'}\gamma_\nu\\ &\quad-2\gamma_\mu\not{x}\gamma^{\mu'}\gamma_{\mu'}\not{x}\gamma_\nu-4x_\nu\gamma^{\mu'}\gamma_\mu\not{x}\gamma_{\mu'}+4g_{\mu'\nu}\not{x}\gamma_\mu\not{x}\gamma^{\mu'}\\ &\quad+4g_{\mu\mu'}\gamma^{\mu'}\not{x}\gamma_\nu\not{x}-4x_\mu\gamma^{\mu'}\gamma_{\mu'}\not{x}\gamma_\nu\}\\ &\quad=\mathrm{tr}\{-8x^2\gamma_\mu\gamma_\nu-8\not{x}\gamma_\mu\not{x}\gamma_\nu-8\gamma_\mu\gamma_\nu x^2\\ &\qquad-16x_\nu x_\mu+8\gamma_\mu\not{x}\gamma_\nu\not{x}-16x_\mu\not{x}\gamma_\nu\}\\ &\quad=-16x^2\cdot 4g_{\mu\nu}-32\cdot 4x_\nu x_\mu=-64\left(x^2g_{\mu\nu}+2x_\mu x_\nu\right) \ ,\\ &T_{\mu\nu}=\frac{1}{96\pi^2}\left\langle 0\left|\frac{\alpha}{\pi}G^a_{\alpha\beta}G^{a\alpha\beta}\right|0\right\rangle\frac{x^2g_{\mu\nu}+2x_\mu x_\nu}{\left(x^2\right)^2} \ .\end{aligned} \tag{7.266}$$

This leads to the same integral we encountered in (7.255) repeating the same calculations therefore gives

$$\begin{aligned}\Pi_{\mu\nu}&=\frac{1}{96\pi^2}\left\langle 0\left|\frac{\alpha}{\pi}G^a_{\alpha\beta}G^{a\alpha\beta}\right|0\right\rangle\frac{4\pi^2}{\left(q^2\right)^2}\left(q^2g_{\mu\nu}-q_\mu q_\nu\right)\cdot 2\\ &=\frac{1}{2}(q^2g_{\mu\nu}-q_\mu q_\nu)\frac{1}{(q^2+\mathrm{i}\varepsilon)^2}\left\langle\frac{\alpha_s}{\pi}G^a_{\alpha\beta}G^{a\alpha\beta}\right\rangle \ .\end{aligned} \tag{7.267}$$

It should now be obvious how to proceed further. Taking into account the following graphs we get[17]

$$\Pi_{\mu\nu}(q) = (q^2 g_{\mu\nu} - q_\mu q_\nu)\left\{-\frac{1}{4\pi^2}\left(1+\frac{\alpha_s}{\pi}\right)\ln\frac{-q^2}{\mu^2} - \frac{6m^2}{q^2+\mathrm{i}\varepsilon}\right.$$

$$+\frac{2}{(q^2+\mathrm{i}\varepsilon)^2} m\langle\bar{q}q\rangle + \frac{1}{12}\frac{1}{(q^2+\mathrm{i}\varepsilon)^2}\left\langle\frac{\alpha_s}{\pi}G^a_{\alpha\beta}G^{a\alpha\beta}\right\rangle$$

$$\left. +\frac{112}{91}\frac{\pi\alpha_s}{(q^2+\mathrm{i}\varepsilon)^3}\langle\bar{q}q\rangle^2\right\} . \tag{7.268}$$

$$\begin{aligned}\frac{1}{3q^2}\mathrm{Im}\,\Pi_{\mu\nu}(q)g^{\mu\nu} = &\left\{\frac{1}{4\pi}\left(1+\frac{\alpha_s}{\pi}\right) - m\langle\bar{q}q\rangle\pi\left(-\frac{\partial}{\partial q^2}\delta\left(q^2\right)\right)\right.\\ &-\frac{\pi}{12}\left(-\frac{\partial}{\partial q^2}\delta\left(q^2\right)\right)\left\langle\frac{\alpha_s}{\pi}G^a_{\alpha\beta}G^{a\alpha\beta}\right\rangle\\ &\left.-\left[\frac{1}{2}\left(\frac{\partial}{\partial q^2}\right)^2\delta\left(q^2\right)\right]2\pi\cdot\frac{112}{91}\langle\bar{q}q\rangle^2\pi\alpha_s\right\} .\end{aligned} \tag{7.269}$$

We have so far calculated the partonic $\Pi_{\mu\nu}(q^2)$ tensor. Next we have to relate our result to the hadron description of $\Pi_{\mu\nu}(q^2)$. The quark description used a current with the quantum numbers of the rho. However, several resonances exist with such quantum numbers as well as continuum states. The unique property of the rho is that it is the lightest state and this property can be used to project it out

[17] L.J. Reinders, H. Rubinstein and S. Yazaki: Phys. Rep. **127**, 1 (1985).
V.A. Novikov, M.A. Shifman, A.I. Vainshtein and V.I. Zakharov: Fortsch. Phys. **32**, 585 (1985).

from both the hadron and the parton descriptions. There are various methods to do this; we wish to discuss here only one, the Borel transformation, defined by

$$I = \frac{1}{\pi M^2} \int_0^\infty \mathrm{e}^{-s/M^2} \mathrm{Im}\,[\Pi(s)]\, \mathrm{d}s \ . \tag{7.270}$$

Inserting (7.206) into this expression and replacing the upper integration bound by s_0 gives

$$\begin{aligned} I &= \frac{1}{\pi M^2} \int_0^{s_0} \mathrm{e}^{-s/M^2} \mathrm{Im}\,(\Pi(s))\, \mathrm{d}s \\ &= \frac{1}{\pi M^2} \sum_{\mathrm{f}} \frac{3\pi}{\mathrm{e}_{\mathrm{f}}^2} \sum_R \frac{m_R^2}{g_R^2} \mathrm{e}^{-m_R^2/M^2} \ . \end{aligned} \tag{7.271}$$

Obviously for small enough M^2 only the lowest-mass state survives. Here, however, we run into problems, since M^2 cannot be arbitrarily small, or otherwise the highest-twist contributions to the partonic description would become arbitrarily large. To understand this one has to know that the Borel transformation is equivalent to the following mathematical operation:

$$\begin{aligned} &\frac{1}{\pi M^2} \int_0^{s_0} \mathrm{e}^{-s/M^2} \mathrm{Im}\,[\Pi(s)]\, \mathrm{d}s \\ &= \lim_{Q^2,n\to\infty,\ Q^2/n=M^2=\mathrm{const}} \frac{1}{(n-1)!} Q^{2n} \left(-\frac{\mathrm{d}}{\mathrm{d}Q^2}\right)^n \Pi(Q^2) \\ &= \lim_{Q^2,n\to\infty,\ Q^2/n=M^2} \frac{nQ^{2n}}{\pi} \int_0^{s_0} \frac{\mathrm{Im}\,\Pi(s)}{\left(s+Q^2\right)^{n+1}}\, \mathrm{d}s \\ &= \lim_{Q^2,n\to\infty,\ Q^2/n=M^2} \frac{n}{\pi Q^2} \int_0^{s_0} \frac{\mathrm{Im}\,\Pi(s)}{\left(1+\frac{s}{Q^2}\right)^{n+1}}\, \mathrm{d}s \\ &= \lim_{Q^2,n\to\infty,\ Q^2/n=M^2} \frac{1}{\pi M^2} \int_0^{s_0} \mathrm{e}^{-(n+1)\ln(1+s/Q^2)} \mathrm{Im}\,\Pi(s)\, \mathrm{d}s \\ &= \frac{1}{\pi M^2} \int_0^{s_0} \mathrm{e}^{-s/M^2} \mathrm{Im}\,\Pi(s)\, \mathrm{d}s \ , \end{aligned} \tag{7.272}$$

where we have started from the well-known dispersion relation

$$\Pi\left(q^2\right) = \frac{1}{\pi} \int_0^{s_0} \frac{\mathrm{Im}\,\Pi(s)}{\left(s+Q^2\right)}\, \mathrm{d}s \ . \tag{7.273}$$

Now the higher twist corrections to the partonic calculations, or the higher-dimension condensates that parametrize them, are suppressed by powers of Q^2. Here Q^2 must be large in order to allow for perturbative calculations in α. Thus making M^2 small would require very large values of n. But such high derivatives increase the importance of the higher twist effects as

$$\frac{Q^{2n}}{(n-1)!}\left(-\frac{\mathrm{d}}{\mathrm{d}Q^2}\right)^n \frac{1}{(Q^2)^N} = \frac{(N+n-1)!}{(N-1)!}\frac{1}{(n-1)!}\frac{1}{(Q^2)^N}$$
$$\sim \frac{(N+n-1)!}{(n-1)!} \tag{7.274}$$

becomes very large with increasing N. Thus QCD sum rules are for most applications an expansion with a limited domain of applicability. If M^2 becomes too small the higher twist effects on the partonic side become uncontrollable; if it becomes too large, the hadronic side becomes a complicated mixture of hadronic states. Luckily for most problems one finds an intermediate domain around $M = 1$ GeV in which the result depends only slightly on M, indicating that the approximations made were acceptable. Inserting (7.206), we obtain for the Borel transform of the hadronic part simply

$$\frac{1}{\pi M}\int_0^{s_0} \mathrm{e}^{-s/M^2} \mathrm{Im}\, \Pi(s)\, \mathrm{d}s = \mathrm{e}^{-m_\varrho^2/M^2}\frac{3\pi m_\varrho^2}{e_\mathrm{f}^2 g_R^2}\frac{1}{\pi M^2} \; . \tag{7.275}$$

The integral (7.273) is now easy, yielding

$$\frac{1}{\pi M^2}\int_0^{s_0} \mathrm{e}^{-s/M^2} \Pi_{\mu\nu}(q')g^{\mu\nu}\, \mathrm{d}s$$
$$= \frac{1}{4\pi^2}\left[\left(1+\frac{\alpha_s}{\pi}\right)\left(1-\mathrm{e}^{-s_0/M^2}\right) + \frac{8\pi^2}{M^4}\langle 0|m\bar{q}q|0\rangle \right.$$
$$\left. + \frac{\pi^2}{3M^4}\left\langle 0\left|\frac{\alpha_s}{\pi}G^a_{\alpha\beta}\, G^{a\alpha\beta}\right|0\right\rangle - \frac{448}{81}\frac{\pi^2\alpha_s}{M^6}|\langle 0|\bar{q}q|0\rangle|^2\right] \; . \tag{7.276}$$

This expression can now be equated with (7.275). However, there is still one last problem to be solved. The current substituted in (7.242) to describe vector mesons can still be arbitrarily normalized. Thus we introduce a factor λ_ϱ and write

$$\frac{1}{4\pi^2}\left[\left(1+\frac{\alpha_s(\lambda)}{\pi}\right)\left(1-\mathrm{e}^{-s_0/M^2}\right) + \frac{8\pi^2}{M^4}\langle 0|m\bar{q}q|0\rangle \right.$$
$$\left. + \frac{\pi^2}{3M^4}\left\langle 0\left|\frac{\alpha_s}{\pi}G^a_{\alpha\beta}\, G^{a\alpha\beta}\right|0\right\rangle - \frac{448}{81}\frac{\pi^2\alpha_s}{M^6}|\langle 0|\bar{q}q|0\rangle|^2\right]$$
$$= \lambda_\varrho \frac{1}{\pi M^2}\frac{3\pi m_\varrho^2}{e_\mathrm{f}^2 g_R^2}\mathrm{e}^{-m_\varrho^2/M^2} = \frac{1}{\pi M^2}J \; . \tag{7.277}$$

We get rid of λ_ϱ and all other constants by dividing $\mathrm{d}J/\mathrm{d}M^2$ by J:

$$\begin{aligned}\frac{\mathrm{d}J}{\mathrm{d}M^2}/J = &\Big[\left(1+\frac{\alpha_s}{\pi}\right)\left(1-\mathrm{e}^{-s_0/M^2}\right) - M^2\left(1+\frac{\alpha_s}{\pi}\right)\frac{s_0}{M^2}\mathrm{e}^{-s_0/M^2} \\ &-\frac{8\pi^2}{M^4}\langle 0|m\bar{q}q|0\rangle - \frac{\pi^2}{3M^4}\left\langle 0\left|\frac{\alpha_s}{\pi}G^a_{\alpha\beta}G^{a\alpha\beta}\right|0\right\rangle \\ &+2\cdot\frac{448}{81}\frac{\pi^2\alpha_s}{M^6}\langle 0|\bar{q}q|0\rangle^2\Big] \\ &\times\Big[M^2\left(1+\frac{\alpha_s}{\pi}\right)\left(1-\mathrm{e}^{-s_0/M^2}\right)+\frac{8\pi^2}{M^2}\langle 0|m\bar{q}q|0\rangle \\ &+\frac{\pi^3}{3M^2}\left\langle 0\left|\frac{\alpha_s}{\pi}G^a_{\alpha\beta}G^{a\alpha\beta}\right|0\right\rangle - \frac{448}{81}\frac{\pi^2\alpha_s}{M^4}\langle 0|\bar{q}q|0\rangle^2\Big]^{-1} = \frac{m_\varrho^2}{M^4} ,\end{aligned} \tag{7.278}$$

$$\begin{aligned}m_\varrho^2 = M^2&\left[\left(1+\frac{\alpha_s}{\pi}\right)\left(1-\left(1+\frac{s_0}{M^2}\right)\mathrm{e}^{-s_0/M^2}\right)-\frac{1}{M^4}A+\frac{2}{M^6}B\right] \\ &\times\left[\left(1+\frac{\alpha_s}{\pi}\right)\left(1-\mathrm{e}^{-s_0/M^2}\right)+\frac{1}{M^4}A-\frac{1}{M^6}B\right]^{-1} .\end{aligned} \tag{7.279}$$

With the condensates from (7.207) the numerical constants are

$$\begin{aligned}A &= -8\pi^2(0.10)^4\ \mathrm{GeV}^4 + \frac{\pi^3}{3}(0.36)^4\ \mathrm{GeV}^4 \\ &= (-0.007+0.055)\ \mathrm{GeV}^4 = 0.05\ \mathrm{GeV}^4 , \\ B &= (0.225)^8\cdot\pi^2\cdot 0.25\cdot\frac{448}{81}\ \mathrm{GeV}^6 = 0.035\ \mathrm{GeV}^6 .\end{aligned} \tag{7.280}$$

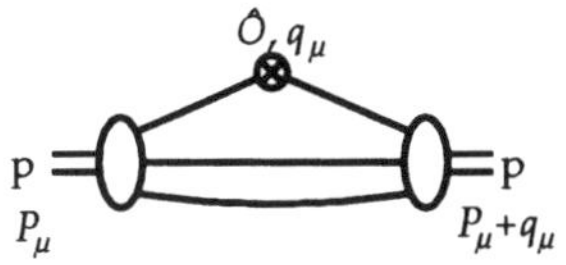

Fig. 7.18. The general form of a proton vertex $\hat{O}$ at which momentum q_μ is transferred

The resulting form of m_ϱ^2 as a function of M^2 is plotted in Fig. 7.19.

Similar calculations can be repeated for any set of parameters, and the resulting masses agree quite well with the real values, except for the pion. However, the QCD sum-rule technique is not limited to the calculation of masses. It is also possible to calculate vertices, such as any coupling to the quark in the proton, according to Fig. 7.18.

The proton current can be written in many different forms. The simplest is the so-called Ioffe current

$$\eta_I(x) = \varepsilon_{abc}\left(u^{ta}(x)\gamma_0\gamma_2\gamma_\mu u^b(x)\right)\gamma_5\gamma^\mu d^c(x) . \tag{7.281}$$

Here, the first $u(x)$ field is transposed so that the bracket is just a number as far as the spinor indices are concerned and a Lorentz vector with respect to the index of γ_μ. It is easy to show that this combination indeed has isospin 1/2 and not 3/2, which is the second possibility in the coupling of 3 isospin-1/2 objects

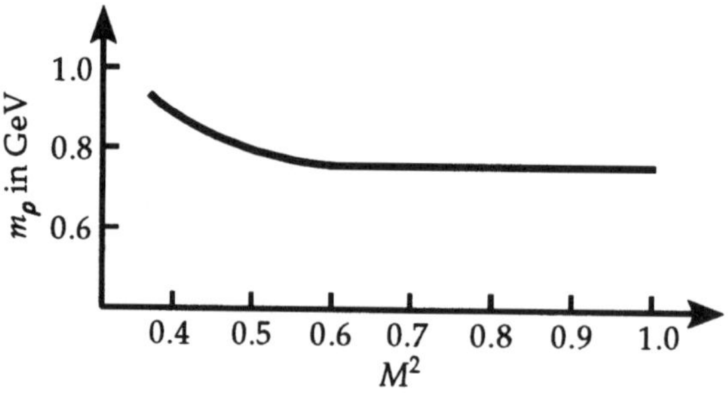

Fig. 7.19. The rho mass as a function of M^2, as given by (7.227) for $s_0 = 1.5\,\mathrm{GeV}^2$. The convergence is clearly convincing in the range $M^2 = 0.5 - 0.8\,\mathrm{GeV}^2$

$$
\begin{aligned}
I^2 &= I_3^2 + \frac{1}{2}(I_+I_- + I_-I_+) \ , \\
I_j &= I_j(1) + I_j(2) + I_j(3) \ ,
\end{aligned}
\tag{7.282}
$$

where the $I_j(k)$ act on quark number k. To show this, one has to use the so-called Fierz transformation:

$$
\begin{aligned}
I^2\eta_I(x) = {} & \frac{1}{4}\eta_I(x) + \frac{1}{2}I_+\left[\varepsilon_{abc}\left(d^{ta}(x)\gamma_0\gamma_2\gamma_\mu u^b(x)\right)\gamma_5\gamma^\mu d^c(x)\right. \\
& \left. +\varepsilon_{abc}\left(u^{ta}(x)\gamma_0\gamma_2\gamma_\mu d^b(x)\right)\gamma_5\gamma^\mu d^c(x)\right] \\
& +\frac{1}{2}I_-\left[\varepsilon_{abc}\left(u^{ta}(x)\gamma_0\gamma_2\gamma_\mu u^b(x)\right)\gamma_5\gamma^\mu u^c(x)\right] \\
= {} & \frac{1}{4}\eta_I(x) + \frac{3}{2}\eta_I(x) + \varepsilon_{abc}\left(d^{ta}(x)\gamma_0\gamma_2\gamma_\mu u^b(x)\right)\gamma_5\gamma^\mu u^c(x) \\
& +\varepsilon_{abc}\left(u^{ta}(x)\gamma_0\gamma_2\gamma_\mu d^b(x)\right)\gamma_5\gamma^\mu u^c(x) \ .
\end{aligned}
\tag{7.283}
$$

The last two terms are equal (please remember that the bracket is just a number).

$$
\begin{aligned}
d^{ta}(1)\gamma_0\gamma_2\gamma_\mu u^b(2) &= \left[d^{ta}(1)\gamma_0\gamma_2\gamma_\mu u^b(2)\right]^t \\
&= u^{bt}(2)\gamma_\mu^t\gamma_2^t\gamma_0^t d^a(1) = -u^{bt}(2)\gamma_2\gamma_0\gamma_\mu d^a(1) \\
&= u^{bt}(2)\gamma_0\gamma_2\gamma_\mu d^a(1) \to u^{at}(1)\gamma_0\gamma_2\gamma_\mu d^b(2) \ .
\end{aligned}
\tag{7.284}
$$

In the last step we have used the antisymmetry of ε_{abc} and the fact that the total wave function is odd under the exchange of any two particles. The Fierz transformation in our case is

$$
\begin{aligned}
& \gamma_5\Big(\left(u^{at}\gamma_0\gamma_2\right)\gamma_\mu d^b\Big)\gamma^\mu u^c \\
& \quad = -\frac{1}{4}\gamma_5\Big\{\left(u^{at}\gamma_0\gamma_2 u^c\right)d^b - 2\left(u^{at}\gamma_0\gamma_2\gamma_\mu u^c\right)\gamma^\mu d^b \\
& \qquad -2\left(u^{at}\gamma_0\gamma_2\gamma_\mu\gamma_5 u^c\right)\gamma^\mu\gamma_5 d^b - \left(u^{at}\gamma_0\gamma_2\gamma_5 u^c\right)\gamma_5 d^b\Big\} \ .
\end{aligned}
\tag{7.285}
$$

Now one can see from (7.284) that all combinations for which the transposed bracket has the opposite sign vanish, owing to the symmetry operation ($a \to c$, $1 \to 2$). But

$$\begin{aligned}\left(u^{at}\gamma_0\gamma_2 u^c\right)^t &= u^{ct}\gamma_2^t\gamma_0^t u^a = -u^{ct}\gamma_0\gamma_2 u^a \ ,\\ \left(u^{at}\gamma_0\gamma_2\gamma_\mu\gamma_5 u^c\right)^t &= u^{ct}\gamma_5^t\gamma_0\gamma_2\gamma_\mu u^a = -u^{ct}\gamma_0\gamma_2\gamma_\mu\gamma_5 u^a \ ,\\ \left(u^{at}\gamma_0\gamma_2\gamma_5 u^c\right)^t &= u^{ct}\gamma_5\gamma_2\gamma_0 u^a = -u^{ct}\gamma_0\gamma_2\gamma_5 u^a \ .\end{aligned} \tag{7.286}$$

Thus only the second term in (7.285) survives:

$$\begin{aligned}\left(u^{at}\gamma_0\gamma_2\gamma_\mu d^b\right)\gamma_5\gamma^\mu u^c &= -\frac{1}{2}\left(u^{at}(1)\gamma_0\gamma_2\gamma_\mu u^c(3)\right)\gamma^\mu\gamma_5 d^b(2)\\ &\Rightarrow -\frac{1}{2}\left(u^{at}(1)\gamma_0\gamma_2\gamma_\mu u^b(2)\right)\gamma^\mu\gamma_5 d^c(3) \ ,\end{aligned} \tag{7.287}$$

and we have

$$I^2\eta_I(x) = \frac{1}{4}\eta_I(x) + \frac{3}{2}\eta_I(x) - 2\cdot\frac{1}{2}\eta_I(x) = \frac{3}{4}\eta_I(x) = \frac{1}{2}\left(\frac{1}{2}+1\right)\eta_I(x) \ , \tag{7.288}$$

which demonstrates that $\eta_I(x)$ indeed has isospin 1/2. In calculating the triangle graph in Fig. 7.18, basically all techniques are the same except that now we have to perform a double Borel transformation to make sure that both the incoming and the outgoing baryons are projected onto the proton state:

$$\begin{aligned}&s_1 = p^2 \ , \quad s_2 = (p+q)^2 \ ,\\ &\int_0^{s_0} ds_1\ e^{-s_1/2M^2} \int_0^{s_0} ds_2\ e^{-s_2/2M^2} \ldots \ .\end{aligned} \tag{7.289}$$

In fact the applicability of QCD sum rules for calculations of quantities other than masses is not trivial. The Borel transformation projects out the lowest-mass eigenstate, which is the correct state only if it is unique, i.e., if any state is uniquely determined by the quantum numbers contained in the current and by the mass. For fully relativistic problems this is ascertained by general theorems. With these remarks we wish to end our introduction to QCD sum rules. Let us summarize what we have done by noting that with the few parameters for the condensates, one is able to calculate with a typical accuracy of 20%

- the masses of ϱ, ϕ, δ, a_1, N, Σ, Λ, Ξ, Δ, ...,
- the vertex constants such as $g_{\pi NN}$, $g_{\pi N\Delta}$, $g_{\omega\varrho\pi}$, ...,
- the heavier quarkonia states ${}^1S_0, {}^3S_1, {}^3P_2, \ldots$ for $\bar{c}c$ and $\bar{b}b$,
- quantities such as the total momentum carried by quarks or gluons at the scale M^2, which is $\sum_q \int_0^1 xq(x)\ dx$,
- the form of the 3-quark component of the nucleon wave function and of the quark–antiquark component of the pion wave function (the so-called Chernyak–Zhitnitsky wave functions), and many more quantities.

8. Phenomenological Models for Nonperturbative QCD Problems

As the complete calculation of many, especially dynamical, nonperturbative problems, for example, using lattice calculations is still impossible, theoreticants have tried to develop simple physical models for these problems. Using suitable assumptions and parameter choices, one then attempts to reproduce as many properties of QCD as possible. We shall consider two such problems here: the ground state of QCD and the quark–gluon plasma. In both cases we shall restrict ourselves to a few remarks only.

8.1 The Ground State of QCD

Many semiphenomenological models have been developed for the QCD ground state. We shall consider here only the so-called "Spaghetti vacuum" which is one of the most promising candidates and has the advantage of being easily visualized. As a starting point, let us ask what the fate is of a hypothetical monopole–antimonopole pair in a superconductor. The Meissner–Ochsenfeld effect teaches us that the flux lines of the magnetic field cannot enter any superconducting region. Consequently a region of normal conduction is created in between the pair (see Fig. 8.1).

Thus a string is created, and, since the energy density in the region of normal conduction is larger than in the superconductivity region, the magnetic charges are confined.

Our model system therefore shows the same properties as QCD, with magnetic charge and magnetic field playing the role of color charges and color electric fields. This leads us to the dual superconductor picture, i.e., the assump-

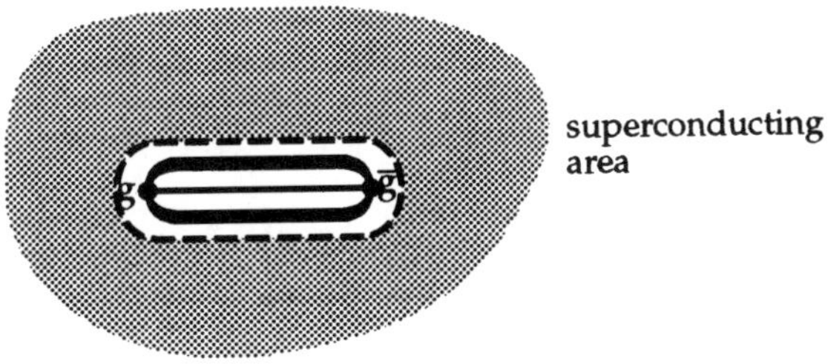

Fig. 8.1. A magnetic monopole–antimonopole ($g\bar{g}$) pair creates a string of normal conductance in a superconductor

tion that the QCD ground state has the properties of a superconductor in which the roles of magnetic and electric fields are interchanged. To put this on a firm foundation, one must show that QCD contains some objects that can play the role of Cooper pairs. As Cooper pairs are bound by electric field, one expects that their dual partners in QCD are bound by color magnetic fields.

From the many ideas about what these objects might be, we shall discuss only the "Spaghetti vacuum", in which the Cooper pairs are strings of color magnetic fields. To obtain this picture we start with the simplest question and ask how spontaneous color magnetization can arise in a pure gauge theory. To this end, we introduce a color magnetic background field H, determine the corresponding vacuum energy, and calculate the total energy with respect to H. The color electric and color magnetic fields are defined as

$$E^{aj} = F^a_{0j} = \nabla_0 A^a_j - \nabla_j A^a_0 + g f^{abc} A^b_0 A^c_j \ , \tag{8.1}$$

$$H^{aj} = -\varepsilon_{jik} F^a_{ik} = -\varepsilon_{jik}\left(\nabla_i A^a_k - \nabla_k A^a_i + g f^{abc} A^b_i A^c_k\right) \ . \tag{8.2}$$

To see whether a homogeneous magnetic field can lower the total energy, we set

$$H^{aj} = H\delta_{a3}\delta_{j3} \Rightarrow A^{0a}_\mu = -Hx_1\delta_{\mu 2}\delta_{a3} \ . \tag{8.3}$$

For simplicity we shall keep to an SU(2) gauge theory. We consider a small variation A^a_ν around A^{0a}_μ:

$$\begin{aligned}
F^a_{\mu\nu}F^{a\mu\nu} = &\Big\{\nabla_\mu A^1_\nu - \nabla_\nu A^1_\mu \\
&\quad + g\left[A^2_\mu\left(A^3_\nu - Hx_1\delta_{\nu 2}\right) - \left(A^3_\mu - Hx_1\delta_{\mu 2}\right)A^2_\nu\right]\Big\}^2 \\
&+ \Big\{\nabla_\mu A^2_\nu - \nabla_\nu A^2_\mu \\
&\quad + g\left[\left(A^3_\mu - Hx_1\delta_{\mu 2}\right)A^1_\nu - A^1_\mu\left(A^3_\nu - Hx_1\delta_{\nu 2}\right)\right]\Big\}^2 \\
&+ 2H^2 + \left[\nabla_\mu A^3_\nu - \nabla_\nu A^3_\mu + g\left(A^1_\mu A^2_\nu - A^2_\mu A^1_\nu\right)\right]^2 \\
&+ 2\left[\nabla_\mu A^3_\nu - \nabla_\nu A^3_\mu + g\left(A^1_\mu A^2_\nu - A^2_\mu A^1_\nu\right)\right] \\
&\times (-H)\left(\delta_{\mu 1}\delta_{\nu 2} - \delta_{\mu 2}\delta_{\nu 1}\right) \ .
\end{aligned} \tag{8.4}$$

We consider small variations around $H^{a\mu}$ and therefore neglect higher powers in A:

$$\begin{aligned}
F^a_{\mu\nu}F^{a\mu\nu} \approx &\left[\nabla_\mu A^1_\nu - \nabla_\nu A^1_\mu - gHx_1\left(A^2_\mu\delta_{\nu 2} - A^2_\nu\delta_{\mu 2}\right)\right]^2 \\
&+ \left[\nabla_\mu A^2_\nu - \nabla_\nu A^2_\mu - gHx_1\left(A^1_\nu\delta_{\mu 2} - A^1_\mu\delta_{\nu 1}\right)\right]^2 \\
&+ 2H^2 + \left(\nabla_\mu A^3_\nu - \nabla_\nu A^3_\mu\right)^2 \\
&- 4H\left[\nabla_1 A^3_2 - \nabla_2 A^3_1 + g\left(A^1_1 A^2_2 - A^2_1 A^1_2\right)\right] \ .
\end{aligned} \tag{8.5}$$

The source terms for A^3_μ that appear in the last term vanish by partial integration of the action integral if we demand that $A^3_2(x_1 \to \pm\infty) = A^3_1(x_2 \to \pm\infty) = 0$. This is equivalent to demanding a fixed background field. Indeed the appearance of this source term is an indication that A^3_μ is not a physical degree of freedom, and it can be shown that A^1_μ and A^2_μ are the only physical degrees of freedom. We shall content ourselves here with the observation that the source terms vanish by partial integration and that the only remaining term in A^3_μ is a free Lagrange density and is therefore not of interest to us. Dropping these terms we have

$$\begin{aligned} F^a_{\mu\nu} F^{a\,\mu\nu} = &\Big|\nabla_\mu \left(A^1_\nu + \mathrm{i}A^2_\nu\right) - \nabla_\nu \left(A^1_\mu + \mathrm{i}A^2_\mu\right) - gHx_1 \Big[\left(A^2_\mu - \mathrm{i}A^1_\mu\right)\delta_{\nu 2} \\ &- \left(A^2_\nu - \mathrm{i}A^1_\nu\right)\delta_{\mu 2}\Big]\Big|^2 + 2H^2 - \mathrm{i}2gH\Big[\left(A^1_1 + \mathrm{i}A^2_1\right)\left(A^1_2 - \mathrm{i}A^2_2\right) \\ &- \left(A^1_2 + \mathrm{i}A^2_2\right)\left(A^1_1 - \mathrm{i}A^2_1\right)\Big] \ . \end{aligned} \tag{8.6}$$

In the last step, we have rewritten $F^a_{\mu\nu} F^{a\mu\nu}$ in a way to show that it is useful to replace A^1_μ and A^2_μ by

$$W_\mu = \frac{1}{\sqrt{2}}\left(A^1_\mu + \mathrm{i}A^2_\mu\right) \ , \quad W^+_\mu = \frac{1}{\sqrt{2}}\left(A^1_\mu - \mathrm{i}A^2_\mu\right) \ . \tag{8.7}$$

With these we obtain

$$\begin{aligned} F^a_{\mu\nu}\, F^{a\mu\nu} = &\, 2\left|\left(\nabla_\mu - \mathrm{i}gHx_1\delta_{\mu 2}\right) W_\nu - \left(\nabla_\nu - \mathrm{i}gHx_1\delta_{\nu 2}\right) W_\mu\right|^2 \\ &+ 4\mathrm{i}gH\left(W^+_1 W_2 - W^+_2 W_1\right) + 2H^2 \ . \end{aligned} \tag{8.8}$$

The Lagrange density to be investigated is thus

$$\begin{aligned} L = -\frac{1}{4} F^a_{\mu\nu}\, F^{a\mu\nu} = &-\frac{1}{2}\left|\left(\nabla_\mu - \mathrm{i}gHx_1\delta_{\mu 2}\right) W_\nu - \left(\nabla_\nu - \mathrm{i}gHx_1\delta_{\nu 2}\right) W_\mu\right|^2 \\ &- \mathrm{i}gH\left(W^+_1 W_2 - W^+_2 W_1\right) - \frac{1}{2}H^2 \ , \end{aligned} \tag{8.9}$$

and the resulting equation of motion is

$$\begin{aligned} &-\left(-\nabla^\mu - \mathrm{i}gHx_1\delta_{\mu 2}\right)\left[\left(\nabla_\mu - \mathrm{i}gHx_1\delta_{\mu 2}\right) W_\nu - \left(\nabla_\nu - \mathrm{i}gHx_1\delta_{\nu 2}\right) W_\mu\right] \\ &- \mathrm{i}gH\left(\delta_{\nu 1} W_2 - \delta_{\nu 2} W_1\right) = 0 \ . \end{aligned} \tag{8.10}$$

To make sure that W^+_μ, W_μ are indeed the physical degrees of freedom, it is useful to choose the so-called background gauge:

$$\left(\partial^\mu - \mathrm{i}gHx_1\delta_{\mu 2}\right) W_\mu = 0 \ . \tag{8.11}$$

(For $H \to 0$, this is the usual transverse gauge.)

That this gauge singles out the physical degrees of freedom is formally demonstrated by choosing the gauge (8.11), determining the corresponding ghost fields, and then showing that they cancel the contribution of A^3_μ and those components of W_μ for which $\left(\partial^\mu - \mathrm{i}gHx_1\delta_{\mu 2}\right) W_\mu \neq 0$.

From (8.11) it follows that

$$\begin{aligned}
&\left(\nabla^{\mu}+\mathrm{i}gHx_1\delta_{\mu 2}\right)\left(\nabla_{\mu}-\mathrm{i}gHx_1\delta_{\mu 2}\right)W_{\nu}\\
&\quad=\left(\nabla^{\mu}-\mathrm{i}gHx_1\delta^{\mu 2}\right)\left(\nabla_{\mu}-\mathrm{i}gHx_1\delta_{\mu 2}\right)W_{\nu}\\
&\quad=\left(\nabla_{\nu}-\mathrm{i}gHx_1\delta_{\nu 2}\right)\left(\nabla^{\mu}+\mathrm{i}gHx_1\delta_{\mu 2}\right)W_{\mu}\\
&\qquad-\mathrm{i}gH(-\delta_{\mu 1})\delta_{\nu 2}W_{\mu}-\mathrm{i}gH\delta_{\nu 1}\delta_{\mu 2}W_{\mu}\\
&\quad=-\mathrm{i}gH(W_2\delta_{\nu 1}-W_1\delta_{\nu 2})\ ,
\end{aligned} \tag{8.12}$$

or

$$\left(\nabla^{\mu}-\mathrm{i}gHx_1\delta^{\mu 2}\right)^2 W_{\nu}-2gH\begin{pmatrix}0&0&0&0\\0&0&\mathrm{i}&0\\0&-\mathrm{i}&0&0\\0&0&0&0\end{pmatrix}_{\nu\mu} W_{\mu}=0\ . \tag{8.13}$$

The matrix in (8.13) has the eigenvalues ± 1 and a double eigenvalue at zero. Its eigenvectors are $(0, 1, \mp\mathrm{i}, 0)$, $(0, 0, 0, 1)$, $(1, 0, 0, 0)$. The last is excluded by the gauge condition (8.11), while the others correspond to spin ± 1 and 0. The matrix in (8.13) contains the submatrix

$$(S_3)_{jk}=\mathrm{i}\varepsilon_{3jk}=\begin{pmatrix}0&\mathrm{i}&0\\-\mathrm{i}&0&0\\0&0&0\end{pmatrix}\ , \tag{8.14}$$

i.e., the third component of the spin matrix for spin 1. The factor 2 is analogous to the g factor. Since gluons are massless, $S=0$ is excluded (formally one finds that ghost fields cancel the contribution of $S=0$ gluons). Thus the two physical fields are those with $S=\pm 1$, and they satisfy

$$\begin{aligned}
&\left(\nabla^{\mu}-\mathrm{i}gHx_1\delta^{\mu 2}\right)^2 W_{\nu}-2gH\cdot(\pm 1)W_{\nu}\\
&\quad=\left\{-\frac{\partial^2}{\partial x_0^2}+\frac{\partial^2}{\partial x_1^2}-\left(\mathrm{i}\frac{\partial}{\partial x_2}+gHx_1\right)^2+\frac{\partial^2}{\partial x_3^2}\pm 2gH\right\}W_{\nu}=0\ .
\end{aligned} \tag{8.15}$$

Continuing to the Fourier transform $\tilde{W}_{\nu}(E, x_1, k_2, k_3)$ we have

$$\left(E^2+\frac{\partial^2}{\partial x_1^2}-(k_2+gHx_1)^2-k_3^2\mp 2gH\right)\tilde{W}_{\nu}=0\ , \tag{8.16}$$

and we substitute $x_1=y-k_2/gH$:

$$\left(E^2-k_3^2\mp 2gH+\frac{\partial^2}{\partial y^2}-g^2H^2y^2\right)\tilde{W}_{\nu}(E, y, k_2, k_3)=0\ . \tag{8.17}$$

The last two terms correspond to the equation of the harmonic oscillator:

$$\omega\Psi(y)=-\left(\frac{1}{2}\frac{\partial^2}{\partial y^2}-\frac{1}{2}g^2H^2y^2\right)\Psi(y) \tag{8.18}$$

with energy eigenvalues

$$\omega_n = \left(n + \frac{1}{2}\right) gH \ . \tag{8.19}$$

With the ansatz

$$\tilde{W}_\nu = \Psi(y)\,\phi(E, k_2, k_3) \tag{8.20}$$

it now follows that

$$\left(E^2 - k_3^2 \mp 2gH - 2\omega_n\right) \tilde{W}_\nu = 0 \tag{8.21}$$

$$\Rightarrow \quad E = \sqrt{2gH\left(n + \frac{1}{2} \pm 1\right) + k_3^2} = E' + \mathrm{i}\Gamma \ . \tag{8.22}$$

Note that E becomes imaginary for $n = 0$ and $k_3^2 < gH$. Since an imaginary energy corresponds to a decay width, this state is unstable (see Fig. 8.2).

The appearance of an unstable mode indicates that the model assumptions of a homogeneous color magnetic field are unrealistic. In spite of this, we shall investigate what result would be obtained if this state were not unstable, since this will help us to understand the role of the unstable mode.

From the single-particle energies (8.22) follows the energy density

$$\varepsilon = 2\,\frac{1}{2}\,C \int \frac{\mathrm{d}k_3}{2\pi} \left(\sum_{n=0}^{\infty} \sqrt{2gH\left(n - \frac{1}{2}\right) + k_3^2} + \sum_{n=0}^{\infty} \sqrt{2gH\left(n + \frac{3}{2}\right) + k_3^2} \right) . \tag{8.23}$$

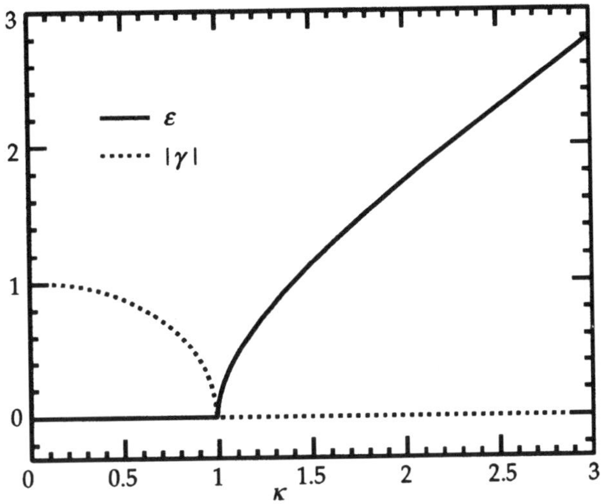

Fig. 8.2. The unstable mode, with $\varepsilon = E_1'/\sqrt{gH}$, $\gamma = \Gamma/\sqrt{gH}$, and $\kappa = k_3/\sqrt{gH}$. For $\kappa < 1$ the energy is zero and the state decays or grows exponentially depending on the sign of γ

The factor 2 stands for the two degrees of freedom W_μ and W_μ^+; the factor 1/2 appears because we are calculating the vacuum energy. The factor C comes from changing $\int \mathrm{d}k_1\,\mathrm{d}k_2$ to $\sum_{n=0}^{\infty}$. Since for the nth Landau state it holds that

$$k_1^2+k_2^2 \sim ngH \ , \tag{8.24}$$

we expect that

$$\frac{\int \mathrm{d}k_1\,\mathrm{d}k_2}{(2\pi)^2} \Rightarrow \frac{\int k\,\mathrm{d}k}{2\pi} \Rightarrow \frac{gH}{4\pi}\sum_n \ . \tag{8.25}$$

A precise calculation yields

$$C = \frac{gH}{2\pi} \ , \tag{8.26}$$

$$\begin{aligned}\varepsilon - \varepsilon(H=0) = {} & \frac{gH}{2\pi}\int \frac{\mathrm{d}k_3}{2\pi}\left[\sum_{n=1}^{\infty}\sqrt{2gH\left(n-\frac{1}{2}\right)+k_3^2}\right.\\ & \left. + \sum_{n=0}^{\infty}\sqrt{2gH\left(n+\frac{3}{2}\right)+k_3^2}\right]\\ & -2\times 2\times\frac{1}{2}\int\frac{\mathrm{d}^3k}{(2\pi)^3}\sqrt{k^2}+\frac{gH}{2\pi}\int\frac{\mathrm{d}k_3}{2\pi}\sqrt{k_3^2-gH} \ . \end{aligned} \tag{8.27}$$

The factors in front of the last-but-one term are spin degrees of freedom × color degrees of freedom $\times\frac{1}{2}$. To perform the sums we introduce parameters δ and η:

$$\begin{aligned}\varepsilon - \varepsilon(H=0) = {} & \lim_{\eta,\delta\to 0}\left[\frac{gH}{\pi}\int_0^\infty\frac{\mathrm{d}k_3}{2\pi}\left\{\sum_{n=1}^{\infty}\left[2gH\left(n-\frac{1}{2}\right)+k_3^2-\mathrm{i}\delta\right]^{1/2+\eta}\right.\right.\\ & \left. + \left[2gH\left(n+\frac{3}{2}\right)+k_3^2-\mathrm{i}\delta\right]^{1/2-\eta}\right\}\\ & \left. -\frac{1}{\pi^2}\int_0^\infty \mathrm{d}k\,k^2\left(k^2-\mathrm{i}\delta\right)^{1/2-\eta}+\frac{gH}{2\pi^2}\int_0^\infty \mathrm{d}k_3(k_3^2-gH-\mathrm{i}\delta)^{1/2-\eta}\right]\end{aligned} \tag{8.28}$$

and use

$$(\omega^2 - \mathrm{i}\delta)^{\nu-\mu} = \frac{\mathrm{i}^{-\nu+\eta}}{\Gamma(-\nu+\eta)} \int\limits_0^\infty \mathrm{d}\tau\; \tau^{-\nu+\eta-1}\; \mathrm{e}^{-\mathrm{i}\tau(\omega^2-\mathrm{i}\delta)} \tag{8.29}$$

$$\begin{aligned}\varepsilon - \varepsilon(H=0) = \lim_{\eta,\delta\to 0} \Bigg[& \frac{gH}{2\pi^2} \frac{\mathrm{i}^{-1/2+\eta}}{\Gamma\left(-\frac{1}{2}+\eta\right)} \int\limits_0^\infty \mathrm{d}k \int\limits_0^\infty \mathrm{d}\tau\tau^{-3/2+\eta} \\ & \times \mathrm{e}^{-\mathrm{i}\tau(k^2-\mathrm{i}\delta)} \left\{ \sum_{n=1}^\infty \mathrm{e}^{-\mathrm{i}\tau 2gH\left(n-\frac{1}{2}\right)} + \sum_{n=0}^\infty \mathrm{e}^{-\mathrm{i}\tau 2gH\left(n+\frac{3}{2}\right)} \right\} \\ & - \frac{\mathrm{i}^{-\frac{1}{2}+\eta}}{\Gamma\left(-\frac{1}{2}+\eta\right)} \frac{1}{\pi^2} \int\limits_0^\infty \mathrm{d}k \int\limits_0^\infty \mathrm{d}\tau k^2 \tau^{-\frac{3}{2}+\eta}\; \mathrm{e}^{-\mathrm{i}\tau(k^2-\mathrm{i}\delta)} \\ & + \frac{gH\mathrm{i}^{-\frac{1}{2}+\eta}}{2\pi^2\Gamma\left(-\frac{1}{2}+\eta\right)} \int\limits_0^\infty \mathrm{d}k \int\limits_0^\infty \mathrm{d}\tau\; \tau^{-\frac{3}{2}+\eta}\; \mathrm{e}^{-\mathrm{i}\tau(k^2-\mathrm{i}\delta-gH)} \Bigg] . \end{aligned} \tag{8.30}$$

The calculation of the integrals is demonstrated in Exercise 8.1. The result is

$$\varepsilon - \varepsilon(H=0) = \frac{11(gH)^2}{48\pi^2} \ln \frac{gH}{\mu^2} - \mathrm{i}\frac{(gH)^2}{8\pi} . \tag{8.31}$$

EXERCISE

8.1 The QCD Vacuum Energy Density

Problem. Derive (8.31) from (8.30).

Solution. The sums in (8.30) can be directly evaluated:

$$\begin{aligned} & \sum_{n=1}^\infty \mathrm{e}^{-\mathrm{i}\tau 2gH\left(n-\frac{1}{2}\right)} + \sum_{n=0}^\infty \mathrm{e}^{-\mathrm{i}\tau 2gH\left(n+\frac{3}{2}\right)} \\ & = \frac{\mathrm{e}^{\mathrm{i}\tau gH}}{1-\mathrm{e}^{-\mathrm{i}\tau 2gH}} - \mathrm{e}^{\mathrm{i}\tau gH} + \frac{\mathrm{e}^{-3\mathrm{i}\tau gH}}{1-\mathrm{e}^{-2\mathrm{i}\tau gH}} \\ & = \frac{\mathrm{e}^{2\mathrm{i}\tau gH} + \mathrm{e}^{-2\mathrm{i}\tau gH}}{\mathrm{e}^{\mathrm{i}\tau gH} - \mathrm{e}^{-\mathrm{i}\tau gH}} - \mathrm{e}^{\mathrm{i}\tau gH} = \frac{\cos(2gH\tau)}{\mathrm{i}\sin(gH\tau)} - \mathrm{e}^{\mathrm{i}\tau gH} . \end{aligned} \tag{1}$$

Exercise 8.1

Also the k integrations can be directly performed:

$$\int_0^\infty \mathrm{d}k\, \mathrm{e}^{-\mathrm{i}\tau k^2} = \frac{1}{2}\sqrt{\frac{\pi}{\tau}}\, \mathrm{e}^{-\mathrm{i}\frac{\pi}{4}} \;,$$

$$\int_0^\infty \mathrm{d}k\, k^2\, \mathrm{e}^{-\mathrm{i}\tau k^2} = \mathrm{i}\frac{\partial}{\partial\tau}\int_0^\infty \mathrm{d}k\, \mathrm{e}^{-\mathrm{i}\tau k^2} = -\frac{\mathrm{i}}{4}\frac{\sqrt{\pi}}{\tau^{\frac{3}{2}}}\, \mathrm{e}^{-\mathrm{i}\frac{\pi}{4}} \;. \tag{2}$$

Therefore we get

$$\begin{aligned}
&\varepsilon - \varepsilon(H=0)\\
&= \lim_{\eta,\delta\to 0}\left[\frac{gH}{4\pi^{\frac{3}{2}}}\frac{\mathrm{e}^{-\mathrm{i}\frac{\pi}{2}+\mathrm{i}\frac{\pi}{2}\eta}}{\Gamma\left(-\frac{1}{2}+\eta\right)}\int_0^\infty \mathrm{d}\tau\, \tau^{-2+\eta}\mathrm{e}^{-\delta\tau}\left(\frac{\cos(2gH\tau)}{\mathrm{i}\sin(gH\tau)} - \mathrm{e}^{\mathrm{i}\tau gH}\right)\right.\\
&\quad + \frac{\mathrm{i}}{4}\frac{\mathrm{e}^{-\mathrm{i}\frac{\pi}{2}+\mathrm{i}\frac{\pi}{2}\eta}}{\pi^{\frac{3}{2}}\Gamma\left(-\frac{1}{2}+\eta\right)}\int_0^\infty \mathrm{d}\tau\, \tau^{-3+\eta}\mathrm{e}^{-\delta\tau}\\
&\quad \left. + \frac{gH}{4}\frac{\mathrm{e}^{-\mathrm{i}\frac{\pi}{2}+\mathrm{i}\frac{\pi}{2}\eta}}{\pi^{\frac{3}{2}}\Gamma\left(-\frac{1}{2}+\eta\right)}\int_0^\infty \mathrm{d}\tau\, \tau^{-2+\eta}\mathrm{e}^{\mathrm{i}\tau gH-\delta\tau}\right] \;.
\end{aligned} \tag{3}$$

Now we employ the relation

$$\cos(2gH\tau) = 1 - 2\sin^2(gH\tau) \;, \tag{4}$$

which yields

$$\begin{aligned}
&\frac{\cos(2gH\tau)}{\mathrm{i}\sin(gH\tau)} - \mathrm{e}^{\mathrm{i}\tau gH}\\
&\quad = \frac{1}{\mathrm{i}\sin(gH\tau)} - \frac{2}{\mathrm{i}}\frac{1}{2\mathrm{i}}\left(\mathrm{e}^{\mathrm{i}\tau gH} - \mathrm{e}^{-\mathrm{i}\tau gH}\right) - \mathrm{e}^{\mathrm{i}\tau gH}\\
&\quad = \frac{1}{\mathrm{i}\sin(gH\tau)} - \mathrm{e}^{-\mathrm{i}\tau gH}\\
&\quad \to \frac{1}{\mathrm{i}\sin(gH\tau - \mathrm{i}\tilde{\eta})} - \mathrm{e}^{-\mathrm{i}\tau gH} \;.
\end{aligned} \tag{5}$$

Here we have defined the singularities by inserting $\mathrm{i}\tilde{\eta}$ ($\tilde{\eta} > 0$). The integral exists provided $\eta > 2$. We calculate it in this region and determine its value for $\eta \to 0$ by analytic continuation. This very procedure has already been used in dimensional regularization. Identity (5) ensures that the integrand vanishes for $\tau = R\,\mathrm{e}^{\mathrm{i}\phi}$, $-\frac{\pi}{2} < \phi < 0\,, R \to \infty$. Therefore the integral surrounding the fourth quadrant vanishes:

$$\int \mathrm{d}\tau \cdots = \int_0^\infty \mathrm{d}\tau \cdots + \int_{-\mathrm{i}\infty}^0 \mathrm{d}\tau \cdots = 0 \;, \tag{6}$$

Exercise 8.1

Since the integrand has no singularities, we substitute $\tau = -\mathrm{i}s$ and evaluate the last integral with the help of (8.29):

$$\begin{aligned}
&\varepsilon - \varepsilon(H=0) \\
&= \lim_{\eta,\delta\to 0} \left[\frac{gH}{4\pi^{\frac{3}{2}}} \frac{-\mathrm{i}\,\mathrm{e}^{\mathrm{i}\frac{\pi}{2}\eta}}{\Gamma\left(-\frac{1}{2}+\eta\right)} (-\mathrm{i})^{-1+\eta} \right. \\
&\quad \times \int_0^\infty \mathrm{d}s\, s^{-2+\eta}\, \mathrm{e}^{-\delta r} \left(\frac{1}{\sinh(gHs)} - \mathrm{e}^{-sgH} \right) \\
&\quad + \frac{\mathrm{e}^{\mathrm{i}\frac{\pi}{2}\eta}(-\mathrm{i})^{-2+\eta}}{4\pi^{\frac{3}{2}}\Gamma\left(-\frac{1}{2}+\eta\right)} \int_0^\infty \mathrm{d}s\, s^{-3+\eta} \frac{gH}{4} \frac{\mathrm{e}^{-\mathrm{i}\frac{\pi}{2}+\mathrm{i}\frac{\pi}{2}\eta}}{\pi^{\frac{3}{2}}\Gamma\left(-\frac{1}{2}+\eta\right)} \\
&\quad \left. \times \frac{(-1+\eta)}{\mathrm{e}^{\mathrm{i}\frac{\pi}{2}(-1+\eta)}} (-gH+\mathrm{i}\delta)^{1-\eta} \right] .
\end{aligned} \tag{7}$$

Finally we substitute $s = v/gH$:

$$\begin{aligned}
\varepsilon - \varepsilon(H=0) &= \lim_{\eta\to 0} \frac{(gH)^{2-\eta}}{4\pi^{\frac{3}{2}}} \frac{1}{\Gamma\left(-\frac{1}{2}+\eta\right)} \\
&\quad \times \left[\int_0^\infty \mathrm{d}v\, v^{-2+\eta} \left(\frac{1}{\sinh(v)} - \mathrm{e}^{-v} - \frac{1}{v} \right) - (-1+\eta)(-)^n \right] .
\end{aligned} \tag{8}$$

The integral of e^{-v} leads to a gamma function:

$$\int_0^\infty \mathrm{d}v\, v^{-2+\eta}\, \mathrm{e}^{-v} = \Gamma(-1+\eta) , \tag{9}$$

while the integral over $1/\sinh v - 1/v$ can be split into a finite part and one that diverges in the limit $\eta \to 0$:

$$\begin{aligned}
&\int_0^\infty \mathrm{d}v\, v^{-2+\eta} \left(\frac{1}{\sinh(v)} - \frac{1}{v} \right) \\
&\quad = \int_0^\infty \mathrm{d}v\, v^{-2+\eta} \left(\frac{1}{\sinh(v)} - \frac{1}{v} + \frac{1}{6}v \right) - \frac{1}{6} \int_0^\infty \mathrm{d}v\, v^{-1+\eta} \\
&\quad = C - \frac{1}{6}\frac{1}{\eta} ,
\end{aligned} \tag{10}$$

Exercise 8.1

$$\varepsilon-\varepsilon(H=0)=\lim_{\eta\to 0}\frac{(gH)^{2-\eta}}{4\pi^{\frac{3}{2}}\Gamma\left(-\frac{1}{2}+\eta\right)} \times\left\{C-\frac{1}{6}\frac{1}{\eta}-\Gamma[(-1+\eta)\left(1+(-)^{\eta}\right)]\right\} . \tag{11}$$

With

$$\Gamma(-1+\eta)=\frac{1}{-1+\eta}\Gamma(\eta)=\frac{\Gamma(1+\eta)}{(-1+\eta)\eta} \tag{12}$$

we are now able to calculate the limiting case $\eta\to 0$:

$$\begin{aligned}
(gH)^{2-\eta}&=(gH)^2(1-\eta\ln(gH)+\cdots) ,\\
\Gamma\left(-\frac{1}{2}+\eta\right)&=\Gamma\left(-\frac{1}{2}\right)\left[1+\eta\Psi\left(-\frac{1}{2}\right)\right]+\cdots\\
&=-2\sqrt{\pi}\left[1+\eta\Psi\left(-\frac{1}{2}\right)\right]+\cdots ,\\
\Gamma(-1+\eta)&=\frac{-1-\eta}{\eta}[1+\eta\Psi(1)]+\cdots ,\\
(-)^{\eta}&=1+\eta\mathrm{i}\pi ,
\end{aligned} \tag{13}$$

$$\Rightarrow \varepsilon-\varepsilon(H=0)=\lim_{\eta\to 0}\left\{\frac{(gH)^2}{8\pi^2\eta}+\frac{(gH)^2}{8\pi^2}\left[\frac{11}{6}\ln(gH)+C'-\mathrm{i}\pi\right]\right\} . \tag{14}$$

Here all constants have been absorbed into C', which does not depend on H. The divergent first part can be renormalized. The renormalized, i.e., the physical, energy density is then

$$\varepsilon-\varepsilon(H=0)=\frac{11}{6}\frac{(gH)^2}{8\pi^2}\ln(gH)+C'-\mathrm{i}\frac{(gH)^2}{8\pi} . \tag{15}$$

In the transition to (8.28), however, we should have introduced a factor $m^{2\eta}$ in order to conserve the dimension (m is supposed to be an energy). Then we obtain

$$\frac{11(gH)^2}{48\pi^2}\left[\ln\left(\frac{gH}{m^2}\right)+\frac{6}{\pi}C'\right]=\frac{11(gH)^2}{48\pi^2}\ln\left(\frac{gH}{\mu^2}\right) , \tag{16}$$

where C' has been absorbed by the definition of μ

$$\mu^2=m^2\mathrm{e}^{-6C'/11} , \tag{17}$$

$$\varepsilon-\varepsilon(H=0)=\frac{11(gH)^2}{48\pi^2}\ln\left(\frac{gH}{\mu^2}\right)-\mathrm{i}\frac{(gH)^2}{8\pi} . \tag{18}$$

It should be noticed that the techniques used in this exercise to calculate (8.31) are practically the same as those introduced systematically in Sect. 4.3 in the context of dimensional regularization.

The presence of the imaginary part is again an indication of the instability. If we minimize the real part of the vacuum energy $H^2/2$ (see the last term in (8.9)), we obtain a minimum at $H \neq 0$:

$$\frac{\partial}{\partial H}\left[\frac{1}{2}H^2 + \frac{11(gH)^2}{48\pi^2}\ln\left(\frac{gH}{\mu^2}\right)\right] = H + \frac{11g^2H}{24\pi^2}\left[\ln\left(\frac{gH}{\mu^2}\right)+\frac{1}{2}\right] = 0$$

$$\Rightarrow \quad gH = \mu^2 \exp\left[-\left(\frac{24\pi^2}{11g^2}+\frac{1}{2}\right)\right] , \tag{8.32}$$

which does not fix the value of gH since μ^2 was arbitrary up to now. As we have effectively performed a dimensional regularization, μ^2 is proportional to the renormalization scale. Indeed, the real part of $\varepsilon - \varepsilon(H=0)$ can be deduced from renormalization group properties in a very elegant fashion. The prefactor then comes from the β function:

$$\frac{1}{2g}\beta \longleftrightarrow \frac{g^2}{32\pi^2}\frac{11\times 2}{3} = \frac{11g^2}{48\pi^2} \qquad \text{for SU(2)} . \tag{8.33}$$

We demonstrate this in Example 8.2 to show the strength of renormalization group arguments. However, this example will also show the weakness of these techniques, since one tacitly uses a number of assumptions without being able to check their validity. So the renormalization group treatment gives no indication of the existence of unstable states and predicts a constant color magnetic field; but the presence of unstable modes shows that this is not a physical solution.

In order to have any hope of obtaining a realistic ground state a consistent treatment of the unstable modes has to be developed. We shall not perform these calculations explicitly but only illustrate the ideas. There is a well-known method of treating unstable modes from the problem of spontaneous symmetry breaking. Here, too, the usual vacuum modes are not stable, and the field drifts into a finite vacuum expectation value. If the Higgs field is expanded instead around this vacuum expectation value, only stable modes are seen. By analogy to this, one is led to the following procedure for treating unstable modes.

1. Isolate the unstable modes and rewrite the Lagrange density to have them appear in the same way as Higgs fields.
2. Insert nonvanishing vacuum expectation values and determine the energetically optimal gauge field configuration.

The first step can indeed be performed. The second step is very difficult and only possible in the framework of certain ansätze. We therefore show just one of the results (see Fig. 8.3).[1] Other ansätze yield slightly different results but a domain structure at a length scale $\sqrt{gH_0}$ in the xy plane is always found, and compensating positive and negative fields in H in large spatial regions. One can immediately imagine why no unstable modes appear in these solutions.

[1] See J. Ambjørn and P. Olesen: Nucl. Phys. B **170**, 60 (1980).

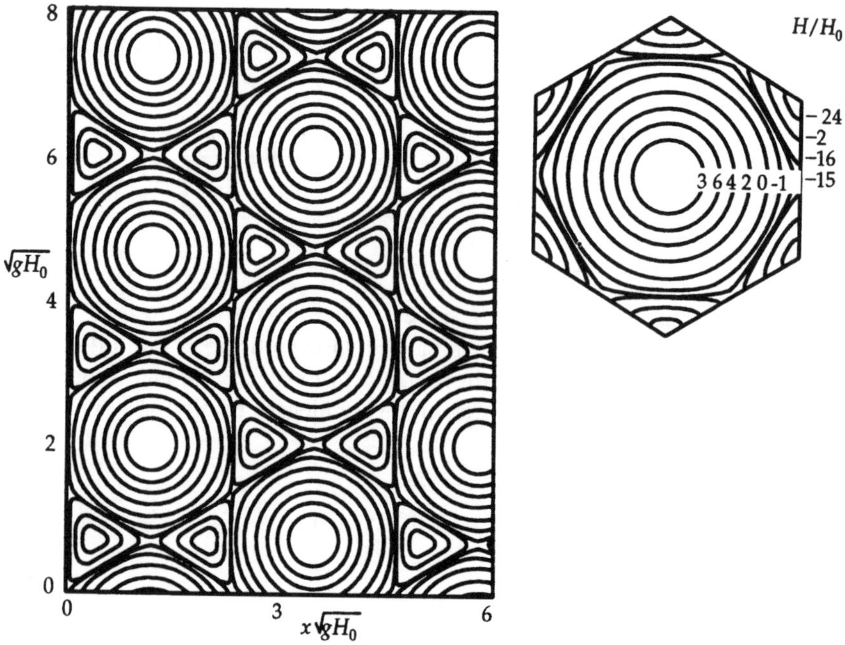

Fig. 8.3. The Ambjørn–Olesen solution for the QCD ground state. The H field is parallel to the z axis. Contour lines are at $0.8H_0$, $0.6H_0, 0.4H_0, 0.2H_0, 0.0H_0$, $-0.1H_0$, $-0.15H_0$, $-0.16H_0$, $-0.2H_0$, $-0.24H_0$ (J. Ambjørn and P. Olesen: Nucl. Phys. B **170**, 60 (1980))

The lowest Landau state ($n = 0$) extends over large spatial domains. Therefore the H fields average out and the term $\pm 2g\langle H\rangle \approx 0$ no longer leads to instabilities. The higher Landau states are localized in the xy plane up to $\sqrt{gM_0}/n$ and thus experience a more or less constant H field, which, as has just been calculated, leads to a lowering of the energy. Figure 8.3 shows a system of parallel tubes made of color magnetic fields that are completely analogous to vortices in type II superconductors. This parallel supports the picture of the QCD vacuum as a dual superconductor. Figure 8.3 has been calculated as a solution of the classical field equations. In quantum mechanics, fields will oscillate around this configuration. In particular, the magnetic flux tubes will no longer be strictly parallel but change their orientation over large spatial regions. This property

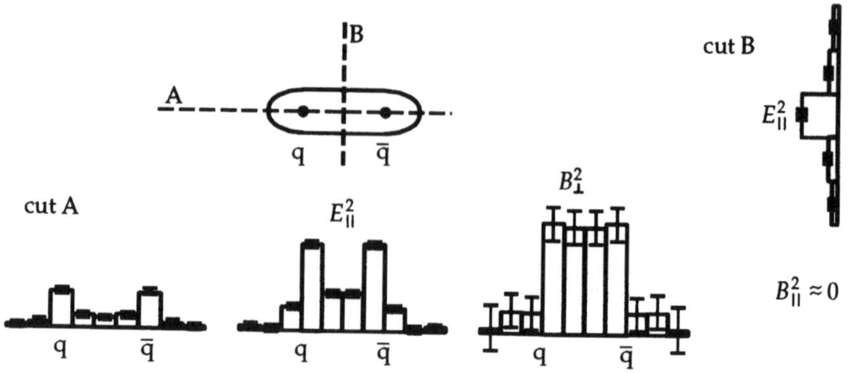

Fig. 8.4. The result of a Monte Carlo calculation for color electric and color magnetic fields around a static quark–antiquark pair. || denotes the direction parallel to the A axis. (From J.W. Flower and S.W. Otto: Phys. Lett. B **160**, 128 (1985).) The color magnetic fields are enlarged by a factor of 10

caused the model to be baptized the "Spaghetti vacuum" and guarantees Lorentz invariance after averaging over sufficiently large spatial domains.

The picture of a dual superconductor is further supported by lattice calculations for some simple systems. One can, for example, calculate the field distribution around a static quark–antiquark pair. If the dual-superconductor picture is correct, the color electric field cannot extend into the QCD vacuum, so that a string is created outside of which $\boldsymbol{E}$ vanishes. Figure 8.4 shows this schematically as well as the results of a Monte Carlo calculation.

In conclusion there are detailed and quasiphenomenological models for the QCD ground state. The "Spaghetti vacuum" is just one of them. All have in common that they lead to highly complicated nonperturbative field configurations and that these have a constant negative energy density compared to the perturbative vacuum. This energy density cannot simply be identified with the bag constant. The fine structure of the QCD ground state should exhibit a length scale of $1/\lambda_{\rm QCD} \approx 1$ fm. Thus the average energy density is not relevant for a hadron; and the precise microscopic structure must be known.

EXAMPLE

8.2 The QCD Ground State and the Renormalization Group

In this example we wish to show how (8.31) (or more accurately the real part of (8.31)) can be derived from renormalization arguments in a very elegant manner. We want to investigate the effective Lagrangian of an SU(2) gauge theory for a constant color magnetic field. In lowest order this is simply $-1/2H^2$. Consequently the next order can be written as

$$L_{\rm eff} = F(H, \mu, g) \; , \tag{1}$$

where μ denotes the renormalization in such a way that $L_{\rm eff}$ becomes the free Lagrangian for $H = \mu^2$. Since the effective Lagrangian must not depend on μ, the renormalization group equation

$$\left[\mu^2 \frac{\partial}{\partial \mu^2} + \beta(g)\frac{\partial}{\partial g} + 2\gamma(g) H^2 \frac{\partial}{\partial H^2}\right] F(H, \mu, \gamma) = 0 \tag{2}$$

must be fulfilled. The derivative of (2) with respect to H^2 is

$$\left[\mu^2 \frac{\partial}{\partial \mu^2} + \beta(g)\frac{\partial}{\partial g} + 2\gamma(g) + 2\gamma(g) H^2 \frac{\partial}{\partial H^2}\right] \frac{\partial}{\partial H^2} F(H, \mu, g) = 0 \; . \tag{3}$$

Since $\partial F/\partial H^2$ is a dimensionless quantity, it can only depend on H/μ^2. Therefore we define

$$t = \ln \frac{H}{\mu^2} \; , \quad \frac{\partial}{\partial H^2} F(H, \mu, g) = G(t, g) \; , \tag{4}$$

$$\left[-(1-\gamma)\frac{\partial}{\partial t} + \beta \frac{\partial}{\partial g} + 2\gamma\right] G(t, g) = 0 \; . \tag{5}$$

Example 8.2

This is the typical form of a renormalization group equation. Taking into account the boundary condition

$$G\big|_{H=\mu^2} = G\big|_{t=0} = -\frac{1}{2} \tag{6}$$

yields

$$G(g,t) = \exp\left(2\int_0^t \bar{\gamma}(g,x)\,\mathrm{d}x\right) G'(g,t)\ , \tag{7}$$

$$G'(g,0) = -\frac{1}{2}\ , \tag{8}$$

and

$$\left[-(1-\gamma)\frac{\partial}{\partial t} + \beta\frac{\partial}{\partial g}\right] G'(g,t) = 0\ . \tag{9}$$

We assign the new name g_0 to the constant g, because we want to introduce a function $g(t, g_0)$ in the following:

$$G(g_0,t) = \exp\left(2\int_0^t \bar{\gamma}(g_0)\,\mathrm{d}x\right) G'(g_0,t)\ , \tag{10}$$

$$G'(G_0,0) = -\frac{1}{2}\ , \tag{11}$$

and

$$\left[-(1-\gamma)\frac{\partial}{\partial t} + \beta\frac{\partial}{\partial g_0}\right] G'(g_0,t) = 0\ . \tag{12}$$

Clearly (12) is solved by every function $G'\big(g(g_0,t)\big)$ if

$$\frac{\partial}{\partial t} g(g_0,t) = \frac{\beta}{1-\gamma}\frac{\partial}{\partial g_0} g(g_0,t)\ . \tag{13}$$

It is always possible to define $G'\big(g(g_0,t)\big)$ and $g(g_0,t)$ in such a way that $\partial g(g_0,t)/\partial g_0 = 1$ holds, such that (13) simplifies to

$$\frac{\partial}{\partial t} g(g_0,t) = \frac{\beta}{1-\gamma}\ . \tag{14}$$

Now (7) is evaluated by perturbative methods. In the case of small t we have $G'(g) \approx -1/2$, and only the result for the *anomalous dimension* γ in the limit of $t \to 0$ is needed. Since we have not discussed the renormalization of gauge theories in this volume, we can only cite the result. It turns out for a pure SU(2) gauge theory that $\overline{\gamma}$ and β are in lowest order proportional to each other:

$$\begin{aligned}\overline{\gamma} &= \frac{\beta}{2g_0} = \frac{\mu^2}{g_0}\frac{\partial g}{\partial \mu^2}\bigg|_{\mu^2=H} \\ &= -\frac{1}{g_0}\frac{\partial g}{\partial t}\bigg|_{t\to 0} = -\frac{\partial \ln g}{\partial t}\bigg|_{t\to 0}\ .\end{aligned} \tag{15}$$

Example 8.2

This relation allows us to evaluate (7):

$$G \approx \frac{1}{2}\exp\left(-2\{\ln[g(g_0,t)]-\ln(g_0)\}\right) = -\frac{1}{2}\left(\frac{g(g_0,t)}{g_0}\right)^{-2}$$

$$\approx -\frac{1}{2}\left(1+\frac{\beta}{g_0}t\right) = -\frac{1}{2}-\frac{1}{2}\frac{g_0^2}{(4\pi)^2}\frac{11\times 2}{3}t\,, \qquad (16)$$

$$G \approx -\frac{1}{2}-\frac{11g_0^2}{48\pi^2}\ln\left(\frac{H}{\mu^2}\right)+O(g^4)\,. \qquad (17)$$

Therefore F assumes the form

$$F = -\frac{11g^2H^2}{48\pi^2}\left[\ln\left(\frac{H}{\mu^2}\right)-\frac{1}{2}\right] = -\frac{11g^2H^2}{48\pi^2}\ln\left(\frac{H}{H_0}\right)\,. \qquad (18)$$

This derivation was first found by Savvidy.[2] Later Nielsen and Olesen discovered in a lengthier calculation the nonstable states, which are overlooked by the renormalization group method. This indicates that one has to be very cautious when employing such abstract principles.

Without color electric fields, the energy density is

$$\varepsilon-\varepsilon(H=0) = -F = \frac{11g^2H^2}{48\pi^2}\ln\frac{H}{H_0}\,.$$

EXAMPLE

8.3 The QGP as a Free Gas

In order to allow for simple calculations the QGP is usually described as a free gas consisting of quarks and gluons. As we have already discussed, this is theoretically not well founded at $T \approx T_c$. However, those calculations frequently yield results which are qualitatively correct. Thus we simply add the gas pressures of a free gluon gas

$$p_G = 16\times\frac{\pi^2}{90}T^4 \qquad (1)$$

and of a free quark gas

$$p_Q = 12\times\frac{T^4}{12}\left[\frac{7\pi^2}{30}+\left(\frac{\mu}{T}\right)^2+\frac{1}{2\pi^2}\left(\frac{\mu}{T}\right)^4\right] \qquad (2)$$

[2] See S.G. Matinyan and G.K. Savvidy: Yad. Fiz. **25**, 218 (1977).

Example 8.3

and identify the result with the bag pressure B. In the case $p_G + p_Q > B$ the QGP region is supposed to expand and one should be able to derive the critical temperature from

$$B = p_G + p_Q = T_{cr}^4 \left[\frac{37\pi^2}{90} + \left(\frac{\mu}{T_{cr}}\right)^2 + \frac{1}{2\pi^2}\left(\frac{\mu}{T_{cr}}\right)^4 \right] \tag{3}$$

for every value of μ. Assuming $B = (145\,\text{MeV})^4$ yields the result shown in Fig. 8.9. This now has to be compared with Fig. 8.5. Apparenty there is a rough qualitative but no quantitative agreement for the phase boundaries. One has

$$\frac{T(\mu = 0)}{T\left(\mu = \frac{4}{3}T_{\max}\right)} = \begin{cases} \frac{155}{113} = 1.4 & \text{for lattice calculations} \\ \frac{102\,\text{MeV}}{91\,\text{MeV}} = 1.1 & \text{for the free gas}\ . \end{cases} \tag{4}$$

Since the position of the phase boundaries is quite insensitive to theoretical subtleties, one can expect that the free gas treatment of QCD leads for more sensitive quantities, to results which are wrong by much more than 30%. Examples of such quantities are the total number of kaons or lambda particles created.

8.2 The Quark–Gluon Plasma

In Sect. 7.1.16 we discussed that lattice calculations show a phase transition at a critical temperature $T_c \sim 100 - 200\,\text{MeV}$ (Figs. 7.10 and 7.12). Such a phase transition is typical for non-Abelian gauge theories. It has been studied extensively for SU(2) and SU(3) but should exist for all SU(N) groups. If it could be experimentally investigated in detail, such studies would definitely improve our understanding of some of the basic properties of QCD. The hope is that in high-energy collisions of heavy ions this new high-temperature phase can indeed be produced for sufficiently long times and in a sufficiently large volume to allow experimental studies. While it is generally agreed that at high enough energies the new phase will be reached in the center of the collision system, the interpretation of possible experimental signals is still very much debated.

The interest in the quark–gluon plasma (QGP) phase transition is further increased by the fact that it is assumed to have played a crucial role in the early universe. As the universe cooled it was the last phase transition to occur and might therefore have left recognizable traces in the present day structure of the universe.

Another interesting point is that the QCD vacuum is a highly nontrivial state, as is the vacuum of the standard model in general. In fact, the existence of a complex vacuum state postulated by modern field theory with an energy density tens of orders of magnitude larger than any observable energy densities is probably one of the most interesting and least-tested features of modern particle theory. To understand some of the problems, let us return to the concept of parton–hadron duality discussed in Example 5.10. We have argued that the description of hadronic reactions is in principle possible on the quark level as well as on the hadron level. These two descriptions are just based on a different set of basis states. Some processes like deep inelastic scattering can be described very easily on the quark level but are extremely hard to treat on the hadron level and such processes therefore allow direct tests of QCD. On the other hand, properties which are easily described in terms of hadrons and their interactions are usually not a good test of QCD. Let us therefore start our detailed discussion by reviewing what hadron models can tell us about the high-temperature phase transition.

As early as in 1965, Hagedorn[3] observed that the experimentally known mass density of hadron states grows exponentially,

$$\frac{\mathrm{d}\varrho}{\mathrm{d}m} \sim m^a\, \mathrm{e}^{m/m_0} \;, \qquad m_0 \approx 200\,\mathrm{MeV} \;, \tag{8.34}$$

and that this fact implies a limiting temperature. With $p^2\,\mathrm{d}p = p\,E\,\mathrm{d}E$, the number of states with energy between E and $E+\mathrm{d}E$ can be written as

$$\mathrm{d}n(E) \sim \mathrm{d}E \int\limits_0^E \mathrm{d}m\, \frac{\mathrm{d}\varrho}{\mathrm{d}m}\, \mathrm{e}^{-E/kT}\; p\,E \;, \qquad p = \sqrt{E^2-m^2} \;, \tag{8.35}$$

implying for (8.34) that

$$\begin{aligned} \mathrm{d}n(E) &\sim \mathrm{d}E \int\limits_0^E m^a\, \mathrm{e}^{m/m_0}\, \mathrm{e}^{-E/kT} \sqrt{E^2-m^2}\; E\, \mathrm{d}m \\ &= \mathrm{d}E\; E^{a+3} \int\limits_0^1 z^a \sqrt{1-z^2}\; \mathrm{e}^{Ez/m_0}\, \mathrm{d}z\; \mathrm{e}^{-E/kT} \;, \end{aligned} \tag{8.36}$$

with $m = Ez$. Substituting $z = \cos(\varphi)$ we get

$$\mathrm{d}E\; E^{a+3}\, \mathrm{e}^{-E/kT} \int\limits_0^{\pi/2} \cos^a(\varphi)\, \sin^2(\varphi)\; \mathrm{e}^{E\cos(\varphi)/m_0}\, \mathrm{d}\varphi \;. \tag{8.37}$$

[3] See Hagedorn: Suppl. Nuov. Cim. **3**, 147 (1965).

Now we assume $E/m_0 \gg 1$. (This is the relevant limit for our discussion.) We can then approximate

$$\int_0^{\pi/2} \cos^a(\varphi)\sin^2(\varphi)\ \mathrm{e}^{E\cos(\varphi)/m_0}\ \mathrm{d}\varphi \approx \int_0^{\pi} \cos^a(\varphi)\sin^2(\varphi)\ \mathrm{e}^{E\cos(\varphi)/m_0}\ \mathrm{d}\varphi$$

$$\approx \int_0^{\pi} \sin^2(\varphi)\ \mathrm{e}^{E\cos(\varphi)/m_0}\ \mathrm{d}\varphi = \sqrt{\pi}\ \frac{2m_0}{E}\ \Gamma\left(\frac{3}{2}\right)\ I_1\left(\frac{E}{m_0}\right)$$

$$\approx \sqrt{\pi}\ \frac{2m_0}{E}\ \frac{\sqrt{\pi}}{2}\ \frac{\mathrm{e}^{E/m_0}}{\sqrt{2\pi\ E/m_0}} = \sqrt{\frac{\pi m_0^3}{2E^3}}\ \mathrm{e}^{E/m_0}\ . \tag{8.38}$$

Obviously the total energy density $\int_0^\infty E\ \mathrm{d}n(E)$ diverges for $kT > kT_0 = m_0 \approx 200$ MeV, because the exponential factor $\exp(E/m_0 - E/kT)$ grows with energy, implying that either no higher temeratures are possible or that some new physics must become relevant. As in this treatment the finite size of the hadrons was neglected, one expects the latter to be the case. Thus we have linked the existence of a phase transition to the fact that the density of states grows exponentially with energy. This behavior is, however, nearly universal to all composite models of the nucleon. It is found, for example, for the MIT bag model and is simply due to the fact that more and more angular momentum states become available with increasing energy. The phase transition temperature is simply determined by the level spacing between hadrons, and one has to conclude that the observation of a phase transition alone is not a very sensitive test of QCD.

In terms of quarks and gluons the phase transition can be linked to a reduction of the effective running coupling constant with increasing temperature as shown in Fig. 8.6. Finite temperature implies that the state in which a process has to be calculated is not the field theoretical vacuum state, but a state containing a thermal distribution of particles. This adds the graphs shown in Fig. 8.5 to the vacuum polarization tensor. It turns out that the resulting effective coupling constant α_{eff} becomes small enough for $T \sim \Lambda_{\mathrm{QCD}}$ to suggest the insignificance of nonperturbative effects.

Another interesting fact is that theoretical studies show that one is really dealing with two phase transitions. The deconfinement phase transition is characterized by the fact that energy density ε and pressure P approach (although not

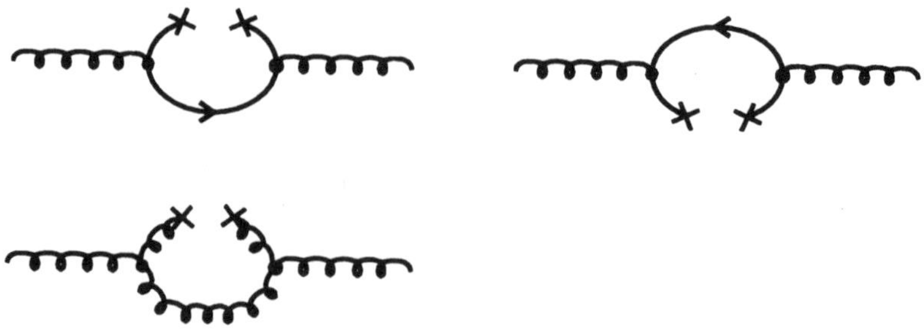

Fig. 8.5. Additional contributions to vacuum polarization in a thermal distribution of particles. Crosses mark a coupling to a thermal field

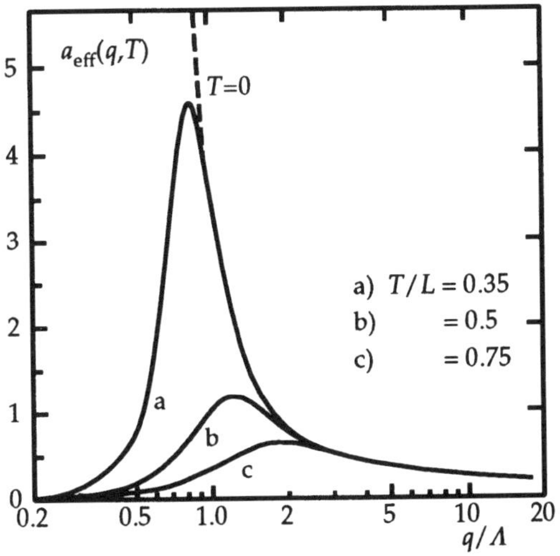

Fig. 8.6. Running coupling constant in a pure SU(2) gauge theory at finite temperature (See B. Müller: *The Physics of the Quark–Gluon Plasma* (Springer, Berlin, Heidelberg 1985), p. 81)

very rapidly) the corresponding Stefan–Boltzmann limits for a free gas of quarks and gluons above the phase transition (see Fig. 7.10). Simultaneously one observes the chiral phase transition which describes the fact that chiral symmetry, which is violated at low temperatures, is conserved for $T > T_c$. According to our discussion in Sect. 7.2 this implies $\langle \bar{q}q \rangle \to 0$ (cf. Fig. 8.7), which is in fact observed in lattice gauge calculations. The physical reason for the coupling of these two phase transitions is still disputed. The link between the two effects is the observation that the condensate can be expanded in terms of the density of "zero modes" (i.e. modes with zero energy), which in turn depends on the long-range properties of the interaction and thus on the presence or absence of confinement:

$$\langle \bar{q}q \rangle = -\frac{\pi}{V}\,\nu(0)\ . \tag{8.39}$$

Here $\nu(\Lambda)$ is the density of states which have the eigenvalue Λ with respect to the Dirac operator $-\mathrm{i}\gamma_\mu(\partial^\mu - \mathrm{i}gA^\mu)$. $\nu(0)$ is the density of states at $\Lambda = 0$. V is a normalization factor, namely the total space–time volume considered.

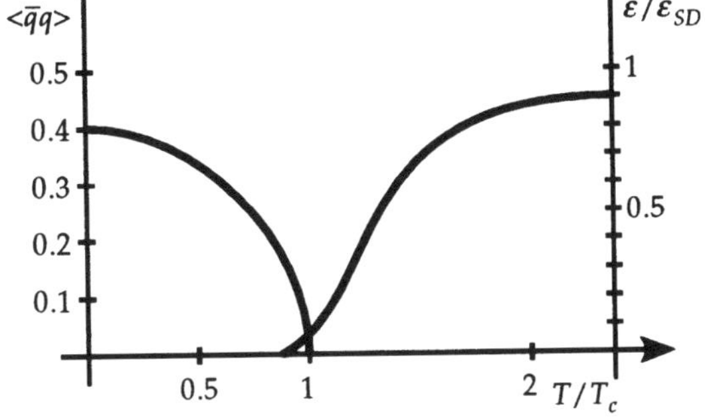

Fig. 8.7. Chiral phase transition $\langle \bar{q}q \rangle \to 0$ and confinement–deconfinement phase transition occur at the same temperature in lattice gauge simulations

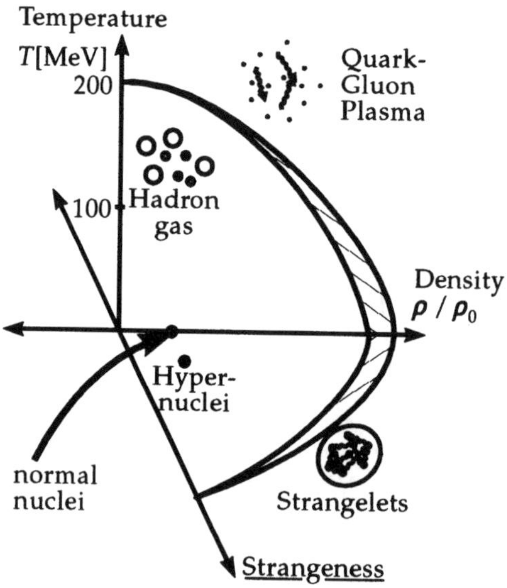

Fig. 8.8. The surface characterizing the transition between normal hadronic matter and the quark–gluon plasma as a function of μ, T, and $\bar{s}s$ content

The properties of the QGP phase transition are further modified by the finite baryon density in a heavy ion collision, characterized by the chemical potential μ. Thus the phase transition is characterized by a line in the μ–T diagram, or even a surface in the μ–T–$\bar{s}s$ diagram if the further modifications due to a finite strangeness–antistrangeness content are added; see Fig. 8.8.

Let us briefly discuss how the μ and T dependence of the phase transition can be understood. The energy density of free quarks and gluons is

$$\begin{aligned}\varepsilon = {} & 8 \times 2 \times \int_0^\infty \frac{\mathrm{d}^3 p}{(2\pi)^3}\, p \left(\mathrm{e}^{p/kT} - 1\right)^{-1} \\ & + 3 \times 2 \times N_\mathrm{f} \times \int_0^\infty \frac{\mathrm{d}^3 p}{(2\pi)^3}\, p \left[\frac{1}{\mathrm{e}^{(p-\mu)/kT} + 1} + \frac{1}{\mathrm{e}^{(p+\mu)/kT} + 1}\right] , \\ & p = E\,, \quad m_q = 0\,. \end{aligned} \tag{8.40}$$

The first term is the gluon contribution with 8 color states and 2 physical spin states. The second term is the quark and antiquark contribution for 3 color states, 2 spin states, and N_f massless flavors.

Making use of the integral formula

$$\begin{aligned} & \int_0^\infty \frac{x^{\nu-1}\,\mathrm{d}x}{\mathrm{e}^{\mu x} - 1} = \frac{\Gamma(\nu)}{\mu^\nu}\,\xi(\nu) \qquad \left(\mathrm{Re}\,\mu > 0,\ \mathrm{Re}\,\nu > 0\right), \\ & \int_0^\infty \frac{x^{\nu-1}\,\mathrm{d}x}{\mathrm{e}^{\mu x} + 1} = \frac{1 - 2^{1-\nu}}{\mu^\nu}\,\Gamma(\nu)\,\xi(\nu)\,, \end{aligned} \tag{8.41}$$

the integrals can be performed analytically:

$$
\begin{aligned}
\varepsilon = & \frac{16\times 4\pi}{(2\pi)^3}\,(kT)^4\,\Gamma(4)\,\xi(4) + \frac{6N_\mathrm{f}}{(2\pi)^3}\left[\int\limits_{-\mu}^{\infty}\mathrm{d}p\,4\pi\,(p+\mu)^3\frac{1}{\mathrm{e}^{\,p/kT}+1}\right.\\
& \left. + \int\limits_{\mu}^{\infty}\mathrm{d}p\,4\pi\,(p-\mu)^3\frac{1}{\mathrm{e}^{\,p/kT}+1}\right]\\
= & \frac{8}{\pi^2}\,(kT)^4\,6\,\frac{\pi^4}{90} + \frac{3N_\mathrm{f}}{\pi^2}\left\{\int\limits_0^{\infty}\mathrm{d}p\left[(p+\mu)^3+(p-\mu)^3\right]\frac{1}{\mathrm{e}^{\,p/kT}+1}\right.\\
& \left. + \int\limits_{-\mu}^{0}\mathrm{d}p\,(p+\mu)^3\frac{1}{\mathrm{e}^{\,p/kT}+1} - \int\limits_0^{\mu}\mathrm{d}p\,(p-\mu)^3\frac{1}{\mathrm{e}^{\,p/kT}+1}\right\}\\
= & \frac{8\pi^2}{15}\,(kT)^4 + \frac{3N_\mathrm{f}}{\pi^2}\left[2(1-2^{-3})\,(kT)^4\,6\,\frac{\pi^4}{90} + 2\cdot 3\,(1-2^{-1})\right.\\
& \left. \times (kT)^2\,\mu^2\cdot 1\cdot\frac{\pi^2}{6} - \int\limits_0^{\mu}\mathrm{d}p\,(p-\mu)^3\left(\frac{1}{\mathrm{e}^{-p/kT}+1}+\frac{1}{\mathrm{e}^{\,p/kT}+1}\right)\right]\\
= & \frac{8\pi^2}{15}\,(kT)^4 + \frac{3N_\mathrm{f}}{\pi^2}\left[\frac{7}{8}\,\frac{2\pi^4}{15}\,(kT)^4 + \frac{\pi^2}{2}\,\mu^2\,(kT)^2\right.\\
& \left. - \int\limits_0^{\mu}\mathrm{d}p\,(p-\mu)^3\,\frac{2+\mathrm{e}^{\,p/kT}+\mathrm{e}^{-p/kT}}{2+\mathrm{e}^{\,p/kT}+\mathrm{e}^{-p/kT}}\right]\\
= & \frac{8\pi^2}{15}\,(kT)^4 + 6N_\mathrm{f}\left[\frac{7\pi^2}{120}\,(kT)^4 + \frac{\mu^2}{4}\,(kT)^2\right.\\
& \left. - \frac{1}{8\pi^2}\,\mu^4 + \frac{3}{2\pi^2}\,\mu\,\frac{\mu^3}{3} - \frac{3}{2\pi^2}\,\mu^2\,\frac{\mu^2}{2} + \frac{1}{2\pi^2}\,\mu^3\cdot\mu\right]\\
= & \frac{8\pi^2}{15}\,(kT)^4 + 6N_\mathrm{f}\left[\frac{7\pi^2}{120}\,(kT)^4 + \frac{\mu^2}{4}\,(kT)^2 + \frac{1}{8\pi^2}\,\mu^4\right]\,.
\end{aligned}
\tag{8.42}
$$

The pressure $P = \varepsilon/3$ can now simply be equated to the bag constant to obtain an estimate for the phase-transition line:

$$
B = \frac{8\pi^2}{45}(kT)^4 + 2N_\mathrm{f}\left[\frac{7\pi^2}{120}(kT)^4 + \frac{\mu^2}{4}(kT)^2 + \frac{1}{8\pi^2}\mu^4\right]\,. \tag{8.43}
$$

In deriving this estimate, all interaction effects were neglected such that the numerical values obtained from (8.43) cannot be trusted. However, the general form shown in Fig. 8.8 should be correct.

Having specified the transition region to the QGP, let us now turn to the most important question, namely whether the new phase can be reached in

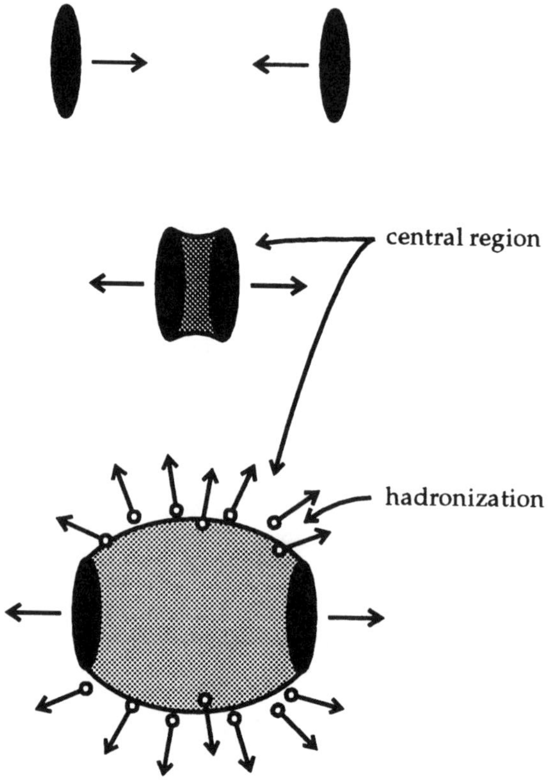

Fig. 8.9. Schematic description of an ultra-relativistic heavy-ion collision

a heavy-ion collision. To this end, heavy-ion collisions were studied in great detail within many different models.[4] After much controversy it is now established that very high densities and enormous rates of particle production are reached. The CERN experiments with 200 GeV/A beams have produced a large amount of data, testing the existing theoretical models and thus allowing a more trustworthy extrapolation to higher energies. The most important result of these and other studies is that the originally often advocated Bjorken picture of the two nuclei passing through one another with only very little interaction is wrong. Instead enormous numbers of particles are produced, absorbing a large fraction of the original energy. Schematically the correct picture is given in Fig. 8.9.

Because so many particles take part in the interactions, it turns out that a hydroydnamic description is applicable except at the highest collision energies. This description was primarily developed by W. Greiner and collaborators in Frankfurt. In its language, the crucial mechanism leading to the large energy deposition is the creation of nuclear shock waves, which had been predicted in a seminal paper by W. Scheid, H. Müller, and W. Greiner.[5] Other effects pre-

[4] For a review see, e.g., W. Greiner, H. Stöcker, and A. Gallmann (Eds.): *Hot and Dense Nuclear Matter*, NATO-ASI SERIES B–Physics (Plenum Press, 1994).

[5] See W. Scheid, H. Müller and W. Greiner: Phys. Lett. **13**, 741 (1974).

dicted by the same model have also been observed, namely the fact that some of the projectiles are deflected for suitable impact parameters (sideways flow) and that a greater than average number of particles are emitted perpendicular to the reaction plane (squeeze-out).

Based on all these investigations, one can be optimistic that a phase transition can indeed be reached, for example, in a heavy-ion collision with 200 GeV/A at RHIC. The still unsolved problem, however, is whether there is any signal that survives the subsequent hadronization. Until now, no undisputed signal has been found. The hope rests therefore on very exotic objects which one cannot imagine being produced in a normal nuclear surrounding. Probably the most interesting of these proposed signals are multibaryon states with a very large strangeness content. B. Witten[6] and E. Farhi and R.L. Jaffe[7] have proposed that droplets of QGP could be stable or at least metastable if they contain a sufficient number of s quarks. The question is thus whether the creation of such droplets, which are termed "strangelets", could be aided in a QGP. Greiner, Koch and Stöcker[8] have shown that this is indeed possible. Under the assumption that a mixed phase of QGP and ordinary nuclear matter is created in a sufficiently large region and lives long enough (10^{-22} s), an unmixing of s and $\bar{s}$ should happen where the s quarks gather in the QGP while the $\bar{s}$ dominate the hadronic phase as kaons. If this mechanism is effective enough to collect sufficiently many s quarks in the QGP, it would not decay and could then remain as a (experimentally easily detected) strangelet. Let us end this chapter by discussing some other rigorous results of QCD besides those of lattice-gauge calculations.

As sketched in Fig. 8.6, at high temperature the coupling to the thermal bath of particles reduces the effective coupling constant. For sufficiently large temperature the resulting theory should therefore again be tractable by perturbation theory. The most advanced formalism along these lines was developed by Braaten and Pisarski. It is based on earlier work by Kapusta. Let us just sketch some of the results obtained. By calculating the lowest-order contributions to the gluon propagator in a thermal background, Kapusta obtained an effective propagator which can be written in the form

$$D_{\mathrm{L}}(\omega, k) = \frac{1}{\varepsilon_{\mathrm{L}}(\omega, k) k^2} \,, \tag{8.44}$$

$$D_{\mathrm{T}}(\omega, k) = \frac{1}{\varepsilon_{\mathrm{T}}(\omega, k)\omega^2 - k^2} \,, \tag{8.45}$$

$$\varepsilon_{\mathrm{L}} = 1 + \frac{g^2 T^2}{k^2}\left[1 - \frac{\omega}{2k}\ln\left|\frac{\omega+k}{\omega-k}\right|\right] \,, \tag{8.46}$$

$$\varepsilon_{\mathrm{T}} = 1 - \frac{g^2 T^2}{2k^2}\left[1 - \left(1 - \frac{k^2}{\omega^2}\right)\frac{\omega}{2k}\ln\left|\frac{\omega+k}{\omega-k}\right|\right] \,, \tag{8.47}$$

[6] See B. Witten: Phys. Rev. D **30**, 272 (1984).

[7] See E. Farhi and R.L. Jaffe: Phys. Rev. D **30**, 2319 (1984).

[8] See C. Greiner, P. Koch, and H. Stöcker: Phys. Rev. Lett. **58**, 1825 (1987).

with transverse and longitudinal color-dielectric functions. For a static source ($\omega \to 0$) these equations imply

$$\varepsilon_{\mathrm{L}} \to 1 + \frac{g^2T^2}{k^2}\left[1 - \frac{\omega}{2k}\frac{2\omega}{k}\right] \to 1 + \frac{g^2T^2}{k^2} ,$$
$$D_{\mathrm{L}} \to \frac{-1}{(-k^2) - g^2T^2} , \tag{8.48}$$
$$\varepsilon_{\mathrm{T}} \to 1 - \frac{g^2T^2}{2k^2}\left[1 - \left(1 - \frac{k^2}{\omega^2}\right)\frac{\omega^2}{k^2}\right] = 1 - \frac{g^2T^2}{2k^2}\, 2 ,$$
$$D_{\mathrm{T}} \to \frac{1}{-k^2} . \tag{8.49}$$

Thus only the longitudinal gluon field is screened, with a screening mass gT, and corresponding Debye length $\lambda_{\mathrm{D}} = 1/gT$. The fact that the transverse propagator is not screened is one of the problems of this approach. This result of lowest-order calculations implies the existence of relevant higher-order terms, e.g. $m_{\mathrm{T}} \sim g^2T$. Great effort is currently being invested in calculating these terms reliably.

An interesting point is that the poles of D_{L} and D_{T} are the effective particles of the thermal state, the "plasmons". Some of their properties can easily be read off from (8.44)–(8.47). For the longitudinal case $\varepsilon_{\mathrm{L}} = 0$ implies

$$0 = k^2 + g^2T^2\left[1 - \frac{\omega}{2k}\ln\left|\frac{\omega+k}{\omega-k}\right|\right] . \tag{8.50}$$

For the transverse case $\varepsilon_{\mathrm{T}} = k^2/\omega^2$ leads to

$$\frac{k^2}{\omega^2} = 1 - \frac{g^2T^2}{2k^2}\left[1 - \left(1 - \frac{k^2}{\omega^2}\right)\frac{\omega}{2k}\ln\left|\frac{\omega+k}{\omega-k}\right|\right] . \tag{8.51}$$

To test whether this high-energy phase is really deconfined one has to check the low k limit, i.e., the properties of the plasmon for large distances. For the longitudinal case one finds

$$0 = k^2 + g^2T^2\left[1 - \frac{\omega}{2k}\left(\frac{2k}{\omega} + \frac{2k^3}{3\omega^3}\right)\right]$$
$$= k^2 + g^2T^2\left(-\frac{k^2}{3\omega^2}\right)$$
$$\Rightarrow \omega^2 \approx \omega^2 - k^2 = \frac{g^2T^2}{3} \tag{8.52}$$

$$m^{\mathrm{L}}_{\mathrm{plasmon}} = \frac{gT}{\sqrt{3}} \tag{8.53}$$

and for the transverse case

$$\frac{k^2}{\omega^2} = 1 - \frac{g^2T^2}{2k^2}\left[1 - \left(1 - \frac{k^2}{\omega^2}\right)\left(1 + \frac{k^2}{3\omega^2}\right)\right]$$
$$= 1 - \frac{g^2T^2}{2k^2}\left(\frac{2k^2}{3\omega^2}\right) = 1 - \frac{g^2T^2}{3\omega^2} \tag{8.54}$$

$$\omega^2 - k^2 = \frac{g^2T^2}{3}$$
$$m^{\mathrm{T}}_{\mathrm{plasmon}} = \frac{gT}{\sqrt{3}} \ . \tag{8.55}$$

Thus both the longitudinal and the transverse gluons have, for large distances, the same effective mass $gt/\sqrt{3}$, implying that the color potentials are screened and confinement is no longer effective. Let us note that for realistically attainable temperatures $T \sim 250$ MeV, $g \sim 2$ one gets

$$m_{\mathrm{plasmon}} \sim 300\ \mathrm{MeV} \ . \tag{8.56}$$

Obviously the simple idea of a free gas of massless gluons is completely ruled out by high-temperature QCD, too.

Braaten and Pisarski pushed this kind of calculation much further. They distinguish three scales, namely $g^2T \ll gT \ll T$. Momenta of order gT can be treated by usual perturbative QCD, those of order T require a careful resummation of contributions of arbitrarily high order in perturbation theory. (This is achieved by deriving effective vertices and propagators.) Momenta of order g^2T cannot be treated. Within this approach a large number of properties were calculated, but the problem with all of this is that it applies only for $g \ll 1$, implying a temperature which cannot be reached in any realistic experiment. As stated above g is even larger than 1. The same criticism applies to approaches in which the QCD field theory is treated as classical field theory. The classical Yang–Mills equations show chaotic behavior and B. Müller and collaborators showed that the corresponding leading Lyapunov exponent can actually be related to the gluon damping rate as obtained by Braaten and Pisarski, implying that the regime of asymptotically high temperatures seems to be really well understood. If one were to simply extrapolate the results of these two descriptions to realistic temperatures, they would imply very short thermalization times and consequently very good chances for quark–gluon plasma formation in heavy-ion collisions, but, as discussed above, such extrapolations are very problematic.

Here we have only been able to sketch some relevant ideas and developments in this very active field. At present it seems very probable that current heavy-ion experiments at RHIC and future ones at LHC will study hadronic matter under conditions where it will show exotic properties. Is is, however, still not clear whether it will be possible to link these experimental observations to basic properties of QCD in an undisputable manner.

Appendix

In this appendix we summarize some useful formulas needed to perform QCD calculations.

A The Group SU(N)

SU(N) denotes the group of unitary matrices of order N with determinant det $= +1$. Every $N \times N$ matrix U of the group can be represented through the generators t^A ($A = 1, \dots, N^2 - 1$) as

$$U(\Lambda_1, \Lambda_2, \dots, \Lambda_{N^2-1}) = \exp(\mathrm{i}\Lambda_A t^A) \ , \tag{A.1}$$

where we have summed over equal indices A. Since the matrix should be unitary, the generators have to be Hermitian. Since, further, det $U = 1$, the generators t^A have to be traceless. The generators obey the algebra

$$\left[t^A, t^B\right] = \mathrm{i} f^{ABC} t^C \tag{A.2}$$

and are normalized as

$$\mathrm{tr}\, t^A t^B = \frac{1}{2}\delta^{AB} \ . \tag{A.3}$$

Here the f^{ABC} are the antisymmetric structure constants of the group. One can also define an anticommutator as

$$\left\{t^A, t^B\right\} = \frac{1}{N}\delta^{AB} + d^{ABC} t^C \ , \tag{A.4}$$

with symmetric structure constants d^{ABC}. The completeness relation is

$$t^A_{ab} t^A_{cd} = \frac{1}{2}\delta_{ad}\delta_{bc} - \frac{1}{2N}\delta_{ab}\delta_{cd} \tag{A.5}$$

and the Fierz identity

$$t^A_{ab} t^A_{cd} = \frac{N^2-1}{2N^2}\delta_{ad}\delta_{bc} - \frac{1}{N} t^A_{ad} t^A_{bc} \ . \tag{A.6}$$

The definitions (A.2) and (A.3) as well as the completeness relation can be used to derive useful identities:

$$t^A_{ab}t^A_{bd} = \frac{N^2-1}{2N}\delta_{ad} \overset{N=3}{=} \frac{4}{3}\delta_{ad} \ , \tag{A.7}$$

$$\mathrm{tr}\, t^A t^A = \frac{N^2-1}{2} \overset{N=3}{=} 4 \ , \tag{A.8}$$

$$\mathrm{tr}\, t^A t^A t^B t^B = \frac{(N^2-1)^2}{4N} \overset{N=3}{=} \frac{16}{3} \ , \tag{A.9}$$

$$\mathrm{tr}\, t^A t^A t^B t^B = \frac{1-N^2}{4N} \overset{N=3}{=} -\frac{2}{3} \ , \tag{A.10}$$

$$f^{ABC}\ \mathrm{tr}\, t^A t^B t^C = \mathrm{i}\frac{N(N^2-1)}{4} \overset{N=3}{=} 6\mathrm{i} \ , \tag{A.11}$$

where we also have given the numerically important case of $N = 3$. Useful relations can also be given for the structure constants:

$$f^{ACD} f^{BCD} = N\delta^{AB} \ , \tag{A.12}$$

$$f^{ABC} f^{ABC} = N(N^2-1) \ . \tag{A.13}$$

For the special case of $N = 3$ the generators t^A are given by the Gell-Mann matrices,

$$\mathrm{SU(3)}: \quad t^A = \frac{\lambda^A}{2} \ , \tag{A.14}$$

and for the case of $N = 2$ we get the Pauli matrices,

$$\mathrm{SU(2)}: \quad t^a = \frac{\tau^a}{2} \ . \tag{A.15}$$

The structure constants of the SU(2) are given by the antisymmetric tensor $\varepsilon^{abc} = -\varepsilon^{acb} = 1$ $(a = 1, 2, 3)$:

$$\mathrm{SU(2)}: \quad \left[\frac{\tau^a}{2}, \frac{\tau^b}{2}\right] = \mathrm{i}\varepsilon^{abc}\frac{\tau^c}{2} \tag{A.16}$$

and the anticommutator in this case reads

$$\mathrm{SU(2)}: \quad \left\{\frac{\tau^a}{2}, \frac{\tau^b}{2}\right\} = \frac{1}{2}\delta^{ab} \ . \tag{A.17}$$

For SU(2) structure constants we have in addition

$$\varepsilon^{abc}\varepsilon^{adf} = \delta^{bd}\delta^{cf} - \delta^{bf}\delta^{cd} \ , \tag{A.18}$$

$$\varepsilon^{acd}\varepsilon^{bcd} = 2\delta^{ab} \ , \tag{A.19}$$

$$\varepsilon^{abc}\varepsilon^{abc} = 6 \ . \tag{A.20}$$

B Dirac Algebra in Dimension *d*

The procedure of dimensional regularization continues the four-dimensional space–time into d dimensions. Accordingly, also the four-dimensional γ matrices have to be continued to d dimensions. Therefore we have to define d γ matrices

$$\gamma_0, \gamma_{\mu_1}, \ldots, \gamma_{\mu_d} . \tag{B.21}$$

γ^0 is Hermitian, γ^i, $i > 0$ anti-Hermitian:

$$\gamma^{\mu\dagger} = \gamma^0 \gamma^\mu \gamma^0 . \tag{B.22}$$

The algebra is defined through the usual anticommutation relations

$$\left\{\gamma^\mu, \gamma^\nu\right\} = 2g^{\mu\nu}\hat{1} . \tag{B.23}$$

The trace of the metric tensor is given by

$$g_\mu^\mu = d \tag{B.24}$$

and therefore

$$\gamma_\mu \gamma^\mu = d\hat{1} , \tag{B.25}$$

while $\mathrm{tr}\,\hat{1} = 4$. Every other choice, for instance $\mathrm{tr}\hat{1} = d$, leads only to a finite renormalization which is irrelevant. By convention one sticks to the first choice. Using the relations above, one can derive some useful identities:

$$\gamma^\mu \gamma_\alpha \gamma_\mu = (2-d)\gamma_\alpha , \tag{B.26}$$
$$\gamma^\mu \gamma_\alpha \gamma_\beta \gamma_\mu = (d-4)\gamma_\alpha \gamma_\beta + 4g_{\alpha\beta} , \tag{B.27}$$
$$\gamma^\mu \gamma_\alpha \gamma_\beta \gamma_\delta \gamma_\mu = -2\gamma_\delta \gamma_\beta \gamma_\alpha - (d-4)\gamma_\alpha \gamma_\beta \gamma_\delta , \tag{B.28}$$
$$\gamma^\mu \sigma_{\alpha\beta} \gamma_\mu = (d-4)\sigma_{\alpha\beta} , \tag{B.29}$$
$$\gamma^\mu \sigma_{\mu\alpha} = \mathrm{i}(d-1)\gamma_\alpha , \tag{B.30}$$
$$\sigma_{\mu\alpha} \gamma^\mu = \mathrm{i}(1-d)\gamma_\alpha , \tag{B.31}$$

where we have in the usual way

$$\sigma^{\mu\nu} = \frac{\mathrm{i}}{2}\left[\gamma^\mu, \gamma^\nu\right] , \tag{B.32}$$
$$\gamma_\mu \gamma_\nu = g_{\mu\nu} - i\sigma_{\mu\nu} . \tag{B.33}$$

The trace over an odd number of γ matrices vanishes:

$$\mathrm{tr}\,\gamma^\mu = 0 , \tag{B.34}$$
$$\mathrm{tr}\,\gamma^\mu \gamma^\nu = 4g^{\mu\nu} , \tag{B.35}$$
$$\mathrm{tr}\,\gamma^\mu \gamma^\nu \gamma^\alpha \gamma^\beta = 4\left(g^{\mu\nu} g^{\alpha\beta} - g^{\mu\alpha} g^{\nu\beta} + g^{\mu\beta} g^{\nu\alpha}\right) . \tag{B.36}$$

The definition of the matrix γ_5 in d dimensions is more complicated. In $d=4$ dimensions γ_5 is defined as

$$\gamma_5 = \gamma^5 = \mathrm{i}\gamma^0\gamma^1\gamma^2\gamma^3 = -\mathrm{i}\gamma_0\gamma_1\gamma_2\gamma_3 = \gamma_5^\dagger \\ = -\frac{\mathrm{i}}{4!}\varepsilon_{\mu\nu\varrho\sigma}\gamma^\mu\gamma^\nu\gamma^\varrho\gamma^\sigma \tag{B.37}$$

with the completely antisymmetric tensor $\varepsilon_{\mu\nu\varrho\sigma}$, $\varepsilon^{0123} = -\varepsilon_{0123} = 1$. In $d=4$ dimensions we have

$$d=4: \quad \{\gamma_5, \gamma^\mu\} = 0 \ . \tag{B.38}$$

For arbitrary dimensions these definitions cannot be carried through consistently. To demonstrate the inconsistencies arising we calculate the following traces, $\mathrm{tr}\,(\gamma_5)$, $\mathrm{tr}\,(\gamma_\mu\gamma_\nu\gamma_5)$, and $\mathrm{tr}\,(\gamma_\mu\gamma_\nu\gamma_\varrho\gamma_\sigma\gamma_5)$ in d dimensions, making use of the anticommutation relation (B.38):

$$d\ \mathrm{tr}\,\gamma_5 = \ \mathrm{tr}\,\gamma_\mu\gamma^\mu\gamma_5 = -\ \mathrm{tr}\,\gamma_\mu\gamma_5\gamma^\mu = -d\ \mathrm{tr}\,\gamma_5 \tag{B.39}$$

from which it follows that $\mathrm{tr}\,\gamma_5 = 0$ for all $d \neq 0$. In the same way we have

$$d\ \mathrm{tr}\,\gamma_\mu\gamma_\nu\gamma_5 = \ \mathrm{tr}\,\gamma_\alpha\gamma^\alpha\gamma_\mu\gamma_\nu\gamma_5 = -\ \mathrm{tr}\,\gamma^\alpha\gamma_\mu\gamma_\nu\gamma_\alpha\gamma_5 \\ = (4-d)\ \mathrm{tr}\,\gamma_\mu\gamma_\nu\gamma_5 - 4g_{\mu\nu}\ \mathrm{tr}\,\gamma_5 = (4-d)\ \mathrm{tr}\,\gamma_\mu\gamma_\nu\gamma_5\ , \tag{B.40}$$

so that

$$(2-d)\ \mathrm{tr}\,\gamma_\mu\gamma_\nu\gamma_5 = 0 \tag{B.41}$$

and therefore $\mathrm{tr}\,\gamma_\mu\gamma_\nu\gamma_5 = 0$ for all $d \neq 2$. Also, it is easy to see that

$$(4-d)\ \mathrm{tr}\,\gamma_\mu\gamma_\nu\gamma_\varrho\gamma_\sigma\gamma_5 = 0\ . \tag{B.42}$$

These relations allow us to define a trace that is different from zero for $d=4$:

$$d=4: \quad \mathrm{tr}\,\gamma_\mu\gamma_\nu\gamma_\varrho\gamma_\sigma\gamma_5 = -4\mathrm{i}\varepsilon_{\mu\nu\varrho\sigma}\ . \tag{B.43}$$

However, it still requires a vanishing trace for arbitrary d. Obviously, the consistent application of the anticommutation relation (B.38) requires a vanishing trace over γ_5 multiplied with any string of γ matrices. Therefore, it was concluded, that either dimensional regularization has to be abandoned for γ_5 or the anticommutation relation has to be changed. In the 't Hooft–Veltman[1] scheme full Lorentz covariance in d dimensions will be abandoned, i.e. γ_5 in 4 and in $d-4$ dimensions is treated differently. The idea is to define γ_5 also in d dimensions with the help of the $\varepsilon_{\mu\nu\varrho\sigma}$, which lives only in 4 dimensions as in (B.37). Each contraction with the $\varepsilon_{\mu\nu\varrho\sigma}$ tensor therefore projects onto 4 dimensions. The definition of γ_5 in the 't Hooft–Veltman scheme therefore is

$$\gamma_5 = \mathrm{i}\gamma^0\gamma^1\gamma^2\gamma^3 = -\frac{\mathrm{i}}{4!}\varepsilon_{\mu\nu\varrho\sigma}\gamma^\mu\gamma^\nu\gamma^\varrho\gamma^\sigma \tag{B.44}$$

[1] G. 't Hooft and M. Veltman: Nucl. Phys. B **44**, 189 (1977)

with the $\varepsilon^{\mu\nu\varrho\sigma}$ tensor

$$\varepsilon^{\mu\nu\varrho\sigma} = \begin{cases} 1 & (\mu\nu\varrho\sigma) \text{ even permutation of } (0123) \\ -1 & (\mu\nu\varrho\sigma) \text{ odd permutation of } (0123) \\ 0 & \text{otherwise} \end{cases} \tag{B.45}$$

From this one gets the following commutation relations:

$$\begin{aligned} \{\gamma_5, \gamma^\mu\} &= 0 \ : \mu = 0, 1, 2, 3 \ , \\ [\gamma_5, \gamma^\mu] &= 0 \ : \text{ otherwise } . \end{aligned} \tag{B.46}$$

γ_5 anticommutes with the other γ matrices in the physical 4 dimensions and commutes with them in $d-4$ dimensions. This procedure may appear unsatisfactory, but still it works in the sense that it reproduces the correct anomalous divergence of the axial current operator. To understand the appearance of a potential contribution to the anomaly we introduce the following notations for γ matrices in d, $d-4$, and 4 dimensions:

$$\gamma_\mu, \quad \hat{\gamma}_\mu, \quad \tilde{\gamma}_\mu . \tag{B.47}$$

From (B.46) we find that the anticommutator with γ_5 does not vanish, instead a contributions is retained which exists in $d-4$ dimensions

$$\{\gamma_5, \gamma^\mu\} = 2\hat{\gamma}^\mu \gamma_5 . \tag{B.48}$$

In the limit $d \to 4$ the original expression is recovered. This, however, is no longer true when in a calculation of some diagram divergencies, which manifest as poles at $d = 4$, appear. This has as consequence that after renormalization, contributions proportional to $d-4$ no longer vanish and give a finite contribution

$$\lim_{d\to 4} \left(\frac{1}{d-4} \gamma_\mu \{\gamma_5, \gamma^\mu\} \right) = \lim_{d\to 4} \left(\frac{1}{d-4} \cdot 2(d-4)\gamma_5 \right) = 2\gamma_5 . \tag{B.49}$$

For massless quarks, the divergence of the axial current in the 't Hooft–Veltman scheme is given by

$$\partial_\mu j_5^\mu = \bar{\Psi} \gamma_5 D_\mu \hat{\gamma}^\mu \Psi , \tag{B.50}$$

which means that, as in (B.49), this term will contribute only anomalously.

As basis for the gamma matrices in $d = 4$ dimensions one can choose the 16 matrices $\hat{1}$, γ_5, γ_μ, $\gamma_5\gamma_\mu$, $\sigma_{\mu\nu}$. For $d > 4$ that basis has to be correspondingly larger. The following relations therefore are valid only in $d = 4$ dimensions:

$$d = 4 : \quad \gamma_\alpha \gamma_\beta \gamma_\varrho = \left(g_{\alpha\beta} g_{\varrho\delta} + g_{\alpha\delta} g_{\varrho\beta} - g_{\alpha\varrho} g_{\beta\delta} \right) \gamma^\delta - \mathrm{i}\varepsilon_{\alpha\beta\varrho\delta} \gamma_5 \gamma^\delta . \tag{B.51}$$

Also, it holds that

$$d = 4 : \quad \gamma_5 \sigma^{\varrho\sigma} = \mathrm{i}\sigma_{\alpha\beta} \frac{1}{2} \varepsilon^{\alpha\beta\varrho\sigma} . \tag{B.52}$$

As another important relation we give the Fierz transformation of the direct product of two quark field operators:

$$d=4: \quad q^a_\alpha \bar{q}^{\,b}_{\ \beta} = -\frac{1}{4}\bar{q}^{\,b} q^a \delta_{\alpha\beta} - \frac{1}{4}\bar{q}^{\,b}\gamma_\mu q^a (\gamma^\mu)_{\alpha\beta} - \frac{1}{8}\bar{q}^{\,b}\sigma_{\mu\nu} q^a (\sigma^{\mu\nu})_{\alpha\beta} - \frac{1}{4}\bar{q}^{\,b}\gamma_5\gamma_\mu q^a (\gamma^\mu\gamma_5)_{\alpha\beta} - \frac{1}{4}\bar{q}^{\,b}\gamma_5 q^a (\gamma_5)_{\alpha\beta} \,. \tag{B.53}$$

C Some Useful Integrals

The master formula for Feynman integrals is

$$\int \frac{\mathrm{d}^d k}{(2\pi)^d} \frac{\left(k^2\right)^a}{(k^2-M^2)^m} = \mathrm{i}\frac{(-1)^{a-m}}{(4\pi)^{d/2}} \frac{\Gamma(a+d/2)\Gamma(m-a-d/2)}{\Gamma(d/2)\Gamma(m)(M^2)^{m-a-d/2}} \,, \tag{C.54}$$

where k is a d-dimensional vector in Minkowski space. For symmetric integration we have

$$\int \mathrm{d}^d k\, k^\mu k^\nu f(k^2) = \frac{g^{\mu\nu}}{d} \int \mathrm{d}^d k\, k^2 f(k^2) \,, \tag{C.55}$$

$$\int \mathrm{d}^d k\, k^\mu k^\nu k^\lambda k^\sigma f(k^2) = \frac{g^{\mu\nu}g^{\lambda\sigma} + g^{\mu\lambda}g^{\nu\sigma} + g^{\mu\sigma}g^{\nu\lambda}}{d(d+2)} \int \mathrm{d}^d k\, k^4 f(k^2) \,. \tag{C.56}$$

For an odd number of k_μ (antisymmetric integration) the corresponding integrals vanish.

The Feynman parameters are

$$\prod_{i=1}^n \frac{1}{a_i^{A_i}} = \frac{\Gamma(A)}{\prod\limits_{i=1}^n \Gamma(A_i)} \int_0^1 \left(\prod_{i=1}^n \mathrm{d}x_i x_i^{A_i-1}\right) \frac{\delta(1-\sum\limits_{i=1}^n x_i)}{\left[\sum\limits_{i=1}^n x_i a_i\right]^A} \,,$$

where $A = \sum\limits_{i=1}^n A_i$. In the simplest case this reduces to

$$\frac{1}{a^A b^B} = \frac{\Gamma(A+B)}{\Gamma(A)\Gamma(B)} \int_0^1 \mathrm{d}u \frac{u^{A-1}\bar{u}^{B-1}}{(ua+\bar{u}b)^{A+B}} \quad ; \quad u+\bar{u}=1 \,. \tag{C.57}$$

For integration over Feynman parameters, the following representation of the beta function is useful:

$$\int_0^1 \mathrm{d}u\, u^{A-1}\bar{u}^{B-1} = \frac{\Gamma(A)\Gamma(B)}{\Gamma(A+B)} \,. \tag{C.58}$$

By suitable substitution one recovers a different representation

$$\int_0^\infty \frac{t^{A-1}}{(1+t)^{A+B}}\,\mathrm{d}t = \frac{\Gamma(A)\Gamma(B)}{\Gamma(A+B)}\,. \tag{C.59}$$

For evaluations of Fourier transforms the Schwinger parameterization might be useful:

$$\frac{1}{(-x^2)^\alpha} = \frac{1}{\Gamma(\alpha)}\int_0^\infty \mathrm{d}u\, u^{\alpha-1}\,\mathrm{e}^{-u(-x^2)}\,. \tag{C.60}$$

Knowledge of the d-dimensional Gauss integral

$$\int_{-\infty}^{+\infty} \mathrm{d}^d\xi e^{-a\xi^2} = \left(\frac{\pi}{a}\right)^{d/2} \tag{C.61}$$

and the definition of the gamma function

$$\Gamma(A) = \int_0^\infty \mathrm{d}t\, t^{A-1}e^{-t} \tag{C.62}$$

then allows us to derive the following formula:

$$\int \mathrm{d}^d x\, \frac{\mathrm{e}^{\mathrm{i}px}}{(-x^2)^\alpha} = -\mathrm{i}2^{d-2\alpha}\pi^{d/2}(-p^2)^{(\alpha-d/2)}\frac{\Gamma(d/2-\alpha)}{\Gamma(\alpha)}\,. \tag{C.63}$$

Note that x is a d-dimensional Minkowski vector in coordinate space, while p is a d-dimensional Minkowski vector in momentum space. The distinction is important since the Wick rotations to Euclidean coordinate or momentum space have different signs:

$$\int \mathrm{d}^d x = -\mathrm{i}\int \mathrm{d}^d x_E\,, \tag{C.64}$$

$$\int \mathrm{d}^d p = +\mathrm{i}\int \mathrm{d}^d p_E\,. \tag{C.65}$$

Here the index E marks an Euclidean vector. This convention leaves the product p^0x_0 invariant under Wick rotation. This has to be kept in mind when using any of the given integrals in either momentum or coordinate space. For instance, (C.63) would read

$$\int \mathrm{d}^d p\, \frac{\mathrm{e}^{-\mathrm{i}px}}{(-p^2)^\alpha} = +\mathrm{i}2^{d-2\alpha}\pi^{d/2}(-x^2)^{(\alpha-d/2)}\frac{\Gamma(d/2-\alpha)}{\Gamma(\alpha)}\,. \tag{C.66}$$

Note that the overall sign has changed compared to (C.63). The Fourier transform (C.63) can be used to derive many more formulas by taking derivatives with respect to p. Differentiating twice, one obtains the following formula:

$$\int \mathrm{d}^d x \; \mathrm{e}^{\mathrm{i}px} \frac{x_\mu x_\nu}{(-x^2)^\alpha} = \frac{\mathrm{i}\pi^{d/2} 2^{d-2\alpha+1}}{\Gamma(\alpha)} (-p^2)^{(\alpha-d/2-2)} \times \Big(2p_\mu p_\nu \Gamma(d/2-\alpha+2) + g_{\mu\nu}(-p^2)\Gamma(d/2-\alpha+1) \Big) \; . \tag{C.67}$$

Subject Index

Printing (Computer to Film): Saladruck Berlin
Binding: Stürtz AG, Würzburg

528